Fundamental constants

Speed of light	c	$2.997\ 924\ 58 \times 10^{8}$ m s^{-1}
Elementary charge	e	$1.602\ 177 \times 10^{-19}$ C
Planck constant	h	$6.626\ 075\ 5 \times 10^{-34}$ J s
	$\hbar = h/2\pi$	$1.054\ 572\ 66 \times 10^{-34}$ J s
Boltzmann constant	k	$1.380\ 658 \times 10^{-23}$ J K^{-1}
Avogadro constant	N_A	$6.022\ 14 \times 10^{23}$ mol^{-1}
Mass of electron	m_e	$9.109\ 39 \times 10^{-31}$ kg
Mass of proton	m_p	$1.672\ 62 \times 10^{-27}$ kg
Mass of neutron	m_n	$1.674\ 93 \times 10^{-27}$ kg
Atomic mass unit	u	$1.660\ 54 \times 10^{-27}$ kg
Vacuum permittivity	$\varepsilon_0 = 1/c^2\mu_0$	$8.854\ 187\ 816 \times 10^{-12}$ J^{-1} C^2 m^{-1}
	$4\pi\varepsilon_0$	$1.112\ 650\ 056 \times 10^{-10}$ J^{-1} C^2 m^{-1}
Vacuum permeability	μ_0	$4\pi \times 10^{-7}$ J s^2 C^{-2} m^{-1}
Bohr magneton	$\mu_B = e\hbar/2m_e$	$9.274\ 015\ 4 \times 10^{-24}$ J T^{-1}
Nuclear magneton	$\mu_N = e\hbar/2m_p$	$5.050\ 786\ 6 \times 10^{-27}$ J T^{-1}
G-value of electron	g_e	$2.002\ 319\ 304$
Bohr radius	$a_0 = 4\pi\varepsilon_0\hbar^2/m_e e^2$	$5.291\ 772\ 49 \times 10^{-11}$ m
Rydberg constant	$\mathcal{R} = m_e e^4/8h^3c\varepsilon_0^2$	$1.097\ 373\ 153 \times 10^{5}$ cm^{-1}
Hartree (unit of energy)	$E_h = 2hc\mathcal{R}$	$4.359\ 748\ 2 \times 10^{-18}$ J
Fine structure constant	$\alpha = \mu_0 e^2 c/2h$ $= e^2/4\pi\hbar c\varepsilon_0$	$7.297\ 353\ 08 \times 10^{-3}$
Compton wavelength of electron	$\lambda_C = h/m_e c$	$2.426\ 309 \times 10^{-12}$ m
Stefan–Boltzmann constant	$\sigma = 2\pi^5 k^4/15h^3c^2$	$5.670\ 51 \times 10^{-8}$ W m^{-2} K^{-4}
First radiation constant	$c_1 = 2\pi hc^2$	$3.741\ 774\ 9 \times 10^{-16}$ J m^2 s^{-1}
Second radiation constant	$c_2 = hc/k$	$1.438\ 769 \times 10^{-2}$ m K

Molecular Quantum Mechanics

P. W. Atkins and R. S. Friedman

Molecular Quantum Mechanics

THIRD EDITION

Oxford New York Tokyo
OXFORD UNIVERSITY PRESS
1997

Oxford University Press, Walton Street, Oxford OX2 6DP

Oxford New York
Athens Auckland Bangkok Bombay
Calcutta Cape Town Dar es Salaam Delhi
Florence Hong Kong Istanbul Karachi
Kuala Lumpur Madras Madrid Melbourne
Mexico City Nairobi Paris Singapore
Taipei Tokyo Toronto

and associated companies in
Berlin Ibadan

Oxford is a trade mark of Oxford University Press

Published in the United States
by Oxford University Press Inc., New York

© *P. W. Atkins and R. S. Friedman, 1997*

First edition 1970
Second edition 1983
Third edition 1997

All rights reserved. No part of this publication may be reproduced, stored in a retrieval system, or transmitted, in any form or by any means, without the prior permission in writing of Oxford University Press. Within the UK, exceptions are allowed in respect of any fair dealing for the purpose of research or private study, or criticism or review, as permitted under the Copyright, Designs and Patents Act, 1988, or in the case of reprographic reproduction in accordance with the terms of licences issued by the Copyright Licensing Agency. Enquiries concerning reproduction outside those terms and in other countries should be sent to the Rights Department, Oxford University Press, at the address above.

This book is sold subject to the condition that it shall not, by way of trade or otherwise, be lent, re-sold, hired out, or otherwise circulated without the publishers prior consent in any form of binding or cover other than that in which it is published and without a similar condition including this condition being imposed on the subsequent purchaser.

A catalogue record for this book is available from the British Library

Library of Congress Cataloguing in Publication Data
ISBN 0 19 855948 8 (Hbk)
ISBN 0 19 855947 X (Pbk)
(Data available)

Typeset by Tradespools Ltd,
Frome, Somerset

Printed in Great Britain by
Bath Press Ltd, Bath, Somerset

Preface

In the universe we inhabit, the pendulum swings with a period of about 13.5 years. The first edition of this book appeared in 1970; the second in 1983; now the third edition is published. Many changes have taken place through the editions, but we hope the essential spirit of the earlier editions has been preserved: there should be as much interpretation as straightforward mathematical presentation in this as in the earlier editions. Another strong feature of the earlier editions has been preserved and extended in this: the emphasis on visualization of often abstract material.

First, the major changes. The whole book has been rewritten. There are two new chapters and a reformulation of other chapters. One entirely new chapter is on the calculation of electronic structure (Chapter 9). No other field in quantum chemistry has made such vigorous strides forwards—into usefulness as well as complexity—as this. It would be hopeless to consider giving a detailed account of this highly technical field, but improper to leave out a discussion of what is now the field of major endeavour in the subject. We have sought to give a survey of the basis of the techniques that are now so widely available in commercial and academic packages so that people can understand the variety of *ab initio* and semiempirical approaches that are now available.

The second entirely new chapter is the one on scattering theory (Chapter 14). We consider that the major analytical advances in molecular quantum mechanics have occurred in this field, under the impetus of advances in experimental procedures (molecular beams and lasers) but also in a shift in attention from structure to the processes of reaction. Once again, here is a topic where a deluge of algebra can readily wash away comprehension. However, we have sought to concentrate on highlights, and to give visualizations of material wherever appropriate. We have drawn the line at a discussion of reactive collisions, for even the slight straying over into this terrain that ends this text shows what a minefield of notation and subtlety it is.

Another feature of this edition is that we have elected to adopt a more systematic approach to quantum mechanics. Now Chapter 1 proceeds axiomatically (but at a highly accessible level, we believe) rather than historically. However, to make this approach accessible, we have given a short historical introduction (in the unnumbered chapter entitled *Introduction and orientation*) so that readers will feel, we hope, that in stepping off into the unknown they are at least stepping off familiar territory.

There are many organizational changes in the text, ranging from the layout of chapters to the choice of words. This edition is in fact a complete rewrite of the second edition. In the rewriting, we have aimed for clarity and precision. Where the depth of the presentation started to seem too great in our judgment for our audience, we have sent material to the back of the book in an extensive and rearranged collection of *Further information* sections.

No text can be, and perhaps even should not aim to be, complete. We know that there are omissions from this text. However, we also believe that no other text is as complete as this. We have aimed to give a balanced account of the electronic structures of atoms and molecules (which other comparable texts also do), but we have also presented a lot of material on spectroscopy of many kinds (which some other texts do too). In addition though, we have presented a lot of material on the electric and magnetic properties of molecules, which is an important zone of quantum chemistry but which is rarely included at this level.

We owe a considerable debt to many people; not least to one another! Our publishers have been helpful and understanding, as always. Our particular thanks are due to a variety of reviewers of the manuscript at several stages of presentation. In particular, we should like to thank Ronald Duchovic (IPFW) and David Schwenke (NASA, Ames) who have been most helpful and generous with their advice. We would also like to thank the readers of the first draft of the text, who were Steven Bernasek (Princeton University), Patrick Fowler (University of Exeter), David Micha (University of Florida), Ian Mills (University of Reading), Robert Sharp (University of Michigan), Charles Trapp (University of Louisville), and Rama Viswanathan (Beloit College). They spotted many points that were best spotted in private, and made valuable suggestions about the organization and content of the book. Over the years, many users have offered advice: they are too numerous to name here, but they will all know how valuable we view their advice and in many cases will see it incorporated into this edition.

PWA, *Oxford*
RSF, *Indiana University Purdue University Fort Wayne*
May 1996

Contents

Introduction and orientation

There are two approaches to quantum mechanics. One is to follow the historical development of the theory from the first indications that the whole fabric of classical mechanics and electrodynamics should be held in doubt to the resolution of the problem in the work of Planck, Einstein, Heisenberg, Schrödinger, and Dirac. The other is to stand back at a point late in the development of the theory and to see its underlying theoretical structure. The first is interesting and compelling because the theory is seen gradually emerging from confusion and dilemma. We see experiment and intuition jointly determining the form of the theory and, above all, we come to appreciate the need for a new theory of matter. The second, more formal approach is exciting and compelling in a different sense: there is logic and elegance in a scheme that starts from only a few postulates, yet reveals as their implications are unfolded, a rich, experimentally verifiable structure.3

This book takes that latter route through the subject. However, to set the scene we shall take a few moments to review the steps that led to the revolutions of the early twentieth century, when some of the most fundamental concepts of the nature of matter and its behaviour were overthrown and replaced by a puzzling but powerful new description.

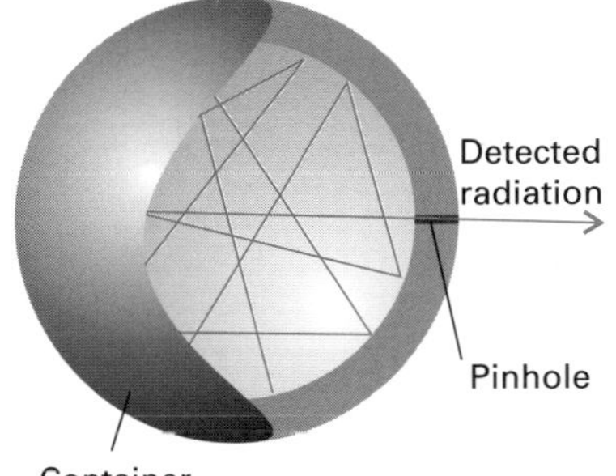

Fig. 0.1 A black-body emitter can be simulated by a heated container with a pinhole in the wall. The electromagnetic radiation is reflected many times inside the container and reaches thermal equilibrium with the walls.

Black-body radiation

In retrospect—and as will become clear—we can now see that theoretical physics hovered on the edge of formulating a quantum mechanical description of matter as it was developed during the nineteenth century. However, it was a series of experimental observations that motivated the revolution. Of these observations, the most important historically was the study of **black-body radiation**, the radiation in thermal equilibrium with a body that absorbs and emits without favouring particular frequencies. A pin-hole in an otherwise sealed container is a good approximation (Fig. 0.1).

Two characteristics of the radiation had been identified by the end of the century and summarized in two laws. According to the **Stefan–Boltzmann law**, the **excitance**, M, the power emitted divided by the area of the emitting region, is proportional to the fourth power of the temperature:

$$M = \sigma T^4 \tag{1}$$

The **Stefan–Boltzmann constant**, σ, is independent of the material from which the body is composed, and its modern value is $5.67 \times 10^{-8}\,\mathrm{W\,m^{-2}\,K^{-4}}$. So, a region of area $1\,\mathrm{cm}^2$ of a black body at 1000 K radiates about 6 W if all frequencies are taken into account.

Not all frequencies (or wavelengths, with $\lambda = c/\nu$), though, are equally represented in the radiation, and the observed peak moves to shorter wavelengths as the temperature is raised. According to **Wien's displacement law**,

$$\lambda_{\max} T = \text{constant} \tag{2}$$

with the constant equal to 2.9 mm K.

One of the most challenging problems in physics at the end of the nineteenth century was to explain these two laws. Lord Rayleigh, with minor help from James Jeans,[1] brought his formidable experience of classical physics to bear on the problem, and formulated the theoretical **Rayleigh–Jeans law** for the energy density ($\mathcal{E}$, the energy divided by the volume) in the wavelength range $\mathrm{d}\lambda$:

$$\mathrm{d}\mathcal{E} = \rho\, \mathrm{d}\lambda \qquad \rho = \frac{8\pi kT}{\lambda^4} \tag{3}$$

where k is Boltzmann's constant ($k = 1.381 \times 10^{-23}\ \mathrm{J\,K^{-1}}$). This formula summarizes the failure of classical physics. It suggests that regardless of the temperature, there should be an infinite energy density at very short wavelengths. This absurd result was termed the **ultraviolet catastrophe** by Ehrenfest.

At this point, Planck made his historic contribution. His suggestion was equivalent to proposing that an oscillation of the electromagnetic field of frequency ν could be excited only in steps of energy of magnitude $h\nu$, where h is a new fundamental constant of nature now known as the **Planck constant**. According to this **quantization** of energy, the oscillator can have the energies $0, h\nu, 2h\nu, \ldots$ and no other energy. Classical physics allowed a *continuous* variation in energy, so even a very high frequency oscillator could be excited with a very small energy: that was the root of the ultraviolet catastrophe. Quantum theory is characterized by discreteness in energies (and, as we shall see, of other properties), and the need for a minimum excitation energy effectively switches off oscillators of very high frequency, and hence eliminates the ultraviolet catastrophe.

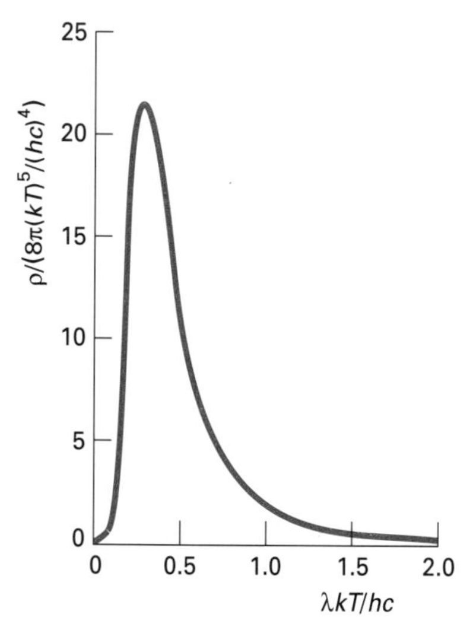

Fig. 0.2 The Planck distribution.

When Planck implemented his suggestion, he derived the following **Planck distribution** for the energy density of a black-body radiator:

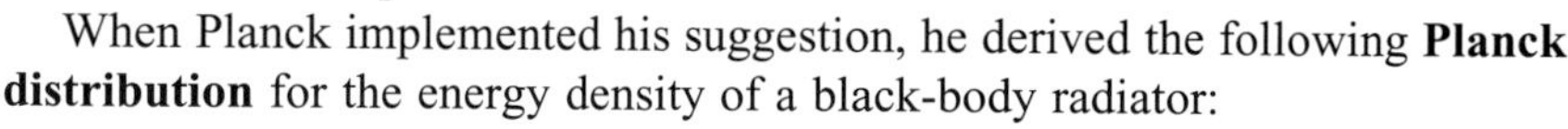

$$\mathrm{d}\mathcal{E} = \rho\, \mathrm{d}\lambda \qquad \rho = \left(\frac{8\pi hc}{\lambda^5}\right) \frac{\mathrm{e}^{-hc/\lambda kT}}{1 - \mathrm{e}^{-hc/\lambda kT}} \tag{4}$$

This expression, which is plotted in Fig. 0.2, avoids the ultraviolet catastrophe, and fits the observed energy distribution extraordinarily well if we take $h = 6.626 \times 10^{-34}\ \mathrm{J\,s}$. Just as the Rayleigh–Jeans law epitomizes the failure of classical physics, the Planck distribution epitomizes the inception of quantum theory. It began the new century as well as a new era, for it was published in 1900.

Heat capacities

In 1819, science had a deceptive simplicity. Dulong and Petit, for example, were able to propose their law that 'the atoms of all simple bodies have

[1] 'It seems to me,' said Jeans, 'that Lord Rayleigh has introduced an unnecessary factor 8 by counting negative as well as positive values of his integers.'

exactly the same heat capacity'. In modern terms, we would phrase the law in terms of the molar isochoric (constant volume) heat capacity, $C_{V,\mathrm{m}}$, and write $C_{V,\mathrm{m}} \approx 3R$ for a solid element, where R is the gas constant ($R = N_\mathrm{A}k$, with N_A the Avogadro constant). Dulong and Petit's rather primitive observations, though, were done at room temperature, and it was unfortunate for them and for classical physics when measurements were extended to lower temperatures. It was found that all elements had heat capacities lower than predicted by Dulong and Petit's law, and the values tended toward zero as $T \to 0$.

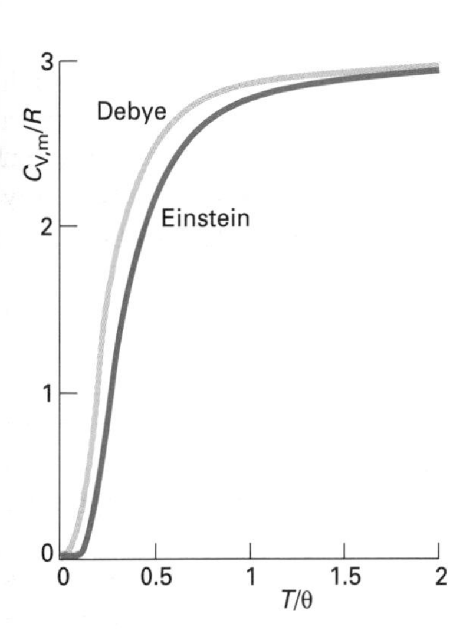

Fig. 0.3 The Einstein and Debye molar heat capacities. The symbol θ denotes the Einstein and Debye temperatures, respectively. Close to $T = 0$ the Debye heat capacity is proportional to T^3.

Dulong and Petit's law was easy to explain in terms of classical physics. All it was necessary to do was to suppose that each atom acted as an oscillator in three dimensions, and then to use classical physics to calculate the corresponding heat capacity. That the heat capacities were smaller than predicted was a serious embarrassment. Einstein recognized the similarity between this problem and black-body radiation, for if each atomic oscillator required a certain minimum energy before it would actively oscillate and hence contribute to the heat capacity, then at low temperatures some would be inactive and the heat capacity would be smaller than expected. He applied Planck's suggestion for electromagnetic oscillators to the material, atomic oscillators of the solid, and deduced the following expression:

$$C_{V,\mathrm{m}} = 3Rf^2 \qquad f = \frac{(\theta_\mathrm{E}/T)\mathrm{e}^{\theta_\mathrm{E}/2T}}{1 - \mathrm{e}^{\theta_\mathrm{E}/T}} \tag{5a}$$

where the **Einstein temperature**, θ_E, is related to the frequency of atomic oscillators by $\theta_\mathrm{E} = h\nu/k$. This function is plotted in Fig. 0.3, and closely reproduces the experimental curve. In fact, the fit is not particularly good at very low temperatures, but that can be traced to Einstein's assumption that all the atoms oscillated with the same frequency. When this restriction was removed by Debye, he obtained

$$C_{V,\mathrm{m}} = 3Rf \qquad f = 3\left(\frac{T}{\theta_\mathrm{D}}\right)^3 \int_0^{\theta_\mathrm{D}/T} \frac{x^4\mathrm{e}^x}{(\mathrm{e}^x - 1)^2}\,\mathrm{d}x \tag{5b}$$

where the **Debye temperature**, θ_D, is related to the maximum frequency of the oscillations that can be supported by the solid. This expression gives a very good fit with observation.

The importance of Einstein's contribution is that it complemented Planck's. Planck had shown that the energy of radiation is quantized; Einstein showed that matter is quantized too. Quantization appears to be universal. Neither was able to justify the form that quantization took (with oscillators excitable in steps of $h\nu$), but that is a problem we shall solve later in the text.

The photoelectric and Compton effects

In those enormously productive months of 1905–6, when Einstein formulated not only his theory of heat capacities but also the special theory of relativity, he found time to make another fundamental contribution to modern physics. His achievement was to relate Planck's quantum hypothesis to the phenomenon of the **photoelectric effect**, the emission of electrons from metals when they are exposed to ultraviolet radiation. The puzzling features of the effect were that the emission was instantaneous when the radiation was applied, however low its intensity, but there was no emission, whatever the intensity of the radiation, unless its frequency exceeded a threshold value typical of each element. It was also known that the kinetic energy of the ejected electrons varied linearly with the frequency of the incident radiation. Einstein pointed out that all the observations fell into place if the electromagnetic field was quantized, and that it consisted of bundles of energy of magnitude $h\nu$.

These bundles were later named **photons** by G.N. Lewis, and we shall use that term from now on. Einstein viewed the photoelectric effect as the outcome of a collision between an incoming projectile, a photon of energy $h\nu$, and an electron buried in the metal. This picture accounts for the instantaneous character of the effect, because even one photon can participate in one collision. It also accounted for the frequency threshold, because a minimum energy (which is normally denoted Φ and called the 'work function' for the metal) must be supplied in a collision before photoejection can occur; hence, only radiation for which $h\nu > \Phi$ can be successful. The linear dependence of the kinetic energy, T, of the photoelectron on the frequency of the radiation is a simple consequence of the conservation of energy, which implies that

$$T = h\nu - \Phi \tag{6}$$

If photons do have a particle-like character, then they should possess a linear momentum, p. The relativistic expression relating a particle's energy to its mass and momentum is

$$E = (m^2c^4 + p^2c^2)^{1/2} \tag{7}$$

where c is the speed of light. In the case of a photon, $E = h\nu$ and $m = 0$, so

$$p = \frac{h\nu}{c} = \frac{h}{\lambda} \tag{8}$$

This linear momentum should be detectable if radiation falls on an electron, for a partial transfer of momentum during the collision should appear as a change in wavelength of the photons. In 1923, A.H. Compton performed the experiment with X-rays scattered from the electrons in a graphite target, and found the results fitted the following formula for the shift in wavelength, $\delta\lambda = \lambda_f - \lambda_i$, when the radiation was scattered through an angle θ:

$$\delta\lambda = 2\lambda_C \sin^2 \tfrac{1}{2}\theta \qquad \lambda_C = \frac{h}{m_e c} \tag{9}$$

where λ_C is called the **Compton wavelength** of the electron. This formula is derived on the supposition that a photon does indeed have a linear momentum h/λ and that the scattering event is like a collision between two particles. There seems little doubt, therefore, that electromagnetic radiation has properties that classically would have been characteristic of particles.

The photon hypothesis seems to be a denial of the extensive accumulation of data that apparently provided unequivocal support for the view that electromagnetic radiation is wavelike. By following the implications of experiments and quantum concepts, we have accounted quantitatively for observations for which classical physics could not supply even a qualitative explanation.

Atomic spectra

There was yet another body of data that classical physics could not elucidate before the introduction of quantum theory. This puzzle was the observation that the radiation emitted by atoms was not continuous but consisted of discrete frequencies, or **spectral lines**. The spectrum of atomic hydrogen had a very simple appearance, and by 1885 J. Balmer had already noticed that their wavenumbers, $\tilde{\nu}$, where $\tilde{\nu} = \nu/c$, fitted the expression

$$\tilde{\nu} = \mathcal{R}_{\mathrm{H}}\left(\frac{1}{2^2} - \frac{1}{n^2}\right) \tag{10}$$

where $\mathcal{R}_{\mathrm{H}}$ has come to be known as the **Rydberg constant** for hydrogen ($\mathcal{R}_{\mathrm{H}} = 1.097 \times 10^5\ \mathrm{cm}^{-1}$) and $n = 3, 4, \ldots$. Rydberg's name is commemorated because he generalized this expression in his **combination principle**, which states that the frequency of any spectral line could be expressed as the difference between two quantities, or **terms**:

$$\tilde{\nu} = T_1 - T_2 \tag{11}$$

This expression strongly suggests that the energy levels of atoms are confined to discrete values, because a transition from one term of energy hcT_1 to another of energy hcT_2 can be expected to release a photon of energy $hc\tilde{\nu}$, or $h\nu$, equal to the difference in energy between the two terms: this argument leads directly to the expression for the wavenumber of the spectroscopic transitions.

But why should the energy of an atom be confined to discrete values? In classical physics, all energies are permissible. The first attempt to weld together Planck's quantization hypothesis and a mechanical model of an atom was made by Niels Bohr in 1913. By arbitrarily assuming that the angular momentum of an electron around a central nucleus (the picture of an atom that had emerged from Rutherford's experiments in 1910) was confined to certain values, he was able to deduce the following expression for the permitted energy levels of an electron in a hydrogen atom:

$$E_n = -\frac{\mu e^4}{8h^2\varepsilon_0^2 n^2} \qquad n = 1, 2, \ldots \tag{12}$$

where $1/\mu = 1/m_{\mathrm{e}} + 1/m_{\mathrm{p}}$ and ε_0 is the vacuum permittivity, a fundamental constant. This formula marks the first appearance in quantum mechanics of a **quantum number**, n, which identifies the state of the system and is used to calculate its energy. Equation 12 is consistent with Balmer's formula and accounted with high precision for all the transitions of hydrogen that were then known.

Bohr's achievement was the union of theories of radiation and models of mechanics. However, it was an arbitrary union, and we now know that it is conceptually untenable (for instance, it is based on the view that an electron travels in a circular path around the nucleus). Nevertheless, the fact that he was able to account quantitatively for the appearance of the spectrum of hydrogen indicated that quantum mechanics was central to any description of atomic phenomena and properties.

The duality of matter

The grand synthesis of these ideas and the demonstration of the deep links that exist between electromagnetic radiation and matter began with Louis de Broglie, who proposed on the basis of relativistic considerations that with any moving body there is 'associated a wave', and that the momentum of the body and the wavelength are related by the **de Broglie relation**:

$$\lambda = \frac{h}{p} \tag{13}$$

We have seen this formula already, in connection with the properties of photons. De Broglie proposed that it is *universally* applicable.

The significance of the de Broglie relation is that it summarizes a fusion of opposites: the momentum is a property of particles; the wavelength is a property of waves. This **duality**, the possession of properties which in classical physics are characteristic of both particles and waves, is a persistent theme in the interpretation of quantum mechanics. It is probably best to regard the terms 'wave' and 'particle' as remnants of a language based on a false (classical) model of the universe, and the term 'duality' as a late attempt to bring the language into line with a current (quantum mechanical) model.

The experimental results that confirmed de Broglie's conjecture are the observation of the diffraction of electrons by the ranks of atoms in a metal crystal acting as a diffraction grating. Davisson and Germer, who performed this experiment in 1925 using a crystal of nickel, found that the diffraction pattern was consistent with the electrons having a wavelength given by the de Broglie relation. Shortly afterwards, G.P. Thomson also succeeded in demonstrating the diffraction of electrons by thin films of celluloid and gold.[2]

If electrons—if *all* particles—have wavelike character, then we should expect there to be observational consequences. In particular, just as a wave of definite wavelength cannot be localized at a point, we should not expect an electron in a state of definite linear momentum (and hence wavelength) to be localized at a single point. It was pursuit of this idea that led Werner Heisenberg to his celebrated uncertainty principle, that it is impossible to specify the location and linear momentum of a particle simultaneously with arbitrary precision. In other words, information about location is at the expense of information about momentum, and vice versa. This **complementarity** of observables, the mutual exclusion of the specification of one property by the specification of another, is also a major theme of quantum mechanics, and almost an icon of the difference between it and classical mechanics, in which the specification of exact trajectories was a central theme.

The consummation of all this faltering progress came in 1926 when Werner Heisenberg and Erwin Schrödinger formulated their seemingly different but equally successful versions of quantum mechanics. These days, we step between the two formalisms as the fancy takes us, for they are mathematically equivalent, and each one has particular advantages in different types of calculation. Although Heisenberg's formulation preceded Schrödinger's by a few

[2] It has been pointed out by M. Jammer that J.J. Thomson was awarded the Nobel Prize for showing that the electron is a particle, and G.P. Thomson, his son, was awarded the Prize for showing that the electron is a wave.

months, it seemed more abstract and was expressed in the then unfamiliar vocabulary of matrices. Still today it is more suited for the more formal manipulations and deductions of the theory, and in the following pages we shall employ it in that manner. Schrödinger's formulation, which was in terms of functions and differential equations, was more familiar in style, but still equally revolutionary in implication. It is more suited to elementary manipulations and to the calculation of numerical results, and we shall employ it in that manner.

'Experiments,' said Planck, 'are the only means of knowledge at our disposal. The rest is poetry, imagination.' It is time for that imagination to unfold.

Problems

0.1 Calculate the size of the quanta involved in the excitation of (a) an electronic motion of period 1.0×10^{-15} s, (b) a molecular vibration of period 1.0×10^{-14} s, and (c) a pendulum of period 1.0 s.

0.2 Find the wavelength corresponding to the maximum in the Planck distribution for a given temperature, and show that the expression reduces to the Wien displacement law at short wavelengths. Determine an expression for the constant in the law in terms of fundamental constants. (This constant is called the second radiation constant.)

0.3 Use the Planck distribution to confirm the Stefan–Boltzmann law and to derive an expression for the Stefan–Boltzmann constant σ.

0.4 The peak in the Sun's emitted energy occurs at about 480 nm. Estimate the temperature of its surface on the basis of it being regarded as a black-body emitter.

0.5 Derive the Einstein formula for the heat capacity of a collection of harmonic oscillators. To do so, use the quantum mechanical result that the energy of a harmonic oscillator of force constant k and mass m is one of the values $(v + \frac{1}{2})h\nu$, with $\nu = (1/2\pi)\sqrt{k/m}$ and $v = 0, 1, 2 \ldots$. *Hint*. Calculate the mean energy, E, of a collection of oscillators by substituting these energies into the Boltzmann distribution, and then evaluate $C = \mathrm{d}E/\mathrm{d}T$.

0.6 Find the (a) low temperature, (b) high temperature forms of the Einstein heat capacity function.

0.7 Show that the Debye expression is proportional to T^3 as $T \to 0$.

0.8 Estimate the molar heat capacities of metallic sodium ($\theta_D = 150\,\mathrm{K}$) and diamond ($\theta_D = 1860\,\mathrm{K}$) at room temperature (300 K).

0.9 Calculate the molar entropy of an Einstein solid at $T = \theta_E$. *Hint*. The entropy is $S = \int_0^T (C_V/T)\,\mathrm{d}T$. Evaluate the integral numerically.

0.10 How many photons would be emitted per second by a sodium lamp rated at 100 W which radiated all its energy with 100 per cent efficiency as yellow light of wavelength 589 nm?

0.11 Calculate the speed of an electron emitted from a clean potassium surface ($\Phi = 2.3\,\mathrm{eV}$) by light of wavelength (a) 300 nm, (b) 600 nm.

0.12 At what wavelength of incident radiation do the relativistic and non-relativistic expressions for the ejection of electrons from potassium differ by 10 per cent. Use $\Phi = 2.3\,\mathrm{eV}$.

0.13 Deduce eqn 9 for the Compton effect on the basis of the conservation of energy and linear momentum. *Hint*. Use the relativistic expressions. Initially the electron is at rest with energy $m_e c^2$. When it is travelling with momentum p its energy is $(p^2c^2 + m_e^2c^4)^{1/2}$. The photon, with initial momentum h/λ_i and energy $h\nu_i$, strikes the stationary electron, is deflected through an angle θ, and emerges with momentum h/λ_f

and energy $h\nu_f$. The electron is initially stationary ($p = 0$) but moves off with an angle θ' to the incident photon. Conserve energy and both components of linear momentum. Eliminate θ', then p, and so arrive at an expression for $\delta\lambda$.

0.14 The first few lines of the visible (Balmer) series in the spectrum of atomic hydrogen lie at $\lambda/\text{nm} = 656.46, 486.27, 434.17, 410.29, \ldots$. Find a value of $\mathcal{R}_H$, the Rydberg constant for hydrogen. The ionization energy, I, is the minimum energy required to remove the electron. Find it from the data and express its value in electron volts. How is I related to $\mathcal{R}_H$? *Hint*. The ionization limit corresponds to $n \to \infty$ for the final state of the electron.

0.15 Calculate the de Broglie wavelength of (a) a mass of 1.0 g travelling at $1.0\,\text{cm}\,\text{s}^{-1}$, (b) the same at 95 per cent of the speed of light, (c) a hydrogen atom at room temperature (300 K); estimate the mean speed from the equipartition principle (which implies that the mean kinetic energy of an atom is equal to $\frac{3}{2}kT$, where k is the Boltzmann constant), (d) an electron accelerated from rest through a potential difference of (i) 1.0 V, (ii) 10 kV. *Hint*. For the momentum in (b) use $p = m_e v/(1 - v^2/c^2)^{1/2}$ and for the speed in (d) use $\frac{1}{2}m_e v^2 = e\mathcal{V}$, where $\mathcal{V}$ is the potential difference.

Further reading

The conceptual development of quantum mechanics. M. Jammer; McGraw-Hill, New York (1966).

Black-body theory and the quantum discontinuity, 1894–1912. T.S. Kuhn; Oxford University Press, New York (1978).

The history of quantum theory. F. Hund; Harrap, London (1974).

The historical development of quantum theory. Vols 1–5. J. Mehra and H. Rechenberg (ed.); Springer, New York (1982 *et seq*).

Physical chemistry. P.W. Atkins; Oxford University Press, Oxford and W.H. Freeman and Co., New York (1994).

Quanta: a handbook of concepts. P.W. Atkins; Oxford University Press, Oxford (1991).

Modern atomic physics. B. Cagnac and J.C. Pebay-Peyroula, Macmillan, London and Wiley, New York (1975).

1 The foundations of quantum mechanics

The whole of quantum mechanics can be expressed in terms of a small set of postulates. When their consequences are developed, they embrace the behaviour of all known forms of matter, including the molecules, atoms, and electrons that will be at the centre of our attention in this book. This chapter introduces the postulates, and illustrates how they are used. The remaining chapters build on them, and show how to apply them to problems of chemical interest, such as atomic and molecular structure and the properties of molecules. We assume that you have already met the concepts of 'hamiltonian' and 'wavefunction' in an elementary introduction, and have seen the Schrödinger equation written in the form

$$H\psi = E\psi$$

This chapter establishes the full significance of this equation, and provides a foundation for its application in the following chapters.

Operators in quantum mechanics

An **observable** is any dynamical variable that can be measured. The principal difference between classical mechanics and quantum mechanics is that whereas in the former physical observables are represented by functions (such as position as a function of time), in quantum mechanics they are represented by mathematical operators. An **operator** is a symbol for an instruction to carry out some action, an operation, on a function. In most of the examples we shall meet, the action will be nothing more complicated than multiplication or differentiation. Thus, one typical operation might be multiplication by x, which is represented by the operator $x\times$. Another operation might be differentiation with respect to x, represented by the operator $\mathrm{d}/\mathrm{d}x$. We shall represent operators by the symbol Ω in general, but use A, $B, \ldots$ when we want to refer to a series of operators. We shall not in general distinguish between the observable and the operator that represents that observable; so the position of a particle along the x-axis will be denoted x and the corresponding operator will also be denoted x (with multiplication implied). We shall always make it clear whether we are referring to the observable or the operator.

We shall need a number of concepts related to operators and functions on which they operate, and this first section introduces some of the more important features.

1.1 Eigenfunctions and eigenvalues

In general, when an operator operates on a function, the outcome is another function. Differentiation of $\sin x$, for instance, gives $\cos x$. However, in certain cases, the outcome of an operation is the same function, multiplied by a constant. Functions of this kind are called 'eigenfunctions' of the operator. More formally, a function f (which may be complex) is an **eigenfunction** of an operator Ω if it satisfies an equation of the form

$$\Omega f = \omega f \tag{1}$$

where ω is a constant. Such an equation is called an **eigenvalue equation**. The function e^{ax} is an eigenfunction of the operator d/dx because $(d/dx)e^{ax} = ae^{ax}$, which is a constant (a) multiplying the original function. In contrast, e^{ax^2} is not an eigenfunction of d/dx, because $(d/dx)e^{ax^2} = 2axe^{ax^2}$ which is a constant ($2a$) times a *different* function of x (the function xe^{ax^2}). The constant ω in an eigenvalue equation is called the **eigenvalue** of the operator Ω.

An important point is that a general function can be expanded in terms of all the eigenfunctions of an operator, a so-called **complete set** of functions. That is, if f_n is an eigenfunction of an operator Ω with eigenvalue ω_n (so $\Omega f_n = \omega_n f_n$), then[1] a general function g can be expressed as the **linear combination**

$$g = \sum_n c_n f_n \tag{2}$$

where the c_n are coefficients and the sum is over a complete set of functions. For instance, the straight line $g = ax$ can be recreated over a certain range by superimposing an infinite number of sine functions, each of which is an eigenfunction of the operator d^2/dx^2. Alternatively, the same function may be constructed from an infinite number of exponential functions, which are eigenfunctions of d/dx. The advantage of expressing a general function as a linear combination of a set of eigenfunctions is that it allows us to deduce the effect of an operator on a function that is not one of its own eigenfunctions. Thus, the effect of Ω on g in eqn 1.2 is simply

$$\Omega g = \Omega \sum_n c_n f_n = \sum_n c_n \Omega f_n = \sum_n c_n \omega_n f_n$$

A special case of these linear combinations is when we have a set of **degenerate** eigenfunctions, which means a set of functions with the same eigenvalue. Thus, suppose that $f_1, f_2, \ldots, f_k$ are all eigenfunctions of the operator Ω, and that they all correspond to the same eigenvalue ω:

$$\Omega f_n = \omega f_n \qquad \text{with } n = 1, 2, \ldots, k \tag{3}$$

Then it is quite easy to show that *any* linear combination of the functions f_n is also an eigenfunction of Ω with the same eigenvalue ω. The proof is as follows. For an arbitrary linear combination g of the degenerate set of functions, we can write

$$\Omega g = \Omega \sum_{n=1}^{k} c_n f_n = \sum_{n=1}^{k} c_n \Omega f_n = \sum_{n=1}^{k} c_n \omega f_n = \omega \left(\sum_{n=1}^{k} c_n f_n \right) = \omega g$$

This expression has the form of an eigenvalue equation ($\Omega g = \omega g$).

A further technical point is that from n basis functions it is possible to construct n linearly independent combinations. A set of functions $g_1, g_2, \ldots, g_n$ is

[1] See P.M. Morse and H. Feschbach, *Methods of theoretical physics*, McGraw-Hill, New York (1953).

said to be **linearly independent** if we cannot find a set of constants $c_1, c_2, \ldots, c_n$ for which

$$\sum_i c_i g_i = 0$$

A set of functions that is not linearly independent is said to be **linearly dependent**. From a set of n linearly independent functions, it is possible to construct an infinite number of sets of linearly independent combinations, but each set can have no more than n members. For example, from three $2p$ orbitals of an atom it is possible to form any number of sets of linearly independent combinations, but each set has no more than three members.

1.2 Representations

The remaining work of this section is to put forward some explicit forms of the operators we shall meet. Much of quantum mechanics can be developed in terms of an abstract set of operators, as we shall see later. However, it is often fruitful to adopt an explicit form for particular operators and to express them in terms of the mathematical operations of multiplication, differentiation, and so on. Different choices of the operators that correspond to a particular observable give rise to the different **representations** of quantum mechanics, because the explicit forms of the operators represent the abstract structure of the theory in terms of actual manipulations.

One of the most common representations is the **position representation**, in which the position operator is represented by multiplication by x (or whatever coordinate is specified) and the linear momentum parallel to x is represented by differentiation with respect to x. Explicitly:

$$\text{position representation:} \qquad x \to x\times \qquad p_x \to \frac{\hbar}{\mathrm{i}}\frac{\partial}{\partial x} \qquad (4)$$

where $\hbar = h/2\pi$. Why the linear momentum should be represented in precisely this manner will be explained in the following section. For the time being, it may be taken to be a basic postulate of quantum mechanics. An alternative choice of operators is the **momentum representation**, in which the linear momentum parallel to x is represented by the operation of multiplication by p_x and the position operator is represented by differentiation with respect to p_x. Explicitly:

$$\text{momentum representation:} \qquad x \to -\frac{\hbar}{\mathrm{i}}\frac{\partial}{\partial p_x} \qquad p_x \to p_x\times \qquad (5)$$

There are other representations. We shall normally use the position representation when the adoption of a representation is appropriate, but we shall also see that many of the calculations in quantum mechanics can be done independently of a representation.

1.3 Commutation and non-commutation

An important feature of operators is that in general the outcome of successive operations (A followed by B, which is denoted BA, or B followed by A, denoted AB) depends on the order in which the operations are carried out. That is, in general $BA \neq AB$. We say that, in general, operators do not **commute**. The quantity $AB - BA$ is called the **commutator** of A and B and is denoted $[A, B]$:

$$[A, B] = AB - BA \tag{6}$$

It is instructive to evaluate the commutator of the position and linear momentum operators in the two representations shown above; the procedure is illustrated in the following example.

Example 1.1 The evaluation of a commutator

Evaluate the commutator $[x, p_x]$ in the position representation.

Method. To evaluate the commutator $[A, B]$ we need to remember that the operators operate on some function, which we shall write f. So, evaluate $[B, A]f$ for an arbitrary function f, and then cancel f at the end of the calculation.

Answer. Substitution of the explicit expressions for the operators into $[x, p_x]$ proceeds as follows:

$$\begin{aligned}[x, p_x]f = (xp_x - p_x x)f &= x \times \frac{\hbar}{\mathrm{i}}\frac{\partial f}{\partial x} - \frac{\hbar}{\mathrm{i}}\frac{\partial}{\partial x} \times xf \\ &= x \times \frac{\hbar}{\mathrm{i}}\frac{\partial f}{\partial x} - \frac{\hbar f}{\mathrm{i}} - x \times \frac{\hbar}{\mathrm{i}}\frac{\partial f}{\partial x} \\ &= \mathrm{i}\hbar \times f\end{aligned}$$

This derivation is true for any function f, so in terms of the operators themselves,

$$[x, p_x] = \mathrm{i}\hbar$$

The right-hand side should be interpreted as the operator 'multiply by the constant $\mathrm{i}\hbar$'.

Exercise 1.1. Evaluate the same commutator in the momentum representation. [Same]

1.4 The construction of operators

Operators for other observables of interest can be constructed from the operators for position and momentum. For example, the kinetic energy operator T can be constructed by noting that kinetic energy is related to linear momentum by $T = p^2/2m$ where m is the mass of the particle. It follows that in one dimension and in the position representation

$$T = \frac{p_x^2}{2m} = \frac{1}{2m}\left(\frac{\hbar}{\mathrm{i}}\frac{\mathrm{d}}{\mathrm{d}x}\right)^2 = -\frac{\hbar^2}{2m}\frac{\mathrm{d}^2}{\mathrm{d}x^2} \tag{7}$$

In three dimensions the operator in the position representation is

$$\begin{aligned} T &= \frac{p_x^2 + p_y^2 + p_z^2}{2m} = \frac{1}{2m}\left\{\left(\frac{\hbar}{\mathrm{i}}\frac{\partial}{\partial x}\right)^2 + \left(\frac{\hbar}{\mathrm{i}}\frac{\partial}{\partial y}\right)^2 + \left(\frac{\hbar}{\mathrm{i}}\frac{\partial}{\partial z}\right)^2\right\} \\ &= -\frac{\hbar^2}{2m}\left\{\frac{\partial^2}{\partial x^2} + \frac{\partial^2}{\partial y^2} + \frac{\partial^2}{\partial z^2}\right\} \\ &= -\frac{\hbar^2}{2m}\nabla^2 \end{aligned} \tag{8}$$

The operator ∇^2, which is read 'del squared' and called the **laplacian**, is the sum of the three second derivatives.

The potential energy of a particle in one dimension, $V(x)$, becomes multiplication by the function $V(x)$ in the position representation. The same is true of the potential energy operator in three dimensions. For example, in the position representation the operator for the Coulomb potential energy of an electron in the field of a nucleus of atomic number Z is the multiplicative operator

$$V = -\frac{Ze^2}{4\pi\varepsilon_0 r}\times \tag{9}$$

where r is the distance from the nucleus to the electron. It is usual to omit the multiplication sign from multiplicative operators, but it should not be forgotten that such expressions are multiplications.

The operator for the total energy of a system is called the **hamiltonian operator** and is denoted H:

$$H = T + V \tag{10}$$

The name commemorates W.R. Hamilton's contribution to the formulation of classical mechanics. To write the explicit form of this operator we simply substitute the appropriate expressions for the kinetic and potential energy operators in the chosen representation. For example, the hamiltonian for a particle of mass m able to move in one dimension is

$$H = -\frac{\hbar^2}{2m}\frac{\mathrm{d}^2}{\mathrm{d}x^2} + V(x) \tag{11}$$

where $V(x)$ is the operator for the potential energy. Similarly, the hamiltonian operator for an electron of mass m_e in a hydrogen atom is

$$H = -\frac{\hbar^2}{2m_e}\nabla^2 - \frac{e^2}{4\pi\varepsilon_0 r} \tag{12}$$

The general prescription for constructing operators in the position representation should be clear from these examples. In short:

1. Write the classical expression for the observable in terms of position coordinates and the linear momentum.
2. Replace x by multiplication by x, and replace p_x by $(\hbar/\mathrm{i})\partial/\partial x$ (and likewise for the other coordinates).

1.5 Linear operators

The operators we meet in quantum mechanics are all linear. A **linear operator** is one for which the following statement is true:

$$\Omega(af + bg) = a\Omega f + b\Omega g$$

where a and b are constants. Multiplication is a linear operation; so is differentiation and integration. An example of a nonlinear operation is that of taking the logarithm of a function, because it is not true, for example, that $\log 2x = 2 \log x$.

1.6 Integrals over operators

When we want to make contact between a calculation done using operators and the actual outcome of an experiment, we need to evaluate certain integrals. These integrals all have the form

$$I = \int f^* \Omega g \, \mathrm{d}\tau \tag{13}$$

where f^* is the complex conjugate of f. In this integral $\mathrm{d}\tau$ is the **volume element**. In one dimension, $\mathrm{d}\tau$ can be identified as $\mathrm{d}x$; in three dimensions it is $\mathrm{d}x\mathrm{d}y\mathrm{d}z$. The integral is taken over the entire space available to the system, which is typically from $x = -\infty$ to $x = +\infty$ (and similarly for the other coordinates). A glance at the later pages of this book will show that many molecular properties are expressed as combinations of integrals of this form (often in a notation which will be explained later). Certain special cases of this type of integral have special names, and we shall introduce them here.

When the operator Ω in eqn 1.13 is simply multiplication by 1, the integral is called an **overlap integral** and commonly denoted S:

$$S = \int f^* g \, \mathrm{d}\tau \tag{14}$$

It is helpful to regard S as a measure of the similarity of two functions: when $S = 0$, the functions are classified as **orthogonal**, rather like two perpendicular vectors. When S is close to 1, the two functions are almost identical. The recognition of mutually orthogonal functions often helps to reduce the amount of calculation considerably, and rules will emerge in later sections and chapters.

The **normalization integral** is the special case of eqn 1.14 for $f = g$. A function f is said to be **normalized** (strictly, normalized to 1) if

$$\int f^* f \, \mathrm{d}\tau = 1 \tag{15}$$

It is almost always easy to ensure that a function is normalized by multiplying it by an appropriate numerical factor, which is called a **normalization factor** and typically denoted N and taken to be real. The procedure is illustrated in the following example.

Example 1.2 How to normalize a function

A certain function is $\sin(\pi x/L)$ between $x = 0$ and $x = L$ and is zero elsewhere. Find the normalized form of the function.

Method. We need to find the factor N such that $N \sin(\pi x/L)$ is normalized to 1. To find N we substitute this expression into eqn 1.15, evaluate the integral, and select N to ensure normalization. Note that 'all space' extends from $x = 0$ to $x = L$.

Answer. The necessary integration is

$$\int f^* f \, d\tau = \int_0^L N^2 \sin^2(\pi x/L) \, dx = \tfrac{1}{2} L N^2$$

For this integral to be equal to 1, we require $N = (2/L)^{1/2}$. The normalized function is therefore $f = (2/L)^{1/2} \sin(\pi x/L)$.

Comment. We shall see later that this function describes the distribution of a particle in a square well, and we shall need its normalized form there.

Exercise 1.2. Normalize the function $f = e^{i\phi}$, where ϕ ranges from 0 to 2π. [$N = 1/(2\pi)^{1/2}$]

A set of functions f_n that are (a) normalized and (b) mutually orthogonal are said to satisfy the **orthonormality condition**:

$$\int f_n^* f_m \, d\tau = \delta_{nm} \tag{16}$$

In this expression, δ_{nm} denotes the **Kronecker delta**, which is 1 when $m = n$ and 0 otherwise.

The postulates of quantum mechanics

Now we turn to an application of the preceding material, and move into the foundations of quantum mechanics.

Postulates in science may be of two kinds. There is the obvious kind, the kind based on direct observation, in which the content is self-evident. An example is the Second Law of thermodynamics in the form 'heat flows spontaneously from a hot to a cold body'. Then there is the subtle kind, from which the content has to be unfolded by a chain of argument. An example is the Second Law in the form 'the entropy of an isolated system increases during a spontaneous change'. The postulates we use as a basis for quantum mechanics are of the second, subtle, sort. They are by no means the most subtle that have been devised, but they are strong enough for what we have to do.

1.7 States and wavefunctions

The first postulate concerns the information we can know about a state:

Postulate 1. The state of a system is fully described by a function $\Psi(\mathbf{r}_1, \mathbf{r}_2, \ldots, t)$.

In this statement, $\mathbf{r}_1, \mathbf{r}_2, \ldots$ are the spatial coordinates of particles $1, 2, \ldots$ that constitute the system and t is the time. The function Ψ plays a central role in quantum mechanics, and is called the **wavefunction** of the system. When we are not interested in how the system changes in time we shall denote the wavefunction $\psi(\mathbf{r}_1, \mathbf{r}_2, \ldots)$. The state of the system may also depend on some internal variable of the particles (their spin states); we ignore that for now and return to it later. By 'describe' we mean that the wavefunction contains information about all the properties of the system that are open to experimental determination.

The wavefunction of a system will turn out to be specified by a set of labels called **quantum numbers**, and may then be written $\psi_{a,b,\ldots}$, where $a, b, \ldots$ are the quantum numbers. The values of these quantum numbers specify the wavefunction and thus allow the values of various physical observables to be calculated. It is often convenient to refer to the **state** of the system without referring to the corresponding wavefunction; the state is specified by listing the values of the quantum numbers that define it.

1.8 The fundamental prescription

The next postulate concerns the selection of operators:

Postulate 2. Observables are represented by operators chosen to satisfy the commutation relations

$$[q, p_{q'}] = i\hbar\delta_{qq'} \qquad [q, q'] = 0 \qquad [p_q, p_{q'}] = 0$$

where q and q' each denote one of the coordinates x, y, z and p_q and $p_{q'}$ the corresponding linear momenta.

This commutation relation is a basic, unprovable, and underivable postulate. It is the basis of the selection of the form of the operators in the position and momentum representations for all observables that depend on the position and the momentum.[2] Thus, if we define the position representation as the representation in which the position operator is multiplication by the position coordinate, then as we saw in Example 1.1, it follows that the momentum operator must be derivation with respect to x, as specified earlier. Similarly, if the momentum representation is defined as the representation in which the linear momentum is represented by multiplication, then the form of the position operator is fixed as a derivative with respect to the linear momentum.

1.9 The outcome of measurements

The next postulate brings together the wavefunction and the operators and establishes the link between formal calculations and experimental observations:

Postulate 3. When a system is described by a wavefunction ψ, the mean value of the observable Ω in a series of measurements is equal to the expectation value of the corresponding operator.

[2]This prescription excludes intrinsic observables, such as spin (Section 4.8).

The **expectation value** of an operator Ω for an arbitrary state ψ is denoted $\langle\Omega\rangle$ and defined as

$$\langle\Omega\rangle = \frac{\int \psi^*\Omega\psi\,\mathrm{d}\tau}{\int \psi^*\psi\,\mathrm{d}\tau} \tag{17}$$

If the wavefunction is chosen to be normalized to 1, then the expectation value is simply

$$\langle\Omega\rangle = \int \psi^*\Omega\psi\,\mathrm{d}\tau \tag{18}$$

Unless we state otherwise, from now on we shall assume that the wavefunction is normalized to 1.

The meaning of Postulate 3 can be unravelled as follows. First, suppose that ψ is an eigenfunction of Ω with eigenvalue ω; then

$$\langle\Omega\rangle = \int \psi^*\Omega\psi\,\mathrm{d}\tau = \int \psi^*\omega\psi\,\mathrm{d}\tau = \omega\int \psi^*\psi\,\mathrm{d}\tau = \omega \tag{19}$$

That is, a series of experiments on identical systems to determine Ω will give the average value ω. Now suppose that although the system is in an eigenstate of the hamiltonian it is not in an eigenstate of Ω. In this case the wavefunction can be expressed as a linear combination of eigenfunctions of Ω:

$$\psi = \sum_n c_n\psi_n \qquad \text{where } \Omega\psi_n = \omega_n\psi_n$$

In this case, the expectation value is

$$\begin{aligned}\langle\Omega\rangle &= \int\left(\sum_m c_m\psi_m\right)^*\Omega\left(\sum_n c_n\psi_n\right)\mathrm{d}\tau = \sum_{m,n} c_m^*c_n\int \psi_m^*\Omega\psi_n\,\mathrm{d}\tau \\ &= \sum_{m,n} c_m^*c_n\omega_n\int \psi_m^*\psi_n\,\mathrm{d}\tau\end{aligned}$$

Because the eigenfunctions form an orthonormal set, the integral in the last expression is zero if $n \neq m$ and the double sum reduces to a single sum:

$$\langle\Omega\rangle = \sum_n c_n^*c_n\omega_n\int \psi_n^*\psi_n\,\mathrm{d}\tau = \sum_n c_n^*c_n\omega_n = \sum_n |c_n|^2\omega_n \tag{20}$$

That is, the expectation value is a weighted sum of the eigenvalues of Ω, the contribution of a particular eigenvalue to the sum being determined by the square modulus of the corresponding coefficient in the expansion of the wavefunction.

We can now interpret the difference between eqns 1.19 and 1.20 in the form of a subsidiary postulate:

> **Postulate 3′.** When ψ is an eigenfunction of the operator Ω, the determination of the property Ω always yields one result, namely the corresponding eigenvalue ω. When ψ is not an eigenfunction of Ω, a single measurement of the property yields a single outcome which is one of the eigenvalues of Ω, and the probability that a particular eigenvalue

n is measured is equal to $|c_n|^2$, where c_n is the coefficient of the eigenfunction ψ_n in the expansion of the wavefunction.

One measurement can give only one result: a pointer can indicate only one value on a dial at any instant. A series of determinations can lead to a series of results with some mean value. The subsidiary postulate asserts that a measurement of the observable Ω always results in the pointer indicating one of the eigenvalues of the corresponding operator. If the function that describes the state of the system is an eigenfunction of Ω, then every pointer reading is precisely ω and the mean value is also ω. If the system has been prepared in a state that is not an eigenfunction of Ω, then different measurements give different values, but every individual measurement is one of the eigenvalues of Ω, and the probability that a particular outcome ω_n is obtained is determined by the value of $|c_n|^2$. In this case, the mean value of all the observations is the weighted average of the eigenvalues.

1.10 The interpretation of the wavefunction

The next postulate concerns the interpretation of the wavefunction itself, and is commonly called the **Born interpretation**:

Postulate 4. The probability that a particle will be found in the volume element $d\tau$ at the point $\mathbf{r}$ is proportional to $|\psi(\mathbf{r})|^2 d\tau$.

As we have already remarked, in one dimension the volume element is dx. In three dimensions the volume element is $dxdydz$. It follows from this interpretation that $|\psi(\mathbf{r})|^2$ is a **probability density**, in the sense that it yields a probability when multiplied by the volume of a region. The wavefunction itself is a **probability amplitude**, and has no direct physical meaning. Note that whereas the probability density is real and non-negative, the wavefunction (amplitude) may be complex and negative. It is usually convenient to use a normalized wavefunction; then the Born interpretation becomes an equality rather than a proportionality. The implication of the Born interpretation is that the wavefunction should be **square integrable**, that is

$$\int |\psi|^2 \, d\tau < \infty$$

because there must be a finite probability of finding the particle somewhere in the whole of space (and that probability is 1 for a normalized wavefunction). This postulate in turn implies that $\psi \rightarrow 0$ as $x \rightarrow \pm\infty$, for otherwise the integral of $|\psi|^2$ would be infinite. We shall make frequent use of this implication throughout the text.

1.11 The equation for the wavefunction

The final postulate concerns the dynamical evolution of the wavefunction with time:

Postulate 5. The wavefunction $\Psi(\mathbf{r}_1, \mathbf{r}_2, \ldots, t)$ evolves in time according to the equation

$$i\hbar \frac{\partial \Psi}{\partial t} = H\Psi \tag{21}$$

This partial differential equation is the celebrated **Schrödinger equation**, which was introduced by Erwin Schrödinger in 1926. At this stage, we are treating the equation as an unsubstantiated postulate. However, in Section 1.23 we shall advance arguments in support of its plausibility. The operator H in the Schrödinger equation is the hamiltonian operator for the system, the operator corresponding to the total energy. For example, by using the expression in eqn 1.11, we obtain the time-dependent Schrödinger equation in one dimension (x) with a time-independent potential energy for a single particle:

$$i\hbar \frac{\partial \Psi}{\partial t} = -\frac{\hbar^2}{2m}\frac{\partial^2 \Psi}{\partial x^2} + V(x)\Psi \tag{22}$$

We shall have a great deal to say about the Schrödinger equation and its solutions in the rest of the text.

1.12 The separation of the Schrödinger equation

The Schrödinger equation can often be separated into equations for the time and space variation of the wavefunction. The separation is possible when the potential energy is independent of the time, and in one dimension the equation has the form

$$H\Psi = -\frac{\hbar^2}{2m}\frac{\partial^2 \Psi}{\partial x^2} + V(x)\Psi = i\hbar \frac{\partial \Psi}{\partial t}$$

Equations of this form can be solved by the technique of **separation of variables**, in which a trial solution takes the form

$$\Psi(x,t) = \psi(x)\theta(t)$$

When this substitution is made, we obtain

$$-\frac{\hbar^2}{2m}\theta\frac{\mathrm{d}^2\psi}{\mathrm{d}x^2} + V(x)\psi\theta = i\hbar\psi\frac{\mathrm{d}\theta}{\mathrm{d}t}$$

Division of both sides of this equation by $\psi\theta$ gives

$$-\frac{\hbar^2}{2m}\frac{1}{\psi}\frac{\mathrm{d}^2\psi}{\mathrm{d}x^2} + V(x) = i\hbar\frac{1}{\theta}\frac{\mathrm{d}\theta}{\mathrm{d}t}$$

Only the left-hand side of this equation is a function of x, so when x changes, only the left can change. But as the left-hand side is equal to the right-hand side, and the latter does not change, the left-hand side must be equal to a constant. Because the dimensions of the constant are those of an energy (the same as those of V), we shall write it E. It follows that the time-dependent equation separates into the following two differential equations:

$$-\frac{\hbar^2}{2m}\frac{\mathrm{d}^2\psi}{\mathrm{d}x^2} + V(x)\psi = E\psi \qquad (23a)$$

$$\mathrm{i}\hbar\frac{\mathrm{d}\theta}{\mathrm{d}t} = E\theta \qquad (23b)$$

The second of these equations has the solution

$$\theta \propto \mathrm{e}^{-\mathrm{i}Et/\hbar} \qquad (24)$$

Therefore, the complete wavefunction ($\Psi = \psi\theta$) has the form

$$\Psi(x, t) = \psi(x)\mathrm{e}^{-\mathrm{i}Et/\hbar} \qquad (25)$$

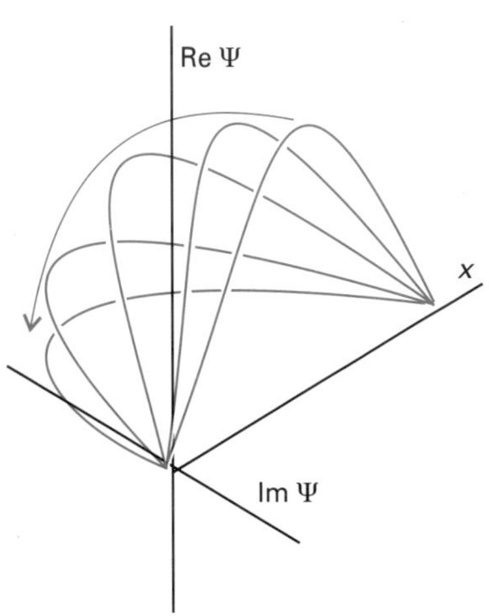

Fig. 1.1. A wavefunction corresponding to an energy E rotates in the complex plane from real to imaginary and back to real at a circular frequency $E/\hbar$.

The constant of proportionality in eqn 1.24 has been absorbed into the normalization constant for ψ. The time-independent wavefunction ψ satisfies eqn 1.23a, which may be written in the form

$$H\psi = E\psi$$

This is the **time-independent Schrödinger equation**, on which much of the following development will be based.

This calculation stimulates several remarks. First, eqn 1.23a has the form of a standing-wave equation. Therefore, so long as we are interested only in the spatial dependence of the wavefunction, it is legitimate to regard the time-independent Schrödinger equation as a wave equation. Second, when the potential energy of the system does not depend on the time, and the system is in a state of energy E, it is a very simple matter to construct the time-dependent wavefunction from the time-independent wavefunction simply by multiplying the latter by $\mathrm{e}^{-\mathrm{i}Et/\hbar}$. The time dependence of such a wavefunction is simply a modulation of its phase, because we can write

$$\mathrm{e}^{-\mathrm{i}Et/\hbar} = \cos(Et/\hbar) - \mathrm{i}\sin(Et/\hbar)$$

It follows that the time-dependent factor oscillates periodically from 1 to $-\mathrm{i}$ to -1 to i and back to 1 with a frequency E/h and period h/E. This behaviour is depicted in Fig. 1.1. Therefore, to imagine the time-variation of a wavefunction of a definite energy, think of it as flickering from positive through imaginary to negative amplitudes with a frequency proportional to the energy.

Although the phase of a wavefunction Ψ with definite energy E oscillates in time, the product $\Psi^*\Psi$ (or $|\Psi|^2$) remains constant:

$$\Psi^*\Psi = \left(\psi^*\mathrm{e}^{\mathrm{i}Et/\hbar}\right)\left(\psi\mathrm{e}^{-\mathrm{i}Et/\hbar}\right) = \psi^*\psi$$

States of this kind are called **stationary states**. From what we have seen so far, it follows that systems with a specific, precise energy and in which the potential energy does not vary with time are in stationary states. Although their wavefunctions flicker from one phase to another in a repetitive manner, the value of $\Psi^*\Psi$ remains constant.

Hermitian operators

With the basic foundation of quantum mechanics laid, we can start to develop some technical points about the operators that are widely encountered. The operators known as **hermitian operators** play a very special role in quantum mechanics because their eigenvalues are real. Hence, it is the hermitian operators that are used to represent observables, because the outcome of an observation must be a real number. We shall meet non-hermitian operators, but as their eigenvalues are not guaranteed to be real, they do not correspond to observables.

1.13 Definitions

An operator is **hermitian** if it satisfies the following relation:

$$\int \psi_m^* \Omega \psi_n \, d\tau = \left\{ \int \psi_n^* \Omega \psi_m \, d\tau \right\}^* \tag{26}$$

for any two wavefunctions ψ_m and ψ_n. An alternative version of this definition is

$$\int \psi_m^* \Omega \psi_n \, d\tau = \int (\Omega \psi_m)^* \psi_n \, d\tau \tag{27}$$

This expression is obtained by taking the complex conjugate of each term on the right-hand side of eqn 1.26.

Example 1.3 How to confirm the hermiticity of operators

Show that the position and momentum operators in the position representation are hermitian.

Method. We need to show that the operators satisfy eqn 1.27 for a general wavefunction. In some cases (the position operator, for instance), the hermiticity is obvious as soon as the integral is written down. When a differential operator is used, it may be necessary to use integration by parts at some stage in the argument to transfer the differentiation from one function to another:

$$\int u \, dv = uv - \int v \, du$$

Answer. That the position operator is hermitian is obvious from inspection:

$$\int \psi_m^* x \psi_n \, d\tau = \int \psi_n x \psi_m^* \, d\tau = \left\{ \int \psi_n^* x \psi_m \, d\tau \right\}^*$$

We have used the facts that $(\psi^*)^* = \psi$ and x is real. The demonstration of the hermiticity of p_x, a differential operator in the position representation, involves an integration by parts:

$$\int \psi_m^* p_x \psi_n \,\mathrm{d}x = \int \psi_m^* \frac{\hbar}{\mathrm{i}} \frac{\mathrm{d}}{\mathrm{d}x} \psi_n \,\mathrm{d}x = \frac{\hbar}{\mathrm{i}} \int \psi_m^* \,\mathrm{d}\psi_n$$
$$= \frac{\hbar}{\mathrm{i}} \left\{ \psi_m^* \psi_n - \int \psi_n \,\mathrm{d}\psi_m^* \right\} \Big|_{x=-\infty}^{x=\infty}$$
$$= \frac{\hbar}{\mathrm{i}} \left\{ \psi_m^* \psi_n \Big|_{x=-\infty}^{x=\infty} - \int_{-\infty}^{\infty} \psi_n \frac{\mathrm{d}}{\mathrm{d}x} \psi_m^* \,\mathrm{d}x \right\}$$

The first term on the right is zero (because when x is infinite, a square-integrable wavefunction is vanishingly small). Therefore,

$$\int \psi_m^* p_x \psi_n \,\mathrm{d}x = -\frac{\hbar}{\mathrm{i}} \int \psi_n \frac{\mathrm{d}}{\mathrm{d}x} \psi_m^* \,\mathrm{d}x$$
$$= \left\{ \int \psi_n^* \frac{\hbar}{\mathrm{i}} \frac{\mathrm{d}}{\mathrm{d}x} \psi_m \,\mathrm{d}x \right\}^* = \left\{ \int \psi_n^* p_x \psi_m \,\mathrm{d}x \right\}^*$$

Hence, the operator is hermitian.

Exercise 1.3. Show that the two operators are hermitian in the momentum representation.

1.14 Dirac bracket notation

We are on the edge of getting lost in a complicated notation. The appearance of many quantum mechanical expressions is greatly simplified by adopting the **Dirac bracket notation** in which integrals are written as follows:

$$\langle m|\Omega|n\rangle = \int \psi_m^* \Omega \psi_n \,\mathrm{d}\tau \tag{28}$$

The symbol $|n\rangle$ is called a **ket**, and denotes the state with wavefunction ψ_n. Similarly, the symbol $\langle n|$ is called a **bra**, and denotes the complex conjugate of the wavefunction, ψ_n^*. When a bra and ket are strung together with an operator between them, as in the **bracket** $\langle m|\Omega|n\rangle$, the integral in eqn 1.28 is to be understood. When the operator is simply multiplication by 1, the 1 is omitted and we use the convention

$$\langle m|n\rangle = \int \psi_m^* \psi_n \,\mathrm{d}\tau \tag{29}$$

This notation is very elegant. For example, the normalization integral becomes $\langle n|n\rangle = 1$ and the orthonormality condition (eqn 1.16) becomes $\langle m|n\rangle = 0$ for $m \neq n$. The combined orthonormality condition is then

$$\langle m|n\rangle = \delta_{mn} \tag{30}$$

A final point is that, as can readily be deduced from the definition of a Dirac bracket,

$$\langle m|n\rangle = \langle n|m\rangle^*$$

1.15 The properties of hermitian operators

In terms of the Dirac notation, the definition of hermiticity is

$$\langle m|\Omega|n\rangle = \langle n|\Omega|m\rangle^* \tag{31}$$

As we shall now see, this property has far-reaching implications. First, we shall establish the following property:

Property 1. The eigenvalues of hermitian operators are real.

To prove this result, we consider the eigenvalue equation

$$\Omega|\omega\rangle = \omega|\omega\rangle$$

The ket $|\omega\rangle$ denotes an **eigenstate** of the operator Ω in the sense that the corresponding wavefunction is an eigenfunction of the operator and we are labelling the eigenstates with the eigenvalue ω of the operator Ω. It is often convenient to use the eigenvalues as labels in this way. Multiplication from the left by $\langle\omega|$ results in the equation

$$\langle\omega|\Omega|\omega\rangle = \omega\langle\omega|\omega\rangle = \omega$$

Now take the complex conjugate of both sides:

$$\langle\omega|\Omega|\omega\rangle^* = \omega^*$$

However, by hermiticity, $\langle\omega|\Omega|\omega\rangle^* = \langle\omega|\Omega|\omega\rangle$. Therefore, it follows that $\omega = \omega^*$, which implies that the eigenvalue ω is real.

The second property we shall prove is as follows:

Property 2. Eigenfunctions corresponding to different eigenvalues of an hermitian operator are orthogonal.

That is, if we have two eigenfunctions of an hermitian operator Ω with eigenvalues ω and ω', with $\omega \neq \omega'$, then $\langle\omega|\omega'\rangle = 0$. For example, it follows at once that all the wavefunctions of a harmonic oscillator are mutually orthogonal, for each one corresponds to a different energy (the eigenvalue of the hamiltonian, an hermitian operator).

The proof of this property runs as follows. Suppose we have the two eigenstates $|\omega\rangle$ and $|\omega'\rangle$ that satisfy the following relations:

$$\Omega|\omega\rangle = \omega|\omega\rangle \text{ and } \Omega|\omega'\rangle = \omega'|\omega'\rangle$$

Then multiplication of the first relation by $\langle\omega'|$ and the second by $\langle\omega|$ gives

$$\langle\omega'|\Omega|\omega\rangle = \omega\langle\omega'|\omega\rangle \text{ and } \langle\omega|\Omega|\omega'\rangle = \omega'\langle\omega|\omega'\rangle$$

Next, we take the complex conjugate of the second relation and subtract it from the first:

$$\langle\omega'|\Omega|\omega\rangle - \langle\omega|\Omega|\omega'\rangle^* = \omega\langle\omega'|\omega\rangle - \omega'\langle\omega|\omega'\rangle^*$$

Because Ω is hermitian, the left-hand side of this expression is zero; so (noting that ω' is real and using $\langle\omega|\omega'\rangle^* = \langle\omega'|\omega\rangle$ as explained earlier) we arrive at

$$(\omega - \omega')\langle\omega'|\omega\rangle = 0$$

However, because the two eigenvalues are different, the only way of satisfying this relation is for $\langle\omega'|\omega\rangle = 0$, as was to be proved.

The specification of states: complementarity

The question that now arises is the following. Let us suppose for the moment that the state of a system can be specified as $|a, b, \ldots\rangle$, where each of the eigenvalues $a, b, \ldots$ corresponds to the operators representing different observables $A, B, \ldots$ of the system. If the system is in the state $|a, b, \ldots\rangle$, then when we measure the property A we shall get exactly a as an outcome, and likewise for the other properties. But can a state be specified *arbitrarily* fully? That is, can it be *simultaneously* an eigenstate of all possible observables $A, B, \ldots$ without restriction? With this question we are moving into the domain of the uncertainty principle.

As a first step, we establish the conditions under which two observables may be specified simultaneously with arbitrary precision. That is, we establish the conditions for a state $|\psi\rangle$ corresponding to the wavefunction ψ to be simultaneously an eigenstate of two operators A and B. In fact, we shall prove the following:

> **Property 3.** If two observables are to have simultaneously precisely defined values, then their corresponding operators must commute.

That is, AB must equal BA, or equivalently, $[A, B] = 0$.

To prove this assertion, we assume that $|\psi\rangle$ *is* an eigenstate of both operators, so we start by knowing that $A|\psi\rangle = a|\psi\rangle$ and $B|\psi\rangle = b|\psi\rangle$. That being so, we can write the following chain of relations:

$$AB|\psi\rangle = Ab|\psi\rangle = bA|\psi\rangle = ba|\psi\rangle = ab|\psi\rangle = aB|\psi\rangle = Ba|\psi\rangle = BA|\psi\rangle$$

Therefore, if $|\psi\rangle$ is an eigenstate of both A and B, and if the same is true for all functions ψ of a complete set, then it is certainly necessary that $[A, B] = 0$. However, does the condition $[A, B] = 0$ actually guarantee that A and B have simultaneous eigenvalues? In other words, if $A|\psi\rangle = a|\psi\rangle$ and $[A, B] = 0$, can we be confident that $|\psi\rangle$ is also an eigenstate of B? We confirm this as follows. Because $A|\psi\rangle = a|\psi\rangle$, we can write

$$BA|\psi\rangle = Ba|\psi\rangle = aB|\psi\rangle$$

Because A and B commute, the first term on the left is equal to $AB|\psi\rangle$. Therefore, this relation has the form

$$A(B|\psi\rangle) = a(B|\psi\rangle)$$

However, on comparison of this eigenvalue equation with $A|\psi\rangle = a|\psi\rangle$, we can conclude that $B|\psi\rangle \propto |\psi\rangle$, or $B|\psi\rangle = b|\psi\rangle$, where b is a coefficient of proportionality. That is, $|\psi\rangle$ is an eigenstate of B, as was to be proved.

It follows from this discussion that we are now in a position to determine which observables may be specified simultaneously. All we need do is to inspect the commutator $[A, B]$: if it is zero, then A and B may be specified simultaneously.

Example 1.4 How to decide whether observables may be specified simultaneously

What restrictions are there on the simultaneous specification of the position and the linear momentum of a particle?

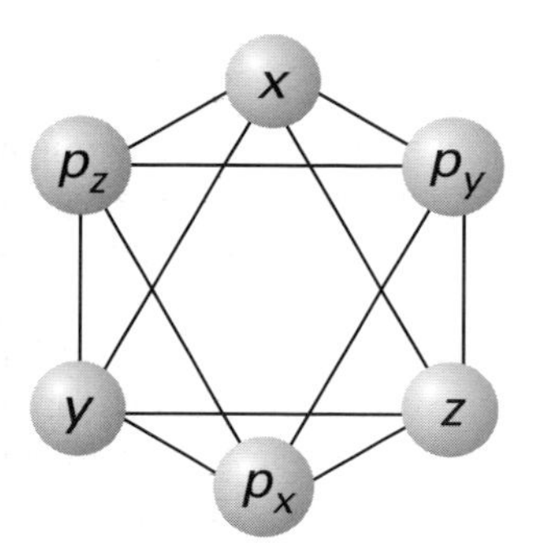

Fig. 1.2. A summary of the position and momentum observables that can be specified simultaneously with arbitrary precision (joined by lines) and those that cannot (not joined).

Method. To answer this question we have to determine whether the position coordinates can be specified simultaneously, whether the momentum components can be specified simultaneously, and whether the position and momentum can be specified simultaneously. The answer is found by examining the commutators of the corresponding operators.

Answer. All three position operators x, y, and z commute with one another, so there is no constraint on the complete specification of position. The same is true of the three operators for the components of linear momenta. So all three components can be determined simultaneously. However, x and p_x do not commute, and so these two observables cannot be specified simultaneously, and likewise for y, p_y and z, p_z. The consequent pattern of permitted simultaneous specifications is illustrated in Fig. 1.2.

Exercise 1.4. Can the kinetic energy and the linear momentum be specified simultaneously? [Yes]

Observables that *cannot* be determined simultaneously are said to be **complementary**. Thus, position along the x-axis and linear momentum parallel to that axis are complementary observables. Classical physics made the mistake of presuming that there was no restriction on the simultaneous determination of observables, that there was no complementarity. Quantum mechanics forces us to choose a selection of all possible observables if we seek to specify a state fully.

The uncertainty principle

Although we cannot specify the eigenvalues of two non-commuting operators simultaneously, it is possible to give up precision in the specification of one property in order to acquire greater precision in the specification of a complementary property. For example, if we know the location of a particle to within a range Δx, then we can specify the linear momentum parallel to x to within a range Δp_x subject to the constraint

$$\Delta x \Delta p_x \geq \tfrac{1}{2}\hbar \tag{32}$$

Thus, as Δx increases (an increased uncertainty in x), the uncertainty in p_x can decrease, and vice versa. This relation between the uncertainties in the specification of two complementary observables is a special case of the **uncertainty principle** proposed by W. Heisenberg in 1927.

1.16 Formal derivation of the principle

A very general form of the uncertainty principle was developed by H.P. Robertson in 1929, and runs as follows. We shall suppose that we are interested in the simultaneous specification of two observables A and B that obey the commutation relation $[A, B] = \mathrm{i}C$. (The i is included for future convenience; for $A = x$ and $B = p_x$ it follows from the fundamental commutation relation that $C = \hbar$.) We shall suppose that the system is prepared in a normalized but otherwise arbitrary state $|\psi\rangle$, which is not necessarily an eigenstate of either operator A or B. The mean results of determining A and B separately are expressed by the expectation values

$$\langle A\rangle = \langle\psi|A|\psi\rangle \text{ and } \langle B\rangle = \langle\psi|B|\psi\rangle$$

The operators for the spread of individual determinations of A and B around their mean values are

$$\delta A = A - \langle A\rangle \text{ and } \delta B = B - \langle B\rangle$$

It is easy to verify that the commutation relation for these deviation operators is

$$[\delta A, \delta B] = [A - \langle A\rangle, B - \langle B\rangle] = [A, B] = \mathrm{i}C$$

because the expectation values $\langle A\rangle$ and $\langle B\rangle$ are simple numbers and commute with operators.

When Robertson developed his version of the uncertainty principle, he started by considering the properties of the following integral, where α is a real but otherwise arbitrary number:

$$I = \int |(\alpha\delta A - \mathrm{i}\delta B)\psi|^2 \,\mathrm{d}\tau$$

The integral I is clearly non-negative as the integrand is positive everywhere. The integral can be developed as follows:

$$\begin{aligned} I &= \int \{(\alpha\delta A - \mathrm{i}\delta B)\psi\}^* \{(\alpha\delta A - \mathrm{i}\delta B)\psi\} \,\mathrm{d}\tau \\ &= \int \psi^* (\alpha\delta A + \mathrm{i}\delta B)(\alpha\delta A - \mathrm{i}\delta B)\psi \,\mathrm{d}\tau \end{aligned}$$

In the second step we have used the hermitian character of the two operators (as expressed in eqn 1.27). At this point it is convenient to recognize that the final expression is an expectation value, and to write it in the form

$$I = \langle(\alpha\delta A + \mathrm{i}\delta B)(\alpha\delta A - \mathrm{i}\delta B)\rangle$$

This expression expands to

$$I = \alpha^2\langle(\delta A)^2\rangle + \langle(\delta B)^2\rangle - \mathrm{i}\alpha\langle\delta A\delta B - \delta B\delta A\rangle = \alpha^2\langle(\delta A)^2\rangle + \langle(\delta B)^2\rangle + \alpha\langle C\rangle$$

In the second step we have recognized the presence of the commutator. The integral is still non-negative, even though that is no longer obvious. At this point we recognize that I has the general form of a quadratic expression in α, and so express it as a square:

$$I = \langle(\delta A)^2\rangle\left(\alpha + \frac{\langle C\rangle}{2\langle(\delta A)^2\rangle}\right)^2 + \langle(\delta B)^2\rangle - \frac{\langle C\rangle^2}{4\langle(\delta A)^2\rangle}$$

(We have 'completed the square' for the first term.) This expression is still non-negative whatever the value of α, and remains non-negative even if we choose a value for α that corresponds to the minimum value of I. That value of α is the value that ensures that the first term on the right is zero (because that term always supplies a positive contribution to I). Therefore, with that choice of α, we obtain

$$I = \langle(\delta B)^2\rangle - \frac{\langle C\rangle^2}{4\langle(\delta A)^2\rangle} \geq 0$$

The inequality rearranges to

$$\langle(\delta A)^2\rangle\langle(\delta B)^2\rangle \geq \tfrac{1}{4}\langle C\rangle^2$$

and a familiar expression is starting to emerge.

The expectation values on the left can be put into a simpler form by writing them as follows:

$$\begin{aligned}\langle(\delta A)^2\rangle &= \langle(A - \langle A\rangle)^2\rangle \\ &= \langle A^2 - 2A\langle A\rangle + \langle A\rangle^2\rangle = \langle A^2\rangle - 2\langle A\rangle\langle A\rangle + \langle A\rangle^2 \\ &= \langle A^2\rangle - \langle A\rangle^2\end{aligned}$$

We see that $\langle(\delta A)^2\rangle$ is the mean square deviation of A from its mean value (and likewise for B). We shall now define the **root mean square deviation** of A as the square-root of this quantity:

$$\Delta A = \{\langle A^2\rangle - \langle A\rangle^2\}^{1/2} \tag{33}$$

and likewise for B. Then the inequality becomes

$$\Delta A \Delta B \geq \tfrac{1}{2}|\langle C\rangle|$$

Then, because $[A, B] = \mathrm{i}C$, we obtain the final form of the uncertainty principle:

$$\Delta A \Delta B \geq \tfrac{1}{2}|\langle[A, B]\rangle| \tag{34}$$

This is an exact and precise form of the uncertainty principle: the precise form of the 'uncertainties' ΔA and ΔB are given (they are root mean square deviations) and the right-hand side gives a precise lower bound on the value of the product of uncertainties.

1.17 Consequences of the uncertainty principle

The first point to note is that the uncertainty principle is consistent with Property 3, for if A and B commute, then C is zero and there is no constraint on the uncertainties: there is no inconsistency in having both $\Delta A = 0$ and $\Delta B = 0$. On the other hand, when A and B do not commute, the values of ΔA and ΔB are related. For instance, while it may be possible to prepare a system in a state in which $\Delta A = 0$, the uncertainty then implies that ΔB must be infinite in order to ensure that $\Delta A \Delta B$ is not less than $\frac{1}{2}|\langle[A, B]\rangle|$. In the particular case of the simultaneous specification of x and p_x, as we have seen, $[x, p_x] = \mathrm{i}\hbar$, so the lower bound on the simultaneous specification of these two complementary observables is $\frac{1}{2}\hbar$.

Example 1.5 How to calculate the joint uncertainty in two observables

A particle was prepared in a state with wavefunction $\psi = N\mathrm{e}^{-x^2/2\Gamma}$, where $N = (1/\pi\Gamma)^{1/4}$. Evaluate Δx and Δp_x, and confirm that the uncertainty principle is satisfied.

Method. We must evaluate the expectation values $\langle x\rangle$, $\langle x^2\rangle$, $\langle p_x\rangle$, and $\langle p_x^2\rangle$ by integration and then combine their values to obtain Δx and Δp_x. There are two short cuts. For $\langle x\rangle$, we note that ψ is symmetrical around $x = 0$, and so $\langle x\rangle = 0$. The value of $\langle p_x\rangle$ can be obtained by noting that p_x is an imaginary hermitian operator and ψ is real. Because hermiticity implies that $\langle p_x\rangle^* = \langle p_x\rangle$ but the imaginary character of p_x implies that $\langle p_x\rangle^* = -\langle p_x\rangle$, we can conclude that $\langle p_x\rangle = 0$. For the remaining integrals we use

$$\int_{-\infty}^{\infty} \mathrm{e}^{-ax^2}\,\mathrm{d}x = \left(\frac{\pi}{a}\right)^{1/2} \qquad \int_{-\infty}^{\infty} x^2\mathrm{e}^{-ax^2}\,\mathrm{d}x = \frac{1}{2a}\left(\frac{\pi}{a}\right)^{1/2}$$

Answer. The following integrals are obtained:

$$\langle x^2\rangle = N^2\int_{-\infty}^{\infty} x^2\mathrm{e}^{-x^2/\Gamma}\,\mathrm{d}x = \tfrac{1}{2}\Gamma$$

$$\langle p_x^2\rangle = N^2\int_{-\infty}^{\infty} \mathrm{e}^{-x^2/2\Gamma}\left(-\hbar^2\frac{\mathrm{d}^2}{\mathrm{d}x^2}\right)\mathrm{e}^{-x^2/2\Gamma}\,\mathrm{d}x$$

$$= \hbar^2N^2\left\{\frac{1}{\Gamma}\int_{-\infty}^{\infty}\mathrm{e}^{-x^2/\Gamma}\,\mathrm{d}x - \frac{1}{\Gamma^2}\int_{-\infty}^{\infty}x^2\mathrm{e}^{-x^2/\Gamma}\,\mathrm{d}x\right\} = \frac{\hbar^2}{2\Gamma}$$

It follows that (because $\langle x\rangle = 0$ and $\langle p_x\rangle = 0$)

$$\Delta x\Delta p_x = \langle x^2\rangle^{1/2}\langle p_x^2\rangle^{1/2} = \tfrac{1}{2}\hbar$$

Comment. In this example, $\Delta x\Delta p_x$ has its minimum permitted value. This is a special feature of gaussian wavefunctions. A gaussian wavefunction is encountered in the ground state of a harmonic oscillator (see Section 2.16).

Exercise 1.5. Calculate the value of $\Delta x\Delta p_x$ for a wavefunction that is zero everywhere except in a region of space of length L, where it has the form $\sqrt{2/L}\sin(\pi x/L)$. [$(\hbar/2\sqrt{3})(\pi^2-6)^{1/2}$]

The uncertainty principle in the form given in eqn 1.34 can be applied to all pairs of complementary observables. We shall see additional examples in later chapters.

1.18 The uncertainty in energy and time

Finally, it is appropriate at this point to make a few remarks about the so-called **energy–time uncertainty relation**, which is often expressed in the form $\Delta E\Delta t \geq \hbar$ and interpreted as implying a complementarity between energy and time. As we have seen, for this relation to be a true uncertainty relation, it would be necessary for there to be a nonzero commutator for energy and time. However, although the energy operator is well defined (it is the hamiltonian for the system), *there is no operator for time in quantum mechanics.* Time is a *parameter*, not an observable. Therefore, strictly speaking, there is no uncertainty relation between energy and time. In Section 6.18 we shall see the true significance of the energy–time 'uncertainty principle', and see that it is a relation between the uncertainty in the energy of a system that has a finite lifetime τ, and is of the form $\delta E \approx \hbar/2\tau$.

Time-evolution and conservation laws

As well as determining which operators are complementary, the commutator of two operators also plays a role in determining the time-evolution of systems and in particular the time-evolution of the expectation values of observables. The precise relation for operators that do not have an intrinsic dependence on the time (in the sense that $\partial\Omega/\partial t = 0$) is

$$\frac{\mathrm{d}\langle\Omega\rangle}{\mathrm{d}t} = \frac{\mathrm{i}}{\hbar}\langle[H, \Omega]\rangle \tag{35}$$

We see that if the operator for the observable commutes with the hamiltonian, then the expectation value of the operator does not change with time. An observable that commutes with the hamiltonian for the system, and which therefore has an expectation value that does not change with time, is called a **constant of the motion**, and its expectation value is said to be **conserved**.

The proof of this relation is as follows. Differentiation of $\langle\Omega\rangle$ with respect to time gives

$$\frac{\mathrm{d}\langle\Omega\rangle}{\mathrm{d}t} = \frac{\mathrm{d}}{\mathrm{d}t}\langle\psi|\Omega|\psi\rangle = \int\left(\frac{\partial\Psi^*}{\partial t}\right)\Omega\Psi\,\mathrm{d}\tau + \int\Psi^*\Omega\left(\frac{\partial\Psi}{\partial t}\right)\mathrm{d}\tau$$

because only the two states (not the operator Ω) depend on the time. The Schrödinger equation lets us write

$$\int\Psi^*\Omega\left(\frac{\partial\Psi}{\partial t}\right)\mathrm{d}\tau = \int\Psi^*\Omega\left(\frac{1}{\mathrm{i}\hbar}\right)H\Psi\,\mathrm{d}\tau = \left(\frac{1}{\mathrm{i}\hbar}\right)\int\Psi^*\Omega H\Psi\,\mathrm{d}\tau$$

$$\int\left(\frac{\partial\Psi^*}{\partial t}\right)\Omega\Psi\mathrm{d}\tau = -\int\left(\frac{1}{\mathrm{i}\hbar}\right)(H\Psi)^*\Omega\Psi\,\mathrm{d}\tau = -\left(\frac{1}{\mathrm{i}\hbar}\right)\int\Psi^*H\Omega\Psi\,\mathrm{d}\tau$$

In the second line we have used the hermiticity of the hamiltonian (in the form of eqn 1.27). It then follows, by combining these two expressions, that

$$\frac{\mathrm{d}\langle\Omega\rangle}{\mathrm{d}t} = -\left(\frac{1}{\mathrm{i}\hbar}\right)(\langle H\Omega\rangle - \langle\Omega H\rangle) = \frac{\mathrm{i}}{\hbar}\langle[H, \Omega]\rangle$$

as was to be proved.

As an important example, consider the rate of change of the expectation value of the linear momentum of a particle in a one-dimensional system. The commutator of H and p_x is

$$[H,p_x] = \left[-\frac{\hbar^2}{2m}\frac{d^2}{dx^2} + V, \frac{\hbar}{i}\frac{d}{dx}\right] = \frac{\hbar}{i}\left[V, \frac{d}{dx}\right]$$

because the derivatives commute. The remaining commutator can be evaluated by remembering that there is an unwritten function on the right on which the operators operate, and writing

$$[H,p_x]\psi = \frac{\hbar}{i}\left\{V\frac{d\psi}{dx} - \frac{d(V\psi)}{dx}\right\} = \frac{\hbar}{i}\left\{V\frac{d\psi}{dx} - V\frac{d\psi}{dx} - \frac{dV}{dx}\psi\right\}$$
$$= -\frac{\hbar}{i}\frac{dV}{dx}\psi$$

This relation is true for all functions ψ; therefore the commutator itself is

$$[H,p_x] = -\frac{\hbar}{i}\left(\frac{dV}{dx}\right) \qquad (36)$$

It follows that the linear momentum is a constant of the motion if the potential energy does not vary with position, when $dV/dx = 0$. Specifically, we can conclude that the rate of change of the expectation value of linear momentum is

$$\frac{d}{dt}\langle p_x\rangle = \frac{i}{\hbar}\langle[H,p_x]\rangle = -\left\langle\frac{dV}{dx}\right\rangle \qquad (37)$$

Then, because the negative slope of the potential energy is by definition the **force** that is acting ($F = -dV/dx$), the rate of change of the expectation value of linear momentum is given by

$$\frac{d}{dt}\langle p_x\rangle = \langle F\rangle \qquad (38)$$

That is, *the rate of change of the expectation value of the linear momentum is equal to the expectation value of the force*. It is also quite easy to prove in the same way that

$$\frac{d}{dt}\langle x\rangle = \frac{\langle p_x\rangle}{m} \qquad (39)$$

which shows that the rate of change of the mean position can be identified with the mean velocity along the x-axis. These two relations jointly constitute **Ehrenfest's theorem**. Ehrenfest's theorem clarifies the relation between classical and quantum mechanics: classical mechanics deals with average values (expectation values); quantum mechanics deals with the underlying details.

Matrices in quantum mechanics

As we have seen, the fundamental commutation relation of quantum mechanics, $[x, p_x] = i\hbar$, implies that x and p_x are to be treated as operators. However, there is an alternative interpretation: that x and p_x should be regarded as matrices, for matrix multiplication is also non-commutative. We shall introduce this approach here as it introduces a language that is widely used throughout quantum mechanics even though matrices are not being used explicitly.

1.19 Matrix elements

A **matrix**, **M**, is an array of numbers (which may be complex), called **matrix elements**. Each element is specified by quoting the row (r) and column (c) that it occupies, and denoting the matrix element as M_{rc}. The rules of matrix algebra are set out in *Further information 23*. For our present purposes it is sufficient to emphasize the rule of **matrix multiplication**: the product of two matrices **M** and **N** is another matrix **P** = **MN** with elements given by the rule

$$P_{rc} = \sum_s M_{rs}N_{sc} \tag{40}$$

The order of matrix multiplication is important, and it is essential to note that **MN** is not necessarily equal to **NM.** Hence, **MN** – **NM** is not in general zero. Heisenberg formulated his version of quantum mechanics, which is called **matrix mechanics**, by representing position and linear momentum by the matrices $\mathbf{x}$ and $\mathbf{p}_x$, and requiring that $\mathbf{x}\mathbf{p}_x - \mathbf{p}_x\mathbf{x} = i\hbar\mathbf{1}$ where $\mathbf{1}$ is the **unit matrix**, a square matrix with all diagonal elements (those for which $r = c$) equal to 1 and all others 0.

Throughout this chapter we have encountered quantities of the form $\langle m|\Omega|n\rangle$. These quantities are commonly abbreviated to Ω_{mn}, which immediately suggests that they are elements of a matrix. For this reason, the Dirac bracket $\langle m|\Omega|n\rangle$ is often called a **matrix element** of the operator Ω. A **diagonal matrix element** Ω_{nn} is then a bracket of the form $\langle n|\Omega|n\rangle$ with the bra and the ket referring to the same state. We shall often encounter sums over products of Dirac brackets that have the form

$$\sum_s \langle r|A|s\rangle\langle s|B|c\rangle$$

If the brackets that appear in this expression are interpreted as matrix elements, then we see that it has the form of a matrix multiplication, and we may write

$$\sum_s \langle r|A|s\rangle\langle s|B|c\rangle = \sum_s A_{rs}B_{sc} = (AB)_{rc} = \langle r|AB|c\rangle \tag{41}$$

That is, the sum is equal to the single matrix element (bracket) of the product of operators AB. Comparison of the first and last terms in this line of equations also allows us to write the symbolic relation

$$\sum_s |s\rangle\langle s| = 1 \tag{42}$$

This **completeness relation** is exceptionally useful for developing quantum mechanical equations. It is often used in reverse: the matrix element $\langle r|AB|c\rangle$ can always be split into a sum of two factors by regarding it as $\langle r|A1B|c\rangle$ and then replacing the 1 by a sum over a complete set of states of the form in eqn 1.42.

Example 1.6 How to make use of the completeness relation

Use the completeness relation to prove that the eigenvalues of the square of an hermitian operator are non-negative.

Method. We have to prove that for $\Omega^2|\omega\rangle = \omega|\omega\rangle$, $\omega > 0$ if Ω is hermitian. If both sides of the eigenvalue equation are multiplied by $\langle\omega|$, converting it to $\langle\omega|\Omega^2|\omega\rangle = \omega$, we see that the proof requires us to show that the expectation value on the left is non-negative. As it has the form $\langle\omega|\Omega\Omega|\omega\rangle$, it suggests that the completeness relation might provide a way forward. The hermiticity of Ω implies that it will be appropriate to use the property $\langle m|\Omega|n\rangle = \langle n|\Omega|m\rangle^*$ at some stage in the argument.

Answer. The diagonal matrix element $\langle\omega|\Omega^2|\omega\rangle$ can be developed as follows:

$$\begin{aligned}\langle\omega|\Omega^2|\omega\rangle = \langle\omega|\Omega\Omega|\omega\rangle &= \sum_s \langle\omega|\Omega|s\rangle\langle s|\Omega|\omega\rangle \\ &= \sum_s \langle\omega|\Omega|s\rangle\langle\omega|\Omega|s\rangle^* = \sum_s |\langle\omega|\Omega|s\rangle|^2 \geq 0\end{aligned}$$

The final inequality follows from the fact that all the terms in the sum are non-negative.

Exercise 1.6. Show that if $(\Omega f)^* = -\Omega f^*$, then $\langle\Omega\rangle = 0$ for any real function.

The origin of the completeness relation, which is also known as the **closure relation**, can be demonstrated by the following argument. Suppose we have a complete set of orthonormal states $|s_i\rangle$. Then, by definition of complete, we can expand an arbitrary function ψ as a linear combination:

$$|\psi\rangle = \sum_i c_i|s_i\rangle$$

Multiplication from the left by the bra $\langle s_j|$ and use of the orthonormality of the complete basis set gives

$$c_j = \langle s_j|\psi\rangle$$

Thus

$$|\psi\rangle = \sum_i \langle s_i|\psi\rangle|s_i\rangle = \sum_i |s_i\rangle\langle s_i|\psi\rangle$$

which immediately implies the completeness relation.

1.20 The diagonalization of the hamiltonian

The time-independent form of the Schrödinger equation, $H\psi = E\psi$, can be given a matrix interpretation. First, we express $|\psi\rangle$ as a linear combination of a complete set of states $|n\rangle$:

$$H|\psi\rangle = H\sum_n c_n|n\rangle = \sum_n c_nH|n\rangle$$

$$E|\psi\rangle = E\sum_n c_n|n\rangle$$

These two lines are equal to one another. Next, multiply from the left by an arbitrary bra $\langle m|$ and use the orthonormality of the states to obtain

$$\sum_n c_n\langle m|H|n\rangle = E\sum_n c_n\langle m|n\rangle = Ec_m$$

In matrix notation this is

$$\sum_n H_{mn}c_n = Ec_m \tag{43}$$

Now suppose that we can find the set of states such that $H_{mn} = 0$ unless $m = n$; that is, when using this set, the hamiltonian has a diagonal matrix. Then this expression becomes

$$H_{mm}c_m = Ec_m \tag{44}$$

and the energy E is seen to be the diagonal element of the hamiltonian. In other words, *solving the Schrödinger equation is equivalent to diagonalizing the hamiltonian matrix* (see *Further information 23*). This is yet another link between the Schrödinger and Heisenberg formulations of quantum mechanics. Indeed, it was reported that when Heisenberg was looking for ways of diagonalizing his matrices, the mathematician David Hilbert suggested to him that he should look for the corresponding differential equation instead. Had he done so, Schrödinger's wave mechanics would have been Heisenberg's too.

Example 1.7 How to diagonalize a simple hamiltonian

In a system that consists of only two states (such as an electron spin in a magnetic field, when the electron spin can be in one of two orientations), the hamiltonian has the following matrix elements: $H_{11} = a$, $H_{22} = b$, $H_{12} = H_{21} = d$. Find the energy levels and the eigenstates of the system.

Method. The energy levels are the eigenvalues of the matrix. Use the procedure explained in *Further information 23* to find them (by solving the secular determinant $\det|\mathbf{H} - E\mathbf{S}| = 0$). Find the eigenstates in the form $c_1|1\rangle + c_2|2\rangle$ by solving the secular equations for each eigenvalue in turn. The best procedure is to express the eigenstates as $|a\rangle = |1\rangle\cos\zeta - |2\rangle\sin\zeta$ and $|b\rangle = |1\rangle\sin\zeta + |2\rangle\cos\zeta$, where ζ is a parameter, for this parametrization ensures that the two states are orthonormal for all values of ζ. These forms of the eigenstates let us establish the transformation matrix $\mathbf{T}$ such that $\mathbf{T}^{-1}\mathbf{HT}$ is diagonal, its diagonal elements being the required eigenvalues. Therefore, we form $\mathbf{T}^{-1}\mathbf{HT}$, equate it to the matrix $E\mathbf{1}$, and then solve for ζ.

Answer. Because the states $|1\rangle$ and $|2\rangle$ are orthonormal,

$$\det|\mathbf{H} - E\mathbf{S}| = \begin{vmatrix} a-E & d \\ d & b-E \end{vmatrix} = (a-E)(b-E) - d^2 = 0$$

This quadratic equation for E has the roots

$$E_{\pm} = \tfrac{1}{2}(a+b) \pm \tfrac{1}{2}\{(a-b)^2 + 4d^2\}^{1/2} = \tfrac{1}{2}(a+b) \pm \Delta$$

where $\Delta = \frac{1}{2}\{(a-b)^2 + 4d^2\}^{1/2}$. These are the eigenvalues, and hence they are the energy levels. According to *Further information 23*, the matrix that diagonalizes $\mathbf{H}$ is formed from the coefficients of the states $|1\rangle$ and $|2\rangle$ as follows. First, form the transformation matrix and its reciprocal:

$$\mathbf{T} = \begin{pmatrix} \cos\zeta & -\sin\zeta \\ \sin\zeta & \cos\zeta \end{pmatrix} \qquad \mathbf{T}^{-1} = \begin{pmatrix} \cos\zeta & \sin\zeta \\ -\sin\zeta & \cos\zeta \end{pmatrix}$$

Then construct the following matrix equation:

$$\begin{pmatrix} E_2 & 0 \\ 0 & E_1 \end{pmatrix} = \mathbf{T}^{-1}\mathbf{H}\mathbf{T} = \begin{pmatrix} \cos\zeta & \sin\zeta \\ -\sin\zeta & \cos\zeta \end{pmatrix}\begin{pmatrix} a & d \\ d & b \end{pmatrix}\begin{pmatrix} \cos\zeta & -\sin\zeta \\ \sin\zeta & \cos\zeta \end{pmatrix}$$

$$= \begin{pmatrix} a\cos^2\zeta + b\sin^2\zeta + 2d\cos\zeta\sin\zeta & d(\cos^2\zeta - \sin^2\zeta) + (b-a)\cos\zeta\sin\zeta \\ d(\cos^2\zeta - \sin^2\zeta) + (b-a)\cos\zeta\sin\zeta & b\cos^2\zeta + a\sin^2\zeta - 2d\cos\zeta\sin\zeta \end{pmatrix}$$

Consequently, by equating matching off-diagonal elements, we obtain

$$d(\cos^2\zeta - \sin^2\zeta) + (b-a)\cos\zeta\sin\zeta = 0$$

which solves to

$$\zeta = -\tfrac{1}{2}\arctan\left(\frac{2d}{b-a}\right)$$

Comment. The two-level system occurs widely in quantum mechanics, and we shall return to it in Chapter 6. The parametrization of the states in terms of the angle ζ is a very useful device, and we shall encounter it again.

The plausibility of the Schrödinger equation

The Schrödinger equation is properly regarded as a postulate of quantum mechanics, and hence we should not ask for a deeper justification. However, it is often more satisfying to set postulates in the framework of the familiar. In this section we shall see that the Schrödinger equation is a plausible description of the behaviour of matter by going back to the formulation of classical mechanics devised by W.R. Hamilton in the nineteenth century. We shall concentrate on the qualitative aspects of the approach: the calculations supporting these remarks will be found in *Further information 1*.

1.21 The propagation of light

In geometrical optics light travels in straight lines in a uniform medium, and we know that the physical nature of light is a wave motion. In classical mechanics particles travel in straight lines unless a force is present. Moreover, we know from the experiments performed at the end of the nineteenth century and the start of the twentieth that particles have a wave character. There are clearly deep analogies here. We shall therefore first establish how, in optics, wave motion can result in straight-line motion, and then argue by analogy about the wave nature of particles.

The basic rule governing light propagation in geometrical optics is **Fermat's principle of least time**. A simple form of the principle is that the path taken by a ray of light through a medium is such that its time of passage is a minimum. As an illustration, consider the relation between the angles of incidence and reflection for light falling on a mirror (Fig. 1.3). The briefest path between source, mirror, and observer is clearly the one corresponding to equal angles of incidence and reflection. In the case of refraction, it is necessary to take into account the different speeds of propagation in the two media. In Fig. 1.4, the geometrically straight path is not necessarily the briefest, because the light travels relatively slowly through the denser medium. The briefest path is in fact easily shown to be the one in which the angles of incidence θ_i and refraction θ_r are related by Snell's law, that $\sin\theta_r/\sin\theta_i = n_1/n_2$. (The refractive indexes n_1 and n_2 enter because the speed of light in a medium of refractive index n is c/n, where c is the speed of light in a vacuum.)

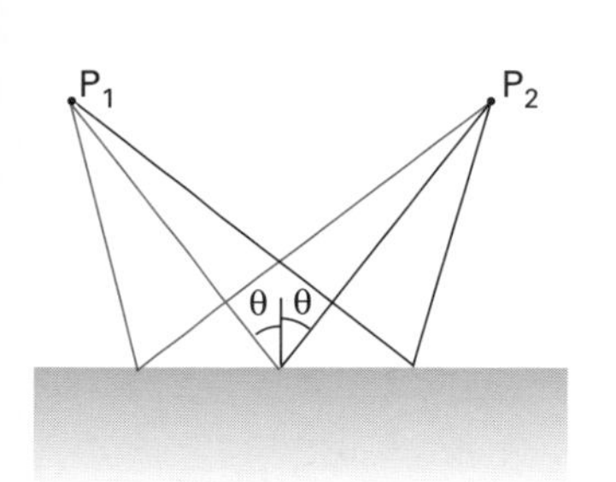

Fig. 1.3. When light reflects from a surface, the angle of reflection is equal to the angle of incidence.

How can the wave nature of light account for this behaviour? Consider the case illustrated in Fig. 1.5, where we are interested in the propagation of light between two fixed points P_1 and P_2. A wave of electromagnetic radiation travelling along some general path A arrives at P_2 with a particular phase that depends on its path length. A wave travelling along a neighbouring path A′ travels a different distance and arrives with a different phase. Path A has very many neighbouring paths, and there is destructive interference between the waves. Hence, an observer concludes that the light does not travel along a path like A. The same argument applies to every path between the two points, with one exception: the straight line path B. The neighbours of B do not interfere destructively with B itself, and it survives. The mathematical reason for this exceptional behaviour can be seen as follows.

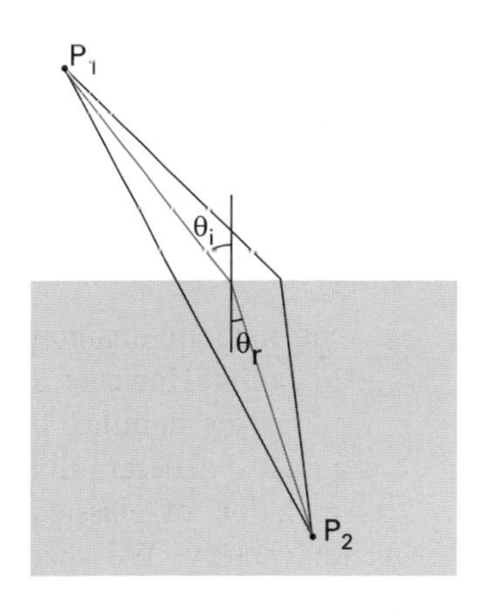

Fig. 1.4. When light is refracted at the interface of two transparent media, the angle of refraction, θ_r, and the angle of incidence, θ_i, are related by Snell's law.

The amplitude of a wave at some point x can be written $ae^{2\pi ix/\lambda}$, where λ is the wavelength. It follows that the amplitude at P_1 is $ae^{2\pi ix_1/\lambda}$ and that at P_2 it is $ae^{2\pi ix_2/\lambda}$. The two amplitudes are therefore related as follows:

$$\Psi(P_2) = ae^{2\pi ix_2/\lambda} = e^{2\pi i(x_2-x_1)/\lambda}ae^{2\pi ix_1/\lambda} = e^{2\pi i(x_2-x_1)/\lambda}\Psi(P_1)$$

This relation between the two amplitudes can be written more simply as

$$\Psi(P_2) = e^{i\phi}\Psi(P_1) \qquad \text{with } \phi = \frac{2\pi(x_2 - x_1)}{\lambda} \tag{45}$$

The function ϕ is the **phase length** of the straight line path. The relative phases at P_2 and P_1 for waves that travel by curved paths are related by an expression of the same kind, but with the phase length determined by the length, L, of the path:

$$\phi = \frac{2\pi L}{\lambda} \tag{46}$$

Now we consider how the path length varies with the distortion of the path from a straight line. If we distort the path from B to A in Fig. 1.5, ϕ changes as depicted in Fig. 1.6. Obviously, ϕ goes through a minimum at B. Now we arrive at the crux of the argument. Consider the phase length of the paths in the vicinity of A. The phase length of A′ is related to the phase length at A by the following

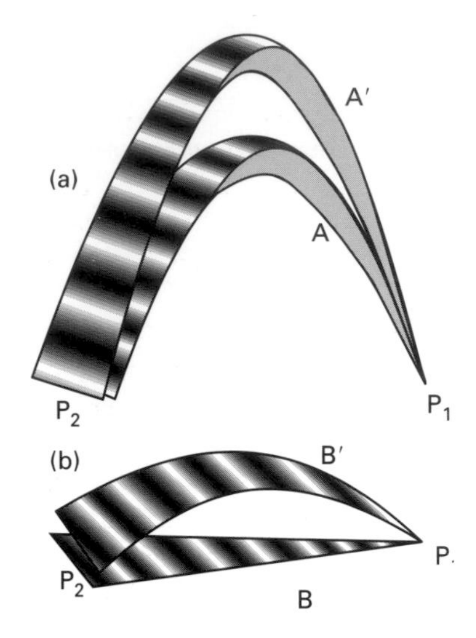

Fig. 1.5. (a) A curved path through a uniform medium has neighbours with significantly different phases at the destination point, and there is destructive interference between them. (b) A straight path between two points has neighbours with almost the same phase, and these paths do not interfere destructively.

Taylor expansion:

$$\phi(\mathrm{A}') = \phi(\mathrm{A}) + \left(\frac{\mathrm{d}\phi}{\mathrm{d}s}\right)_{\mathrm{A}} \delta s + \tfrac{1}{2}\left(\frac{\mathrm{d}^2\phi}{\mathrm{d}s^2}\right)_{\mathrm{A}} \delta s^2 + \dots \tag{47}$$

where δs is a measure of the distortion of the path. This expression should be compared with the similar expression for the path lengths of B and its neighbours:

$$\begin{aligned}\phi(\mathrm{B}') &= \phi(\mathrm{B}) + \left(\frac{\mathrm{d}\phi}{\mathrm{d}s}\right)_{\mathrm{B}} \delta s + \tfrac{1}{2}\left(\frac{\mathrm{d}^2\phi}{\mathrm{d}s^2}\right)_{\mathrm{B}} \delta s^2 + \dots \\ &= \phi(\mathrm{B}) + \tfrac{1}{2}\left(\frac{\mathrm{d}^2\phi}{\mathrm{d}s^2}\right)_{\mathrm{B}} \delta s^2 + \dots\end{aligned} \tag{48}$$

The term in δs is zero because the first derivative is zero at the minimum of the curve. In other words, to first order in the displacement, *straight line paths have neighbours with the same phase length*. On the other hand, *curved paths have neighbours with different phase lengths*. This difference is the reason why straight line propagation survives whereas curved paths do not: the latter have annihilating neighbours.

Two further points now need to be made. When the medium is not uniform, the wavelength of a wave varies with position. Because $\lambda = v/\nu$, and v, the speed of propagation is equal to c/n, where the refractive index n varies with position, a more general form of the phase length is

$$\phi = 2\pi \int_{\mathrm{P}_1}^{\mathrm{P}_2} \frac{\mathrm{d}x}{\lambda(x)} = \frac{2\pi\nu}{c} \int_{\mathrm{P}_1}^{\mathrm{P}_2} n(x)\,\mathrm{d}x \tag{49}$$

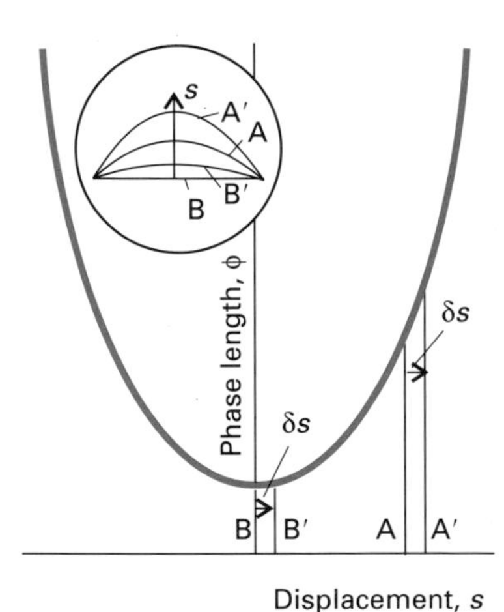

Fig. 1.6. The variation of phase length with displacement from a straight line path. The phase length at A′ differs from that at A by a first-order term; the phase lengths at B and B′ differ only to second order in the displacement.

The same argument applies, but because of the dependence of the refractive index on position, *a curved or kinked path may turn out to correspond to the minimum phase length*, and therefore have, to first order at least, no destructive neighbours. Hence, the path adopted by the light will be curved or kinked. The focusing caused by a lens is a manifestation of this effect.

The second point concerns the stringency of the conclusion that the minimum-phase-length paths have non-destructive neighbours. Because the wavelength of the radiation occurs in the denominator of the expression defining the phase length, waves of short wavelength will have larger phase lengths for a given path than radiation of long wavelength. The variation of phase length with wavelength is indicated in Fig. 1.7. It should be clear that neighbours annihilate themselves much more strongly when the light has a short wavelength than when it is long. Therefore, the rule that light (or any other form of wave motion) propagates itself in straight lines becomes more stringent as its wavelength shortens. Sound waves travel only in approximately straight lines; light waves travel in almost exactly straight lines. Geometrical optics is the limit of infinitely short wavelengths, where the annihilation by neighbours is so effective that the light appears to travel in perfectly straight lines.

1.22 The propagation of particles

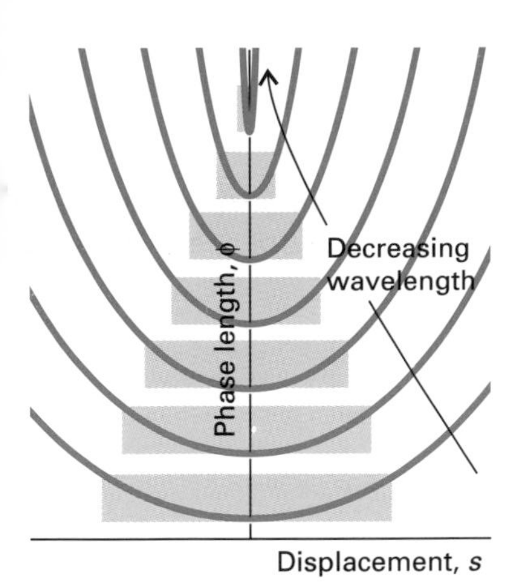

Fig. 1.7. The variation of phase length with wavelength. Interference between neighbours is most acute for short wavelengths. The geometrical limit corresponds to zero wavelength, where even infinitesimal neighbours interfere destructively and completely.

The path taken by a particle in classical mechanics is determined by Newton's laws. However, it turns out that these laws are equivalent to **Hamilton's principle**, which states that particles adopt paths between two given points such that the **action** S associated with the path is a minimum. There is clearly a striking analogy between Fermat's principle of least time and Hamilton's principle of least action.

The formal definition of action is given in *Further information 1*, where it is seen to be an integral taken along the path of the particle, just like the phase length in optics. When we turn to the question of why particles adopt the path of least action, we can hardly avoid the conclusion that the reason must be the same as why light adopts the path of least phase length. But to apply that argument to particles, we have to suppose that particles have an associated wave character. You can see that this attempt to 'explain' classical mechanics leads almost unavoidably to the heart of quantum mechanics and the duality of matter. We have the experimental evidence to encourage us to pursue the analogy; Hamilton did not.

1.23 The transition to quantum mechanics

The hypothesis we now make is that a particle is described by some kind of amplitude Ψ, and that amplitudes at different points are related by an expression of the form $\Psi(\mathrm{P}_2) = \mathrm{e}^{\mathrm{i}\phi}\Psi(\mathrm{P}_1)$. By analogy with optics, we say that the wave is propagated along the path that makes ϕ a minimum. But we also know that in the classical limit, the particle propagates along a path that corresponds to least action. As ϕ is dimensionless (because it appears as an exponent), the constant of proportionality between ϕ and S must have the dimensions of 1/action. Furthermore, we have seen that geometrical optics, the classical form of optics, corresponds to the limit of short wavelengths and very large phase lengths. In classical mechanics, particles travel along 'geometrical' trajectories, corresponding to large ϕ. Hence, the constant with the dimensions of action must be very small. The natural quantity to introduce is Planck's constant, or some small multiple of it. It turns out that agreement with experiment (that is, the correct form of the Schrödinger equation) is obtained if we use $\hbar$; we therefore conclude that we should write $\phi = S/\hbar$.

You should notice the relation between this approach and Heisenberg's. In his, a 0 was replaced by $\hbar$ (in the commutator $[x, p_x]$), and classical mechanics evolved into quantum mechanics. In the approach we are presenting here, a 0 has also been replaced by $\hbar$, for had we wanted *precise* geometrical trajectories, then we would have divided S by 0.

We have arrived at the stage where the amplitude associated with a particle is described by a relation of the form

$$\Psi(\mathrm{P}_2) = \mathrm{e}^{\mathrm{i}S/\hbar}\Psi(\mathrm{P}_1) \tag{50}$$

where S is the action associated with the path from P_1 (at x_1, t_1) to P_2 (at x_2, t_2). This expression lets us develop an equation of motion, because we can differentiate Ψ with respect to the time t_2:

$$\left(\frac{\partial\Psi(\mathrm{P}_2)}{\partial t_2}\right) = \frac{\mathrm{i}}{\hbar}\left(\frac{\partial S}{\partial t_2}\right)\mathrm{e}^{\mathrm{i}S/\hbar}\Psi(\mathrm{P}_1) = \frac{\mathrm{i}}{\hbar}\left(\frac{\partial S}{\partial t_2}\right)\Psi(\mathrm{P}_2)$$

One of the results derived in *Further information 1* is that the rate of change of the action is equal to $-E$, where E is the total energy, $T+V$:

$$\frac{\partial S}{\partial t} = -E \tag{51}$$

Therefore, the equation of motion at all points of a trajectory is

$$\frac{\partial\Psi}{\partial t} = -\frac{\mathrm{i}}{\hbar}E\Psi$$

The final step involves replacing E by its corresponding operator H, which then results in the time-dependent Schrödinger equation, eqn 1.21.

There are a few points that are worth noting about this justification. First, we have argued by analogy with classical optics, and have sought to formulate equations that are consistent with classical mechanics. It should therefore not be surprising that the approach might not generate some purely quantum mechanical properties. Indeed, we shall see later that the property of electron spin has been missed, for despite its evocative name, it has no classical counterpart. A related point is that the derivation has been entirely non-relativistic: at no point have we tried to ensure that space and time are treated on an equal footing. The alignment of relativity and quantum mechanics was achieved by P.A.M. Dirac, who found a way of treating space and time symmetrically, and in the process accounted for the existence of electron spin. Finally, it should be noted that the time-dependent Schrödinger equation is not a wave equation. A wave equation has a second derivative with respect to time, whereas the Schrödinger equation has a first derivative. We have to conclude that the time-dependent Schrödinger equation is therefore a kind of diffusion equation, an equation of the form

$$\frac{\partial f}{\partial t} = D\nabla^2 f \tag{52}$$

where f is a probability density and D is a diffusion coefficient. There is perhaps an intuitive satisfaction in the notion that the solutions of the basic equation of quantum mechanics evolve by some kind of diffusion.

Problems

1.1 Which of the following operations are linear and which are nonlinear? (a) integration, (b) extraction of a square root, (c) translation (replacement of x by $x + a$, where a is a constant), (d) inversion (replacement of x by $-x$).

1.2 Find the operator for position x if the operator for momentum p is taken to be $(\hbar/2m)^{1/2}(A + B)$, with $[A, B] = 1$ and all other commutators zero.

1.3 Which of the following functions are eigenfunctions of (a) $\mathrm{d}/\mathrm{d}x$, (b) $\mathrm{d}^2/\mathrm{d}x^2$: (i) e^{ax}, (ii) e^{ax^2}, (iii) x, (iv) x^2, (v) $ax + b$, (vi) $\sin x$?

1.4 Construct quantum mechanical operators in the position representation for the following observables: (a) kinetic energy in one and in three dimensions, (b) the inverse separation, $1/x$, (c) electric dipole moment, (d) z-component of angular momentum, (e) the mean square deviations of the position and momentum of a particle from the mean values.

1.5 Repeat Problem 1.4, but find operators in the momentum representation. *Hint*. The observable $1/x$ should be regarded as x^{-1}; hence the operator required is the inverse of the operator for x.

1.6 In relativistic mechanics, energy and momentum are related by the expression $E^2 = p^2c^2 + m^2c^4$. Show that when $p^2c^2 \ll m^2c^4$ this reduces to $E = p^2/2m + mc^2$. Construct the relativistic analogue of the Schrödinger equation from the relativistic expression. What can be said about the conservation of probability?

1.7 Confirm that the operators (a) $T = -(\hbar^2/2m)(\mathrm{d}^2/\mathrm{d}x^2)$ and (b) $l_z = (\hbar/\mathrm{i})(\mathrm{d}/\mathrm{d}\phi)$ are hermitian. *Hint*. Consider the integrals $\int_0^L \psi^* T\psi\,\mathrm{d}x$ and $\int_0^{2\pi} \psi^* l_z\psi\,\mathrm{d}\phi$ and integrate by parts.

1.8 Demonstrate that the linear combinations $A + \mathrm{i}B$ and $A - \mathrm{i}B$ are not hermitian if A and B are hermitian operators.

1.9 Evaluate the expectation values of the operators p_x and p_x^2 for a particle with wavefunction $\sqrt{2/L}\sin(\pi x/L)$ in the range 0 to L.

1.10 Are the linear combinations $2x - y - z$, $2y - x - z$, $2z - x - y$ linearly independent or not?

1.11 Evaluate the commutators (a) $[x, y]$, (b) $[p_x, p_y]$, (c) $[x, p_x]$, (d) $[x^2, p_x]$, (e) $[x^n, p]$.

1.12 Evaluate the commutators (a) $[(1/x), p_x]$, (b) $[(1/x), p_x^2]$, (c) $[xp_y - yp_x, yp_z - zp_y]$, (d) $[x^2(\mathrm{d}^2/\mathrm{d}y^2), y(\mathrm{d}/\mathrm{d}x)]$.

1.13 Show that (a) $[A, B] = -[B, A]$, (b) $[A^m, A^n] = 0$ for all m, n, (c) $[A^2, B] = A[A, B] + [A, B]A$, (d) $[A, [B, C]] + [B, [C, A]] + [C, [A, B]] = 0$.

1.14 Evaluate the commutator $[l_y, [l_y, l_z]]$ given that $[l_x, l_y] = \mathrm{i}\hbar l_z$, $[l_y, l_z] = \mathrm{i}\hbar l_x$, and $[l_z, l_x] = \mathrm{i}\hbar l_y$.

1.15 The operator e^A has a meaning if it is expanded as a power series: $\mathrm{e}^A = \sum_n (1/n!)A^n$. Show that if $|a\rangle$ is an eigenstate of A with eigenvalue a, then it is also an eigenstate of e^A. Find the latter's eigenvalue.

1.16 (a) Show that $\mathrm{e}^A\mathrm{e}^B = \mathrm{e}^{A+B}$ only if $[A, B] = 0$. (b) If $[A, B] \neq 0$ but $[A, [A, B]] = [B, [A, B]] = 0$, show that $\mathrm{e}^A\mathrm{e}^B = \mathrm{e}^{A+B}\mathrm{e}^f$, where f is a simple function of $[A, B]$. *Hint*. This is another example of the differences between operators (q-numbers) and ordinary numbers (c-numbers). The simplest approach is to expand the exponentials and to collect and compare terms on both sides of the equality. Note that $\mathrm{e}^A\mathrm{e}^B$ will give terms like $2AB$ while e^{A+B} will give $AB + BA$. Be careful with order.

1.17 Evaluate the commutators (a) $[H, p_x]$ and (b) $[H, x]$, where $H = p^2/2m + V(x)$. Choose (i) $V(x) = V$, a constant, (ii) $V(x) = \frac{1}{2}kx^2$, (iii) $V(x) \rightarrow V(r) = e^2/4\pi\varepsilon_0 r$.

1.18 Evaluate (via eqn 1.34) the limitation on the simultaneous specification of the following observables: (a) the position and momentum of a particle, (b) the three components of linear momentum of a particle, (c) the kinetic energy and potential energy of a particle, (d) the electric dipole moment and the total energy of a one-dimensional system, (e) the kinetic energy and the position of a particle in one dimension.

1.19 Use eqn 1.35 to find expressions for the rate of change of the expectation values of position and momentum of a harmonic oscillator; solve the pair of differential equations, and show that the expectation values change in time in the same way as for a classical oscillator.

1.20 Confirm that the z-component of angular momentum, $l_z = (\hbar/\mathrm{i})\mathrm{d}/\mathrm{d}\phi$, is a constant of the motion for a particle on a ring with uniform potential energy $V(\phi) = V$.

1.21 The only nonzero matrix elements of x and p_x for a harmonic oscillator are

$$\langle v+1|x|v\rangle = \left(\frac{\hbar}{2m\omega}\right)^{1/2}(v+1)^{1/2} \qquad \langle v-1|x|v\rangle = \left(\frac{\hbar}{2m\omega}\right)^{1/2} v^{1/2}$$

$$\langle v+1|p_x|v\rangle = \mathrm{i}\left(\frac{\hbar m\omega}{2}\right)^{1/2}(v+1)^{1/2} \qquad \langle v-1|p_x|v\rangle = -\mathrm{i}\left(\frac{\hbar m\omega}{2}\right)^{1/2} v^{1/2}$$

(and their hermitian conjugates). Write out the matrices of x and p_x explicitly (label the rows and columns $v = 0, 1, 2, \ldots$) up to $v = 4$, and confirm by matrix multiplication that they satisfy the commutation rule. Construct the hamiltonian matrix by forming $p_x^2/2m + \frac{1}{2}kx^2$ by matrix multiplication and addition, and infer the eigenvalues.

1.22 Use the completeness relation, eqn 1.42, and the information in Problem 1.21 to deduce the value of the matrix element $\langle v|xp_x^2x|v\rangle$.

1.23 Write the time-independent Schrödinger equations for (a) the hydrogen atom, (b) the helium atom, (c) the hydrogen molecule, (d) a free particle, (e) a particle subjected to a constant, uniform force.

1.24 The time-dependent Schrödinger equation is separable when V is independent of time. (a) Show that it is also separable when V is a function only of time and uniform in space. (b) Solve the pair of equations. Let $V(t) = V\cos\omega t$; find an expression for $\Psi(x,t)$ in terms of $\Psi(x,0)$. (c) Is $\Psi(x,t)$ stationary in the sense specified in Section 1.12?

1.25 The ground-state wavefunction of a hydrogen atom has the form $\psi(r) = N\mathrm{e}^{-ar}$, a being a collection of fundamental constants with the magnitude 53 pm. Normalize this spherically symmetrical function. *Hint.* The volume element is $\mathrm{d}\tau = \sin\theta\,\mathrm{d}\theta\,\mathrm{d}\phi\,r^2\mathrm{d}r$, with $0 \le \theta \le \pi$, $0 \le \phi \le 2\pi$, and $0 \le r < \infty$. 'Normalize' always means 'normalize to unity' in this text.

1.26 A particle in an infinite one-dimensional system was described by the wavefunction $\psi(x) = N\mathrm{e}^{-x^2/2\Gamma^2}$. Normalize this function. Calculate the probability of finding the particle in the range $-\Gamma \le x \le \Gamma$. *Hint.* The integral encountered in the second part is the error function. It is defined and tabulated in M. Abramowitz and I.A. Stegun, *Handbook of mathematical functions*, Dover (1965).

1.27 An excited state of the system in the previous problem is described by the wavefunction $\psi(x) = Nx\mathrm{e}^{-x^2/2\Gamma^2}$. Where is the most probable location of the particle?

1.28 On the basis of the information in Problem 1.25, calculate the probability density of finding the electron (a) at the nucleus, (b) at a point in space 53 pm from the nucleus. Calculate the probabilities of finding the electron inside a region of volume 1.0 pm^3 located at these points.

1.29 (a) Calculate the probability of the electron being found anywhere within a sphere of radius 53 pm for the atom defined in Problem 1.25. (b) If the radius of the atom is defined as the radius of the sphere inside which there is a 90 per cent probability of finding the electron, what is the atom's radius?

1.30 Explore the concept of phase length as follows. First, consider two points P_1 and P_2 separated by a distance l, and let the paths taken by waves of wavelength λ be a straight line from P_1 to a point a distance d above the midpoint of the line P_1P_2, and then on to P_2. Find an expression for the phase length and sketch it as a function of d for various values of λ. Confirm explicitly that $\phi' = 0$ at $d = 0$.

1.31 Confirm that the path of minimum phase length for light passing from one medium to another corresponds to the refraction of light at their interface as given by Snell's law (Section 1.21).

1.32 Show that if the Schrödinger equation had the form of a true wave equation, then the integrated probability would be time-dependent. *Hint*. A wave equation has $\kappa\partial^2/\partial t^2$ in place of $\partial/\partial t$, where κ is a constant with the appropriate dimensions (what are they?). Solve the time component of the separable equation and investigate the behaviour of $\int \Psi^*\Psi\, d\tau$.

Further reading

Introduction to quantum mechanics. B.H. Bransden and C.J. Joachain; Longman, London and Wiley, New York (1989).
Introduction to quantum mechanics. D.J. Griffiths; Prentice-Hall, Englewood Cliffs (1995).
Quantum mechanics. L.E. Ballentine; Prentice-Hall, Englewood Cliffs (1990).
Quantum mechanics. W. Greiner; Springer, New York (1994).
Quantum mechanics. P.J.E. Peebles; Princeton University Press, Princeton (1992).
Quantum mechanics. A.I.M. Rae; Institute of Physics, Bristol (1992).
Quantum mechanics: a first course. B.C. Reed; Wuerz Publishing, Winnipeg (1994).
Quantum mechanics II: a second course in quantum theory. R.H. Landau; Wiley, New York (1995).
Quantum mechanics. A.S. Davydov; Pergamon Press, Oxford (1976).
Foundations of quantum mechanics. J.M. Jauch; Addison-Wesley, Reading, Mass. (1968).
The principles of quantum mechanics. P.A.M. Dirac; Clarendon Press, Oxford (1958).
Mathematical foundations of quantum mechanics. J. von Neumann; Princeton University Press, Princeton (1955).
Matrix mechanics. H.S. Green; Noordhoff, Groningen (1965).
Linear operators for quantum mechanics. T.F. Jordan; Wiley, New York (1969).
Quantum mechanics in simple matrix form. T.F. Jordan; Wiley, New York (1986).
The uncertainty principle and foundations of quantum mechanics. W.C. Price and S.S. Chissick (ed.); Wiley, London (1977).
Mathematical methods in the physical sciences. M.L. Boas; Wiley, New York (1983).
Mathematics in chemistry: an introduction to modern methods. H.G. Hecht; Prentice-Hall, Englewood Cliffs (1990).

2 Linear motion and the harmonic oscillator

In this chapter we consider the quantum mechanics of translation and vibration. Both types of motion can be solved exactly in certain cases, and both are important not only in their own right but also because they form a basis for the description of the more complicated types of motion encountered in quantum chemistry. Translational motion also has the advantage of introducing in a simple way many of the striking features of quantum mechanics. However, there are certain features of wavefunctions that are common to all the problems we shall encounter, and we start by considering them. As we shall see, it is the combination of these features with the solution of the Schrödinger equation that results in one of the most characteristic features of quantum mechanics, the quantization of energy.

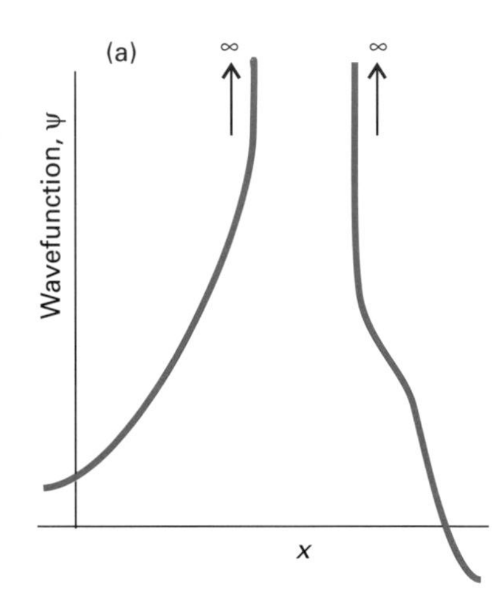

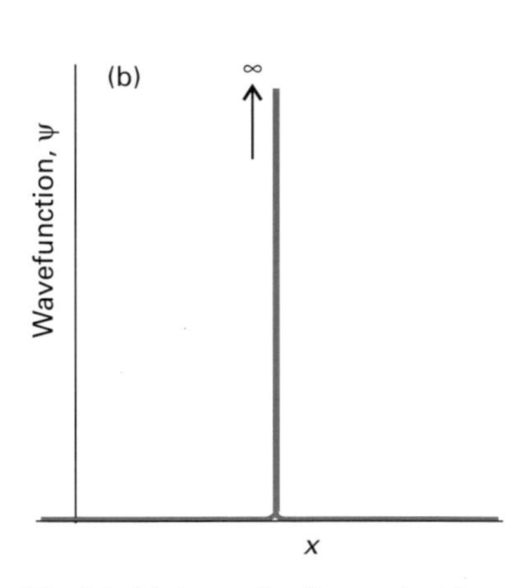

Fig. 2.1. (a) A wavefunction must not be infinite over a finite range because it is then not square-integrable. (b) However, it may be infinite over an infinitesimal range for such a function is square-integrable (it corresponds to a Dirac δ-function).

The characteristics of acceptable wavefunctions

We have seen that the Born interpretation of the wavefunction ψ, like that of its time-dependent version Ψ, is that $\psi^*\psi$ is a probability density. It must therefore be square-integrable (Section 1.10), and specifically the wavefunction must satisfy the normalization condition

$$\int \psi^*\psi \, \mathrm{d}\tau = 1 \tag{1}$$

The implication of this condition is that the wavefunction cannot become infinite over a finite region of space, as in Fig. 2.1. If it did become infinite, the integral would be infinite, and the Born interpretation would be untenable. This restriction does not rule out the possibility that the wavefunction could be infinite over an *infinitesimal* region of space because then its integral may remain finite (the integral is the area under the curve of $\psi^*\psi$, and infinitely high $\times$ infinitely narrow may result in a finite area). Such a wavefunction corresponds to the localization of a particle at a single, precise point, like the centre of mass of a speck of dust on a table at absolute zero. By the uncertainty principle, we know that a particle described by a wavefunction of this kind would have an infinitely uncertain linear momentum.

Another implication of the Born interpretation is that for $\psi^*\psi$ to be a valid probability density, it must be *single valued*; that is, have one value at each point. The Born interpretation would be untenable if $\psi^*\psi$ could take more than one value at each point of space. In simple applications, the single-valued character of $\psi^*\psi$ implies that ψ itself must be single valued, and we shall normally impose that condition on the wavefunction. (The exceptions arise when electron spin is taken into account.)

There are two other conditions on the form of the wavefunction that stem

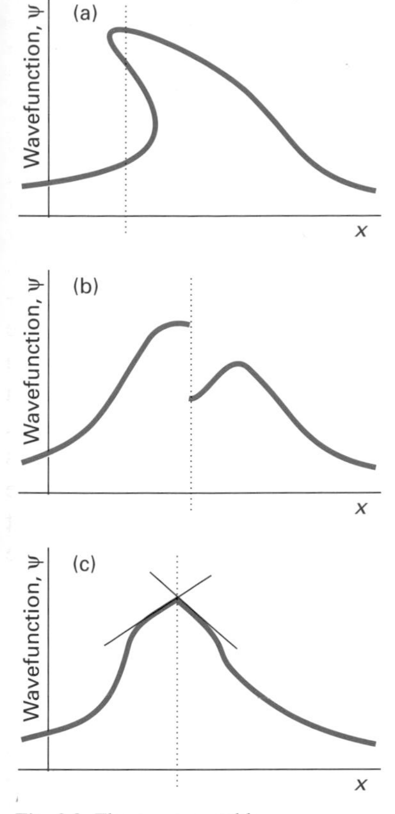

Fig. 2.2. Three unacceptable wavefunctions. (a) A wavefunction that is not single valued everywhere. (b) A discontinuous wavefunction. (c) A wavefunction with a discontinuous slope.

from the requirement that ψ is a solution of a second-order differential equation, and therefore that its second derivative should exist. In the first place, in order to define a second derivative of a function, it is necessary that the function itself should be continuous (Fig. 2.2). A weaker requirement is that the first derivative should also be continuous. This condition is weaker because there are systems—those with certain ill-behaved potential energies—where the restriction is too severe. For example, when we deal with a particle in a box, we shall encounter a potential energy that is excessively ill-behaved because it jumps from zero to infinity in an infinitesimal distance (when the particle touches the wall of the box). In such a case there is no need for the particle to have a continuous first derivative.

In summary, in general a wavefunction must satisfy the following conditions:

1. It must be single valued (strictly, $\psi^*\psi$ should be single valued).
2. It must not be infinite over a finite range.
3. It must be continuous everywhere.
4. It must have a continuous first derivative, except at ill-behaved regions of the potential.

Some general remarks on the Schrödinger equation

The time-independent Schrödinger equation is an equation for the *curvature* of the wavefunction. With this idea established, it is possible to guess the form of its solutions even when the form of the potential energy is complicated.

The curvature of a function is its second derivative.[1] A function with positive curvature looks like $\smile$ and one with negative cuvature looks like $\frown$. The one-dimensional Schrödinger equation expresses the curvature of the wavefunction as

$$\frac{d^2\psi}{dx^2} = \frac{2m}{\hbar^2}(V - E)\psi \tag{2}$$

Therefore, if we know the values of $V - E$ and ψ at a particular point, we can state the curvature of the wavefunction there. In this section, we concentrate on the qualitative features of the equations, because they show us how to unfold the qualitative features of the solutions without the clutter of detail.

2.1 The curvature of the wavefunction

First, we should note that *the curvature of ψ is proportional to the amplitude of ψ*. Therefore, for a given value of $V - E$ when ψ is large, the curvature is large. Where ψ falls towards zero its curvature decreases (Fig. 2.3). Where ψ is zero,

[1]This use of the term curvature is colloquial. In fact, in mathematics, curvature is a precisely defined concept in the theory of surfaces: in one dimension the curvature of a function f is equal to

$$\text{curvature of } f = \frac{(d^2f/dx^2)}{\{1 + (df/dx)^2\}^{3/2}}$$

We shall invariably adopt the colloquial meaning, and identify curvature with the second derivative d^2f/dx^2.

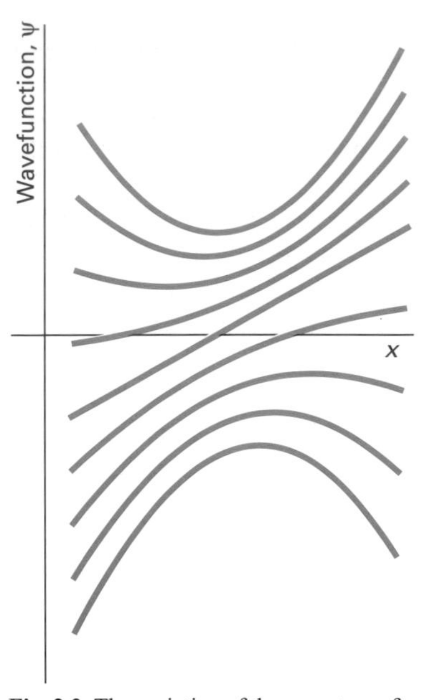

Fig. 2.3. The variation of the curvature of a wavefunction with its amplitude, for constant energy.

its curvature is zero.

Next, note that where E is greater than V, the factor $V - E$ is negative, and so the sign of the curvature of ψ is opposite to the sign of ψ itself. That is, if $E > V$ and $\psi > 0$, then ψ has negative curvature and looks like ⌒. On the other hand, where E is smaller than V, $V - E$ is positive, and the curvature of ψ has the same sign as its amplitude. A wavefunction with positive amplitude would then have a positive curvature, and look like ⌣. Finally, the curvature is proportional to the difference $|V - E|$. If the total energy is greatly in excess of the potential energy, then the curvature is large. These features are summarized in Fig. 2.4, which contains all the information we need to solve the Schrödinger equation qualitatively for a one-particle, one-dimensional system.

2.2 Qualitative solutions

Consider a system in which the potential energy depends on position as depicted in Fig. 2.5. Suppose that at x'' the wavefunction has the amplitude and slope as shown as A, and that the energy of the particle is E. Note that $E < V$ for positions to the right of x' but that $E > V$ to the left of x': the sign of $E - V$ therefore changes at x'. Because ψ_A is positive at x'' and $V < E$, the curvature of ψ_A is negative. The wavefunction remains positive at x', but to the right of that point $V > E$. Its curvature therefore becomes positive, and it bends away from the x-axis and rises to infinity as x increases. Therefore, according to the Born interpretation, ψ is an inadmissable wavefunction.

With this failure in mind, we select a function ψ_B that has a different slope at x'' but the same amplitude. This function has a negative curvature (because $E > V$). Its curvature becomes positive to the right of x' because its amplitude is positive but now $E < V$. The change in curvature is insufficient to stop ψ_B falling through zero to a negative value, and as it does so its curvature changes sign. This negative curvature forces ψ_B to a negatively infinite value as x increases, and it is therefore an inadmissable wavefunction.

Learning from our mistakes, we now select a wavefunction ψ_C that has a slope intermediate between those of ψ_A and ψ_B. Its curvature changes sign at x' but it does so in such a way that ψ_C approaches zero asymptotically as x increases. As it does so, its curvature lessens (because the curvature is proportional to the amplitude) and it curls off to neither positive nor negative infinity. Such a wavefunction is acceptable. Note that for the potential shown in Fig. 2.5, a well-behaved wavefunction can be found for *any* value of E simply by adjusting the amplitude or slope of the function at x''. Therefore, the energies of such systems are not quantized.

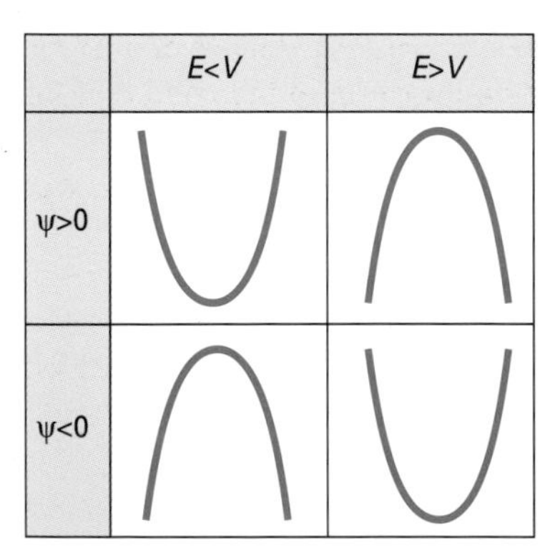

Fig. 2.4. The variation of the curvature of a wavefunction with the sign of the wavefunction at the point in question and the relative size of the energy and potential energy at the point.

2.3 The emergence of quantization

Now that we have seen the sensitivity of the wavefunction to a potential that rises to a large value only on one side, it should be easy to appreciate the difficulty of fitting a function to a system in which the potential confines the particle on both sides (Fig. 2.6). The function ψ_C that was acceptable in the system shown in Fig. 2.5 has been traced to x''' to the left of which V rises above E again. We see that its behaviour at this boundary means that ψ_C is unacceptable. In fact, in general it is impossible to find an acceptable solution for an arbitrary value of E. *Only for some values of E is it possible to construct a well-behaved*

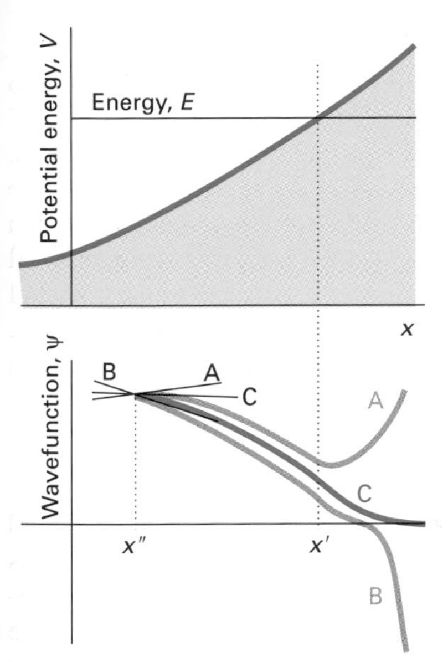

Fig. 2.5. The acceptability of a wavefunction is determined by the amplitude and slope at a particular point and the consequent implications on the behaviour of the wavefunction at the boundary. Only C is acceptable.

function. One such function is ψ_D in Fig. 2.6. In other words, *the energy is quantized in a system with a boundary on each side.*

The considerable importance of this conclusion cannot be overemphasized. The Schrödinger equation, being a differential equation, has an infinite number of solutions. It has *mathematically acceptable* solutions for any value of E. However, the Born interpretation imposes restrictions on the solutions. When the system has boundaries that confine the particle to a finite region, almost all the solutions are unacceptable: acceptable solutions occur only for special values of E. That is, *energy quantization is a consequence of boundary conditions.*

The diagram in Fig. 2.7 depicts the effect of boundaries on the quantization of the energy of a particle. Quantization occurs only when the particle is confined to a finite region of space. When its energy exceeds E' the particle can escape to positive values of x, and when its energy exceeds E'' the particle can travel indefinitely to positive and negative values of x. Furthermore, as the potential becomes less confining (that is, when the region for which $E > V$ becomes larger), the separation between neighbouring quantized levels is reduced because it gets progressively easier to find energies that give well-behaved functions. The region of quantized energy is generally taken to signify that we are dealing with **bound states** of a system, in which the wavefunction is localized in a definite region (like an electron in a hydrogen atom). The region of non-quantized energy is typically associated with **scattering problems** in which projectiles collide and then travel off to infinity. We introduce both types of solution in this chapter, but delay the complications of scattering problems until Chapter 14 at the end of the book.

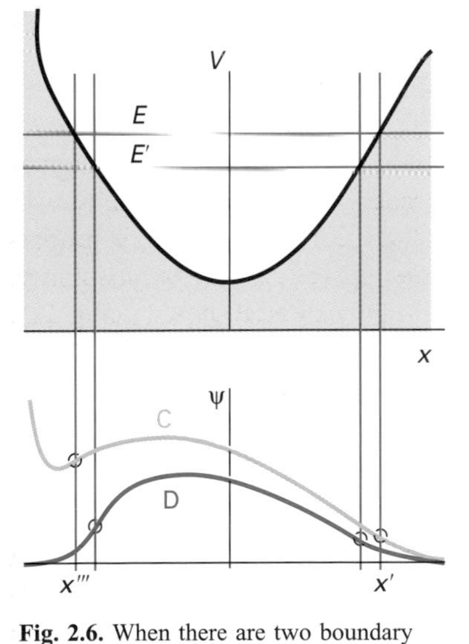

Fig. 2.6. When there are two boundary conditions to satisfy (in the sense that the particle is bounded), then it is possible to find acceptable solutions only for certain values of E. That is, the need to satisfy boundary conditions implies the quantization of the energy of the system.

2.4 Penetration into nonclassical regions

A glance at Fig. 2.6 shows that a wavefunction may be nonzero even where $E < V$; that is, ψ need not vanish where the kinetic energy is negative. A negative kinetic energy is forbidden classically, and the fact that a particle may be found in a region where the kinetic energy is negative is an example of quantum mechanical 'penetration'. We shall elaborate on this term in the course of this chapter.

The penetration of a particle into a region where the kinetic energy is negative is no particular cause for alarm. We have seen that observed energies are the expectation values of operators, and the expectation value of the kinetic energy operator is invariably positive (it is proportional to the square of an hermitian operator, p_x). Secondly, because the eigenvalues of the squares of hermitian operators are always positive (Example 1.6), each individual determination of the kinetic energy will have a positive outcome. Finally, any attempt to confine a particle within a nonclassical region, and then to measure its kinetic energy, will be doomed by the uncertainty principle. The confinement would have to be to such a small region that the corresponding uncertainty in momentum, and hence in kinetic energy, would be so great that we would be unable to conclude that the kinetic energy was indeed negative.

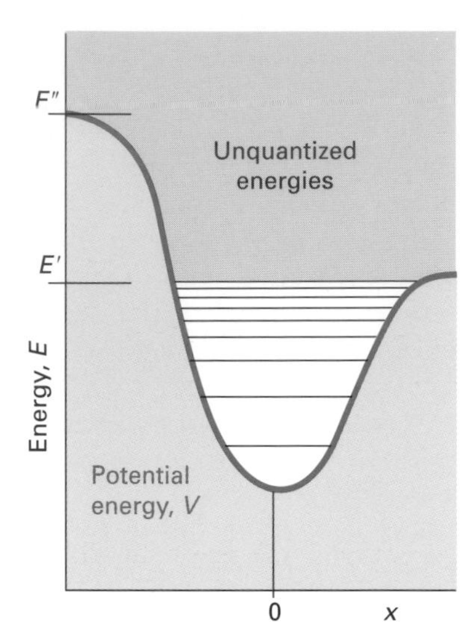

Fig. 2.7. A general summary of the role of boundaries: the system is quantized only if it is confined to a finite region of space. A single boundary does not entail quantization.

Translational motion

The easiest type of motion to consider is that of a completely free particle travelling in an unbounded one-dimensional region. Because the potential is constant, and may be chosen to be zero, the hamiltonian for the system is

$$H = -\frac{\hbar^2}{2m}\frac{\mathrm{d}^2}{\mathrm{d}x^2} \tag{3}$$

The time-independent Schrödinger equation, $H\psi = E\psi$, therefore has the form

$$-\frac{\hbar^2}{2m}\frac{\mathrm{d}^2\psi}{\mathrm{d}x^2} = E\psi \tag{4}$$

The general solutions of this equation are

$$\psi = A\mathrm{e}^{\mathrm{i}kx} + B\mathrm{e}^{-\mathrm{i}kx} \qquad k = \left(\frac{2mE}{\hbar^2}\right)^{1/2} \tag{5}$$

as may readily be checked by substitution. Because $\mathrm{e}^{\mathrm{i}kx} = \cos kx + \mathrm{i}\sin kx$, an alternative form of this solution is

$$\psi = C\cos kx + D\sin kx \tag{6}$$

In both forms, the solutions of the coefficients A, B, C, and D are to be found by considering the boundary conditions (see below). However, an important point is that functions of the form $\mathrm{e}^{\pm\mathrm{i}kx}$ are not square-integrable (Section 1.10), and so care needs to be taken with their interpretation. Indeed, because they correspond to a uniform probability distribution throughout space, they cannot be a description of real physical systems. To cope with this problem we need the concept of wavepacket (Section 2.8).

2.5 Energy and momentum

The first point to note about the solutions is that, as the motion is completely unconfined, the energy of the particle is not quantized. An acceptable solution exists for any value of E: we simply use the appropriate value of k in eqn 2.5.

The relation between the energy of a free particle and its linear momentum is $E = p^2/2m$. According to eqn 2.5, the energy is related to the parameter k by $E = k^2\hbar^2/2m$. It follows that the magnitude of the linear momentum of a particle described by the wavefunctions in eqn 2.5 is

$$p = k\hbar \tag{7}$$

This expression can be developed in a number of ways. For example, we can turn it round, and say that the form of the wavefunction of a particle with linear momentum of magnitude p is given by eqn 2.5 with $k = p/\hbar$. A second point is that the wavefunctions in eqn 2.5 have a definite wavelength. This may be easier to see in the case of eqn 2.6, because a wave of wavelength λ is commonly written as $\cos(2\pi x/\lambda)$ or as $\sin(2\pi x/\lambda)$. It follows that the wavelength of the wavefunction in eqn 2.6 is $\lambda = 2\pi/k$. That is, the wavefunction for a particle with linear momentum $p = k\hbar$ has a wavelength $\lambda = 2\pi/k$. It follows that the wavelength and linear momentum are related by

$$p = \frac{2\pi}{\lambda} \times \hbar = \frac{h}{\lambda} \tag{8}$$

This is the **de Broglie relation**.

2.6 The significance of the coefficients

The significance of the coefficients in the wavefunction can be determined by considering the effect of the linear momentum operator in the position representation, $p = (\hbar/\mathrm{i})\mathrm{d}/\mathrm{d}x$. Suppose initially that $B = 0$, then

$$p\psi = \frac{\hbar}{\mathrm{i}}\frac{\mathrm{d}\psi}{\mathrm{d}x} = k\hbar A\mathrm{e}^{\mathrm{i}kx} = k\hbar\psi \tag{9}$$

We see that the wavefunction is an eigenfunction of the linear momentum operator, and that its eigenvalue is $k\hbar$. Alternatively, if $A = 0$, then

$$p\psi = \frac{\hbar}{\mathrm{i}}\frac{\mathrm{d}\psi}{\mathrm{d}x} = -k\hbar B\mathrm{e}^{-\mathrm{i}kx} = -k\hbar\psi$$

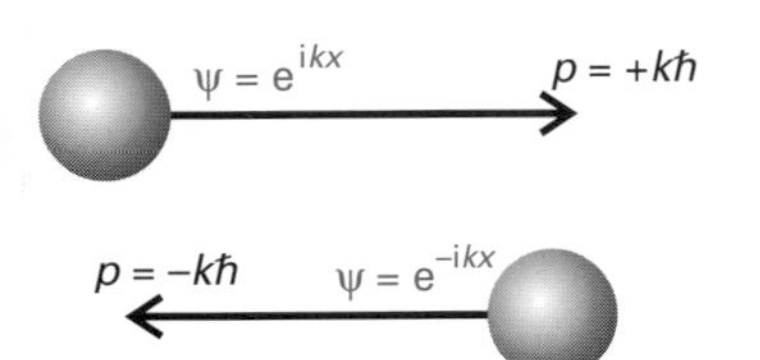

Fig. 2.8. Wavefunctions for a particle travelling to the right (positive x) and left (negative x) with a given magnitude of linear momentum ($k\hbar$) are each other's complex conjugate.

The distinction between the two solutions is the sign of the eigenvalue. Because linear momentum is a vector quantity, we are immediately led to the conclusion that *the two wavefunctions correspond to states of the particle with the same magnitude of linear momentum but in opposite directions*. This is a very important point, for it lets us write down the wavefunctions for particles that not only have a definite kinetic energy and therefore magnitude of linear momentum, but for which we can also specify directions of travel (Fig. 2.8).

The significance of the coefficients A and B should now be clearer: they depend on how the state of the particle was prepared. If it was shot from a gun in the direction of positive x, then $B = 0$. If it had been shot in the opposite direction (by the duelling partner), then its state would be described by a wavefunction with $A = 0$.

Now we turn to the significance of the coefficients C and D in the alternative form of the wavefunction. Suppose $D = 0$, so that the particle is described by the wavefunction $C\cos kx$. When we examine the effect of the momentum operator we find

$$p\psi = \frac{\hbar}{\mathrm{i}}\frac{\mathrm{d}}{\mathrm{d}x}C\cos kx = \mathrm{i}k\hbar C\sin kx$$

We see that the wavefunction is not an eigenfunction of the linear momentum operator. However, by writing

$$\psi = \tfrac{1}{2}C\mathrm{e}^{\mathrm{i}kx} + \tfrac{1}{2}C\mathrm{e}^{-\mathrm{i}kx}$$

we see that the wavefunction is a superposition of the two linear momentum eigenstates with equal coefficients. From the general considerations set out in Section 1.9, we can conclude that in a series of observations, we would obtain the linear momentum $+k\hbar$ half the time and $-k\hbar$ half the time, but we would not be able to predict which direction we would detect in any given observation. The expectation value of the linear momentum, its average value, is zero if its wavefunction is a sine (or a cosine) function.

An important general point illustrated by this discussion is that a complex

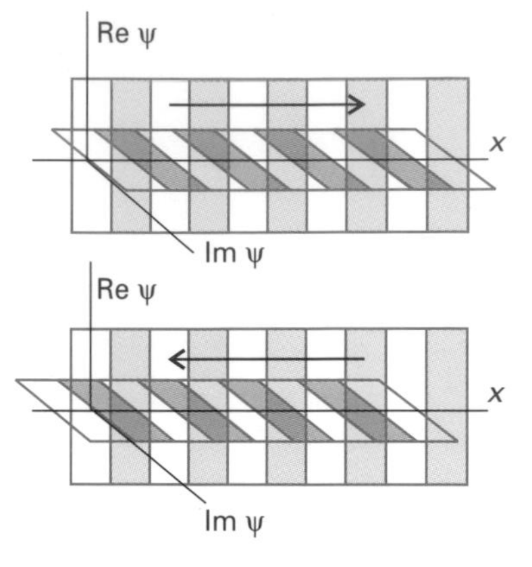

Fig. 2.9. The relative phase of the imaginary and real components of a wavefunction determine the direction of propagation of the particle: the real component seems to chase the imaginary component. The sign of the function is depicted by different shading.

wavefunction (such as e^{ikx}), or any function that cannot be made real simply by multiplication by a constant, corresponds to a definite state of linear momentum (in direction as well as in magnitude), whereas a real function (such as $\cos kx$) does not (Fig. 2.9; see also Exercise 1.6).

2.7 The flux density

Further insight into the form of the general solutions of the Schrödinger equation for free particles can be obtained by introducing a quantity called the **flux density**, J_x. The full usefulness of this quantity will become clear in later chapters where we are interested in the flow of charge in a molecule and the impact of beams of molecules on one another. The flux density in the x-direction is defined as follows:

$$J_x = \frac{\hbar}{2mi}\left(\Psi^* \frac{\partial \Psi}{\partial x} - \Psi \frac{\partial \Psi^*}{\partial x}\right) \tag{10}$$

For a state with a definite energy, the time-dependent phase factors in Ψ cancel, and the flux density is

$$J_x = \frac{\hbar}{2mi}\left(\psi^* \frac{d\psi}{dx} - \psi \frac{d\psi^*}{dx}\right) \tag{11}$$

To see its significance, we shall calculate the flux density for a system that is described by the wavefunction in eqn 2.5 with $B = 0$:

$$J_x = \frac{\hbar |A|^2}{2mi}\{e^{-ikx}(ik)e^{ikx} - e^{ikx}(-ik)e^{-ikx}\} = \frac{k\hbar |A|^2}{m}$$

For the wavefunction with $A = 0$ we find

$$J_x = -\frac{k\hbar |B|^2}{m}$$

We should now note that $\pm k\hbar/m$ is the classical velocity of the particle, so the flux density is the velocity multiplied by the probability that the particle is in that particular state.

Example 2.1 The uniformity of the flux density

Show that the flux density associated with a wavefunction of definite energy is independent of location.

Method. The question implies that we need to consider the form of J_x given in eqn 2.10 in conjunction with the time-dependent Schrödinger equation. To show that J_x is independent of x we need to establish an equation for $\partial J_x/\partial x$ by using the Schrödinger equation and show that it is zero.

Answer. The derivative of J_x is

$$\frac{\partial J_x}{\partial x} = \frac{\hbar}{2mi}\left(\Psi^* \frac{\partial^2 \Psi}{\partial x^2} - \Psi \frac{\partial^2 \Psi^*}{\partial x^2}\right)$$

(Two terms have cancelled when the derivative was taken.) We can now use the Schrödinger equation in the form

$$-\frac{\hbar^2}{2m}\frac{\partial^2\Psi}{\partial x^2} + V\Psi = \mathrm{i}\hbar\frac{\partial\Psi}{\partial t}$$
$$-\frac{\hbar^2}{2m}\frac{\partial^2\Psi^*}{\partial x^2} + V\Psi^* = -\mathrm{i}\hbar\frac{\partial\Psi^*}{\partial t}$$

This argument leads to

$$\frac{\partial J_x}{\partial x} = -\left(\Psi^*\frac{\partial\Psi}{\partial t} + \Psi\frac{\partial\Psi^*}{\partial t}\right) = -\frac{\partial(\Psi^*\Psi)}{\partial t}$$

However, we know that if the energy is well defined and V is independent of t, then $|\Psi|^2$ is independent of time; consequently,

$$\frac{\partial J_x}{\partial x} = 0$$

This derivation is in fact the justification of the definition of flux density in eqn 2.10.

Comment. The general equation for $\partial J_x/\partial x$ is the same as the continuity equation for the flow of an incompressible fluid of density ρ, which in three dimensions is

$$\frac{\partial\rho}{\partial t} + \nabla\cdot\mathbf{J} = 0$$

Exercise 2.1. Show that the flux density associated with a real wavefunction is zero.

2.8 Wavepackets

So far we have considered a case in which the energy of the particle is specified exactly. But suppose that the particle had been prepared with an imprecisely specified energy. Because the energy is imprecise, the wavefunction that describes the particle must be a superposition of functions corresponding to different energies. Such a superposition is called a **wavepacket**. For example, suppose the particle is a projectile fired towards positive x; then we know that the wavefunction of the projectile must be a superposition of functions of the form e^{ikx} with a range of values of k corresponding to the range of linear momenta (and hence kinetic energies) possessed by the particle.

A wavepacket is a wavefunction that has a nonzero amplitude in a small region of space and is close to zero elsewhere. In general, wavepackets move through space in a manner that resembles the motion of a classical particle. To see both these features, we consider a superposition of time-dependent wavefunctions of the form

$$\Psi_k(x,t) = A\mathrm{e}^{\mathrm{i}kx}\mathrm{e}^{-\mathrm{i}E_k t/} \tag{12}$$

The superposition is a linear combination of such functions, each one of which is weighted by a coefficient $g(k)$ called the **shape function** of the packet.

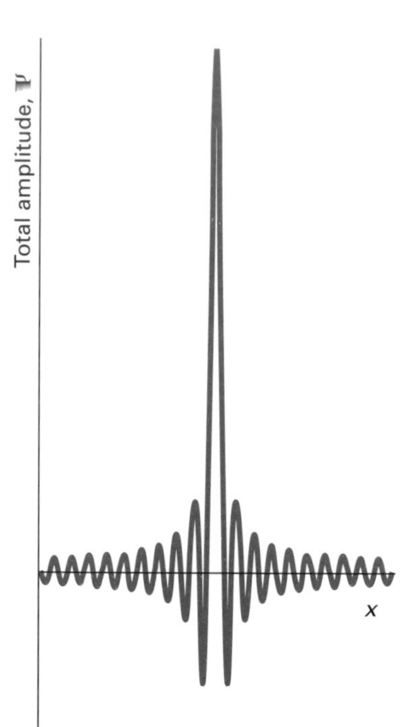

Fig. 2.10. A wavepacket formed by the superposition of many waves with different wavelengths. Twenty waves have been superimposed to produce this figure.

Because k is a continuously variable parameter, the sum is actually an integral over k, and so the wavepacket has the form

$$\Psi(x,t) = \int g(k)\Psi_k(x,t)\,\mathrm{d}k \tag{13}$$

The pictorial form of such a packet is shown in Fig. 2.10. It can be seen that as a result of the interference between the component waves at one instant the wavepacket has a large amplitude at one region of space, but because the time-dependent factor affects the phases of the waves that contribute to the superposition, the region of constructive interference changes with time (Fig. 2.11). It should not be hard to believe that the centre of the packet moves to the right, and this is confirmed by a mathematical analysis of the motion (see *Further information 5*). The classical motion of a projectile is captured by the motion of the wavepacket, and once again we see how classical mechanics emerges from quantum mechanics.

Penetration into and through barriers

A highly instructive extension of the results for free translational motion is to the case where the potential energy of a particle rises sharply to a high, constant value, perhaps to decline to zero again after a finite distance. Classically we know what happens: if a particle approaches the barrier from the left, then it will pass over it only if its initial energy is greater than the potential energy it possesses when it is inside the barrier. If its energy is lower than the height of the barrier, then the particle is reflected. To see what quantum mechanics predicts, we shall consider three types of barrier of increasing difficulty.

2.9 An infinitely thick potential wall

The Schrödinger equation for the problem falls apart into two equations, one for each zone in Fig. 2.12. The hamiltonians for the two zones are

$$\begin{aligned} \text{Zone I } (x<0): \quad & H = -\frac{\hbar^2}{2m}\frac{\mathrm{d}^2}{\mathrm{d}x^2} \\ \text{Zone II } (x>0): \quad & H = -\frac{\hbar^2}{2m}\frac{\mathrm{d}^2}{\mathrm{d}x^2} + V \end{aligned} \tag{14}$$

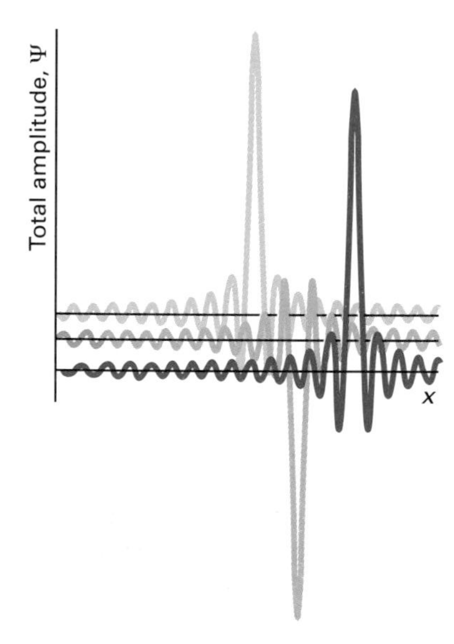

Fig. 2.11. Because each wave in a superposition oscillates with a different frequency, the point of constructive interference moves as time increases.

The corresponding equations are free-particle Schrödinger equations, except for the replacement of E by $E - V$ in Zone II. Therefore, the general solutions can be written down by referring to eqn 2.5:

$$\begin{aligned} \text{Zone I: } \psi &= A\mathrm{e}^{\mathrm{i}kx} + B\mathrm{e}^{-\mathrm{i}kx} \qquad & k\hbar &= (2mE)^{1/2} \\ \text{Zone II: } \psi &= A'\mathrm{e}^{\mathrm{i}k'x} + B'\mathrm{e}^{-\mathrm{i}k'x} \qquad & k'\hbar &= \{2m(E-V)\}^{1/2} \end{aligned} \tag{15}$$

We shall concentrate on the case when E is less than V, so that classically the particle cannot be found at $x > 0$ (inside the wall). The condition $E < V$ implies that k' is imaginary; so we shall write $k' = \mathrm{i}\kappa$, where κ is real. It then follows that

$$\text{Zone II: } \psi = A'\mathrm{e}^{-\kappa x} + B'\mathrm{e}^{\kappa x} \qquad \kappa\hbar = \{2m(V-E)\}^{1/2} \tag{16}$$

This wavefunction is a mixture of decaying and increasing exponentials: we see

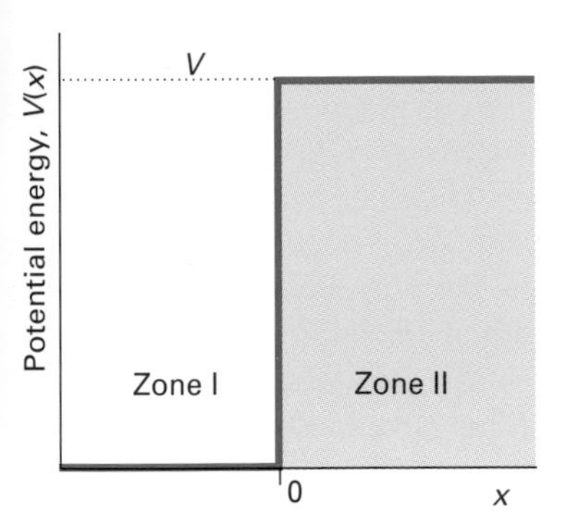

Fig. 2.12. The potential energy of a barrier of finite height but of semi-infinite extent.

that a wavefunction does not oscillate when $E < V$.

Because the barrier is infinitely wide, the increasing exponential must be ruled out because it leads to an infinite amplitude. Therefore, inside a barrier like that shown in Fig. 2.12, the wavefunction must be simply an exponentially decaying function, $e^{-\kappa x}$. One important point about this conclusion is that *the particle may be found inside a classically forbidden region*. This effect is an example of **penetration** into a barrier. The rapidity with which the wavefunction decays to zero is determined by the value of κ, for the amplitude of the wavefunction decreases to $1/e$ of its value at the edge of the barrier in a distance $1/\kappa$, which is called the **penetration depth**. The penetration depth decreases with the mass of the particle and the height of the barrier above the energy of the incident particle. Macroscopic particles have such large masses that their penetration depth is almost zero whatever the height of the barrier, and for all practical purposes they are not found in classically forbidden regions. An electron or a proton, on the other hand, may penetrate into a forbidden zone to an appreciable extent. For example, an electron that has been accelerated through a potential difference of 1.0 V, and which has acquired a kinetic energy of 1.0 eV, incident on a potential barrier equivalent to 2.0 eV, will have a wavefunction that decays to $1/e$ of its initial amplitude after 0.20 nm, which is comparable to the diameter of one atom. Hence, penetration can have very important effects on processes at surfaces, such as electrodes, and for all events on an atomic scale.

2.10 A barrier of finite width

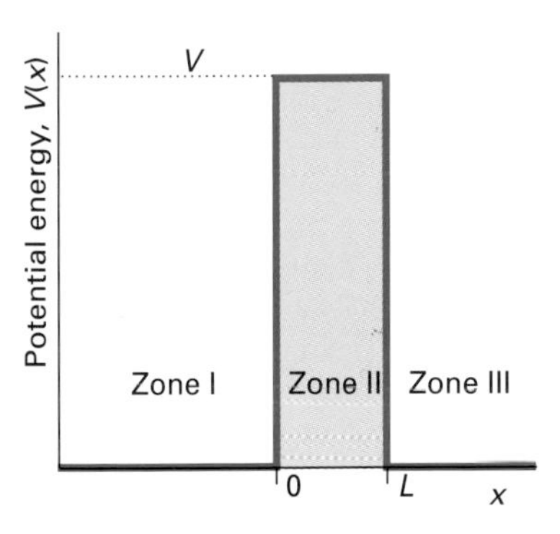

Fig. 2.13. The potential energy of a finite barrier. Particles incident from one side may be found on the opposite side of the barrier. According to classical mechanics, that is possible only if E is not less than V. According to quantum mechanics, however, barrier penetration may occur whatever the energy.

We now consider the case of a barrier of a finite width (Fig. 2.13). In particular, the potential energy, $V(x)$, has the form:

$$\begin{aligned} &\text{Zone I } (x < 0): && V(x) = 0 \\ &\text{Zone II } (0 \le x \le L): && V(x) = V \\ &\text{Zone III } (x > L): && V(x) = 0 \end{aligned} \tag{17}$$

The general solutions of the time-independent Schrödinger equation can immediately be written down:

$$\begin{aligned} &\text{Zone I: } \psi = Ae^{ikx} + Be^{-ikx} \\ &\text{Zone II: } \psi = A'e^{ik'x} + B'e^{-ik'x} \\ &\text{Zone III: } \psi = A''e^{ikx} + B''e^{-ikx} \end{aligned} \tag{18}$$

where k and k' are defined in the previous section. In scattering problems, of which this is a simple example, it is common to distinguish between 'incoming' and 'outgoing' waves. An **incoming wave** is a contribution to the total wavefunction with a component of linear momentum towards the target (from any direction). An **outgoing wave** is a contribution with a component of linear momentum away from the target. Each contribution corresponds to a flux of particles either towards or away from the target. In the problem we are currently considering, in Zone I the component A is the coefficient of the incoming wave and B the coefficient of the outgoing wave. In Zone III, A'' is the coefficient of the outgoing wave and B'' the coefficient of the incoming wave.

In this section we first consider solutions for $E < V$. Classically, the particle does not have enough energy to overcome the potential barrier. Therefore, for a particle incident from the left, the probability is exactly zero that it will be found on the right of the barrier ($x > L$). Quantum mechanically, however, the particle can be found on the right of the barrier even though $E < V$. In Zone II, the wavefunction has the form given in eqn 2.16. We need to note that the increasing exponential function in the wavefunction in this zone will not rise to infinity before the potential has fallen to zero again and oscillations resume. Therefore, the coefficient B' will not be zero. The values of the coefficients are established by using the acceptablity criteria for wavefunctions set out at the beginning of this chapter, and in particular the requirement that they and their slopes must be continuous. The continuity condition lets us match the wavefunction at the points where the zones meet, and therefore to find conditions for the coefficients. For example, the continuity of the amplitude at $x = 0$ and at $x = L$ leads to the two conditions

$$A + B = A' + B'$$
$$A'\mathrm{e}^{-\kappa L} + B'\mathrm{e}^{\kappa L} = A''\mathrm{e}^{\mathrm{i}kL} + B''\mathrm{e}^{-\mathrm{i}kL} \tag{19}$$

Similarly, the continuity of slopes at the same two points leads to the two conditions

$$\mathrm{i}kA - \mathrm{i}kB = -\kappa A' + \kappa B'$$
$$-\kappa A'\mathrm{e}^{-\kappa L} + \kappa B'\mathrm{e}^{\kappa L} = \mathrm{i}kA''\mathrm{e}^{\mathrm{i}kL} - \mathrm{i}kB''\mathrm{e}^{-\mathrm{i}kL} \tag{20}$$

These four equations give four conditions for finding six unknowns. The remaining conditions include a normalization requirement (one more condition) and a statement about the initial state of the particle (such as the fact that it approaches the barrier from the left).

Consider the case where the particles are prepared in Zone I with a linear momentum that carries them to the right. It then follows that the coefficient $B'' = 0$, because the exponential function it multiplies corresponds to particles with linear momentum towards the left on the right-hand side of the barrier, and there can be no such particles. That is, there is no incoming wave, no inward flux of particles, in Zone III. There may be particles travelling to the left on the left of the barrier because reflection can take place at the barrier. We can therefore identify the coefficient B as determining (via $|B|^2$) the flux density of particles reflected from the barrier in Zone I. The **reflection probability**, R, is the ratio of the reflected flux density to the incident flux density:

$$R = \frac{|B|^2}{|A|^2} \tag{21}$$

Similarly, the coefficient A'', the coefficient of the outgoing wave in Zone III, determines (via $|A''|^2$) the flux of particles streaming away from the barrier on the right. The **transmission probability**, P, is the ratio of the transmitted flux density to the incident flux density, and is given by

$$P = \frac{|A''|^2}{|A|^2} \tag{22}$$

The complete calculation of P involves only elementary manipulations of the

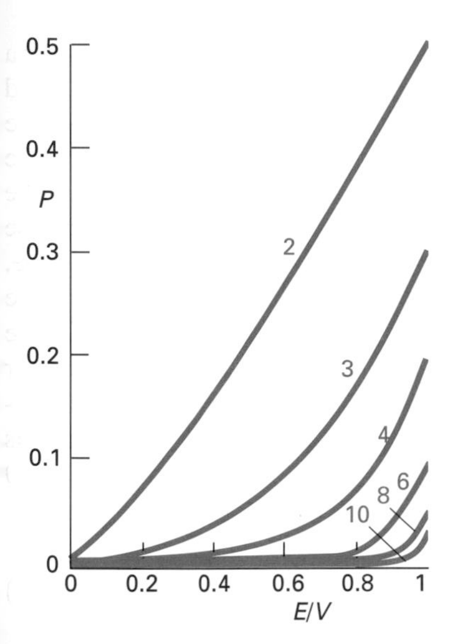

Fig. 2.14. The tunnelling probability through a finite rectangular barrier as a function of incident energy. The curves are labelled with the value of $L(2mV)^{1/2}/\hbar$.

relations given above, and the result is

$$P = \left(1 + \frac{(e^{\kappa L} - e^{-\kappa L})^2}{16(E/V)(1 - E/V)}\right)^{-1} \qquad \hbar\kappa = \{2mV(1 - E/V)\}^{1/2} \qquad (23)$$

and $R = 1 - P$. Because we have been considering energies $E < V$, P represents the probability that a particle incident on one side of the barrier will penetrate the barrier and emerge on the opposite side. That is, P is the probability of **tunnelling**, nonclassical penetration, through the barrier (Fig. 2.14).

We now deal with energies $E > V$. Classically, the particle now has sufficient energy to overcome the potential barrier. A particle incident from the left would have unit probability of being found on the right of the barrier. Once again, though, quantum mechanics gives a different result. To determine the expressions for P and R we could proceed as we did above for energies $E < V$, write down four relations for the six coefficients, and then manipulate them. However, it is considerably easier to take the expression for P given above and replace κ by $k'/i = -ik'$. This procedure[2] gives

$$P = \left(1 + \frac{\sin^2(k'L)}{4(E/V)(E/V - 1)}\right)^{-1} \qquad \hbar k' = \{2mV(E/V - 1)\}^{1/2} \qquad (24)$$

and, again, $R = 1 - P$. This function is plotted in Fig. 2.15.

The transmission coefficient, P, takes on its maximum value of 1 when $\sin(k'L) = 0$. In other words, at energies E corresponding to

$$k' - \frac{n\pi}{L} \qquad (25)$$

where n is a positive integer, there is a maximum in the transmission coefficient and transmission is complete. Furthermore, P has minima near

$$k' = \frac{n\pi}{2L} \qquad (26)$$

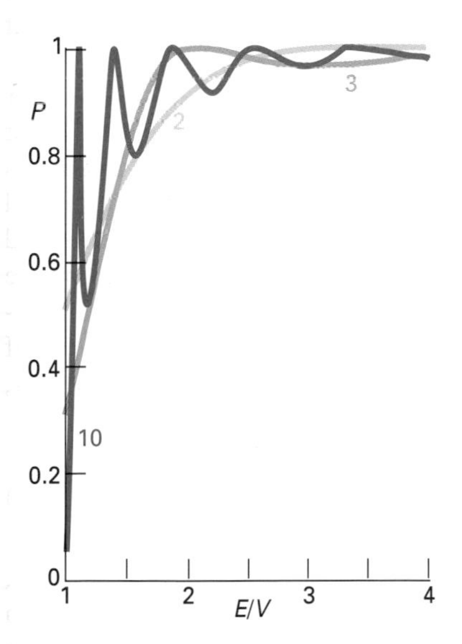

Fig. 2.15. The same as in the preceding illustration, but for $E > V$. Note that according to quantum mechanics, the particle may be reflected back from the barrier (so that $P < 1$) even though classically it has enough energy to pass over it.

At high energies ($E \gg V$), P approaches its classical value of 1. We see in Fig. 2.15 how the transmission coefficient for energies above the barrier height fluctuates between maxima and minima.

You should take note of two striking differences between the quantum mechanical and classical results. First, even when $E > V$, there is still a probability of the particle being reflected by the potential barrier even though classically it has enough energy to travel over the barrier. This phenomenon is known as **antitunnelling** or **nonclassical reflection**. Secondly, the strong variation of P with the energy of the incident particle is a purely quantum mechanical effect. The peaks in the transmssion coefficient for energies above V are examples of **scattering resonances**. We shall have more to say concerning resonances in Chapter 14 when we discuss scattering in general.

[2] We have used the relation $2i\sin x = e^{ix} - e^{-ix}$.

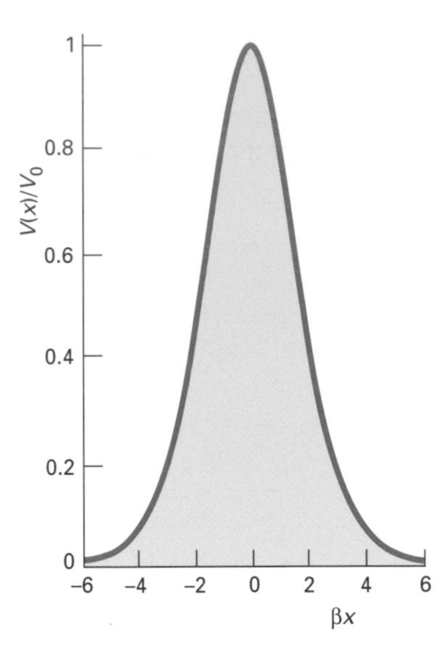

Fig. 2.16. The Eckart potential barrier, as described in the text.

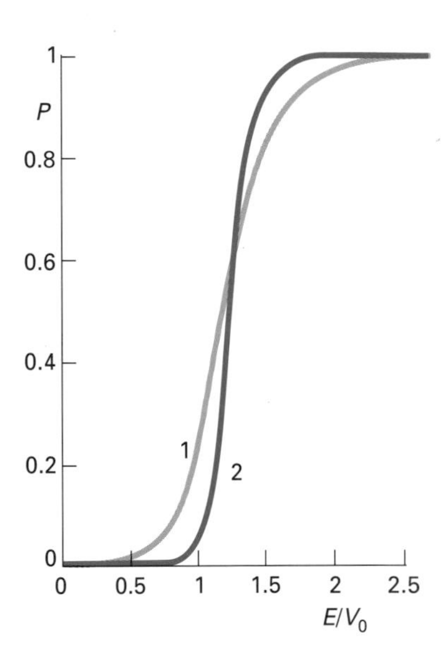

Fig. 2.17. The tunnelling probability for an Eckart barrier and its variation with energy. The curves are labelled with the value of $(2mV_0)^{1/2}/\beta\hbar$.

2.11 The Eckart potential barrier

The rectangular barrier we have been considering is obviously not very realistic, but it does serve to introduce a number of concepts, and it has properties that are found in more realistic models. In fact, there are only a few realistic potentials for which analytical expressions for the reflection and transmission coefficients are available. One such potential is the **Eckart potential barrier**:

$$V(x) = \frac{4V_0 e^{\beta x}}{(1 + e^{\beta x})^2} \tag{27}$$

where V_0 and β are constants with dimensions of energy and inverse length, respectively. This barrier was investigated by C. Eckart in 1930.[3] It is shown in Fig. 2.16; we see that it is symmetric in x with a maximum value of V_0 at $x = 0$, and approaches zero as $|x| \to \infty$. The Schrödinger equation associated with this potential can be solved, but its solutions are the so-called hypergeometric functions, which are beyond the scope of this book. All we shall do is quote the analytical expression for the transmission coefficient P:

$$P = \frac{\cosh(4\pi\sqrt{2mE}/\hbar\beta) - 1}{\cosh(4\pi\sqrt{2mE}/\hbar\beta) + \cosh(2\pi\sqrt{8mV_0 - (\hbar\beta/2)^2}/\hbar\beta)} \tag{28}$$

We have plotted P as a function of energy E in Fig. 2.17. Notice that for energies $E \ll V_0$, $P \approx 0$; as the energy approaches the top of the barrier, the transmission probability increases. This increase corresponds to the tunnelling of the particle through the classically forbidden region of the barrier and its emergence on the other side. As the energy increases beyond V_0, P approaches 1. Notice that $P < 1$ even when $E > V_0$. There is still a probability of the particle being reflected by the barrier even when classically it can pass over the barrier. This behaviour is another example of the antitunnelling displayed by the rectangular barrier. Finally, when $E \gg V_0$, $P \approx 1$ as is expected classically.

Particle in a box

We now turn to a case in which a particle is confined by walls to a region of space of length L. The walls are represented by a potential energy that is zero inside the region and which rises abruptly to infinity at the edges (Fig. 2.18). This system is called a **one-dimensional square well** or a **particle in a box**. The squareness in the former name refers to the steepness with which the potential energy goes to infinity at the ends of the box. Because the particle is confined, its energy is quantized, and the boundary conditions determine which energies are permitted.

2.12 The solutions

The hamiltonian for the system is

$$H = -\frac{\hbar^2}{2m}\frac{d^2}{dx^2} + V(x) \qquad V(x) = \begin{cases} 0 & \text{for } 0 \le x \le L \\ \infty & \text{otherwise} \end{cases} \tag{29}$$

[3] For details, see C. Eckart, *Phys. Rev.*, 1303, **35** (1930).

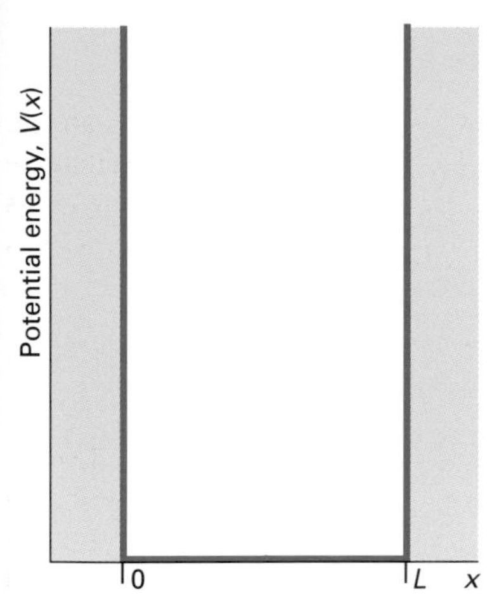

Fig. 2.18. The infinite square-well potential characteristic of a particle in a box.

Because the potential energy of a particle that touches the walls is infinite, the particle cannot in fact penetrate them. This result is justified by the behaviour of the wavefunctions described in Section 2.9. It follows that the hamiltonian for the region where the potential is not infinite, and therefore the only region where the wavefunction is nonzero, is

$$H = -\frac{\hbar^2}{2m}\frac{d^2}{dx^2} \tag{30}$$

This expression is the same as the hamiltonian for free translational motion, and therefore we know at once that the solutions are those given earlier (eqn 2.6). However, in this case there are boundary conditions to satisfy, and they will have the effect of eliminating most of the possible solutions.

The wavefunctions are zero outside the box where $x < 0$ or $x > L$. Wavefunctions are everywhere continuous. Therefore, the wavefunctions must be zero *at* the walls. The boundary conditions are therefore $\psi(0) = 0$ and $\psi(L) = 0$. We now apply each condition in turn to a general solution of the form

$$\psi(x) = C\cos kx + D\sin kx \qquad k\hbar = (2mE)^{1/2}$$

First, at $x = 0$, because $\cos 0 = 1$ and $\sin 0 = 0$, $\psi(0) = C$. Therefore, to satisfy the condition $\psi(0) = 0$ we require $C = 0$. Next, at $x = L$

$$\psi(L) = D\sin kL = 0$$

One way to achieve this condition is to set $D = 0$, but then the wavefunction would be zero everywhere and the particle found nowhere. The alternative is to require that the sine function itself vanishes. It does so if kL is equal to an integral multiple of π. That is, we must require k to take the values

$$k = \frac{n\pi}{L} \qquad n = 1, 2, \ldots \tag{31}$$

The value $n = 0$ is excluded because it would give $\sin kx = 0$ for all x, and the particle would not be found anywhere. The integer n is an example of a **quantum number**, a number that labels a state of the system and which, by the use of an appropriate formula, can be used to calculate the value of an observable of the system. For instance, because $E = k^2\hbar^2/2m$, it follows that the energy is related to n by

$$E_n = \frac{n^2\hbar^2\pi^2}{2mL^2} = \frac{n^2h^2}{8mL^2} \qquad n = 1, 2, \ldots \tag{32}$$

A major conclusion of this calculation at this stage is that the energy of the particle is quantized; that is, confined to a series of discrete values.

There now remains only the constant D to determine before the solution is complete. The probability of finding the particle somewhere within the box must be 1, so the integral of ψ^2 over the region between $x = 0$ and $x = L$ must be equal to 1:

$$1 = \int_0^L \psi^*\psi\, dx = D^2\int_0^L \sin^2\left(\frac{n\pi x}{L}\right) dx = \tfrac{1}{2}LD^2$$

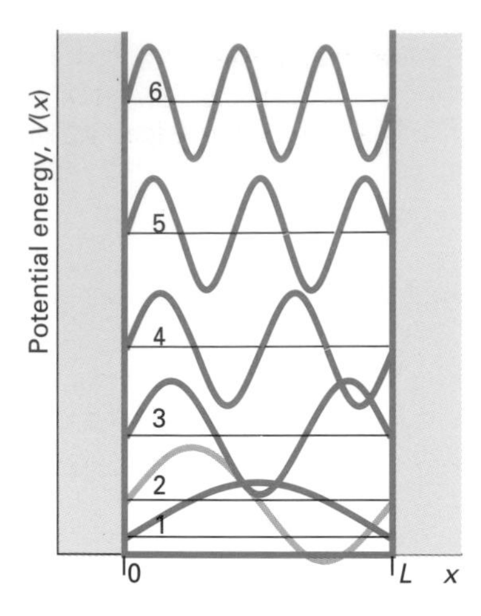

Fig. 2.19. The first six energy levels and the corresponding wavefunctions for a particle in a box. Notice that the levels are more widely separated as the energy increases; the maximum amplitude of the wavefunctions is the same in all cases.

Therefore, as we saw in Example 1.2, $D = (2/L)^{1/2}$. The complete solution is

$$\psi = \left(\frac{2}{L}\right)^{1/2} \sin\left(\frac{n\pi x}{L}\right)$$
$$E_n = \frac{n^2h^2}{8mL^2} \qquad n = 1, 2, \ldots \tag{33}$$

We see that there is a single quantum number, n, which determines the wavefunctions and the energies.

Some of the solutions are illustrated in Fig. 2.19. The squares of the wavefunctions are shown in Fig. 2.20: they represent the probability densities for finding the particle in each state. Note how the particle seems to avoid the walls in the low energy states but becomes increasingly uniformly distributed as n increases. A point where a wavefunction passes through zero (not simply approaches zero without passing through) is called a **node**. We see from Fig. 2.19 that the lowest energy state has no nodes, and that the number increases as n increases (and is equal to $n - 1$). It is a common feature of wavefunctions that the higher the number of nodes, the higher the energy.

2.13 Features of the solutions

The lowest energy that the particle can have is $E_1 = h^2/8mL^2$. This irremovable energy is called the **zero-point energy**. It is a purely quantum mechanical property, and in a hypothetical universe in which $h = 0$ there would be no zero-point energy. The uncertainty principle gives some insight into its origin, because the uncertainty in the position of the particle is finite (it is somewhere between 0 and L), so the uncertainty in the momentum of the particle cannot be zero. Because $\Delta p \neq 0$, it follows that $\langle p^2 \rangle \neq 0$ and consequently that the average kinetic energy, which is proportional to $\langle p^2 \rangle$, also cannot be zero. A more fundamental way of understanding the origin of the zero-point energy, though, is to note that the wavefunction is necessarily curved if it is to be zero at each wall but not zero throughout the interior of the box. We have already seen that the curvature of a wavefunction signifies the possession of kinetic energy, so the particle necessarily possesses nonzero kinetic energy if it is inside the box.

The energy separation of neighbouring states decreases as the walls move back and give the particle more freedom:

$$E_{n+1} - E_n = \left\{(n+1)^2 - n^2\right\} \frac{h^2}{8mL^2} = (2n+1)\frac{h^2}{8mL^2} \tag{34}$$

As the length of the box approaches infinity (corresponding to a box of macroscopic dimensions), the separation of neighbouring levels approaches zero, and the effects of quantization become completely negligible. In effect, the particle becomes unbounded and free, and its state is described by the wavefunctions in eqn 2.5.

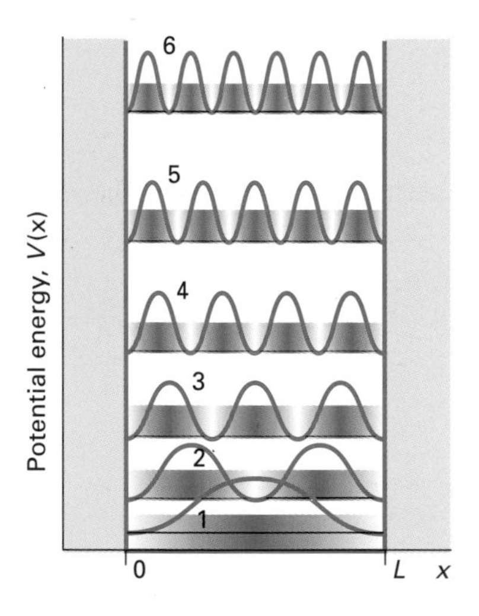

Fig. 2.20. The probability distribution of a particle in a box. Two techniques have been used to display the probability density: the lines show the value of ψ^2, and the ribbons depict the probability density by shading. Note that the distribution becomes more uniform as the energy increases.

2.14 The two-dimensional square well

Interesting new features arise when we consider a particle confined to a rectangular planar surface with linear dimensions L_1 in the x direction and L_2 in the y direction. The potential energy for such an arrangement is illustrated in

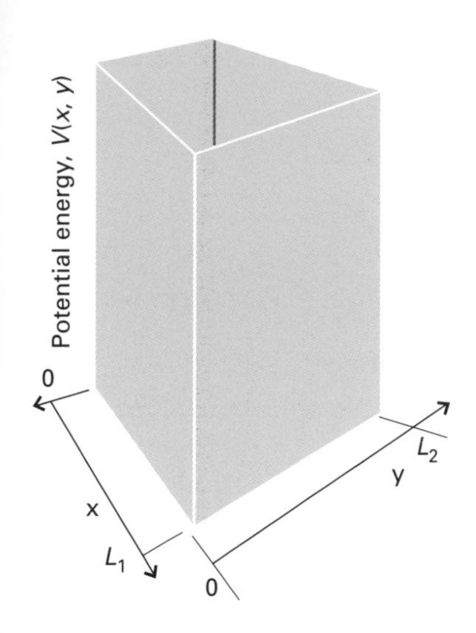

Fig. 2.21. A view of the potential energy of a particle in a two-dimensional square well.

Fig. 2.21. Just as in one dimension, where the wavefunctions look like those of a vibrating string with clamped ends, so in two dimensions they can be expected to correspond to the vibrations of a plate with the edges rigidly clamped.

The hamiltonian for the two-dimensional, infinitely deep square well is

$$H = -\frac{\hbar^2}{2m}\left(\frac{\partial^2}{\partial x^2} + \frac{\partial^2}{\partial y^2}\right) + V(x,y)$$
$$V(x,y) = \begin{cases} 0 & \text{for } 0 \leq x \leq L_1 \text{ and } 0 \leq y \leq L_2 \\ \infty & \text{otherwise} \end{cases} \tag{35}$$

The Schrödinger equation for the particle inside the walls, the only region where the wavefunction is nonzero, is therefore

$$\frac{\partial^2 \psi}{\partial x^2} + \frac{\partial^2 \psi}{\partial y^2} = -\frac{2mE}{\hbar^2}\psi \tag{36}$$

The boundary conditions are that the wavefunction must vanish at all four walls.

To solve this equation in two variables, we try the separation of variables technique described in Section 1.12. The trial solution is written $\psi(x,y) = XY$, where X is a function of only x and Y is a function only of y. Inserting the trial solution into the Schrödinger equation and dividing through by XY, we get

$$\frac{X''}{X} + \frac{Y''}{Y} = -\frac{2mE}{\hbar^2}$$

where $X'' = \mathrm{d}^2X/\mathrm{d}x^2$ and $Y'' = \mathrm{d}^2Y/\mathrm{d}y^2$. We now use the same argument as in Section 1.12, and conclude that the original equation can be separated into two parts:

$$X'' = -\left(\frac{2mE^X}{\hbar^2}\right)X \qquad Y'' = -\left(\frac{2mE^Y}{\hbar^2}\right)Y$$

with $E^X + E^Y = E$. Both equations have the same form as the equation for a one-dimensional system, and the boundary conditions are the same. Therefore, we may write the solutions immediately (using $\psi = XY$):

$$\psi_{n_1 n_2}(x,y) = \frac{2}{(L_1 L_2)^{1/2}}\sin\left(\frac{n_1 \pi x}{L_1}\right)\sin\left(\frac{n_2 \pi y}{L_2}\right)$$
$$E_{n_1 n_2} = \frac{h^2}{8m}\left(\frac{n_1^2}{L_1^2} + \frac{n_2^2}{L_2^2}\right) \qquad n_1 = 1, 2, \ldots;\ n_2 = 1, 2, \ldots \tag{37}$$

Note that to define the state of a particle in a two-dimensional system, we need to specify the values of *two* quantum numbers; n_1 and n_2 can take any integer values in their range independently of each other.

Many of the features of the one-dimensional system are reproduced in higher dimensions. There is a zero-point energy (E_{11}), and the energy separations decrease as the walls move apart and become less confining. The energy is quantized as a consequence of the boundary conditions. The shapes of some of the low-energy wavefunctions are illustrated in Fig. 2.22 and the corresponding probability densities are shown in Fig. 2.23. As in the one-dimensional case, the particle is distributed more uniformly at high energies than at low.

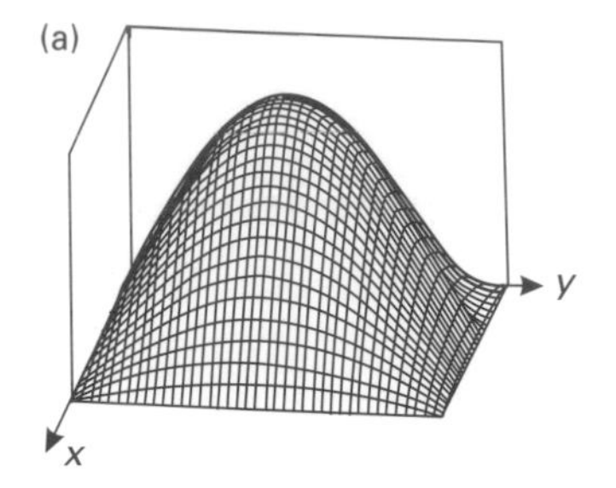

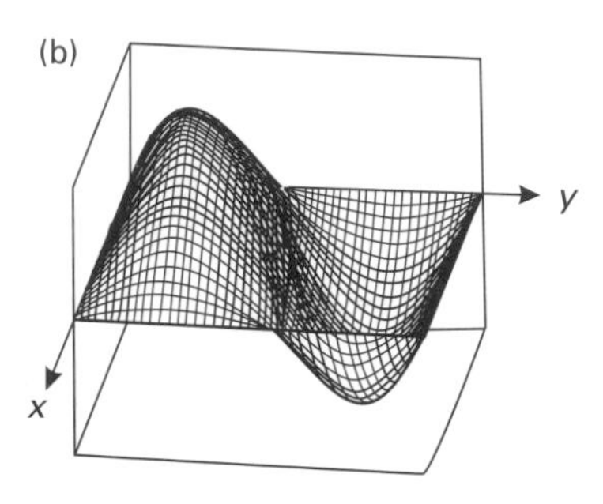

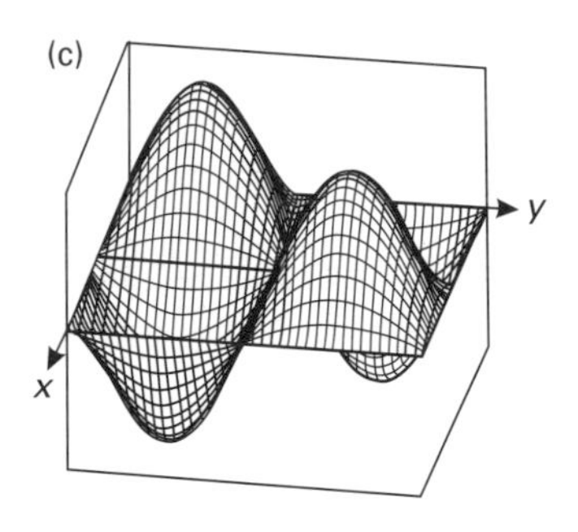

Fig. 2.22. Three wavefunctions for a particle in a two-dimensional square well: (a) $n_1 = 1, n_2 = 1$, (b) $n_1 = 1, n_2 = 2$, and (c) $n_1 = 2, n_2 = 2$.

2.15 Degeneracy

One feature found in two dimensions but not in one dimension is apparent when the box is geometrically square. Then $L_1 = L_2 = L$ and the energies are given by

$$E_{n_1 n_2} = \frac{h^2}{8mL^2}\left(n_1^2 + n_2^2\right) \tag{38}$$

This expression implies that a state with the quantum numbers $n_1 = a$ and $n_2 = b$ (which we could denote $|a, b\rangle$) has exactly the same energy as one with $n_1 = b$ and $n_2 = a$ (the state $|b, a\rangle$) even when $a \neq b$. This is an example of the degeneracy of states mentioned in Section 1.1. For example, the two states $|1, 2\rangle$ and $|2, 1\rangle$ both have the energy $5h^2/8mL^2$ but their two wavefunctions are different:

$$\psi_{12} = \left(\frac{2}{L}\right)\sin\left(\frac{\pi x}{L}\right)\sin\left(\frac{2\pi y}{L}\right) \qquad \psi_{21} = \left(\frac{2}{L}\right)\sin\left(\frac{2\pi x}{L}\right)\sin\left(\frac{\pi y}{L}\right)$$

Inspection of Fig. 2.24 shows the origin of this degeneracy: one wavefunction can be transformed into the other by rotation of the box through 90°. We should always expect degeneracies to be present in systems that have a high degree of symmetry, as we shall see in more detail in Chapter 5.

In the case of a rectangular but not square box, the symmetry and the degeneracy are lost. However, sometimes degeneracy is encountered where there is no rotation that transforms one wavefunction into another; it is then called **accidental degeneracy**. In certain cases, accidental degeneracy is known to arise when the full symmetry of the system has not been recognized, and a deeper analysis of the system shows the presence of a **hidden symmetry** that does interrelate the degenerate functions. It may be the case that all accidental degeneracies can be traced to the existence of hidden symmetries. Accidental degeneracy occurs in the hydrogen atom, and we shall continue the discussion there.

Example 2.2 Hidden symmetry and accidental degeneracy

Show that in a rectangular box with sides $L_1 = L$ and $L_2 = 2L$ there is an accidental degeneracy between the states $|1, 4\rangle$ and $|2, 2\rangle$.

Method. To confirm the degeneracy, all we need do is to substitute the data into the expression for the energy, eqn 2.38.

Answer. The two states have the following energies:

$$E_{14} = \frac{h^2}{8m}\left(\frac{1^2}{L^2} + \frac{4^2}{(2L)^2}\right) = \frac{5h^2}{8mL^2}$$

$$E_{22} = \frac{h^2}{8m}\left(\frac{2^2}{L^2} + \frac{2^2}{(2L)^2}\right) = \frac{5h^2}{8mL^2}$$

The energies are the same, despite the lack of symmetry.

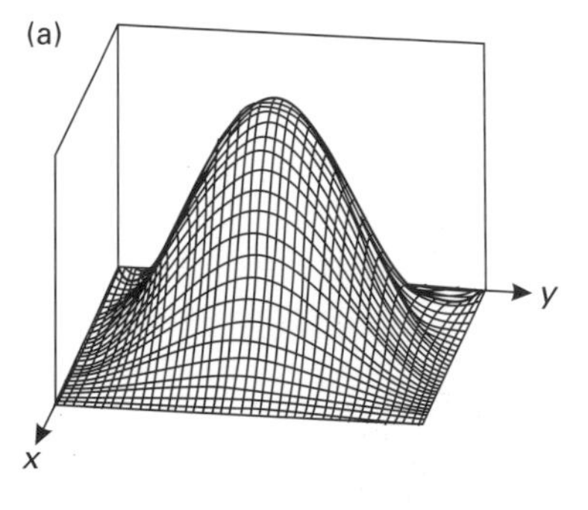

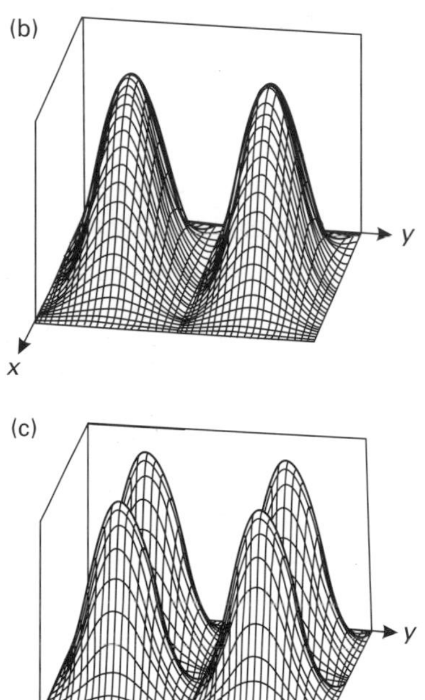

Fig. 2.23. Three probability distributions for a particle in a two-dimensional square well: (a) $n_1 = 1, n_2 = 1$, (b) $n_1 = 1$, $n_2 = 2$, and (c) $n_1 = 2, n_2 = 2$ (as in the previous illustration).

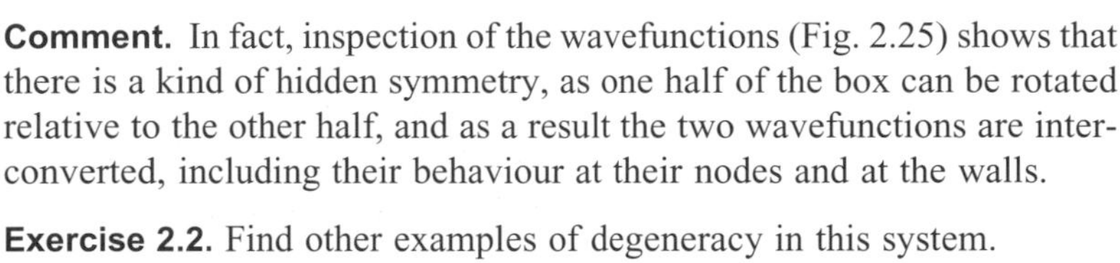

Comment. In fact, inspection of the wavefunctions (Fig. 2.25) shows that there is a kind of hidden symmetry, as one half of the box can be rotated relative to the other half, and as a result the two wavefunctions are interconverted, including their behaviour at their nodes and at the walls.

Exercise 2.2. Find other examples of degeneracy in this system.

[For instance, $|2, 8\rangle, |4, 4\rangle$]

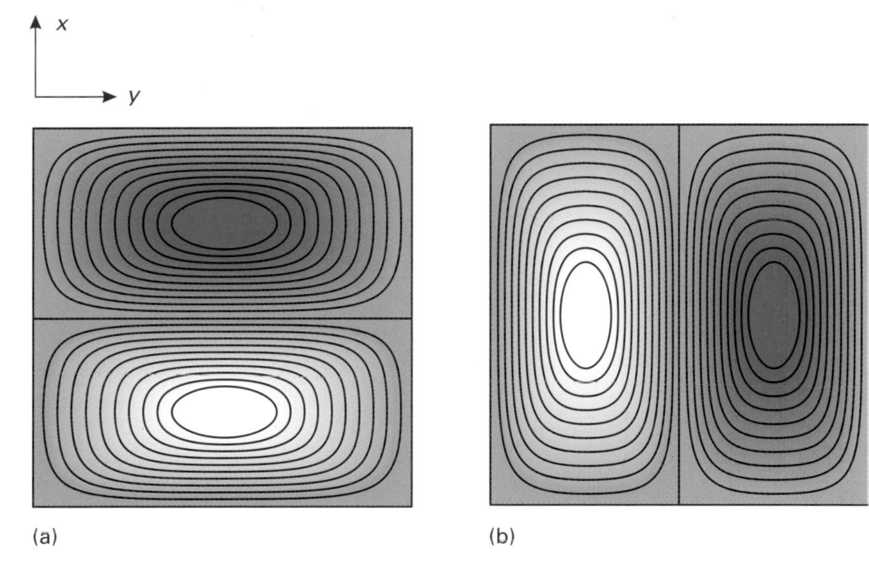

Fig. 2.24. A contour representation of the two degenerate states (a) $n_1 = 2, n_2 = 1$ and (b) $n_1 = 1, n_2 = 2$ for a particle in a square square well. Note that one wavefunction is rotated into the other by a symmetry transformation of the box (its rotation through 90° about a perpendicular axis).

The harmonic oscillator

We now turn to one of the most important individual topics in quantum mechanics, the harmonic oscillator. Harmonic oscillations occur when a system contains a part that experiences a restoring force proportional to the displacement from equilibrium. Pendulums and vibrating strings are familiar examples. An example of chemical importance is the vibration of atoms in a molecule. Another example is the electromagnetic field, which can be treated as a collection of harmonic oscillators, one for each frequency of radiation present. The importance of the harmonic oscillator also lies in the way that the same algebra occurs in a variety of different problems; for example, it also occurs in the treatment of rotational motion.

The restoring force in a one-dimensional harmonic oscillator is $-kx$, where the constant of proportionality k is called the **force constant**. Because the force acting on a particle is the negative gradient of the potential energy ($F = -\mathrm{d}V/\mathrm{d}x$), it follows that the potential energy of the oscillator varies

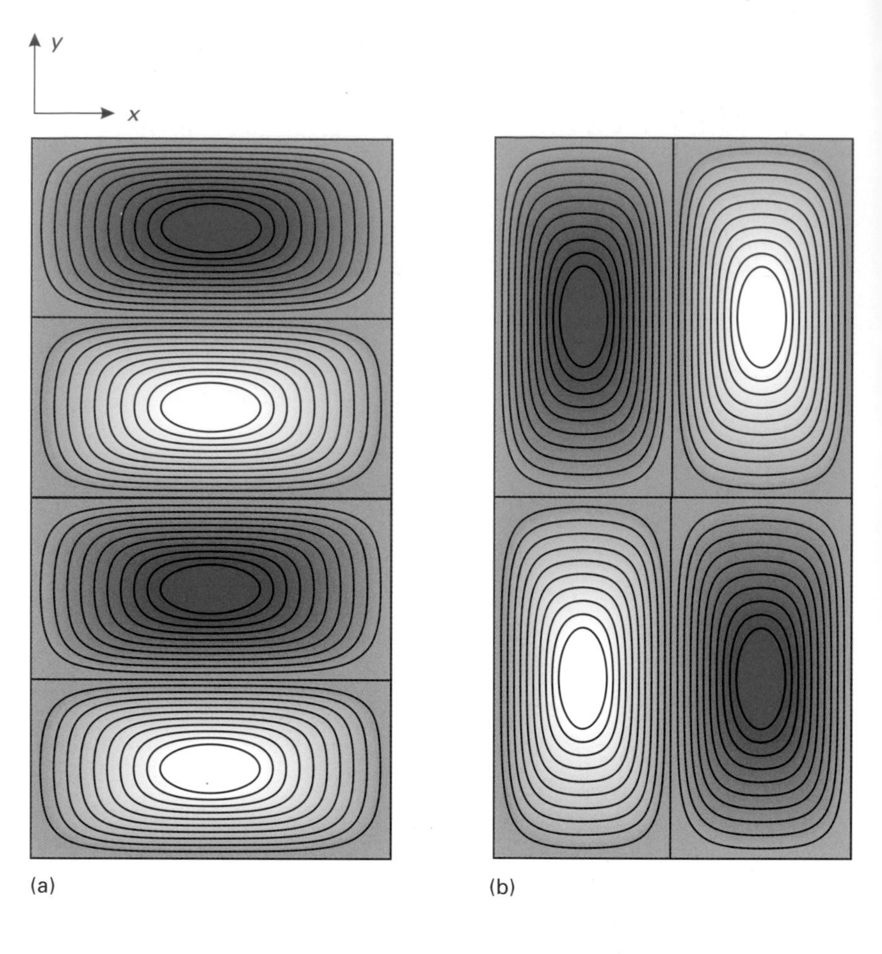

Fig. 2.25. An example of accidental degeneracy: the two functions shown here are degenerate even though one cannot be transformed into the other by a symmetry transformation of the system. Note, however, that a hidden symmetry (the separate rotation of the two halves of the box) does interconvert them.

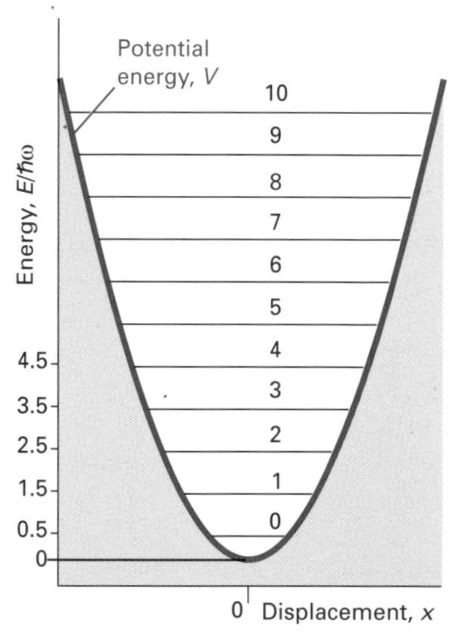

Fig. 2.26. The parabolic potential energy characteristic of a harmonic oscillator and the evenly spaced ladder of allowed energies (which continues up to infinity).

with displacement as

$$V(x) = \tfrac{1}{2}kx^2 \qquad (39)$$

This **parabolic potential** is illustrated in Fig. 2.26; it is zero at $x = 0$ because we are interested only in the potential energy with respect to zero displacement, not its absolute value. The difference between this potential and the square-well potential is the rapidity with which it rises to infinity: the 'walls' of the oscillator are much softer, and so we should expect the wavefunctions of the oscillator to penetrate them slightly. In other respects the two potentials are similar, and we can imagine the slow deformation of the square well into the smooth parabola of the oscillator. The wavefunctions of one system should change slowly into those of the other: they will have the same general form, but will penetrate into classically disallowed displacements.

Another point about the harmonic oscillator is that it is really much too

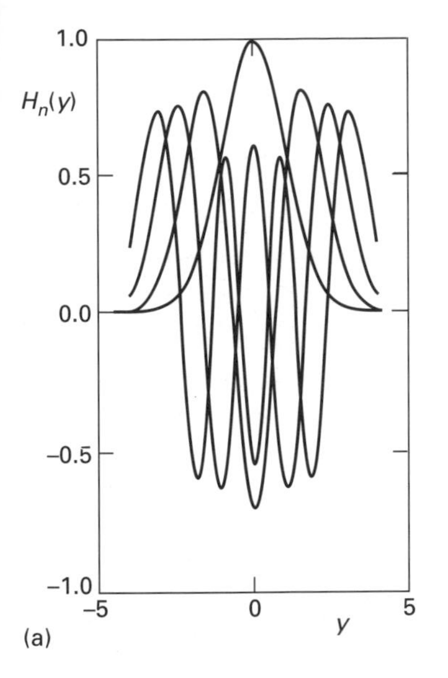

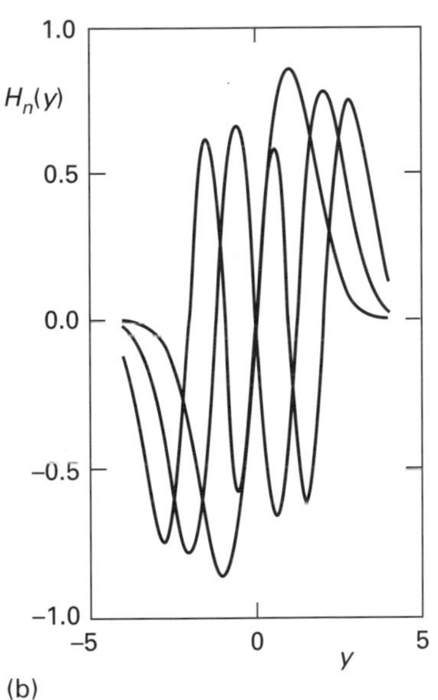

Fig. 2.27. The wavefunctions of a harmonic oscillator for v up to 6: (a) even values, (b) odd values. Note that the number of nodes increases with v, and that even v functions are symmetric whereas odd v functions are antisymmetric about $x = 0$.

simple. Its simplicity arises from the symmetrical occurrence of momentum and displacement in the expression for the total energy. Classically, the energy is $E = p^2/2m + kx^2/2$, and both p and x occur as their squares. This hidden symmetry has important implications, one being that if one has a new theory that can be applied to the harmonic oscillator and solved, then it may still be unsolvable for other systems. Another implication involves the uncertainty principle, for in the ground state of the harmonic oscillator, the product of the uncertainties Δp and Δx is *equal* to $\frac{1}{2}\hbar$ (see Problem 2.29).

2.16 The solutions

Because the potential energy is $V = \frac{1}{2}kx^2$, the hamiltonian operator for the harmonic oscillator is

$$H = -\frac{\hbar^2}{2m}\frac{\mathrm{d}^2}{\mathrm{d}x^2} + \tfrac{1}{2}kx^2 \tag{40}$$

The Schrödinger equation is therefore

$$-\frac{\hbar^2}{2m}\frac{\mathrm{d}^2\psi}{\mathrm{d}x^2} + \tfrac{1}{2}kx^2\psi = E\psi \tag{41}$$

The best method for solving this equation—a method that also works for rotational motion and the hydrogen atom—is set out in *Further information 6*. This method depends on looking for a way of factorizing the hamiltonian and introduces the concepts of 'creation and annihilation operators'. The conventional solution, which involves expressing the solutions as polynomials in the displacement, is described in *Further information 7*. That algebra, however, need not deflect us from the main thread of this chapter, the discussions of the solutions themselves. It turns out that their properties are remarkably simple.

The energy of a harmonic oscillator is quantized (as expected from the shape of the potential) and limited to the values

$$E_v = \left(v + \tfrac{1}{2}\right)\hbar\omega \qquad \text{where } \omega = \left(\frac{k}{m}\right)^{1/2} \qquad v = 0,\ 1,\ 2,\ldots \tag{42}$$

These energy levels are illustrated in Fig. 2.26. The wavefunctions are no longer the simple sine functions of the square well, but do show a family resemblance to them. They can be pictured as sine waves that collapse towards zero at large displacements (Fig. 2.27). Their precise form is that of a bell-shaped **gaussian function**, a function of the form e^{-x^2}, multiplied by a polynomial in the displacement:

$$\psi_v = N_v H_v(y)\mathrm{e}^{-\frac{1}{2}y^2} \qquad y = \frac{x}{\alpha} \qquad \alpha = \left(\frac{\hbar^2}{mk}\right)^{1/4} \tag{43}$$

where N_v is a normalization constant:

$$N_v = \left(\frac{1}{2^v v! \pi^{1/2}\alpha}\right)^{1/2} \tag{44}$$

The $H_v(y)$ are **Hermite polynomials**. Because $H_0(y) = 1$, the wavefunction for the state with $v = 0$ is proportional to the gaussian function $\mathrm{e}^{-y^2/2}$. When $v = 1$, because $H_1(y) = 2y$, the wavefunction is the same gaussian function multiplied by $2y$. The Hermite polynomials get progressively more complicated as v increases, and the first few are set out in Table 2.1.

Table 2.1 Hermite polynomials

v	$H_v(y)$
0	1
1	$2y$
2	$4y^2 - 2$
3	$8y^3 - 12y$
4	$16y^4 - 48y^2 + 12$
5	$32y^5 - 160y^3 + 120y$
6	$64y^6 - 480y^4 + 720y^2 - 120$
7	$128y^7 - 1344y^5 + 3360y^3 - 1680y$
8	$256y^8 - 3584y^6 + 13440y^4 - 13440y^2 + 1680$

Differential equation: $H_v'' - 2yH_v' + 2vH_v = 0$
Recursion relation: $H_{v+1} = 2yH_v - 2vH_{v-1}$
Orthogonality relation: $\int_{-\infty}^{\infty} H_v H_{v'} e^{-y^2}\, dy = 0$ for $v \neq v'$
Normalization: $\int_{-\infty}^{\infty} H_v^2 e^{-y^2}\, dy = \pi^{1/2} 2^v v!$

Example 2.3 The nodes of harmonic oscillator wavefunctions

Locate the nodes of the harmonic oscillator wavefunction with $v = 4$.

Method. The gaussian function has no nodes, so we need to determine the nodes of the Hermite polynomials by determining the values of y at which they pass through 0. The polynomials are listed in Table 2.1.

Answer. Because $H_4(y) = 16y^4 - 48y^2 + 12$, we need to solve

$$16y^4 - 48y^2 + 12 = 0$$

This is a quadratic equation in $z = y^2$ with roots

$$z = \frac{48 \pm (48^2 - 4 \times 16 \times 12)^{1/2}}{2 \times 16} = 2.7247 \text{ and } 0.2753$$

The nodes are therefore at ± 1.6507 and ± 0.5246 (see Fig. 2.27).

Comment. For more complicated polynomials it is best and sometimes essential to use numerical methods (the root extracting program of a mathematics package). The graph in Fig. 2.28 shows the pattern of nodes: note how they spread away from the origin but become more uniformly distributed as v increases.

Exercise 2.3. Identify the location of the five nodes of H_5.
[At $y = 0, \pm 0.959, \pm 2.020$]

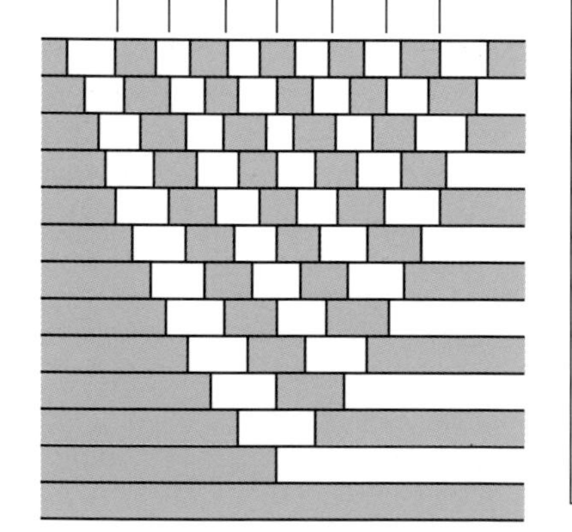

Fig. 2.28. The distribution of nodes in the first 13 states of a harmonic oscillator (up to $v = 12$). The white regions show where the wavefunction is negative and the shaded regions where it is positive.

2.17 Properties of the solutions

The properties of the harmonic oscillator are summarized in Table 2.2. The most significant point about the energy levels is that they form a ladder with

Table 2.2 Properties of the harmonic oscillator

Energies:	$E_v = (v + \frac{1}{2})\hbar\omega$ with $\omega = (k/m)^{1/2}$ $v = 0, 1, 2, \ldots$
Wavefunctions:	$\psi_v(x) = N_v H_v(y) e^{-\frac{1}{2}y^2}$, with $y = x/\alpha$, $\alpha = (\hbar^2/mk)^{1/4}$ $N_v = (2^v v! \pi^{1/2}\alpha)^{-1/2}$ for $\int_{-\infty}^{\infty} \psi_v^2(x)\,dx = 1$
Other relations:	$\langle T \rangle = \langle V \rangle$

$\langle v+1|x|v\rangle = (\hbar/2m\omega)^{1/2}(v+1)^{1/2}$ $\quad\langle v-1|x|v\rangle = (\hbar/2m\omega)^{1/2}v^{1/2}$
$\langle v+1|p_x|v\rangle = \mathrm{i}(m\omega/2)^{1/2}(v+1)^{1/2}$ $\quad\langle v-1|p_x|v\rangle = -\mathrm{i}(m\omega/2)^{1/2}v^{1/2}$

equal spacing. The energy separation between neighbours is $E_{v+1} - E_v = \hbar\omega$ regardless of the value of v. The equal spacing of the energy levels is another consequence of the hidden x^2, p^2 symmetry of the harmonic oscillator.

As the force constant k increases, so the separation between neighbouring levels also increases ($\omega \propto \sqrt{k}$). As k decreases or the mass increases, so ω decreases, and the separation between neighbouring levels decreases too. In the limit of zero force constant the parabolic potential fails to confine the particle (it corresponds to an infinitely weak spring) and the energy can vary continuously. There is no quantization in this limit of an unconstrained, free particle.

When thinking about the contributions to the total energy of a harmonic oscillator we have to take into account both the kinetic energy (which depends on the curvature of the wavefunction) and the potential energy (which depends on the probability of the particle being found at large displacements from equilibrium). The greater the force constant, the more confining the potential, and therefore the more sharply curved the wavefunctions. But whereas in the square well the walls are stringent limitations on the distribution of the particle, in the harmonic oscillator they can be penetrated. Therefore, as the force constant increases, the curvature need not follow the confinement so obediently, and some of the sharpness of the curvature can be lost by the wavefunction spreading further out into greater displacements, but at the cost of acquiring a greater average potential energy.

The discussion of the balance between the kinetic and potential contributions to the total energy is greatly simplified by the **virial theorem**, which although originally derived from classical mechanics has a quantum mechanical counterpart (see *Further information 3*). The virial theorem implies that if the potential energy can be expressed in the form $V = ax^s$, where a is a constant, then the mean kinetic and potential energies are related by

$$2\langle T \rangle = s\langle V \rangle \tag{45}$$

It follows that the total mean energy is

$$E = \langle T \rangle + \langle V \rangle = \left(1 + \frac{2}{s}\right)\langle T \rangle \tag{46}$$

For the harmonic oscillator, $s = 2$, so $\langle T \rangle = \langle V \rangle$ (hidden symmetry again), and therefore $E = 2\langle T \rangle$. Therefore, as the total energy increases (as it does as k increases for a given quantum state), both the kinetic and the potential energy increase. Not only does the curvature of the wavefunction increase, but it also spreads into regions of higher potential energy. In classical terms, this behaviour corresponds to a pendulum swinging more rapidly and with greater amplitude as its energy is increased.

A harmonic oscillator has a zero-point energy of magnitude $E_0 = \frac{1}{2}\hbar\omega$. The classical interpretation of such a conclusion is that the oscillator never stops fluctuating about its equilibrium position. The reason for its existence is the same as for a particle in a box: the wavefunctions must be zero at large displacements in either direction (because the potential energy is confining), nonzero in between (because the particle must be somewhere), and continuous (as for all wavefunctions). These conditions can be satisfied only if the wavefunction has curvature; hence the oscillator must possess kinetic energy in all its states. By the virial theorem, it must also possess the same potential energy; hence it must possess a nonzero energy even in its lowest state. This argument can also be turned round: if $E = 0$, then for an oscillator $\langle T \rangle = \langle V \rangle = 0$, which implies that both $\langle p^2 \rangle = 0$ and $\langle x^2 \rangle = 0$. For these to be possible mean values, both p and x must be zero, which is contrary to the uncertainty principle.

2.18 The classical limit

The shapes of the wavefunctions have already been drawn in Fig. 2.27. Their similarities to the square-well wavefunctions should be noted. The major difference between the two is the penetration of the harmonic oscillator wavefunctions into classically forbidden regions where $E < V$. In the same way as for the square well, the particle clusters away from the walls (and stays close to $x = 0$) in its lowest energy states. This is the behaviour to be expected classically of a stationary particle, for such a particle will be found at zero displacement and nowhere else. When the oscillator is moving, the classical prediction is that it has the highest probability of being found at its maximum displacement, the **turning points** of its classical trajectory, where it is briefly stationary. The behaviour of the quantum oscillator is quite different for low energy levels, but the two descriptions become increasingly similar as it is excited into higher levels. We see from Fig. 2.29 that at high v, the wavefunctions have their dominant maxima close to the classical turning points and resemble the classical distribution.

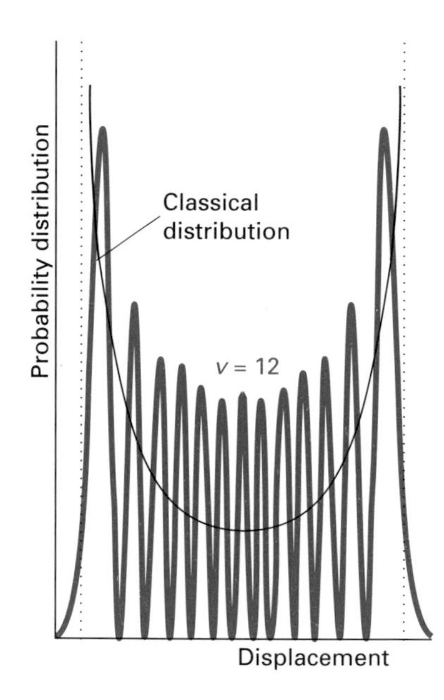

Fig. 2.29. A comparison of the probability distribution for a highly excited state of a harmonic oscillator ($v = 12$) and that of a classical oscillator with the same energy. Note how the former is starting to resemble the latter.

When the energy levels of the oscillator are close in comparison with the precision with which its state can be prepared (for example, when the parabola is so broad or the mass so great that the levels lie close together), the state of the oscillator must be expressed as a superposition of the wavefunctions considered so far. For example, because the energy levels are only about 10^{-34} J apart for a pendulum of period 1 s, we cannot hope to set it swinging with such precision that we can be confident that only one level is occupied. Setting the pendulum swinging results in its being described by a superposition of wavefunctions, and the interference between the components of the superposition results in the formation of a wavepacket. The time-dependence of the components results in a region of constructive interference that moves from one side of the potential to the other with an angular frequency ω. That is, for coarse preparations of initial states, there is a sharply defined wavepacket which oscillates in the potential with the angular frequency $\omega = (k/m)^{1/2}$. This is precisely the classical behaviour of an oscillator, with the wavepacket denoting the location of the classical particle. In other words, when we see a pendulum swing, we are seeing a display of the separation of its quantized energy levels.

Example 2.4 The construction and motion of a wavepacket

Show that whatever superposition of harmonic oscillator states is used to construct a wavepacket, it is localized at the same place at the times $0, T, 2T, \ldots$, where T is the classical period of the oscillator.

Method. The classical period is $T = 2\pi/\omega$. We need to form a time-dependent wavepacket by superimposing the Ψ_v for the oscillator, and then evaluate it at $t = nT$, with $n = 0, 1, 2, \ldots$.

Answer. The wavepacket has the following form:

$$\Psi(x, t) = \sum_v c_v \Psi_v(x, t) = \sum_v c_v \psi_v(x) \mathrm{e}^{-\mathrm{i}E_v t/\hbar}$$
$$= \sum_v c_v \psi_v(x) \mathrm{e}^{-\mathrm{i}(v+\frac{1}{2})\omega t}$$

It follows that

$$\Psi(x, nT) = \sum_v c_v \psi_v(x) \mathrm{e}^{-2\mathrm{i}\pi n(v+\frac{1}{2})} = \sum_v c_v \psi_v(x)(-1)^n = (-1)^n \Psi(x, 0)$$

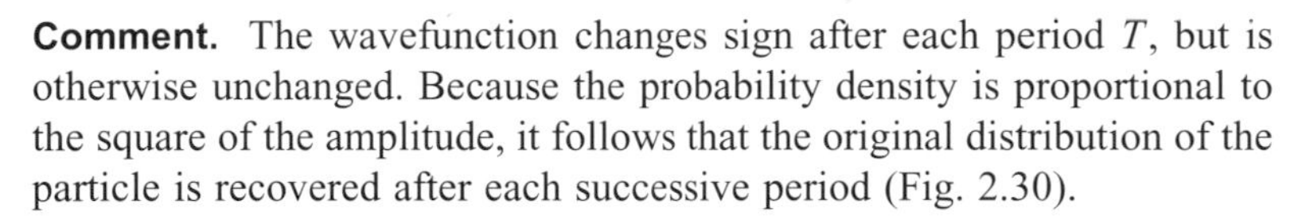

because $\mathrm{e}^{2\pi\mathrm{i}} = 1$ and $\mathrm{e}^{\mathrm{i}\pi} = -1$.

Comment. The wavefunction changes sign after each period T, but is otherwise unchanged. Because the probability density is proportional to the square of the amplitude, it follows that the original distribution of the particle is recovered after each successive period (Fig. 2.30).

Exercise 2.4. Construct the explicit form of Ψ at $x = 0$ and discuss its time behaviour.

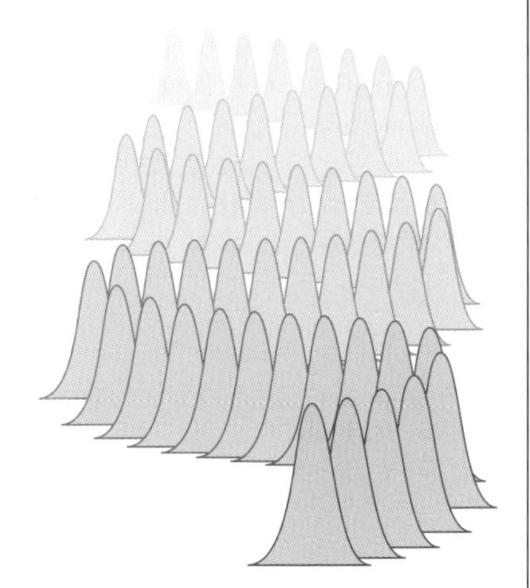

Fig. 2.30. The trajectory of a wavepacket. This particular wavepacket oscillates backwards and forwards with the classical frequency, and although it spreads and contracts with time, at the end of each period it has its initial shape and location. (The spreading and contraction of the wavepacket are hard to see in this illustration.)

Translation revisited: the scattering matrix

In this concluding brief section of the chapter we return to the discussion of unbound translational motion and show that it can be expressed more succinctly. The aim of this section is to introduce one of the most important concepts in scattering theory, namely the scattering matrix. To do so, we shall redevelop the finite-barrier problem treated in Section 2.10, and express it in a way that utilizes this concept. The material here will be developed further in Chapter 14 and could be ignored at this stage.

We pick up the finite-barrier story at eqn 2.19 and express the relations between the coefficients in forms of matrices.[4] The coefficients will be written as follows:

$$\mathbf{C} = \begin{pmatrix} A \\ B \end{pmatrix} \qquad \mathbf{C}' = \begin{pmatrix} A' \\ B' \end{pmatrix} \qquad \mathbf{C}'' = \begin{pmatrix} A'' \\ B'' \end{pmatrix} \tag{47}$$

for Zones I, II, and III, respectively. The two equations relating the coefficients

[4] The properties of matrices are reviewed in *Further information 23*; in this section we deal only with 2×2 matrices, and the manipulations required are very straightforward. Only matrix multiplication is required.

A, B, A', and B' for the wavefunction in Zones I and II can now be expressed in matrix form as:

$$\mathbf{C}' = \mathbf{MC} \qquad \mathbf{M} = \tfrac{1}{2}\begin{pmatrix} 1 - \mathrm{i}k/\kappa & 1 + \mathrm{i}k/\kappa \\ 1 + \mathrm{i}k/\kappa & 1 - \mathrm{i}k/\kappa \end{pmatrix} \tag{48}$$

Likewise, the relations between the coefficients A', B', A'', and B'' in Zones II and III can be expressed as another matrix expression:

$$\mathbf{C}'' = \mathbf{QC}' \quad \mathbf{Q} = \frac{\mathrm{e}^{-(\kappa+\mathrm{i}k)L}}{2\mathrm{i}k}\begin{pmatrix} -\kappa + \mathrm{i}k & \mathrm{e}^{2\kappa L}(\kappa + \mathrm{i}k) \\ \mathrm{e}^{2\mathrm{i}kL}(\kappa + \mathrm{i}k) & \mathrm{e}^{2\kappa L}\mathrm{e}^{2\mathrm{i}kL}(-\kappa + \mathrm{i}k) \end{pmatrix} \tag{49}$$

We have the connection between Zone I and Zone II and between Zone II and Zone III in matrix form. The connection between the coefficients in Zones III and I is now easy to deduce by combining the two relations:

$$\mathbf{C}'' = \mathbf{TC} \qquad \mathbf{T} = \mathbf{QM} \tag{50}$$

In exactly the same way, we can set up a matrix relation between the coefficients of the outgoing and incoming waves. First we write

$$\mathbf{C}_{\text{in}} = \begin{pmatrix} B'' \\ A \end{pmatrix} \qquad \mathbf{C}_{\text{out}} = \begin{pmatrix} A'' \\ B \end{pmatrix} \tag{51}$$

Then the two are related by

$$\mathbf{C}_{\text{out}} = \mathbf{SC}_{\text{in}} \tag{52}$$

Some straightforward algebra shows that the matrices $\mathbf{S}$ and $\mathbf{T}$ are related by

$$\begin{pmatrix} S_{11} & S_{12} \\ S_{21} & S_{22} \end{pmatrix} = \begin{pmatrix} T_{21}/T_{22} & T_{11} - T_{21}T_{12}/T_{22} \\ 1/T_{22} & -T_{12}/T_{22} \end{pmatrix} \tag{53}$$

The matrix $\mathbf{S}$ is called the **scattering matrix**, or ***S* matrix**. It will play a central role in the discussion of scattering in Chapter 14.

One of the many advantages of introducing the scattering matrix is that reflection and transmission coefficients can be easily expressed in terms of its elements. For example, if the particle is incident from the left, so that $B'' = 0$, then it follows from eqn 2.52 that

$$A'' = S_{12}A \qquad B = S_{22}A$$

Therefore, the reflection and transmission probabilities are

$$R = |S_{22}|^2 \qquad P = |S_{12}|^2 \tag{54}$$

Example 2.5 Properties of the *S* matrix

A property of the *S* matrix is that it is unitary (see below). Show that the unitarity of the *S* matrix implies that $P + R = 1$.

Method. The unitarity of the *S* matrix means that

$$\mathbf{S}^\dagger\mathbf{S} = \mathbf{S}(\mathbf{S}^\dagger) = \mathbf{1}$$

where $\mathbf{S}^\dagger$ is the adjoint of $\mathbf{S}$ (the complex conjugate of its transpose). This property of the *S* matrix is established in *Further information 13*. The condition $P + R = 1$ can be expressed in terms of the elements of the *S* matrix by using eqn 2.54. We should inspect the relation and see if it is implied by the unitarity condition by writing the latter out in terms of the elements of **S**.

Answer. In terms of the elements of the S matrix, the condition $P + R = 1$ is

$$|S_{12}|^2 + |S_{22}|^2 = 1$$

The condition $\mathbf{S}^{\dagger}\mathbf{S} = \mathbf{1}$, when written out in full, is

$$\begin{pmatrix} S_{11}^* & S_{21}^* \\ S_{12}^* & S_{22}^* \end{pmatrix} \begin{pmatrix} S_{11} & S_{12} \\ S_{21} & S_{22} \end{pmatrix} = \begin{pmatrix} S_{11}^*S_{11} + S_{21}^*S_{21} & S_{11}^*S_{12} + S_{21}^*S_{22} \\ S_{12}^*S_{11} + S_{22}^*S_{21} & S_{12}^*S_{12} + S_{22}^*S_{22} \end{pmatrix}$$
$$= \begin{pmatrix} 1 & 0 \\ 0 & 1 \end{pmatrix}$$

Comparison of the (2,2)-elements implies that

$$S_{12}^*S_{12} + S_{22}^*S_{22} = 1$$

which is the same as $P + R = 1$.

Comment. As this calculation suggests, the unitarity of the S matrix is essentially a way of saying that the number of particles is preserved during the scattering event, because $P + R = 1$ expresses the fact that the sum of the probabilities of transmission and reflection is 1. Whenever you see 'unitarity' referred to, you should think of it as implying the conservation of probability. Conversely, if you want to ensure that probability is conserved, then you should impose the property of unitarity on the matrices you are using.

Exercise 2.5. Suppose the particle flux is incident from the right of the barrier. Define P and R in terms of the appropriate S matrix elements and confirm that $P + R = 1$.

Problems

2.1 Write the wavefunctions for (a) an electron travelling to the right ($x > 0$) after being accelerated from rest through a potential difference of (i) 1.0 V, (ii) 10 kV, (b) a particle of mass 1.0 g travelling to the right at $10\,\mathrm{m\,s^{-1}}$.

2.2 Find expressions for the probability densities of the particles in the preceding problem.

2.3 Use the qualitative 'wavefunction generator', Fig. 2.4, to sketch the wavefunctions for (a) a particle with a potential energy that decreases linearly to the right, (b) a particle with a potential energy that is constant to $x = 0$, then falls in the shape of a semicircle to a low value to climb back to its original constant value at $x = L$, (c) the same as in (b), but with the dip replaced by a hump.

2.4 Express the coefficients C and D in eqn 2.6 in terms of the coefficients A and B in eqn 2.5.

2.5 Calculate the flux density (eqn 2.11) for a particle with a wavefunction with coefficients $A = A_0 \cos\zeta$ and $B = A_0 \sin\zeta$, for a particle undergoing free motion in one dimension, with ζ a parameter, and plot J_x as a function of ζ.

2.6 A particle was prepared travelling to the right with all momenta between $(k-\frac{1}{2}\Delta k)\hbar$ and $(k+\frac{1}{2}\Delta k)\hbar$ contributing equally to the wavepacket. Find the explicit form of the wavepacket at $t=0$, normalize it, and estimate the range of positions, Δx, within which the particle is likely to be found. Compare the last conclusion with a prediction based on the uncertainty principle. *Hint.* Use eqn 2.13 with $g=B$, a constant, inside the range $k-\frac{1}{2}\Delta k$ to $k+\frac{1}{2}\Delta k$ and zero elsewhere, and eqn 2.12 with $t=0$ for Ψ_k. To evaluate $\int|\Psi_k|^2\,\mathrm{d}\tau$ (for the normalization step) use the integral $\int_{-\infty}^{\infty}(\sin^2 x/x^2)\,\mathrm{d}x=\pi$. Take Δx to be determined by the locations where $|\Psi|^2$ falls to half its value at $x=0$. For the last part use $\Delta p_x\approx\hbar\Delta k$.

2.7 Sketch the form of the wavepacket constructed in Problem 2.6. Sketch its form a short time after, when t is nonzero but still small. *Hint.* For the second part use eqn 2.13, but with $\mathrm{e}^{-\mathrm{i}k^2t/2m}\approx 1-\mathrm{i}k^2t/2m$. Use a computer to draw the wavepacket at longer times, evaluating the appropriate integrals numerically.

2.8 Repeat the evaluation leading to eqn 2.23 for the case $E>V$ and calculate the tunnelling probability.

2.9 A particle of mass m is incident from the left on a wall of infinite thickness and which may be represented by a potential energy V. Calculate the reflection coefficient for (a) $E\le V$, (b) $E>V$. For electrons incident on a metal surface $V=10\,\mathrm{eV}$. Evaluate and plot the reflection coefficient. *Hint.* Proceed as in Problem 2.8 but consider only two domains, inside the barrier and outside it. The reflection coefficient is the ratio $|B|^2/|A|^2$ in the notation of eqn 2.21.

2.10 A particle of mass m is confined to a one-dimensional box of length L. Calculate the probability of finding it in the following regions: (a) $0\le x\le\frac{1}{2}L$, (b) $0\le x\le\frac{1}{4}L$, (c) $\frac{1}{2}L-\delta x\le x\le\frac{1}{2}L+\delta x$. Deduce the expressions for general values of n, and then specialize to $n=1$.

2.11 An electron is confined to a one-dimensional box of length L. What should be the length of the box in order for its zero-point energy to be equal to its rest mass energy $(m_{\mathrm{e}}c^2)$? Express the result in terms of the Compton wavelength, $\lambda_{\mathrm{C}}=h/m_{\mathrm{e}}c$.

2.12 Energy is required to compress the box when a particle is inside: this suggests that the particle exerts a force on the walls. (a) On the basis that when the length of the box changes by $\mathrm{d}L$ the energy changes by $\mathrm{d}E=F\mathrm{d}L$, find an expression for the force. (b) At what length does $F=1\,\mathrm{N}$ when an electron is in the state $n=1$?

2.13 The mean position $\langle x\rangle$ of a particle in a one-dimensional well can be calculated by weighting its position x by the probability that it will be found in the region $\mathrm{d}x$ at x, which is $\psi^2(x)\,\mathrm{d}x$, and then summing (i.e. integrating) these values. Show that $\langle x\rangle=\frac{1}{2}L$ for all values of n. *Hint.* Evaluate $\int_0^L x\psi^2(x)\,\mathrm{d}x$.

2.14 The root mean square deviation of the particle from its mean position is$\Delta x=\{\langle x^2\rangle-\langle x\rangle^2\}^{1/2}$. Evaluate this quantity for a particle in a well and show that it approaches its classical value as $n\to\infty$. *Hint.* Evaluate $\langle x^2\rangle=\int_0^L x^2\psi^2(x)\,\mathrm{d}x$. In the classical case the distribution is uniform across the box, and so in effect $\psi(x)=1/\sqrt{L}$.

2.15 The mean value and mean square value of the linear momentum are given by $\int_0^L\psi^*p\psi\,\mathrm{d}x$ and $\int_0^L\psi^*p^2\psi\,\mathrm{d}x$, respectively. Evaluate these quantities, form the r.m.s. deviation $\Delta p=\{\langle p^2\rangle-\langle p\rangle^2\}^{1/2}$ and investigate the consistency of the outcome with the uncertainty principle. *Hint.* Use $p=(\hbar/\mathrm{i})\mathrm{d}/\mathrm{d}x$. For $\langle p^2\rangle$ notice that $E=p^2/2m$ and we already know E for each n. For the last part, form $\Delta x\Delta p$ and show that $\Delta x\Delta p\ge\frac{1}{2}\hbar$, the precise form of the principle, for all n; evaluate $\Delta x\Delta p$ for $n=1$.

2.16 Calculate the energies and wavefunctions for a particle in a one-dimensional square well in which the potential energy rises to a finite value V at each end, and is zero inside the well. Show that for any V and L there is always at least one bound level, and that as $V\to\infty$ the solutions coincide with those in eqn 2.32. *Hint.* This is a difficult problem. Divide space into three zones, solve the Schrödinger equations, and impose the boundary conditions (finiteness of ψ and its continuity and the continuity of $\mathrm{d}\psi/\mathrm{d}x$

across the zone boundaries: combine the latter requirements into the continuity of the logarithmic derivatives $(1/\psi)(\mathrm{d}\psi/\mathrm{d}x)$). After some algebra arrive at

$$kL + 2\arcsin\left\{\frac{k\hbar}{(2mV)^{1/2}}\right\} = n\pi, \qquad k\hbar = (2mE)^{1/2}$$

Solve this expression graphically for k and hence find the energies for each value of the integer n.

2.17 (a) Confirm eqn 2.23 and eqn 2.24 for the one-dimensional transmission probability. (b) Demonstrate that the two expressions coincide at $E = V$ and identify the value of P at that energy.

2.18 Identify the locations of the nodes in the wavefunction with $n = 4$ for a particle in a one-dimensional square well.

2.19 A very simple model of polyenes is the free electron molecular orbital model. Regard a chain of N conjugated carbon atoms, bond length R_{CC}, as forming a box of length $L = (N-1)R_{\mathrm{CC}}$. Find the wavefunctions and their energies. Suppose that the electrons enter the states in pairs so that the lowest $\frac{1}{2}N$ states are occupied. Estimate the wavelength of the lowest energy transition. Sometimes the length of the chain is taken to be $(N+1)R_{\mathrm{CC}}$, allowing for electrons to spill over the ends slightly.

2.20 (a) Show that the variables in the Schrödinger equation for a cubic box may be separated and the overall wavefunctions expressed as $X(x)Y(y)Z(z)$. (b) Deduce the energy levels and wavefunctions. (c) Show that the functions are orthonormal. (d) What is the degeneracy of the level with $E = 14(h^2/8mL^2)$?

2.21 (a) Demonstrate that accidental degeneracies can exist in a rectangular infinite square-well potential provided that the lengths of the sides are in a rational proportion. (b) What are the degeneracies when $L_1 = \lambda L_2$, with λ an integer?

2.22 Find the form of the ground-state wavefunction of a particle of mass m in an infinitely deep circular square well of radius R. *Hint*. Separate the Schrödinger equation for the system; the radial wavefunctions are related to Bessel functions.

2.23 The Hermite polynomials $H_v(y)$ satisfy the differential equation

$$H_v''(y) - 2yH_v'(y) + 2vH_v(y) = 0$$

Confirm that the wavefunctions in eqn 2.43 are solutions of the harmonic oscillator Schrödinger equation.

2.24 Locate the nodes of the harmonic oscillator wavefunction for the state with $v = 6$.

2.25 Confirm the expression for the normalization factor of a harmonic oscillator wavefunction, eqn 2.44.

2.26 Evaluate the matrix elements (a) $\langle v+1|x|v\rangle$ and (b) $\langle v+2|x^2|v\rangle$ of a harmonic oscillator by using the recursion relations of the Hermite polynomials.

2.27 The oscillation of the atoms around their equilibrium positions in the molecule HI can be modelled as a harmonic oscillator of mass $m \approx m_{\mathrm{H}}$ (the iodine atom is almost stationary) and force constant $k = 313.8\,\mathrm{N\,m^{-1}}$. Evaluate the separation of the energy levels and predict the wavelength of the light needed to induce a transition between neighbouring levels.

2.28 What is the relative probability of finding the HI molecule with its bond length 10 per cent greater than its equilibrium value (161 pm) when it is in (a) the $v = 0$ state, (b) the $v = 4$ state? Use the information in Problem 2.27.

2.29 Calculate the values of (a) $\langle x \rangle$, (b) $\langle x^2 \rangle$, (c) $\langle p_x \rangle$, (d) $\langle p_x^2 \rangle$ for a harmonic oscillator in its ground state and examine the value of $\Delta x \Delta p_x$ in the light of the uncertainty principle. *Hint*. Use the integrals

$$\int_{-\infty}^{\infty} e^{-\alpha x^2}\,dx = \left(\frac{\pi}{\alpha}\right)^{1/2} \qquad \int_{0}^{\infty} x e^{-\alpha x^2}\,dx = \frac{1}{2\alpha} \qquad \int_{-\infty}^{\infty} x^2 e^{-\alpha x^2}\,dx = \tfrac{1}{2}\left(\frac{\pi}{\alpha^3}\right)^{1/2}$$

2.30 Use the information in eqns 2.47–2.52 to write down the form of the S matrix for a one-dimensional system in which a particle is scattered from a finite rectangular region of nonzero potential energy.

Further reading

Introduction to quantum mechanics with applications to chemistry. L. Pauling and E.B. Wilson; McGraw-Hill, New York (1935).
The tunnel effect in chemistry. R.P. Bell; Chapman and Hall, London (1980).
Solvable models in quantum mechanics. S. Albeverio; Springer, New York (1988).
Introduction to quantum mechanics. B.H. Bransden and C.J. Joachain; Longman, London and Wiley, New York (1989).
Introduction to quantum mechanics. D.J. Griffiths; Prentice-Hall, Englewood Cliffs (1995).
Quantum mechanics. L.E. Ballentine; Prentice-Hall, Englewood Cliffs (1990).
Quantum mechanics. W. Greiner; Springer, New York (1994).
Quantum mechanics. P.J.E. Peebles; Princeton University Press, Princeton (1992).
Quantum mechanics. A.I.M. Rae; Institute of Physics, Bristol (1992).
Quantum mechanics: a first course. B.C. Reed; Wuerz Publishing, Winnipeg (1994).
Quantum mechanics II: a second course in quantum theory. R.H. Landau; Wiley, New York (1995).

3 Rotational motion and the hydrogen atom

The second class of motion we consider is rotational motion, the motion of an object around a fixed point. With this problem we encounter 'angular momentum', which is one of the most important topics in quantum mechanics. In this chapter we discuss rotational motion and angular momentum in terms of solutions of the Schrödinger equation, but we return to the topic in the next chapter and see how its properties emerge from the operators for angular momentum. This is a chapter for pictures; the next provides the algebra beneath the pictures.

The material we describe here occurs throughout quantum mechanics. In particular, it crops up wherever we are interested in the motion of a particle in a **central potential**, in which the potential energy depends only on the distance from a single point. One example is the central potential experienced by an electron in a hydrogen atom. That problem is also exactly solvable, and we shall consider it in this chapter too.

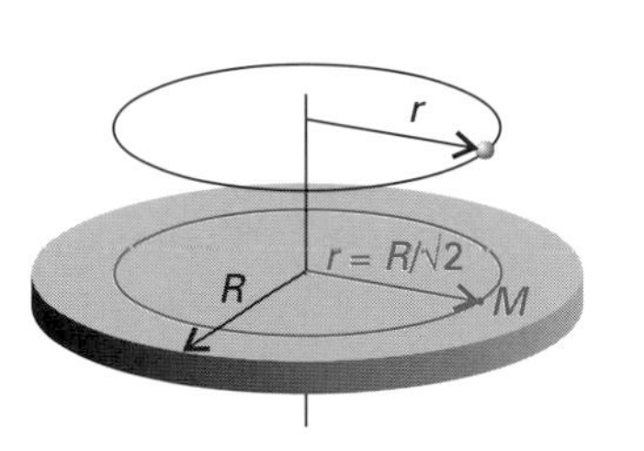

Fig. 3.1 The rotational characteristics of a uniform disk are represented by the motion of a single mass point at its radius of gyration.

Particle on a ring

As a first step, we consider the quantum mechanical description of a particle travelling on a circular ring. This problem is more general than it might seem, for as well as applying to the motion of a bead on a circle of wire, it also applies to any body rotating in a plane (e.g. a compact disk, Fig. 3.1). This generality stems from the fact that any such body can be represented by a mass point moving in a circle of radius r, its **radius of gyration** about the centre of mass. We shall see, in fact, that the property that determines the characteristics of the rotational motion of a body is the **moment of inertia**, $I = mr^2$, and it is not necessary to enquire into whether the value of I for a body is that of an actual particle moving on a ring of radius r or is that of a body of mass m and radius of gyration r rotating about its own centre of mass.

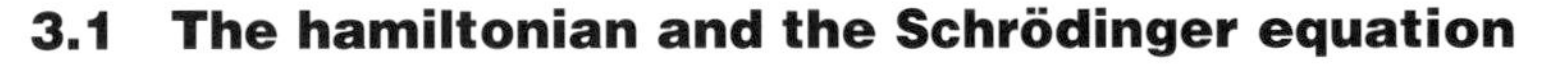

3.1 The hamiltonian and the Schrödinger equation

The particle of mass m travels on a circle of radius r in the xy-plane. Its potential energy is constant, and may be taken to be zero. The hamiltonian is therefore

$$H = -\frac{\hbar^2}{2m}\left(\frac{\partial^2}{\partial x^2} + \frac{\partial^2}{\partial y^2}\right) \tag{1}$$

Because the motion is confined to a circle, a simpler expression is obtained by adopting cylindrical coordinates and writing $x = r\cos\phi$ and $y = r\sin\phi$. The laplacian in two dimensions is

$$\frac{\partial^2}{\partial x^2} + \frac{\partial^2}{\partial y^2} = \frac{\partial^2}{\partial r^2} + \frac{1}{r}\frac{\partial}{\partial r} + \frac{1}{r^2}\frac{\partial^2}{\partial \phi^2} \tag{2}$$

Then, with r constant so that derivatives with respect to r can be discarded, the hamiltonian is

$$H = -\frac{\hbar^2}{2mr^2}\frac{\mathrm{d}^2}{\mathrm{d}\phi^2} = -\frac{\hbar^2}{2I}\frac{\mathrm{d}^2}{\mathrm{d}\phi^2} \tag{3}$$

The wavefunction depends only on the angle ϕ, and so we denote it Φ. The Schrödinger equation is therefore

$$\frac{\mathrm{d}^2\Phi}{\mathrm{d}\phi^2} = -\frac{2IE}{\hbar^2}\Phi \tag{4}$$

The general solutions are

$$\Phi = A\mathrm{e}^{\mathrm{i}m_l\phi} + B\mathrm{e}^{-\mathrm{i}m_l\phi} \qquad m_l = \left(\frac{2IE}{\hbar^2}\right)^{1/2} \tag{5}$$

The quantity m_l is a dimensionless number, and at this stage it is completely unrestricted in value; the significance of the subscript l will become apparent later.

Example 3.1 The separation of the Schrödinger equation

The wavefunctions for a particle on a ring also arise in connection with a particle confined to a circular region of zero potential energy by potential walls of infinite height (a 'circular square well'). Show that the Schrödinger equation is separable, and find equations for the radial and angular components.

Method. We try to separate the equation by proposing a solution in the form $\psi(r, \phi) = R(r)\Phi(\phi)$. The hamiltonian for the problem has only a kinetic energy contribution in the region where the particle may be found. It follows from the symmetry of the problem that it is sensible to express the hamiltonian in cylindrical coordinates. The laplacian in two dimensions, which is needed to write the hamiltonian, is given in eqn 3.2.

Answer. It follows from eqn 3.2 that the Schrödinger equation inside the well is

$$-\frac{\hbar^2}{2m}\left\{\frac{\partial^2\psi}{\partial r^2} + \frac{1}{r}\frac{\partial\psi}{\partial r} + \frac{1}{r^2}\frac{\partial^2\psi}{\partial\phi^2}\right\} = E\psi$$

Substitution of $\psi = R\Phi$ and then division of both sides by $R\Phi$ gives

$$-\left(\frac{\hbar^2}{2m}\right)\frac{1}{R}\{R'' + \frac{1}{r}R'\} - \frac{\hbar^2}{2mr^2}\frac{\Phi''}{\Phi} = E$$

where R' and R'' are first and second derivatives with respect to r and Φ'' is a second derivative with respect to ϕ. The $1/r^2$ in the second term can be eliminated by multiplication through by r^2, and after a little rearrangement the equation becomes

$$-\left(\frac{\hbar^2}{2m}\right)\frac{1}{R}\{r^2R''+rR'\}-Er^2=\left(\frac{\hbar^2}{2m}\right)\frac{\Phi''}{\Phi}$$

This equation is separable, because the left is a function only of r and the right is a function only of ϕ. We therefore write

$$\Phi''=-m_l^2\Phi$$

which implies that

$$r^2R''+rR'+\left(\frac{2mE}{\hbar^2}\right)r^2R=m_l^2R$$

Exercise 3.1. Go on to solve the radial equation by identifying the form of the equation by reference to Chapter 9 of M. Abramowitz and I.A. Stegun, *Handbook of mathematical functions*, Dover (1965), or a similar source.

Now we introduce the boundary conditions. There are no barriers to the particle's motion so long as it remains on the ring, and so there is no requirement for the wavefunctions to vanish at any point. However, wavefunctions must be single-valued (Chapter 2), and so it follows that $\Phi(\phi+2\pi)=\Phi(\phi)$. This requirement is an example of a **cyclic boundary condition**. It follows that

$$Ae^{im_l\phi}e^{2\pi im_l}+Be^{-im_l\phi}e^{-2\pi im_l}=Ae^{im_l\phi}+Be^{-im_l\phi}$$

This relation is satisfied only if m_l is an integer, for then $e^{2\pi im_l}=1$. The boundary conditions therefore imply that

$$m_l=0,\pm1,\pm2,\ldots$$

It follows that the allowed energies are

$$E_{m_l}=\frac{m_l^2\hbar^2}{2I}\qquad\text{with } m_l=0,\pm1,\pm2,\ldots\tag{6}$$

3.2 The angular momentum

The significance of the different signs of m_l can be discovered in exactly the same way as for linear motion, and by analogy with the discussion of wavefunctions for linear momenta $p=k\hbar$ with opposite signs of k, it can be anticipated that opposite signs correspond to opposite direction of travel. To confirm that this is so, we take the classical expression for the z-component of angular momentum, which is derived from the classical expression[1]

$$\mathbf{l}=\mathbf{r}\times\mathbf{p}=\begin{vmatrix}\hat{\mathbf{i}} & \hat{\mathbf{j}} & \hat{\mathbf{k}}\\ x & y & z\\ p_x & p_y & p_z\end{vmatrix}\tag{7}$$

The z-component (the coefficient of $\hat{\mathbf{k}}$) can be picked out by expanding the determinant, and is

$$l_z=xp_y-yp_x\tag{8}$$

[1] This relation is discussed in detail in Chapter 4.

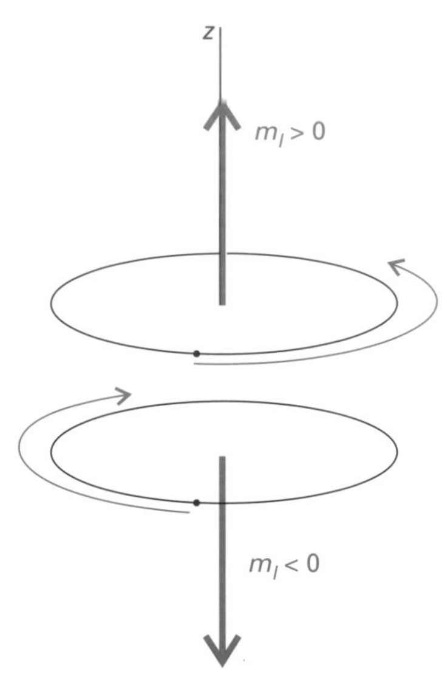

Fig. 3.2 The vector representation of angular momentum of a particle (or an effective particle) confined to a plane. Note the right-hand screw convention for the orientation of the vector.

At this point we express the classical observable as an operator in the position representation:

$$l_z \to x\left(\frac{\hbar}{\mathrm{i}}\right)\frac{\partial}{\partial y} - y\left(\frac{\hbar}{\mathrm{i}}\right)\frac{\partial}{\partial x}$$

Substitution of the polar coordinates defined above results in the expression

$$l_z = \frac{\hbar}{\mathrm{i}}\frac{\partial}{\partial \phi} \tag{9}$$

Now consider the effect of this operator on the wavefunction with $B = 0$:

$$l_z \Phi_{m_l} = \frac{\hbar}{\mathrm{i}}\frac{\partial}{\partial \phi} A\mathrm{e}^{\mathrm{i}m_l\phi} = m_l\hbar A\mathrm{e}^{\mathrm{i}m_l\phi} = m_l\hbar\Phi_{m_l} \tag{10}$$

This is an eigenvalue equation, and we see that the wavefunction corresponds to an angular momentum $m_l\hbar$. If $m_l > 0$, then the angular momentum is positive, and if $m_l < 0$, then the angular momentum is negative (Fig. 3.2).

The remaining task is to normalize the wavefunctions. For the function with $B = 0$, we write

$$\int_0^{2\pi} \Phi^*\Phi\,\mathrm{d}\phi = |A|^2\int_0^{2\pi} \mathrm{e}^{-\mathrm{i}m_l\phi}\mathrm{e}^{\mathrm{i}m_l\phi}\,\mathrm{d}\phi = |A|^2\int_0^{2\pi}\mathrm{d}\phi = 2\pi|A|^2 = 1$$

It follows that $|A| = 1/(2\pi)^{1/2}$, and A is conventionally chosen to be real (so the modulus bars can be dropped from this relation). It is easy to go on to show that the wavefunctions with different values of m_l are mutually orthogonal (see Problem 3.4).

3.3 The shapes of the wavefunctions

The physical basis of the quantization of rotation becomes clear when we inspect the shapes of the wavefunctions. The wavefunction corresponding to a state of definite angular momentum $m_l\hbar$ is

$$\Phi_{m_l} = \left(\frac{1}{2\pi}\right)^{1/2}\mathrm{e}^{\mathrm{i}m_l\phi} = \left(\frac{1}{2\pi}\right)^{1/2}\{\cos m_l\phi + \mathrm{i}\sin m_l\phi\} \tag{11}$$

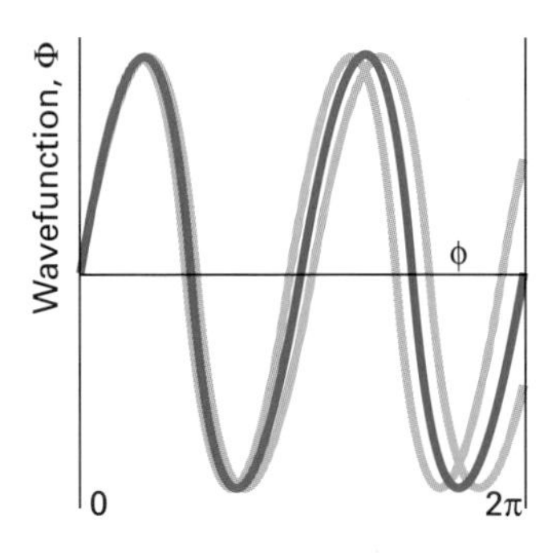

Fig. 3.3 The wavefunction must satisfy cyclic boundary conditions; only the dark curve of these three is acceptable. The horizontal coordinate corresponds to an entire circumference of the ring, and the end points should be considered to be joined.

Note that the wavefunction is complex (for $m_l \neq 0$), which is another illustration of the fact that wavefunctions corresponding to definite states of motion (other than being stationary in the sense that $m_l = 0$) are complex. We shall consider explicitly only the cosine component of the function, but similar remarks apply to the sine component too: the two components are 90° out of phase.

When m_l is an integer, the cosine functions form a wave with an integral number of wavelengths wrapped round the circular ring. The 'ends' of the wave join at ϕ and $\phi + 2\pi$, and the function reproduces itself on the next circuit (Fig. 3.3). When m_l is not an integer (for one of the disallowed solutions), the wavefunction has an incomplete number of wavelengths between 0 and 2π, and does not reproduce itself on the next circuit of the ring. At any point, it is double-valued, and hence must be rejected.

A glance at the expression for the energy shows that all the levels except the lowest ($m_l = 0$) are doubly degenerate: because $E_{m_l} \propto m_l^2$, the states $+|m_l|$ and

$-|m_l|$ have the same energy. This degeneracy stems from the fact that the particle can travel in either direction around the ring with the same magnitude of angular momentum, and hence with the same kinetic energy. The ground state is non-degenerate because when $m_l = 0$ the particle is stationary and the question of alternative directions of travel does not arise.

There are several ways of depicting the wavefunctions. The simplest procedure is to plot the real part of Φ on the perimeter of the ring (Fig. 3.4). It should be noted that in general the wavefunction is complex, and so it has real and imaginary components displaced by 90°. It is therefore easier to unwrap the ring into a straight line in the range $0 \le \phi \le 2\pi$ and to plot the wavefunctions on this line (Fig. 3.5). Drawing the two components helps to remind us that although the amplitude varies from point to point, the probability density is uniform:

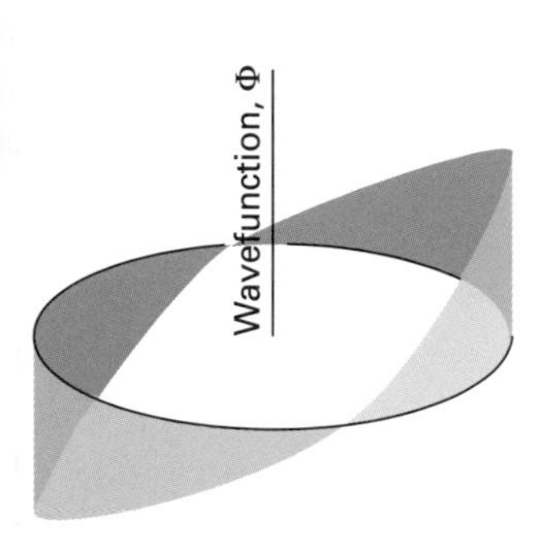

Fig. 3.4 One wavefunction for a particle on a ring (with $m_l = \pm 1$). Only the real part is shown.

$$|\Phi|^2 = \left\{\left(\frac{1}{2\pi}\right)^{1/2} e^{-im_l\phi}\right\}\left\{\left(\frac{1}{2\pi}\right)^{1/2} e^{im_l\phi}\right\} = \frac{1}{2\pi} \tag{12}$$

In a state of definite angular momentum, the particle is distributed uniformly round the ring: certainty in the value of the angular momentum implies total uncertainty in the location of the particle. A second point is that as the energy and the angular momentum increase, so the number of nodes in the real and imaginary components of the wavefunction increases too. This is an example of the behaviour we have already discussed: as the number of nodes is increased, the wavefunction is buckled backwards and forwards more sharply to fit into the perimeter of the ring, and consequently the kinetic energy of the particle increases. A further point that will prove to be of significance later is that the wavefunctions have the following symmetry properties:

$$\begin{aligned}\Phi_{m_l}(\phi + \pi) &= \left(\frac{1}{2\pi}\right)^{1/2} e^{im_l(\phi+\pi)} \\ &= \left(\frac{1}{2\pi}\right)^{1/2} e^{im_l\phi}\left(e^{i\pi}\right)^{m_l} = (-1)^{m_l}\Phi_{m_l}(\phi)\end{aligned} \tag{13}$$

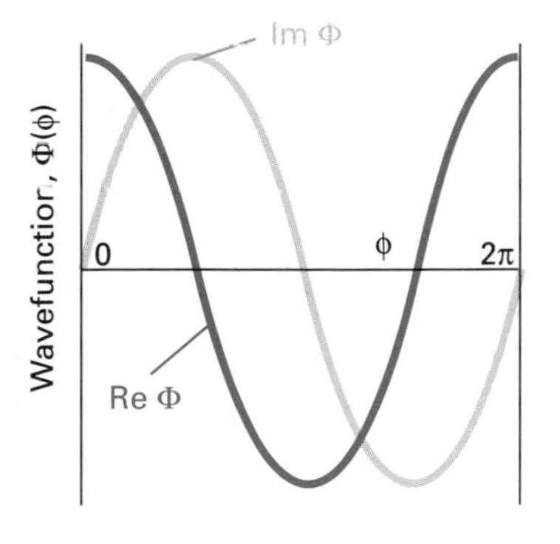

Fig. 3.5 A wavefunction corresponding to a definite state of motion is complex. The real and imaginary components shown here correspond to $m_l = +1$. Note that the real component seems to chase the imaginary one. The state with $m_l = -1$ has the imaginary component shifted in phase by π (that is, the component is multiplied by -1).

That is, points separated by 180° across the diameter of the ring have the same amplitude but differ in sign if m_l is odd.

A particle on a ring has no zero-point energy ($E_0 = 0$). The particle can satisfy the cyclic boundary conditions without its wavefunction needing to be curved, so one possible state has zero kinetic energy. The same argument is sometimes expressed in terms of the uncertainty principle in the form that as the particle may be anywhere in an infinite range of angles, its angular momentum can be specified precisely, and may be zero. However, great care must be taken when applying the uncertainty principle to periodic variables. In such cases it is appropriate to use more elaborate forms of the observables than simply ϕ itself, and then[2]

$$\Delta l_z \Delta \sin\phi \ge \tfrac{1}{2}\hbar|\langle\cos\phi\rangle| \tag{14}$$

[2] See P. Carruthers and M.M. Nieto in *Rev. Mod. Phys.*, 411, **40** (1968).

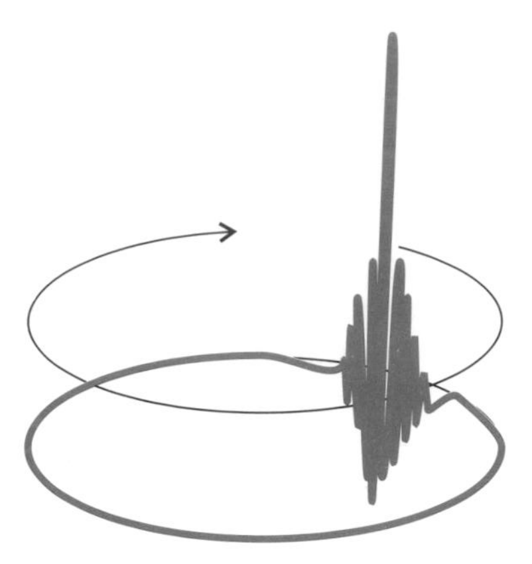

Fig. 3.6 A wavepacket formed from the superposition of many angular momentum eigenfunctions moves round the ring like the location of a classical particle. However, it also spreads with time.

3.4 The classical limit

When a particle is prepared with an energy that is imprecise in comparison with the energy-level separations, as when a macroscopic disk is set spinning, the correct description of the system is as a superposition of angular momentum eigenfunctions. The superposition results in a wavepacket. The amplitude of the wavepacket may represent the location of the actual particle or of a point representing the mass of the spinning disk. Because each component has the form

$$\Psi_{m_l}(\phi, t) = \left(\frac{1}{2\pi}\right)^{1/2} \mathrm{e}^{\mathrm{i}m_l\phi - \mathrm{i}m_l^2\hbar t/2I} \tag{15}$$

the point of maximum interference rotates around the ring (Fig. 3.6). This motion corresponds to the classical description of a rotating body.

Rotating motion in classical physics is normally denoted by a vector that represents the state of angular momentum of the body. For motion confined to the xy-plane, the vector lies parallel to the z-axis. The length of the vector represents the magnitude of the angular momentum, and its direction indicates the direction of motion. The right-hand screw convention is adopted: a vector pointing towards positive z represents clockwise rotation seen from below (as in Fig. 3.2). A vector pointing towards negative z represents motion in a counter-clockwise sense seen from below. The same representation can be used in quantum mechanics, the only difference being that in this case the length of the vector is confined to discrete values corresponding to the allowed values of m_l whereas in classical physics the length is continuously variable.

Particle on a sphere

Now we consider the case of a particle free to move over the surface of a sphere. The mass point can be an actual particle or a point in a solid body that represents the motion of the whole body. For example, a solid uniform sphere of mass m and radius R can be represented by the motion of a single point of mass m at a distance $r = (2/5)^{1/2}R$ (the radius of gyration) from the centre of the sphere. This problem will build on the material done in the previous section and prove to be the foundation for many applications in later chapters.

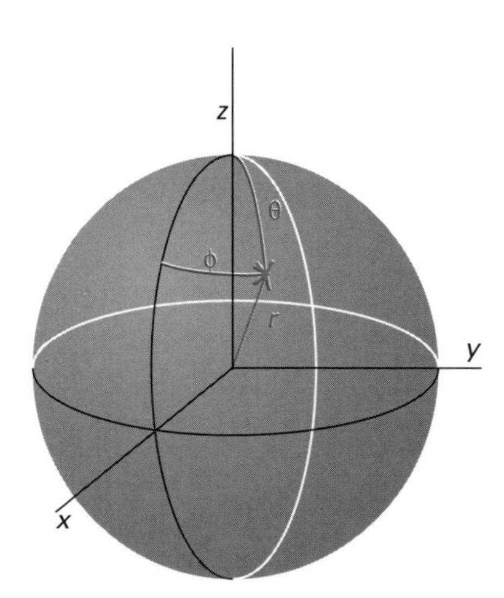

Fig. 3.7 Spherical polar coordinates. The angle θ is called the colatitude and the angle ϕ is the azimuth.

3.5 The Schrödinger equation and its solution

The potential energy of the particle is a constant taken to be zero, so the hamiltonian for the problem is simply

$$H = -\frac{\hbar^2}{2m}\nabla^2 \tag{16}$$

It is convenient to mirror the spherical symmetry of the problem by expressing the derivatives in terms of **spherical polar coordinates** (Fig. 3.7):

$$x = r\sin\theta\cos\phi \qquad y = r\sin\theta\sin\phi \qquad z = r\cos\theta \tag{17}$$

Standard manipulation of the differentials leads to the following expression for the laplacian operator:

$$\nabla^2 = \frac{1}{r}\frac{\partial^2}{\partial r^2}r + \frac{1}{r^2}\Lambda^2 \qquad (18)$$

where the **legendrian**, Λ^2, is defined as

$$\Lambda^2 = \frac{1}{\sin^2\theta}\frac{\partial^2}{\partial\phi^2} + \frac{1}{\sin\theta}\frac{\partial}{\partial\theta}\sin\theta\frac{\partial}{\partial\theta} \qquad (19)$$

The legendrian is the angular part of the laplacian. The condition that the particle is confined to the surface of fixed radius is equivalent to ignoring the radial derivatives, so we retain only the legendrian part of the laplacian and treat r as a constant. The hamiltonian is therefore

$$H = -\frac{\hbar^2}{2mr^2}\Lambda^2 \qquad (20)$$

Then, because the moment of inertia is $I = mr^2$, the Schrödinger equation we have to solve is

$$\Lambda^2\psi = -\left(\frac{2IE}{\hbar^2}\right)\psi \qquad (21)$$

where ψ is a function of the angles θ and ϕ.

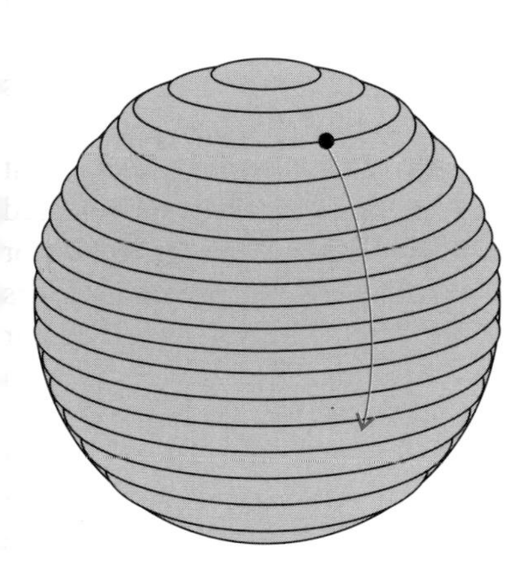

Fig. 3.8 The motion of a particle on the surface of a sphere is like its motion on a stack of rings with the ability to pass between the rings.

There are three ways of solving this second-order partial differential equation. One is to realize that the functions should resemble the solutions we have already found for the particle on a ring, for from one point of view (from any point of view, in fact) a sphere can be regarded as a stack of rings (Fig. 3.8). The difference is that for a sphere, the particle can travel from ring to ring. This view suggests that the wavefunction ought to be separable and of the form $\psi(\theta,\phi) = (\theta)\Phi(\phi)$. Indeed, it is easy to verify that the Schrödinger equation does separate, and that the component for Φ is

$$\frac{d^2\Phi}{d\phi^2} = \text{constant} \times \Phi$$

This equation is the same as the one for a particle on a ring, and the cyclic boundary conditions are the same. The solutions are therefore the same as before, and are specified by the quantum number m_l, with integral values. The equation for is much more involved and its solution by elementary techniques is cumbersome (it is given in *Further information 9*). The second method of solution is to avoid dealing with the Schrödinger equation directly, and to use the properties of the angular momentum operators themselves. The latter is a succinct and powerful approach, and will be described in Chapter 4. The third method of solution is to make the straighforward claim that we recognize eqn 3.21 as a well-known equation in mathematics, so that we can simply refer to tables for its solutions.[3] Indeed, solution by recognition is in fact the way that many differential equations are tackled by professional theoreticians, and it is a method not to be scorned!

[3] This is in practice a common way of solving differential equations, and the *Handbook of mathematical functions* mentioned in Exercise 3.1 is an excellent source of the appropriate information. It is an ideal desert-island book for shipwrecked quantum chemists.

As we show in *Further information 9*, the solutions of eqn 3.21 are the functions called **spherical harmonics**, $Y_{lm_l}(\theta, \phi)$. These highly important functions satisfy the equation

$$\Lambda^2 Y_{lm_l} = -l(l+1)Y_{lm_l} \tag{22}$$

where the labels l and m_l have the following values:

$$l = 0, 1, 2, \ldots \qquad m_l = l, l-1, \ldots, -l$$

Equation 3.22 has the same form as eqn 3.21, and so the wavefunctions ψ are proportional to the spherical harmonics. The spherical harmonics are composed of two factors:

$$Y_{lm_l}(\theta, \phi) = {}_{lm_l}(\theta)\Phi_{m_l}(\phi) \tag{23}$$

in accord with the separability of the Schrödinger equation. The functions Φ are the same as those already described for a particle on a ring. The functions are called **associated Legendre functions**. The first few spherical harmonics are listed in Table 3.1.

Table 3.1 Spherical harmonics

l	m_l	$Y_{lm_l}(\theta, \phi)$
0	0	$\frac{1}{2\pi^{1/2}}$
1	0	$\frac{1}{2}\left(\frac{3}{\pi}\right)^{1/2}\cos\theta$
	±1	$\mp\frac{1}{2}\left(\frac{3}{2\pi}\right)^{1/2}\sin\theta\, e^{\pm i\phi}$
2	0	$\frac{1}{4}\left(\frac{5}{\pi}\right)^{1/2}(3\cos^2\theta - 1)$
	±1	$\mp\frac{1}{2}\left(\frac{15}{2\pi}\right)^{1/2}\cos\theta\sin\theta\, e^{\pm i\phi}$
	±2	$\frac{1}{4}\left(\frac{15}{2\pi}\right)^{1/2}\sin^2\theta\, e^{\pm 2i\phi}$
3	0	$\frac{1}{4}\left(\frac{7}{\pi}\right)^{1/2}(2 - 5\sin^2\theta)\cos\theta$
	±1	$\mp\frac{1}{8}\left(\frac{21}{\pi}\right)^{1/2}(5\cos^2\theta - 1)\sin\theta\, e^{\pm i\phi}$
	±2	$\frac{1}{4}\left(\frac{105}{2\pi}\right)^{1/2}\cos\theta\sin^2\theta\, e^{\pm 2i\phi}$
	±3	$\mp\frac{1}{8}\left(\frac{35}{\pi}\right)^{1/2}\sin^3\theta\, e^{\pm 3i\phi}$

Example 3.2 How to confirm that a spherical harmonic is a solution

Confirm that the spherical harmonic Y_{10} is a solution of eqn 3.22.

Method. The direct method is to substitute the explicit expression for the spherical harmonic, taken from Table 3.1, into the left-hand side of eqn 3.22 and to verify that it is equal to the expression given on the right-hand side. The expression for the legendrian operator is given in eqn 3.19; because Y_{10} is independent of ϕ (see Table 3.1), the partial derivatives with respect to ϕ are zero, and we need consider only the derivatives with respect to θ.

Answer. It follows from Table 3.1 (writing N for the normalization constant) that

$$\Lambda^2 Y_{10} = \frac{1}{\sin\theta}\frac{\partial}{\partial\theta}\sin\theta\frac{\partial}{\partial\theta}N\cos\theta = -N\frac{1}{\sin\theta}\frac{\mathrm{d}}{\mathrm{d}\theta}\sin^2\theta$$
$$= -2N\frac{1}{\sin\theta}\sin\theta\cos\theta = -2Y_{10}$$

This result is consistent with eqn 3.22 when $l = 1$.

Exercise 3.2. Confirm that Y_{21} is a solution.

Comparison of eqns 3.21 and 3.22 shows that the energies of the particle are confined to the values

$$E_{lm_l} = l(l+1)\left(\frac{\hbar^2}{2I}\right) \tag{24}$$

The quantum number l is a label for the energy of the particle. Notice that E_{lm_l} is independent of the value of m_l. Therefore, because for a given value of l there are $2l+1$ values of m_l, we conclude that each energy level is $(2l+1)$-fold degenerate.

3.6 The angular momentum of the particle

The quantum numbers l and m_l have a further significance. The rotational energy of a spherical body of moment of inertia I and angular velocity ω is given by classical physics as $E = \frac{1}{2}I\omega^2$. Because the magnitude of the angular momentum is related to the angular velocity by $l = I\omega$, this energy can be expressed as $E = l^2/2I$. Comparison of this expression with the one in eqn 3.24 shows that

$$\text{magnitude of the angular momentum} = \{l(l+1)\}^{1/2}\hbar \tag{25}$$

Thus, the magnitude of the angular momentum is quantized in quantum mechanics. Indeed, l is called the **angular momentum quantum number**. This result will be confirmed formally in Chapter 4.

The spherical harmonics are also eigenfunctions of l_z:

$$l_z Y_{lm_l} = \frac{\hbar}{\mathrm{i}}\frac{\partial}{\partial\phi}\left(\Theta_{lm_l}\frac{\mathrm{e}^{\mathrm{i}m_l\phi}}{\sqrt{2\pi}}\right) = m_l\hbar Y_{lm_l} \tag{26}$$

This result too will be derived more formally in Chapter 4. We see from it that m_l specifies the component of the angular momentum around the z-axis, the contribution to the total angular momentum that can be ascribed to rotation around that axis. However, because m_l is restricted to certain values, the z-component of the angular momentum is also restricted to $2l+1$ discrete values for a given value of l. This restriction of the component of angular momentum is called **space quantization**. The name stems from the **vector representation** of angular momentum in which the angular momentum is represented by a vector of length $\{l(l+1)\}^{1/2}$ orientated so that its component on the z-axis is of length m_l. The vector can adopt only $2l+1$ orientations (Fig. 3.9), in contrast to the classical description in which the orientation of the rotating body is con-

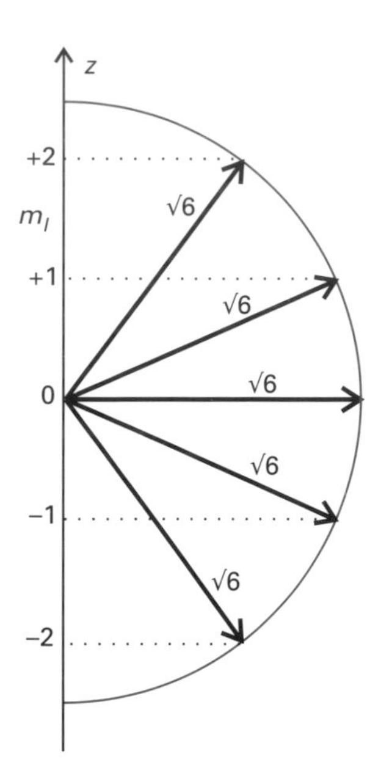

Fig. 3.9 The five (that is, $2l+1$) allowed orientations of the angular momentum with $l=2$. The length of the vector is $\sqrt{l(l+1)}$, which in this case is $\sqrt{6}$.

tinuously variable.

The quantum numbers l and m_l do not enable us to specify the x- and y-components of the angular momentum. Indeed, as we shall see later (Section 4.1), the operators corresponding to these components do not commute with the operator for the z-component, and so these components cannot in general be specified if the z-component is known. Therefore, a better representation of the states of angular momentum of a body is in terms of the cones shown in Fig. 3.10, in which no attempt is made to display any components other than the z-component. At this stage you should not think of the angular momentum vector as sweeping around the cones but simply as lying at some unspecified position on them.

It is a feature of space quantization that the angular momentum vector cannot lie exactly parallel to an arbitrarily specified z-axis. Its maximum z-component is $l\hbar$, which in general is less than its magnitude, $\{l(l+1)\}^{1/2}\hbar$. Only for very large values of l (in the classical limit) is $\{l(l+1)\}^{1/2} \approx l$, and then rotation can take place around a single axis.

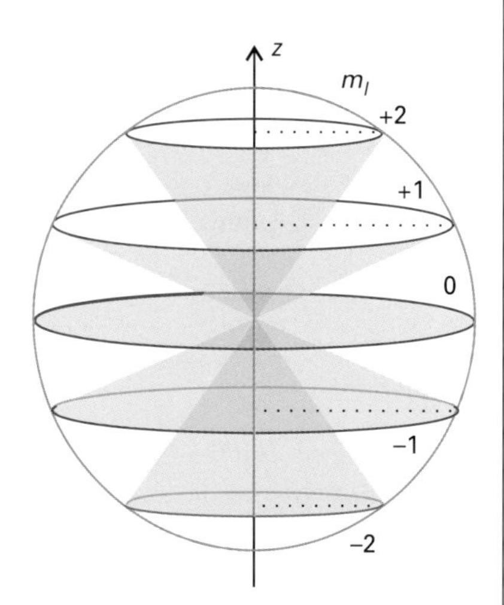

Fig. 3.10 To represent the fact that if the z-component of angular momentum is specified, the x- and y-components cannot in general be specified, the angular momentum vector is supposed to lie at an indeterminate position on one of the cones shown here (for $l=2$).

Example 3.3 The quantization of angular momentum for a macroscopic body

A solid ball of mass 250 g and radius 4.0 cm is spinning at 5.0 revolutions per second. Estimate the value of l and the minimum angle its angular momentum vector can make with respect to a selected axis.

Method. We need to calculate first its angular momentum, by using $l = I\omega$ and the expression for $I = mr^2$, with r the radius of gyration given in the text for a solid sphere of radius R, which is $r = (2/5)^{1/2}R$. Then identify l by setting the calculated value of angular momentum equal to the expression in eqn 3.25. The minimum angle can be obtained by trigonometry for a general value of m_l and then setting $m_l = l$.

Answer. The angular velocity of the ball is $\omega = 2\pi \times (5.0\,\mathrm{s}^{-1})$. Its moment of inertia is $I = \frac{2}{5}mR^2$, so its angular momentum is the product of these two quantities:

$$\{l(l+1)\}^{1/2}\hbar = \tfrac{2}{5}mR^2 \times (2\pi \times 5.0\,\mathrm{s}^{-1}) \approx 5.0 \times 10^{-3}\,\mathrm{kg\,m^2\,s^{-1}}$$

It follows that $l \approx 4.7 \times 10^{31}$. For general values of l and m_l, it follows from the vector representation that

$$\cos\theta = \frac{m_l}{\{l(l+1)\}^{1/2}}$$

When $m_l = l$ and $l \gg 1$ we can write

$$\cos\theta = \frac{l}{\{l(l+1)\}^{1/2}} = \frac{1}{\{1+\frac{1}{l}\}^{1/2}} = \frac{1}{\{1+\frac{1}{2l}+\ldots\}} = 1 - \frac{1}{2l} + \ldots$$

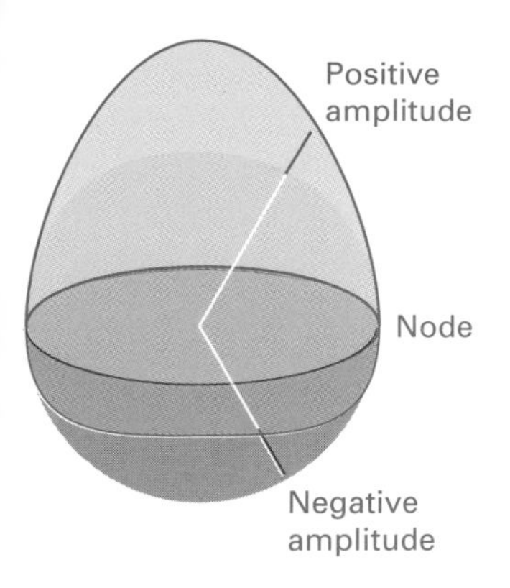

Fig. 3.11 One representation of the wavefunction of a particle on a sphere (with $l = 1, m_l = 0$) plots the function in terms of a height above or below the surface of the sphere.

and, because $\theta \ll 1$, equate this expression to $\cos\theta = 1 - \frac{1}{2}\theta^2 + \ldots$. It follows that

$$\theta \approx \frac{1}{l^{1/2}} = 1.5 \times 10^{-16}\,\text{rad}$$

Comment. This angle is virtually zero. Hence a macroscopic object can rotate effectively solely around a single specified axis.

Exercise 3.3. Show that the difference between the angles θ for the vectors with $m_l = l$ and $m_l = l - 1$ becomes zero as l becomes infinite.

3.7 Properties of the solutions

The wavefunctions for a particle on a sphere—the spherical harmonics—can be represented diagramatically in a variety of ways. The most cumbersome method is to plot the amplitude of the function relative to the surface of the sphere, by analogy with the wavefunctions for a particle on a ring (Fig. 3.11). It is more convenient, however, to plot the amplitudes of the spherical harmonics as a surface, the distance from the origin indicating the amplitude at that orientation (Fig. 3.12). The spherical harmonics are complex functions for $m_l \neq 0$, and the diagrams show only their real components. As for the particle on a ring, the complex function consists of a real and an imaginary component, the latter being the same shape as the former but rotated by 90° around the z-axis. An example is shown in Fig. 3.13. This illustration is included to emphasize the point that if m_l is specified, then the azimuthal distribution of the particle (the distribution with respect to the azimuth ϕ) is uniform: it is impossible to specify the azimuthal location of a particle with a well-defined component of angular momentum around the z-axis.

The probability densities $|Y_{lm_l}|^2$ for $l = 0$, 1, and 2 are illustrated in Fig. 3.14, and the azimuthal uniformity is clearly apparent. Notice too how the distribution shifts towards the equator as $|m_l|$ approaches l. This change corresponds to a reduced tilt in the plane of classical rotation.

It should be noticed that there is no zero-point energy ($E_{00} = 0$) because the wavefunction need not be curved (relative to the surface of the sphere); indeed, Y_{00} is a constant and all its derivatives are zero. The classical description of a rotating particle is achieved when the particle is set rotating with an imprecisely defined energy. In that case, its wavefunction is a wavepacket formed from a superposition of the spherical harmonics. This wavepacket moves in accord with the predictions of classical physics and migrates through all angles, but spreads with time (Fig. 3.15).

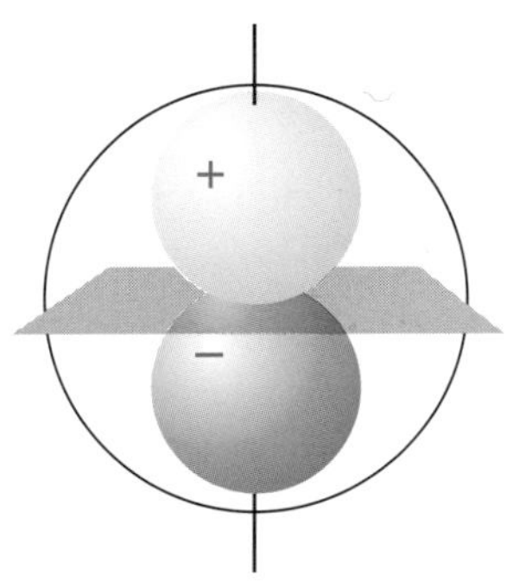

Fig. 3.12 In another representation of the same wavefunction as in the preceding illustration, the function is plotted along a radius to the point in question. In this case, the resulting surface consists of two touching spheres.

3.8 The rigid rotor

It is convenient at this point to introduce a variation on the topic of a particle on a sphere, to see how the same results apply to a body made up of two masses m_1 and m_2 at a fixed separation R. We have seen that any rigid object will be described by the same equations as for a single effective particle, but it is appro-

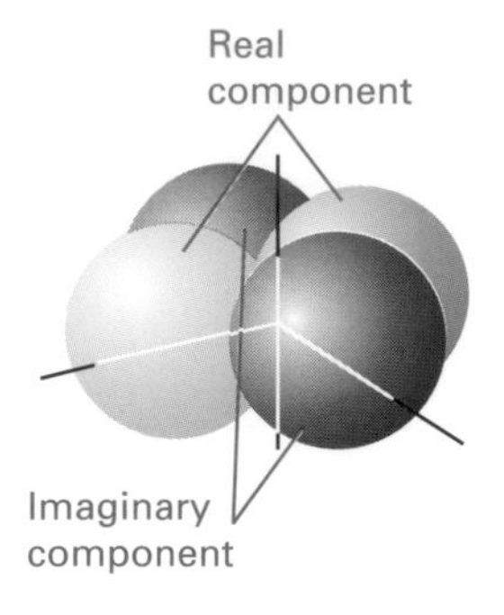

Fig. 3.13 The wavefunctions corresponding to $l = 1, m_l = \pm 1$ are complex, with real and imaginary components like those shown here. The direction of motion is determined by the relative phases of the two components: the real chases the imaginary.

priate to present the argument more formally. As we shall see, the separation of variables technique is the key.

The hamiltonian for two particles moving in free space is

$$H = -\frac{\hbar^2}{2m_1}\nabla_1^2 - \frac{\hbar^2}{2m_2}\nabla_2^2 \tag{27}$$

where ∇_i^2 differentiates with respect to the coordinates of particle i. As we show in *Further information 4*, this expression may be transformed by using

$$\frac{1}{m_1}\nabla_1^2 + \frac{1}{m_2}\nabla_2^2 = \frac{1}{m}\nabla_{\mathrm{cm}}^2 + \frac{1}{\mu}\nabla^2 \tag{28}$$

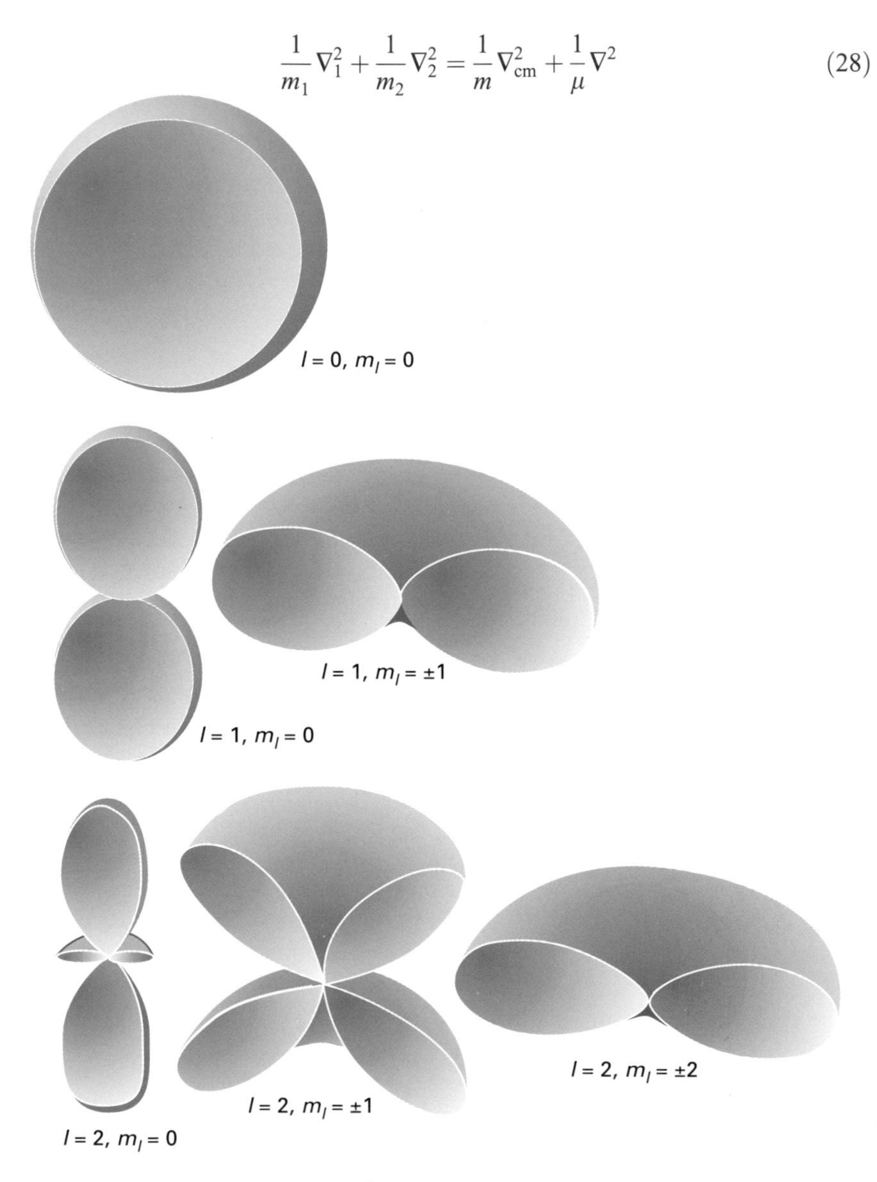

Fig. 3.14 The boundary surfaces for $|\psi|^2$ corresponding to $l = 0, 1, 2$ and the allowed values of $|m_l|$ in each case.

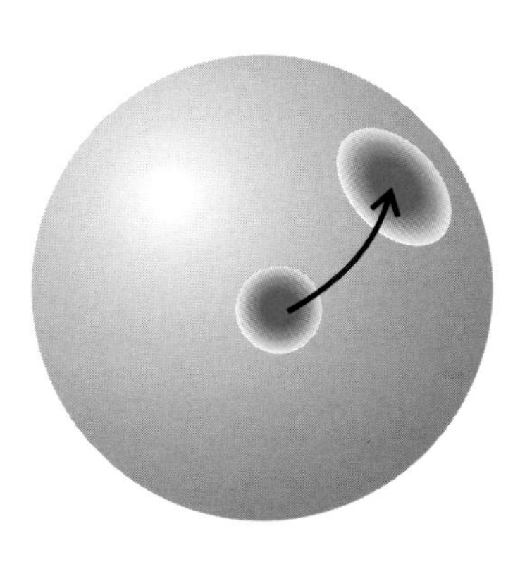

Fig. 3.15 The motion of a wavepacket on the surface of a sphere. As the wavepacket traces out the path like that of a classical particle, it also spreads.

where $m = m_1 + m_2$ and

$$\frac{1}{\mu} = \frac{1}{m_1} + \frac{1}{m_2} \tag{29}$$

The quantity μ is called the **reduced mass** of the system; the subscript 'cm' on the first laplacian on the right indicates that the derivatives are with respect to the centre of mass coordinates of the joint system, and the absence of subscripts on the second laplacian indicates that it is composed of derivatives with respect to the relative coordinates of the pair.

At this stage, the Schrödinger equation has become

$$-\frac{\hbar^2}{2m}\nabla_{\text{cm}}^2\Psi - \frac{\hbar^2}{2\mu}\nabla^2\Psi = E_{\text{total}}\Psi \tag{30}$$

This equation can be separated into equations for the motion of the centre of mass and for the relative motion of the particles. To do so we write $\Psi = \psi_{\text{cm}}\psi$, and by the same arguments as we have used several times before, find that the two factors separately satisfy the equations

$$\begin{aligned} -\frac{\hbar^2}{2m}\nabla_{\text{cm}}^2\psi_{\text{cm}} &= E_{\text{cm}}\psi_{\text{cm}} \\ -\frac{\hbar^2}{2\mu}\nabla^2\psi &= E\psi \end{aligned} \tag{31}$$

with $E_{\text{total}} = E_{\text{cm}} + E$. The first of these equations should be recognized as the translational motion of a free particle of mass m, which we solved in Chapter 2.

The second equation needs a little more work, for although it looks as simple as the first equation, the fact that r is a constant must be taken into account by working in spherical coordinates. Because the separation r of the two particles is constant (for a rigid rotor), the derivative with respect to r plays no role in eqn 3.18. Consequently, only the legendrian component need be retained, and we obtain

$$-\frac{\hbar^2}{2\mu r^2}\Lambda^2\psi = E\psi \tag{32}$$

At this stage we write

$$I = \mu r^2 \tag{33}$$

and obtain exactly the equation we have already considered (eqn 3.21). The solutions of this equation require two quantum numbers playing the role of l and m_l, and for the rigid rotor it is common to use J and M_J. The wavefunctions of the diatomic rigid rotor are the spherical harmonics Y_{JM_J}, and the energy levels are

$$E_{JM_J} = J(J+1)\frac{\hbar^2}{2I} \tag{34}$$

with $J = 0, 1, 2, \ldots$ and $M_J = 0, \pm1, \ldots, \pm J$. Note that each energy level is $(2J+1)$-fold degenerate (because the energy is independent of M_J and there are $2J+1$ values of M_J for each value of J). All the other features of the particle on a sphere apply equally to the rigid diatomic rotor, including the quantization of the angular momentum and space quantization.

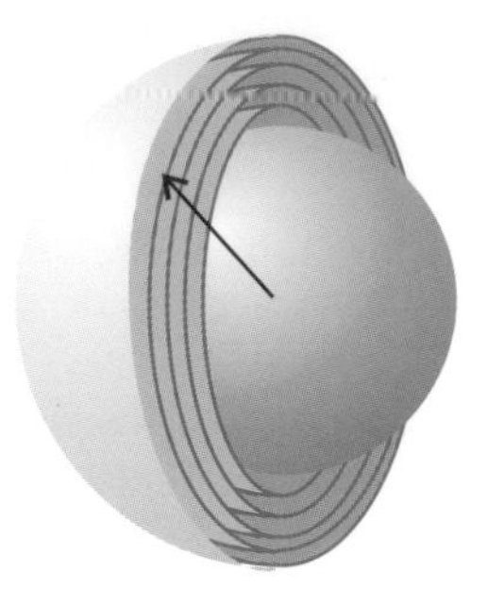

Fig. 3.16 The motion of a particle in a central field of force is like its motion on a stack of spheres with the ability to pass between the spheres.

Motion in a Coulombic field

The motion of an electron in a Coulombic field, one in which the potential varies as $1/r$, is of central importance in chemistry because it includes the structure of **hydrogenic atoms**, or one-electron species with arbitrary atomic number Z ($Z = 1$ for hydrogen itself). Most of the work of solving the Schrödinger equation has in fact already been done, for the motion can be regarded as that of an electron on a series of concentric spheres (Fig. 3.16). It follows that the wavefunctions can be expected to contain factors that correspond to the motion of a particle on a sphere. The additional work we must do is to account for the radial dependence of the motion, the extra degree of freedom that allows the electron to travel between the nested spherical surfaces.

3.9 The Schrödinger equation for hydrogenic atoms

The hamiltonian for the two-particle electron–nucleus system is

$$H = -\frac{\hbar^2}{2m_e}\nabla_e^2 - \frac{\hbar^2}{2m_N}\nabla_N^2 - \frac{Ze^2}{4\pi\varepsilon_0 r} \tag{35}$$

where m_e is the mass of the electron, m_N is the mass of the nucleus, and ∇_e^2 and ∇_N^2 are the laplacian operators that act on the electron and nuclear coordinates, respectively. The quantity ε_0 is the vacuum permittivity, a fundamental constant (see inside front cover). Apart from the Coulombic potential energy term, this hamiltonian is the same as we considered for the two-particle rotor. When we convert to centre-of-mass and relative coordinates, the potential energy term remains unchanged because it depends only on the separation of the particles. Therefore, we can use the work in *Further information 4* to write

$$H = -\frac{\hbar^2}{2m}\nabla_{cm}^2 - \frac{\hbar^2}{2\mu}\nabla^2 - \frac{Ze^2}{4\pi\varepsilon_0 r} \tag{36}$$

The resulting Schrödinger equation is separable on account of the dependence of the potential energy on the particle separation alone, and by the same argument as above, the Schrödinger equation for the relative motion of the electron and nucleus is

$$-\frac{\hbar^2}{2\mu}\nabla^2\psi - \frac{Ze^2}{4\pi\varepsilon_0 r}\psi = E\psi \tag{37}$$

The other component of the Schrödinger equation is that for the translational motion of the atom as a whole, and we do not need to consider it further.

Unlike the rigid rotor, the electron and nucleus are not constrained to have a fixed separation. We have to include the radial derivative in the laplacian, and so write the Schrödinger equation as

$$\frac{1}{r}\frac{\partial^2}{\partial r^2}r\psi + \frac{1}{r^2}\Lambda^2\psi + \frac{Ze^2\mu}{2\pi\varepsilon_0\hbar^2 r}\psi = -\left(\frac{2\mu E}{\hbar^2}\right)\psi \tag{38}$$

3.10 The separation of the relative coordinates

We have anticipated that the Schrödinger equation for the relative motion will be separable into angular and radial components, with the former being the equation for a particle on a sphere. We therefore attempt a solution of the form

$$\psi(r,\theta,\phi) = R(r)Y(\theta,\phi) \tag{39}$$

When this trial solution is substituted into the Schrödinger equation and we use $\Lambda^2 Y = -l(l+1)Y$, it turns into

$$\frac{1}{r}\frac{\partial^2}{\partial r^2}rRY - \frac{l(l+1)}{r^2}RY + \left(\frac{Ze^2\mu}{2\pi\varepsilon_0\hbar^2 r}\right)RY = -\left(\frac{2\mu E}{\hbar^2}\right)RY$$

The function Y may be cancelled throughout, and that leaves an equation for the **radial wavefunction**, R:

$$\frac{1}{r}\frac{\mathrm{d}^2(rR)}{\mathrm{d}r^2} + \left\{\left(\frac{Ze^2\mu}{2\pi\varepsilon_0\hbar^2 r}\right) - \frac{l(l+1)}{r^2}\right\}R = -\left(\frac{2\mu E}{\hbar^2}\right)R$$

At this stage we write $\Pi = rR$, and so obtain

$$\frac{\mathrm{d}^2\Pi}{\mathrm{d}r^2} + \left(\frac{2\mu}{\hbar^2}\right)\left\{\left(\frac{Ze^2}{4\pi\varepsilon_0 r}\right) - \frac{l(l+1)\hbar^2}{2\mu r^2}\right\}\Pi = -\left(\frac{2\mu E}{\hbar^2}\right)\Pi \tag{40}$$

This is the one-dimensional Schrödinger equation in the coordinate r that would have been obtained if, instead of the Coulomb potential, we had used an **effective potential energy** V_{eff}:

$$V_{\text{eff}} = -\frac{Ze^2}{4\pi\varepsilon_0 r} + \frac{l(l+1)\hbar^2}{2\mu r^2} \tag{41}$$

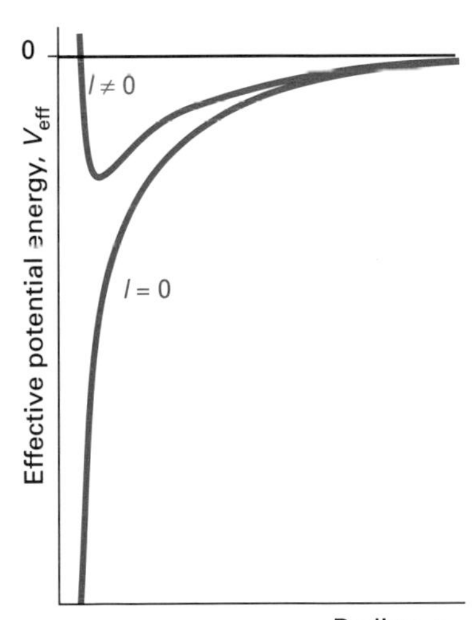

Fig. 3.17 The effective potential experienced by an electron in a hydrogen atom. When $l \neq 0$ there is a centrifugal contribution to the potential that prevents the close approach of the electron to the nucleus, as it increases more rapidly (as $1/r^2$) than the Coulomb attraction (which varies as $-1/r$).

3.11 The radial Schrödinger equation

The effective potential energy may be given a simple physical interpretation. The first part is the attractive Coulomb potential energy. The second part is a repulsive contribution that corresponds to the existence of a centrifugal force that impels the particle away from the nucleus by virtue of its motion. When $l = 0$ the electron has no orbital angular momentum and the force—now solely the Coulombic force—is everywhere attractive. The potential energy for this special case is everywhere negative (Fig. 3.17). When $l > 0$ the electron possesses an orbital angular momentum which tends to fling it away from the vicinity of the nucleus, and there is a competition between this effect and the attractive part of the potential. At very short distances from the nucleus, the repulsive component tends more strongly to infinity (as r^2) than the attractive part (which varies as $1/r$), and the former dominates. The two effective potentials (for $l = 0$ and $l \neq 0$) are qualitatively quite different near $r = 0$, and we shall investigate them separately.

When $l = 0$, the repulsive part of the effective potential energy is absent and the potential is everywhere attractive, even close to $r = 0$. When r is close to zero, the potential energy is locally so much larger than E that the latter may be neglected in eqn 3.40. The equation then becomes

$$\frac{\mathrm{d}^2\Pi}{\mathrm{d}r^2} + \left(\frac{2\mu}{\hbar^2}\right)\left(\frac{Ze^2}{4\pi\varepsilon_0 r}\right)\Pi \approx 0 \qquad \text{for } l = 0 \text{ and } r \approx 0$$

A solution of this equation is

$$\Pi \approx Ar + Br^2 + \text{higher-order terms}$$

as can be verified by substitution of the solution and taking the limit $r \to 0$. Therefore, close to $r = 0$ the radial wavefunction itself has the form $R = \Pi/r \approx A$, which may be nonzero. Therefore, when $l = 0$, there may be a nonzero probability of finding the electron at the nucleus.

When $l \neq 0$, the large repulsive component of the effective potential energy of the electron when it is close to the nucleus has the effect of excluding it from that region. In classical terms, the centrifugal force on an electron with nonzero angular momentum is strong enough at short distances to overcome the attractive Coulomb force. When $l \neq 0$ and r is close to zero, eqn 3.40 becomes

$$\frac{\mathrm{d}^2\Pi}{\mathrm{d}r^2} - \left(\frac{l(l+1)}{r^2}\right)\Pi \approx 0 \qquad (42)$$

because $1/r^2$ is the dominant term. The solution has the form

$$\Pi \approx Ar^{l+1} + \frac{B}{r^l} \qquad \text{for } l \neq 0 \text{ and } r \approx 0$$

Because $\Pi = rR$, at $r = 0$ we know that $\Pi = 0$; so it follows that $B = 0$. Therefore, the radial wavefunction has the form

$$R = \frac{\Pi}{r} \approx Ar^l \qquad \text{for } l \neq 0 \text{ and } r \approx 0$$

This function implies that the amplitude is zero at $r = 0$ for all wavefunctions with $l \neq 0$, and that the electron described by such a wavefunction will not be found at the nucleus.[4]

Example 3.4 The asymptotic form of atomic wavefunctions at large distances

Show that at large distances from the nucleus, bound-state atomic wavefunctions decay exponentially towards zero.

Method. We need to identify the terms in eqn 3.40 that survive as $r \to \infty$, and then solve the resulting equation. When solving such asymptotic equations, the solutions should also be tested in the limit $r \to \infty$.

Answer. When $r \to \infty$, eqn 3.40 reduces to

$$\frac{\mathrm{d}^2\Pi}{\mathrm{d}r^2} \simeq -\left(\frac{2\mu E}{\hbar^2}\right)\Pi$$

(The sign $\simeq$ means 'asymptotically equal to'.) However, because $\Pi = rR$, in the same limit this equation becomes

[4] Note that the radial wavefunction does not have a node at $r = 0$: a node is a point where a function passes through *zero*.

$$\frac{d^2R}{dr^2} = \frac{d^2}{dr^2}\frac{\Pi}{r} = \frac{\Pi''}{r} - \frac{2\Pi'}{r^2} + \frac{2\Pi}{r^3}$$
$$\simeq \frac{\Pi''}{r} \simeq -\left(\frac{2\mu E}{\hbar^2}\right)R = +\left(\frac{2\mu|E|}{\hbar^2}\right)R$$

where we have made use of the fact that $E < 0$ for bound states. This equation is satisfied (asymptotically) by

$$R \simeq e^{-(2\mu|E|/\hbar^2)^{1/2}r}$$

The alternative solution, with a positive exponent, is not square-integrable and so can be rejected. Hence, we can conclude that the wavefunction decays exponentially at large distances.

Comment. All atomic wavefunctions, even those for many-electron atoms, decay exponentially at large distances.

Exercise 3.4. Show that the unbound states (for which $E > 0$) are travelling waves at large distances from the nucleus.

$[R \simeq e^{\pm i(2\mu|E|/\hbar^2)^{1/2}r}]$

Explicit solutions of the radial wave equation can be found in a variety of ways. The most elementary method of solution is given in *Further information 8*. As explained there, the acceptable solutions are the **associated Laguerre functions**; the solutions are acceptable in the sense of being well-behaved and corresponding to states of negative energy (bound states of the atom). The first few hydrogenic wavefunctions are listed in Table 3.2.[5] They consist of a decaying exponential function multiplied by a simple polynomial in r. Each one is specified by the labels n and l, with

$$n = 1, 2, \ldots \qquad l = 0, 1, \ldots, n-1$$

Table 3.2 Hydrogenic radial wavefunctions

n	l	$R_{nl}(r)$
1	0 (1s)	$\left(\frac{Z}{a}\right)^{3/2} 2e^{-\rho/2}$
2	0 (2s)	$\left(\frac{Z}{a}\right)^{3/2}\frac{1}{2\sqrt{2}}(2-\rho)e^{-\rho/2}$
	1 (2p)	$\left(\frac{Z}{a}\right)^{3/2}\frac{1}{2\sqrt{6}}\rho e^{-\rho/2}$
3	0 (3s)	$\left(\frac{Z}{a}\right)^{3/2}\frac{1}{9\sqrt{3}}(6-6\rho+\rho^2)e^{-\rho/2}$
	1 (3p)	$\left(\frac{Z}{a}\right)^{3/2}\frac{1}{9\sqrt{6}}(4-\rho)\rho e^{-\rho/2}$
	2 (3d)	$\left(\frac{Z}{a}\right)^{3/2}\frac{1}{9\sqrt{30}}\rho^2 e^{-\rho/2}$

$\rho = (2Z/na)r$ with $a = 4\pi\varepsilon_0\hbar^2/\mu e^2$.

For an infinitely heavy nucleus, $\mu = m_e$ and $a = a_0$, the Bohr radius.

Relation to associated Laguerre functions:

$$R_{nl}(r) = -\left\{\left(\frac{2Z}{na}\right)^3 \frac{(n-l-1)!}{2n[(n+l)!]^3}\right\}\rho^l L_{n+l}^{2l+1}(\rho)e^{-\rho/2}$$

[5] See M. Abramowitz and I.A. Stegun, *Handbook of mathematical functions*, Dover, New York (1965), Chapter 22.

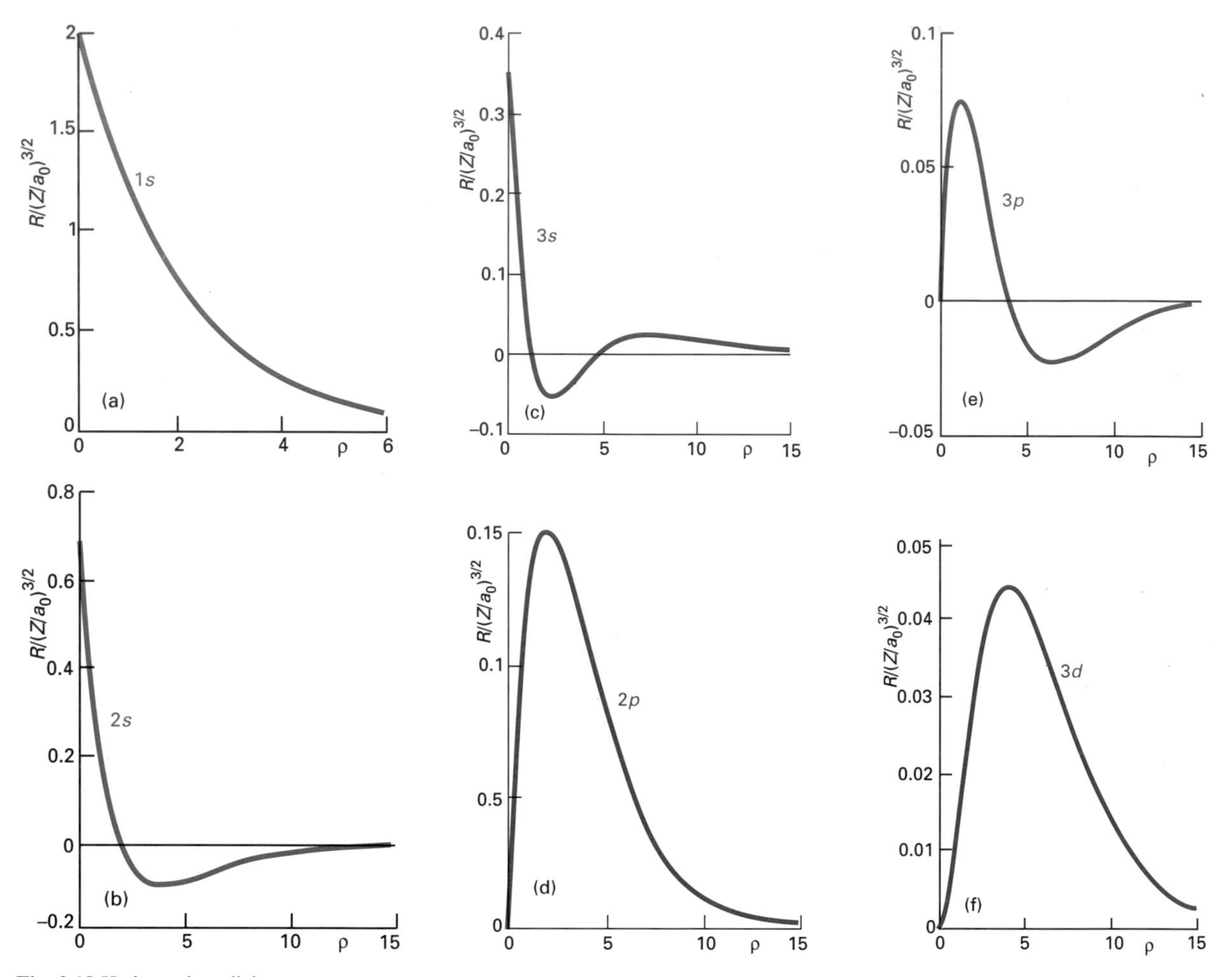

Fig. 3.18 Hydrogenic radial wavefunctions: (a) 1*s*, (b) 2*s*, (c) 3*s*, (d) 2*p*, (e) 3*p*, (f) 3*d*.

Some of the radial wavefunctions are plotted as functions of $\rho = 2Zr/na_0$ in Fig. 3.18, where a_0 is the **Bohr radius**:[6]

$$a_0 = \frac{4\pi\varepsilon_0\hbar^2}{m_e e^2} \tag{43}$$

The numerical value of a_0 is approximately 52.9 pm. Note that the functions with $l = 0$ are nonzero (and finite) at $r = 0$, whereas all the functions with $l \neq 0$ are zero at $r = 0$.

Each radial wavefunction has $n - l - 1$ nodes (the zero amplitude at $r = 0$ for functions with $l \neq 0$ are not nodes; recall the definition in Section 2.12). The locations of these nodes are found by determining where the polynomial in the associated Laguerre function is equal to zero. For example, the zeros of the

[6] In a precise calculation, the Bohr radius a_0, which depends on the mass of the electron, should be replaced by a, in which the reduced mass μ appears instead. Very little error is introduced by using a_0 in place of a in this and the other equations in this chapter.

function with $n = 3$ and $l = 0$ occur where

$$6 - 6\rho + \rho^2 = 0 \qquad \text{with } \rho = \left(\frac{2Z}{3a_0}\right)r$$

The zeros of this polynomial occur at $\rho = 3 \pm \sqrt{3}$, which corresponds to $r = (3 \pm \sqrt{3})(3a_0/2Z)$.

Insertion of the radial wavefunctions into eqn 3.40 gives the following expression for the energy:

$$E_n = -\left(\frac{Z^2\mu e^4}{32\pi^2\varepsilon_0^2\hbar^2}\right)\frac{1}{n^2} \qquad n = 1, 2, \ldots \tag{44}$$

The same values are obtained whatever the value of l or m_l. Therefore, in hydrogenic atoms (but not in any other kind of atom) the energy depends only on the **principal quantum number** n and is independent of the values of l and m_l, and each level is n^2-fold degenerate (that being the total number of wavefunctions for a given n). This degeneracy is peculiar to the Coulomb potential in a non-relativistic system, and we shall return to it shortly.

The roles of the quantum numbers in the hydrogen atom should now be clear, but may be summarized as follows:

1. The principal quantum number, n, specifies the energy through eqn 3.44 and controls the range of values of $l = 0, 1, \ldots, n - 1$; it also gives the total number of orbitals with the specified value of n as n^2 and gives the total number of radial and angular nodes as $n - 1$.
2. The orbital angular momentum quantum number, l, specifies the orbital angular momentum of the electron through eqn 3.25, and determines the number of orbitals with a given n and l as $2l + 1$. There are l angular nodes in the wavefunction.
3. The magnetic quantum number, m_l, specifies the component of orbital angular momentum of an electron through $m_l\hbar$ (see eqn 3.26) and, for a given n and l, specifies an individual orbital.

3.12 Probabilities and the radial distribution function

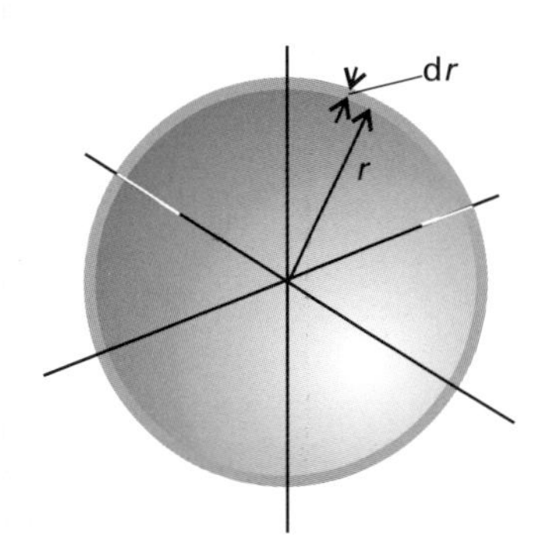

Fig. 3.19 The radial distribution function gives the probability that an electron will be found anywhere between two concentric spheres with radii that differ by dr.

The complete wavefunctions of the electron in a hydrogenic atom have the form $\psi_{nlm_l} = R_{nl}Y_{lm_l}$, where the R_{nl} are related to the (real) associated Laguerre functions and the Y_{lm_l} are the (in general, complex) spherical harmonics. The probability of finding an electron in a volume element $\mathrm{d}\tau = r^2 \sin\theta\, \mathrm{d}\theta\mathrm{d}\phi\mathrm{d}r$ at a point specified by the polar coordinates r, θ, ϕ when it is described by the wavefunction ψ_{nlm_l} is $|\psi_{nlm_l}(r, \theta, \phi)|^2\mathrm{d}\tau$.

Although the wavefunction gives the probability of finding an electron at a specified location, it is sometimes more helpful to know the probability of finding the particle at a given radius regardless of the direction. This probability is obtained by integration over the volume contained between two concentric spheres of radii r and $r + \mathrm{d}r$ (Fig. 3.19):

$$P(r)\mathrm{d}r = \int_{\text{surface}} |\psi_{nlm_l}|^2\, \mathrm{d}\tau = \int_0^{\pi}\int_0^{2\pi} R^2|Y|^2 r^2 \sin\theta\, \mathrm{d}r\mathrm{d}\theta\mathrm{d}\phi \tag{45}$$

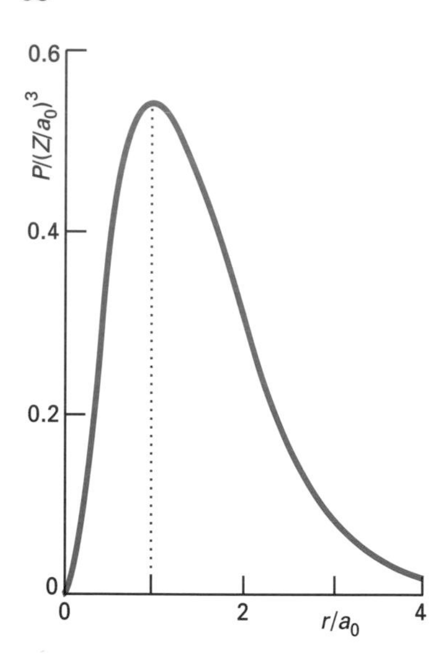

Fig. 3.20 The radial distribution function for a 1*s*-electron. The function passes through a maximum at the Bohr radius, a_0.

The spherical harmonics are normalized to 1 in the sense that

$$\int_0^{\pi}\int_0^{2\pi} |Y|^2 \sin\theta \, d\theta d\phi = 1$$

Therefore,

$$P(r)dr = R^2 r^2 \, dr \tag{46}$$

The quantity $P(r) = R^2(r)r^2$ is the **radial distribution function**: when multiplied by dr it gives the probability that the electron will be found between r and $r + dr$.[7] For an orbital with $n = 1$ and $l = 0$, it follows from Table 3.2 that

$$P(r) = 4\left(\frac{Z}{a_0}\right)^3 r^2 e^{-2Zr/a_0} \tag{47}$$

This function is illustrated in Fig. 3.20. Note that it is zero at $r = 0$ (on account of the factor r^2) and approaches zero as $r \to \infty$ on account of the exponential factor. By differentiation with respect to r and setting $dP/dr = 0$ it is easy to show that it goes through a maximum at

$$r_{max} = \frac{a_0}{Z}$$

For a hydrogen atom ($Z = 1$), $r_{max} = a_0$. Therefore, the radius that Bohr calculated for the orbital of lowest energy in a hydrogen atom in his early pre-quantum mechanical model of the atom is in fact the most probable distance of the electron from the nucleus in the quantum mechanical model. Note that this most probable radius decreases in hydrogenic atoms as Z increases, because the electron is drawn closer to the nucleus as the charge of the latter increases.

3.13 Atomic orbitals

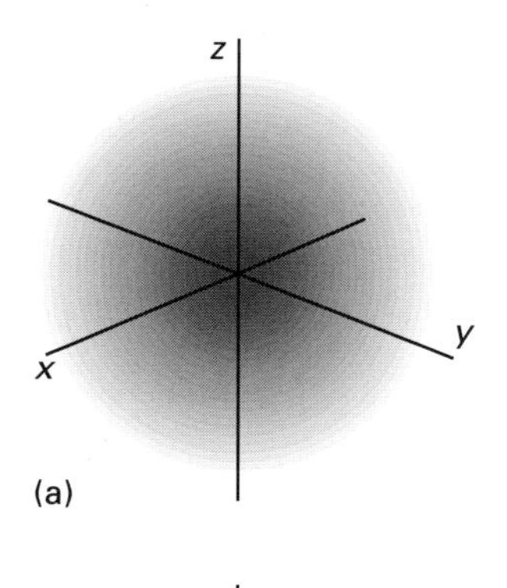

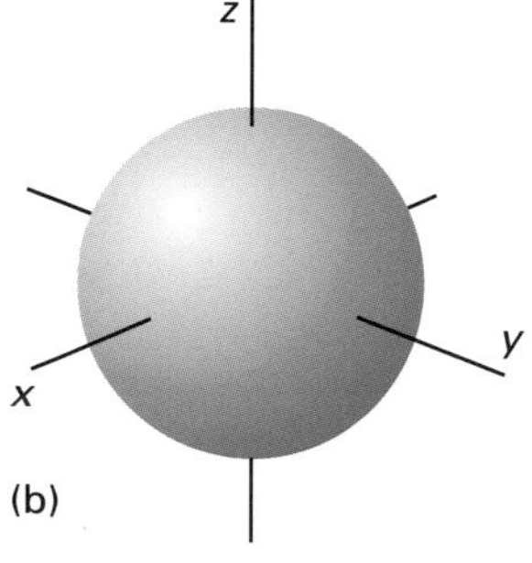

Fig. 3.21 Two representations of the probability density corresponding to a 1*s*-orbital: (a) the density represented by the darkness of shading, (b) the boundary surface of the orbital.

One-electron wavefunctions in atoms are called **atomic orbitals**; this name was chosen because it conveys a sense of less certainty than the term 'orbit' of classical theory. For historical reasons, atomic orbitals with $l = 0$ are called ***s*-orbitals**, those with $l = 1$ are called ***p*-orbitals**, those with $l = 2$ are called ***d*-orbitals**, and those with $l = 3$ are called ***f*-orbitals**. When an electron is described by the wavefunction ψ_{nlm_l} we say that that electron **occupies** the orbital. An electron that occupies an *s*-orbital is called an ***s*-electron**, and similarly for electrons that occupy other types of orbitals.

The shapes of atomic orbitals can be expressed in a number of ways. One way is to denote the probability of finding an electron in a region by the density of shading there (Fig. 3.21). A simpler and generally adequate procedure is to draw the **boundary surface**, the surface of constant probability within which there is a specified proportion of the probability density (typically 90 per cent). For real forms of the orbitals, the sign of the wavefunction itself is often indicated either by tinting the positive amplitude part of the boundary surface or by attaching + and − signs to the relevant lobes of the orbitals. There are few occasions when a precise portrayal of either the amplitude or the probability density is required, and the qualitative boundary surfaces shown in Fig. 3.22

[7] For an *s*-orbital, R^2r^2 is equivalent to $4\pi r^2|\psi|^2$.

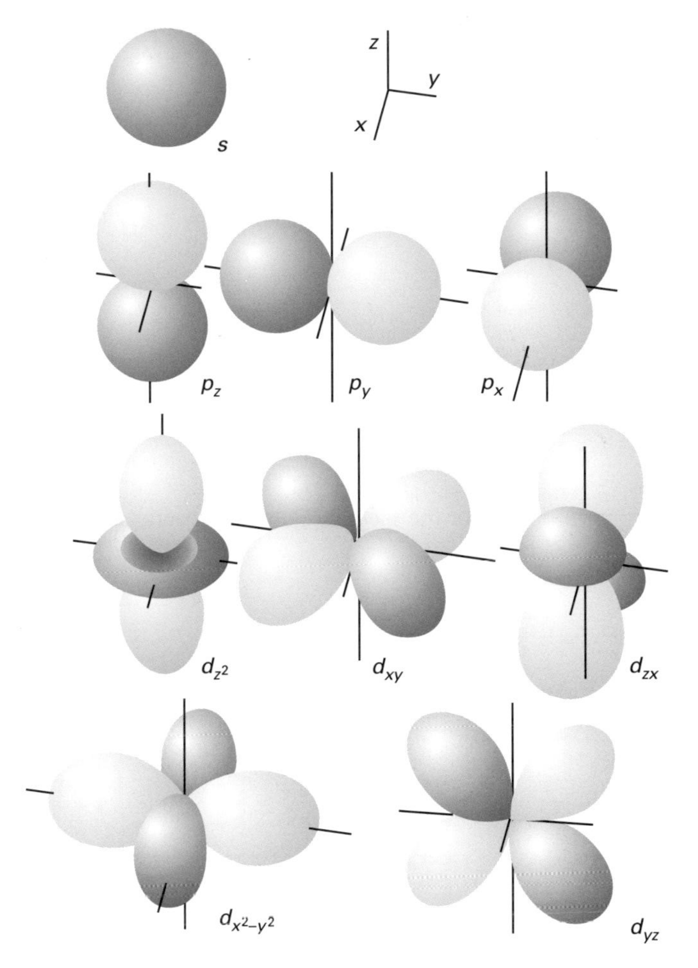

Fig. 3.22 Boundary surfaces for *s*-, *p*-, and *d*-orbitals.

are generally adequate.

The boundary surfaces in Fig. 3.22 show that *s*-orbitals are spherically symmetrical with Y_{00} a constant independent of angle; we have also already seen that *s*-orbitals differ from other types of orbitals insofar as they have a nonzero amplitude at the nucleus. This feature stems from their lack of orbital angular momentum. It may be puzzling why, with no orbital angular momentum, an *s*-orbital can exist, because a classical electron without angular momentum would plunge into the nucleus as a result of the nuclear attraction. The answer is found in a quantum mechanical competition between kinetic and potential energies. For an *s*-electron to cluster close to the nucleus and hence minimize its potential energy, it needs a wavefunction that peaks strongly at the nucleus and is zero elsewhere. However, such a wavefunction is sharply curved, and hence corresponds to a very high kinetic energy for the electron. If, instead, the wavefunction spreads over a very wide region with a gentle curvature, then although its kinetic energy will be low, its potential energy will be high because it spends so much time far from the nucleus. The lowest total energy is obtained when the wavefunction is a compromise between confined-but-curved and dispersed-but-gently-curved.

The three p-orbitals with a given value of n correspond to the three values that m_l may have, namely 0 and ± 1. The orbital with $m_l = 0$ is real and has zero component of angular momentum around the z-axis; it is called a $\boldsymbol{p_z}$**-orbital**. The other two orbitals, p_+ and p_-, are complex, and have their maximum amplitude in the xy-plane (recall Fig. 3.14):

$$p_z = \left(\frac{3}{4\pi}\right)^{1/2} R_{n1}(r)\cos\theta$$

$$p_+ = -\left(\frac{3}{8\pi}\right)^{1/2} R_{n1}(r)\sin\theta\, e^{i\phi} \qquad (48)$$

$$p_- = \left(\frac{3}{8\pi}\right)^{1/2} R_{n1}(r)\sin\theta\, e^{-i\phi}$$

It is usual to depict the real and imaginary components, and to call these orbitals p_x and p_y:

$$p_x = \frac{1}{\sqrt{2}}(p_- - p_+) = \left(\frac{3}{4\pi}\right)^{1/2} R_{n1}(r)\sin\theta\cos\phi$$

$$p_y = \frac{i}{\sqrt{2}}(p_- + p_+) = \left(\frac{3}{4\pi}\right)^{1/2} R_{n1}(r)\sin\theta\sin\phi \qquad (49)$$

The complex orbitals are the appropriate forms to use in atoms and linear molecules where there are no well defined x- and y-axes; the real forms are more appropriate when x- and y-axes are well defined, such as in nonlinear molecules. All three real orbitals (p_x, p_y, and p_z) have the same double-lobed shape, but aligned along the x-, y-, and z-axes, respectively.

Example 3.5 How to analyse the probability distribution of an electron

What is the most probable point at which a hydrogenic $2p_z$-electron will be found, and what is the probability of finding the electron inside a sphere of radius R centred on the nucleus?

Method. For the first part, we need to inspect the form of the wavefunction and identify the location of the maximum amplitude by considering the maxima in r, θ, and ϕ separately. The wavefunction itself is given by combining the information in Tables 3.1 and 3.2, and using $n = 2$, $l = 1$, and $m_l = 0$. For the second part, we integrate $|\psi|^2$ over a sphere of radius R. To do so, make use of the fact that the spherical harmonics are normalized (to 1) under integration over the surface of a sphere.

Answer. The wavefunction we require is $\psi_{210} = R_{21}Y_{10}$. The spherical harmonic is proportional to $\cos\theta$, and its maximum amplitude therefore lies at $\theta = 0$ or π, which is along the z-axis. The wavefunction is constant with respect to the azimuth ϕ. The radial wavefunction is proportional to $\rho e^{-\rho/2}$ with $\rho = (Z/a_0)r$. To find the location of the maximum of this function we differentiate with respect to ρ (which is proportional to r) and set the result equal to zero:

$$\frac{d}{d\rho}\rho e^{-\rho/2} = \left(1 - \frac{\rho}{2}\right)e^{-\rho/2} = 0$$

It follows that the maximum occurs at $\rho = 2$, or at $r = 2a_0/Z$. There are two points at which the probability reaches a maximum, at $\rho = 2, \theta = 0$ on the positive z-axis and at $\rho = 2, \theta = \pi$ on the negative z-axis.

For the second part of the question, we need to integrate:

$$P(R) = \int_{\text{Sphere of radius } R} R_{21}^2 |Y_{10}|^2 \, d\tau = \int_0^R R_{21}^2 r^2 \, dr$$

We have used the fact that the spherical harmonics are normalized to 1 when integrated over the surface of a sphere. It then follows from Table 3.2 that

$$P(R) = \tfrac{1}{24}\left(\frac{Z}{a_0}\right)^3 \int_0^R (\rho^2 e^{-\rho}) r^2 \, dr$$

with $\rho = Zr/a_0$. Therefore,

$$P(R) = \tfrac{1}{24}\left(\frac{Z}{a_0}\right)^5 \int_0^R r^4 e^{-Zr/a_0} dr = \tfrac{1}{24}\int_0^{ZR/a_0} x^4 e^{-x} \, dx$$

$$= 1 - \left\{1 + \left(\frac{ZR}{a_0}\right) + \tfrac{1}{2}\left(\frac{ZR}{a_0}\right)^2 + \tfrac{1}{6}\left(\frac{ZR}{a_0}\right)^3 + \tfrac{1}{24}\left(\frac{ZR}{a_0}\right)^4\right\} e^{-ZR/a_0}$$

For a hydrogen atom with $Z = 1$, we find that the probability of the electron being within a sphere of radius $2a_0$ is

$$P(2a_0) = 1 - 7e^{-2} = 0.053$$

Exercise 3.5. Repeat the calculation for a $2s$-electron in a hydrogenic atom and evaluate $P(2a_0)$ for a hydrogen atom.

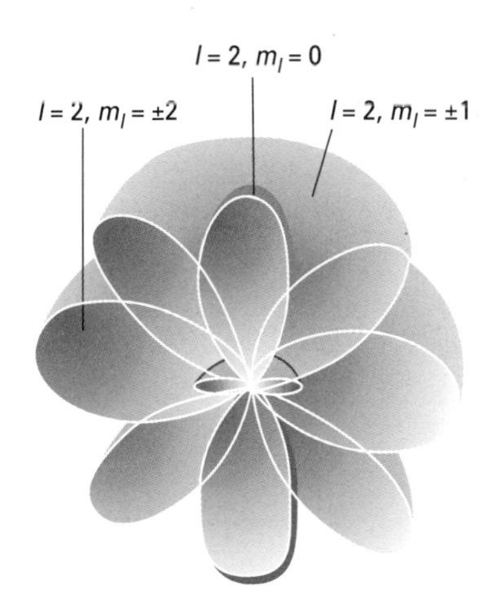

Fig. 3.23 The real parts of the wavefunctions for the five atomic orbitals with $l = 2$. Note that depicted in this way the unique form of the wavefunction with $m_l = 0$ is seen to be a part of a family of cylindrically symmetrical functions.

There are five d-orbitals ($l = 2$) for each $n \geq 3$. All except the orbital with $m_l = 0$ are complex, and correspond to definite states of orbital angular momentum around the z-axis. These complex orbitals have cylindrical symmetry around the z-axis, and can be depicted as shown in Fig. 3.23. However, it is more common to display them as their real components, as in Fig. 3.22:

$$d_{z^2} = d_0 = \left(\frac{5}{16\pi}\right)^{1/2} R_{n2}(r)(3\cos^2\theta - 1)$$

$$= \left(\frac{5}{16\pi}\right)^{1/2} R_{n2}(r)(3z^2 - r^2)r^{-2}$$

$$d_{x^2-y^2} = \frac{1}{\sqrt{2}}(d_{+2} + d_{-2}) = \left(\frac{15}{16\pi}\right)^{1/2} R_{n2}(r)(x^2 - y^2)r^{-2}$$

$$d_{xy} = \frac{1}{i\sqrt{2}}(d_{+2} - d_{-2}) = \left(\frac{15}{4\pi}\right)^{1/2} R_{n2}(r)xyr^{-2}$$

$$d_{yz} = \frac{1}{\mathrm{i}\sqrt{2}}(d_{+1} + d_{-1}) = -\left(\frac{15}{4\pi}\right)^{1/2} R_{n2}(r) yzr^{-2}$$

$$d_{zx} = \frac{1}{\sqrt{2}}(d_{+1} - d_{-1}) = -\left(\frac{15}{4\pi}\right)^{1/2} R_{n2}(r) zxr^{-2}$$

The notation stems from the identification of the angular dependence of the orbitals with the relations $x = r \sin\theta \cos\phi$ and so on (eqn 3.17). In deriving these results, we have used the phases of the spherical harmonics specified in Table 3.1.[8]

Once the wavefunctions of orbitals are available it is a simple matter to calculate various properties of the electron distributions they represent. For example, the mean radius of an orbital can be evaluated by calculating the expectation value of r by using one of the radial wavefunctions given in Table 3.2. However, it is usually easier to use one of the following general expressions that are obtained by using the general properties of associated Laguerre functions to evaluate the expectation values:

$$\begin{aligned} \langle r \rangle_{nlm_l} &= \frac{n^2 a_0}{Z}\left\{1 + \tfrac{1}{2}\left(1 - \frac{l(l+1)}{n^2}\right)\right\} \\ \left\langle \frac{1}{r} \right\rangle_{nlm_l} &= \frac{Z}{a_0 n^2} \end{aligned} \tag{50}$$

Note that the first of these expressions shows that the mean radius of an *ns*-orbital is *greater* than that of an *np*-orbital, which is contrary to what one might expect on the basis of the centrifugal effect of orbital angular momentum. It is due to the existence of an additional *radial* node in the *ns*-orbital, which tends to extend its radial distribution function out to greater distances. The fact that the average value of $1/r$ is independent of l is in line with the degeneracy of hydrogenic atoms, for the Coulomb potential energy of the electron is proportional to the mean value of $1/r$, and the result implies that all orbitals with a given value of n have the same potential energy.

3.14 The degeneracy of hydrogenic atoms

We have already seen that the energies of hydrogenic orbitals depend only on the principal quantum number n. To appreciate this conclusion, we can note that the virial theorem (Section 2.17) for a system in which the potential is Coulombic ($s = -1$) implies that

$$\langle T \rangle = -\tfrac{1}{2}\langle V \rangle \tag{51}$$

However, we have just seen that the mean value of $1/r$ is independent of l; therefore both the average potential energy and (by the virial theorem) the average kinetic energy are independent of l. Hence the total energy is independent of l, and all orbitals with a given value of n have the same energy. Because the permissible values of l are $l = 0, 1, \ldots, n-1$, and for each value of l there

[8] We have also used the trigonometric relations $\cos 2\phi = \cos^2\phi - \sin^2\phi$ and $\sin 2\phi = 2\sin\phi\cos\phi$.

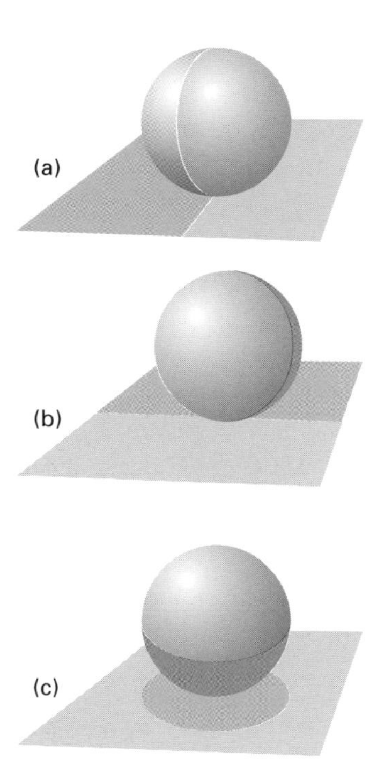

Fig. 3.24 A representation of the origin of the degeneracy of $2s$- and $2p$-orbitals in hydrogenic atoms. The object (a) can be rotated into (b), corresponding (when the projection on the two-dimensional plane is inspected) to the rotation of a $2p_y$-orbital into a $2p_x$-orbital. However, rotation about another axis results in a projection that corresponds to a $2s$-orbital (c). Thus, in a space of higher dimension, rotations can interconvert $2s$- and $2p$-orbitals.

are $2l + 1$ orbitals, the degeneracy of a level with quantum number n is

$$g_n = \sum_{l=0}^{n-1}(2l+1) = n^2 \tag{52}$$

The degeneracy of orbitals with the same value of n but different l is unique to hydrogenic atoms, and is lost in the presence of more than one electron. However, the degeneracy of the orbitals with different values of m_l but the same values of n and l remains even in the presence of many electrons because orbitals with different m_l differ only in the orientation of their angular momentum relative to an arbitrary axis.

The high degeneracy of a hydrogenic atom is an example of an accidental degeneracy, because there is no obvious rotation of the atom that allows an s-orbital to be transformed into a p-orbital, or some other orbital (recall Section 2.15). However, the Coulomb potential does have a hidden symmetry, a symmetry that is not immediately apparent. This hidden symmetry shows up in spaces of dimension higher than 3. It implies that a four-dimensional being would be able to see at a glance that a $2s$-orbital can be rotated into a $2p$-orbital, and would therefore not be surprised at their degeneracy, any more than we are surprised at the degeneracy of the three $2p$-orbitals. A way of illustrating this hidden symmetry is shown in Fig. 3.24, where we have imagined how a two-dimensional being might experience the projection of a patterned sphere. It is quite easy for us to see that one of the rotations of the sphere results in a change in the projection of the sphere which would lead a Flatlander to think that a p-orbital has been transformed into an s-orbital. We, in our three dimensions, can easily see that the orbitals are related by symmetry; the low-dimensional being, however, might not, and would remain puzzled about the degeneracy. The hydrogen atom has exactly the same kind of higher-dimensional symmetry.

Problems

3.1 The rotation of the HI molecule can be pictured as an orbiting of the hydrogen atom at a radius of 160 pm about a virtually stationary I atom. If the rotation is thought of as taking place in a plane (a restriction removed later), what are the rotational energy levels? What wavelength radiation is emitted in the transition $m_l = +1 \rightarrow m_l = 0$?

3.2 Confirm eqn 3.2 for the laplacian in two dimensions.

3.3 Show that $l_z = (\hbar/\mathrm{i})\mathrm{d}/\mathrm{d}\phi$ (that is, confirm eqn 3.9) for a particle confined to a planar surface.

3.4 Show that the wavefunctions in eqn 3.11 are mutually orthogonal.

3.5 Calculate the rotational energy levels of a compact disk of radius 10 cm and mass 50 g free to rotate in a plane. To what value of m_l does a rotation rate of 100 Hz correspond?

3.6 Construct the analogues of Fig. 3.4 for the states of a rotor with $m_l = +3$ and +4.

3.7 (a) Construct a wavepacket $\Psi = N\sum_{m_l=0}^{\infty}(1/m_l!)\mathrm{e}^{\mathrm{i}m_l\phi}$ and normalize it to unity. Sketch the form of $|\Psi|^2$ for $0 \le \phi \le 2\pi$. (b) Calculate $\langle\phi\rangle$, $\langle\sin\phi\rangle$, and $\langle l_z\rangle$. (c) Why is $\langle l_z\rangle \le \hbar$? *Hint*. Draw on a variety of pieces of information, including $\sum_{n=0}^{\infty}x^n/n! = \mathrm{e}^x$, and the following integrals:

$$\int_0^{2\pi}\mathrm{e}^{z\cos\phi}\,\mathrm{d}\phi = 2\pi I_0(z) \qquad \int_0^{2\pi}\cos\phi\,\mathrm{e}^{z\cos\phi}\,\mathrm{d}\phi = 2\pi I_1(z)$$

with $I_0(2) = 2.280\ldots$, $I_1(2) = 1.591\ldots$; the $I(z)$ are modified Bessel functions.

3.8 Investigate the properties of the wavepacket $\Psi = N\sum_{m_l=0}^{\infty}(\alpha^{m_l}/m_l!)\mathrm{e}^{\mathrm{i}m_l\phi}$, and show that when α is large $\langle l_z\rangle \approx \alpha\hbar$. *Hint*. Proceed as in Problem 3.7. The large-value expansions of $I_0(z)$ and $I_1(z)$ are $I_0(z) \simeq I_1(z) \simeq \mathrm{e}^z/\sqrt{2\pi z}$.

3.9 Confirm that the wavefunctions for a particle on a sphere may be written $\psi(\theta, \phi) = (\theta)\Phi(\phi)$ by the method of separation of variables, and find the equation for .

3.10 Confirm eqn 3.18 for the laplacian in three dimensions.

3.11 Confirm that the Schrödinger equation for a particle free to rotate in three dimensions does indeed separate into equations for the variation with θ and ϕ.

3.12 (a) Confirm that $Y_{1,+1}$ and $Y_{2,0}$ as listed in Table 3.1 are solutions of the Schrödinger equation for a particle on a sphere. (b) Confirm by explicit integration that $Y_{1,+1}$ and $Y_{2,0}$ are normalized and mutually orthogonal. *Hint*. The volume element for the integration is $\sin\theta\,\mathrm{d}\theta\,\mathrm{d}\phi$, with $0 \le \phi \le 2\pi$ and $0 \le \theta \le \pi$.

3.13 (a) Confirm that the radius of gyration of a solid uniform sphere of radius R is $r = \sqrt{2/5}R$. (b) What is the radius of gyration of a solid uniform cylinder of radius R and length l?

3.14 Modify Problem 3.1 so that the molecule is free to rotate in three dimensions. Calculate the energies and degeneracies of the lowest three rotational levels, and predict the wavelength of radiation involved in the $l = 1 \rightarrow 0$ transition.

3.15 Calculate the angle that the angular momentum vector makes with the z-axis when the system is described by the wavefunction ψ_{lm_l}. Show that the minimum angle approaches zero as l approaches infinity. Calculate the allowed angles when l is 1, 2, and 3.

3.16 Draw the analogues of Fig. 3.23 for $l = 3$. Observe how the maxima of $|Y|^2$ migrate into the equatorial plane as $|m_l|$ increases.

3.17 Calculate the mean kinetic and potential energies of an electron in the ground state of the hydrogen atom, and confirm that the virial theorem is satisfied. *Hint*. Evaluate $\langle T\rangle = -(\hbar^2/2\mu)\int\psi_{1s}^*\nabla^2\psi_{1s}\,\mathrm{d}\tau$ and $\langle V\rangle = -(e^2/4\pi\varepsilon_0)\int\psi_{1s}^*(1/r)\psi_{1s}\,\mathrm{d}\tau$. The laplacian is given in eqn 3.18, and the virial theorem is dealt with in *Further information 3*.

3.18 Confirm that the radial wavefunctions R_{10}, R_{20}, and R_{31} satisfy the radial wave equation, eqn 3.40. Use Table 3.2.

3.19 Locate the radial nodes of the (a) $2s$-orbital, (b) $3s$-orbital.

3.20 Calculate (a) the mean radius, (b) the mean square radius, and (c) the most probable radius of the $1s$-, $2s$-, and $3s$-orbitals of a hydrogenic atom of atomic number Z. *Hint*. For the most probable radius look for the principal maximum of the radial distribution function.

3.21 Calculate the probability of finding an electron within a sphere of radius a_0 for (a) a $3s$-orbital, (b) a $3p$-orbital.

3.22 Calculate the values of (a) $\langle r\rangle$ and (b) $\langle 1/r\rangle$ for a $3s$- and a $3p$-orbital.

3.23 Confirm that ψ_{1s} and ψ_{2s} are mutually orthogonal.

3.24 A quantity important in some branches of spectroscopy (Section 13.16) is the probability of an electron being found at the same location as the nucleus. Evaluate this probability density for an electron in the $1s$-, $2s$-, and $3s$-orbitals of a hydrogenic atom.

3.25 Another quantity of interest in spectroscopy is the average value of $1/r^3$ (for example, the average magnetic dipole interaction between the electron and nuclear magnetic moments depends on it). Evaluate $\langle 1/r^3\rangle$ for an electron in a $2p$-orbital of a hydrogenic atom.

3.26 Calculate the difference in ionization energies of ^{1}H and ^{2}H on the basis of differences in their reduced masses.

3.27 For a given principal quantum number n, l takes the values $0, 1, \ldots, n-1$ and for each l, m_l takes the values $l, l-1, \ldots, -l$. Confirm that the degeneracy of the term with principal quantum number n is equal to n^2 in a hydrogenic atom.

3.28 Confirm, by drawing pictures like those in Fig. 3.24, that a whimsical Flatlander might be shown that $3s$-, $3p$-, and $3d$-orbitals are degenerate.

Further reading

Atomic structure. E.U. Condon and H. Odabaşi; Cambridge University Press, Cambridge (1980).
Angular momentum: an illustrated guide to rotational symmetries for physical systems. W.J. Thompson; Wiley, New York (1994).
Symmetry and the hydrogen atom. H.V. MacIntosh; *Group theory and its applications*, II, (ed. E.M. Loebl), Academic Press, New York (1971).
Group theory and the hydrogen atom. M. Bander and C. Itzykson; *Rev. mod. Phys.*, **38**, 330 and 346 (1966).
Group theory and the Coulomb problem. M.J. Englefield; Wiley, New York (1972).
Quantum mechanics, Vol. I. A. Messiah; North-Holland, Amsterdam (1961).
Quantum mechanics. E. Merzbacher; Wiley, New York (1970).

4 Angular momentum

In this chapter, we develop the material introduced in Chapter 3 by showing that many of the results obtained there can be inferred from the properties of operators, as introduced in Chapter 1. For instance, although we have seen that solving the Schrödinger equation leads to the conclusion that orbital angular momentum is quantized, the same conclusion can in fact be reached from the form of the angular momentum operators directly, without solving the Schrödinger equation. A further point is that as the development will be based solely on the commutation properties of the angular momentum operators, it follows that the same conclusions apply to observables that are described by operators with the same commutation properties. Therefore, whenever we meet a set of operators with the angular momentum commutation rules, we will immediately know *all* the properties of the corresponding observables. This generality is one of the reasons why angular momentum is of such central importance in quantum mechanics.

Angular momentum has many more mundane applications. It is central to the discussion of the structures of atoms (we have already caught a glimpse of that in the discussion of hydrogenic atoms), to the discussion of the rotation of molecules, as well as to virtually all forms of spectroscopy. We shall draw heavily on this material when we turn to the applications of quantum mechanics in Chapter 7 onwards.

The angular momentum operators

It follows from the general introduction to quantum mechanics in Chapter 1, that the quantum mechanical operators for angular momentum can be constructed by replacing the position, q, and linear momentum, p_q, variables by operators that satisfy the commutation relation

$$[q, p_{q'}] = \mathrm{i}\hbar\delta_{qq'} \tag{1}$$

We shall set up these operators and then show how to determine their commutation relations.

4.1 The operators and their commutation relations

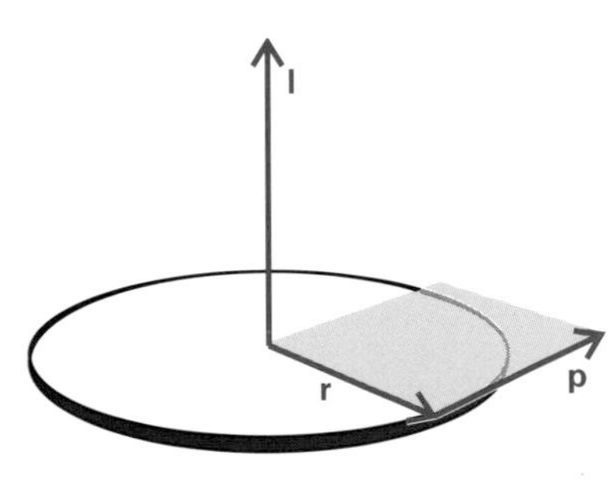

Fig. 4.1 The definition of orbital angular momentum as $\mathbf{l} = \mathbf{r} \times \mathbf{p}$. Note that the angular momentum vector $\mathbf{l}$ stands perpendicular to the plane of the motion of the particle.

In classical mechanics, the angular momentum $\mathbf{l}$ is defined as the vector product $\mathbf{l} = \mathbf{r} \times \mathbf{p}$ (Fig. 4.1). Note that $\mathbf{l}$ displays the sense of rotation according to the right-hand screw rule: it points in the direction a right-hand (conventional) screw travels when it is turned in the same sense as the rotation.

If the position of the particle is expressed in terms of the components of the vector $\mathbf{r} = x\,\hat{\mathbf{i}} + y\,\hat{\mathbf{j}} + z\,\hat{\mathbf{k}}$ where $\hat{\mathbf{i}}$, $\hat{\mathbf{j}}$, and $\hat{\mathbf{k}}$ are orthogonal unit vectors, and the linear momentum is expressed in terms of its components, $\mathbf{p} = p_x\,\hat{\mathbf{i}} + p_y\,\hat{\mathbf{j}} + p_z\,\hat{\mathbf{k}}$, then it follows that the angular momentum can be expressed in terms of its components $\mathbf{l} = l_x\,\hat{\mathbf{i}} + l_y\,\hat{\mathbf{j}} + l_z\,\hat{\mathbf{k}}$ by means of the standard definition of the vector product as a determinant:

$$\mathbf{l} = \mathbf{r} \times p = \begin{vmatrix} \hat{\mathbf{i}} & \hat{\mathbf{j}} & \hat{\mathbf{k}} \\ x & y & z \\ p_x & p_y & p_z \end{vmatrix} \\ = (yp_z - zp_y)\,\hat{\mathbf{i}} + (zp_x - xp_z)\,\hat{\mathbf{j}} + (xp_y - yp_x)\,\hat{\mathbf{k}} \tag{2}$$

We can therefore identify the three components of the angular momentum as

$$l_x = yp_z - zp_y \qquad l_y = zp_x - xp_z \qquad l_z = xp_y - yp_x \tag{3}$$

Note how each one can be generated from its predecessor by cyclic permutation of x, y, and z.

The magnitude of the angular momentum, l, is related to its components by the normal expression for constructing the magnitude of a vector:

$$l^2 = l_x^2 + l_y^2 + l_z^2 \tag{4}$$

Classical mechanics puts no constraints on the magnitude of angular momentum, which is consistent with the energy $E = l^2/2I$ being unconstrained too. Nor does it put any constraints on the components of angular momentum about the three axes, other than the requirement that the components do not exceed the magnitude.

The definitions of the components and the magnitude carry over into quantum mechanics, with the q and p_q in the definitions of the l_q interpreted as operators. The operators l_q in the position representation are obtained, as explained in Section 1.4, by replacing q by $q\times$ and p_q by $(\hbar/\mathrm{i})\partial/\partial q$:

$$\begin{aligned} l_x &= \frac{\hbar}{\mathrm{i}}\left(y\frac{\partial}{\partial z} - z\frac{\partial}{\partial y}\right) \\ l_y &= \frac{\hbar}{\mathrm{i}}\left(z\frac{\partial}{\partial x} - x\frac{\partial}{\partial z}\right) \\ l_z &= \frac{\hbar}{\mathrm{i}}\left(x\frac{\partial}{\partial y} - y\frac{\partial}{\partial x}\right) \end{aligned} \tag{5}$$

However, instead of developing the properties of angular momentum in a specific representation, it is more general, more powerful, and more time-saving to develop them without selecting a representation. Later in the chapter we shall make use of the fact that because the operators l_q and l^2 correspond to observables, they must be hermitian (Section 1.13). The property of hermiticity could be demonstrated explicitly in the position representation, as was illustrated in Example 1.3; but it must be true in any representation if the operators are to stand for observables.

To make progress, we need to establish the commutation relations of the l_q operators. Consider first the commutator of l_x and l_y:

$$\begin{aligned} [l_x, l_y] &= [(yp_z - zp_y), (zp_x - xp_z)] \\ &= [yp_z, zp_x] - [yp_z, xp_z] - [zp_y, zp_x] + [zp_y, xp_z] \\ &= y[p_z, z]p_x - 0 - 0 + xp_y[z, p_z] = \mathrm{i}\hbar(-yp_x + xp_y) \\ &= \mathrm{i}\hbar l_z \end{aligned}$$

In line 1 we have inserted the definitions. In line 2 we have expanded the commutators term by term. In line 3 we have used the fact that y and p_x commute with each other and also with z and p_z. The same is true of x and p_y. The remaining commutators can be derived in the same way, but it is more efficient to note that because the three operators l_q are obtained from one another by cyclic permutation, the commutators can be obtained in the same way. We therefore conclude that

$$[l_x, l_y] = \mathrm{i}\hbar l_z \qquad [l_y, l_z] = \mathrm{i}\hbar l_x \qquad [l_z, l_x] = \mathrm{i}\hbar l_y \tag{6}$$

The remaining operator is l^2, the operator corresponding to the square of the magnitude of the angular momentum. We need its commutator with the operators l_q, and proceed as follows. First, we write

$$[l^2, l_z] = [l_x^2 + l_y^2 + l_z^2, l_z] = [l_x^2, l_z] + [l_y^2, l_z]$$

We have used the fact that the commutator of l_z^2 and l_z is zero:

$$[l_z^2, l_z] = l_z^2 l_z - l_z l_z^2 = l_z^3 - l_z^3 = 0$$

Next, consider the following commutator, which we develop by drawing on the three fundamental relations derived above:

$$\begin{aligned}[l_x^2, l_z] &= l_x l_x l_z - l_z l_x l_x \\ &= l_x l_x l_z - l_x l_z l_x + l_x l_z l_x - l_z l_x l_x \\ &= l_x[l_x, l_z] + [l_x, l_z]l_x = -\mathrm{i}\hbar(l_x l_y + l_y l_x)\end{aligned}$$

Similarly,

$$[l_y^2, l_z] = \mathrm{i}\hbar(l_x l_y + l_y l_x)$$

The sum of the two terms is zero, so we can conclude that the commutator of l^2 with l_z is zero. However, because l_x, l_y, and l_z occur symmetrically in l^2, all three operators must commute with l^2 if any one of them does. That is,

$$[l^2, l_q] = 0 \tag{7}$$

The commutation relations in eqns 4.6 and 4.7 are the foundations for the entire theory of angular momentum. Whenever we encounter four operators having these commutation relations, we know that the properties of the observables they represent are identical to the properties we are about to derive. Therefore, we shall say that an observable *is* an angular momentum if its operators satisfy these commutation relations.[1]

4.2 Angular momentum observables

The immediate conclusion from the commutation relations is that only one component of angular momentum can be specified precisely. We saw in

[1] Because all the properties of the observables are the same, this seems to be an appropriate course of action. However, the procedure does capture some strange bed-fellows. The electric charge of fundamental particles is described by operators that satisfy the same set of commutation relations, but should we regard it—or imagine it—as an angular momentum? Electron spin is also described by the same set of commutation relations, but should we regard it—or imagine it—as an angular momentum?

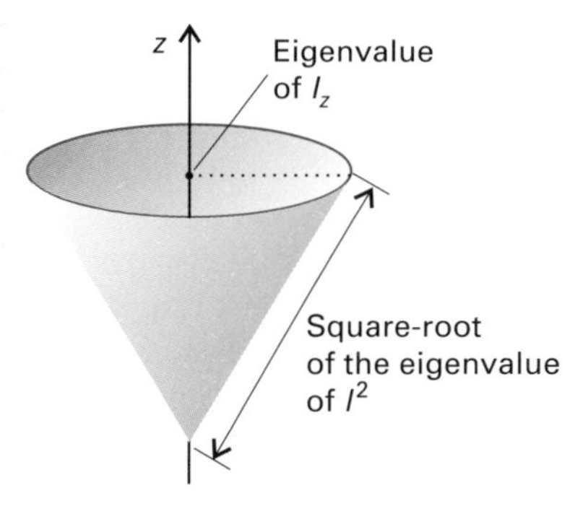

Fig. 4.2 The cone used to represent a state of angular momentum with specified magnitude and z-component.

Section 1.16 that observables are complementary and restricted by the uncertainty relation if their operators do not commute, and l_z does not commute with either l_x or l_y. However, l^2 does commute with all three components, so the magnitude of the angular momentum may be specified simultaneously with any of its components. These conclusions are the quantum mechanical basis of the vector model of angular momentum introduced in Section 3.6, where we represent an angular momentum state by a vector of indeterminate orientation on a cone of given side (the magnitude of the momentum) and height (the eigenvalue of l_z, Fig. 4.2).

At this point, though, we can begin to see that the vector model must be regarded with caution. The commutation relations in eqn 4.6 can be written in a compact fashion as follows:

$$\mathbf{l} \times \mathbf{l} = i\hbar\mathbf{l} \tag{8}$$

To confirm that this is so, write the left-hand side as a determinant and expand it; then compare it term by term with the expression on the right-hand side: this procedure reproduces the three commutation relations (see Problem 4.3). However, it is an elementary feature of vector algebra that the vector product of a vector with itself is zero (the magnitude of $\mathbf{a} \times \mathbf{b}$ is $ab \sin\theta$, where θ is the angle between the vectors **a** and **b**; but when the two vectors are identical that angle is zero). Therefore, because the vector product of **l** with itself is not zero, we have to conclude that **l** is not a vector. The vector model is useful only if we realize that it is not the whole truth, and note that **l** is a vector *operator*, not a classical vector.

4.3 The shift operators

It will prove expedient to introduce two new operators, called the **shift operators**. One operator, l_+, is called the **raising operator**; the other, l_-, is called the **lowering operator**. They are defined as follows:

$$l_+ = l_x + \mathrm{i}l_y \qquad l_- = l_x - \mathrm{i}l_y \tag{9}$$

The inverse relations

$$l_x = \frac{l_+ + l_-}{2} \qquad l_y = \frac{l_+ - l_-}{2\mathrm{i}} \tag{10}$$

are also sometimes useful.

We shall require the commutators of the shift operators. They are easily derived from the fundamental commutation relations. For example,

$$[l_z, l_+] = [l_z, l_x] + \mathrm{i}[l_z, l_y] = \mathrm{i}\hbar l_y + \hbar l_x = \hbar l_+$$

The other commutation relations are obtained similarly, and all three are

$$[l_z, l_+] = \hbar l_+ \qquad [l_z, l_-] = -\hbar l_- \qquad [l_+, l_-] = 2\hbar l_z \tag{11}$$

Furthermore, because l^2 commutes with each of its components, it also commutes with $l_\pm$. Therefore, we can add to these relations the rule

$$[l^2, l_\pm] = 0 \tag{12}$$

The definition of the states

The next task is to see how the commutation relations govern the values of the permitted eigenvalues of l^2 and *one* of the components l_q. It is conventional to call the selected component l_z, but that is entirely arbitrary (as is the choice of the direction denoted z). In the course of this development we shall discover that the solutions found in Chapter 3 are incomplete in a very important respect. We shall also set up an elegant way of constructing the spherical harmonics, and find a simple way of evaluating the matrix elements of angular momentum operators.

4.4 The effect of the shift operators

We shall suppose that the simultaneous eigenstates of l^2 and l_z are distinguished by two quantum numbers, which for the time being we shall denote l' and m_l. The eigenstates are therefore denoted $|l', m_l\rangle$. We shall define m_l through the relation

$$l_z|l', m_l\rangle = m_l\hbar|l', m_l\rangle \tag{13}$$

This relation must be true, because $\hbar$ has the same dimensions as an angular momentum ($\mathrm{M\,L^2\,T^{-1}}$), and so the eigenvalue of l_z must be a numerical multiple of $\hbar$; we are not presupposing that m_l is restricted to discrete values, but that will emerge in due course. All we know is that m_l is a real number: that follows from the hermiticity of l_z. Because l^2 commutes with l_z, the state $|l', m_l\rangle$ is also an eigenstate of l^2. At this stage we shall allow for the possibility that the eigenvalues of l^2 depend on both quantum numbers, and write

$$l^2|l', m_l\rangle = f(l', m_l)\hbar^2|l', m_l\rangle$$

where f is a function that we need to determine: from the work we did in Chapter 3 we know that it will turn out to be equal to $l(l+1)$ where l is the maximum value of $|m_l|$, but that is something we shall derive. All we know at this stage is that because l^2 is hermitian, f is real. Moreover, because l^2 is the sum of squares of hermitian operators, we also know (recall Example 1.6) that its eigenvalues are non-negative.

Because $l^2 - l_z^2 = l_x^2 + l_y^2$, it follows that the eigenvalues of the operator $l^2 - l_z^2$ are non-negative:

$$(l^2 - l_z^2)|l', m_l\rangle = (l_x^2 + l_y^2)|l', m_l\rangle \geq 0$$

However, we also know from the definitions of the effects of l^2 and l_z^2 that

$$(l^2 - l_z^2)|l', m_l\rangle = \{f(l', m_l) - m_l^2\}\hbar^2|l', m_l\rangle$$

For these two relations to be consistent, it follows that

$$f(l', m_l) \geq m_l^2 \tag{14}$$

To take the next step we use the commutation relations to establish the effect of the shift operators (and see why they are so-called). Consider the effect of the operator l_+ on $|l', m_l\rangle$. Because $|l', m_l\rangle$ is an eigenstate of neither l_x nor l_y, when l_+ acts on it, it generates a new state. First, we show that $l_+|l', m_l\rangle$ is an eigenstate of l^2 with the same value of f; that is $|l', m_l\rangle$ and $l_+|l', m_l\rangle$ share the same eigenvalue of l^2. To do so, consider the effect of l^2 on the state obtained by acting with l_+:

$$l^2 l_+|l', m_l\rangle = l_+ l^2|l', m_l\rangle = l_+ f(l', m_l)\hbar^2|l', m_l\rangle = f(l', m_l)\hbar^2 l_+|l', m_l\rangle$$

where the first equality follows from the commutability of l^2 and l_+. It follows, because the eigenvalue of l^2 for the state $l_+|l', m_l\rangle$ is the same as that for the original state $|l', m_l\rangle$, that l_+ leaves the magnitude of the angular momentum unchanged when it acts.

Now consider the same argument applied to $|l', m_l\rangle$ treated as an eigenstate of l_z. The conclusion will be different, because l_+ and l_z do not commute. Instead, we must use the following string of equalities to find the effect of l_z on $l_+|l', m_l\rangle$:

$$\begin{aligned} l_z l_+|l', m_l\rangle &= (l_+ l_z + [l_z, l_+])|l', m_l\rangle \\ &= (l_+ l_z + \hbar l_+)|l', m_l\rangle = (l_+ m_l \hbar + \hbar l_+)|l', m_l\rangle \\ &= (m_l + 1)\hbar l_+|l', m_l\rangle \end{aligned}$$

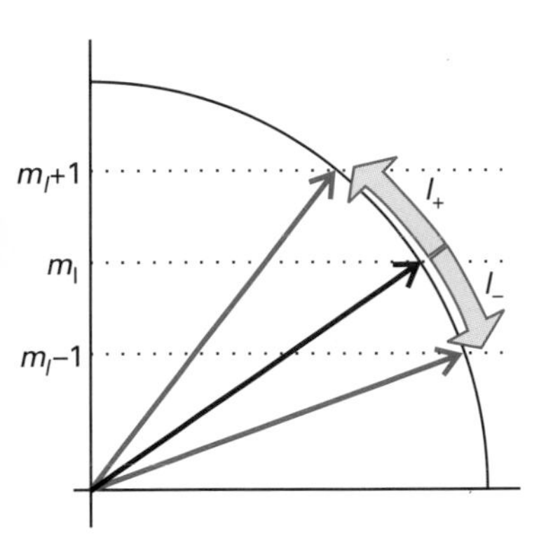

Fig. 4.3 The effect of the shift operators l_+ and l_-.

However, we know from eqn 4.13 that

$$l_z|l', m_l + 1\rangle = (m_l + 1)\hbar|l', m_l + 1\rangle$$

Therefore, the state $l_+|l', m_l\rangle$ must be proportional to the state $|l', m_l + 1\rangle$ and we can write

$$l_+|l', m_l\rangle = c_+(l', m_l)\hbar|l', m_l + 1\rangle \tag{15}$$

where $c_+(l', m_l)$ is a dimensionless numerical coefficient. We now see why l_+ is called a raising operator: when it operates on a state with z-component $m_l\hbar$, it generates from it a state with the same magnitude of angular momentum but with a z-component one unit greater, $(m_l + 1)\hbar$ (Fig. 4.3). In exactly the same way, the effect of the operator l_- can be shown to lower the z-component from $m_l h$ to $(m_l - 1)\hbar$:

$$l_-|l', m_l\rangle = c_-(l', m_l)\hbar|l', m_l - 1\rangle \tag{16}$$

4.5 The eigenvalues of the angular momentum

The shift operators step m_l by ± 1 each time they operate. However, we have already established from the hermiticity of the operators that m_l^2 cannot exceed $f(l', m_l)$; it follows that m_l must have a maximum value, which we shall denote l. When we operate with l_+ on a state in which $m_l = l$, we generate nothing, because there is no state with a larger value of m_l:

$$l_+|l', l\rangle = 0$$

This relation will give us the value of the unknown function f. When acted on by l_-, it gives

$$l_- l_+|l', l\rangle = 0$$

However, the product $l_- l_+$ can be expanded as follows:

$$\begin{aligned} l_- l_+ &= (l_x - \mathrm{i}l_y)(l_x + \mathrm{i}l_y) = l_x^2 + l_y^2 + \mathrm{i}l_x l_y - \mathrm{i}l_y l_x \\ &= l_x^2 + l_y^2 + \mathrm{i}[l_x, l_y] = l^2 - l_z^2 + \mathrm{i}(\mathrm{i}\hbar l_z) \end{aligned} \tag{17}$$

Therefore, the last equation can be written

$$(l^2 - l_z^2 - \hbar l_z)|l', l\rangle = 0$$

When we rearrange this expression and use the definition of the effect of l_z on state, we obtain

$$l^2|l', l\rangle = (l_z^2 + \hbar l_z)|l', l\rangle = (l^2 + l)\hbar^2|l', l\rangle$$

It follows that

$$f(l', l) = l(l+1) \quad (18$$

We have already established that when l_- acts on a state, it leaves the eigenvalue of l^2 unchanged. Therefore, all the states $|l', l\rangle$, $|l', l-1\rangle$, etc. have the same eigenvalue of l^2. Therefore,

$$f(l', m_l) = l(l+1) \qquad \text{for } m_l = l, l-1, \ldots$$

We know that there is a lower bound on m_l because the eigenvalue of l cannot exceed the eigenvalue of l^2, and for the moment we denote this lower bound by k. It is quite easy to show that $k = -l$. To see that this is the case, we start from $l_-|l', k\rangle = 0$, and by a similar argument but using $l_+l_-|l', k\rangle = 0$, conclude that $f(l', k) = k(k-1)$. However, because $f(l', m_l)$ is independent of m_l we must have $l(l+1) = k(k-1)$. Of the two solutions $k = -l$ and $k = l+1$ only the former is acceptable (the lower bound must be below the upper bound!). Therefore,

$$f(l', m_l) = l(l+1) \qquad \text{for } m_l = l, l-1, \ldots, -l$$

At this point we can put the spare quantum number l' to work, and identify it as l, the maximum value of $|m_l|$. Then,

$$f(l, m_l) = l(l+1) \qquad \text{for } m_l = l, l-1, \ldots, -l \quad (19$$

That is, we now know that

$$l^2|l, m_l\rangle = l(l+1)\hbar^2|l, m_l\rangle \quad (20$$

and we now see that the value of l (the maximum value of m_l) determines the magnitude of the angular momentum. We already know that

$$l_z|l, m_l\rangle = m_l\hbar|l, m_l\rangle \quad (21$$

and so we have an effectively complete description of angular momentum.

Finally, we need to decide on the allowed values of l and m_l. As we have seen, the shift operators step the states $|l, m_l\rangle$ from $|l, +l\rangle$ to $|l, -l\rangle$ in unit steps. The symmetry of this ladder of states allows for only two types of value for l: it may be integral or half-integral. For example, we can have $l = 2$, to give the ladder $m_l = +2, +1, 0, -1, -2$, or we could have $l = \frac{3}{2}$, to give $m_l = +\frac{3}{2}, +\frac{1}{2}, -\frac{1}{2}, -\frac{3}{2}$. We cannot obtain a symmetrical ladder with any other type of value ($l = \frac{3}{4}$, for instance, would give the unsymmetrical ladder $m_l = +\frac{3}{4}, -\frac{1}{4}$).

We can summarize the conclusions so far. On the basis of the hermiticity of the angular momentum operators and their commutation relations, we have shown that the magnitude of the angular momentum is confined to the values $\{l(l+1)\}^{1/2}\hbar$, with $l = 0, \frac{1}{2}, 1, \ldots$, and its component on an arbitrary z-axis is limited to the $2l+1$ values $m_l\hbar$ with $m_l = l, l-1, \ldots, -l$. These conclusions differ in one detail from those obtained by solving the Schrödinger equation in Chapter 3. There we saw that l was confined to the *integral* values $l = 0, 1, 2, \ldots$. In that analysis, we obtained the permitted values of l by imposing cyclic boundary conditions. What the present analysis does is to show that angular momentum *may* be described by half-integral quantum numbers, but such quantum numbers do not necessarily apply to a particular physical situation. For orbital angular momentum, where the Born interpretation requires cyclic boundary conditions to be satisfied, only integral values are admissible. Where cyclic boundary conditions are not relevant, as for the intrinsic angular momentum known as spin, the half-integral values may be appropriate.

To emphasize that there is a distinction between angular momenta according to the boundary conditions that have to be satisfied, we shall use the following notation. For orbital angular momenta, when the boundary conditions on the wavefunctions allow only integral quantum numbers, we shall use the notation l and m_l and write states as $|l, m_l\rangle$. When internal angular momentum (spin) is being considered, we shall use the notation s and m_s for the (possibly half-integral) quantum numbers and write the states $|s, m_s\rangle$. When the discussion is general and applicable to either kind of angular momentum, we shall use the quantum numbers j and m_j, and write the states as $|j, m_j\rangle$. The expressions we have deduced so far may therefore be written in this general notation as

$$\begin{aligned} j^2|j, m_j\rangle &= j(j+1)\hbar^2|j, m_j\rangle \\ j_z|j, m_j\rangle &= m_j\hbar|j, m_j\rangle \end{aligned} \tag{22}$$

with $m_j = j, j-1, \ldots, -j$.

4.6 The matrix elements of the angular momentum

One outstanding problem at this point is the value of the coefficient $c_\pm$ introduced in connection with the effect of the shift operators:

$$j_\pm|j, m_j\rangle = c_\pm(j, m_j)\hbar|j, m_j \pm 1\rangle \tag{23}$$

Because the states $|j, m_j\rangle$ form an orthonormal set, multiplication from the left by the bra $\langle j, m_j \pm 1|$ gives

$$\langle j, m_j \pm 1|j_\pm|j, m_j\rangle = c_\pm(j, m_j)\hbar \tag{24}$$

So, we need to know the coefficients if we want to know the values of these matrix elements. Matrix elements of this kind occur in connection with the calculation of magnetic properties and the intensities of transitions in magnetic resonance (Chapter 13).

The first step involves finding two expressions for the matrix elements of the operator j_-j_+. First, we can use eqn 4.17 to write

$$\begin{aligned} j_-j_+|j, m_j\rangle &= (j^2 - j_z^2 - \hbar j_z)|j, m_j\rangle \\ &= \{j(j+1) - m_j(m_j+1)\}\hbar^2|j, m_j\rangle \end{aligned}$$

Alternatively, we can use eqn 4.23 to write

$$\begin{aligned} j_- j_+ |j, m_j\rangle &= j_- c_+(j, m_j)\hbar |j, m_j + 1\rangle \\ &= c_+(j, m_j) c_-(j, m_j + 1)\hbar^2 |j, m_j\rangle \end{aligned}$$

Comparison of the two expressions shows that

$$c_+(j, m_j) c_-(j, m_j + 1) = j(j+1) - m_j(m_j + 1) \tag{25}$$

The next step is to find a relation between the two coefficients that occur in the last expression. We shall base the calculation on the matrix element

$$\langle j, m_j | j_- | j, m_j + 1\rangle = c_-(j, m_j + 1)\hbar$$

and the hermiticity of j_x and j_y. Consider the following string of changes:

$$\begin{aligned} \langle j, m_j | j_- | j, m_j + 1\rangle &= \langle j, m_j | j_x - \mathrm{i} j_y | j, m_j + 1\rangle \\ &= \langle j, m_j | j_x | j, m_j + 1\rangle - \mathrm{i}\langle j, m_j | j_y | j, m_j + 1\rangle \\ &\overset{\mathcal{H}}{=} \langle j, m_j + 1 | j_x | j, m_j\rangle^* - \mathrm{i}\langle j, m_j + 1 | j_y | j, m_j\rangle^* \\ &= \{\langle j, m_j + 1 | j_x | j, m_j\rangle + \mathrm{i}\langle j, m_j + 1 | j_y | j, m_j\rangle\}^* \\ &= \langle j, m_j + 1 | j_+ | j, m_j\rangle^* \end{aligned} \tag{26}$$

The symbol $\mathcal{H}$ means that we have invoked the property of hermiticity. The relation

$$\langle j, m_j | j_- | j, m_j + 1\rangle = \langle j, m_j + 1 | j_+ | j, m_j\rangle^* \tag{27}$$

shows that j_- and j_+ are each other's **hermitian conjugate**. Neither operator is hermitian, and so neither operator corresponds to a physical observable.

The relation we have just derived implies a relation between the coefficients $c_\pm$. Because the matrix element on the left is equal to $c_-(j, m_j + 1)\hbar$ and that on the right is equal to $c_+^*(j, m_j)\hbar$, it follows that

$$c_-(j, m_j + 1) = c_+^*(j, m_j) \tag{28}$$

It then follows from eqn 4.25 that

$$|c_+(j, m_j)|^2 = j(j+1) - m_j(m_j + 1)$$

If we make a convenient choice of phase (choosing c_+ to be real and positive), it follows that

$$c_+(j, m_j) = \{j(j+1) - m_j(m_j + 1)\}^{1/2} \tag{29}$$

Moreover, because $c_-(j, m_j) = c_+^*(j, m_j - 1)$, we can also write

$$c_-(j, m_j) = \{j(j+1) - m_j(m_j - 1)\}^{1/2} \tag{30}$$

With these matrix elements established, we can calculate a wide range of other quantities, as illustrated in the following example.

Example 4.1 How to evaluate matrix elements of the angular momentum

Evaluate the matrix elements (a) $\langle j, m_j + 1|j_x|j, m_j\rangle$, (b) $\langle j, m_j + 2|j_x|j, m_j\rangle$, and (c) $\langle j, m_j + 2|j_x^2|j, m_j\rangle$.

Method. Because we know the matrix elements of the shift operators, one approach is to express all the operators in the questions in terms of them and then to use eqns 4.24, 4.29, and 4.30. Note that $j_x^2 = j_x j_x$ and $\langle j', m_j'|j, m_j\rangle = \delta_{j'j}\delta_{m_j'm_j}$.

Answer.

$$\begin{aligned}\text{(a) } \langle j, m_j + 1|j_x|j, m_j\rangle &= \tfrac{1}{2}\langle j, m_j + 1|j_+ + j_-|j, m_j\rangle \\ &= \tfrac{1}{2}\langle j, m_j + 1|j_+|j, m_j\rangle + \tfrac{1}{2}\langle j, m_j + 1|j_-|j, m_j\rangle \\ &= \tfrac{1}{2}c_+(j, m_j)\hbar\end{aligned}$$

because $\langle j, m_j + 1|j_-|j, m_j\rangle \propto \langle j, m_j + 1|j, m_j - 1\rangle = 0$.

(b) $\langle j, m_j + 2|j_x|j, m_j\rangle = 0$

because $j_\pm$ step m_j only by one unit, and the resulting states are orthogonal to the state $|j, m_j + 2\rangle$.

$$\begin{aligned}\text{(c) } \langle j, m_j + 2|j_x^2|j, m_j\rangle &= \tfrac{1}{4}\langle j, m_j + 2|j_+^2 + j_-^2 + j_+j_- + j_-j_+|j, m_j\rangle \\ &= \tfrac{1}{4}\langle j, m_j + 2|j_+^2|j, m_j\rangle \\ &= \tfrac{1}{4}c_+(j, m_j + 1)c_+(j, m_j)\hbar^2 \\ &= \tfrac{1}{4}\{j(j+1) - (m_j + 1)(m_j + 2)\}^{1/2}\{j(j+1) \\ &\qquad - m_j(m_j + 1)\}^{1/2}\hbar^2\end{aligned}$$

Comment. Note that it is quite easy to spot short-cuts, as in (c), where it should be obvious that only j_+^2 can contribute to the matrix element.

Exercise 4.1. Evaluate the matrix element $\langle j, m_j + 1|j_x^3|j, m_j\rangle$.

4.7 The angular momentum eigenfunctions

Now we shall consider orbital angular momentum explicitly. This version of the general theory refers to the angular momentum arising from the distribution of a particle in space, and so it is subject to cyclic boundary conditions on the wavefunctions. These conditions limit the angular momentum quantum numbers to integral values, and we denote them l and m_l. In Chapter 3 we saw that the wavefunctions are solutions of a second-order differential equation, and we asserted (and proved in *Further information 9*) that they were the spherical harmonics. With the work done in this chapter, we can show that they can also be obtained by solving a first-order differential equation, which is a much simpler task.

We begin by finding the wavefunction for the state $|l, l\rangle$ (for which $m_l = l$). Once this wavefunction has been determined, the wavefunctions for the states $|l, m_l\rangle$ can be generated by acting on $|l, l\rangle$ with l_- the appropriate number of times. The equation we have to solve is

$$l_+|l,l\rangle = 0$$

To express this equation as a differential equation, we must adopt a representation for the operators. In the position representation, the orbital angular momentum operators are

$$\begin{aligned} l_x &= -\frac{\hbar}{\mathrm{i}}\left(\sin\phi\frac{\partial}{\partial\theta} + \cot\theta\cos\phi\frac{\partial}{\partial\phi}\right) \\ l_y &= \frac{\hbar}{\mathrm{i}}\left(\cos\phi\frac{\partial}{\partial\theta} - \cot\theta\sin\phi\frac{\partial}{\partial\phi}\right) \\ l_z &= \frac{\hbar}{\mathrm{i}}\frac{\partial}{\partial\phi} \end{aligned} \tag{31}$$

These operators are obtained from the cartesian forms given in eqn 4.5 by expressing them in terms of spherical polar coordinates. It follows that the shift operators in the position representation are

$$\begin{aligned} l_+ &= \hbar \mathrm{e}^{\mathrm{i}\phi}\left(\frac{\partial}{\partial\theta} + \mathrm{i}\cot\theta\frac{\partial}{\partial\phi}\right) \\ l_- &= -\hbar \mathrm{e}^{-\mathrm{i}\phi}\left(\frac{\partial}{\partial\theta} - \mathrm{i}\cot\theta\frac{\partial}{\partial\phi}\right) \end{aligned} \tag{32}$$

To obtain these expressions we have used the relation $\mathrm{e}^{\pm\mathrm{i}\phi} = \cos\phi \pm \mathrm{i}\sin\phi$.

It follows from the equation $l_+|l,l\rangle = 0$ that

$$\hbar \mathrm{e}^{\mathrm{i}\phi}\left(\frac{\partial}{\partial\theta} + \mathrm{i}\cot\theta\frac{\partial}{\partial\phi}\right)\psi_{l,l}(\theta,\phi) = 0 \tag{33}$$

This partial differential equation can be separated by writing $\psi = \Theta\Phi$, for in the normal way we then obtain

$$\frac{\tan\theta}{\Theta}\frac{\mathrm{d}\Theta}{\mathrm{d}\theta} = -\frac{\mathrm{i}}{\Phi}\frac{\mathrm{d}\Phi}{\mathrm{d}\phi}$$

According to the usual separation of variables argument, both sides are equal to a constant, which we shall denote c. The equation therefore separates into the following two first-order ordinary differential equations:

$$\tan\theta\frac{\mathrm{d}\Theta}{\mathrm{d}\theta} = c\Theta \qquad \frac{\mathrm{d}\Phi}{\mathrm{d}\phi} = \mathrm{i}c\Phi$$

The two equations integrate immediately to

$$\Theta \propto \sin^c\theta \qquad \Phi \propto \mathrm{e}^{\mathrm{i}c\phi}$$

The value of c is found to be l by requiring that $l_z\psi_{l,l} = l\hbar\psi_{l,l}$. Therefore, the complete solution is

$$\psi_{l,l} = N\sin^l\theta\,\mathrm{e}^{\mathrm{i}l\phi} \tag{34}$$

where N is a normalization constant. This is the explicit form of the spherical harmonic Y_{ll} given in Table 3.1, apart from the normalization constant, which can be obtained by integration over the surface of a sphere. With this function found, it is a straightforward matter to apply the operator l_- to obtain the rest of the functions with a given value of l.

Example 4.2 How to construct wavefunctions for states with $m_l < l$

Construct the wavefunction for the state $|l, l-1\rangle$.

Method. We know that $l_-|l,l\rangle = c_-\hbar|l,l-1\rangle$. We also know the position representation form of l_- (eqn 4.32). We need to combine the two expressions.

Answer. In the position representation we have

$$\begin{aligned} l_-\psi_{l,l} &= -\hbar e^{-i\phi}\left(\frac{\partial}{\partial\theta} - i\cot\theta\frac{\partial}{\partial\phi}\right)N\sin^l\theta\, e^{il\phi} \\ &= -N\hbar e^{-i\phi}\{l\sin^{l-1}\theta\cos\theta - i(il)\cot\theta\sin^l\theta\}e^{il\phi} \\ &= -2Nl\hbar\sin^{l-1}\theta\cos\theta\, e^{i(l-1)\phi} \end{aligned}$$

However, we also know that

$$l_-|l,l\rangle = \{l(l+1) - l(l-1)\}^{1/2}\hbar|l,l-1\rangle = (2l)^{1/2}\hbar|l,l-1\rangle$$

Therefore,

$$\psi_{l,l-1} = -(2l)^{1/2}N\sin^{l-1}\theta\cos\theta\, e^{i(l-1)\phi}$$

Comment. If $\psi_{l,l}$ is normalized to unity, then so is $\psi_{l,l-1}$ and all the other states that can be generated in this way. The normalization constant is

$$N = \frac{1}{(2^l l)!}\left\{\frac{(2l+1)!}{4\pi}\right\}^{1/2}$$

Exercise 4.2. Derive an expression for the wavefunction with $m_l = l - 2$ in the same way.

4.8 Spin

The Dutch physicists George Uhlenbeck and Samuel Goudsmit realized in 1925 that a great simplification of the description of atomic spectra could be obtained if it was assumed that an electron possessed an intrinsic angular momentum with quantum number $s = \frac{1}{2}$, and which could exist in two states with $m_s = +\frac{1}{2}$, denoted α or $\uparrow$, and $m_s = -\frac{1}{2}$, denoted β or $\downarrow$. This intrinsic angular momentum is called the **spin** of the electron, but footnote 1 of this chapter should be recalled.

Spin is a purely quantum mechanical phenomenon in the sense that in a universe in which $h \to 0$ the spin angular momentum would be zero. Orbital angular momentum survives in a classical world, because l can be allowed to approach infinity as $h \to 0$ and the quantity $\{l(l+1)\}^{1/2}\hbar \simeq l\hbar$ can be finite. Uhlenbeck and Goudsmit's proposal was no more than a hypothesis, but when Dirac showed how to combine quantum mechanics and special relativity, the existence of particles with half-integral angular momentum quantum numbers appeared automatically.

The angular momentum operators describe spin, but for $s = \frac{1}{2}$ they do so in a very simple way. If we denote the state $|\frac{1}{2},+\frac{1}{2}\rangle$ by α and the state $|\frac{1}{2},-\frac{1}{2}\rangle$ by β, then the general expressions given earlier become

$$s_z\alpha = +\tfrac{1}{2}\hbar\alpha \qquad s_z\beta = -\tfrac{1}{2}\hbar\beta \qquad s^2\alpha = \tfrac{3}{4}\hbar^2\alpha \qquad s^2\beta = \tfrac{3}{4}\hbar^2\beta \tag{35}$$

and the effects of the shift operators are

$$s_+\alpha = 0 \qquad s_+\beta = \hbar\alpha \qquad s_-\alpha = \hbar\beta \qquad s_-\beta = 0 \tag{36}$$

It follows that the only nonzero matrix elements of the shift operators are

$$\langle\alpha|s_+|\beta\rangle = \hbar \qquad \langle\beta|s_-|\alpha\rangle = \hbar \tag{37}$$

The angular momenta of composite systems

We now consider a system in which there are two sources of angular momentum, which we denote $\mathbf{j}_1$ and $\mathbf{j}_2$. The system might be a single particle that possesses both spin and orbital angular momentum, or it might consist of two particles with spin or orbital momentum. The question we investigate here is what the commutation rules imply for the total angular momentum $\mathbf{j}$ of the system.

4.9 The specification of coupled states

The state of particle 1 is fully specified by reporting the quantum numbers j_1 and m_{j1}, and the same is true of particle 2 in terms of its quantum numbers j_2 and m_{j2}. If we are to be able to specify the overall state as $|j_1m_{j1};j_2m_{j2}\rangle$, we need to know whether all the corresponding operators commute with one another. In fact, operators for independent sources of angular momentum do commute with one another, and we can write

$$[j_{1q}, j_{2q'}] = 0 \tag{38}$$

for all components q and q'. One way to see that this is so is to note that in the position representation the operators are expressed in terms of the coordinates and derivatives of each particle separately, and the derivatives for one particle treat the coordinates of the other particle as constants. *Operators that refer to independent components of a system always commute with one another.* Because the operators j_1^2 and j_2^2 are defined in terms of their components, which commute, so too do these two operators. Hence, all four operators j_1^2, j_{1z}, j_2^2, and j_{2z} commute with one another, and it is permissible to express the state as $|j_1m_{j1};j_2m_{j2}\rangle$.

We now explore whether the **total angular momentum**, $\mathbf{j} = \mathbf{j}_1 + \mathbf{j}_2$, can also be specified. First, we investigate whether $\mathbf{j}$ is indeed an angular momentum. To do so, we evaluate the commutators of its components, such as

$$\begin{aligned}[j_x, j_y] &= [j_{1x}+j_{2x}, j_{1y}+j_{2y}] \\ &= [j_{1x}, j_{1y}] + [j_{2x}, j_{2y}] + [j_{1x}, j_{2y}] + [j_{2x}, j_{1y}] \\ &= \mathrm{i}\hbar j_{1z} + \mathrm{i}\hbar j_{2z} + 0 + 0 = \mathrm{i}\hbar j_z\end{aligned} \tag{39}$$

This commutation relation, and the other two that can be derived from it by cyclic permutation of the coordinate labels, is characteristic of angular momentum, so $\mathbf{j}$ is an angular momentum ($\mathbf{j}_1 - \mathbf{j}_2$, on the other hand, is not). Because $\mathbf{j}$ is an angular momentum, we can conclude without further work that its magnitude is $\{j(j+1)\}^{1/2}\hbar$ with j integral or half-integral, and its z component has the values $m_j\hbar$ with $m_j = j, j-1, \ldots, -j$.

We now need to work towards discovering which values of j can exist in the system. The initial question is whether we can actually specify j if j_1 and j_2 have been specified. Because j_1^2 commutes with all its components, and j_2^2 does likewise, and because j^2 can be expressed in terms of those same components, it follows that

$$[j^2, j_1^2] = [j^2, j_2^2] = 0 \tag{40}$$

Therefore, we can conclude that the eigenvalues of j_1^2, j_2^2, and j^2 can be specified simultaneously. For instance, a p-electron (for which $l = 1$ and $s = \frac{1}{2}$) can be regarded as having a well-defined total angular momentum with a magnitude given by some value of j (the actual permitted values of which we have yet to find).

Because j^2 commutes with its own components, in particular it commutes with $j_z = j_{1z} + j_{2z}$. Therefore, we know that we can specify the value of m_j as well as j. At this point, we have established that a state of coupled angular momentum can be denoted $|j_1j_2;jm_j\rangle$. Note, however, that we have not yet established that it can be specified more fully as $|j_1m_{j1}j_2m_{j2};jm_j\rangle$ because we have not yet established whether j^2 commutes with j_{1z} and j_{2z}.

To explore this point we proceed as follows:

$$\begin{aligned}
[j_{1z}, j^2] &= [j_{1z}, j_x^2] + [j_{1z}, j_y^2] + [j_{1z}, j_z^2] \\
&= [j_{1z}, (j_{1x} + j_{2x})^2] + [j_{1z}, (j_{1y} + j_{2y})^2] + [j_{1z}, (j_{1z} + j_{2z})^2] \\
&= [j_{1z}, j_{1x}^2 + 2j_{1x}j_{2x}] + [j_{1z}, j_{1y}^2 + 2j_{1y}j_{2y}] \\
&= [j_{1z}, j_{1x}^2 + j_{1y}^2] + 2[j_{1z}, j_{1x}]j_{2x} + 2[j_{1z}, j_{1y}]j_{2y} \\
&= [j_{1z}, j_1^2 - j_{1z}^2] + 2i\hbar j_{1y}j_{2x} - 2i\hbar j_{1x}j_{2y} \\
&= 2i\hbar(j_{1y}j_{2x} - j_{1x}j_{2y})
\end{aligned} \tag{41}$$

The commutator is *not* zero, and so we *cannot* specify m_{j1} (or m_{j2}) if we specify the value of j.

It follows from this analysis that we have to make a choice when specifying the system. Either we use the **uncoupled picture** $|j_1m_{j1}; j_2m_{j2}\rangle$, which leaves the total angular momentum unspecified and therefore, in effect, says nothing about the relative orientation of the two momenta, or we use the **coupled picture** $|j_1 j_2;jm_j\rangle$ which leaves the individual components unspecified. At this stage, which choice we make is arbitrary. Later, when we consider the energy of interaction between different angular momenta we shall see that one picture is more natural than the other. At this stage, the two pictures are simply alternative ways of specifying a composite system.

4.10 The permitted values of the total angular momentum

If we decide to use the coupled picture, the question arises as to the permissible values of j and m_j. We know that the commutation relations permit j to have any positive integral or half-integral values, but we need to determine which of

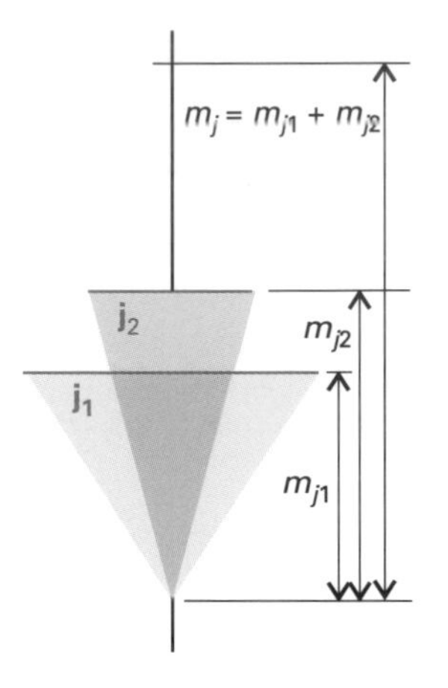

Fig. 4.4 A representation of the requirement that $m_j = m_{j1} + m_{j2}$.

these many values actually occur for a given j_1 and j_2. For example, the total angular momentum of a p-electron ($l = 1$ and $s = \frac{1}{2}$) is unlikely to exceed $j = \frac{3}{2}$.

The allowed values of m_j follow immediately from the relation $j_z = j_{1z} + j_{2z}$, and are

$$m_j = m_{j1} + m_{j2} \tag{42}$$

That is, the total component of angular momentum about an axis is the sum of the components of the two contributing momenta (Fig. 4.4).

To determine the allowed values of j, we first note that the total number of states in the uncoupled picture is $(2j_1 + 1)(2j_2 + 1) = 4j_1j_2 + 2j_1 + 2j_2 + 1$. There is only one state in which both components have their maximum values, $m_{j1} = j_1$ and $m_{j2} = j_2$, and this state corresponds to $m_j = j_1 + j_2$. However, the maximum value of m_j is by definition j, so the maximum value of j is $j = j_1 + j_2$. There are $2j + 1 = 2j_1 + 2j_2 + 1$ states corresponding to this value of j, and so there are a further $4j_1j_2$ states to find.

Although the state with $m_j = j_1 + j_2$ can arise in only one way, the state with $m_j = j_1 + j_2 - 1$ can arise in two ways, from $m_{j1} = j_1 - 1$ and $m_{j2} = j_2$ and from $m_{j1} = j_1$ and $m_{j2} = j_2 - 1$. The state with $j = j_1 + j_2$ accounts for only one of these states (or for one of their two linear combinations), and so there must be another coupled state for which the maximum value of m_j is $m_j = j_1 + j_2 - 1$. This state corresponds to a state with $j = j_1 + j_2 - 1$. A system with this value of j accounts for a further $2j + 1 = 2j_1 + 2j_2 - 1$ states. The process can be continued by considering the next lower value of m_j, which is $m_j = j_1 + j_2 - 2$, and which can be produced in three ways. The two states with $j = j_1 + j_2$ and $j = j_1 + j_2 - 1$ account for two of them; the third (or the third linear combination) must arise from the state with $j = j_1 + j_2 - 2$. This argument can be continued, and all the states are accounted for by the time we have reached $j = |j_1 - j_2|$ (j is a positive number, hence the modulus signs). Therefore, the permitted states of angular momentum that can arise from a system composed of two sources of angular momentum are given by the **Clebsch–Gordan series**:

$$j = j_1 + j_2, j_1 + j_2 - 1, \ldots, |j_1 - j_2| \tag{43}$$

To verify that this series does indeed account for all $4j_1j_2 + 2j_1 + 2j_2 + 1$ states, we sum the number of states $(2j + 1)$ for each permitted value of j. For $j_1 \geq j_2$ the sum is

$$\sum_{j=j_1-j_2}^{j_1+j_2} (2j + 1) = 4j_1j_2 + 2j_1 + 2j_2 + 1 = (2j_1 + 1)(2j_2 + 1) \tag{44}$$

as required.[2]

[2] The sum of an arithmetical progression with n terms starting at a_1 and terminating at a_n is $\frac{1}{2}n(a_1 + a_n)$; the number of terms in the sum is $2j_2 + 1$ when $j_1 \geq j_2$.

Example 4.3 Using the Clebsch–Gordan series

What angular momentum states can arise from a system with two sources of angular momentum, one with $j_1 = \frac{1}{2}$ and the other with $j_2 = \frac{3}{2}$? Specify the states.

Method. Use the Clebsch–Gordan series in eqn 4.43 to find the highest and lowest values of j first, and then complete the series. The composite system has $(2j_1 + 1)(2j_2 + 1)$ states, which may either be specified as $|j_1 m_{j1};\ j_2 m_{j2}\rangle$ or as $|j_1 j_2; j m_j\rangle$.

Answer. The highest and lowest values of j are $\frac{1}{2} + \frac{3}{2} = 2$ and $|\frac{1}{2} - \frac{3}{2}| = 1$, respectively. So the complete Clebsch–Gordan series is

$$j = 2, 1$$

A specification of the $4 \times 2 = 8$ states in the uncoupled representation is:

$$\begin{array}{cccc} |\frac{1}{2},+\frac{1}{2};\frac{3}{2},+\frac{3}{2}\rangle & |\frac{1}{2},+\frac{1}{2};\frac{3}{2},+\frac{1}{2}\rangle & |\frac{1}{2},+\frac{1}{2};\frac{3}{2},-\frac{1}{2}\rangle & |\frac{1}{2},+\frac{1}{2};\frac{3}{2},-\frac{3}{2}\rangle \\ |\frac{1}{2},-\frac{1}{2};\frac{3}{2},+\frac{3}{2}\rangle & |\frac{1}{2},-\frac{1}{2};\frac{3}{2},+\frac{1}{2}\rangle & |\frac{1}{2},-\frac{1}{2};\frac{3}{2},-\frac{1}{2}\rangle & |\frac{1}{2},-\frac{1}{2};\frac{3}{2},-\frac{3}{2}\rangle \end{array}$$

The alternative specification, in the coupled representation, is

$$\begin{array}{cccc} |\frac{1}{2},\frac{3}{2};2,+2\rangle & |\frac{1}{2},\frac{3}{2};2,+1\rangle & |\frac{1}{2},\frac{3}{2};2,0\rangle & |\frac{1}{2},\frac{3}{2};2,-1\rangle \\ |\frac{1}{2},\frac{3}{2};2,-2\rangle & |\frac{1}{2},\frac{3}{2};1,+1\rangle & |\frac{1}{2},\frac{3}{2};1,0\rangle & |\frac{1}{2},\frac{3}{2};1,-1\rangle \end{array}$$

Comment. The eight states in the coupled representation are linear combinations of the eight states in the uncoupled representation. We explore the relation between them in Section 4.13.

Exercise 4.3. Repeat the question for $j_1 = 1$ and $j_2 = 2$.

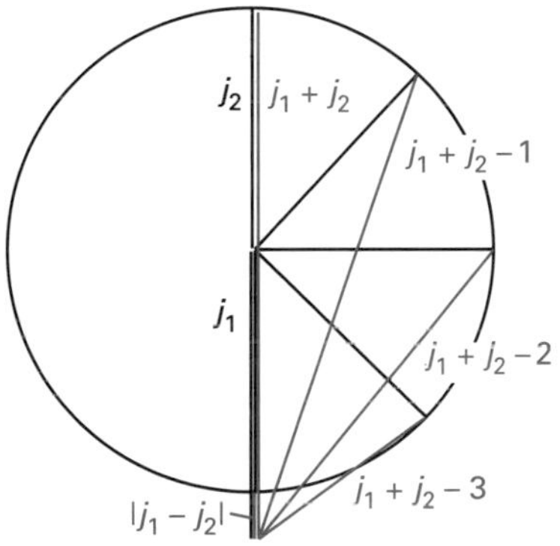

The Clebsch–Gordan series can be expressed in a simple pictorial way. Suppose we are given rods of lengths j_1 and j_2 and are asked for the lengths j of the third side of a triangle that can be formed using these two rods (with all three lengths integers or half-integers). Then the answer would be precisely those given by the Clebsch–Gordan series (Fig. 4.5). For example, $j_1 = 1$ and $j_2 = 1$ require rods of lengths $j = 2, 1, 0$ to form a triangle. Although the **triangle condition** is no more than a simple and helpful mnemonic, it does suggest that angular momenta in quantum mechanics do in some respects behave like vectors and that the total angular momentum can be regarded as the resultant of the contributing momenta. The exploration of this point leads to the vector model of coupled angular momenta.

Fig. 4.5 The triangle condition corresponding to the Clebsch–Gordan series. The allowed values of j are those for which lines of length j, j_1, and j_2 can be used to form a triangle.

4.11 The vector model of coupled angular momenta

The vector model of coupled angular momentum is an attempt to represent pictorially the features of coupled angular momenta that we have deduced from the commutation relations. The approach gives insight into the significance of various coupling schemes and is often a helpful guide to the imagination: it puts visual flesh on the operator bones.

The features that the vector diagrams of coupled momenta must express are as follows:

1. The length of the vector representing the total angular momentum is $\{j(j+1)\}^{1/2}$, with j one of the values permitted by the Clebsch–Gordan series.
2. This vector must lie at an indeterminate angle on a cone about the z-axis (because j_x and j_y cannot be specified if j_z has been specified).
3. The lengths of the contributing angular momentum vectors are $\{j_1(j_1+1)\}^{1/2}$ and $\{j_2(j_2+1)\}^{1/2}$. These lengths have definite values even when j is specified.
4. The projection of the total angular momentum on the z-axis is m_j; in the coupled picture (in which j is specified), the values of m_{j1} and m_{j2} are indefinite, but their sum is equal to m_j.
5. In the uncoupled picture (in which j is not specified), the individual components m_{j1} and m_{j2} may be specified, and their sum is equal to m_j.

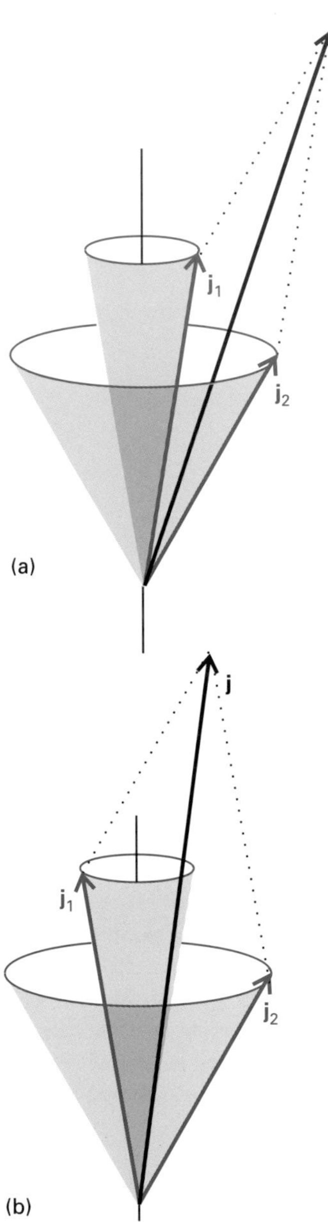

Fig. 4.6 Two possible states of total angular momentum that can arise from two specified contributing momenta with quantum numbers j_1 and j_2. The relative orientations of the contributing momenta on their cones determine the total magnitude.

The diagrams in Fig. 4.6 and Fig. 4.7 capture these points. In Fig. 4.6(a) is shown one of the states of the uncoupled picture: both m_{j1} and m_{j2} are specified, but there is no indication of the relative orientation of j_1 and j_2 apart from the fact that they lie on their respective cones. The total angular momentum is therefore indeterminate, for it could be either of the resultants shown in (a) or (b) or anything in between. In Fig. 4.7 is shown one of the states of the coupled picture. Now the resultant, the total angular momentum, has a well-defined magnitude and resultant on the z-axis, but the individual components m_{j1} and m_{j2} are indeterminate. It is important not to think of the vectors as precessing around their cones: at this stage the vector model is a display of possible but unspecifiable orientations.

An important example, and one that we shall encounter many times in later chapters, is the case of two particles with spin $s = \frac{1}{2}$, such as two electrons or two protons. For each particle, $s = \frac{1}{2}$ and $m_s = \pm\frac{1}{2}$. In the uncoupled picture, the electrons may be in any of the four states

$$\alpha_1\alpha_2 \qquad \alpha_1\beta_2 \qquad \beta_1\alpha_2 \qquad \beta_1\beta_2$$

These four states are illustrated in Fig. 4.8. The individual angular momenta lie at unspecified positions on their cones and the total angular momentum is indeterminate.

Now consider the coupled picture. The triangle condition (or the Clebsch–Gordan series) tells us that the total spin S (upper-case letters are used to denote the angular momenta of collections of particles) can take the values 1 and 0. When $S = 0$, there is only one possible value of its z-component, namely 0, corresponding to $M_S = 0$. Such a coupled state is called a **singlet**. When $S = 1$, $M_S = +1, 0, -1$, and so this coupled arrangement is called a **triplet**.

The vector model of the triplet is shown in Fig. 4.9. The cones have been drawn to scale, and several points should be apparent. One is that to arrive at a resultant corresponding to $S = 1$ (of length $\sqrt{2}$) using component vectors corresponding to $s = \frac{1}{2}$ (of length $\frac{1}{2}\sqrt{3}$), the vectors must lie at a definite angle relative to one another. In fact, they must lie in the same plane, as shown in the illustration, for only that orientation results in a vector of the correct length. Note that although spins are said to be 'parallel' in a triplet state (and represented $\uparrow\uparrow$), they are in fact at an acute angle (of close to 70°). The two spins

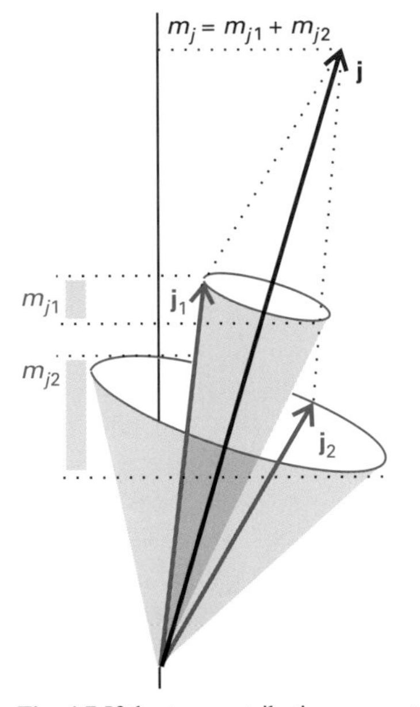

Fig. 4.7 If the two contributing momenta are locked together so that they give rise to a specified total, the projections of the contributing momenta span a range (as depicted by the vertical bars) and although their sum can be specified, their individual values cannot be specified.

make the same angle to one another in the states with $M_S = \pm 1$; that is necessary if they are to have the same resultant.

The vector model of the singlet must represent a state in which the spin angular momentum vectors sum to give a zero resultant (Fig. 4.10). It is clear from the illustration that the two spins are truly antiparallel ($\uparrow\downarrow$) in this state. As in the triplet states, only the relative orientation of the vectors is fixed; the absolute orientation is completely indeterminate.

4.12 The relation between schemes

The state $|j_1j_2;jm_j\rangle$ is built from all values of m_{j1} and m_{j2} such that $m_{j1} + m_{j2} = m_j$. This remark suggests that it should be possible to express the coupled state as a sum over all the uncoupled states $|j_1m_{j1};j_2m_{j2}\rangle$ that conform to $m_{j1} + m_{j2} = m_j$. It follows that we should be able to write

$$|j_1j_2;jm_j\rangle = \sum_{m_{j1},m_{j2}} C_{m_{j1}m_{j2}}|j_1m_{j1};j_2m_{j2}\rangle \tag{45}$$

The coefficients $C_{m_{j1}m_{j2}}$ are called **vector coupling coefficients**. Alternative names are 'Clebsch–Gordan coefficients', 'Wigner coefficients', and (in a slightly modified form), the '3j-symbols'.

We shall illustrate the use of vector coupling coefficients by considering the singlet and triplet states of two spin-$\frac{1}{2}$ particles. The values are set out in Table 4.1 (more values for other cases will be found in Appendix 2). The values in the

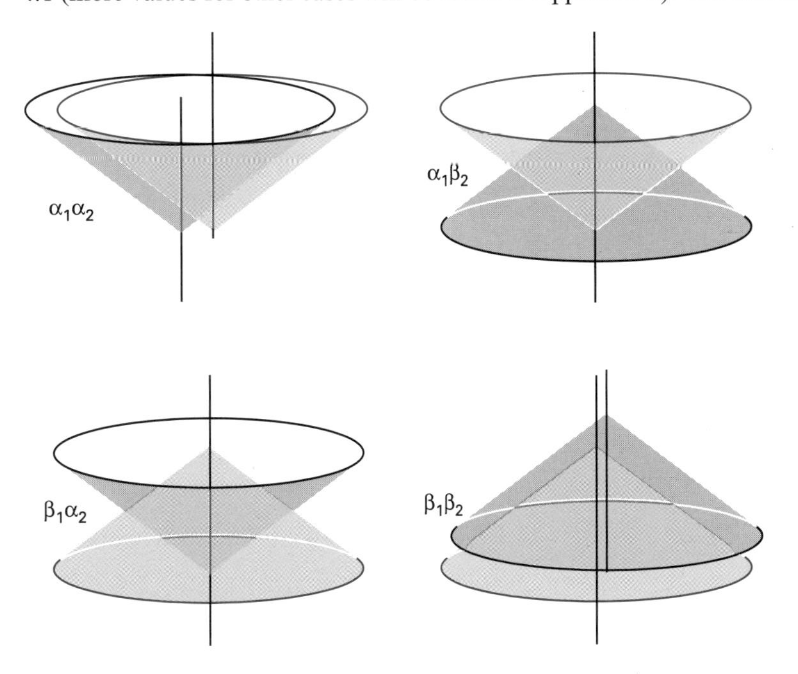

Fig. 4.8 The four uncoupled states of a system consisting of two spin-$\frac{1}{2}$ particles (such as electrons).

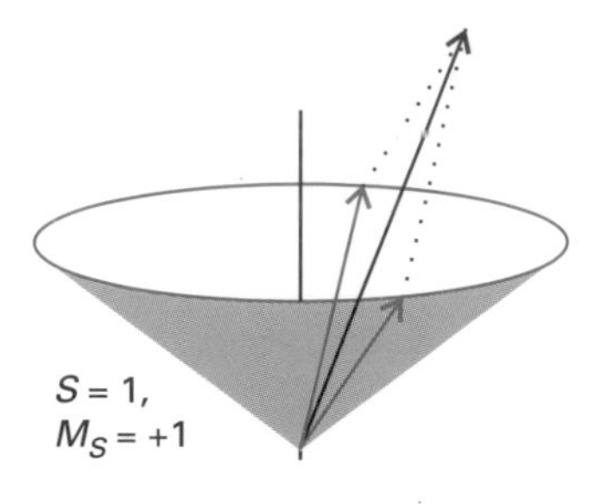

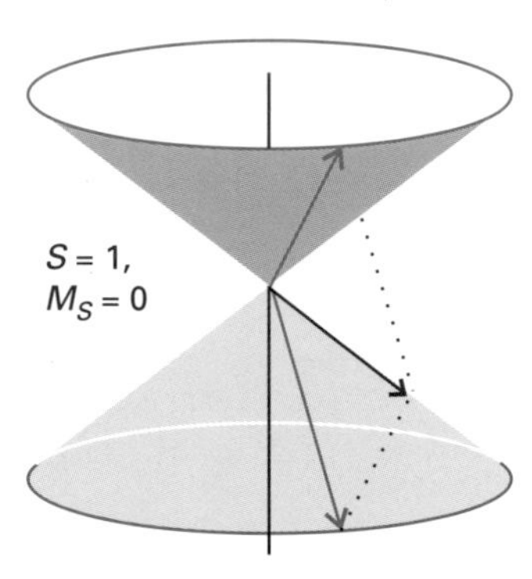

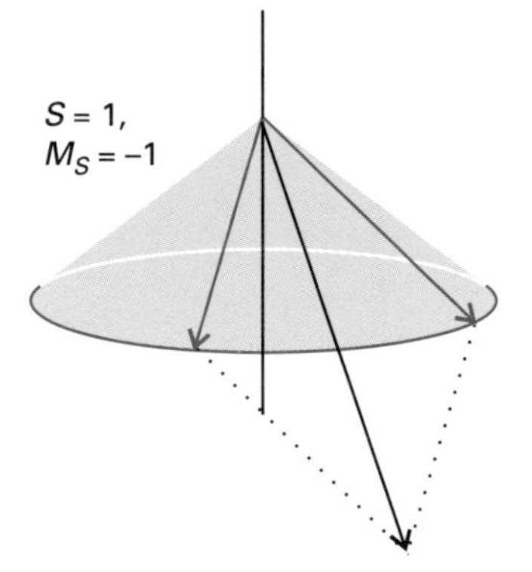

Fig. 4.9 Three of the four coupled states of a system consisting of two spin-$\frac{1}{2}$ particles. These states all correspond to $S = 1$. The relative orientations of the individual angular momenta are the same in each case (the angle is about 70°).

table imply that, using the notation $|S, M_S\rangle$,

$$\begin{aligned}
|1, +1\rangle &= \alpha_1\alpha_2 \\
|1, 0\rangle &= \frac{1}{\sqrt{2}}\alpha_1\beta_2 + \frac{1}{\sqrt{2}}\beta_1\alpha_2 \\
|1, -1\rangle &= \beta_1\beta_2 \\
|0, 0\rangle &= \frac{1}{\sqrt{2}}\alpha_1\beta_2 - \frac{1}{\sqrt{2}}\beta_1\alpha_2
\end{aligned}$$

There are two points to note. One is that even a 'spin-parallel' triplet state, can be composed of 'opposite' spins (see the composition of $|1, 0\rangle$). Secondly, the + in $|1, 0\rangle$ is taken to signify that the α and β spins from which it is built are in phase with one another (as suggested by the vector diagram for this state), whereas the $-$ sign in $|0, 0\rangle$ signifies that they are out of phase. This too is captured by the antiparallel arrangement of vectors in the vector diagram.

General expressions for the vector coupling coefficients can be derived, but they are very complicated and it is usually simplest to use tables of numerical values. These values can be derived quite simply in special cases, and we shall indicate the procedure for the values in Table 4.1. The general point to note is that the coefficients are in fact the overlap integrals for coupled states with uncoupled states. To see that this is so, multiply both sides of eqn 4.45 from the left by $\langle j_1 m'_{j1}; j_2 m'_{j2}|$: the only term that survives on the right is the one with $m_{j1} = m'_{j1}$ and $m_{j2} = m'_{j2}$ (by the orthogonality of the states), so

Table 4.1 Vector coupling coefficients for $s_1 = \frac{1}{2}, s_2 = \frac{1}{2}$

m_{s1}	m_{s2}	$\|1, +1\rangle$	$\|1, 0\rangle$	$\|0, 0\rangle$	$\|1, -1\rangle$
$+\frac{1}{2}$	$+\frac{1}{2}$	1	0	0	0
$+\frac{1}{2}$	$-\frac{1}{2}$	0	$\frac{1}{\sqrt{2}}$	$\frac{1}{\sqrt{2}}$	0
$-\frac{1}{2}$	$+\frac{1}{2}$	0	$\frac{1}{\sqrt{2}}$	$-\frac{1}{\sqrt{2}}$	0
$-\frac{1}{2}$	$-\frac{1}{2}$	0	0	0	1

$$\langle j_1 m'_{j1}; j_2 m'_{j2} | j_1 j_2; j m_j\rangle = C_{m'_{j1} m'_{j2}} \tag{46}$$

Thus, the coefficient $C_{m_{j1} m_{j2}}$ can be interpreted as the extent to which the coupled state $|j_1 j_2; j m_j\rangle$ resembles the uncoupled state $|j_1 m_{j1}; j_2 m_{j2}\rangle$.

The state $|1, +1\rangle$ must be composed of $\alpha_1\alpha_2$, because only this state corresponds to $M_S = +1$. It follows that

$$|1, +1\rangle = \alpha_1\alpha_2 \tag{47}$$

The effect of the lowering operator S_- on $|1, +1\rangle$ is given by eqns 4.23 and 4.30, which in the current notation reads

$$S_-|S, M_S\rangle = \{S(S+1) - M_S(M_S - 1)\}^{1/2}\hbar|S, M_S - 1\rangle \tag{48}$$

and so

$$S_-|1, +1\rangle = \sqrt{2}\hbar|1, 0\rangle$$

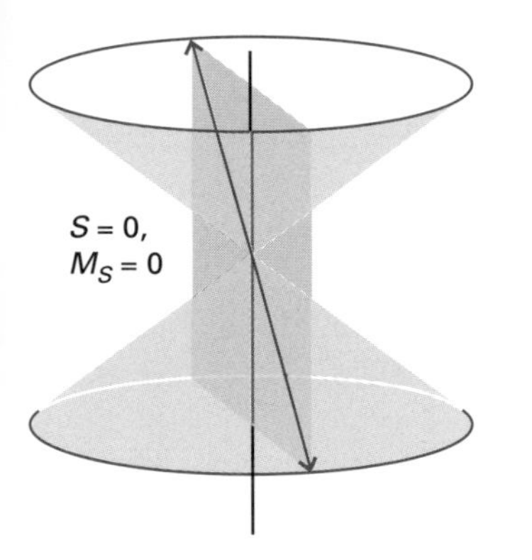

Fig. 4.10 The remaining coupled state of two spin-$\frac{1}{2}$ particles. This state corresponds to $S = 0$. Note that the two contributing momenta are perfectly antiparallel.

However, because $S_- = s_{1-} + s_{2-}$, the effect of S_- can also be written

$$S_-|S, M_S\rangle = (s_{1-} + s_{2-})\alpha_1\alpha_2 = \hbar(\alpha_1\beta_2 + \beta_1\alpha_2)$$

Comparison of these two expressions results in

$$|1, 0\rangle = \frac{1}{\sqrt{2}}(\alpha_1\beta_2 + \beta_1\alpha_2) \tag{49}$$

as found from Table 4.1. The third state of the triplet is obtained by repeating the procedure:

$$\begin{aligned} S_-|1, 0\rangle = \sqrt{2}\hbar|1, -1\rangle &= (s_{1-} + s_{2-})\frac{1}{\sqrt{2}}(\alpha_1\beta_2 + \beta_1\alpha_2) \\ &= \sqrt{2}\hbar\beta_1\beta_2 \end{aligned}$$

It follows that

$$|1, -1\rangle = \beta_1\beta_2 \tag{50}$$

as we found from the table and exactly as would be expected on physical grounds (namely, that there is only one way of achieving a state with $M_S = -1$ from two spin-$\frac{1}{2}$ systems).

Only the singlet state remains to be found. Because it necessarily has $M_S = 0$ and $M_S = m_{s1} + m_{s2}$, it is constructed from $\alpha_1\beta_2$ and $\beta_1\alpha_2$. However, it must (by the hermiticity of S^2) be orthogonal to the state $|1, 0\rangle$. Therefore, we can write immediately that

$$|0, 0\rangle = \frac{1}{\sqrt{2}}(\alpha_1\beta_2 - \beta_1\alpha_2) \tag{51}$$

as was given by the use of Table 4.1.

Example 4.4 How to confirm that a state is correctly labelled

Confirm that the state $|0, 0\rangle = \frac{1}{\sqrt{2}}(\alpha_1\beta_2 - \beta_1\alpha_2)$ does in fact correspond to $S = 0$ and $M_S = 0$.

Method. To determine the value of M_S we determine the eigenvalue of $S_z = s_{1z} + s_{2z}$. To evaluate the eigenvalue of S^2, we can make use of the relation

$$S^2 = (\mathbf{s}_1 + \mathbf{s}_2) \cdot (\mathbf{s}_1 + \mathbf{s}_2) = s_1^2 + s_2^2 + 2\mathbf{s}_1 \cdot \mathbf{s}_2$$

to express S^2 in terms of the individual spin operators. We then need to recognize that

$$\mathbf{s}_1 \cdot \mathbf{s}_2 = s_{1z}s_{2z} + \tfrac{1}{2}(s_{1+}s_{2-} + s_{1-}s_{2+})$$

Answer. First, we use

$$\begin{aligned} S_z|0, 0\rangle &= \frac{1}{\sqrt{2}}(s_{1z} + s_{2z})(\alpha_1\beta_2 - \beta_1\alpha_2) \\ &= \frac{1}{\sqrt{2}}(s_{1z}\alpha_1\beta_2 - s_{1z}\beta_1\alpha_2 + \alpha_1 s_{2z}\beta_2 - \beta_1 s_{2z}\alpha_2) \\ &= \frac{\hbar}{\sqrt{2}}(\tfrac{1}{2}\alpha_1\beta_2 + \tfrac{1}{2}\beta_1\alpha_2 - \tfrac{1}{2}\alpha_1\beta_2 - \tfrac{1}{2}\beta_1\alpha_2) = 0 \end{aligned}$$

It follows that $M_S = 0$. For the eigenstate of S^2, we proceed as follows:

$$S^2|0,0\rangle = \frac{1}{\sqrt{2}}(s_1^2\alpha_1\beta_2 - s_1^2\beta_1\alpha_2 + \alpha_1 s_2^2\beta_2 - \beta_1 s_2^2\alpha_2 + 2s_{1z}\alpha_1 s_{2z}\beta_2 - 2s_{1z}\beta_1 s_{2z}\alpha_2 + s_{1+}\alpha_1 s_{2-}\beta_2 + s_{1-}\alpha_1 s_{2+}\beta_2 - s_{1+}\beta_1 s_{2-}\alpha_2 - s_{1-}\beta_1 s_{2+}\alpha_2)$$

$$= \frac{\hbar^2}{\sqrt{2}}(\tfrac{3}{4}\alpha_1\beta_2 - \tfrac{3}{4}\beta_1\alpha_2 + \tfrac{3}{4}\alpha_1\beta_2 - \tfrac{3}{4}\beta_1\alpha_2 - \tfrac{1}{2}\alpha_1\beta_2 + \tfrac{1}{2}\beta_1\alpha_2 - \alpha_1\beta_2 + \beta_1\alpha_2)$$

$$= 0$$

Hence, $S = 0$.

Exercise 4.4. Apply the same analysis to the state $|1,0\rangle$.

As a second illustration, consider two d-electrons. The Clebsch–Gordan series gives the total orbital angular momentum, L, as

$$L = 4, 3, 2, 1, 0$$

With these states there are associated 25 states, so the problem is somewhat larger than before. The state with $L = 4$ must have $M_L = +4$ as one of its components, and this state can be obtained in only one way, when $m_{l1} = +2$ and $m_{l2} = +2$. It follows that

$$|4,+4\rangle = |+2,+2\rangle$$

where the notation on the left is $|L, M_L\rangle$ and that on the right is $|m_{l1}, m_{l2}\rangle$. To avoid this rather confusing symbolism, we shall denote the states with $L = 0, 1, \ldots, 4$ by the letters S, P, D, F, G (by analogy with the labels for atomic orbitals). Then instead of the line above we can write

$$|\mathrm{G},+4\rangle = |+2,+2\rangle$$

We may now proceed to generate the remaining eight states with $L = 4$ by applying the operator $L_- = l_{1-} + l_{2-}$. From L_- applied to the left of the last equation we get

$$L_-|\mathrm{G},+4\rangle = \sqrt{8}\hbar|\mathrm{G},+3\rangle$$

and from $l_{1-} + l_{2-}$ applied to the right we get

$$(l_{1-} + l_{2-})|+2,+2\rangle = \sqrt{4}\hbar(|+1,+2\rangle + |+2,+1\rangle)$$

from which it follows that

$$|\mathrm{G},+3\rangle = \frac{1}{\sqrt{2}}(|+1,+2\rangle + |+2,+1\rangle)$$

The remaining seven states of this set may be generated similarly. The state $|\mathrm{F},+3\rangle$ also arises from the states $|+1,+2\rangle$ and $|+2,+1\rangle$ and must be orthogonal to $|\mathrm{G},+3\rangle$. Therefore, we can immediately write

$$|\mathrm{F},+3\rangle = \frac{1}{\sqrt{2}}(|+1,+2\rangle - |+2,+1\rangle)$$

The remaining six states of this set can now be generated. The same argument may then be applied to generate the D, P, and S states and the table of coefficients given in Appendix 2 compiled.

Example 4.5 How to use vector coupling coefficients

Construct the state with $j = \frac{3}{2}$ and $m_j = -\frac{1}{2}$ for a p-electron.

Method. For a p-electron, $l = 1$ and $s = \frac{1}{2}$. The state with $j = \frac{3}{2}$ is a linear combination of the states $|1, m_l; \frac{1}{2}, m_s\rangle$ with $m_l + m_s = -\frac{1}{2}$. Use Appendix 2 for the vector coupling coefficients.

Answer. We write the coupled state in the form

$$|\tfrac{3}{2}, -\tfrac{1}{2}\rangle = \left(\tfrac{2}{3}\right)^{1/2}|1, 0; \tfrac{1}{2}, -\tfrac{1}{2}\rangle + \left(\tfrac{1}{3}\right)^{1/2}|1, -1; \tfrac{1}{2}, +\tfrac{1}{2}\rangle$$

Exercise 4.5. Find the expression for the state $|\mathrm{D}, 0\rangle$ arising from the orbital angular momenta of two p-electrons. Use the tables in Appendix 2.

4.13 The coupling of several angular momenta

The final point we need to make in this section concerns the case where three or more momenta are coupled together. In the case of three momenta, we have the choice of first coupling j_1 to j_2 to form j_{12} and then coupling j_3 to that to give the overall resultant j. As an illustration, consider the total orbital angular momenta of three p-electrons. The coupling of one pair gives $l_{12} = 2, 1, 0$. Then the third couples with each of these resultants in turn: $l_{12} = 2$ gives rise to $L = 3, 2, 1$; $l_{12} = 1$ gives rise to $L = 2, 1, 0$; and $l_{12} = 0$ gives rise to only $L = 1$. The angular momentum states are therefore F + 2D + 3P + S

When there are more than two sources of angular momentum, the overall states may be formed in different ways. Thus, instead of the scheme described above, j_1 and j_3 can first be coupled to form j_{13}, and then j_2 coupled to j_{13} to form j. The triangle condition applies to each step in the coupling procedure, but the compositions of the states obtained are different. The states obtained by the first coupling procedure can be expressed as linear combinations of the states obtained by the second procedure, and the expansion coefficients are known as **Racah coefficients** or, in slightly modified form, as '6j-symbols'. The question of alternative coupling schemes, and how to select the most appropriate ones, arises in discussions of atomic and molecular spectra, and we shall meet it again there.

Problems

4.1 Evaluate the commutator $[l_x, l_y]$ in (a) the position representation, (b) the momentum representation.

4.2 Evaluate the commutators (a) $[l_y^2, l_x]$, (b) $[l_y^2, l_x^2]$, and (c) $[l_x, [l_x, l_y]]$. *Hint.* Use the basic commutators in eqn 4.6.

4.3 Verify that eqn 4.8 expresses the basic angular momentum commutation rules. *Hint.* Expand the left of eqn 4.8 and compare coefficients of the unit vectors. Be careful with the ordering of the vector components when expanding the determinant: the operators in the second row always precede those in the third.

4.4 (a) Confirm that the Pauli matrices

$$\sigma_x = \begin{pmatrix} 0 & 1 \\ 1 & 0 \end{pmatrix} \qquad \sigma_y = \begin{pmatrix} 0 & -\mathrm{i} \\ \mathrm{i} & 0 \end{pmatrix} \qquad \sigma_z = \begin{pmatrix} 1 & 0 \\ 0 & -1 \end{pmatrix}$$

satisfy the angular momentum commutation relations when we write $s_q = \frac{1}{2}\hbar\sigma_q$, and hence provide a matrix representation of angular momentum. (b) Why does the representation correspond to $s = \frac{1}{2}$? *Hint.* For the second part, form the matrix representing s^2 and establish its eigenvalues.

4.5 Using the Pauli matrix representation, reduce each of the operators (a) $s_x s_y$, (b) $s_x s_y^2 s_z^2$, and (c) $s_x^2 s_y^2 s_z^2$ to a single spin operator.

4.6 Evaluate the effect of (a) $e^{\mathrm{i}s_x/\hbar}$, (b) $e^{\mathrm{i}s_y/\hbar}$, and (c) $e^{\mathrm{i}s_z/\hbar}$ on an α spin state. *Hint.* Expand the exponential operators as in Problem 1.15 and use arguments like those in Problem 4.5.

4.7 Suppose that in place of the actual angular momentum commutation rules, the operators obeyed $[l_x, l_y] = -i\hbar l_z$. What would be the roles of $l_\pm$?

4.8 Calculate the matrix elements (a) $\langle 0,0|l_z|0,0\rangle$, (b) $\langle 2,1|l_+|2,0\rangle$, (c) $\langle 2,2|l_+^2|2,0\rangle$, (d) $\langle 2,0|l_+l_-|2,0\rangle$, (e) $\langle 2,0|l_-l_+|2,0\rangle$, and (f) $\langle 2,0|l_-^2 l_z l_+^2|2,0\rangle$.

4.9 Demonstrate that $\mathbf{j}_1 - \mathbf{j}_2$ is not an angular momentum.

4.10 Calculate the values of the following matrix elements between *p*-orbitals: (a) $\langle p_x|l_z|p_y\rangle$, (b) $\langle p_x|l_+|p_y\rangle$, (c) $\langle p_z|l_y|p_x\rangle$, (d) $\langle p_z|l_x|p_y\rangle$, and (e) $\langle p_z|l_x|p_x\rangle$.

4.11 Evaluate the matrix elements (a) $\langle j, m_j+1|j_x^3|j, m_j\rangle$ and (b) $\langle j, m_j+3|j_x^3|j, m_j\rangle$.

4.12 Confirm that the spherical polar forms of the orbital angular momentum operators in eqn 4.31 satisfy the angular momentum commutation relation $[l_x, l_y] = \mathrm{i}\hbar l_z$ and that the shift operators in eqn 4.32 satisfy $[l_+, l_-] = 2\hbar l_z$.

4.13 Verify that successive application of l_- to ψ_{ll} with $l = 2$ generates the five normalized spherical harmonics Y_{2m_l} as set out in Table 3.1.

4.14 (a) Demonstrate that if $[j_{1q}, j_{2q'}] = 0$ for all q, q', then $\mathbf{j}_1 \times \mathbf{j}_2 = -\mathbf{j}_2 \times \mathbf{j}_1$. (b) Go on to show that if $\mathbf{j}_1 \times \mathbf{j}_1 = \mathrm{i}\hbar\mathbf{j}_1$ and $\mathbf{j}_2 \times \mathbf{j}_2 = \mathrm{i}\hbar\mathbf{j}_2$, then $\mathbf{j} \times \mathbf{j} = \mathrm{i}\hbar\mathbf{j}$ where $\mathbf{j} = \mathbf{j}_1 + \mathbf{j}_2$.

4.15 In some cases m_{j1} and m_{j2} may be specified at the same time as j because although $[j^2, j_{1z}]$ is nonzero, the effect of $[j^2, j_{1z}]$ on the state with $m_{j1} = j_1$, $m_{j2} = j_2$ is zero. Confirm that $[j^2, j_{1z}]|j_1 j_1; j_2 j_2\rangle = 0$ and $[j^2, j_{1z}]|j_1, -j_1; j_2, -j_2\rangle = 0$.

4.16 Determine what total angular momenta may arise in the following composite systems: (a) $j_1 = 3$, $j_2 = 4$; (b) the orbital momenta of two electrons (i) both in *p*-orbitals, (ii) both in *d*-orbitals, (iii) in the configuration p^1d^1; (c) the spin angular momenta of four electrons. *Hint.* Use the Clebsch–Gordan series, eqn 4.43; apply it successively in (c).

4.17 Construct the vector coupling coefficients for a system with $j_1 = 1$ and $j_2 = \frac{1}{2}$ and evaluate the matrix elements $\langle j'm_j'|j_{1z}|jm_j\rangle$. *Hint.* Proceed as in Section 4.12 and check the answer against the values in Appendix 2. For the matrix element, express the coupled states in the uncoupled representation, and then operate with j_{1z}.

4.18 Use the vector model of angular momentum to derive the value of the angle between the vectors representing (a) two α spins, (b) an α and a β spin in a state with $S = 1$ and $M_S = +1$ and $M_S = 0$, respectively.

4.19 Set up a quantum mechanical expression that can be used to derive the same result as in Problem 4.18. *Hint.* Consider the expectation value of $\mathbf{s}_1 \cdot \mathbf{s}_2$.

4.20 Apply both procedures (of the preceding two problems) to calculate the angle between α spins in the $\alpha\alpha\alpha$ state with $S = \frac{3}{2}$.

4.21 Consider a system of two electrons that can have either paired or unpaired spins (e.g. a biradical). The energy of the system depends on the relative orientation of their spins. Show that the operator $(hJ/\hbar^2)\mathbf{s}_1 \cdot \mathbf{s}_2$ distinguishes between singlet and triplet states. The system is now exposed to a magnetic field in the z-direction. Because the two electrons are in different environments, they experience different local fields and their interaction energy can be written $(\mu_B/\hbar)B(g_1 s_{1z} + g_2 s_{2z})$ with $g_1 \neq g_2$; μ_B is the Bohr magneton and g is the electron g-value; these two quantities are discussed in detail in Chapter 13. Establish the matrix of the total hamiltonian, and demonstrate that when $hJ \gg \mu_B B$, the coupled representation is 'better', but that when $\mu_B B \gg hJ$ the uncoupled representation is 'better'. Find the eigenvalues and eigenstates of the system in each case.

4.22 What is the expectation value of the z-component of orbital angular momentum of electron 1 in the $|\mathrm{G}, M_L\rangle$ state of the configuration d^2? *Hint*. Express the coupled state in terms of the uncoupled states, find $\langle \mathrm{G}, M_L | l_{1z} | \mathrm{G}, M_L\rangle$ in terms of the vector coupling coefficients, and evaluate it for $M_L = +4, +3, \ldots, -4$.

4.23 Prove that $\sum_{m_{j1}, m_{j2}} |C_{m_{j1} m_{j2}}|^2 = 1$ for a given j_1, j_2, and j. *Hint*. Use eqn 4.45 and form $\langle j_1 j_2; j m_j | j_1 j_2; j m_j\rangle$.

Further reading

Elementary theory of angular momentum. M.E. Rose; Wiley, New York (1975).
Angular momentum. D.M. Brink and G.R. Satchler; Clarendon Press, Oxford (1993).
Angular momentum in quantum mechanics. A.R. Edmonds; Princeton University Press, Princeton (1960).
Atomic structure. E.U. Condon and H. Odabaşi; Cambridge University Press, Cambridge (1980).
Operator techniques in atomic spectroscopy. B.R. Judd; McGraw-Hill, New York (1963).
Angular momentum for diatomic molecules. B.R. Judd; Academic Press, New York (1975).
Angular momentum in quantum physics: theory and application. L.C. Biedenharn and J.D. Louck; Addison-Wesley, New York (1981).
Quantum theory of angular momentum. L.C. Biedenharn and H. van Dam (ed.); Academic Press, New York (1965).
Angular momentum: Understanding spatial aspects in chemistry and physics. R.N. Zare; Wiley, New York (1988).
Quantum mechanics, Vol. 1. C. Cohen-Tannoudji, B. Diu, and F. Laloë; Wiley, New York (1977).
The 3j- and 6j-symbols. M. Rotenberg, R. Bivins, N. Metropolis, and J.R. Wooten; MIT Press, Cambridge, Mass. (1959).
Quantum theory of angular momentum: irreducible tensors, spherical harmonics, vector coupling coefficients, 3 j-symbols. D.A. Varshalovich, A.N. Moskalev, and V.K. Khersonskii; World Scientific, Singapore (1988).

5 Group theory

The subject of this chapter—the mathematical theory of symmetry—is one of the most remarkable in quantum mechanics. Not only does it simplify calculations, but it also reveals unexpected connections between apparently disparate phenomena. Whole regions of study are brought together in terms of its concepts. Angular momentum is a part of group theory; so too are the properties of the harmonic oscillator. The conservation of energy and of momentum can be discussed in terms of group theory. Group theory is used to classify the fundamental particles, to discuss the selection rules that govern what spectroscopic transitions are allowed, and to formulate molecular orbitals. The subject simply glitters with power and achievements.

What are the capabilities of group theory within quantum chemistry? We shall see that group theory is particularly helpful for deciding whether an integral is zero. Integrals occur throughout quantum chemistry, for they include expectation values, overlap integrals, and matrix elements. It is particularly helpful to know, with minimum effort, whether these integrals are necessarily zero. A limitation of group theory, though, is that it cannot give the magnitude of integrals that it cannot show to be necessarily zero. The values of nonzero integrals typically depend on a variety of fundamental constants, and group theory is silent on them. One particular type of matrix element is the 'transition dipole moment' between two states. This quantity determines the intensities of spectroscopic transitions, and if we know that they are necessarily zero, then we have established a selection rule for the transition. In Chapter 2 we encountered the phenomenon of degeneracy and saw qualitatively at least that it is related to the symmetry of the system; group theory lets us anticipate the occurrence and degree of degeneracy that may exist in a system. Finally, we shall see that group theory, by making use of the full symmetry of a system, provides a very powerful way of constructing and classifying molecular orbitals.

The symmetries of objects

We shall begin by establishing the qualitative aspects of the symmetries of objects. This will enable us to classify molecules according to their symmetry. Once molecules have been classified, many properties follow immediately. Moreover, this is a first step to the mathematical formulation of the theory, from which its full power flows.

5.1 Symmetry operations and elements

An **operation** applied to an object is an act of doing something to it, such as rotating it through some angle. A **symmetry operation** is an operation that leaves an object apparently unchanged. For example, the rotation of a sphere through any angle around its centre leaves it apparently unchanged, and is thus a symmetry operation. The translation of the function $\sin x$ through an interval 2π

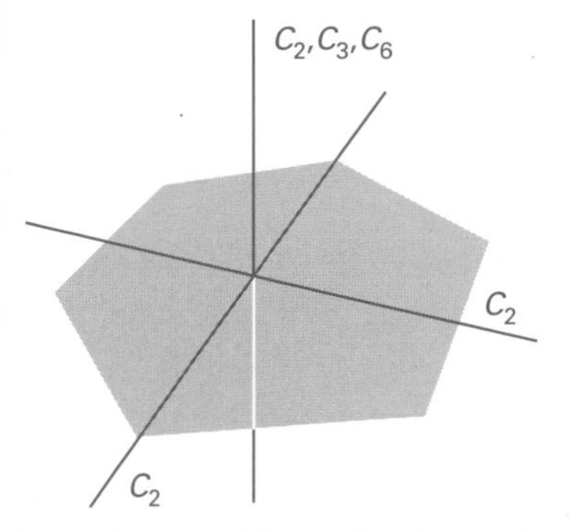

Fig. 5.1 Some of the rotational axes of a regular hexagon, such as a benzene molecule.

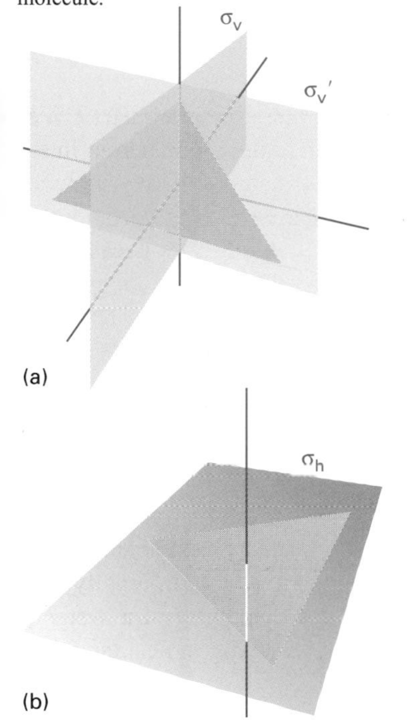

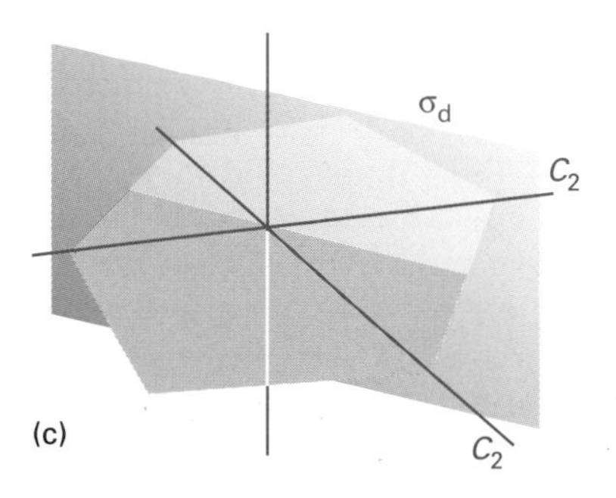

Fig. 5.2 (a) Two vertical mirror planes, (b) a horizontal mirror plane, and (c) a dihedral mirror plane.

leaves it apparently unchanged, and so it is a symmetry operation of the function. Not all operations are symmetry operations. The rotation of a rectangle through 90° is only a symmetry operation if the rectangle happens to be a square. Every object has at least one symmetry operation: the **identity**, the operation of doing nothing.

To each symmetry operation there corresponds a **symmetry element**, the point, line, or plane with respect to which the operation is carried out. For example, a rotation is carried out with respect to a line called an 'axis of symmetry', and a reflection is carried out with respect to a plane called a 'mirror plane'. If we disregard translational symmetry operations, then there are five types of symmetry operations that leave the object apparently unchanged, and five corresponding types of symmetry element:

E The **identity** operation, the act of doing nothing. The corresponding symmetry element is the object itself.

C_n An *n*-**fold rotation**, the operation, a rotation by $2\pi/n$ around an **axis of symmetry**, the element.

A hexagon, or a hexagonal molecule such as benzene, has two-, three-, and six-fold axes (C_2, C_3, and C_6) perpendicular to the plane and several two-fold axes (C_2) in the plane (Fig. 5.1). For $n > 2$ the direction of rotation is significant, and the n orientations of the object are visited in a different order depending on whether the rotation is clockwise as seen from below (C_n^+) or counter-clockwise (C_n^-). Therefore, for $n > 2$, there are two rotations associated with each symmetry axis. If an object (such as a hexagon) has several axes of rotation, then the one with the largest value of n is called the **principal axis**, provided it is unique.

σ A **reflection**, the operation, in a **mirror plane**, the element.

When the mirror plane includes the principal axis of symmetry, it is termed a **vertical plane** and denoted σ_V. If the principal axis is perpendicular to the mirror plane, then the latter symmetry element is called a **horizontal plane** and denoted σ_h. A **dihedral plane**, σ_d, is a vertical plane that bisects the angle between two C_2 axes that lie perpendicular to the principal axis (Fig. 5.2).

i An **inversion**, the operation, through a **centre of symmetry**, the element.

The inversion operation is a hypothetical operation which consists of taking each point of an object through its centre and out to an equal distance on the other side (Fig. 5.3).

S_n An *n*-**fold improper rotation**, the operation (which is also called a 'rotary-reflection') occurs about an **axis of improper rotation**, the symmetry element (or 'rotary-reflection axis').

An improper rotation is a composite operation consisting of an n-fold rotation followed by a horizontal reflection in a plane perpendicular to the n-fold axis. Neither operation alone is in general a symmetry operation, but the overall outcome is. A methane molecule, for example, has three S_4 axes (Fig. 5.4). Care should be taken to recognize improper rotations in disguised form. Thus, S_1 is equivalent to a reflection, and S_2 is equivalent to an inversion.

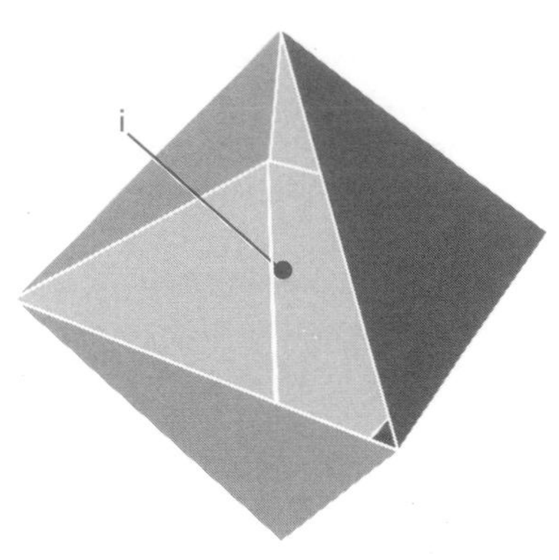

Fig. 5.3 The centre of inversion of a regular octahedron.

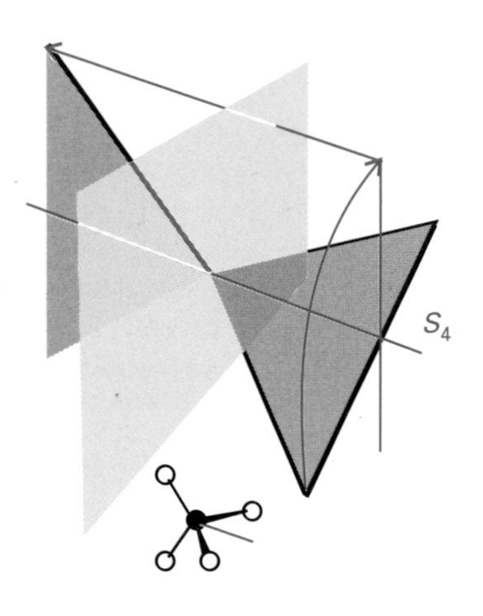

Fig. 5.4 An axis of improper rotation in a tetrahedral molecule (such as methane).

5.2 The classification of molecules

To classify a molecule according to its symmetry, we list all its symmetry operations, and then ascribe a label based on the list of those operations. In other words, we use the list of symmetry operations to identify the **point group** of the molecule. The term 'point' indicates that we are considering only the operations corresponding to symmetry elements that intersect in at least one point. That point is not moved by any operation. To classify crystals, we would also need to consider translational symmetry, which would lead us to classify them according to their **space group**.

The name of the point group is expressed using either the **Schoenflies system** or the **International system** (which is also called the 'Hermann–Mauguin system'). It is common to use the former for individual molecules and the latter when considering species in solids. We shall describe and use the Schoenflies system here, but a translation table is given in Table 5.1. In the

Table 5.1 Schoenflies and International notation for point groups

$C_i : \bar{1}$	$C_s : m$			
$C_1 : 1$	$C_2 : 2$	$C_3 : 3$	$C_4 : 4$	$C_6 : 6$
	$C_{2v} : 2mm$	$C_{3v} : 3m$	$C_{4v} : 4mm$	$C_{6v} : 6mm$
	$C_{2h} : 2/m$	$C_{3h} : \bar{6}$	$C_{4h} : 4/m$	$C_{6h} : 6/m$
	$D_2 : 222$	$D_3 : 32$	$D_4 : 422$	$D_6 : 622$
	$D_{2h} : mmm$	$D_{3h} : \bar{6}2m$	$D_{4h} : 4/mmm$	$D_{6h} : 6/mmm$
	$D_{2d} : \bar{4}2m$	$D_{3d} : \bar{3}m$	$S_4 : \bar{4}$	$S_6 : \bar{3}$
$T : 23$	$T_d : \bar{4}3m$	$T_h : m3$	$O : 432$	$O_h : m3m$

The entries in the table are in the form Schoenflies:International. The International system is also known as the Hermann–Mauguin system. The group $D_2 : 222$ is sometimes denoted V and called the *Vierer group* (the group of four).

Schoenflies system, the name of the point group is based on a dominant feature of the symmetry of the molecule, and the label given to the group is in some cases the same as the label of that feature. This double use of a symbol is actually quite helpful, and rarely leads to confusion.

1. The groups C_1, C_s, and C_i. These groups consist of the identity alone (C_1), the identity and a reflection (C_s), and the identity and an inversion (C_i) (Fig. 5.5).
2. The groups C_n. These groups consist of the identity and an n-fold rotation (Fig. 5.6).
3. The groups C_{nv}. In addition to the operations of the groups C_n, these groups also contain n vertical reflections (Fig. 5.7). An important example is the group $C_{\infty v}$, the group to which a cone and a heteronuclear diatomic molecule belong.
4. The groups C_{nh}. In addition to the operations of the groups C_n, these groups contain a horizontal reflection together with whatever operations the presence of these operations implies (Fig. 5.8).

It is important to note, as remarked in the last definition, that the presence of a particular set of operations may imply the presence of other operations that are

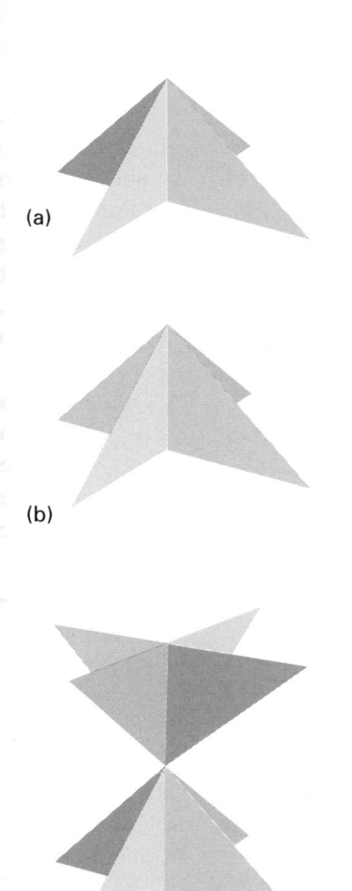

Fig. 5.5 Objects belonging to the groups (a) C_1, (b) C_s, and (c) C_i.

not mentioned explicitly in the definition. For example, C_{2h} automatically possesses an inversion, because rotation by 180° followed by a horizontal reflection is equivalent to inversion. The full set of operations in each group can be found by referring to the tables (the 'character tables') listed in Appendix 1. These tables contain a mass of additional information, and they will gradually move to centre stage as the chapter progresses.

5. The groups D_n. In addition to the operations of the groups C_n, these groups possess n two-fold rotations perpendicular to the n-fold (principal) axis, together with whatever operations the presence of these operations implies (Fig. 5.9).
6. The groups D_{nh}. These groups consist of the operations present in D_n together with a horizontal reflection, together with whatever operations the presence of these operations implies (Fig. 5.10). An important example is $D_{\infty h}$, the group to which a uniform cylinder and a homonuclear diatomic molecule belong.
7. The groups D_{nd}. These groups contain the operations of the groups D_n and n dihedral reflections, together with whatever operations the presence of these operations implies (Fig. 5.11).
8. The groups S_n, with n even. These groups contain the identity and an n-fold improper rotation, together with whatever operations the presence of these operations implies (Fig. 5.12).

Only the even values of n need be considered, because groups with odd n are identical to the groups C_{nh}, which have already been classified. Note also that the group S_2 is equivalent to the group C_i.

9. The cubic and icosahedral groups. These groups contain more than one n-fold rotation with $n \geq 3$. The cubic groups are labelled T (for tetrahedral) and O for octahedral; the icosahedral group is labelled I. The group T_d is the group of the regular tetrahedron; T is the same group but without the reflections of the tetrahedron; T_h is a tetrahedral group with an inversion. The group of the regular octahedron is called O_h; if it lacks reflections it is called O. The group of the regular icosahedron is called I_h; if it lacks inversion it is called I. Some of these groups are summarized in Fig. 5.13, Fig. 5.14, and Fig. 5.15.
10. The full rotation group, R_3. This group consists of all rotations through any angle and in any orientation. It is the symmetry group of the sphere.

Atoms belong to R_3, but no molecule does. The properties of R_3 turn out to be the properties of angular momentum. This is the deep link between this chapter and Chapter 4, and we explore it later.

There are two simple ways of determining to what point group a molecule belongs. One way is to work through the decision tree illustrated in Fig. 5.16. The other is to recognize the group by comparing the molecule with the objects in Fig. 5.17.

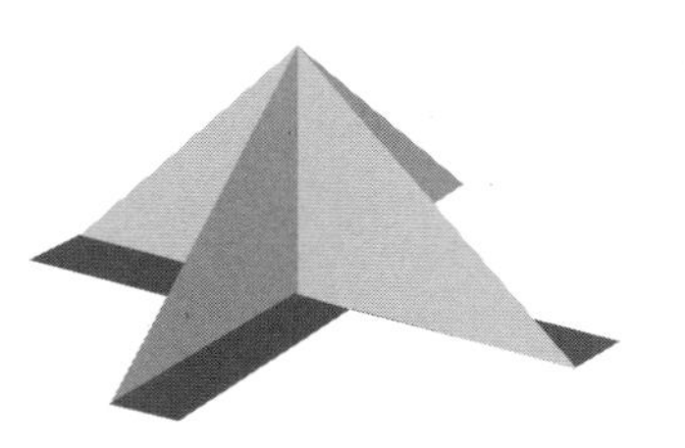

Fig. 5.6 An object belonging to the group C_4. In this and the following illustrations (up to Fig. 5.15), the shading should not be taken into account when considering the symmetry operations.

The calculus of symmetry

Power comes to group theory from its mathematical structure. We shall present the material in two stages. The first considers the symmetry operations them-

Fig. 5.7 An object belonging to the group C_{4v}.

selves, and shows how they may be combined together. The second stage shows how to associate matrices with each symmetry operation and to draw on the properties of matrices to establish several important results.

5.3 The definition of a group

Symmetry operations can be performed consecutively. We shall use the convention that the operation R followed by the operation S is denoted SR. The order of operations is important because in general the outcome of the operation SR is not the same as the outcome of the operation RS. However, a general feature of symmetry operations is that the outcome of a joint symmetry operation is *always* equivalent to a single symmetry operation. We have already seen this property when we saw that a two-fold rotation followed by a reflection in a plane perpendicular to the two-fold axis is equivalent to an inversion:

$$\sigma_h C_2 = i$$

Fig. 5.8 An object belonging to the group C_{4h}.

In general, it is true that for all the symmetry operations of an object, we can write

$$RS = T \tag{1}$$

where T is an operation of the group.

A further point about symmetry operations is that there is no difference between the outcomes of the operations $(RS)T$ and $R(ST)$, where (RS) is the outcome of the joint operation S followed by R and (ST) is the outcome of the joint operation T followed by S. In other words, $(RS)T = R(ST)$ and the multiplication of symmetry operations is **associative**.

These observations, together with two others which are true by inspection, can be summarized as follows:

1. The identity is a symmetry operation.
2. Symmetry operations combine in accord with the associative law of multiplication.
3. If R and S are symmetry operations, then RS is also a symmetry operation.
4. The inverse of each symmetry operation is also a symmetry operation.

The third observation implies that R^2 (which is shorthand for RR) is a symmetry operation. In observation 4, the inverse of an operation R is generally denoted R^{-1} and is defined such that

$$RR^{-1} = R^{-1}R = E \tag{2}$$

The remarkable point to notice is that in mathematics a set of entities called **elements** form a **group** if they satisfy the following conditions:

1. The identity is an element of the set.
2. The elements multiply associatively.
3. If R and S are elements, then RS is also an element of the set.
4. The inverse of each element is a member of the set.

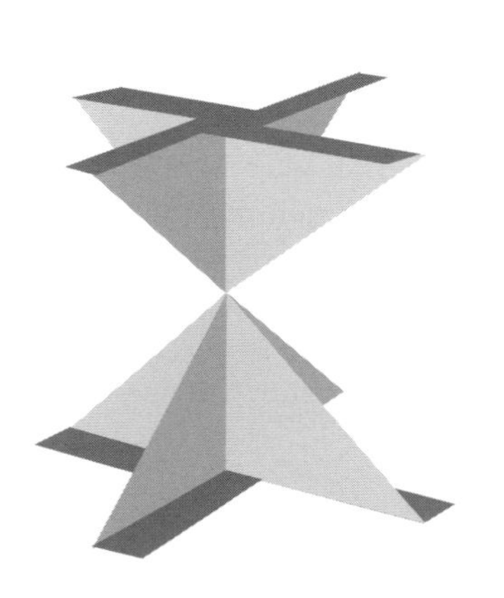

Fig. 5.9 An object belonging to the group D_4.

That is, the set of symmetry operations of an object fulfill conditions that ensure they form a group in the mathematical sense. Consequently, the mathematical theory of groups, which is called **group theory**, may be applied

Fig. 5.10 An object belonging to the group D_{4h}.

Fig. 5.11 An object belonging to the group D_{4d}.

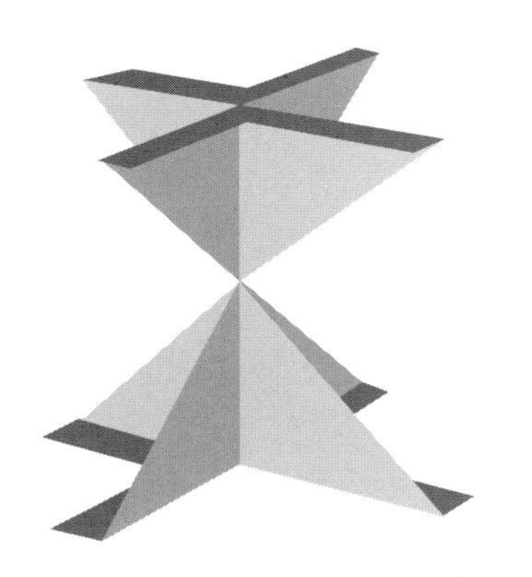

Fig. 5.12 An object belonging to the group S_4.

to the study of the symmetry of molecules. This is the justification for the title of this chapter.[1]

5.4 Group multiplication tables

A table showing the outcome of forming the products RS for all symmetry operations in a group is called a **group multiplication table**. The procedure used to construct such tables can be illustrated by the group C_{3v}. The symmetry operations for this group are illustrated in Fig. 5.18. We see that there are six members of the group, so it is said to have **order** 6, which we write $h = 6$.

To determine the outcome of a sequence of symmetry operations, we consider diagrams like those in Fig. 5.19. You should note that the sequence of changes takes place against a constant background of symmetry elements, in the sense that if a C_3^+ operation is performed, the line representing the σ_v plane remains in the same position on the page and is not rotated through 120° by the C_3^+ operation. Thus it follows that

$$C_3^- C_3^+ = E \qquad \sigma_v C_3^+ = \sigma_v'' \qquad \sigma_v' \sigma_v = C_3^+$$

The complete set of 36 (in general, h^2) products is shown in Table 5.2. As can be seen, each product is equivalent to a *single* element of the group. Note that RS is not always the same as SR. Similar tables can be constructed for all the point groups.

Table 5.2 The C_{3v} group multiplication table

First:	E	C_3^+	C_3^-	σ_v	σ_v'	σ_v''
Second						
E	E	C_3^+	C_3^-	σ_v	σ_v'	σ_v''
C_3^+	C_3^+	C_3^-	E	σ_v'	σ_v''	σ_v
C_3^-	C_3^-	E	C_3^+	σ_v''	σ_v	σ_v'
σ_v	σ_v	σ_v''	σ_v'	E	C_3^-	C_3^+
σ_v'	σ_v'	σ_v	σ_v''	C_3^+	E	C_3^-
σ_v''	σ_v''	σ_v'	σ_v	C_3^-	C_3^+	E

Example 5.1 How to construct a group multiplication table

Construct the group multiplication table for the group C_{2v}, the elements of which are shown in Fig. 5.20.

Method. Consider a single point, and the effect on the point of each pair of symmetry operations (RS). Identify the single operation that reproduces the effect of the joint application ($RS = T$), and enter it into the table. Note that $ER = R$ for all R, where E is the identity operation. The orientation on the page of the symmetry elements is unchanged by all the operations.

[1] The unfortunate double meaning of the term 'element' should be noted. It is important to distinguish 'element', in the sense of a member of a group, from 'symmetry element', as defined earlier. The symmetry operations are the elements that comprise the group.

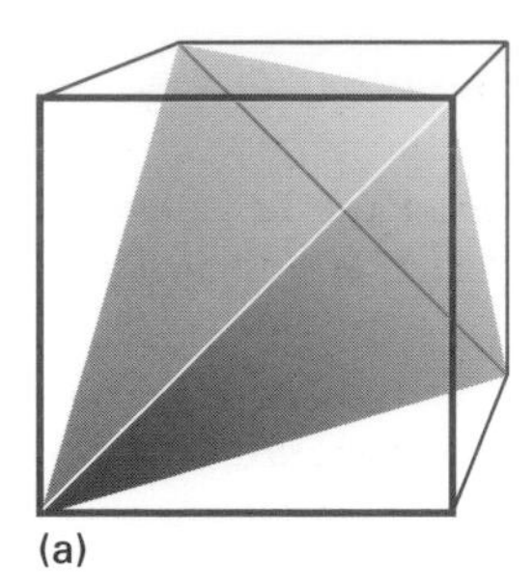

(a)

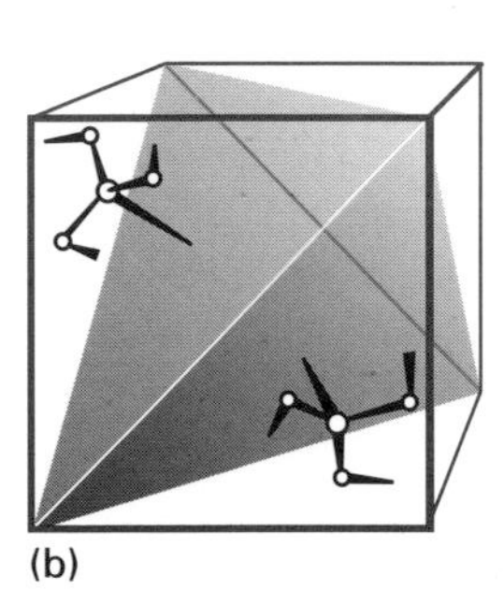

(b)

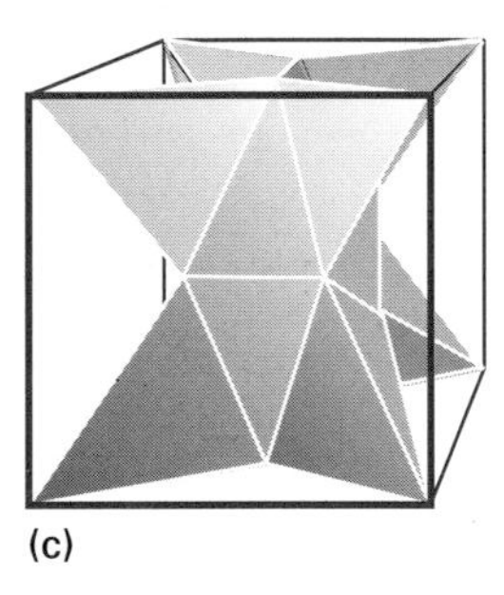

(c)

Fig. 5.13 Objects belonging to the groups (a) T_d, (b) T, and (c) T_h.

Answer. The group multiplication table is as follows:

	E	C_2	σ_v	σ'_v
E	E	C_2	σ_v	σ'_v
$\mathbf{C_2}$	C_2	E	σ'_v	σ_v
$\boldsymbol{\sigma_v}$	σ_v	σ'_v	E	C_2
$\boldsymbol{\sigma'_v}$	σ'_v	σ_v	C_2	E

Comment. Note that in this group $RS = SR$ for all entries in the table. Groups of this kind, in which the elements commute, are called 'Abelian'. The group C_{3v} is an example of a 'non-Abelian group'.

Exercise 5.1. Construct the group multiplication table for the group D_3, with the elements shown in Fig. 5.21.

5.5 Matrix representations

Relations such as $RS = T$ are symbolic summaries of the effect of actions carried out on objects. We can enrich this symbolic representation of symmetry operations by representing the operations by entities that can be manipulated just like ordinary algebra. However, because symmetry operations are in general non-commutative (that is, their outcome depends on the order in which they are applied), we should expect to need to use matrices rather than simple numbers, for matrix multiplication is also non-commutative in general. The **matrix representative** of a symmetry operation is a matrix that reproduces the effect of the symmetry operation (in a manner we describe below). A **matrix representation** is a set of representatives, one for each element of the group, which multiply together as summarized by the group multiplication table.

To set up a matrix representative for a particular operation of a group, we need to choose a **basis**, a set of functions on which the operation takes place. To illustrate the procedure, we shall consider the set of *s*-orbitals s_A, s_B, s_C, and s_N on an NH_3 molecule (Fig. 5.22), which belongs to the group C_{3v}. We have chosen this basis partly because it is simple enough to illustrate a number of points in a straightforward fashion but also because it will be used in the discussion of the electronic structure of an ammonia molecule when we construct molecular orbitals in Chapter 8. The **dimension** of this basis, the number of members, is 4. We can write the basis as a four-component vector (s_N, s_A, s_B, s_C). In general, a basis of dimension d can be written as the vector **f**, where

$$\mathbf{f} = (f_1, f_2, \ldots, f_d)$$

Under the operation σ_v, the vector changes from (s_N, s_A, s_B, s_C) to $\sigma_v(s_N, s_A, s_B, s_C) = (s_N, s_A, s_C, s_B)$. This transformation can be represented by a matrix multiplication:

$$\sigma_v(s_N, s_A, s_B, s_C) = (s_N, s_A, s_B, s_C)\begin{bmatrix} 1 & 0 & 0 & 0 \\ 0 & 1 & 0 & 0 \\ 0 & 0 & 0 & 1 \\ 0 & 0 & 1 & 0 \end{bmatrix} \quad (3)$$

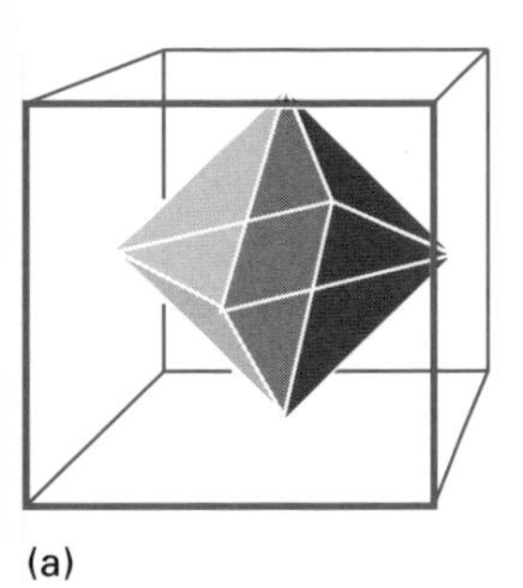

(a)

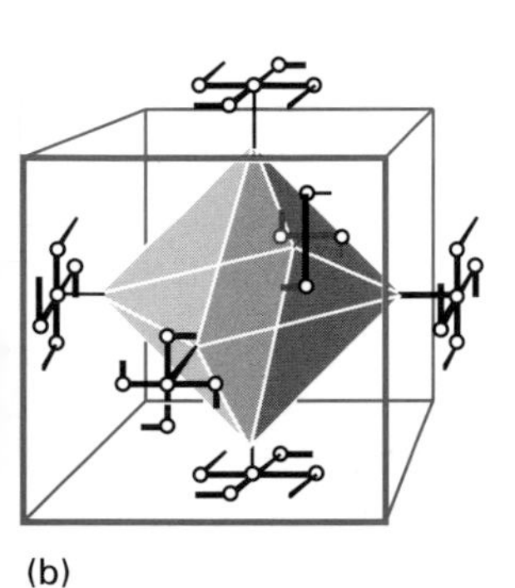

(b)

Fig. 5.14 Objects belonging to the groups (a) O_h and (b) O.

This portrayal of the effect of the symmetry operation can be verified by carrying out the matrix multiplication. (For information on matrices, see *Further information 23*.) The matrix in this expression is the representative of the operation σ_v for the chosen basis, and is denoted $\mathbf{D}(\sigma_v)$. Note that a four-dimensional basis gives rise to a 4×4-dimensional representative, and that in general a d-dimensional basis gives rise to a $d \times d$-dimensional representative. In terms of the explicit rules for matrix multiplication, the effect of an operation R on the general basis $\mathbf{f}$ is to convert the component f_i into

$$Rf_i = \sum_j f_j D_{ji}(R) \tag{4}$$

where $\mathbf{D}(\mathrm{R})$ is the representative of the operation R. For example,

$$\sigma_v s_B = s_N \times 0 + s_A \times 0 + s_B \times 0 + s_C \times 1 = s_C$$

The representatives of the other operations of the group can be found in the same way. Note that because $E\mathbf{f} = \mathbf{f}$, the representative of the identity operation is always the unit matrix.

Example 5.2 How to formulate a matrix representative

Find the matrix representative for the operation C_3^+ in the group C_{3v} for the s-orbital basis used above.

Method. Examine Fig. 5.22 to decide how each member of the basis is transformed under the operation, and write it in the form $R\mathbf{f} = \mathbf{f}'$. Then construct a $d \times d$ matrix $\mathbf{D}(R)$ which generates $\mathbf{f}'$ when $\mathbf{fD}(R)$ is formed and multiplied out.

Answer. Inspection of the illustration shows that under the operation,

$$C_3^+(s_N, s_A, s_B, s_C) = (s_N, s_B, s_C, s_A)$$

This transformation can be expressed as the matrix product

$$C_3^+(s_N, s_A, s_B, s_C) = (s_N, s_A, s_B, s_C)\begin{bmatrix} 1 & 0 & 0 & 0 \\ 0 & 0 & 0 & 1 \\ 0 & 1 & 0 & 0 \\ 0 & 0 & 1 & 0 \end{bmatrix}$$

Therefore, the 4×4 matrix in this expression is the representative of the operation C_3^+ in the basis.

Exercise 5.2. Find the matrix representative of the operation C_3^- in the same basis.

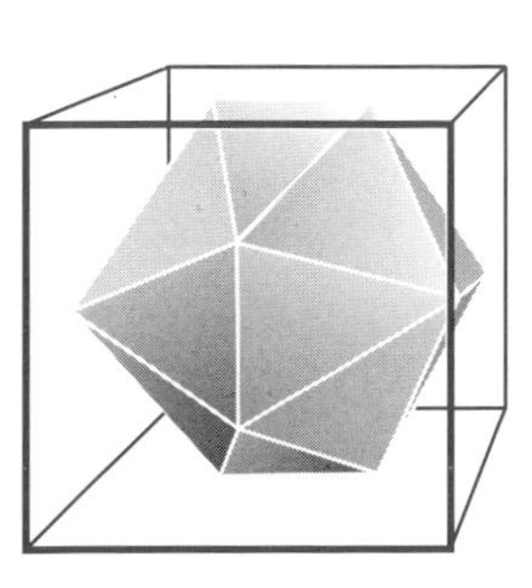

Fig. 5.15 An object belonging to the group I.

The complete set of representatives for this basis are displayed in Table 5.3.

We now arrive at a centrally important point. Consider the effect of the consecutive operations C_3^+ followed by σ_v. From the group multiplication table we know that the effect of the joint operation $\sigma_v C_3^+$ is the same as the effect of the reflection σ_v''. That is,

$$\sigma_v C_3^+ = \sigma_v''$$

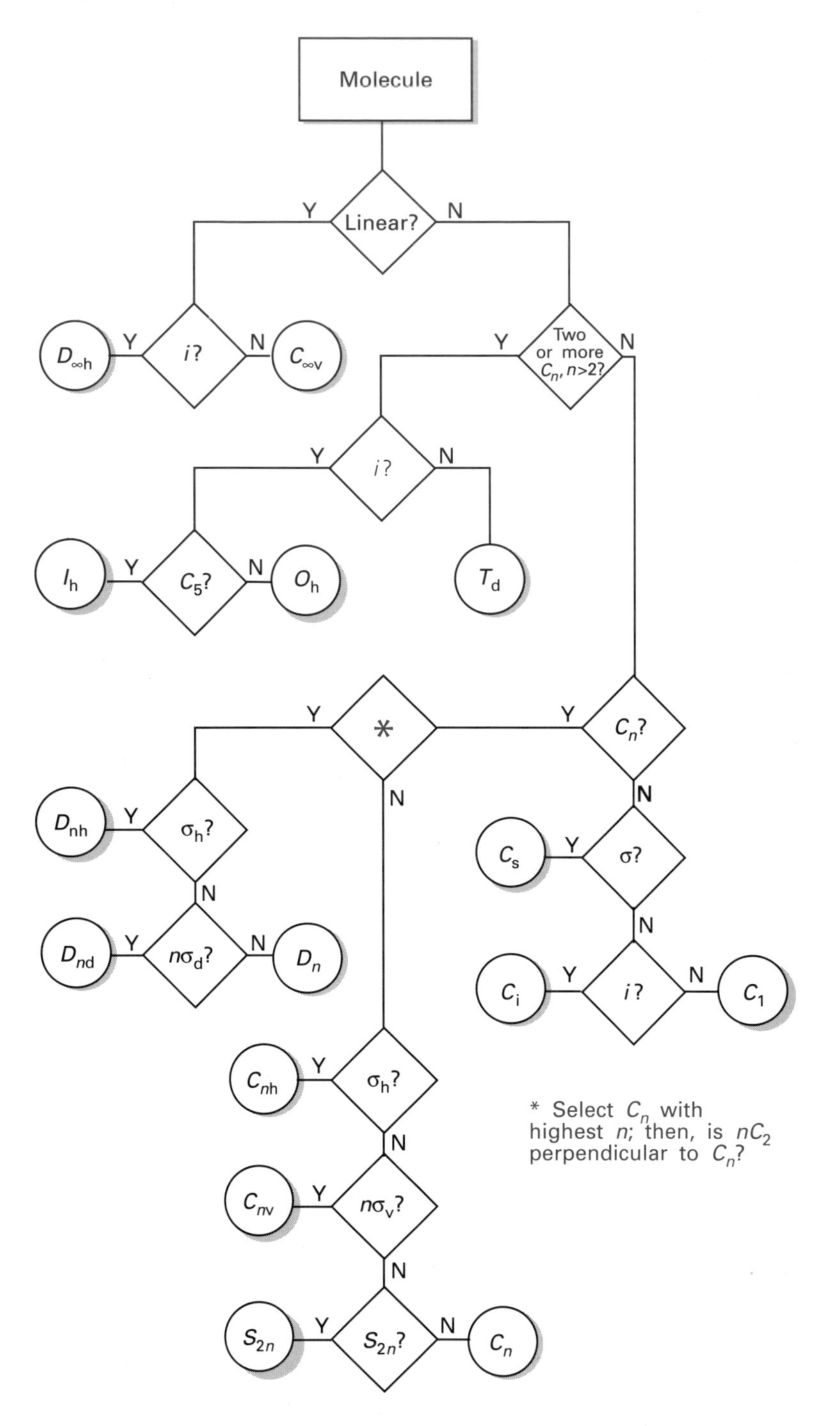

Fig. 5.16 A flow chart for deciding on the name of a point group to which an object belongs.

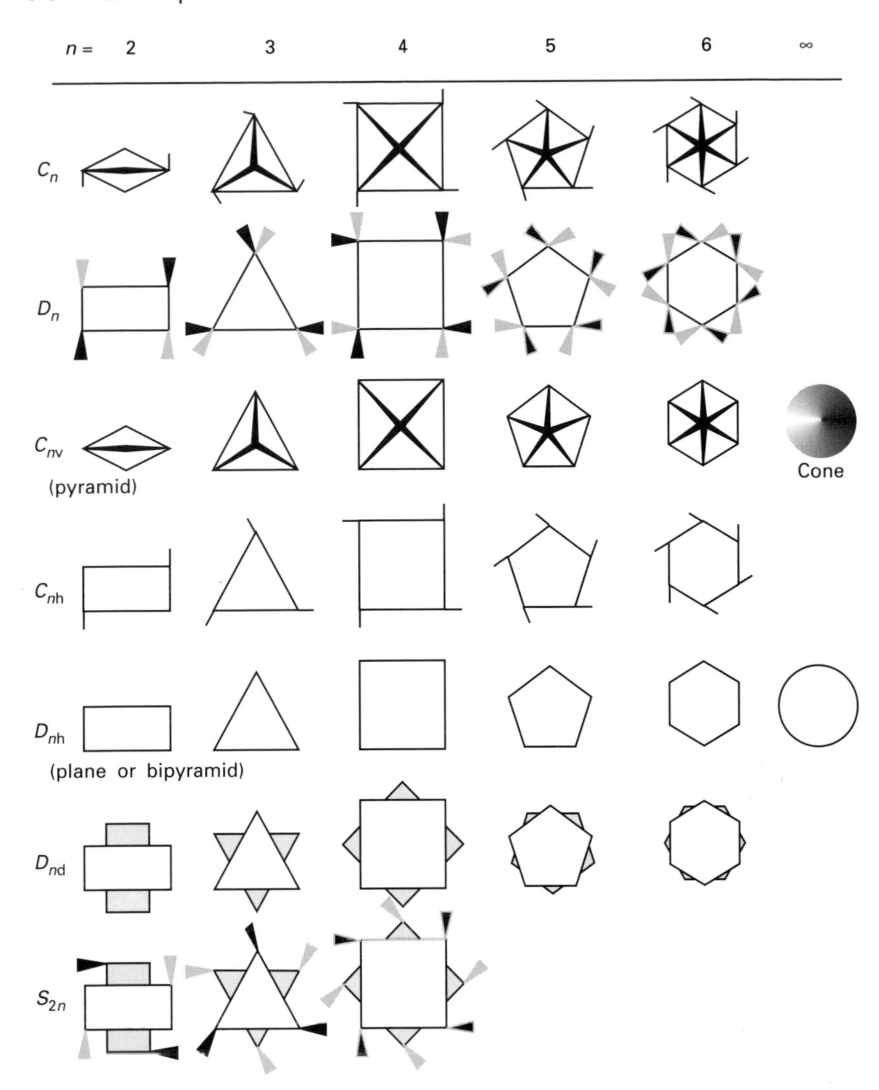

Fig. 5.17 Representative shapes for a variety of point groups.

Table 5.3 The matrix representation of C_{3v} in the basis (s_N, s_A, s_B, s_C)

$$\mathbf{D}(E)=\begin{bmatrix}1&0&0&0\\0&1&0&0\\0&0&1&0\\0&0&0&1\end{bmatrix}\quad \mathbf{D}(C_3^+)=\begin{bmatrix}1&0&0&0\\0&0&0&1\\0&1&0&0\\0&0&1&0\end{bmatrix}\quad \mathbf{D}(C_3^-)=\begin{bmatrix}1&0&0&0\\0&0&1&0\\0&0&0&1\\0&1&0&0\end{bmatrix}$$

$$\chi(E)=4 \qquad \chi(C_3^+)=1 \qquad \chi(C_3^-)=1$$

$$\mathbf{D}(\sigma_v)=\begin{bmatrix}1&0&0&0\\0&1&0&0\\0&0&0&1\\0&0&1&0\end{bmatrix}\quad \mathbf{D}(\sigma_v')=\begin{bmatrix}1&0&0&0\\0&0&1&0\\0&1&0&0\\0&0&0&1\end{bmatrix}\quad \mathbf{D}(\sigma_v'')=\begin{bmatrix}1&0&0&0\\0&0&0&1\\0&0&1&0\\0&1&0&0\end{bmatrix}$$

$$\chi(\sigma_v)=2 \qquad \chi(\sigma_v')=2 \qquad \chi(\sigma_v'')=2$$

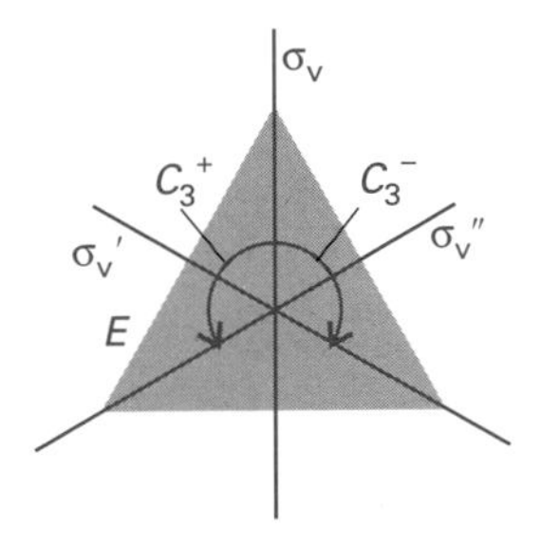

Fig. 5.18 The symmetry elements of the group C_{3v}.

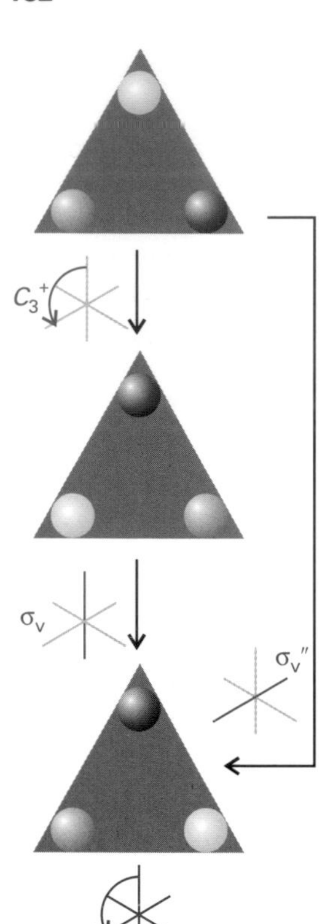

Fig. 5.19 The effect of the operation C_3^+ followed by σ_v is equivalent to the single operation σ_v''.

Now consider this joint operation in terms of the matrix representatives.

$$\mathbf{D}(\sigma_v)\mathbf{D}(C_3^+) = \begin{bmatrix} 1 & 0 & 0 & 0 \\ 0 & 1 & 0 & 0 \\ 0 & 0 & 0 & 1 \\ 0 & 0 & 1 & 0 \end{bmatrix} \begin{bmatrix} 1 & 0 & 0 & 0 \\ 0 & 0 & 0 & 1 \\ 0 & 1 & 0 & 0 \\ 0 & 0 & 1 & 0 \end{bmatrix} = \begin{bmatrix} 1 & 0 & 0 & 0 \\ 0 & 0 & 0 & 1 \\ 0 & 0 & 1 & 0 \\ 0 & 1 & 0 & 0 \end{bmatrix}$$
$$= \mathbf{D}(\sigma_v'')$$

That is, the matrix representatives multiply together in exactly the same way as the operations of the group. This is true whichever operations are considered, and so the set of six 4×4 matrices in Table 5.3 form a matrix representation of the group for the selected basis in the sense that

$$\text{if } RS = T, \text{ then } \mathbf{D}(R)\mathbf{D}(S) = \mathbf{D}(T) \qquad (5)$$

for all members of the group.

The formal proof that the representatives multiply in the same way as the symmetry operations gives a taste of the kind of manipulation that will be needed later. Once again we consider two elements R and S which multiply together to give the element T. It follows from eqn 5.4 that for the general basis $\mathbf{f}$,

$$RSf_i = R\sum_j f_j D_{ji}(S) = \sum_{j,k} f_k D_{kj}(R) D_{ji}(S)$$

The sum over j of $D_{kj}(R)D_{ji}(S)$ is the definition of a matrix product, and so

$$RSf_i = \sum_k f_k \{D(R)D(S)\}_{ki}$$

However, we also know that $RS = T$, so we can also write

$$RSf_i = Tf_i = \sum_k f_k \{D(T)\}_{ki}$$

By comparing the two equations we see that

$$\{D(R)D(S)\}_{ki} = \{D(T)\}_{ki}$$

for all elements k and i. Therefore,

$$\mathbf{D}(R)\mathbf{D}(S) = \mathbf{D}(T)$$

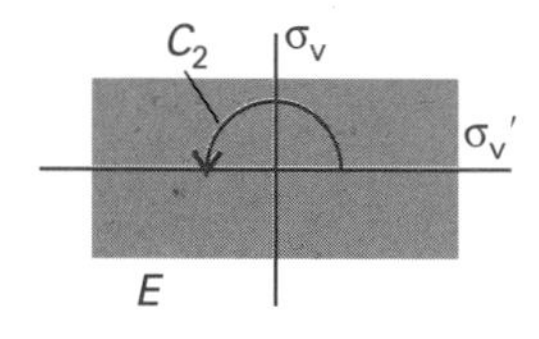

Fig. 5.20 The symmetry elements of the group C_{2v}.

This shows that the representatives do indeed multiply like the group elements, as we set out to prove. In particular, the representatives of an operation R and its inverse R^{-1} are related by

$$\mathbf{D}(R^{-1}) = \mathbf{D}(R)^{-1} \qquad (6)$$

where $\mathbf{D}^{-1}$ denotes the inverse of the matrix $\mathbf{D}$. For instance, because $RR^{-1} = E$, it follows that

$$\mathbf{D}(R)\mathbf{D}(R^{-1}) = \mathbf{D}(R)\mathbf{D}(R)^{-1} = \mathbf{1} = \mathbf{D}(E)$$

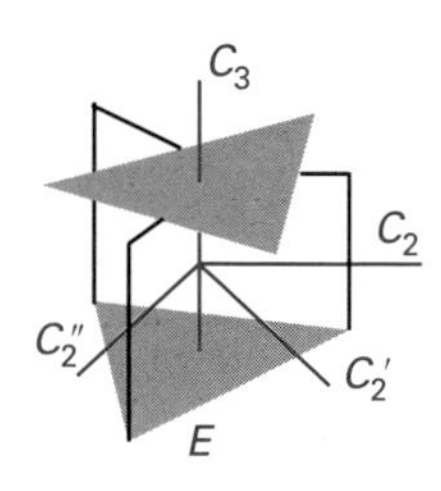

Fig. 5.21 The symmetry elements of the group D_3.

5.6 The properties of matrix representations

To develop the content of matrix representations, we need to introduce some of their properties. In each case we shall introduce the concept using the s-orbital

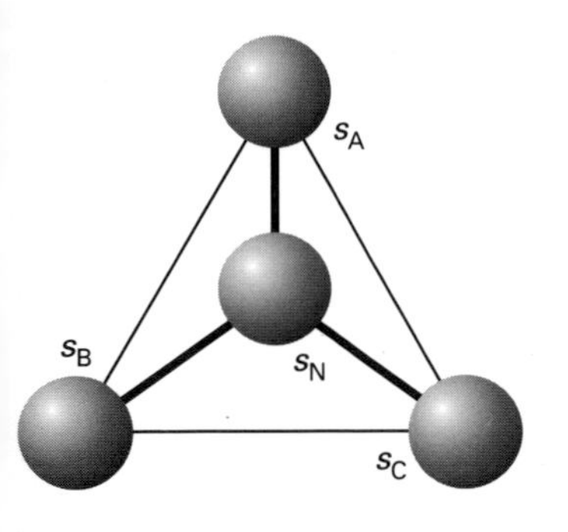

Fig. 5.22 One basis for a discussion of the representation of the group C_{3v}; each sphere can be regarded as an s-orbital centred on an atom.

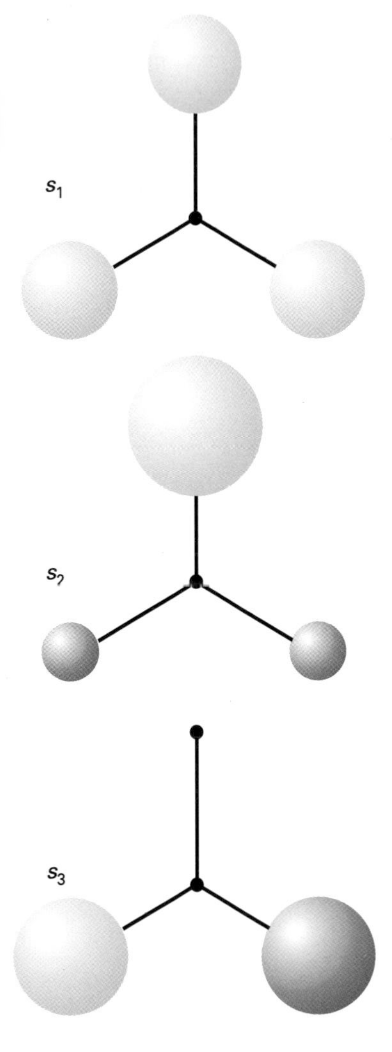

Fig. 5.23 The symmetry-adapted linear combinations of the peripheral atom orbitals in a C_{3v} molecule.

basis for C_{3v} to fix our ideas, and then we shall generalize the concept to any basis for any group.

To begin, we introduce the concept of 'similarity transformation'. Suppose that instead of the s-orbital basis, we select a linear combination of these orbitals. One such set might be (s_N, s_1, s_2, s_3), where $s_1 = s_A + s_B + s_C$, $s_2 = 2s_A - s_B - s_C$, and $s_3 = s_B - s_C$ (apart from the requirement that the combinations are linearly independent, the choice is arbitrary, but later we shall see that this set has a special significance). The combinations are illustrated in Fig. 5.23. We should expect the matrix representation in this basis to be similar to that in the original basis. This similarity can be given a formal definition as follows.

Because the new basis $\mathbf{f}' = (f_1', f_2', \ldots, f_d')$ is a linear combination of the original basis $\mathbf{f} = (f_1, f_2, \ldots, f_d)$, we can express any member as

$$f_i' = \sum_j f_j c_{ji} \tag{7}$$

where the c_{ji} are coefficients. This expansion can be expressed as a matrix product by writing

$$\mathbf{f}' = \mathbf{fc} \tag{8}$$

where $\mathbf{c}$ is the matrix formed of the coefficients c_{ji}. Now suppose that in the original basis the representative of the element R is $\mathbf{D}(R)$ in the sense that

$$Rf_i = \sum_k f_k D_{ki}(R), \qquad \text{or } R\mathbf{f} = \mathbf{fD}(R) \tag{9}$$

Likewise, the effect of the same operation on a member of the transformed basis set is

$$Rf_i' = \sum_k f_k' D_{ki}'(R), \qquad \text{or } R\mathbf{f}' = \mathbf{f}'\mathbf{D}'(R)$$

The relation between the two 'similar' representatives can be found by substituting $\mathbf{f}' = \mathbf{fc}$ into the last equation, which then becomes

$$R\mathbf{fc} = \mathbf{fcD}'(R)$$

If we then multiply through from the right by $\mathbf{c}^{-1}$, the reciprocal of the matrix $\mathbf{c}$ (in the sense that $\mathbf{cc}^{-1} = \mathbf{c}^{-1}\mathbf{c} = \mathbf{1}$), then we obtain

$$R\mathbf{f} = \mathbf{fcD}'(R)\mathbf{c}^{-1}$$

Comparison of this expression with eqn 5.9 shows that the representatives for the two bases are related by the **similarity transformation**

$$\mathbf{D}(R) = \mathbf{cD}'(R)\mathbf{c}^{-1} \tag{10}$$

The inverse relation is obtained by multiplication from the left by $\mathbf{c}^{-1}$ and from the right by $\mathbf{c}$:

$$\mathbf{D}'(R) = \mathbf{c}^{-1}\mathbf{D}(R)\mathbf{c} \tag{11}$$

Example 5.3 How to construct a similarity transformation

The representative of the operation C_3^+ in C_{3v} for the s-orbital basis is given in Table 5.3. Derive an expression for the representative in the transformed basis given at the start of this subsection.

Method. To implement the recipe in eqn 5.11, we need to construct the matrices $\mathbf{c}$ and $\mathbf{c}^{-1}$. Therefore, begin by expressing the relation between the two bases in matrix form (as $\mathbf{f}' = \mathbf{fc}$), and find the reciprocal of $\mathbf{c}$ by the methods described in *Further information 23*. Finally, evaluate the matrix product $\mathbf{c}^{-1}\mathbf{D}(R)\mathbf{c}$.

Answer. The relation between the two bases,

$$\begin{aligned} s_N &= s_N \\ s_1 &= s_A + s_B + s_C \\ s_2 &= 2s_A - s_B - s_C \\ s_3 &= s_B - s_C \end{aligned}$$

can be expressed as the following matrix relation:

$$(s_N, s_1, s_2, s_3) = (s_N, s_A, s_B, s_C)\begin{bmatrix} 1 & 0 & 0 & 0 \\ 0 & 1 & 2 & 0 \\ 0 & 1 & -1 & 1 \\ 0 & 1 & -1 & -1 \end{bmatrix}$$

which lets us identify the matrix $\mathbf{c}$. The reciprocal of this matrix is

$$\mathbf{c}^{-1} = \frac{1}{6}\begin{bmatrix} 6 & 0 & 0 & 0 \\ 0 & 2 & 2 & 2 \\ 0 & 2 & -1 & -1 \\ 0 & 0 & 3 & -3 \end{bmatrix}$$

The representative of C_3^+ in the new basis is therefore

$$\begin{aligned} \mathbf{D}'(C_3^+) &= \mathbf{c}^{-1}\mathbf{D}(C_3^+)\mathbf{c} \\ &= \frac{1}{6}\begin{bmatrix} 6 & 0 & 0 & 0 \\ 0 & 2 & 2 & 2 \\ 0 & 2 & -1 & -1 \\ 0 & 0 & 3 & -3 \end{bmatrix}\begin{bmatrix} 1 & 0 & 0 & 0 \\ 0 & 0 & 0 & 1 \\ 0 & 1 & 0 & 0 \\ 0 & 0 & 1 & 0 \end{bmatrix}\begin{bmatrix} 1 & 0 & 0 & 0 \\ 0 & 1 & 2 & 0 \\ 0 & 1 & -1 & 1 \\ 0 & 1 & -1 & -1 \end{bmatrix} \\ &= \frac{1}{6}\begin{bmatrix} 6 & 0 & 0 & 0 \\ 0 & 6 & 0 & 0 \\ 0 & 0 & -3 & -3 \\ 0 & 0 & 9 & -3 \end{bmatrix} = \begin{bmatrix} 1 & 0 & 0 & 0 \\ 0 & 1 & 0 & 0 \\ 0 & 0 & -\frac{1}{2} & -\frac{1}{2} \\ 0 & 0 & \frac{3}{2} & -\frac{1}{2} \end{bmatrix} \end{aligned}$$

Exercise 5.3. Find the representative for the operation σ_v in the transformed basis. [See Table 5.4]

The same technique as that illustrated in the example may be applied to the other representatives, and the results are collected in Table 5.4.

Table 5.4 The matrix representation of C_{3v} in the basis (s_N, s_1, s_2, s_3)

$\mathbf{D}(E)$	$\mathbf{D}(C_3^+)$	$\mathbf{D}(C_3^-)$
$\begin{bmatrix} 1 & 0 & 0 & 0 \\ 0 & 1 & 0 & 0 \\ 0 & 0 & 1 & 0 \\ 0 & 0 & 0 & 1 \end{bmatrix}$	$\begin{bmatrix} 1 & 0 & 0 & 0 \\ 0 & 1 & 0 & 0 \\ 0 & 0 & -\frac{1}{2} & -\frac{1}{2} \\ 0 & 0 & \frac{3}{2} & -\frac{1}{2} \end{bmatrix}$	$\begin{bmatrix} 1 & 0 & 0 & 0 \\ 0 & 1 & 0 & 0 \\ 0 & 0 & -\frac{1}{2} & \frac{1}{2} \\ 0 & 0 & -\frac{3}{2} & -\frac{1}{2} \end{bmatrix}$
$\chi(E) = 4$	$\chi(C_3^+) = 1$	$\chi(C_3^-) = 1$
$\mathbf{D}(\sigma_v)$	$\mathbf{D}(\sigma_v')$	$\mathbf{D}(\sigma_v'')$
$\begin{bmatrix} 1 & 0 & 0 & 0 \\ 0 & 1 & 0 & 0 \\ 0 & 0 & 1 & 0 \\ 0 & 0 & 0 & -1 \end{bmatrix}$	$\begin{bmatrix} 1 & 0 & 0 & 0 \\ 0 & 1 & 0 & 0 \\ 0 & 0 & -\frac{1}{2} & \frac{1}{2} \\ 0 & 0 & \frac{3}{2} & \frac{1}{2} \end{bmatrix}$	$\begin{bmatrix} 1 & 0 & 0 & 0 \\ 0 & 1 & 0 & 0 \\ 0 & 0 & -\frac{1}{2} & -\frac{1}{2} \\ 0 & 0 & -\frac{3}{2} & \frac{1}{2} \end{bmatrix}$
$\chi(\sigma_v) = 2$	$\chi(\sigma_v') = 2$	$\chi(\sigma_v'') = 2$

5.7 The characters of representations

There is one striking feature of the two representations in Tables 5.3 and 5.4. Although the matrices differ for the two bases, for a given operation the sum of the diagonal elements of the representative is the same in the two bases. The diagonal sum of matrix elements is called the **character** of the matrix, and is denoted by the symbol $\chi(R)$. We see, therefore, that *the character of an operation is invariant under a similarity transformation of the basis.*

The definition of the character of an operation in a basis is formally

$$\chi(R) = \sum_i D_{ii}(R) \tag{12}$$

In matrix algebra, the sum of diagonal elements is called the **trace** of the matrix, and denoted tr. So, a succinct definition of the character of the operation R is

$$\chi(R) = \operatorname{tr} \mathbf{D}(R) \tag{13}$$

The general proof that the character of an operation is invariant under a similarity transformation makes use of the fact that the trace of a product of matrices is invariant under cyclic permutation of their order:

$$\operatorname{tr} \mathbf{ABC} = \operatorname{tr} \mathbf{CAB} = \operatorname{tr} \mathbf{BCA} \tag{14}$$

The proof of this invariance (which we shall use several times in the following discussion) runs as follows. First, we express the trace as a diagonal sum:

$$\operatorname{tr} \mathbf{ABC} = \sum_i (ABC)_{ii}$$

Then we expand the matrix product by the rules of matrix multiplication:

$$\operatorname{tr} \mathbf{ABC} = \sum_{ijk} A_{ij} B_{jk} C_{ki}$$

Matrix elements are simple numbers that may be multiplied in any order. If they are permuted cyclically in this expression, neighbouring subscripts continue to match, and so the matrix product may be reformulated with the matrices in a permuted order:

$$\operatorname{tr} \mathbf{ABC} = \sum_{ijk} B_{jk} C_{ki} A_{ij} = \sum_j (BCA)_{jj} = \operatorname{tr} \mathbf{BCA}$$

as required.

Now we apply this general result to establish the invariance of the characte under a similarity transformation brought about by the matrix **c**:

$$\begin{aligned}\chi(R) &= \operatorname{tr}\mathbf{D}(R) = \operatorname{tr}\mathbf{c}\mathbf{D}'(R)\mathbf{c}^{-1} \\ &= \operatorname{tr}\mathbf{D}'(R)\mathbf{c}^{-1}\mathbf{c} = \operatorname{tr}\mathbf{D}'(R) = \chi'(R)\end{aligned} \quad (15$$

That is, the characters of R in the two representations, $\chi(R)$ and $\chi'(R)$, are equal as we wanted to prove.

5.8 Characters and classes

One feature of the characters shown in Tables 5.3 and 5.4 is that the character of the two rotations are the same, as are the characters of the three reflections This suggests that the operations fall into various classes that can be distin guished by their characters.

The formal definition of the **class** of a symmetry operation is that two opera tions R and R' belong to the same class if there is some symmetry operation S o the group such that

$$R' = S^{-1}RS \quad (16$$

The elements R and R' are said to be **conjugate**. Conjugate members belong t the same class. The physical interpretation of conjugacy and membership of a class is that R and R' are the same kind of operation (such as a rotation) bu performed with respect to symmetry elements that are related by a symmetr operation.

Example 5.4 How to show that two symmetry operations are conjugate

Show that the symmetry operations C_3^+ and C_3^- are conjugate in the group $C_{3\text{v}}$.

Method. We need to show that there is a symmetry transformation of the group that transforms C_3^+ into C_3^-. Intuitively, we know that the reflection of a rotation in a vertical plane reverses the sense of the rotation, so we can suspect that a reflection is the necessary operation. To work out the effect of a succession of operations, we use the information in the group multiplication table (Table 5.2); to find the reciprocal of an operation, we look for the element that produces the identity E in the group multiplication table.

Answer. We consider the joint operation $\sigma_\text{v}^{-1}C_3^+\sigma_\text{v}$. According to Table 5.2, the reciprocal of σ_v is σ_v itself. Therefore, from the group multiplication table we can write

$$\sigma_\text{v}^{-1}C_3^+\sigma_\text{v} = \sigma_\text{v}(C_3^+\sigma_\text{v}) = \sigma_\text{v}\sigma_\text{v}' = C_3^-$$

Hence, the two rotations belong to the same class.

Exercise 5.4. Show that σ_v and σ_v' are members of the same class in $C_{3\text{v}}$.

With the concept of conjugacy established, it is now straightforward to demonstrate that symmetry operations in the same class have the same character in a given representation. To do so, we use the cyclic invariance of the trace of the product of representatives (eqn 5.14):

$$\begin{aligned}\chi(R') &= \operatorname{tr}\mathbf{D}(R') = \operatorname{tr}\mathbf{D}^{-1}(S)\mathbf{D}(R)\mathbf{D}(S)\\ &= \operatorname{tr}\mathbf{D}(R)\mathbf{D}(S)\mathbf{D}^{-1}(S) = \operatorname{tr}\mathbf{D}(R)\\ &= \chi(R)\end{aligned}$$

A word of warning: although it is true that all members of the same class have the same character in a given representation, the characters of different classes may be the same as one another. For example, as we shall see, one matrix representation of a group consists of 1×1 matrices each with the single element 1. Such a representation certainly reproduces the group multiplication table, but does so in a trivial way, and hence is called the **unfaithful representation** of the group. We shall see later that this representation is in fact one of the most important of all possible representations. The characters of all the operations of the group are 1 in the unfaithful representation, and although it is true that members of the same class have the same character (1 in each case), different classes also share that character.

5.9 Irreducible representations

Inspection of the representation of the group C_{3v} in Table 5.3 for the original s-orbital basis shows that all the matrices have a **block-diagonal form**:

$$\begin{bmatrix} 1 & 0 & 0 & 0 \\ 0 & \blacksquare & \blacksquare & \blacksquare \\ 0 & \blacksquare & \blacksquare & \blacksquare \\ 0 & \blacksquare & \blacksquare & \blacksquare \end{bmatrix}$$

As a consequence, we see that the original four-dimensional basis may be broken into two, one consisting of s_N alone and the other of the three-dimensional basis (s_A, s_B, s_C):

$$\begin{array}{ccc} E & C_3^+ & C_3^- \\ 1 & 1 & 1 \\ \begin{bmatrix} 1 & 0 & 0 \\ 0 & 1 & 0 \\ 0 & 0 & 1 \end{bmatrix} & \begin{bmatrix} 0 & 0 & 1 \\ 1 & 0 & 0 \\ 0 & 1 & 0 \end{bmatrix} & \begin{bmatrix} 0 & 1 & 0 \\ 0 & 0 & 1 \\ 1 & 0 & 0 \end{bmatrix} \\ \sigma_v & \sigma_v' & \sigma_v'' \\ 1 & 1 & 1 \\ \begin{bmatrix} 1 & 0 & 0 \\ 0 & 0 & 1 \\ 0 & 1 & 0 \end{bmatrix} & \begin{bmatrix} 0 & 1 & 0 \\ 1 & 0 & 0 \\ 0 & 0 & 1 \end{bmatrix} & \begin{bmatrix} 0 & 0 & 1 \\ 0 & 1 & 0 \\ 1 & 0 & 0 \end{bmatrix} \end{array}$$

The first row in each case is the one-dimensional representation spanned by s_N and the 3×3 matrices form the three-dimensional representation spanned by the three-dimensional basis (s_A, s_B, s_C).

The separation of the representation into sets of matrices of lower dimension is called the **reduction** of the representation. In this case, we write

$$\mathbf{D}^{(4)} = \mathbf{D}^{(3)} \oplus \mathbf{D}^{(1)} \tag{17}$$

and say that the four-dimensional representation has been reduced to a **direct sum** (the significance of the sign $\oplus$) of a three-dimensional and a one-dimensional representation. The term 'direct sum' is used because we are not simply adding together matrices in the normal way but creating a matrix of high dimension from matrices of lower dimension.

There are several points that should be noted about the reduction. First, we see that one of the representations obtained is the unfaithful representation mentioned earlier, in which all the representatives are 1×1 matrices with the same single element, 1, in each case. Another point is that the characters of the representatives of symmetry operations of the same class are the same, as we proved earlier. That is true of $\mathbf{D}^{(4)}$, $\mathbf{D}^{(3)}$, and $\mathbf{D}^{(1)}$ (although the characters do have different values for each representation).

The question that we now confront is whether $\mathbf{D}^{(3)}$ is reducible. A glance at the representation in Table 5.4 shows that the similarity transformation we discussed earlier converts $\mathbf{D}^{(4)}$ to a block-diagonal form of structure

$$\begin{bmatrix} 1 & 0 & 0 & 0 \\ 0 & 1 & 0 & 0 \\ 0 & 0 & \blacksquare & \blacksquare \\ 0 & 0 & \blacksquare & \blacksquare \end{bmatrix}$$

which corresponds to the reduction

$$\mathbf{D}^{(4)} = \mathbf{D}^{(1)} \oplus \mathbf{D}^{(1)} \oplus \mathbf{D}^{(2)}$$

The two one-dimensional representations in this expression are the same as the single one-dimensional (and unfaithful) representation introduced above, so in effect the new feature we have achieved is the reduction of the 3-dimensional representation:

$$\mathbf{D}^{(3)} = \mathbf{D}^{(1)} \oplus \mathbf{D}^{(2)}$$

In this case, $\mathbf{D}^{(1)}$ is spanned by the linear combination s_1 whereas before it was spanned by the single orbital s_{N}. A glance at Fig. 5.23 shows the physical basis of this analogy: the orbital s_{N} has the 'same symmetry' as s_1. However, we are now moving to a position where we can say what we mean by the colloquial term 'same symmetry': we mean *act as a basis of the same matrix representation*.

The question that immediately arises is whether the two-dimensional representation can be reduced to the direct sum of two one-dimensional representations by another choice of similarity transformation. As we shall see shortly, group theory can be used to confirm that $\mathbf{D}^{(2)}$ is an **irreducible representation**, or 'irrep', of the molecular point group in the sense that no similarity transformation (that is, linear combination of basis functions) can be found that *simultaneously* converts the representatives to block-diagonal form. The unfaithful $\mathbf{D}^{(1)}$ representation is another example of an irreducible representation.

Each irreducible representation of a group has a label called a **symmetry species**. The symmetry species is ascribed on the basis of the list of characters of the representation. Thus, the unfaithful representation has the list of characters $(1,1,1,1,1,1)$ and belongs to the symmetry species named A_1. The two-dimensional irreducible representation has characters $(2,-1,-1,0,0,0)$, and its label is E. The letters A and B are used for the symmetry species of one-dimensional irreducible representations, E is used for two-dimensional irreducible representations, and T is used for three-dimensional irreducible representations. The irreducible representations labelled A_1 and E are also labelled $\Gamma^{(1)}$ and $\Gamma^{(3)}$, respectively (we meet $\Gamma^{(2)}$ shortly: the numbers on Γ do not refer to the dimension of the irreducible representation, they are just labels). We shall use the Γ notation for general expressions and the A, B,... labels in particular cases. If a particular set of functions is a basis for an irreducible representation Γ, then we say that the basis **spans** that irreducible representation. The complete list of characters of all possible irreducible representations of a group is called a **character table**. As we shall shortly show, there are only a finite number of irreducible representations for groups of finite order, and we shall see that these tables are of enormous importance and usefulness.

We are now left with three tasks. One is to determine which symmetry species of irreducible representation may occur in a group and establish their characters. The second is to determine to what direct sum of irreducible representations an arbitrary matrix representation can be reduced—that is equivalent to deciding which irreducible representations an arbitrary basis spans. The third is to construct the linear combinations of members of an arbitrary basis that span a particular irreducible representation. This work requires some powerful machinery, which the next subsection provides.

5.10 The great and little orthogonality theorems

The quantitative development of group theory is based on the **Great Orthogonality Theorem** (GOT), which states the following. Consider a group of order h, and let $\mathbf{D}^{(l)}(R)$ be the representative of the operation R in a d_l-dimensional irreducible representation of symmetry species $\Gamma^{(l)}$ of the group. Then

$$\sum_R D_{ij}^{(l)}(R)^* D_{i'j'}^{(l')}(R) = \frac{h}{d_l}\delta_{ll'}\delta_{ii'}\delta_{jj'} \tag{18}$$

Note that this form of the theorem allows for the possibility that the representatives have complex elements; in the applications in this chapter, however, they will in fact be real and complex conjugation has no effect. Although this expression may look fearsome, it is simple to apply. In words, it states that if you select any location in a matrix of one irreducible representation, and any location in a matrix of the same or different irreducible representation of the group, multiply together the numbers found in those two locations, and then sum the products over all the operations of the group, then the answer is zero unless the locations of the elements are the same in both sets of matrices, and indeed the same set of matrices (the same irreducible representations) are chosen. If the locations are the same, and the two irreducible representations are the same, then the result of the calculation is h/d_l.

Example 5.5 How to use the Great Orthogonality Theorem

Illustrate the validity of the GOT by choosing two examples from Table 5.3 that (a) give a nonzero and (b) a zero value according to the theorem.

Method. For a nonzero outcome, we must choose the same location in the same matrix representation: a simple example would be to use the one-dimensional unfaithful representation A_1. For the outcome zero, we can choose either different locations in a single irreducible representation or arbitrary locations in two different irreducible representations. Refer to Table 5.3 for the specific values of the matrix elements.

Answer. (a) For C_{3v}, for which $h = 6$, take the irreducible representation A_1 (which has $d = 1$), in which the matrices are 1, 1, 1, 1, 1, 1. The sum on the left of the GOT with each matrix element taken with itself is

$$\sum_R D_{11}^{(A_1)}(R)^* D_{11}^{(A_1)}(R) = 1 \times 1 + 1 \times 1 + 1 \times 1 \\ + 1 \times 1 + 1 \times 1 + 1 \times 1 = 6$$

which is equal to $6/1 = 6$, as required by the theorem. (b) Consider two different locations in the two-dimensional irreducible representation E. For example, take the 34 and 33 elements of the matrices in Table 5.3:

$$\begin{aligned}\sum_R D_{34}^{(E)}(R)^* D_{33}^{(E)}(R) &= D_{34}^{(E)}(E)^* D_{33}^{(E)}(E) + D_{34}^{(E)}(C_3^+)^* D_{33}^{(E)}(C_3^+) + \ldots \\ &= 0 \times 1 + 0 \times 0 + 1 \times 0 + 1 \times 0 \\ &\quad + 0 \times 0 + 0 \times 1 \\ &= 0\end{aligned}$$

which is also in accord with the theorem.

Exercise 5.5. Confirm the validity of the GOT by using the irreducible representation A_1 and any element of the irreducible representation E for the matrices in Table 5.4.

The Great Orthogonality Theorem is too great for most of our purposes, and it is possible to derive from it a weaker statement in terms of the characters of irreducible representations. To derive this version, we set $i = j$ and $j' = i'$, to obtain diagonal elements on the left of eqn 5.18, and then sum over all these diagonal elements. The left of eqn 5.18 becomes

$$\begin{aligned}\sum_{i,i'} \sum_R D_{ii}^{(l)}(R)^* D_{i'i'}^{(l')}(R) &= \sum_R \left\{ \sum_i D_{ii}^{(l)}(R)^* \right\} \left\{ \sum_{i'} D_{i'i'}^{(l')}(R) \right\} \\ &= \sum_R \chi^{(l)}(R)^* \chi^{(l')}(R)\end{aligned}$$

Under the same manipulations, the right-hand side of eqn 5.18 becomes

$$\sum_{i,i'} \left(\frac{h}{d_l} \right) \delta_{ll'} \delta_{ii'} \delta_{ii'} = \frac{h}{d_l} \delta_{ll'} \sum_i \delta_{ii}$$

There are d_l values of the index i in a matrix of dimension d_l, and so the sum on the right is the sum of 1 taken d_l times, or d_l itself. Hence, on combining the two halves of the equation, we arrive at the **little orthogonality theorem**:

$$\sum_R \chi^{(l)}(R)^* \chi^{(l')}(R) = h\delta_{ll'} \tag{19}$$

The little orthogonality theorem (LOT) can be expressed slightly more simply by making use of the fact that all operations of the same class have the same character. Suppose that the number of symmetry operations in a class c is $g(c)$, so that $g(C_3) = 2$ and $g(\sigma_v) = 3$ in the group C_{3v}. Then

$$\sum_c g(c)\chi^{(l)}(c)^* \chi^{(l')}(c) = h\delta_{ll'} \tag{20}$$

where the sum is now over the classes. When $l' = l$, this expression becomes

$$\sum_c g(c)\left|\chi^{(l)}(c)\right|^2 = h \tag{21}$$

which signifies that the sum of the squares of the characters of any irreducible representation of a group is equal to the order of the group.

The form of the LOT suggests the following analogy. Suppose we interpret the quantity $\sqrt{g(c)}\chi_c^{(l)}$ as a component $v_c^{(l)}$ of a vector $\mathbf{v}^{(l)}$, with each component distinguished by the index c, then the LOT can be written

$$\sum_c v_c^{(l)*} v_c^{(l')} = \mathbf{v}^{(l)*} \cdot \mathbf{v}^{(l')} = h\delta_{ll'} \tag{22}$$

This expression shows that the LOT is equivalent to the statement that two vectors are orthogonal unless $l' = l$. However, the number of orthogonal vectors in a space of dimension N cannot exceed N (think of the three orthogonal vectors in ordinary space). In the present case, the dimensionality of the 'space' occupied by the vectors is equal to the number of classes of the group. Therefore, the number of values of l which distinguish the different orthogonal vectors cannot exceed the number of classes of the group. Because l labels the symmetry species of the irreducible representations of the group, it follows that the number of symmetry species cannot exceed the number of classes of the group. In fact, it follows from a more detailed analysis of the GOT (as distinct from the LOT) that these two numbers are equal. Hence, we arrive at the following restriction on the structure of a group:

The number of symmetry species is equal to the number of classes.

The vector interpretation can be applied to the GOT itself. To do so, we identify $D_{ij}^{(l)}(R)$ as the Rth component of a vector $\mathbf{v}$ identified by the three indices l, i, and j. The orthogonality condition is then

$$\mathbf{v}^{(l,i,j)*} \cdot \mathbf{v}^{(l',i',j')} = \frac{h}{d_l}\delta_{ll'}\delta_{ii'}\delta_{jj'} \tag{23}$$

This condition implies that any pair of vectors with different labels are orthogonal. The orthogonality condition is expressed in terms of a sum over all h elements of a group, so the vectors are h-dimensional. The total number of vectors of a given irreducible representation is d_l^2 because the labels i and j can each take d_l values in a $d_l \times d_l$ matrix. The total dimensionality of the space is therefore the sum of d_l^2 over all the symmetry species. The resulting number ($\sum_l d_l^2$) cannot exceed the dimension h of the space the vectors inhabit, and it may be shown that the two numbers are in fact equal. Therefore, we have the following further restriction on the structure of the group:

$$\sum_l d_l^2 = h \tag{24}$$

Example 5.6 How to construct a character table

Use the restrictions derived above and the LOT to complete the C_{3v} character table.

Method. We have identified two of the irreducible representations of the six-dimensional group, A_1 and E. The restriction given above will give us the number of symmetry species to look for, and we can use eqn 5.24 to determine their dimensions. The characters themselves can be found from the LOT by ensuring that they are orthogonal to the two irreducible representations we have already found.

Answer. The order of the group is $h = 6$ and there are three classes of operation; therefore, we expect there to be three symmetry species of irreducible representation. The dimensionality, d, of the unidentified irreducible representation must be such as to satisfy

$$1^2 + 2^2 + d^2 = 6$$

Hence, $d = 1$, and the missing irreducible representation is one-dimensional. We shall call it A_2. At this stage we can use the LOT to construct three equations for the three unknown characters. With $l = l' = A_2$, eqn 5.21 is

$$\chi^{(A_2)}(E)^2 + 2\chi^{(A_2)}(C_3)^2 + 3\chi^{(A_2)}(\sigma_v)^2 = 6$$

With $l = A_2$ and $l' = A_1$ we obtain

$$\chi^{(A_2)}(E)\chi^{(A_1)}(E) + 2\chi^{(A_2)}(C_3)\chi^{(A_1)}(C_3) + 3\chi^{(A_2)}(\sigma_v)\chi^{(A_1)}(\sigma_v) = 0$$

and with $l = A_2$ and $l' = E$

$$\chi^{(A_2)}(E)\chi^{(E)}(E) + 2\chi^{(A_2)}(C_3)\chi^{(E)}(C_3) + 3\chi^{(A_2)}(\sigma_v)\chi^{(E)}(\sigma_v) = 0$$

When the known values of the characters of A_1 and E are substituted, these two equations become

$$\chi^{(A_2)}(E) + 2\chi^{(A_2)}(C_3) + 3\chi^{(A_2)}(\sigma_v) = 0$$

and

$$2\chi^{(A_2)}(E) - 2\chi^{(A_2)}(C_3) = 0$$

The three equations are enough to determine the three unknown characters, and we find $\chi^{(A_2)}(E) = 1$, $\chi^{(A_2)}(C_3) = 1$, and $\chi^{(A_2)}(\sigma_v) = -1$. The complete set of characters is displayed in Table 5.5.

Comment. The character of the identity in a one-dimensional irreducible representation is 1, so that value could have been obtained without any calculation.

Exercise 5.6. Construct the character table for the group C_{2v}.

[See Table 5.6]

Table 5.5 The C_{3v} character table

C_{3v}	E	$2C_3$	$3\sigma_v$
A_1	1	1	1
A_2	1	1	−1
E	2	−1	0

Table 5.6 The C_{2v} character table

C_{2v}	E	C_2	σ_v	σ_v'
A_1	1	1	1	1
A_2	1	1	−1	−1
B_1	1	−1	1	−1
B_2	1	−1	−1	1

The character table for any symmetry group can be constructed as we have illustrated, and a selection is given in Appendix 1.

Reduced representations

A great deal depends on being able to establish what irreducible representations are spanned by a given basis. This problem leads us into the applications of group theory that we shall use throughout the text.

5.11 The reduction of representations

The question we now tackle is, given a general set of basis functions, how do we find the symmetry species of the irreducible representations they span? Often, as we shall see, we are interested more in the symmetry species and its characters than in the actual irreducible representation (the set of matrices). We have seen that a representation may be expressed as a direct sum of irreducible representations

$$\mathbf{D}(R) = \mathbf{D}^{(\Gamma_1)}(R) \oplus \mathbf{D}^{(\Gamma_2)}(R) \oplus \dots \qquad (25)$$

by finding a similarity transformation that *simultaneously* converts the matrix representatives to block-diagonal form. It is notationally simpler to express this reduction in terms of the symmetry species of the irreducible representations that occur in the reduction:

$$\Gamma = \sum_l a_l \Gamma^{(l)} \qquad (26)$$

where a_l is the number of times the irreducible representation of symmetry species $\Gamma^{(l)}$ appears in the direct sum. For example, the reduction of the *s*-orbital basis we have been considering would be written $\Gamma = 2A_1 + E$.

Our task is to find the coefficients a_l. To do so, we make use of the fact that because the character of an operation is invariant under a similarity transformation, the character of the original representative is the sum of the characters of

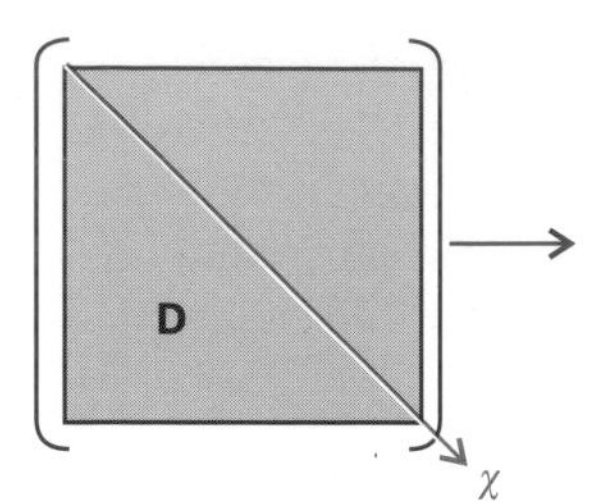

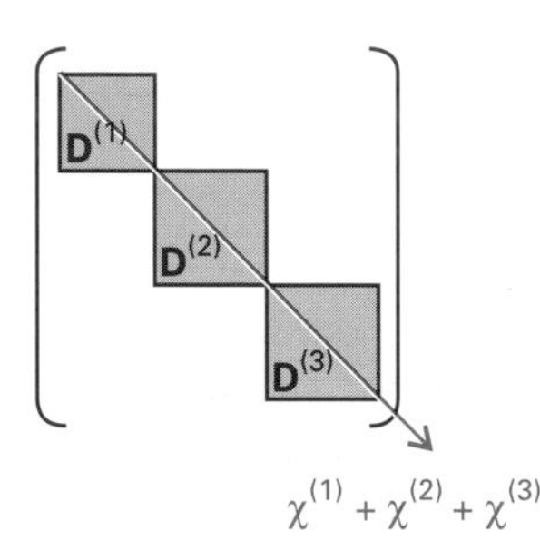

Fig. 5.24 A diagrammatic representation of the reduction of a matrix to block-diagonal form. The sum of the diagonal elements remains unchanged by the reduction.

the irreducible representations into which it is reduced (Fig. 5.24). Therefore,

$$\chi(R)=\sum_{l}a_l\chi^{(l)}(R) \qquad (27)$$

Now we use the LOT to determine the coefficients. To do so, we multiply both sides of this equation by $\chi^{(l')}(R)^*$ and sum over all the elements of the group:

$$\sum_{R}\chi^{(l')}(R)^*\chi(R)=\sum_{R}\sum_{l}a_l\chi^{(l')}(R)^*\chi^{(l)}(R)$$
$$=h\sum_{l}a_l\delta_{ll'}=ha_{l'}$$

That is, the coefficients are given by the rule

$$a_l=\frac{1}{h}\sum_{R}\chi^{(l)}(R)^*\chi(R) \qquad (28)$$

Because the characters of members of the same class of operation are the same, we can express this equation in terms of the characters of the classes:

$$a_l=\frac{1}{h}\sum_{c}g(c)\chi^{(l)}(c)^*\chi(c) \qquad (29)$$

Although the last two expressions provide a formal procedure for finding the reduction coefficients, in many cases it is possible to find them by inspection. For example, in the *s*-orbital basis for C_{3v}, the characters are (4, 1, 2) for the classes (E, $2C_3$, $3\sigma_v$). By inspection of the character table (Table 5.5), it is immediately clear that the reduction is $2A_1 + E$. However, in more complicated cases, the formal procedure is almost essential.

Example 5.7 How to determine the reduction of a representation

What symmetry species do the four H1*s*-orbitals of methane span?

Method. Methane belongs to the group T_d; the character table will be found in Appendix 1. The character of each operation in the four-dimensional basis (H_a, H_b, H_c, H_d) can be determined by noting the number (N) of members left in their original location after the application of each operation: a 1 occurs in the diagonal of the representative in each case, and so the character is the sum of 1 taken N times. Only one operation from each class need be considered because the characters are the same for all members of a class. With the characters $\chi(c)$ established, apply eqn 5.29 to determine the reduction.

Answer. Refer to Fig. 5.25. The numbers of unchanged basis members under the operations E, C_3, C_2, S_4, σ_d are 4, 1, 0, 0, 2, respectively. The order of the group is $h = 24$. It follows from eqn 5.29 that

$$a(A_1)=\tfrac{1}{24}\{(4\times 1)+8(1\times 1)+3(0\times 1)+6(0\times 1)+6(2\times 1)\}=1$$
$$a(A_2)=\tfrac{1}{24}\{(4\times 1)+8(1\times 1)+3(0\times 1)-6(0\times 1)-6(2\times 1)\}=0$$
$$a(E)=\tfrac{1}{24}\{(4\times 2)-8(1\times 1)+3(0\times 2)+6(0\times 0)+6(2\times 0)\}=0$$
$$a(T_1)=\tfrac{1}{24}\{(4\times 3)+8(1\times 0)-3(0\times 1)+6(0\times 1)-6(2\times 1)\}=0$$
$$a(T_2)=\tfrac{1}{24}\{(4\times 3)+8(1\times 0)-3(0\times 1)-6(0\times 1)+6(2\times 1)\}=1$$

Hence, the four orbitals span $A_1 + T_2$.

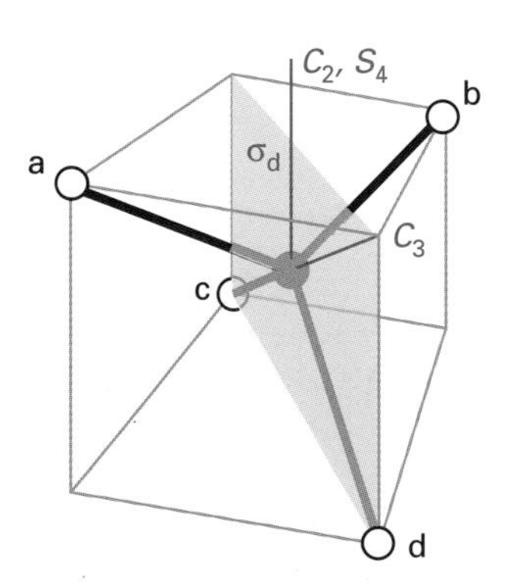

Fig. 5.25 The symmetry elements of the group T_d used in Example 5.7.

Comment. In some cases, an operation changes the sign of a member of the basis without moving its location (an example is the O2p_x-orbital in H_2O under the operation C_2). This results in -1 appearing on the diagonal. In other cases, such as for the basis (p_x, p_y) on the central atom in a molecule belonging to the group C_{3v}, a fractional value appears on the diagonal: see Section 5.13.

Exercise 5.7. What symmetry species do the five Cl3s-orbitals of PCl_5, a trigonal bipyramidal molecule in the gas phase, span?

5.12 Symmetry-adapted bases

We now establish how to find the linear combinations of the members of a basis that span an irreducible representation of a given symmetry species. This procedure is called finding a **symmetry-adapted basis**. The next couple of pages will bristle with subscripts; if you do not wish to pick your way through the thicket, you will be able to use the final result (eqn 5.39).

Consider the set of functions $\mathbf{f}^{(l')} = (f_1^{(l')}, f_2^{(l')}, \ldots, f_d^{(l')})$ that form a basis for a $d_{l'}$-dimensional irreducible representation $\mathbf{D}^{(l')}$ of symmetry species $\Gamma^{(l')}$ of a group of order h. We can express the effect of any operation of the group as

$$Rf_{j'}^{(l')} = \sum_{i'} f_{i'}^{(l')} D_{i'j'}^{(l')}(R) \tag{30}$$

The GOT may now be invoked. First we multiply by the complex conjugate of an element $D_{ij}^{(l)}(R)$ of a representative of the same operation, and then sum over the elements, using the GOT to simplify the outcome:

$$\begin{aligned}
\sum_R D_{ij}^{(l)}(R)^* Rf_{j'}^{(l')} &= \sum_R \sum_{i'} D_{ij}^{(l)}(R)^* f_{i'}^{(l')} D_{i'j'}^{(l')}(R) \\
&= \sum_{i'} f_{i'}^{(l')} \left\{ \sum_R D_{ij}^{(l)}(R)^* D_{i'j'}^{(l')}(R) \right\} \\
&= \sum_{i'} f_{i'}^{(l')} \left(\frac{h}{d_{l'}} \right) \delta_{ll'} \delta_{ii'} \delta_{jj'} \\
&= \left(\frac{h}{d_{l'}} \right) \delta_{ll'} \delta_{jj'} f_i^{(l')} = \left(\frac{h}{d_l} \right) \delta_{ll'} \delta_{jj'} f_i^{(l)}
\end{aligned} \tag{31}$$

We then define a **projection operator**:

$$P_{ij}^{(l)} = \frac{d_l}{h} \sum_R D_{ij}^{(l)}(R)^* R \tag{32}$$

This operator can be thought of as a mixture of the operations of the group, with a weight given by the value of the matrix elements of the representation. It follows from eqns 5.31 and 5.32 that the effect of the projection operator is as follows:

$$P_{ij}^{(l)} f_{j'}^{(l')} = f_i^{(l)} \delta_{ll'} \delta_{jj'} \tag{33}$$

The reason why P is called a projection operator can now be made clear. In the first case, suppose that either $l \neq l'$ or $j \neq j'$; then when $P_{ij}^{(l)}$ acts on some member $f_{j'}^{(l')}$, it gives zero. That is, when $P_{ij}^{(l)}$ acts on a function that is not a

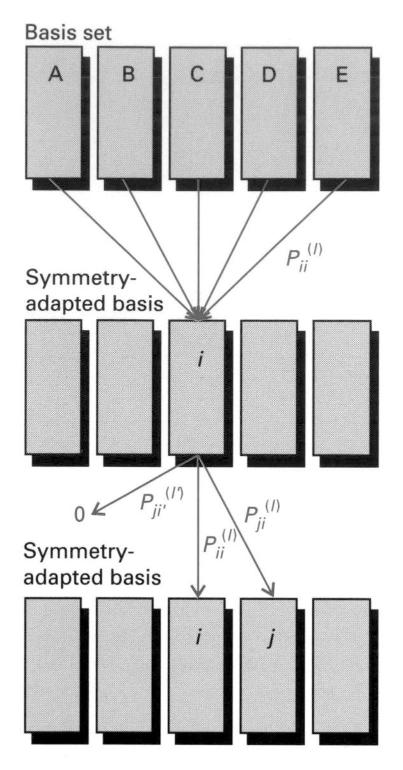

Fig. 5.26 A schematic diagram to illustrate the effect of the various projection operators.

member of the set that spans $\Gamma^{(l)}$, or—if it is a member—is not at the location j in the set, then it gives zero. On the other hand, if the member is at the location j of the set that does span $\Gamma^{(l)}$, then it converts the function standing at the location j into the function standing at the location i. That is, P *projects* a member from one location to another location (Fig. 5.26). The importance of this result is that if we know only one member of a basis of a representation, then we can project all the other members out of it.

In the special case of $l' = l$ and $i = j$, the effect of the projection operator on some member of the basis is

$$P_{ii}^{(l)} f_{j'}^{(l)} = f_i^{(l)} \delta_{ij'} \tag{34}$$

That is, P then generates either 0 (if $i \neq j'$) or regenerates the original function (if $i = j'$). The significance of this special case will be apparent soon.

Now suppose that we are given a linearly independent but otherwise arbitrary set of functions $\mathbf{f} = (f_1, f_2, \ldots)$. An example might be the *s*-orbital basis we considered earlier. What is the effect of the projection operator $P_{ii}^{(l)}$ on any one member? Just as any member of the symmetry-adapted basis can be expressed as the appropriate linear combination of the members of the arbitrary basis, we can express any f_j as a linear combination of all the $f_{j'}^{(l')}$:

$$f_j = \sum_{l'j'} f_{j'}^{(l')} \tag{35}$$

(The expansion coefficients have been absorbed into the $f_{j'}^{(l')}$.) If we now operate on this expression with the projection operator $P_{ii}^{(l)}$, we obtain

$$P_{ii}^{(l)} f_j = \sum_{l'j'} P_{ii}^{(l)} f_{j'}^{(l')} = \sum_{l'j'} \delta_{ll'} \delta_{ij'} f_{j'}^{(l')} = f_i^{(l)} \tag{36}$$

That is, when $P_{ii}^{(l)}$ operates on *any* member of the arbitrary initial basis, it generates the ith member of the basis for the irreducible representation of symmetry species $\Gamma^{(l)}$. With that member obtained, we can act on it with $P_{ji}^{(l)}$ to construct the jth member of the set. This solves the problem of finding a symmetry-adapted basis.

The problem with the method is that to set up the projection operators we need to know the elements of all the representatives of the irreducible representation. It is normally the case that only the characters (the sums of the diagonal elements) are available. However, even that limited information can be useful. Consider the projection operator $p^{(l)}$ formed by summing $P^{(l)}$ over its diagonal elements:

$$p^{(l)} = \sum_i P_{ii}^{(l)} = \frac{d_l}{h} \sum_{i,R} D_{ii}^{(l)}(R)^* R \tag{37}$$

The sum over the diagonal elements of a representative is the character of the corresponding operation, so

$$p^{(l)} = \frac{d_l}{h} \sum_R \chi^{(l)}(R)^* R \tag{38}$$

This operator can therefore be constructed from the character tables alone. Its effect is to generate a sum of the members of a basis spanning an irreducible

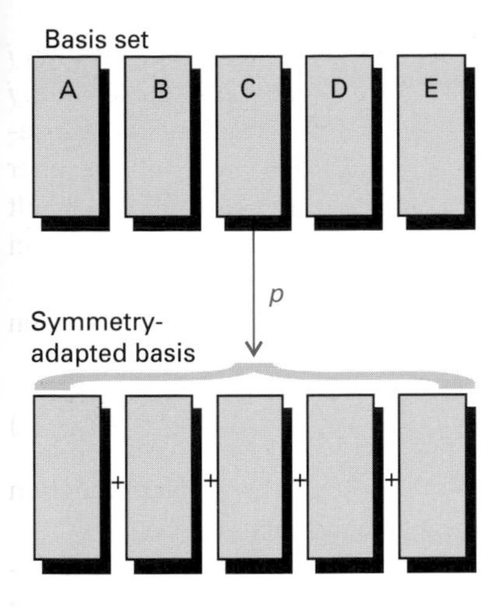

Fig. 5.27 The projection operator p generates a sum of the symmetry-adapted basis functions when it is applied to any member of the original basis.

representation (Fig. 5.27):

$$p^{(l)}f_j = \sum_i P_{ii}^{(l)}f_j = \sum_i f_i^{(l)} \tag{39}$$

The fact that a sum is generated is of no consequence for one-dimensional irreducible representations because in such cases there is only one member of the basis set. However, for two- and higher-dimensional irreducible representations the projection operator gives a sum of two or more members of the basis. Nevertheless, because we are generally concerned only with low-dimensional irreducible representations, this is rarely a severe complication, and the following example shows how any ambiguity can be resolved.

Example 5.8 How to use projection operators

Construct the symmetry-adapated bases for the group C_{3v} using the s-orbital basis.

Method. We have already established that the s-orbital basis spans $2A_1 + E$, so we can use eqn 5.39 to construct the appropriate symmetry-adapted bases by projection. We shall take all the characters to be real. The simplest way to use eqn 5.39 is to follow this recipe:

1. Draw up a table headed by the basis and showing in the columns the effect of the operations. (A given column is headed by f_j and an entry in the table shows Rf_j.)
2. Multiply each member of the column by the character of the corresponding operation. (This step produces $\chi(R)Rf_j$ at each location.)
3. Add the entries within each column. (This produces $\sum_R \chi(R)Rf_j$ for a given f_j.)
4. Multiply by dimension/order. (This produces pf_j.)

For the group C_{3v}, $h = 6$.

Answer. The table to construct is as follows:

Original set:	s_N	s_A	s_B	s_C
Under E	s_N	s_A	s_B	s_C
C_3^+	s_N	s_B	s_C	s_A
C_3^-	s_N	s_C	s_A	s_B
σ_v	s_N	s_A	s_C	s_B
σ_v'	s_N	s_B	s_A	s_C
σ_v''	s_N	s_C	s_B	s_A

For the irreducible representation of symmetry species A_1, $d = 1$ and all $\chi(R) = 1$. Hence, the first column gives

$$\tfrac{1}{6}(s_N + s_N + s_N + s_N + s_N + s_N) = s_N$$

The second column gives

$$\tfrac{1}{6}(s_A + s_B + s_C + s_A + s_B + s_C) = \tfrac{1}{3}(s_A + s_B + s_C)$$

The remaining two columns give the same outcome. For E, $d = 2$ and for the six operations $\chi = (2, -1, -1, 0, 0, 0)$ for the six operations. The first column gives

$$\tfrac{2}{6}(2s_{\mathrm{N}} - s_{\mathrm{N}} - s_{\mathrm{N}} + 0 + 0 + 0) = 0$$

The second column gives

$$\tfrac{2}{6}(2s_{\mathrm{A}} - s_{\mathrm{B}} - s_{\mathrm{C}} + 0 + 0 + 0) = \tfrac{1}{3}(2s_{\mathrm{A}} - s_{\mathrm{B}} - s_{\mathrm{C}})$$

The remaining columns produce $\frac{1}{3}(2s_{\mathrm{B}} - s_{\mathrm{C}} - s_{\mathrm{A}})$ and $\frac{1}{3}(2s_{\mathrm{C}} - s_{\mathrm{A}} - s_{\mathrm{B}})$. These three linear combinations are not linearly independent (the sum of them is zero), so we can form a linear combination of the second two combinations that is orthogonal to the first. The combination

$$s_3 = \tfrac{1}{3}(2s_{\mathrm{B}} - s_{\mathrm{C}} - s_{\mathrm{A}}) - \tfrac{1}{3}(2s_{\mathrm{C}} - s_{\mathrm{A}} - s_{\mathrm{B}}) = s_{\mathrm{B}} - s_{\mathrm{C}}$$

is orthogonal to $s_2 = \frac{1}{3}(2s_{\mathrm{A}} - s_{\mathrm{B}} - s_{\mathrm{C}})$. Note that the two orbitals s_2 and s_3 have a different character under σ_{v} (+1 and −1, respectively).

Exercise 5.8. Find the symmetry-adapted linear combinations of the *p*-orbitals in NO_2.

The symmetry properties of functions

We now turn to a consideration of the transformation properties of functions in general. To set the scene, we shall investigate how the three *p*-orbitals of the nitrogen atom in NH_3 transform under the operations of the group $C_{3\mathrm{v}}$. The basis for the representation we shall develop is (p_x, p_y, p_z). Intuitively, we can expect the representation to reduce to an irreducible representation spanned by p_z because $p_z \rightarrow p_z$ under all operations of the group (but is it of symmetry species A_1 or A_2?) and a two-dimensional irreducible representation spanned by (p_x, p_y), because these orbitals are mixed by the symmetry operations, which is of symmetry species E. But suppose the basis was extended to include *d*-orbitals on the central atom—what irreducible representations would then be spanned? To answer questions like that, we need a systematic procedure that can be applied even when—especially when—the conclusions are not obvious. The systematic approach is set out below. The procedures are essentially the same as we have already described, but they are more generally applicable than the calculations done above.

5.13 The transformation of *p*-orbitals

Consider the basis (p_x, p_y, p_z) for $C_{3\mathrm{v}}$. We know from Section 3.13 that the orbitals have the form

$$p_x = xf(r) \qquad p_y = yf(r) \qquad p_z = zf(r)$$

where r is the distance from the nucleus. All operations of a point group leave r unchanged, and so the orbitals transform in the same way as the basis (x, y, z). Some of the transformations of this basis are illustrated in Fig. 5.28.

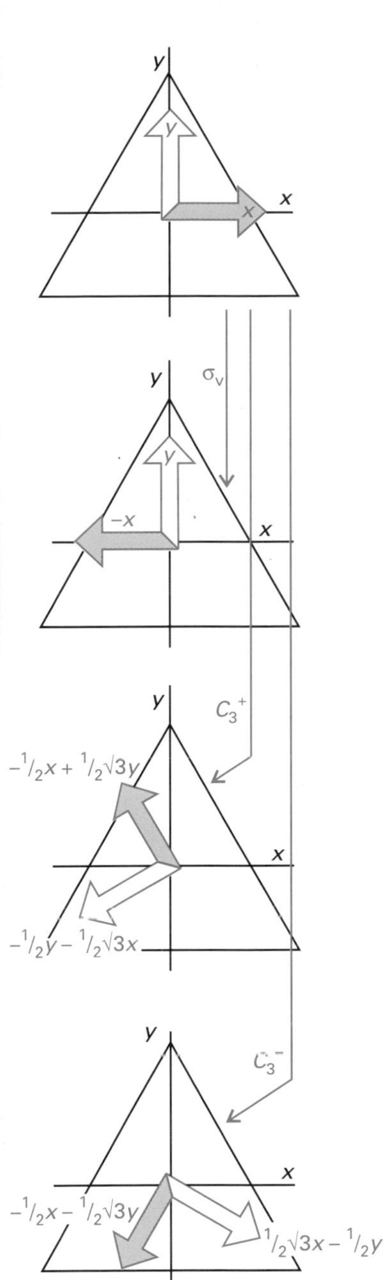

Fig. 5.28 The effect of certain symmetry operations of the group C_{3v} on the functions x and y.

The effect of σ_v on the basis is

$$\sigma_v(x, y, z) = (-x, y, z) = (x, y, z)\begin{bmatrix} -1 & 0 & 0 \\ 0 & 1 & 0 \\ 0 & 0 & 1 \end{bmatrix}$$

This relation identifies $\mathbf{D}(\sigma_v)$ in this basis. Under the rotation C_3^+ we have

$$\begin{aligned} C_3^+(x, y, z) &= (-\tfrac{1}{2}x + \tfrac{1}{2}\sqrt{3}y, -\tfrac{1}{2}\sqrt{3}x - \tfrac{1}{2}y, z) \\ &= (x, y, z)\begin{bmatrix} -\frac{1}{2} & -\frac{1}{2}\sqrt{3} & 0 \\ \frac{1}{2}\sqrt{3} & -\frac{1}{2} & 0 \\ 0 & 0 & 1 \end{bmatrix} \end{aligned}$$

and we can identify $\mathbf{D}(C_3^+)$ for the basis. The complete representation can be established in this way, and is set out in Table 5.7, together with the characters.

Table 5.7 The matrix representation of C_{3v} in the basis (x, y, z)

$\mathbf{D}(E)$	$\mathbf{D}(C_3^+)$	$\mathbf{D}(C_3^-)$
$\begin{bmatrix} 1 & 0 & 0 \\ 0 & 1 & 0 \\ 0 & 0 & 1 \end{bmatrix}$	$\begin{bmatrix} -\frac{1}{2} & -\frac{1}{2}\sqrt{3} & 0 \\ \frac{1}{2}\sqrt{3} & -\frac{1}{2} & 0 \\ 0 & 0 & 1 \end{bmatrix}$	$\begin{bmatrix} -\frac{1}{2} & \frac{1}{2}\sqrt{3} & 0 \\ -\frac{1}{2}\sqrt{3} & -\frac{1}{2} & 0 \\ 0 & 0 & 1 \end{bmatrix}$
$\chi(E) = 3$	$\chi(C_3^+) = 0$	$\chi(C_3^-) = 0$
$\mathbf{D}(\sigma_v)$	$\mathbf{D}(\sigma_v')$	$\mathbf{D}(\sigma_v'')$
$\begin{bmatrix} -1 & 0 & 0 \\ 0 & 1 & 0 \\ 0 & 0 & 1 \end{bmatrix}$	$\begin{bmatrix} \frac{1}{2} & -\frac{1}{2}\sqrt{3} & 0 \\ -\frac{1}{2}\sqrt{3} & -\frac{1}{2} & 0 \\ 0 & 0 & 1 \end{bmatrix}$	$\begin{bmatrix} \frac{1}{2} & \frac{1}{2}\sqrt{3} & 0 \\ \frac{1}{2}\sqrt{3} & -\frac{1}{2} & 0 \\ 0 & 0 & 1 \end{bmatrix}$
$\chi(\sigma_v) = 1$	$\chi(\sigma_v') = 1$	$\chi(\sigma_v'') = 1$

The characters of the operations E, $2C_3$, σ_v in the basis (x, y, z) are 3, 0, and 1, respectively. This corresponds to the reduction $A_1 + E$. The function z is a basis for A_1, and the pair (x, y) span E. We therefore now also know that the three p-orbitals also span $A_1 + E$, and that p_z is a basis for A_1 and (p_x, p_y) is a basis for E.

The identities of the symmetry species of the irreducible representations spanned by x, y, and z are so important that they are normally given specifically in the character tables (see Appendix 1). Exactly the same procedure may be applied to the quadratic forms x^2, xy, etc. that arise when the d-orbitals are expressed in cartesian coordinates (Section 3.13):

$$d_{xy} = xyf(r) \qquad d_{yz} = yzf(r) \qquad d_{zx} = zxf(r)$$

$$d_{x^2-y^2} = (x^2 - y^2)f(r) \qquad d_{z^2} = (3z^2 - r^2)f(r)$$

and the symmetry species these functions span are also normally reported: in C_{3v} the five functions span $A_1 + 2E$.

5.14 The decomposition of direct-product bases

The question that now arises is stimulated by noticing that the quadratic forms that govern the symmetry properties of the d-orbitals are expressed as products of the linear terms that govern the symmetry properties of p-orbitals. We can now explore whether it is possible to find the symmetry species of the quadratic forms such as xy, for instance, directly from the properties of x and y without having to go through the business of setting up the symmetry transformations and their representatives all over again. In more general terms, if we know what symmetry species are spanned by a basis $(f_1, f_2, \ldots)$, can we state the symmetry species spanned by their products, such as $(f_1^2, f_1 f_2, \ldots)$? We shall now show that this information is carried by the character tables.

Suppose that $f_i^{(l)}$ is a member of a basis for an irreducible representation of symmetry species $\Gamma^{(l)}$ of dimension d_l, and that $f_{i'}^{(l')}$ is a member of a basis for an irreducible representation of symmetry species $\Gamma^{(l')}$ of dimension $d_{l'}$. That being so, under an operation R of a group they transform as follows:

$$Rf_i^{(l)} = \sum_j f_j^{(l)} D_{ji}^{(l)}(R) \qquad Rf_{i'}^{(l')} = \sum_{j'} f_{j'}^{(l')} D_{j'i'}^{(l')}(R) \tag{40}$$

It follows that their product transforms as

$$R\left(f_i^{(l)} f_{i'}^{(l')}\right) = \left(Rf_i^{(l)}\right)\left(Rf_{i'}^{(l')}\right) = \sum_{jj'} f_j^{(l)} f_{j'}^{(l')} D_{ji}^{(l)}(R) D_{j'i'}^{(l')}(R) \tag{41}$$

which is a linear combination of the products $f_j^{(l)} f_{j'}^{(l')}$. It follows that the products also form a basis for a representation, which is called a **direct-product representation**. Its dimension is $d_l d_{l'}$.

To discover whether the direct-product representation is reducible, we need to work out its characters. The matrix representative of the operation R in the direct-product basis is $D_{ji}^{(l)}(R) D_{j'i'}^{(l')}(R)$, where the pair of indices jj' now label the row of the matrix and the indices ii' label the column. The diagonal elements are the elements with $j = i$ and $j' = i'$. It follows that the character of the operation R is

$$\begin{aligned} \chi(R) &= \sum_{ii'} D_{ii}^{(l)}(R) D_{i'i'}^{(l')}(R) \\ &= \left\{\sum_i D_{ii}^{(l)}(R)\right\}\left\{\sum_{i'} D_{i'i'}^{(l')}(R)\right\} = \chi^{(l)}(R)\chi^{(l')}(R) \end{aligned} \tag{42}$$

This is a very simple and useful result: it states that the characters of the operations in the direct-product basis are the products of the corresponding characters for the original bases. With the characters of the representation established, we can then use the standard techniques described above to decide on the reduction of the representation. This procedure is illustrated in the following example.

Example 5.9 The reduction of a direct-product representation

Determine the symmetry species of the irreducible representations spanned by (a) the quadratic forms and (b) the basis (xz, yz) in the group C_{3v}.

Method. For both parts of the problem we use the result set out in eqn 5.42 to establish the characters of the direct-product representation, and then reconstruct that set of characters as a linear combination of the characters of the irreducible representations of the group. If the decomposition of the characters is not obvious, use the procedure set out in Example 5.7.

Answer. (a) The basis (x, y, z) spans a (reducible) representation with characters 3, 0, 1 (in the usual order). The direct-product basis therefore spans a representation with characters 9, 0, 1. This set of characters corresponds to $2A_1 + A_2 + 3E$. (b) The basis (xz, yz) is the direct product of the bases z and (x, y) which span A_1 and E, respectively. The direct-product basis therefore has characters

$$(1\ 1\ 1) \times (2\ -1\ 0) = (2\ -1\ 0)$$

which we recognize as the characters of E itself. Therefore, (xz, yz) is a basis for E.

Comment. The fact that the direct product of bases that span A_1 and E spans E is normally written

$$A_1 \times E = E$$

Exercise 5.9. What irreducible representations are spanned by the direct product of (x, y) with itself in the group C_{3v}?

In the example we have shown that $A_1 \times E = E$, which is a formal way of expressing the fact that the direct-product basis (xz, yz) spans E. In the same way, the direct product of (x,y) with itself, which consists of the basis (x^2, xy, yx, y^2), spans

$$E \times E = A_1 + A_2 + E$$

(The significance of the appearance of both xy and yx is discussed below.) Tables of decompositions of direct products like these are called **direct-product tables**. They can be worked out once and for all, and some are listed in Appendix 1. We shall see that they are often as important as the character tables themselves! A particularly important point to note from the tables is that the product $\Gamma^{(l)} \times \Gamma^{(l')}$ contains the totally symmetric irreducible representation (A_1 in many groups) only if $l' = l$.

Finally, we need to account for the presence of both xy and yx in the direct-product basis. We need to note that the **symmetrized direct product**

$$f_{ij}^{(+)} = \tfrac{1}{2}\{f_i^{(l)} f_j^{(l)} + f_j^{(l)} f_i^{(l)}\} \tag{43}$$

and the **antisymmetrized direct product**

$$f_{ij}^{(-)} = \tfrac{1}{2}\{f_i^{(l)} f_j^{(l)} - f_j^{(l)} f_i^{(l)}\} \tag{44}$$

of a basis taken with itself also form bases for the group. Clearly, the latter vanish identically in this case because $xy - yx = 0$. We need to establish which irreducible representations are spanned by the antisymmetrized direct product and discard them from the decomposition. The characters of the products are given by the following expressions:[2]

$$\chi^{+}(R) = \frac{1}{2}\{\chi^{(l)}(R)^2 + \chi^{(l)}(R^2)\} \qquad \chi^{-}(R) = \frac{1}{2}\{\chi^{(l)}(R)^2 - \chi^{(l)}(R^2)\} \tag{45}$$

[2] For a derivation, see M. Hamermesh, *Group theory and its applications to physical problems*, Addison-Wesley, Reading, Mass. (1962).

In the direct-product tables the symmetry species of the antisymmetrized product is denoted $[\Gamma]$. The fact that it is reported at all signifies that it has some use: we shall see what it is in Section 7.16. In the present case

$$\mathrm{E} \times \mathrm{E} = \mathrm{A}_1 + [\mathrm{A}_2] + \mathrm{E}$$

and so we now know that (x^2, xy, y^2) spans $\mathrm{A}_1 + \mathrm{E}$. One of the most important applications of this type of procedure is to the determination of selection rules (see below, Section 5.16).

5.15 Direct-product groups

We can now consider another example of using group theory to build up information from existing results. Here we shall show how to build up the properties of larger groups by cementing together the character tables for smaller groups.

Suppose there exists a group G of order h with elements $R_1, R_2, \ldots, R_h$ and another group G' of order h' with elements $R'_1, R'_2, \ldots, R'_{h'}$. Let the groups satisfy the following two conditions:

1. The only element in common is the identity.
2. The elements of group G commute with the elements of group G'.

'Commutation' here has the same meaning as in the chapter on operators: $RR' = R'R$. Examples of two such groups are C_s and $C_{3\mathrm{v}}$. Then the products RR' of each element of G with each element of G' form a group called the **direct-product group**:

$$G'' = G \otimes G' \tag{46}$$

That G'' is in fact a group can be verified by checking that the group property is obeyed for all pairs of elements. Then, because $R_iR_j = R_k$ (because G is a group) and $R'_rR'_s = R'_t$ (for a similar reason), in G'' with elements $R_iR'_r$:

$$(R_iR'_r)(R_jR'_s) = R_iR'_rR_jR'_s = R_iR_jR'_rR'_s = R_kR'_t$$

and the element so generated is a member of G''. The order of the direct-product group is hh' (so the order of $C_\mathrm{s} \otimes C_{3\mathrm{v}}$ is $2 \times 6 = 12$).

The direct-product group can be identified by constructing its elements ($C_\mathrm{s} \otimes C_{3\mathrm{v}}$ will turn out to be $D_{3\mathrm{h}}$), and the character table can be constructed from the character tables of the component groups. To do so, we proceed as follows. Let $(f_1, f_2, \ldots)$ be a basis for an irreducible representation of G and $(f'_1, f'_2, \ldots)$ be a basis for an irreducible representation of G'. It follows that we can write

$$Rf_i = \sum_j f_j D_{ji}(R) \qquad R'f'_r = \sum_s f'_s D_{sr}(R') \tag{47}$$

Then the effect of RR' on the direct-product basis is

$$RR'f_if'_r = (Rf_i)(R'f'_r) = \sum_{js} f_jf'_s D_{ji}(R) D_{sr}(R')$$

The character of the operation RR' is the sum of the diagonal elements:

$$\chi(RR') = \sum_{ir} D_{ii}(R) D_{rr}(R') = \chi(R)\chi(R') \tag{48}$$

Therefore, the character table of the direct-product group can be written down simply by multiplying together the appropriate characters of the two contributing groups.

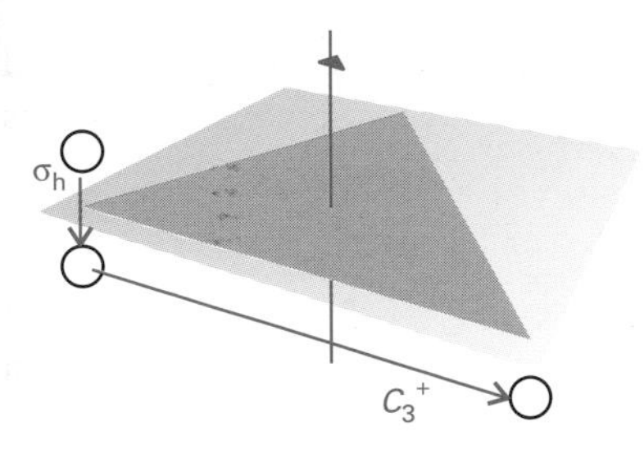

Fig. 5.29 A combination of the operations σ_h and C_3^+ is equivalent to the operation S_3^+.

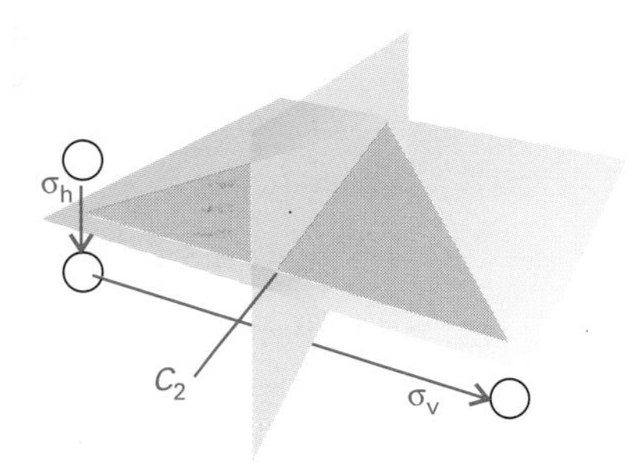

Fig. 5.30 A combination of the operations σ_h and σ_v is equivalent to the operation C_2.

C_{3v}	E	$2C_3$	$3\sigma_v$
A_1	1	1	1
A_2	1	1	−1
E	2	−1	0

C_s	E	σ_h
A′	1	1
A″	1	−1

Example 5.10 How to construct the character table of a direct-product group

Construct the direct-product group $C_s \otimes C_{3v}$, identify it, and build its character table from the constituent groups.

Method. To construct the direct-product group, we form elements by combining each element of one group with each element of the other group in turn. It is often sufficient to deal with the products of *classes* of operation rather than each individual operation. The resulting group is recognized by noting its composition and referring to Fig. 5.16. The characters are constructed by multiplying together the characters contributing to each operation.

Answer. The groups C_s and C_{3v} have, respectively, 2 and 3 classes, so the direct-product group has $2 \times 3 = 6$ classes. It follows that it also has six symmetry species of irreducible representations. The classes of C_s are (E, σ_h) and those of C_{3v} are $(E, 2C_3, 3\sigma_v)$. When each class of C_{3v} is multiplied by the identity operation of C_s, the same three classes, $(E, 2C_3, 3\sigma_v)$, are reproduced. Each of these classes is also multiplied by σ_h. The operation $E\sigma_h$ is the same as σ_h itself. The operations $C_3^+\sigma_h$ and $C_3^-\sigma_h$ are the improper rotations S_3^+ and S_3^-, respectively (see Fig. 5.29). The operations $\sigma_v\sigma_h$ are the same as two-fold rotations about the bisectors of the angles of the triangular object (Fig. 5.30) and are denoted C_2. The direct-product group is therefore formed as follows:

C_{3v}:	E		$2C_3$		$3\sigma_v$	
C_s:	E	σ_h	E	σ_h	E	σ_h
$C_{3v} \otimes C_s$:	E	σ_h	$2C_3$	$2S_3$	$3\sigma_v$	$3C_2$

According to the system of nomenclature described in Section 5.2, this set of operations corresponds to the group D_{3h}. At this point, we use the rule about characters to construct the character table. The two component group character tables are shown in the margin. On taking all the appropriate products we obtain the following table:

	$E = EE$	$\sigma_h = E\sigma_h$	$2C_3 = E(2C_3)$	$2S_3 = \sigma_h(2C_3)$	$3\sigma_v = E(3\sigma_v)$	$3C_2 = \sigma_h(3\sigma_v)$
$A_1'(= A_1A')$	1	1	1	1	1	1
$A_1''(= A_1A'')$	1	−1	1	−1	−1	1
$A_2'(= A_2A')$	1	1	1	1	−1	−1
$A_2''(= A_2A'')$	1	−1	1	−1	1	−1
$E'(= EA')$	2	2	−1	−1	0	0
$E''(= EA'')$	2	−2	−1	1	0	0

This is the table for this group given in Appendix 1.

Comment. The procedure described here is an important and easy way of constructing the character tables for more complex groups, such as $D_{6h} = D_6 \otimes C_i$ and $O_h = O \otimes C_i$.

Exercise 5.10. Construct the character table for the group $D_{6h} = D_6 \otimes C_i$.

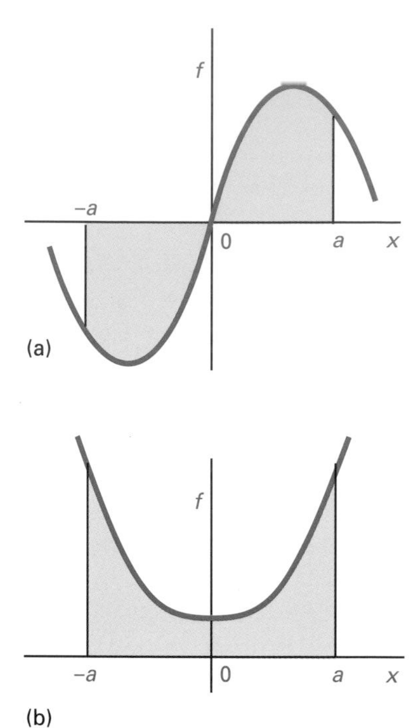

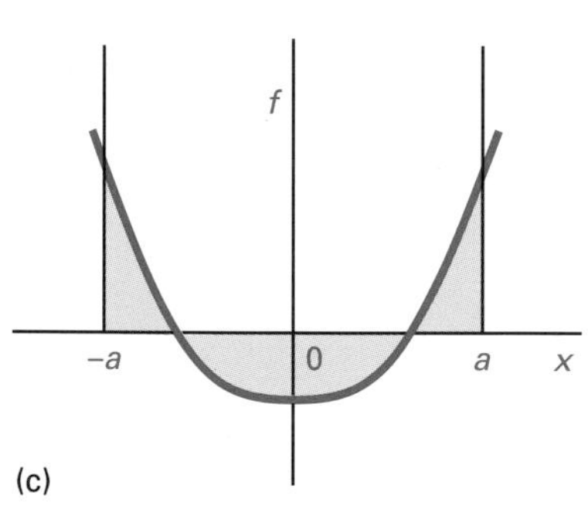

Fig. 5.31 (a) An antisymmetric function with necessarily zero integral over a symmetric range about the origin. (b) A symmetric function with nonzero integral over a symmetric range. (c) The integral of this symmetric function, however, is zero.

5.16 Vanishing integrals

One of the more important applications of group theory is to the problem of deciding when integrals are necessarily zero on account of the symmetry of the system. This application can be illustrated quite simply by considering two functions $f(x)$ and $g(x)$, and the integral over a symmetrical range around $x = 0$.

Let $f(x)$ be a function that is antisymmetric with respect to $x \to -x$, so $f(-x) = -f(x)$. The integral of this function over a range from $x = -a$ to $x = +a$ is zero (Fig. 5.31). On the other hand, if $g(x)$ is a symmetrical function in the sense that $g(-x) = g(x)$, then its integral over the same range is not necessarily zero. Note that the integral of g may, by accident, be zero, whereas the integral of f is *necessarily* zero. Now consider another way of looking at the two functions. The range $(-a, a)$ is an 'object' with two symmetry elements: the identity and a mirror plane perpendicular to the x-axis (Fig. 5.32). Such an object belongs to the point group C_s. The function f spans the irreducible representation of symmetry species A″ because $Ef = f$ and $\sigma_h f = -f$. On the other hand, g spans A′ because $Eg = g$ and $\sigma_h g = g$. That is, *if the integrand is not a basis for the totally symmetric irreducible representation of the group, then the integral is necessarily zero*. If the integrand is a basis for the totally symmetric irreducible representation, then the integral is not necessarily zero (but may, by accident, be zero).

This simple example also introduces a further point that generalizes to all groups. The integrals of f^2 and g^2 are not zero, but the integral of fg is necessarily zero. This feature is consistent with the discussion above, because f^2 is a basis for A″ × A″ = A′, which is the totally symmetric irreducible representation; likewise g^2 is a basis for A′ × A′ = A′, which is also the totally symmetric irreducible representation. However fg is a basis for A″ × A′ = A″, which is not totally symmetric, so the integral necessarily vanishes. Another way of looking at this result is to note that f spans one species of irreducible representation, g spans another. Then, *basis functions that span irreducible representations of different symmetry species are orthogonal*.

More formally: if $f_i^{(l)}$ is the ith member of a basis that spans the irreducible representation of symmetry species $\Gamma^{(l)}$ of a group, and $f_j^{(l')}$ is the jth member of a basis that spans the irreducible representation of symmetry species $\Gamma^{(l')}$ of the same group, then for a symmetric range of integration:

$$\int f_i^{(l)*} f_j^{(l')}\, d\tau \propto \delta_{ll'}\delta_{ij} \tag{49}$$

The proof of this conclusion is based on the GOT, and is given in *Further information 14*. Note that the integral *may* be zero even though $l' = l$ and $i = j$, because the theorem is silent on the value of the proportionality constant.

We have now arrived at one of the most important results of group theory. The conclusion can be summarized as follows:

> An integral $\int f^{(l)*} f^{(l')}\, d\tau$ over a symmetric range is necessarily zero unless the integrand is a basis for the totally symmetric irreducible representation of the group. This is the case only if $\Gamma^{(l)} = \Gamma^{(l')}$.

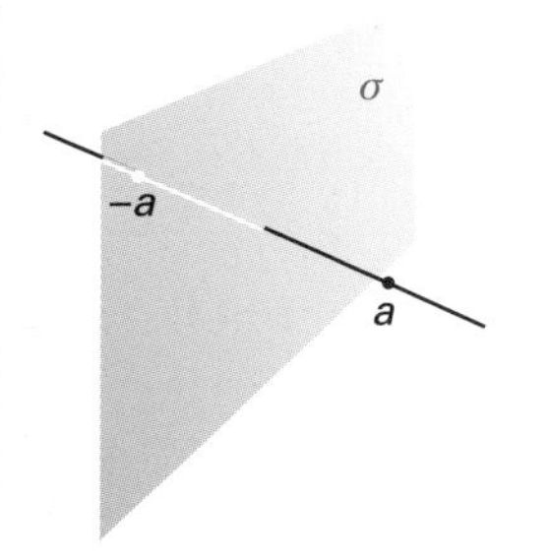

Fig. 5.32 The symmetry element of a symmetric integration range.

Example 5.11 The identification of zero integrals

Determine which orbitals of nitrogen in ammonia may have non-vanishing overlap with the symmetry-adapted linear combination of hydrogen 1*s*-orbitals.

Method. The overlap integral has the form $\int \psi_i^* \psi_j \, d\tau$; hence it is non-vanishing only if $\Gamma_i \times \Gamma_j$ includes A_1. Begin by identifying the symmetry species of the N2*s*- and N2*p*-orbitals by using the character table in Appendix 1 and noting that p_x transforms as *x*, etc., and decide which can have non-vanishing overlap with the symmetry-adapted linear combinations of the H1*s*-orbitals. Use the direct-product tables in Appendix 1.

Answer. In C_{3v}, the N2*p*-orbitals span $A_1(p_z)$ and $E(p_x, p_y)$. Because $A_1 \times A_1 = A_1$ and $E \times E = A_1 + A_2 + E$, the p_z orbital can have nonzero overlap with the combination s_1, and the p_x and p_y orbitals can have nonzero overlap with s_2 and s_3. The N2*s*-orbital also spans A_1, and so may also overlap with s_1.

Comment. Note that whether the s_1 combination has nonzero overlap with N2p_z depends on the bond angle: when the molecule is flat, s_1 lies in the nodal plane of N2p_z and the overlap is zero.

Exercise 5.11. Show group theoretically that the overlap of s_1 and N2p_z is necessarily zero when the molecule is planar.

An integral of the form

$$I = \int f^{(l)*} f^{(l')} f^{(l'')} \, d\tau \tag{50}$$

over all space is also necessarily zero unless the integrand is a basis for the totally symmetric irreducible representation (such as A_1). To determine whether that is so, we first form $\Gamma^{(l)} \times \Gamma^{(l')}$ and expand it in the normal way. Then we take each $\Gamma^{(k)}$ in the expansion and form the direct product $\Gamma^{(k)} \times \Gamma^{(l'')}$. If A_1 (or its equivalent) occurs nowhere in the resulting expression, then the integral *I* is necessarily zero. This conclusion is of the greatest importance in quantum mechanics because we often encounter integrals of the form

$$\langle a|\Omega|b\rangle = \int \psi_a^* \Omega \psi_b \, d\tau$$

Therefore, we can use group theory to decide when matrix elements are necessarily zero. This often results in an immense simplification of the construction of molecular orbitals, the interpretation of spectra, and the calculation of molecular properties.

Example 5.12 The identification of vanishing matrix elements

Do the integrals (a) $\langle d_{xy}|z|d_{x^2-y^2}\rangle$ and (b) $\langle d_{xy}|l_z|d_{x^2-y^2}\rangle$ vanish in a C_{4v} molecule?

Method. We need to assess whether $\Gamma^{(l)} \times \Gamma^{(l')} \times \Gamma^{(l'')}$ contains A_1. To do so, we use the character tables in Appendix 1 to identify the symmetry species of each function in the integral. Angular momenta transform as rotations (so l_z transforms as the rotation R_z which is listed in the tables). Use Appendix 1 for the direct-product decomposition.

Answer. In C_{4v}, d_{xy} and $d_{z^2-y^2}$ span B_2 and B_1, respectively, z spans A_1, and l_2 spans A_2. (a) The integral spans

$$B_2 \times A_1 \times B_1 = B_2 \times B_1 = A_2$$

and hence the matrix element must vanish. (b) The integrand spans

$$B_2 \times A_2 \times B_1 = B_2 \times B_2 = A_1$$

and hence the integral is not necessarily zero.

Comment. Matrix elements of this kind are particularly important for discussing electronic spectra: we shall see that they occur in the formulation of selection rules.

5.17 Symmetry and degeneracy

We have already mentioned (in Section 2.15) that the presence of degeneracy is a consequence of the symmetry of a system. We are now in a position to discuss the relation. To do so, we note that the hamiltonian of a system must be invariant under every operation of the relevant point group:

$$(RH) = H \tag{51}$$

A qualitative interpretation of this result is that the hamiltonian is the operator for the energy, and energy does not change under a symmetry operation. An example is the hamiltonian for the harmonic oscillator: the kinetic energy operator is proportional to $\mathrm{d}^2/\mathrm{d}x^2$ and the potential energy operator is proportional to x^2. Both terms are invariant under the replacement $x \rightarrow -x$, and so the hamiltonian spans the totally symmetric irreducible representation of the point group C_s. Because H is invariant under a similarity transformation of the group (that is, any symmetry operation leaves it unchanged), we can write

$$RHR^{-1} = H$$

Multiplication from the right by R gives $RH = HR$, so we can conclude that symmetry operations must commute with the hamiltonian.

Now consider an eigenfunction ψ_i of H with eigenvalue E. That is, $H\psi_i = E\psi_i$. We can multiply this equation from the left by R, giving $RH\psi_i = ER\psi_i$, and insert $R^{-1}R$ for the identity, to obtain

$$RHR^{-1}R\psi_i = ER\psi_i$$

From the invariance of H it then follows that

$$HR\psi_i = ER\psi_i$$

Therefore, ψ_i and $R\psi_i$ correspond to the same energy E. Thus we conclude that functions that can be generated from one another by any symmetry operation of the system, have the same energy. That is:

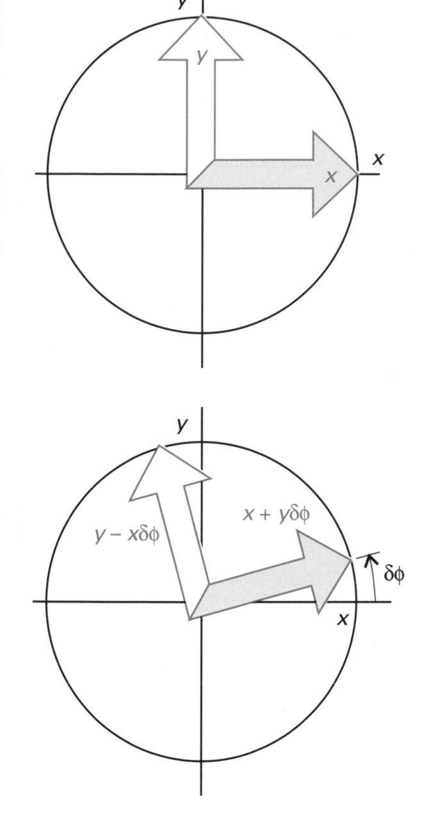

Fig. 5.33 The effect on the functions x and y of an infinitesimal rotation $\delta\phi$ about the z-axis.

> Eigenfunctions that are related by symmetry transformations of the system are degenerate.

We have already seen an example in the discussion of the geometrically square two-dimensional square-well eigenfunctions in Section 2.15.

We can go on to formulate a rule for the maximum degree of degeneracy that can occur in a system of given symmetry. Consider a member ψ_j of a basis for an irreducible representation of dimension d of the point group for the system, and suppose it has an energy E. We have already seen that all the other members of the basis can be generated by acting on this function with the projection operator P_{ij} defined in eqn 5.32. However, because P_{ij} is a linear combination of the symmetry operations of the group, it commutes with the hamiltonian. Therefore,

$$P_{ij}H\psi_j = HP_{ij}\psi_j = H\psi_i \qquad \text{and } P_{ij}H\psi_j = P_{ij}E\psi_j = E\psi_i$$

and hence $H\psi_i = E\psi_i$, and ψ_i has the same eigenvalue as ψ_j. But we can generate all d members of the d-dimensional basis by choosing the index i appropriately, and so all d basis functions have the same energy. We can conclude that:

> The degree of degeneracy of a set of functions is equal to the dimension of the irreducible representation they span.

This dimension is always given by $\chi(E)$, the character of the identity.

In the harmonic oscillator, with point group C_s, the only irreducible representations are one-dimensional, and so all the eigenfunctions are nondegenerate. For a geometrically square two-dimensional square-well potential, with point group C_{4v}, two-dimensional irreducible representations are allowed, and so some levels can be doubly degenerate. Triply degenerate levels occur in systems with cubic point-group symmetry, and five-fold degeneracy is encountered in icosahedral systems. The full rotation group, R_3, has irreducible representations of arbitrarily high dimension, and so degeneracies of any degree can occur.

The full rotation group

We shall now consider the full rotation groups in two and three dimensions (R_2 and R_3) and discover the deep connection between group theory and the quantum mechanics of angular momentum. The techniques are no different in principle from those introduced earlier in the chapter, but there are some interesting points of detail.

5.18 The generators of rotations

Consider first the full rotation group in two dimensions, the point group of a circular system (Fig. 5.33). You should bear in mind the analogous illustration for the equilateral triangle (Fig. 5.28) to see the analogies between the treatment of finite and infinite rotation groups. The difference is that in R_2 (which is a synonym of $C_{\infty v}$) a rotation through any angle is a symmetry operation. In particular, rotations through infinitesimal angles are symmetry operations.

We shall first establish the effect of an infinitesimal counter-clockwise rotation through an angle $\delta\phi$ about the z-axis on the basis (x, y). It will be convenient to work in polar coordinates and to write the basis as $(r\cos\phi, r\sin\phi)$, with r a constant under all operations of the group. Under the infinitesimal rotation $\delta\phi$, which we denote $C_{\delta\phi}$, the basis transforms as follows:

$$\begin{aligned} C_{\delta\phi}(x,y) &= \{r\cos(\phi-\delta\phi), r\sin(\phi-\delta\phi)\} \\ &= \{r\cos\phi\cos\delta\phi + r\sin\phi\sin\delta\phi, r\sin\phi\cos\delta\phi - r\cos\phi\sin\delta\phi\} \\ &= \{r\cos\phi + r\delta\phi\sin\phi + \ldots, r\sin\phi - r\delta\phi\cos\phi + \ldots\} \\ &= (x + y\delta\phi + \ldots, y - x\delta\phi + \ldots) = (x,y) - (-y,x)\delta\phi + \ldots \end{aligned}$$

We have used the expansions $\sin x = x - \frac{1}{6}x^3 + \ldots$ and $\cos x = 1 - \frac{1}{2}x^2 + \ldots$ and have kept only lowest-order terms in the infinitesimal angle $\delta\phi$. That is:

$$C_{\delta\phi}(x,y) = (x,y) - (-y,x)\delta\phi + \ldots$$

Now we identify an important fact. Consider the effect of the angular momentum operator

$$l_z = \frac{\hbar}{\mathrm{i}}\left(x\frac{\partial}{\partial y} - y\frac{\partial}{\partial x}\right) \tag{52}$$

on the basis:

$$l_z(x, y) = \frac{\hbar}{\mathrm{i}}\left(x\frac{\partial}{\partial y} - y\frac{\partial}{\partial x}\right)(x, y) = \frac{\hbar}{\mathrm{i}}(-y, x)$$

By comparing this result with the effect of $C_{\delta\phi}$, we see that

$$C_{\delta\phi}(x,y) = \left\{1 - \frac{\mathrm{i}}{\hbar}\delta\phi l_z + \ldots\right\}(x,y) \tag{53}$$

and that the operator itself can be written

$$C_{\delta\phi} = 1 - \frac{\mathrm{i}}{\hbar}\delta\phi l_z + \ldots \tag{54}$$

The infinitesimal rotation operator therefore differs from the identity to first order in $\delta\phi$ by a term that is proportional to the operator l_z. The operator $1 - (\mathrm{i}/\hbar)\delta\phi l_z$ is therefore called the **generator** of the infinitesimal rotation about the z-axis. In a similar way, the operators l_x and l_y are the generators for rotations about the x- and y-axes in R_3.

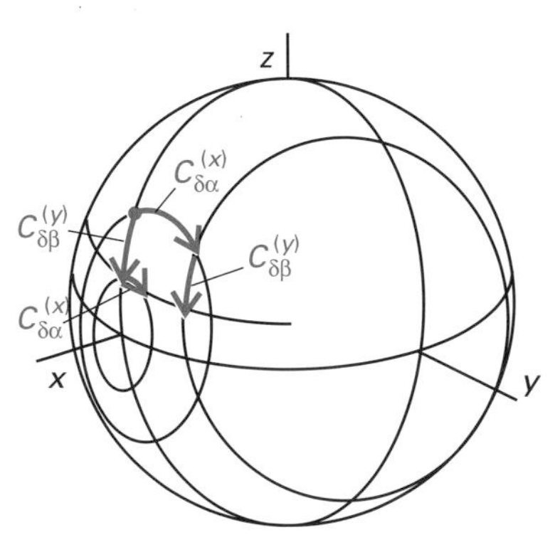

Fig. 5.34 The non-commutation of perpendicular rotations. Notice that the outcome of the combined rotation $C^{(y)}_{\delta\beta}C^{(x)}_{\delta\alpha}$ is different from the outcome of $C^{(x)}_{\delta\alpha}C^{(y)}_{\delta\beta}$.

We know that the angular momentum operators satisfy a set of commutation relations. These can be seen in a different light as follows. The effect of a sequence of rotations about different axes depends on the order in which they are applied (Fig. 5.34). Under a rotation by $\delta\alpha$ about x followed by a rotation by $\delta\beta$ about y, we have

$$\begin{aligned} C^{(y)}_{\delta\beta}C^{(x)}_{\delta\alpha} &= \left(1 - \frac{\mathrm{i}}{\hbar}\delta\beta l_y + \ldots\right)\left(1 - \frac{\mathrm{i}}{\hbar}\delta\alpha l_x + \ldots\right) \\ &= 1 - \frac{\mathrm{i}}{\hbar}(\delta\beta l_y + \delta\alpha l_x) + \left(\frac{\mathrm{i}}{\hbar}\right)^2\delta\beta\delta\alpha l_y l_x + \ldots \end{aligned}$$

However, if the rotations are applied in the opposite order the outcome is

$$\begin{aligned} C_{\delta\alpha}^{(x)} C_{\delta\beta}^{(y)} &= \left(1 - \frac{\mathrm{i}}{\hbar}\delta\alpha l_x + \ldots\right)\left(1 - \frac{\mathrm{i}}{\hbar}\delta\beta l_y + \ldots\right) \\ &= 1 - \frac{\mathrm{i}}{\hbar}(\delta\beta l_y + \delta\alpha l_x) + \left(\frac{\mathrm{i}}{\hbar}\right)^2 \delta\alpha\delta\beta l_x l_y + \ldots \end{aligned}$$

The difference between these two operations to second order is

$$C_{\delta\beta}^{(y)} C_{\delta\alpha}^{(x)} - C_{\delta\alpha}^{(x)} C_{\delta\beta}^{(y)} = \left(\frac{\mathrm{i}}{\hbar}\right)^2 \delta\alpha\delta\beta(l_y l_x - l_x l_y) = \frac{\mathrm{i}}{\hbar}\delta\alpha\delta\beta l_z \tag{55}$$

where the last equality follows from the commutation relation $[l_x, l_y] = \mathrm{i}\hbar l_z$.

The result we have established is that the difference between two infinitesimal rotations is equivalent to a single infinitesimal rotation through the angle $-\delta\alpha\delta\beta$ about the z-axis, which is geometrically plausible (Fig. 5.34). The *reverse* argument, that it is geometrically obvious that the difference is a single rotation therefore implies that $[l_x, l_y] = \mathrm{i}\hbar l_z$. Hence, the angular momentum commutation relations can be regarded as a direct consequence of the geometrical properties of composite rotations.

5.19 The representation of the full rotation group

We shall now look for the irreducible representations of the full rotation group R_3. As a starting point, we note that the spherical harmonics Y_{lm_l} for a given l transform into linear combinations of one another under a rotation. (For example, p-orbitals rotate into one another, d-orbitals do likewise, and so on, but p-orbitals do not rotate into d-orbitals.) Therefore, the functions $Y_{ll}, Y_{l,l-1}, \ldots, Y_{l,-l}$ form a basis for a $(2l+1)$-dimensional (and it turns out, irreducible) representation of the group. Each spherical harmonic has the form $Y_{lm_l} = P(\theta)\mathrm{e}^{\mathrm{i}m_l\phi}$, and so, as a result of a rotation by α around the z-axis, each one transforms into $P(\theta)\mathrm{e}^{\mathrm{i}m_l(\phi-\alpha)}$. The entire basis therefore transforms as follows:

$$\begin{aligned} &C_{\alpha}^{(z)}(Y_{ll}, Y_{l,l-1}, \ldots, Y_{l,-l}) \\ &\quad = \left(P(\theta)\mathrm{e}^{\mathrm{i}l(\phi-\alpha)}, P(\theta)\mathrm{e}^{\mathrm{i}(l-1)(\phi-\alpha)}, \ldots, P(\theta)\mathrm{e}^{-\mathrm{i}l(\phi-\alpha)}\right) \\ &\quad = (Y_{ll}, Y_{l,l-1}, \ldots, Y_{l,-l}) \begin{bmatrix} \mathrm{e}^{-\mathrm{i}l\alpha} & 0 & 0 & \ldots & 0 \\ 0 & \mathrm{e}^{-\mathrm{i}(l-1)\alpha} & 0 & \ldots & 0 \\ 0 & 0 & & & \vdots \\ \vdots & \vdots & & & \vdots \\ 0 & 0 & \ldots & \ldots & \mathrm{e}^{\mathrm{i}l\alpha} \end{bmatrix} \end{aligned} \tag{56}$$

This expression lets us recognize the matrix representative of the rotation in the basis.

The character of a rotation through the angle α about the z-axis (and therefore about any axis, because in R_3 all rotations through a given angle belong to the same class) is the following sum:

$$\begin{aligned}\chi(C_\alpha) &= e^{-il\alpha} + e^{-i(l-1)\alpha} + \ldots + e^{il\alpha} \\ &= 1 + 2\cos\alpha + 2\cos 2\alpha + \ldots + 2\cos(l-1)\alpha + 2\cos l\alpha\end{aligned} \tag{57}$$

To obtain this expression, we have used $e^{ix} + e^{-ix} = 2\cos x$; the leading 1 comes from the term with $m_l = 0$. This simple expression can be used to establish the character of any rotation for a $(2l+1)$-dimensional basis. An even simpler version is obtained by recognizing that the first line is a geometric series.[3] Hence, it is the sum

$$\chi(C_\alpha) = \sum_{m_l=-l}^{l} e^{im_l\alpha} = \frac{e^{-il\alpha}\left(e^{i(2l+1)\alpha} - 1\right)}{e^{i\alpha} - 1} \tag{58}$$

This slightly awkward expression can be manipulated into

$$\chi(C_\alpha) = \frac{\sin(l+\frac{1}{2})\alpha}{\sin\frac{1}{2}\alpha} \tag{59}$$

When $\alpha = 0$, the character is $2l + 1$ (take the limit as $\alpha \to 0$), and so the levels with quantum number l are $(2l+1)$-fold degenerate in a spherical system.

Example 5.13 How to determine the symmetry species of atoms in various environments

An atom has a configuration that gives rise to an F term. What symmetry species would it give rise to in an octahedral environment?

Method. We need to identify the rotations that are common to both R_3 and O, and then to calculate their characters from eqn 5.59 with $l = 3$. Then, by referring to the character table for O in Appendix 1 we can identify the symmetry species spanned by the term in the reduced symmetry environment.

Answer. The rotation angles in O are $\alpha = 0$ for E, $\alpha = \frac{2}{3}\pi(C_3)$, $\pi(C_2)$, $\frac{1}{2}\pi(C_4)$, $\pi(C_2')$. Because

$$\chi(C_\alpha) = \frac{\sin(7\alpha/2)}{\sin(\alpha/2)}$$

we find $\chi = (7, 1, -1, -1, -1)$ for (E, C_3, C_2, C_4, C_2'). Then, use of eqn 5.29 with $h = 24$ gives $a(A_2) = 1$, $a(T_1) = 1$, and $a(T_2) = 1$. Therefore, $F \to A_2 + T_1 + T_2$.

Comment. The step down from a group to its subgroup is called 'descent in symmetry'. It is a particularly important technique in the theory of the structure and spectra of *d*-metal complexes (see Chapter 8).

Exercise 5.13. What irreducible representations does a G term span in tetrahedral symmetry?

[3] The sum of a series $a + ar + ar^2 + \ldots + ar^n$ is $a(r^{n+1} - 1)/(r - 1)$. In the present case, $a = e^{-il\alpha}$, $r = e^{i\alpha}$, and $n = 2l$.

5.20 Coupled angular momenta

We now explore the group-theoretical description of the coupling of two angular momenta. We suppose that we have two sets of functions that are the bases for irreducible representations $\Gamma^{(j_1)}$ and $\Gamma^{(j_2)}$ of the full rotation group. The functions will be denoted $f_{m_{j1}}^{(j_1)}$ and $f_{m_{j2}}^{(j_2)}$, respectively. The products $f_{m_{j1}}^{(j_1)} f_{m_{j2}}^{(j_2)}$ provide a basis for the direct-product representation $\Gamma^{(j_1)} \times \Gamma^{(j_2)}$. This representation is in general reducible, and we can reduce it as explained in Section 5.14.

First, we write

$$\Gamma^{(j_1)} \times \Gamma^{(j_2)} = \sum_j a_j \Gamma^{(j)} \tag{60}$$

To determine the coefficients we consider the characters:

$$\begin{aligned} \chi(C_\alpha) &= \chi^{(j_1)}(C_\alpha)\chi^{(j_2)}(C_\alpha) \\ &= \sum_{m_{j1}=-j_1}^{j_1} \sum_{m_{j2}=-j_2}^{j_2} \mathrm{e}^{\mathrm{i}(m_{j1}+m_{j2})\alpha} \end{aligned} \tag{61}$$

The question we now address is whether the right-hand side of this equation can be expressed as a sum over $\sum_{m_j} \mathrm{e}^{\mathrm{i}m_j\alpha}$ and, if so, how many times each term in the sum appears. We shall now demonstrate that each term appears exactly once, and that j varies from $j_1 + j_2$ down to $|j_1 - j_2|$.

The argument runs as follows. Because $|m_{j1} + m_{j2}| \leq j_1 + j_2$, it follows that $|m_j| \leq j_1 + j_2$, and so $j \leq j_1 + j_2$. Therefore, $a_j = 0$ if $j > j_1 + j_2$. The maximum value of m_j may be obtained from m_{j1} and m_{j2} in only one way: when $m_{j1} = j_1$ and $m_{j2} = j_2$. Therefore, $a_{j_1+j_2} = 1$. The next value of m_j, which is $j - 1$, may be obtained in two ways, namely $m_{j1} = j_1 - 1$ and $m_{j2} = j_2$ or $m_{j1} = j_1$ and $m_{j2} - j_2 - 1$; one of these ways is accounted for by the representation with $j = j_1 + j_2$, and so we can conclude that $a_{j_1+j_2-1} = 1$. This argument can be continued down to $j = |j_1 - j_2|$, and so eqn 5.61 is equivalent to

$$\chi(C_\alpha) = \sum_{j=|j_1-j_2|}^{j_1+j_2} \sum_{m_j=-j}^{j} \mathrm{e}^{\mathrm{i}m_j\alpha} = \sum_{j=|j_1-j_2|}^{j_1+j_2} \chi^{(j)}(C_\alpha) \tag{62}$$

Therefore, we can conclude that the direct product decomposes as follows:

$$\Gamma^{(j_1)} \times \Gamma^{(j_2)} = \Gamma^{(j_1+j_2)} + \Gamma^{(j_1+j_2-1)} + \ldots + \Gamma^{(|j_1-j_2|)} \tag{63}$$

which is nothing other than the Clebsch–Gordan series, eqn 4.43. This result shows, in effect, that the whole of angular momentum theory can be regarded as an aspect of group theory and the symmetry properties of rotations.

Applications

There are numerous applications of group theory, both explicit and implicit. We shall encounter many of them in the following pages. That being so, we shall only indicate here the types of applications that are encountered, and where in the text. The application of the rotation groups (R_3, $D_{\infty h}$, and $C_{\infty v}$) will appear wherever we discuss the angular momentum of atoms and molecules (Chapters

7, 10, and 11). Finite groups play an important role in the discussion of molecular structure and properties, both in the setting up of molecular orbitals (Chapter 8) and in the evaluation of the matrix elements and expectation values that are needed to evaluate molecular properties (Chapter 6). When an atom or ion is embedded in a local environment, as in a crystal or a complex, the degeneracy of its orbitals is removed with important consequences for its spectroscopic features (Chapter 11). Spectroscopy in general relies heavily on group-theoretical arguments in its classification of states, the construction of normal modes of vibration, and the derivation of selection rules. The calculation of the electric and magnetic properties of molecules relies heavily on the evaluation of matrix elements, and group theory helps by eliminating many integrals on the basis of symmetry alone (Chapters 12 and 13). The following chapters will confirm that group theory does indeed pervade the whole of quantum chemistry.

Problems

5.1 Classify the following molecules according to their point symmetry group: (a) H_2O, (b) CO_2, (c) C_2H_4, (d) *cis*-ClHC=CHCl, (e) *trans*-ClCH=CHCl, (f) benzene, (g) naphthalene, (h) CHClFBr, (i) $B(OH)_3$.

5.2 Which of the molecules listed above may possess a permanent electric dipole moment? *Hint*. Decide on the criterion for the non-vanishing of $\langle\boldsymbol{\mu}\rangle = \int \psi^* \boldsymbol{\mu} \psi \, d\tau$ and refer to the tables in Appendix 1; $\boldsymbol{\mu}$ transforms as $\mathbf{r} = (x, y, z)$.

5.3 Find the representatives of the operations of the group C_{2v} using as a basis the valence orbitals of H and O in H_2O (that is, $H1s_A$, $H1s_B$, O2*s*, three O2*p*). *Hint*. The group is of order 4 and so there are four six-dimensional matrices to find.

5.4 Confirm that the representatives established in Problem 5.3 reproduce the group multiplications $C_2^2 = E$, $\sigma_v C_2 = \sigma'_v$.

5.5 Determine which symmetry species are spanned by the six orbitals of H_2O described in Problem 5.3. Find the symmetry-adapted linear combinations, and confirm that the representatives are in block-diagonal form.

5.6 Find the representatives of the operations of the group T_d by using as a basis four 1*s*-orbitals, one at each apex of a regular tetrahedron (as in CH_4). *Hint*. The basis is four-dimensional; the order of the group is 24, and so there are 24 matrices to find.

5.7 Confirm that the representations established in Problem 5.6 reproduce the group multiplications $C_3^+ C_3^- = E$, $S_4 C_3 = S'_4$, and $S_4 C_3 = \sigma_d$.

5.8 Determine which irreducible representations are spanned by the four 1*s*-orbitals in methane. Find the symmetry-adapted linear combinations, and confirm that the representatives for C_3^+ and S_4 are in block-diagonal form. *Hint*. Decompose the representation into irreducible representations by analysing the characters. Use the projection operator in eqn 5.32 to establish the symmetry-adapted bases.

5.9 Analyse the following direct products into the symmetry species they span: (a) C_{2v}: $A_2 \times B_1 \times B_2$, (b) C_{3v}: $A_1 \times A_2 \times E$, (c) C_{6v}: $B_2 \times E_1$, (d) $C_{\infty v}$: E_1^2, (e) O: $T_1 \times T_2 \times E$.

5.10 Show that $3x^2y - y^3$ is a basis for an A_1 irreducible representation of C_{3v}. *Hint*. Use the information in Section 5.13 to show that $C_3^+(3x^2y - y^3) \propto 3x^2y - y^3$; likewise for the other elements of the group.

5.11 A function $f(x, y, z)$ was found to be a basis for a representation of C_{2v}, the characters being $(4, 0, 0, 0)$. What symmetry species of irreducible representation does it span? *Hint*. Proceed by inspection to find the a_l in eqn 5.26 or use eqn 5.28.

5.12 Find the components of the function $f(x,y,z)$ in Problem 5.11 acting as a basis for each irreducible representation it spans. *Hint.* Use eqn 5.28. The basis for A_1, for example, turns out to be $\frac{1}{4}\{f(x,y,z)+f(-x,-y,-z)+f(x,-y,-z)+f(-x,y,z)\}$.

5.13 Regard the naphthalene molecule as having C_{2v} symmetry (with the C_2 axis perpendicular to the plane), which is a subgroup of its full symmetry group. Consider the *p*-orbitals on each carbon as a basis. What symmetry species do they span? Construct the symmetry-adapted bases.

5.14 Repeat the process for benzene, using the subgroup C_{6v} of the full symmetry group. After constructing the symmetry-adapted linear combinations, refer to the D_{6h} character table to label them according to the full group.

5.15 Show that in an octahedral array, hydrogen 1*s*-orbitals span $\mathrm{A}_{1g}+\mathrm{E}_g+\mathrm{T}_{1u}$ of the group O_h.

5.16 Classify the terms that may arise from the following configurations: (a) C_{2v}: $a_1^2b_1^1b_2^1$; (b) C_{3v}: $a_2^1e^1$, e^2; (c) T_d: $a_2^1e^1$, $e^1t_1^1$, $t_1^1t_2^1$, t_1^2, t_2^2; (d) O: e^2, $e^1t_1^1$, t_2^2. *Hint.* Use the direct product tables; triplet terms have antisymmetric spatial functions.

5.17 Construct the character tables for the groups O_h and D_{6h}. *Hint.* Use $D_{6h}=D_6\times C_i$ and $O_h=O\times C_i$ and the procedure in Section 5.15.

5.18 Demonstrate that there are no nonzero integrals of the form $\int\psi' H\psi\,d\tau$ when ψ' and ψ belong to different symmetry species.

5.19 The ground states of the C_{2v} molecules NO_2 and ClO_2 are $^2\mathrm{A}_1$ and $^2\mathrm{B}_1$, respectively; the ground state of O_2 is $^3\Sigma_g^-$. To what states may (a) electric-dipole, (b) magnetic-dipole transitions take place? *Hint.* The electric-dipole operator transforms as translations, the magnetic as rotations.

5.20 What is the maximum degeneracy of the energy levels of a particle confined to the interior of a regular tetrahedron?

5.21 Demonstrate that the linear momentum operator $p=(\hbar/i)(d/dx)$ is the generator of infinitesimal translations. *Hint.* Proceed as in Section 5.18.

5.22 An atom bearing a single *p*-electron is trapped in an environment with C_{3v} symmetry. What symmetry species does it span? *Hint.* Use eqn 5.59 with $\alpha=120°$.

Further reading

Chemical applications of group theory. F.A. Cotton; Wiley, New York (1990).
Symmetry in molecules and crystals. M.F.C. Ladd; Wiley, New York (1989).
Group theory and its applications to physical problems. M. Hamermesh; Addison-Wesley, Reading, Mass. (1962).
Tables for group theory. P.W. Atkins, M.S. Child, and C.S.G. Phillips; Oxford University Press, Oxford (1970).
Symmetry in bonding and spectra: an introduction. B.E. Douglas and C.A. Hollingsworth; Academic Press, Orlando (1985).
Group theory in spectroscopy: with applications to magnetic circular dichroism. S.B. Piepho and P.N. Schatz; Wiley, New York (1983).
Symmetry principles of molecules. G.S. Ezra; Springer, Berlin (1982).
Group theory and its applications in physics. T. Inui, Y. Tanabe, and Y. Onodera; Springer, New York (1990).
Icons and symmetries. S.L. Altmann; Oxford University Press, Oxford (1992).

6 Techniques of approximation

This is a sad but necessary chapter. It is sad because we have reached the point at which the hope of finding exact solutions is set aside and we begin to look for methods of approximation. It is necessary, because most of the problems of quantum chemistry cannot be solved exactly, and so we must learn how to tackle them. There are very few problems for which the Schrödinger equation can be solved exactly, and the examples in previous chapters almost exhaust the list. As soon as the shape of the potential is distorted from the forms already considered, or more than two particles interact with one another (as in a helium atom), the equation cannot be solved exactly.

There are three ways of making progress. The first is to try to guess the shape of the wavefunction of the system. Even people with profound insight need a criterion of success, and this is provided by the *variation principle*, which we specify below. It is useful to be guided to the form of the wavefunction by a knowledge of the distortion of the system induced by the complicating aspects of the potential or the interactions. For example, the exact solutions of the system that resembles the true system may be known and can be used as a guide to the true solutions by noting how the hamiltonians of the two systems differ. This procedure is the province of *perturbation theory*. Perturbation theory is particularly useful when we are interested in the response of atoms and molecules to electric and magnetic fields. When these fields change with time (as in a light wave) we have to deal with *time-dependent perturbation theory*. The third important method of approximation, which is dealt with in detail in Chapters 7 and 9, makes use of *self-consistent field* procedures, which is an iterative method for solving the Schrödinger equation for systems of many particles.

Time-independent perturbation theory

Our first concern is with time-independent perturbation theory. In this technique we make use of the fact that the hamiltonians for the true and simpler model system, H and $H^{(0)}$, respectively, differ by a contribution that is independent of the time:

$$H = H^{(0)} + H^{(1)} \tag{1}$$

We refer to $H^{(1)}$ as the **perturbation**.

6.1 Perturbation of a two-level system

Consider first a system that has only two eigenstates. We suppose that the two eigenstates of $H^{(0)}$ are known, and denote them $|1\rangle$ and $|2\rangle$. The corresponding wavefunctions are $\psi_1^{(0)}$ and $\psi_2^{(0)}$, respectively. These states and functions form a complete orthonormal basis. They correspond to the energies E_1 and E_2:

$$H^{(0)}\psi_m^{(0)} = E_m\psi_m^{(0)} \qquad m = 1, 2$$

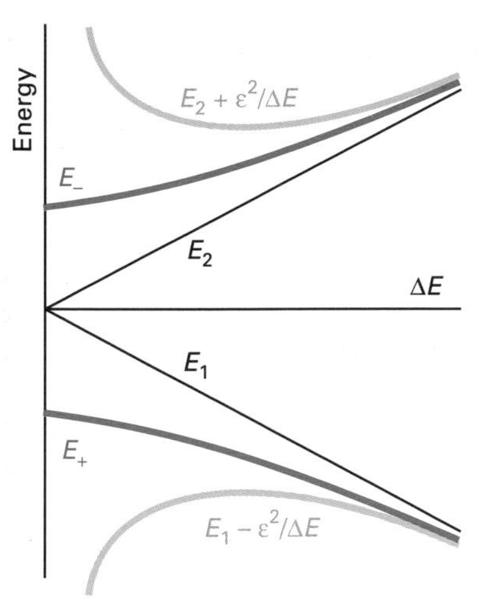

Fig. 6.1 The variation of the energies of a two-level system with a constant perturbation as the separation of the unperturbed levels is increased. The pale lines show the energies according to second-order perturbation theory.

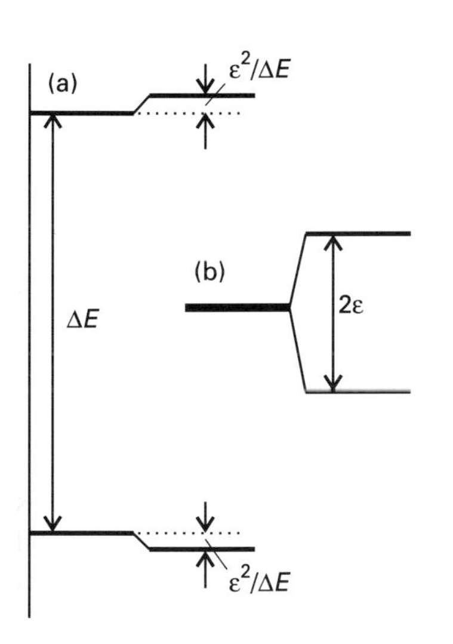

Fig. 6.2 (a) When the unperturbed levels are far apart in energy, the shift in energy caused by a perturbation of strength ε is $\pm\varepsilon^2/\Delta E$. (b) If the levels are initially degenerate, then the shift in energy is much larger, and is equal to $\pm\varepsilon$.

The wavefunctions of the true system differ only slightly from those of the model system, and we can hope to solve the equation

$$H\psi = E\psi \tag{2}$$

in terms of them by writing

$$\psi = a_1\psi_1^{(0)} + a_2\psi_2^{(0)} \tag{3}$$

where a_1 and a_2 are constants.

To find the constants a_m we insert the linear combination into the Schrödinger equation and obtain (using ket notation)

$$a_1(H - E)|1\rangle + a_2(H - E)|2\rangle = 0$$

When this equation is multiplied from the left by the bras $\langle 1|$ and $\langle 2|$ in turn, and use is made of the orthonormality of the two states, we obtain the two equations

$$a_1(H_{11} - E) + a_2H_{12} = 0 \qquad a_1H_{21} + a_2(H_{22} - E) = 0 \tag{4}$$

where $H_{mn} = \langle m|H|n\rangle$.

The condition for the existence of non-trivial solutions of this pair of equations is that the determinant of the coefficients of the constants a_1 and a_2 should disappear:

$$\begin{vmatrix} H_{11} - E & H_{12} \\ H_{21} & H_{22} - E \end{vmatrix} = 0$$

This condition is satisfied for the following values of E:

$$E_\pm = \tfrac{1}{2}(H_{11} + H_{22}) \pm \tfrac{1}{2}\{(H_{11} - H_{22})^2 + 4H_{12}H_{21}\}^{1/2} \tag{5}$$

In the special but common case of a perturbation for which the diagonal matrix elements are zero ($H_{mm}^{(1)} = 0$), this expression simplifies to

$$E_\pm = \tfrac{1}{2}(E_1 + E_2) \pm \tfrac{1}{2}\{(E_1 - E_2)^2 + 4\varepsilon^2\}^{1/2} \tag{6}$$

where $\varepsilon^2 = H_{12}^{(1)}H_{21}^{(1)}$. Because $H^{(1)}$ is hermitian, we can write $\varepsilon^2 = |H_{12}^{(1)}|^2$. When the perturbation is absent, $\varepsilon = 0$ and $E_+ = E_1$, $E_- = E_2$, the two unperturbed energies.

The variation of the energies of the system as the separation of the states of the model system is increased is illustrated in Fig. 6.1. As can be seen, the lower of the two levels is lowered in energy whereas that of the upper level is raised. In other words, the effect of the perturbation is to drive the energy levels apart and to prevent their crossing. This **non-crossing rule** is a common feature of all perturbations. A second general feature can also be seen from the illustration: the effect of the perturbation is greater the smaller the energy separation of the unperturbed levels. For instance, when the two original energies have the same energy ($E_1 = E_2$), then

$$E_+ - E_- = 2\varepsilon$$

Equation 6.6 also shows that the stronger the perturbation, the stronger the effective repulsion of the levels (Fig. 6.2). In summary:

1. When a perturbation is applied, the lower level moves down in energy and the upper level moves up.

2. The closer the unperturbed states are in energy, the greater the effect of a perturbation.
3. The stronger the perturbation, the greater the effect on the energies of the levels.

The effect of the perturbation can be seen in more detail by considering the case of a perturbation that is weak compared with the separation of the energy levels in the sense that $\varepsilon^2 \ll (E_1 - E_2)^2$. When this condition holds, eqn 6.6 can be expanded by making use of $(1+x)^{1/2} = 1 + \frac{1}{2}x + \ldots$, to obtain

$$E_{\pm} = \tfrac{1}{2}(E_1 + E_2) \pm \tfrac{1}{2}(E_1 - E_2)\left(1 + \frac{2\varepsilon^2}{\Delta E^2} + \ldots\right)$$

where $\Delta E = E_2 - E_1$. Then, to second-order in ε we have

$$E_+ \approx E_1 - \frac{\varepsilon^2}{\Delta E} \qquad E_- \approx E_2 + \frac{\varepsilon^2}{\Delta E} \tag{7}$$

These two solutions converge on the exact solutions when $(2\varepsilon/\Delta E)^2 \ll 1$, as shown in Fig. 6.1. A general feature of all perturbation theory calculations is that the shifts in energy are of the order of $\varepsilon^2/\Delta E$.

The perturbed wavefunctions are obtained by solving eqn 6.4 for the coefficients setting in turn $E = E_+$ (to obtain ψ_+) and $E = E_-$ (to obtain ψ_-). A convenient way to express the solutions is to write

$$\psi_+ = \psi_1^{(0)} \cos\zeta + \psi_2^{(0)} \sin\zeta \qquad \psi_- = -\psi_1^{(0)} \sin\zeta + \psi_2^{(0)} \cos\zeta \tag{8}$$

and then it is found that[1]

$$\tan 2\zeta = \frac{2|H_{12}^{(1)}|}{E_1 - E_2} \tag{9}$$

For a degenerate model system ($E_1 = E_2$), we have $\tan 2\zeta = \infty$, corresponding to $\zeta = \pi/4$. In this case the perturbed wavefunctions are

$$\psi_+ = \frac{1}{\sqrt{2}}\left(\psi_1^{(0)} + \psi_2^{(0)}\right) \qquad \psi_- = -\frac{1}{\sqrt{2}}\left(\psi_1^{(0)} - \psi_2^{(0)}\right) \tag{10}$$

It follows that each perturbed state is a 50 per cent mixture of the two model states. In contrast, for a perturbation acting on two widely separated states we can write $\tan 2\zeta \approx 2\zeta = -2|H_{12}^{(1)}|/\Delta E$. Furthermore, because $\sin\zeta \approx \zeta$ and $\cos\zeta \approx 1$, it follows that

$$\psi_+ \approx \psi_1^{(0)} - \frac{|H_{12}^{(1)}|}{\Delta E}\psi_2^{(0)} \qquad \psi_- \approx \psi_2^{(0)} + \frac{|H_{12}^{(1)}|}{\Delta E}\psi_1^{(0)} \tag{11}$$

We see that each model state is slightly contaminated by the other state.

6.2 Many-level systems

Now we generalize these results to a system in which there are numerous (possibly infinite) levels. Special precautions have to be taken if the state of interest is degenerate, and we consider that possibility in Section 6.8.

[1] In general, a complex matrix element $H_{12}^{(1)}$ can be written as $|H_{12}^{(1)}|e^{i\phi}$. In the following, we suppose that $\phi = 0$.

We suppose that we know all the eigenfunctions and eigenvalues of a model system with hamiltonian $H^{(0)}$ that differs from the true system to a small extent. An example might be an anharmonic oscillator or a molecule in a weak electric field: the model systems would then be a harmonic oscillator or a molecule in the absence of a field, respectively. We therefore suppose that we have found the solutions of the equations

$$H^{(0)}|n\rangle = E_n|n\rangle \tag{12}$$

with $n = 0, 1, 2, \ldots$, and $|n\rangle$ a member of an orthonormal basis. We shall suppose that we are calculating the perturbed form of the state $|0\rangle$ of energy E_0, but this state is not necessarily the ground state of the system.

The hamiltonian of the perturbed system will be written

$$H = H^{(0)} + \lambda H^{(1)} + \lambda^2 H^{(2)} + \ldots \tag{13}$$

The only significance of the parameter λ is that it keeps track of the order of the perturbation, and will enable us to identify all first-order terms in the energy, all second-order terms, and so on. At the end of the calculation we set $\lambda = 1$ because by then it will have served its purpose. Similarly, the perturbed wavefunction of the system will be written

$$\psi = \psi_0 + \lambda\psi_0^{(1)} + \lambda^2\psi_0^{(2)} + \ldots \tag{14}$$

which shows how the unperturbed function (ψ_0) is corrected by terms that are of various orders in the perturbation. The energy of the perturbed state also has correction terms of various orders, and we write

$$E = E_0 + \lambda E_0^{(1)} + \lambda^2 E_0^{(2)} + \ldots \tag{15}$$

We shall refer to $E_0^{(1)}$ as the **first-order correction** to the energy, to $E_0^{(2)}$ as the **second-order correction**, and so on.

The equation we need to solve is

$$H\psi = E\psi \tag{16}$$

Insertion of the preceding equations into this equation, followed by collecting terms that have the same power of λ, then results in

$$\begin{aligned}
&\lambda^0\{H^{(0)}\psi_0 - E_0\psi_0\} \\
&\quad + \lambda^1\{H^{(0)}\psi_0^{(1)} + H^{(1)}\psi_0 - E_0\psi_0^{(1)} - E_0^{(1)}\psi_0\} \\
&\quad + \lambda^2\{H^{(0)}\psi_0^{(2)} + H^{(1)}\psi_0^{(1)} + H^{(2)}\psi_0 - E_0\psi_0^{(2)} \\
&\qquad - E_0^{(1)}\psi_0^{(1)} - E_0^{(2)}\psi_0\} + \ldots = 0
\end{aligned}$$

Because λ is an arbitrary parameter, the coefficient of each power of λ must equal zero separately, and so we have the following set of equations:

$$\begin{aligned}
H^{(0)}\psi_0 &= E_0\psi_0 \\
\{H^{(0)} - E_0\}\psi_0^{(1)} &= \{E_0^{(1)} - H^{(1)}\}\psi_0 \\
\{H^{(0)} - E_0\}\psi_0^{(2)} &= \{E_0^{(2)} - H^{(2)}\}\psi_0 + \{E_0^{(1)} - H^{(1)}\}\psi_0^{(1)}
\end{aligned} \tag{17}$$

and so on.

6.3 The first-order correction to the energy

The solution of the first of these equations is assumed known (it is eqn 6.12). The first-order correction to the wavefunction is written as a linear combination of the unperturbed wavefunctions of the system:

$$\psi_0^{(1)} = \sum_n a_n \psi_n^{(0)} \tag{18}$$

The sum is over all states of the model system including those belonging to the continuum, if there is one. When this expansion is inserted into the equation for $\psi_0^{(1)}$, we obtain (in ket notation)

$$\begin{aligned}\sum_n a_n\{H^{(0)} - E_0\}|n\rangle &= \sum_n a_n\{E_n - E_0\}|n\rangle \\ &= \{E_0^{(1)} - H^{(1)}\}|0\rangle\end{aligned} \tag{19}$$

When this expression is multiplied from the left by the bra $\langle 0|$, we obtain

$$\sum_n a_n\{E_n - E_0\}\delta_{n0} = E_0^{(1)} - \langle 0|H^{(1)}|0\rangle$$

The left-hand side of this equation is zero, so we can conclude that the first-order correction to the energy of the state $|0\rangle$ is

$$E_0^{(1)} = \langle 0|H^{(1)}|0\rangle = H_{00}^{(1)} \tag{20}$$

The matrix element $H_{00}^{(1)}$ is the average value of the first-order perturbation over the unperturbed state $|0\rangle$. An analogy is the first-order shift in the frequency of a violin string when small weights are added along its length: those at the nodes have no effect on the frequency, those at the antinodes (the points of maximum amplitude) affect the frequency most strongly, and the overall effect is an average taking into account the displacement of the string at the location of each weight.

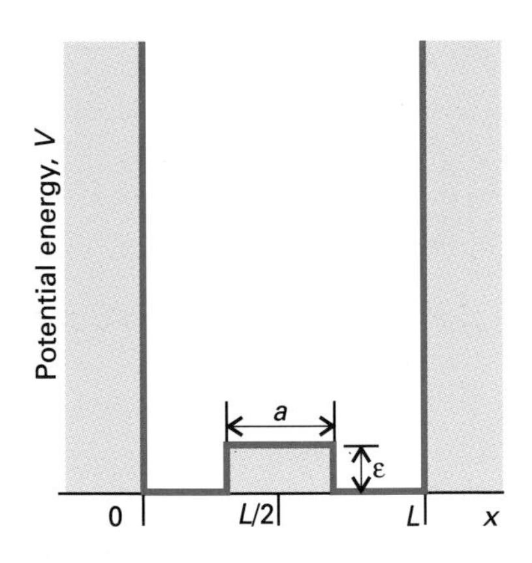

Fig. 6.3 The perturbation to a square-well potential used in Example 6.1.

Example 6.1 How to calculate the first-order correction to the energy

A small step in the potential energy is introduced into the one-dimensional square-well problem as illustrated in Fig. 6.3. Calculate the first-order correction to the energy of a particle confined to the well and evaluate it for $a = L/10$, so the blip in the potential occupies the central 10 per cent of the well, and for (a) $n = 1$, (b) $n = 2$.

Method. We need to evaluate eqn 6.20 by using

$$H^{(1)} = \begin{cases} \varepsilon & \text{if } \tfrac{1}{2}(L-a) \le x \le \tfrac{1}{2}(L+a) \\ 0 & \text{if } x \text{ is outside this region} \end{cases}$$

The wavefunctions are given in eqn 2.33. We should anticipate that the effect of the perturbation will be much smaller for $n = 2$ than for $n = 1$ because in the former it is applied in the vicinity of a node. A useful integral is

$$\int \sin^2 kx \, dx = \tfrac{1}{2}x - \frac{1}{4k}\sin 2kx + \text{constant}$$

Answer. The integral required is

$$E^{(1)} = \frac{2\varepsilon}{L}\int_{\frac{1}{2}(L-a)}^{\frac{1}{2}(L+a)} \sin^2\left(\frac{n\pi x}{L}\right) \mathrm{d}x$$
$$= \varepsilon\left\{\frac{a}{L} - (-1)^n\left(\frac{1}{n\pi}\right)\sin\left(\frac{n\pi a}{L}\right)\right\}$$

(a) For $n = 1$, $E^{(1)} = 0.1984\varepsilon$; (b) for $n = 2$, $E^{(1)} = 0.0065\varepsilon$.

Comment. The relative sizes of the two answers are consistent with the perturbation being close to an antinode and a node, respectively. When n is very large, $E^{(1)} \approx (a/L)\varepsilon$, independent of n. At such high quantum numbers, the probability of finding the particle in the region a is a/L regardless of n. Note that if $\varepsilon > 0$, then the energy of the states is increased from the unperturbed values.

Exercise 6.1. Evaluate the first-order correction to the energy of a particle in a box for a perturbation of the form $\varepsilon \sin x\pi/L$ for $n = 1$ and $n = 2$.

6.4 The first-order correction to the wavefunction

Now we look for the first-order correction to the state of the system. To find it, we multiply eqn 6.19 from the left by the bra $\langle k|$, where $k \neq 0$. The orthonormality of the states again simplifies the resulting expression, and we obtain the following expression:

$$\sum_n a_n\{E_n - E_0\}\delta_{kn} = a_k\{E_k - E_0\} = E_0^{(1)}\langle k|0\rangle - \langle k|H^{(1)}|0\rangle$$
$$= -\langle k|H^{(1)}|0\rangle = -H_{k0}^{(1)}$$

Because the state $|0\rangle$ is non-degenerate, the differences $E_k - E_0$ are all non-zero for $k \neq 0$. Therefore, the coefficients are given by

$$a_k = \frac{H_{k0}^{(1)}}{E_0 - E_k} \tag{21}$$

It follows that the wavefunction of the system corrected to first-order in the perturbation is

$$\psi \approx \psi_0 + \sum_k{}' \left\{\frac{H_{k0}^{(1)}}{E_0 - E_k}\right\}\psi_k^{(0)} \tag{22}$$

where the prime on the sum means that the state with $k = 0$ should be omitted.

The last equation echoes the expression derived for the two-level system in the limit of a weak perturbation and widely separated energy levels. As in that case, perturbation theory guides us towards the form of the perturbed state of the system. In this case, the procedure simulates the distortion of the state by mixing into it the other states of the system. This is expressed by saying that the perturbation induces **virtual transitions** to these other states of the model system. However, that is only a pictorial way of speaking: in fact, the distorted state is being simulated as a linear superposition of the unperturbed states of the system. The equation shows that a particular state k makes no contribution to the superposition if $H_{k0}^{(1)} = 0$, and (for a given magnitude of the matrix element) the contribution of a state is smaller the larger the energy difference $|E_0 - E_k|$.

6.5 The second-order correction to the energy

We use the same technique to extract the second-order correction to the energy from eqn 6.17. The second-order correction to the wavefunction is written as the linear combination

$$\psi_0^{(2)} = \sum_n b_n \psi_n^{(0)} \tag{23}$$

and this expansion is substituted into the third equation in eqn 6.17, which in ket notation becomes

$$\sum_n b_n \{E_n - E_0\}|n\rangle = \{E_0^{(2)} - H^{(2)}\}|0\rangle + \sum_n a_n \{E_0^{(1)} - H^{(1)}\}|n\rangle$$

This equation is multiplied through from the left by $\langle 0|$, which gives

$$\sum_n b_n \{E_n - E_0\}\delta_{0n} = E_0^{(2)} - \langle 0|H^{(2)}|0\rangle + \sum_n a_n \{E_0^{(1)}\delta_{0n} - \langle 0|H^{(1)}|n\rangle\}$$

The left-hand side is zero, and so

$$E_0^{(2)} = \langle 0|H^{(2)}|0\rangle - \sum_n a_n \{E_0^{(1)}\delta_{0n} - \langle 0|H^{(1)}|n\rangle\} = H_{00}^{(2)} + \sum_n{}' a_n H_{0n}^{(1)}$$

because when $n = 0$ in the sum, the term in braces is zero (by eqn 6.20), and when $n \neq 0$ the term $E_0^{(1)}\delta_{0n}$ disappears.

At this point we can import eqn 6.21 for the coefficient a_n, and obtain the following expression for the second-order correction to the energy:

$$E_0^{(2)} = H_{00}^{(2)} + \sum_n{}' \frac{H_{0n}^{(1)} H_{n0}^{(1)}}{E_0 - E_n} \tag{24}$$

The prime on the sum signifies omission of the state with $n = 0$. This is a very important result, and we shall use it frequently. It is a generalization of the approximate form of the solutions for the two-level problem, and consists of two parts. One, $H_{00}^{(2)}$, is the same kind of average as occurs for the first-order correction, and is an average of the second-order perturbation over the unperturbed wavefunction of the system. The second term is more involved, but can be interpreted as the average of the first-order perturbation taking into account the first-order distortion of the original wavefunction. It should be noticed that the sum in eqn 6.24 gives a negative contribution (lowers the energy) if $E_n > E_0$ for all n, which is the case if $|0\rangle$ is the ground state.

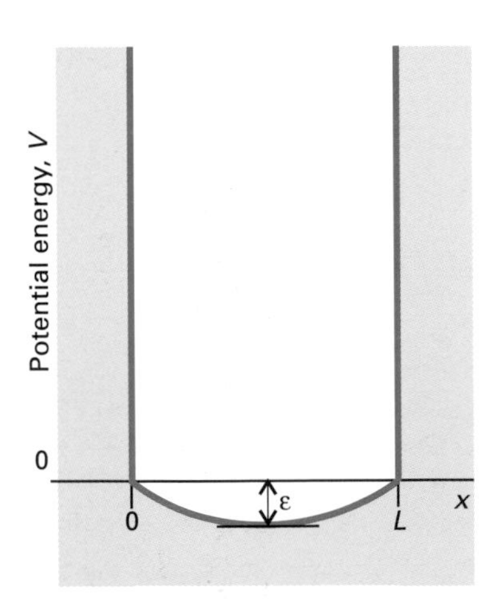

Fig. 6.4 The perturbation to a square-well potential used in Example 6.2.

Example 6.2 How to evaluate a second-order correction to the energy

The square-well potential was modified by the addition of a contribution of the form $-\varepsilon \sin(\pi x/L)$ (Fig. 6.4). Find the second-order correction to the energy of the state with $n = 1$ by numerical evaluation of the perturbation sum.

Method. Evaluate the matrix elements $H_{n0}^{(1)}$ analytically using the wavefunctions given in eqn 2.33. Tabulate their numerical values. The denominator in eqn 6.24 is obtained from the energy expression in eqn 2.33 and is proportional to $1 - n^2$. Evaluate the terms in the perturbation sum up to about $n = 9$. By symmetry, only odd values of n contribute. In this problem, $H^{(2)} = 0$ and $H_{n0}^{(1)}$ is real.

Answer. The matrix elements we require are as follows:

$$H_{n0}^{(1)} = -\frac{2\varepsilon}{L}\int_0^L \sin\left(\frac{n\pi x}{L}\right)\sin\left(\frac{\pi x}{L}\right)\sin\left(\frac{\pi x}{L}\right)\mathrm{d}x$$
$$= \frac{\varepsilon}{\pi}\left\{\frac{1}{n} - \frac{1}{2(n+2)} - \frac{1}{2(n-2)}\right\}\{(-1)^n - 1\}$$
$$E_1 - E_n = (1 - n^2)\frac{h^2}{8mL^2}$$

The following terms are then found:

n	3	5	7	9	...
$H_{n0}^{(1)}/\varepsilon$	0.1698	0.0243	0.0081	0.0037	...
$\dfrac{(H_{n0}^{(1)}/\varepsilon)^2}{n^2 - 1}$	3.62×10^{-3}	2.45×10^{-5}	1.36×10^{-6}	1.69×10^{-7}	...

It follows that

$$E^{(2)} = -3.62 \times 10^{-3} \times \frac{8mL^2\varepsilon^2}{h^2}$$

Comment. The energy of the ground state is lowered by the perturbation: this is a general feature of the $\mathrm{H}^{(1)2}$ term in second-order perturbation theory. The distorted wavefunction can be calculated from eqn 6.22 and is

$$\psi = \psi_1 - \frac{8mL^2\varepsilon}{h^2}\{2.12 \times 10^{-2}\psi_3 + 1.01 \times 10^{-3}\psi_5 + 1.68 \times 10^{-4}\psi_7 + \ldots\}$$

This wavefunction corresponds to a greater accumulation of amplitude in the middle of the well.

Exercise 6.2. Repeat the calculation for a perturbation of the form $\varepsilon \sin(2\pi x/L)$.

6.6 Comments on the perturbation expressions

We could now go on to find the second-order correction to the wavefunction, and use that result to deduce the third-order correction to the energy, and so on. However, such high-order corrections are only rarely needed and more advanced techniques are generally employed. Furthermore, a useful theorem states that to know the energy correct to order $2n + 1$ in the perturbation, it is sufficient to know the wavefunctions only to nth order in the perturbation. Thus, from the first-order wavefunction, we can calculate the energy up to third order. A final technical problem is to know whether the perturbation theory expansion actually converges. This is answered affirmatively for most common cases by a theorem due to Rellich and Kato,[2] but it is normally simply assumed that convergence occurs. The *Further reading* section suggests places where this delicate question can be pursued.

The practical difficulty with eqn 6.24 is that we do not normally have detailed information about the states and energies that occur in the sum. The sum extends, for instance, over all the states of the system, which includes the continuum, if that exists. There are, happily, several aspects of the formulation that alleviate this problem.

In the first place, the contribution can be expected to be small for states that differ by a large energy from the state of interest, and so their contribution will be small on account of the appearance of energy differences in the denominator. Other things being equal, only energetically nearby states contribute appreciably to the sum. The continuum states are generally so high in energy (they correspond, for instance, to ionized states of the system), that they can often safely be ignored.

A further apparent difficulty is that although states that are high in energy make only small individual contributions to the sum, there may be very many of them, so their total contribution may be significant. For the hydrogen atom, the number of states of a given energy increases as n^2, and when $n = 10^3$ there are 10^6 states of the same energy, each one making a small contribution to the sum. However, it often turns out that the matrix elements in the numerators of the perturbation sum vanish identically for many states. For instance, for a hydrogen atom in a uniform electric field in the z-direction, for each n only one of the n^2 states of the same energy has nonvanishing matrix elements to the ground state of the atom. Thus, although there may be 10^6 states lining up to be included, only one of them is selected.

The vanishing of matrix elements that so greatly simplifies the perturbation formulas and helps to guarantee convergence of perturbation expansions depends on the symmetry properties of the system. This is where group theory plays such a striking role. The matrix elements of interest are in fact integrals:

$$H_{0n}^{(1)} = \int \psi_0^* H^{(1)} \psi_n \, d\tau \qquad (25)$$

We saw in Section 5.16 that such integrals are necessarily zero unless the direct product $\Gamma^{(0)} \times \Gamma^{(\text{pert})} \times \Gamma^{(n)}$ contains the totally symmetric irreducible representation (for instance, A_1 or its equivalent). The physical basis of this important conclusion can be understood by considering the distortion of the wavefunction induced by the perturbation. Suppose that the state of interest (the state $|0\rangle$) is totally symmetric (it might be the 1s-orbital of a hydrogenic atom). Then

$$\Gamma^{(0)} \times \Gamma^{(\text{pert})} \times \Gamma^{(n)} = A_1 \times \Gamma^{(\text{pert})} \times \Gamma^{(n)} = \Gamma^{(\text{pert})} \times \Gamma^{(n)}$$

[2] See the volume edited by C.H. Wilcox, *Perturbation theory and its applications in quantum mechanics*, Wiley, New York (1966), for a discussion of these matters.

and this product must contain A_1 (or its equivalent). It does so only if $\Gamma^{(\text{pert})} = \Gamma^{(n)}$. It follows that the only states that are mixed into the ground state by the perturbation are those with the same symmetry as the perturbation. In other words: the distortion impressed on the system has the same symmetry as the perturbation; the perturbation leaves its footprint on the system. For example, if the perturbation is an electric field in the z-direction, then only the p_z-orbitals of the atom have the correct symmetry to mirror the effect of the perturbation and are the only orbitals to be included in the sum.

Example 6.3 How to determine the states to include in a perturbation calculation

What orbitals should be mixed into a d_{zx} orbital when it is perturbed by the application of an electric field in the x-direction?

Method. An electric field of strength $\mathcal{E}$ in the x-direction corresponds to the perturbation $H^{(1)} = -\mu_x\mathcal{E}$, where μ_x is the x-component of the electric dipole moment operator: $\mu_x = -ex$. Therefore, we need to decide which matrix elements $\langle d_{zx}|x|n\rangle$ are nonzero. To do so, we decide on the symmetry species that gives a component A_1 when we evaluate $\Gamma(d_{zx}) \times \Gamma(x) \times \Gamma^{(n)}$. We use the full rotation group, the group to which atoms belong, for the analysis, and refer to Appendix 1 for its characters.

Answer. The function d_{zx} is a component of the basis for $\Gamma^{(2)}$ and x is likewise a component of the basis for $\Gamma^{(1)}$. Because $\Gamma^{(2)} \times \Gamma^{(1)} = \Gamma^{(3)} + \Gamma^{(2)} + \Gamma^{(1)}$, at this stage we can infer that f-, d-, and p-orbitals can be mixed into the d_{zx}-orbital. However, under inversion, d_{zx} is even and x is odd, so the admixed function must be odd, which eliminates d-orbitals. The functional form of the product of d_{zx} and x is zx^2 times a radial function. Reflection in the xy-plane changes the sign of this product, so the admixed function must also change sign under this reflection. The appropriate functions are therefore f_{z^3} and p_z.

Exercise 6.3. Consider the same system. What orbitals would be mixed in for a field applied in the y-direction?

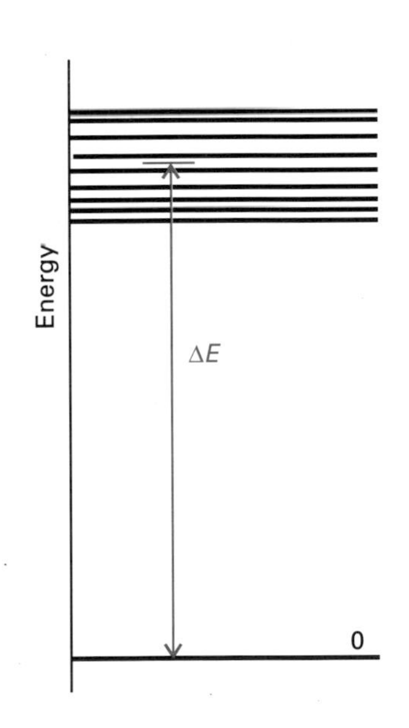

Fig. 6.5 The qualitative basis of the closure approximation, in which it is supposed that the individual excitation energies can all be set equal to a single average value.

6.7 The closure approximation

It is often the case that a quantum chemist wants a 'back-of-the-envelope' assessment of the magnitude of a property without evaluating the perturbation sum in detail. If the spectrum of energy levels of the system resembles that shown in Fig. 6.5, then we can make the approximation that all the energy differences $E_n - E_0$ in the perturbation expression can be replaced by their average value ΔE. Then the expression for the second-order correction to the energy becomes

$$E_0^{(2)} \approx H_{00}^{(2)} - \frac{1}{\Delta E}\sum_n{}' H_{0n}^{(1)} H_{n0}^{(1)}$$

The sum is almost in the form of a matrix product. It would be such a product if the sum extended over all n, including $n = 0$. So, we extend the sum, but cancel the term that should not be present:

$$\begin{aligned} E_0^{(2)} &\approx H_{00}^{(2)} - \frac{1}{\Delta E}\sum_n H_{0n}^{(1)} H_{n0}^{(1)} + \frac{1}{\Delta E} H_{00}^{(1)} H_{00}^{(1)} \\ &\approx H_{00}^{(2)} - \frac{1}{\Delta E}\left(H^{(1)}H^{(1)}\right)_{00} + \frac{1}{\Delta E} H_{00}^{(1)2} \end{aligned} \tag{26}$$

The energy correction is now expressed solely in terms of integrals over the ground state of the system, and so we need no information about excited states other than their average energy above the ground state. Because the approximation effectively 'closes' the sum over matrix elements down into a single term, it is called the **closure approximation**.

The closure approximation for the second-order energy can be expressed succinctly by introducing the term

$$\varepsilon^2 = \langle 0|H^{(1)2}|0\rangle - \langle 0|H^{(1)}|0\rangle^2$$

for then it becomes

$$E_0^{(2)} \approx H_{00}^{(2)} - \frac{\varepsilon^2}{\Delta E} \tag{27}$$

We shall use this expression several times later in the text.

Two comments are in order at this point. One is that the closure approximation is a *very* crude procedure in most instances, because the array of energy levels often differs quite significantly from that supposed in Fig. 6.5. The energy levels of a particle in a box are an example of an array of levels that is much more uniform than the bunching supposed in the approximation. However, an alternative way of regarding the approximation is to identify ΔE not with a mean energy but with the following ratio:

$$\Delta E = \frac{\left(H_{00}^{(1)2}\right) - \left(H^{(1)2}\right)_{00}}{\sum_n' H_{0n}^{(1)} H_{n0}^{(1)}/(E_0 - E_n)} \tag{28}$$

With this definition of ΔE, the closure approximation is exact; but of course the net effect is to create more work, and the formal procedure is only useful in so far as it establishes the significance of ΔE somewhat more precisely.

Example 6.4 How to use the closure approximation

Derive an approximate expression for the ground-state energy of a hydrogen atom in the presence of an electric field of strength $\mathcal{E}$ applied in the z-direction by using the closure approximation.

Method. The perturbation hamiltonian is $H^{(1)} = ez\mathcal{E}$. The first-order correction to the energy is zero because $e\mathcal{E}\langle 0|z|0\rangle = 0$. There is no second-order hamiltonian, so the energy expression is slightly simplified in so far as it has no terms in $H^{(2)}$. Set up the expression for $E_0^{(2)}$ and then apply closure. The resulting expression can be simplified by taking into account the spherical symmetry of the atom in its ground state and relating the expectation value of z^2 to the expectation value of r^2.

Answer. The full perturbation expression is

$$E_0^{(2)} = e^2\mathcal{E}^2 \sum_n{}' \frac{z_{0n}z_{n0}}{E_0 - E_n}$$

We now apply closure, and note that $\langle 0|z|0\rangle = 0$ by symmetry; therefore,

$$\varepsilon^2 = e^2\mathcal{E}^2\langle 0|z^2|0\rangle$$

The expectation value of z^2 in a spherical system is the same as the expectation values of x^2 and y^2, and because $r^2 = x^2 + y^2 + z^2$ it follows that

$$\langle 0|z^2|0\rangle = \tfrac{1}{3}\langle 0|r^2|0\rangle = \tfrac{1}{3}\langle r^2\rangle$$

where $\langle r^2\rangle$ is the mean square radius of the atom in its ground state. It follows from eqn 6.27 that

$$E_0^{(2)} = -\frac{e^2\mathcal{E}^2\langle r^2\rangle}{3\Delta E}$$

Comment. This is a very much simpler expression than the full perturbation formula. The mean excitation energy may be identified with the ionization energy of the atom, which is close to $hc\mathcal{R}_{\text{H}}$, where $\mathcal{R}_{\text{H}}$ is the Rydberg constant of the hydrogen atom (see Section 7.1).

Exercise 6.4. Derive a similar expression for the effect of an electric field on a one-dimensional harmonic oscillator treated as an electric dipole of magnitude eR and force constant k.

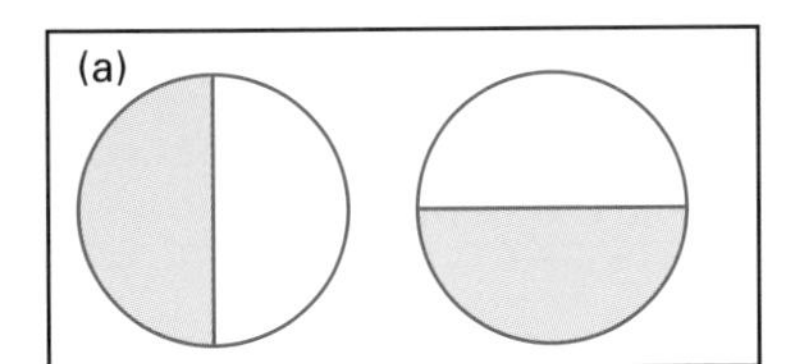

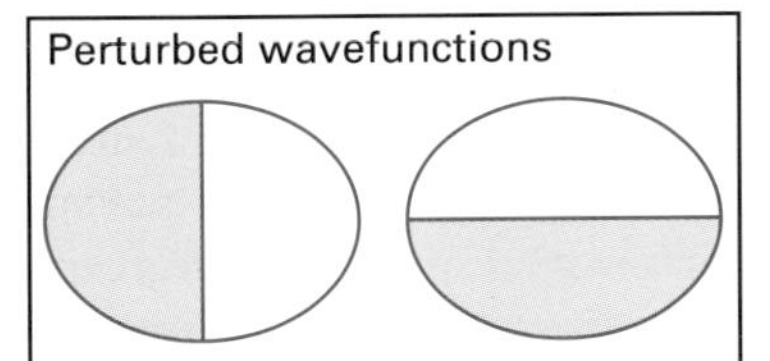

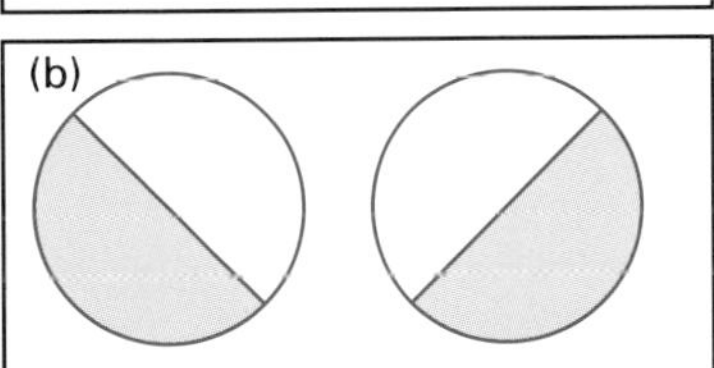

Fig. 6.6 A representation of the importance of making the correct choice of basis when considering the effect of a perturbation on degenerate states. In this diagram, the perturbation is represented by the squashing of the circle in a vertical direction. (a) A good choice of basis, because the wavefunctions undergo least change. (b) A poor choice, because both linear combinations are extensively distorted by the perturbation.

6.8 Perturbation theory for degenerate states

Figure 6.1 warns that the totally wrong result may be obtained for systems in which perturbations are applied to degenerate states, because the denominators $E_n - E_0$ then stand the risk of becoming zero. Another problem with degeneracies is that a small perturbation can induce very large changes in the forms of functions. This point is illustrated schematically in Fig. 6.6, where we see that the perturbation (which is represented by the conversion of a circle to an ellipse) leads to a large change in the initial pair of degenerate states for one particular choice of starting functions, but to a much more modest change for another choice in which the nodes remain in the same locations. The fact that any linear combination of degenerate functions is also an eigenfunction of the hamiltonian means that we have the freedom to select the combination that most closely resembles the final form of the functions once the perturbation has been applied. We shall now show that both these problems—the selection of optimum starting combinations and the avoidance of zeros in the energy denominators—can be solved by a single procedure.

We suppose that the system is r-fold degenerate and that the states corresponding to the energy E_0 are $|0, l\rangle$, with $l = 1, 2, \ldots, r$; the corresponding wavefunctions are $\psi_{0l}^{(0)}$. All r states satisfy

$$H^{(0)}|0, l\rangle = E_0|0, l\rangle \tag{29}$$

The linear combinations of the degenerate states that most closely resemble the perturbed states are

$$\phi_{0i} = \sum_{l=1}^{r} c_{il}\psi_{0l}^{(0)} \tag{30}$$

When the perturbation is applied, we suppose the state ϕ_{0i} is distorted into ψ_i, which it closely resembles, and its energy changes from E_0 to E_i, which has a similar value. The index i is needed on the new energy E_i because the degeneracy may be removed by the perturbation. As in Section 6.2, we write

$$\psi_i = \phi_{0i} + \lambda\psi_{0i}^{(1)} + \ldots$$
$$E_i = E_0 + \lambda E_{0i}^{(1)} + \ldots$$

Substitution of these expansions into $H\psi_i = E_i\psi_i$ and collection of powers of λ, just as for the non-degenerate case, gives (up to first order in λ)

$$\begin{gathered} H^{(0)}\phi_{0i} = E_0\phi_{0i} \\ \{H^{(0)} - E_0\}\psi_{0i}^{(1)} = \{E_{0i}^{(1)} - H^{(1)}\}\phi_{0i} \end{gathered} \tag{31}$$

As before, we attempt to express the first-order correction to the wavefunction as a sum over all functions. The simplest procedure is to divide the sum into two parts, one being a sum over the members of the degenerate set $|0, l\rangle$, and the other the sum over all the other states (which may or may not have degeneracies among themselves):

$$\psi_{0i}^{(1)} = \sum_l a_l\psi_{0l}^{(0)} + \sum_n{}' a_n\psi_n^{(0)}$$

On insertion of this expression into eqn 6.31 and conversion to ket notation, we obtain

$$\sum_l a_l\{E_0 - E_0\}|0, l\rangle + \sum_n{}' a_n\{E_n - E_0\}|n\rangle = \sum_l c_{il}\{E_{0i}^{(1)} - H^{(1)}\}|0, l\rangle$$

The first term is plainly zero. On multiplying the remaining terms from the left by the bra $\langle 0, k|$, we obtain zero on the left (because the states $|n\rangle$ are orthogonal to the states $|0, k\rangle$), and hence we are left with

$$\sum_l c_{il}\{E_{0i}^{(1)}\langle 0, k|0, l\rangle - \langle 0, k|H^{(1)}|0, l\rangle\} = 0$$

The degenerate functions need not be orthogonal, so we introduce the following overlap integral:

$$S_{kl} = \langle 0, k|0, l\rangle \tag{32}$$

and similarly write

$$H_{kl}^{(1)} = \langle 0, k|H^{(1)}|0, l\rangle \tag{33}$$

and so obtain

$$\sum_l c_{il}\{E_{0i}^{(1)}S_{kl} - H_{kl}^{(1)}\} = 0 \tag{34}$$

The linear combinations can always be chosen to be mutually orthogonal, and if that is done, the overlap integral $S_{kl} = \delta_{kl}$. These equations (there is one for each value of i) are called the **secular equations**. They are a set of r simultaneous equations for the coefficients c_{il}, and therefore have non-trivial solutions only if the determinant of the factors of the c_{il} vanish:

$$\det\left|H_{kl}^{(1)} - E_{0i}^{(1)}S_{kl}\right| = 0 \tag{35}$$

This determinant is called the **secular determinant**. The solution of this equation gives the energies $E_{0i}^{(1)}$ that we seek. The solution of the secular equations for each of these values of the energy then gives the coefficients that define the optimum form of the linear combinations to use for any subsequent perturbation distortion.

Example 6.5 The perturbation of degenerate states

What is the first-order correction to the energies of a doubly degenerate pair of states?

Method. We set up the secular determinant and solve it for the energies by expanding it and looking for the roots of the resulting polynomial in E. We suppose that the linear combinations of the degenerate states have been chosen to be orthogonal, which is the case for the combinations $(1/\sqrt{2})\{|0,1\rangle \pm |0,2\rangle\}$.

Answer. The secular determinant is

$$\begin{vmatrix} H_{11}^{(1)} - E_{0i}^{(1)} & H_{12}^{(1)} \\ H_{21}^{(1)} & H_{22}^{(1)} - E_{0i}^{(1)} \end{vmatrix} = 0$$

This equation expands to

$$(H_{11}^{(1)} - E_{0i}^{(1)})(H_{22}^{(1)} - E_{0i}^{(1)}) - H_{12}^{(1)}H_{21}^{(1)} = 0$$

which corresponds to the following polynomial in the energy:

$$E_{0i}^{(1)2} - (H_{11}^{(1)} + H_{22}^{(1)})E_{0i}^{(1)} + (H_{11}^{(1)}H_{22}^{(1)} - H_{12}^{(1)}H_{21}^{(1)}) = 0$$

The roots of this quadratic equation are

$$E_{0i}^{(1)} = \tfrac{1}{2}\{H_{11}^{(1)} + H_{22}^{(1)}\} \pm \tfrac{1}{2}\{(H_{11}^{(1)} + H_{22}^{(1)})^2 - 4(H_{11}^{(1)}H_{22}^{(1)} - H_{12}^{(1)}H_{21}^{(1)})\}^{1/2}$$

Comment. This result is the same as we obtained for the two-level problem in Section 6.1.

Variation theory

Perturbation theory is not the only approximate procedure when the exact energies and wavefunctions are unknown. Another very useful method for estimating the energy and approximating the wavefunction of a known hamiltonian is based on variation theory. Variation theory is a way of assessing and improving guesses about the forms of wavefunctions in complicated systems. The first step is to guess the form of a **trial function**, ψ_{trial}, and then the procedure shows how to optimize it.

6.9 The Rayleigh ratio

We suppose the system is described by a hamiltonian H and denote the lowest eigenvalue of this hamiltonian as E_0. The **Rayleigh ratio**, $\mathcal{E}$, is then defined as

$$\mathcal{E} = \frac{\int \psi_{\text{trial}}^* H \psi_{\text{trial}} \, d\tau}{\int \psi_{\text{trial}}^* \psi_{\text{trial}} \, d\tau} \tag{36}$$

Then the **variation theorem** states that

$$\mathcal{E} \geq E_0 \text{ for any } \psi_{\text{trial}} \tag{37}$$

The equality holds only if the trial function is identical to the true ground state wavefunction of the system.

The proof of the theorem runs as follows. The trial function can be written as a linear combination of the true (but unknown) eigenfunctions of the hamiltonian (which form a complete set):

$$\psi_{\text{trial}} = \sum_n c_n \psi_n \qquad \text{where } H\psi_n = E_n \psi_n$$

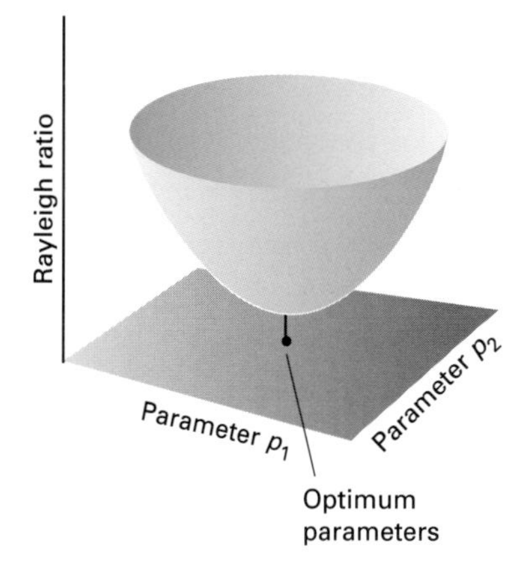

Fig. 6.7 The variation principle seeks the values of the parameters (two are shown here) that minimize the energy. The resulting wavefunction is the optimum wavefunction of the selected form.

Now consider the integral

$$\begin{aligned} I &= \int \psi_{\text{trial}}^* (H - E_0) \psi_{\text{trial}} \, d\tau = \sum_{n,n'} c_n^* c_{n'} \int \psi_n^* (H - E_0) \psi_{n'} \, d\tau \\ &= \sum_{n,n'} c_n^* c_{n'} (E_{n'} - E_0) \int \psi_n^* \psi_{n'} \, d\tau \\ &= \sum_n c_n^* c_n (E_n - E_0) \geq 0 \end{aligned}$$

The final inequality follows from $E_n \geq E_0$ and $|c_n|^2 \geq 0$. It follows that

$$\int \psi_{\text{trial}}^* (H - E_0) \psi_{\text{trial}} \, d\tau \geq 0$$

which rearranges into $\mathcal{E} \geq E_0$, so completing the proof.

The significance of the variation theorem is that it enables us to modify the trial function so as to lower the Rayleigh ratio towards the true energy, and the trial function that gives the lowest energy is the optimum function of that form. We also know that the value of the Rayleigh ratio is not less than the true ground-state energy of the system: that is, we have a way of calculating an **upper bound** to the true energy of the system. Typically, the trial function is expressed in terms of one or more parameters that are varied until the Rayleigh ratio is minimized (Fig. 6.7). The procedure is illustrated in the following example.

Example 6.6 The use of the variation theorem to find an optimized wavefunction

Find the optimum form of a trial function e^{-kr} and the upper bound to the ground-state energy of a hydrogenic atom.

Method. Begin by writing the hamiltonian for the problem and then evaluate the integrals that occur in the expression for the Rayleigh ratio. The ratio will be obtained as a function of the parameter k, so to find the minimum value of the ratio we need to find the value of k that corresponds to $\mathrm{d}\mathcal{E}/\mathrm{d}k = 0$.

Answer. The hamiltonian for the atom is

$$H = -\frac{\hbar^2}{2\mu}\nabla^2 - \frac{Ze^2}{4\pi\varepsilon_0 r}$$

However, because the trial function is independent of angle, we need consider only the radial derivatives in the laplacian (see eqn 3.18). The integrals we require are therefore

$$\int \psi_{\text{trial}}^* \psi_{\text{trial}}\,\mathrm{d}\tau = \int_0^{2\pi}\mathrm{d}\phi\int_0^{\pi}\sin\theta\,\mathrm{d}\theta\int_0^{\infty}\mathrm{e}^{-2kr}r^2\,\mathrm{d}r = \frac{\pi}{k^3}$$

$$\int \psi_{\text{trial}}^*\left(\frac{1}{r}\right)\psi_{\text{trial}}\,\mathrm{d}\tau = \int_0^{2\pi}\mathrm{d}\phi\int_0^{\pi}\sin\theta\,\mathrm{d}\theta\int_0^{\infty}\mathrm{e}^{-2kr}r\,\mathrm{d}r = \frac{\pi}{k^2}$$

$$\begin{aligned}\int \psi_{\text{trial}}^*\nabla^2\psi_{\text{trial}}\,\mathrm{d}\tau &= \int \psi_{\text{trial}}^*\left(\frac{1}{r}\frac{\mathrm{d}^2}{\mathrm{d}r^2}\right)\left(r\mathrm{e}^{-kr}\right)\mathrm{d}\tau \\ &= \int \psi_{\text{trial}}^*\left(k^2 - \frac{2k}{r}\right)\psi_{\text{trial}}\,\mathrm{d}\tau \\ &= \frac{\pi}{k} - \frac{2\pi}{k} = -\frac{\pi}{k}\end{aligned}$$

Therefore,

$$\int \psi_{\text{trial}}^* H\psi_{\text{trial}}\,\mathrm{d}\tau = \frac{\hbar^2\pi}{2\mu k} - \frac{Ze^2}{4\varepsilon_0 k^2}$$

and so the Rayleigh ratio is

$$\mathcal{E} = \frac{(\hbar^2\pi/2\mu k) - (Ze^2/4\varepsilon_0 k^2)}{(\pi/k^3)} = \frac{\hbar^2 k^2}{2\mu} - \frac{Ze^2 k}{4\pi\varepsilon_0}$$

To find the minimum value of this ratio we differentiate with respect to k:

$$\begin{aligned}\frac{\mathrm{d}\mathcal{E}}{\mathrm{d}k} &= \frac{\hbar^2 k}{\mu} - \frac{Ze^2}{4\pi\varepsilon_0} \\ &= 0 \text{ when } k = \frac{Ze^2\mu}{4\pi\varepsilon_0\hbar^2}\end{aligned}$$

The best value of $\mathcal{E}$ is therefore

$$\mathcal{E} = -\frac{Z^2 e^4 \mu}{32\pi^2 \varepsilon_0^2 \hbar^2}$$

and the optimum form of the wavefunction has

$$k = \frac{Ze^2 \mu}{4\pi\varepsilon_0 \hbar^2}$$

Comment. This optimum value of the Rayleigh ratio turns out to be the exact ground-state energy and the corresponding trial function is the true wavefunction for the atom. This special result follows from the fact that the trial function happens to include the exact wavefunction as a special case.

Exercise 6.6. Repeat the calculation for a trial function of the form e^{-kr^2} and confirm that the Rayleigh ratio lies above the true energy of the ground state.

6.10 The Rayleigh–Ritz method

The variation procedure we have described was devised by Lord Rayleigh. A modification called the **Rayleigh–Ritz method** replaces the parametrization of the trial function by a linear combination of fixed basis functions with variable coefficients; these coefficients are treated as the variables to be changed until an optimized set is obtained.

The trial function is taken to be

$$\psi_{\text{trial}} = \sum_i c_i \psi_i \tag{38}$$

with only the coefficients variable; we shall suppose that all coefficients and basis functions are real. The Rayleigh ratio is

$$\begin{aligned}\mathcal{E} &= \frac{\int \psi_{\text{trial}}^* H \psi_{\text{trial}}\, d\tau}{\int \psi_{\text{trial}}^* \psi_{\text{trial}}\, d\tau} = \frac{\sum_{i,j} c_i c_j \int \psi_i H \psi_j\, d\tau}{\sum_{i,j} c_i c_j \int \psi_i \psi_j\, d\tau} \\ &= \frac{\sum_{i,j} c_i c_j H_{ij}}{\sum_{i,j} c_i c_j S_{ij}}\end{aligned} \tag{39}$$

To find the minimum value of this ratio, we differentiate with respect to each coefficient in turn and set $\partial\mathcal{E}/\partial c_k = 0$ in each case:

$$\begin{aligned}\frac{\partial \mathcal{E}}{\partial c_k} &= \frac{\sum_j c_j H_{kj} + \sum_i c_i H_{ik}}{\sum_{i,j} c_i c_j S_{ij}} - \frac{\left(\sum_j c_j S_{kj} + \sum_i c_i S_{ik}\right) \sum_{i,j} c_i c_j H_{ij}}{\left(\sum_{i,j} c_i c_j S_{ij}\right)^2} \\ &= \frac{\sum_j c_j (H_{kj} - \mathcal{E} S_{kj})}{\sum_{i,j} c_i c_j S_{ij}} + \frac{\sum_i c_i (H_{ik} - \mathcal{E} S_{ik})}{\sum_{i,j} c_i c_j S_{ij}} = 0\end{aligned}$$

This expression is satisfied if the numerators vanish, which means that we must solve the **secular equations**

$$\sum_i c_i(H_{ik} - \mathcal{E}S_{ik}) = 0 \tag{40}$$

This is a set of simultaneous equations for the coefficients c_i. The condition for the existence of solutions is that the **secular determinant** should be zero:

$$\det|H_{ik} - \mathcal{E}S_{ik}| = 0 \tag{41}$$

Solution of the last equation leads to a set of values of $\mathcal{E}$ as the roots of the corresponding polynomial, and the lowest value is the best value of the ground state of the system with a basis set of the selected form. The coefficients in the linear combination are then found by solving the set of secular equations with this value of $\mathcal{E}$. The procedure is illustrated in the following example.

Example 6.7 An application of the Rayleigh–Ritz method

Suppose we were investigating the effect of mass of the nucleus on the ground-state wavefunctions of the hydrogen atom. One approach might be to use as a trial function a linear combination of the $1s$- and $2s$-orbitals of a hydrogen atom with an infinitely heavy nucleus but to use the true hamiltonian for the atom. Find the optimum linear combination of these orbitals and the ground-state energy of the atom.

Method. We use the wavefunctions of a hydrogen atom with an infinitely heavy nucleus as the basis, and the hamiltonian of the actual hydrogen atom: neither orbital is an eigenfunction of the hamiltonian, but a linear combination of them can be expected to be a reasonable approximation to an eigenfunction. The first step is to evaluate the matrix elements needed for the secular determinant: these can be expressed in terms of the Rydberg constant $\mathcal{R}$ with a suitable correction for the energy. Then set the secular determinant equal to zero and find the lowest root of the resulting polynomial in $\mathcal{E}$. Use this value in the secular equations for the coefficients.

Answer. The basis functions are

$$\psi_1 = \left(\frac{1}{\pi a_0^3}\right)^{1/2} \mathrm{e}^{-r/a_0} \qquad \psi_2 = \left(\frac{1}{32\pi a_0^3}\right)^{1/2} \left(2 - \frac{r}{a_0}\right) \mathrm{e}^{-r/2a_0}$$

The trial function is then $\psi_{\text{trial}} = c_1\psi_1 + c_2\psi_2$. The basis functions are orthonormal, so $S_{11} = S_{22} = 1$ and $S_{12} = S_{21} = 0$. The hamiltonian is

$$H = -\frac{\hbar^2}{2\mu}\nabla^2 - \frac{e^2}{4\pi\varepsilon_0 r}$$

The basis functions are independent of angles, so only the radial derivatives need be retained. Express the energies in terms of $hc\mathcal{R} = \hbar^2/2a_0^2 m_e$.

The integrals required are quite straightforward to evaluate and are as follows:

$$H_{11} = (\gamma - 1)hc\mathcal{R} \quad H_{22} = \tfrac{1}{4}(\gamma - 1)hc\mathcal{R} \quad H_{12} = H_{21} = \frac{16\gamma}{27\sqrt{2}}hc\mathcal{R}$$

with $\gamma = m_e/m_p$. The secular determinant expands as follows:

$$\begin{vmatrix} H_{11} - \mathcal{E}S_{11} & H_{12} - \mathcal{E}S_{12} \\ H_{21} - \mathcal{E}S_{21} & H_{22} - \mathcal{E}S_{22} \end{vmatrix} = \begin{vmatrix} H_{11} - \mathcal{E} & H_{12} \\ H_{21} & H_{22} - \mathcal{E} \end{vmatrix}$$
$$= \mathcal{E}^2 - (H_{11} + H_{22})\mathcal{E} + (H_{11}H_{22} - H_{12}H_{21})$$
$$= 0$$

Substitution of the matrix elements gives the lower root

$$\mathcal{E} = \tfrac{1}{8}(\gamma - 1)\{5 + 3(1 + 2\Gamma^2)^{1/2}\}hc\mathcal{R} \qquad \text{where } \Gamma = \frac{2^6\gamma}{3^4(\gamma - 1)}$$

Because $\Gamma = -0.000\,43$, it follows that $\mathcal{E} = -0.999\,46hc\mathcal{R}$. The secular equations are

$$c_1(H_{11} - \mathcal{E}) + c_2H_{21} = 0 \qquad c_1H_{12} + c_2(H_{22} - \mathcal{E}) = 0$$

and for the trial function to be normalized we also know that $c_1^2 + c_2^2 = 1$. It follows that with the value of $\mathcal{E}$ found above,

$$c_1 \approx 1.000\,00 \qquad c_2 = -0.000\,54$$

Comment. The wavefunction has a 3.0×10^{-5} per cent admixture of 2*s*-orbital into the 1*s*-orbital, with a negative sign for the coefficient. The latter signifies a small decrease in amplitude of the overall wavefunction at the nucleus. The explanation of this reduction can be traced to the fact that the reduced mass is slightly less than the mass of the electron, and so the 'effective particle' has slightly more freedom than an electron.

It should be noted that the variation principle leads to an upper bound for the energy of the system. It is also possible to use the principle to determine an upper bound for the first excited state by formulating a trial function that is orthogonal to the ground-state function. There are also variational techniques for finding lower bounds, and so the true energy can be sandwiched above and below and hence located reasonably precisely. These calculations, though, are often quite difficult because they involve integrals over the square of the hamiltonian. A further remark is that although the variation principle may give a good value for the energy, there is no guarantee that the optimum trial function will give a good value for some other property of the system, such as its dipole moment.

The Hellmann–Feynman theorem

Whereas the variation theorem takes a fixed hamiltonian and investigates how the energy (more precisely, the Rayleigh ratio) varies as the trial function is modified, the Hellmann–Feynman theorem investigates how the energy varies as the *hamiltonian* varies.

Consider a system characterized by a hamiltonian that depends on a parameter P. This parameter might be the internuclear distance in a molecule or the strength of the electric field to which the molecule is exposed. The exact (not trial) wavefunction for the system is a solution of the Schrödinger equation, and so it and its energy also depend on the parameter P. The question we tackle is how the energy of the system varies as the parameter is varied, and we shall now prove the following relation, which is the **Hellmann–Feynman theorem**:

$$\frac{\mathrm{d}E}{\mathrm{d}P}=\left\langle\frac{\partial H}{\partial P}\right\rangle \tag{42}$$

The proof of the theorem is as follows. We shall suppose that the wavefunction is normalized to 1 for all values of P, in which case

$$E(P)=\int\psi(P)^*H(P)\psi(P)\,\mathrm{d}\tau$$

The derivative of E with respect to P is

$$\begin{aligned}\frac{\mathrm{d}E}{\mathrm{d}P}&=\int\left(\frac{\partial\psi^*}{\partial P}\right)H\psi\,\mathrm{d}\tau+\int\psi^*\left(\frac{\partial H}{\partial P}\right)\psi\,\mathrm{d}\tau+\int\psi^*H\left(\frac{\partial\psi}{\partial P}\right)\mathrm{d}\tau\\&=E\int\left(\frac{\partial\psi^*}{\partial P}\right)\psi\,\mathrm{d}\tau+\int\psi^*\left(\frac{\partial H}{\partial P}\right)\psi\,\mathrm{d}\tau+E\int\psi^*\left(\frac{\partial\psi}{\partial P}\right)\mathrm{d}\tau\\&=E\frac{\mathrm{d}}{\mathrm{d}P}\int\psi^*\psi\,\mathrm{d}\tau+\int\psi^*\left(\frac{\partial H}{\partial P}\right)\psi\,\mathrm{d}\tau\end{aligned}$$

The first term is zero because the integral is equal to 1 for all values of P. The second term is the expectation value of the first-derivative of the hamiltonian. (In the second line of the proof, we have employed the hermiticity of H to let it operate on the function standing to its left.)

The great advantage of the Hellmann–Feynman theorem is that the operator $\partial H/\partial P$ might be very simple. For example, if the total hamiltonian is $H=H^{(0)}+Px$, then $\partial H/\partial P=x$, and there is no mention of $H^{(0)}$, which might be a very complicated operator. In this case,

$$\frac{\mathrm{d}E}{\mathrm{d}P}=\langle x\rangle$$

and the calculation is apparently very simple.

There is, as always, a complication. The proof of the theorem has supposed that the wavefunctions are the exact eigenfunctions of the total hamiltonian. Therefore, to evaluate the expectation value of even a simple operator like x, we need to have solved the Schrödinger equation for the complete, complicated hamiltonian. Nevertheless, we can use the perturbation theory described earlier in the chapter to arrive at successively better approximations to the true wavefunctions, and therefore can calculate successively better approximations to the value of $\mathrm{d}E/\mathrm{d}P$, the response of the system to changes in the hamiltonian. We shall use this technique in Chapters 12 and 13 to calculate the properties of molecules in electric and magnetic fields.

Time-dependent perturbation theory

Just about every perturbation in chemistry is time-dependent, even those that appear to be stationary. Even stationary perturbations have to be turned on: samples are inserted into electric and magnetic fields, the shapes of vessels are changed, and so on. The reason why time-independent perturbation theory can often be applied in these cases is that the response of a molecule is so rapid that for all practical purposes the systems forget that they were ever unperturbed and settle rapidly into their final perturbed states. Nevertheless, if we really want to understand the properties of molecules, we need to see how systems respond to newly imposed perturbations and then settle into stationary states after an interval.

But there is a much more important reason for studying time-dependent perturbations. Many important perturbations never 'settle down' to a constant value. A molecule exposed to electromagnetic radiation, for instance, experiences an electromagnetic field that oscillates for as long as the perturbation is imposed. Time-dependent perturbation theory is essential for such problems, and is used to calculate transition probabilities in spectroscopy and the intensities of spectral lines.

We shall adopt the same approach as for time-dependent perturbation theory. First, we consider a two-level system. Then we generalize that special case to systems of arbitrary complexity.

6.11 The time-dependent behaviour of a two-level system

The total hamiltonian of the system is

$$H = H^{(0)} + H^{(1)}(t) \tag{43}$$

A typical example of a time-dependent perturbation is one that oscillates at an angular frequency ω, in which case

$$H^{(1)}(t) = 2H^{(1)} \cos \omega t \tag{44}$$

where $H^{(1)}$ is a time-independent operator and the 2 is present for future convenience. We need to deal with the time-dependent Schrödinger equation:

$$H\Psi = \mathrm{i}\hbar \frac{\partial \Psi}{\partial t} \tag{45}$$

As in the earlier part of the chapter, we denote the energies of the two states as E_1 and E_2 and the corresponding time-independent wavefunctions as ψ_1 and ψ_2. These wavefunctions are the solutions of

$$H^{(0)}\psi_n = E_n\psi_n \tag{46}$$

and are related to the time-dependent unperturbed wavefunctions by

$$\Psi_n(t) = \psi_n \mathrm{e}^{-\mathrm{i}E_n t/\hbar} \tag{47}$$

In the presence of the perturbation $H^{(1)}(t)$, the state of the system is expressed as a linear combination of the basis functions:

$$\Psi(t) = a_1(t)\Psi_1(t) + a_2(t)\Psi_2(t) \tag{48}$$

Notice that the coefficients are also time-dependent because the composition of the state may evolve with time. The total time-dependence of the wavefunction therefore arises from the oscillation of the basis functions and the evolution of the coefficients. The probability that at any time t the system is in state n is $|a_n(t)|^2$.

Substitution of the linear combination into the Schrödinger equation, eqn 6.45, leads to the following expression:

$$\begin{aligned} H\Psi &= a_1 H^{(0)}\Psi_1 + a_1 H^{(1)}(t)\Psi_1 + a_2 H^{(0)}\Psi_2 + a_2 H^{(1)}(t)\Psi_2 \\ &= \mathrm{i}\hbar\frac{\partial}{\partial t}(a_1\Psi_1 + a_2\Psi_2) \\ &= \mathrm{i}\hbar a_1\frac{\partial \Psi_1}{\partial t} + \mathrm{i}\hbar\Psi_1\frac{\mathrm{d}a_1}{\mathrm{d}t} + \mathrm{i}\hbar a_2\frac{\partial \Psi_2}{\partial t} + \mathrm{i}\hbar\Psi_2\frac{\mathrm{d}a_2}{\mathrm{d}t} \end{aligned}$$

Each basis function satisfies

$$H^{(0)}\Psi_n = \mathrm{i}\hbar\frac{\partial \Psi_n}{\partial t}$$

so the last equation simplifies to

$$a_1 H^{(1)}(t)\Psi_1 + a_2 H^{(1)}(t)\Psi_2 = \mathrm{i}\hbar\dot{a}_1\Psi_1 + \mathrm{i}\hbar\dot{a}_2\Psi_2$$

where $\dot{a} = \mathrm{d}a/\mathrm{d}t$.

The next step is to extract equations for the time-variation of the coefficients. To do so, we write the time-dependence of the wavefunctions explicitly:

$$\begin{aligned} a_1 H^{(1)}(t)|1\rangle \mathrm{e}^{-\mathrm{i}E_1 t/\hbar} &+ a_2 H^{(1)}(t)|2\rangle \mathrm{e}^{-\mathrm{i}E_2 t/\hbar} \\ &= \mathrm{i}\hbar\dot{a}_1|1\rangle \mathrm{e}^{-\mathrm{i}E_1 t/\hbar} + \mathrm{i}\hbar\dot{a}_2|2\rangle \mathrm{e}^{-\mathrm{i}E_2 t/\hbar} \end{aligned} \tag{49}$$

We have also taken this opportunity to express the wavefunctions ψ_n as the kets $|n\rangle$. Now multiply through from the left by $\langle 1|$ and use the orthonormality of the states to obtain

$$a_1 H_{11}^{(1)}(t)\mathrm{e}^{-\mathrm{i}E_1 t/\hbar} + a_2 H_{12}^{(1)}(t)\mathrm{e}^{-\mathrm{i}E_2 t/\hbar} = \mathrm{i}\hbar\dot{a}_1\mathrm{e}^{-\mathrm{i}E_1 t/\hbar} \tag{50}$$

where

$$H_{ij}^{(1)}(t) = \langle i|H^{(1)}(t)|j\rangle \tag{51}$$

The expression we have obtained can be simplified in a number of ways. In the first place, we shall write $\hbar\omega_{21} = E_2 - E_1$, and so obtain

$$a_1 H_{11}^{(1)}(t) + a_2 H_{12}^{(1)}(t)\mathrm{e}^{-\mathrm{i}\omega_{21}t} = \mathrm{i}\hbar\dot{a}_1 \tag{52}$$

Next, it is commonly the case that the perturbation has no diagonal elements, and so we can set $H_{11}^{(1)}(t) = H_{22}^{(1)}(t) = 0$. The equation then reduces to

$$a_2 H_{12}^{(1)}(t)\mathrm{e}^{-\mathrm{i}\omega_{21}t} = \mathrm{i}\hbar\dot{a}_1$$

which may be rearranged to

$$\dot{a}_1 = \frac{1}{\mathrm{i}\hbar}a_2 H_{12}^{(1)}(t)\mathrm{e}^{-\mathrm{i}\omega_{21}t} \tag{53}$$

This differential equation for a_1 depends on a_2, so we need an equation for that coefficient too. The same procedure, but with multiplication by $\langle 2|$, leads to

$$\dot{a}_2 = \frac{1}{i\hbar} a_1 H_{21}^{(1)}(t) e^{i\omega_{21}t} \tag{54}$$

First, suppose the perturbation is absent, so its matrix elements are zero. In that simple case $\dot{a}_1 = 0$ and $\dot{a}_2 = 0$. The coefficients do not change from their initial values and the state is

$$\Psi = a_1 \psi_1^{(0)} e^{-iE_1 t/\hbar} + a_2 \psi_2^{(0)} e^{-iE_2 t/\hbar} \tag{55}$$

with a_1 and a_2 constants. Although Ψ oscillates with time, the probability of finding the system in either of the states is constant, because the square modulus of the coefficients of either a_i is constant. That is, in the absence of a perturbation, the state of the system is frozen at whatever was its initial composition.

Now consider the case of a constant perturbation applied at $t = 0$. We shall write $H_{12}^{(1)}(t) = \hbar V$ and (by hermiticity) $H_{21}^{(1)}(t) = \hbar V^*$ when the perturbation is present. Then

$$\dot{a}_1 = -iVa_2 e^{-i\omega_{21}t} \qquad \dot{a}_2 = -iV^* a_1 e^{i\omega_{21}t} \tag{56}$$

There are several ways of solving coupled differential equations such as these. The most elementary method (which we employ here) is to substitute one equation into the other.[3] On differentiation of $\dot{a}_2$ and then using the expression for $\dot{a}_1$ we obtain

$$\begin{aligned} \ddot{a}_2 &= -i\dot{a}_1 V^* e^{i\omega_{21}t} + \omega_{21} a_1 V^* e^{i\omega_{21}t} \\ &= -a_2 V^* V + i\omega_{21}\dot{a}_2 \end{aligned} \tag{57}$$

The corresponding expression for $\ddot{a}_1$ is obtained by differentiating $\dot{a}_1$. Note that two coupled first-order equations lead to one second-order differential equation for either a_1 or a_2. The general solutions of this second-order differential equation are

$$a_2(t) = \left(Ae^{i\Omega t} + Be^{-i\Omega t}\right) e^{\frac{1}{2}i\omega_{21}t} \qquad \Omega = \tfrac{1}{2}(\omega_{21}^2 + 4|V|^2)^{1/2} \tag{58}$$

where A and B are constants determined by the initial conditions. A similar expression holds for a_1.

Now suppose that at $t = 0$ the system is definitely in state 1. Then $a_1(0) = 1$ and $a_2(0) = 0$. These initial conditions are enough to determine the two constants in the general solution, and after some straightforward algebra we find the following two particular solutions:

$$\begin{aligned} a_1(t) &= \left\{\cos \Omega t + \frac{i\omega_{21}}{2\Omega} \sin \Omega t\right\} e^{-\frac{1}{2}i\omega_{21}t} \\ a_2(t) &= -\left(\frac{i|V|}{\Omega}\right) \sin \Omega t \, e^{\frac{1}{2}i\omega_{21}t} \end{aligned} \tag{59}$$

These are the *exact* solutions for the problem: we have made no approximations in their derivation.

[3] A much more powerful method is to use Laplace transforms.

6.12 The Rabi formula

We are interested in the probability of finding the system in one of the two states as a function of time. These probabilities are $P_1(t) = |a_1(t)|^2$ and $P_2(t) = |a_2(t)|^2$. For state 2, the initially unoccupied state, we find the **Rabi formula**:

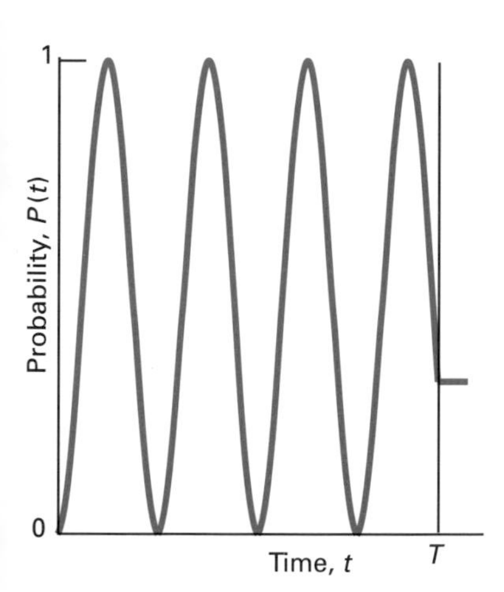

Fig. 6.8 The variation with time of the probability of being in an initially empty state of a two-level degenerate system that is subjected to a constant perturbation turned on at $t = 0$ and extinguished at $t = T$.

$$P_2(t) = \left(\frac{4|V|^2}{\omega_{21}^2 + 4|V|^2}\right) \sin^2 \tfrac{1}{2}\left(\omega_{21}^2 + 4|V|^2\right)^{1/2} t \qquad (60)$$

This expression will be at the centre of the following discussion. The probability of the system being in state 1 is of course $P_1(t) = 1 - P_2(t)$, so we do not need to make a special calculation for its value.

The first case we consider is that of a degenerate pair of states, so $\omega_{21} = 0$. The probability that the system will be found in state 2 if at $t = 0$ it was certainly in state 1 is then

$$P_2(t) = \sin^2 |V|t \qquad (61)$$

This function is plotted in Fig. 6.8. We see that the system oscillates between the two states, and periodically is certainly in state 2. Because the frequency of the oscillation is governed by $|V|$, we also see that strong perturbations drive the system between its two states more rapidly than weak perturbations. However, provided we wait long enough (specifically, for a time $t = \pi/2|V|$), then, whatever the perturbation, in due course the system will be found with certainty in state 2. This responsiveness is a special characteristic of degenerate systems. Degenerate systems are 'loose' in the sense that the populations of their states may be transferred completely even by weak stimuli.

Now consider the other extreme, when the energy levels are widely separated in comparison with the strength of the perturbation, in the sense $\omega_{21}^2 \gg 4|V|^2$. In this case, $4|V|^2$ can be ignored in both the denominator and the argument of the sine function, and we obtain

$$P_2(t) \approx \left(\frac{2|V|}{\omega_{21}}\right)^2 \sin^2 \tfrac{1}{2}\omega_{21} t \qquad (62)$$

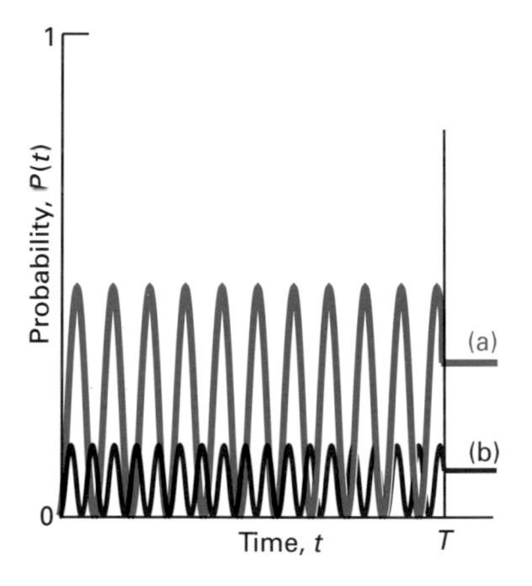

Fig. 6.9 The variation with time of the probability of being in an initially empty state of a two-level non-degenerate system that is subjected to a constant perturbation turned on at $t = 0$ and extinguished at $t = T$. The variation labelled (a) corresponds to a small energy separation and that in (b) corresponds to a large separation. Note that the latter oscillates more rapidly than the former.

The behaviour of the system is now quite different (Fig. 6.9). The populations oscillate, but $P_2(t)$ never rises above $4|V|^2/\omega_{21}^2$, which is very much less than 1. There is now only a very small probability that the perturbation will drive the system from state 1 to state 2. Moreover, the frequency of oscillation of the population is determined solely by the separation of the states and is independent of the strength of the perturbation. That is like the behaviour of a bell that is struck by a hammer: the frequency is largely independent of the strength of the blow. (Indeed, there is a deep connection between the two phenomena.) The only role of the perturbation, other than its role in causing the transitions, is to govern the maximum extent to which population transfer occurs. If the perturbation is strong (but still weak in comparison with the energy separation of the states), then there is a higher probability of finding the system in state 2 than when the perturbation is weak.

Example 6.8 How to prepare systems in specified states

Suggest how you could prepare a degenerate two-level system in a mixed state in which there is equal likelihood of finding it in either state.

Method. We know that a state, once prepared, persists with constant composition in the absence of a perturbation. This suggests that we should use the Rabi formula to find the time for which a perturbation should be applied to result in $P_2(t) = 0.5$, and then immediately extinguish the perturbation.

Answer. The Rabi formula shows that $P_2(t) = 0.5$ when $t = \pi/4|V|$. Therefore, the perturbation should be applied to a system that is known to be in state 1 initially, and removed at $t = \pi/4|V|$. Although the wavefunction of the system will oscillate, the probability of finding the system in either state will remain 0.5 until another perturbation is applied.

Comment. This state preparation procedure is the quantum mechanical basis of pulse techniques in nuclear magnetic resonance.

Exercise 6.8. For how long should the perturbation be applied to the same system to obtain a state with probability 0.25 of being in state 2?

6.13 Many-level systems: the variation of constants

The discussion of the two-level system has revealed two rather depressing features. One is that even very simple systems lead to very complicated differential equations. For a two-level system the problem requires the solution of a second-order differential equation; for an n-level system, the solution requires dealing with an nth-order differential equation, which is largely hopeless. The second point is that even for a two-level system, the differential equation could be solved only for a trivially simple perturbation, one that did not vary with time. The differential equation is very much more complicated to solve when the perturbation has a realistic time-dependence, such as an oscillation in time. Even the case $\cos \omega t$ is very complicated. Clearly, we need to set up an approximation technique for dealing with systems of many levels and which can cope with realistic perturbations.

We shall describe the technique invented by P.A.M. Dirac and known (agreeably paradoxically) as the **variation of constants**. It is a generalization of the two-level problem, and that relationship should be held in mind as we go through the material.

As before, the hamiltonian is taken to be $H = H^{(0)} + H^{(1)}(t)$. The eigenstates of $H^{(0)}$ will be denoted by the ket $|n\rangle$ or by the corresponding wavefunction $\psi_n^{(0)}$ as convenient, where

$$\Psi_n(t) = \psi_n^{(0)} e^{-iE_nt/\hbar} \qquad H^{(0)}\Psi_n = i\hbar \frac{\partial \Psi_n}{\partial t} \tag{63}$$

The state of the perturbed system is Ψ. As before, we express it as a time-dependent linear combination of the time-dependent unperturbed states:

$$\Psi(t) = \sum_n a_n(t)\Psi_n(t) = \sum_n a_n(t)\psi_n^{(0)}\mathrm{e}^{-\mathrm{i}E_n t/\hbar} \qquad H\Psi = \mathrm{i}\hbar\frac{\partial\Psi}{\partial t} \tag{64}$$

Our problem, as for the two-level case, is to find how the linear combination evolves with time. To do so, we set up and then solve the differential equations satisfied by the coefficients a_n.

We proceed as before. Substitution of Ψ into the Schrödinger equation leads to the following expressions:

$$H\Psi = \sum_n a_n(t)H^{(0)}(t)\Psi_n(t) + \sum_n a_n(t)H^{(1)}(t)\Psi_n(t)$$

$$\mathrm{i}\hbar\frac{\partial\Psi}{\partial t} = \sum_n a_n(t)\mathrm{i}\hbar\frac{\partial\Psi_n}{\partial t} + \sum_n \mathrm{i}\hbar\dot{a}_n(t)\Psi_n(t)$$

The terms in the two boxes on the left are equal (by the Schrödinger equation), and those in the two boxes on the right are also equal (by eqn 6.63). Therefore,

$$\sum_n a_n(t)H^{(1)}(t)\Psi_n(t) = \mathrm{i}\hbar\sum_n \dot{a}_n(t)\Psi_n(t)$$

In terms of the time-independent kets, this equation is

$$\sum_n a_n(t)H^{(1)}(t)|n\rangle\mathrm{e}^{-\mathrm{i}E_n t/\hbar} = \mathrm{i}\hbar\sum_n \dot{a}_n(t)|n\rangle\mathrm{e}^{-\mathrm{i}E_n t/\hbar}$$

At this point we have to extract one of the $\dot{a}_n$ on the right. To do so, we make use of the orthonormality of the eigenstates, and multiply through by $\langle k|$:

$$\sum_n a_n(t)\langle k|H^{(1)}(t)|n\rangle\mathrm{e}^{-\mathrm{i}E_n t/\hbar} = \mathrm{i}\hbar\dot{a}_k(t)\mathrm{e}^{-\mathrm{i}E_k t/\hbar}$$

This expression can be simplified by writing $H_{kn}^{(1)}(t) = \langle k|H^{(1)}(t)|n\rangle$ and $\hbar\omega_{kn} = E_k - E_n$, when it becomes

$$\dot{a}_k(t) = \frac{1}{\mathrm{i}\hbar}\sum_n a_n(t)H_{kn}^{(1)}(t)\mathrm{e}^{\mathrm{i}\omega_{kn}t} \tag{65}$$

It is easy to verify that this equation reduces to eqn 6.54 in the case of a two-level system with vanishing diagonal elements of the perturbation.

The last equation is exact. We can move towards finding exact solutions and from this point on the development diverges from the exact two-level calculation described earlier. To solve a first-order differential equation, we integrate it from $t = 0$, when the coefficients had the values $a_n(0)$, to the time t of interest:

$$a_k(t) - a_k(0) = \frac{1}{\mathrm{i}\hbar}\sum_n\int_0^t a_n(t)H_{kn}^{(1)}(t)\mathrm{e}^{\mathrm{i}\omega_{kn}t}\,\mathrm{d}t \tag{66}$$

This equation is still exact.

The trouble with the last equation is that although it appears to give an expression for any coefficient $a_k(t)$, it does so in terms of all the coefficients, including a_k itself. These other coefficients are unknown, and must be determined from equations of a similar form. So, to solve eqn 6.66, it appears that we must already know all the coefficients! A way out of this cyclic problem is to

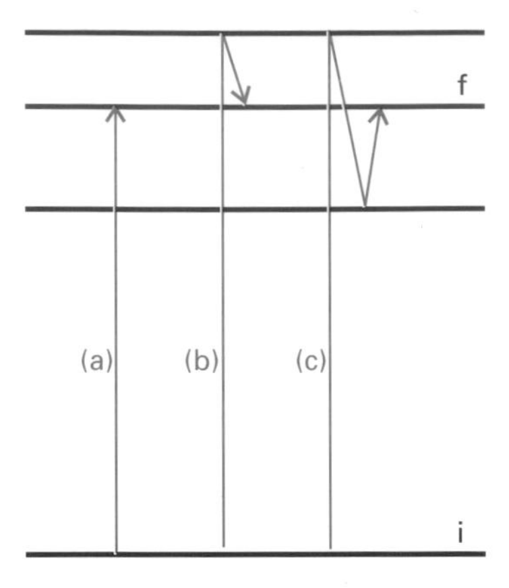

Fig. 6.10 The procedure described in the text corresponds to considering only direct transitions between the initial and final states (as in (a)), and ignoring indirect transitions (as in (b) and (c)), which correspond to higher-order processes.

make an approximation. We shall base the approximation on the supposition that the perturbation is so weak and applied for so short a time that all the coefficients remain close to their initial values. Then, if the system is certainly in state $|i\rangle$ at $t = 0$, all coefficients other than a_i are close to zero throughout the period for which the perturbation is applied, and any single coefficient, such as the coefficient of state $|f\rangle$ that is zero initially, is given by

$$a_f(t) \approx \frac{1}{i\hbar}\int_0^t a_i(t)H_{fi}^{(1)}(t)e^{i\omega_{fi}t}\,dt$$

because all terms in the sum are zero ($a_n(t) \approx 0$) except for the term corresponding to the initial state. We have also made use in the sum of the fact that $a_f(t) \approx a_f(0) = 0$. However, the coefficient of the initial state remains close to 1 for all the time of interest, so we can set $a_i(t) \approx 1$, and obtain

$$a_f(t) \approx \frac{1}{i\hbar}\int_0^t H_{fi}^{(1)}(t)e^{i\omega_{fi}t}\,dt \qquad (67)$$

This is an explicit expression for the value of the coefficient of a state that was initially unoccupied and will be the formula that we employ in the following discussion.

The nature of the approximation can be seen quite simply. It ignores the possibility that the perturbation can take the system from its initial state $|i\rangle$ to some final state $|f\rangle$ by an indirect route in which the perturbation induces a sequence of several transitions (Fig. 6.10). Put another way: the approximation assumes that the perturbation acts only once, and that we are therefore dealing with first-order perturbation theory. In modern books you will often see diagrams like those in Fig. 6.11. The intersection of the sloping and horizontal lines is intended to convey the idea that the perturbation (the sloping line) acts on the molecular states (the horizontal line) only once. This **diagrammatic perturbation theory** has been extensively developed, and many calculations can be performed by analysing the diagrams themselves. In the present case, the upper diagram in Fig. 6.11 can be regarded as a succinct expression for the right-hand side of eqn 6.67. Second-order perturbation theory (which we are not doing here) would give rise to diagrams like the one shown in the lower part of Fig. 6.11. These diagrams are sometimes associated with the name of R.P. Feynman, who introduced similar diagrams in the context of fundamental particle interactions, and are called **Feynman diagrams**.

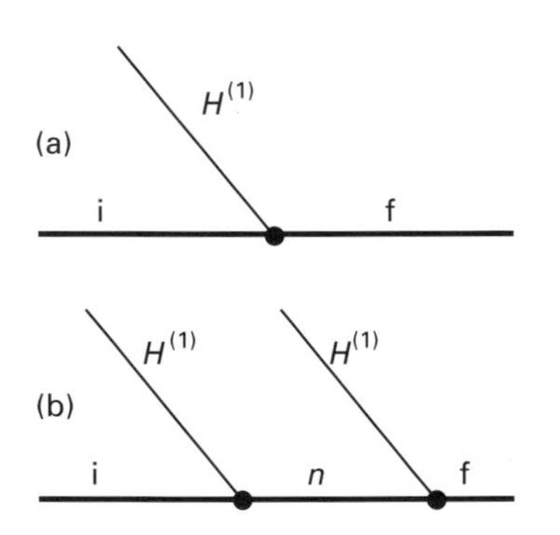

Fig. 6.11 Diagrams for (a) first-order and (b) second-order contributions to the perturbation of a system.

6.14 The effect of a slowly switched constant perturbation

As a first example of the use of eqn 6.67, consider a perturbation that rises slowly from zero to a steady final value (Fig. 6.12). Such a switched perturbation is

$$H^{(1)}(t) = \begin{cases} 0 & \text{for } t < 0 \\ H^{(1)}(1 - e^{-kt}) & \text{for } t \geq 0 \end{cases} \qquad (68)$$

where $H^{(1)}$ is a time-independent operator and, for slow switching, k is small (and positive). The coefficient of an initially unoccupied state is given by

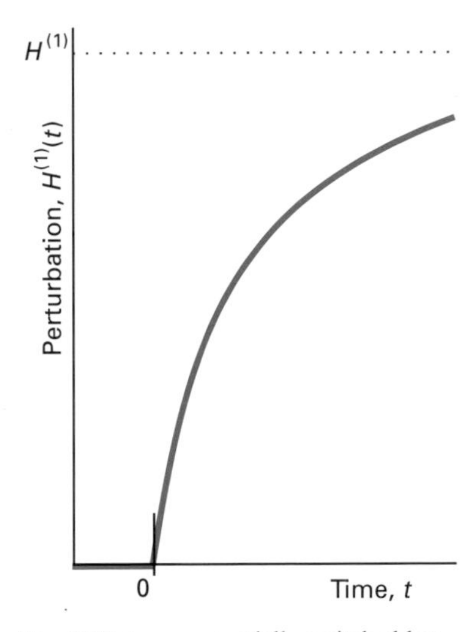

Fig. 6.12 An exponentially switched but otherwise constant perturbation.

eqn 6.67 as

$$\begin{aligned} a_{\mathrm{f}}(t) &= \frac{H_{\mathrm{fi}}^{(1)}}{\mathrm{i}\hbar}\int_0^t \left(1-\mathrm{e}^{-kt}\right)\mathrm{e}^{\mathrm{i}\omega_{\mathrm{fi}}t}\,\mathrm{d}t \\ &= \frac{H_{\mathrm{fi}}^{(1)}}{\mathrm{i}\hbar}\left\{\frac{\mathrm{e}^{\mathrm{i}\omega_{\mathrm{fi}}t}-1}{\mathrm{i}\omega_{\mathrm{fi}}}+\frac{\mathrm{e}^{-(k-\mathrm{i}\omega_{\mathrm{fi}})t}-1}{k-\mathrm{i}\omega_{\mathrm{fi}}}\right\} \end{aligned} \tag{69}$$

This result, which is exact within first-order perturbation theory, can be simplified by supposing that we are interested in times very long after the perturbation has reached its final value, which means $t \gg 1/k$, and that the perturbation is switched slowly in the sense that $k^2 \ll \omega_{\mathrm{fi}}^2$. Then the $|a_{\mathrm{f}}|^2$ in eqn 6.69 simplifies to

$$|a_{\mathrm{f}}(t)|^2 = \frac{|H_{\mathrm{fi}}^{(1)}|^2}{\hbar^2\omega_{\mathrm{fi}}^2} \tag{70}$$

We can now see why time-independent perturbation theory can be used for most problems of chemical interest, except where the perturbation continues to change after it has been applied. When a 'constant' perturbation is switched on, it is done very slowly in comparison with the frequencies associated with the transitions in atoms and molecules ($k \approx 10^3\ \mathrm{s}^{-1}$, $\omega_{\mathrm{fi}} \approx 10^{15}\ \mathrm{s}^{-1}$). Furthermore, we are normally interested in a system's properties at times long after the switching is complete ($t \gg 10^{-3}$ s; and in general $kt \gg 1$). These are the conditions under which time-dependent perturbation theory has effectively settled down into time-independent perturbation theory. All the transients stimulated by the switching have subsided and the populations of states are steady.

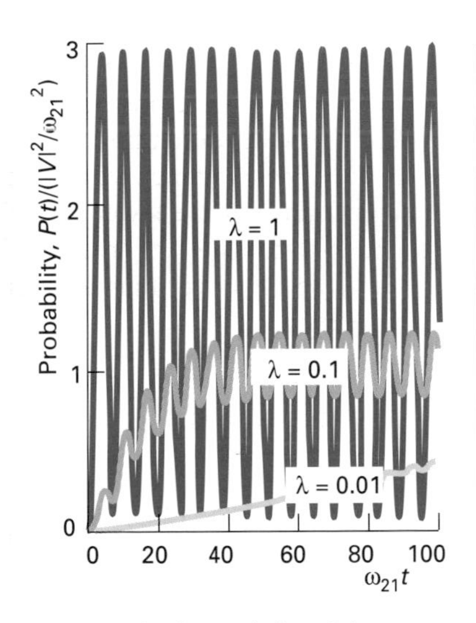

Fig. 6.13 The time variation of the probability of occupying an initially unoccupied state when the perturbation is switched on at different rates; see Example 6.9.

Example 6.9 The effect of a constant perturbation

A constant perturbation was switched on exponentially starting at $t = 0$. Evaluate the probability of finding a system in state 2 given that initially it was in state 1, and illustrate the role of transients.

Method. The perturbation is given by eqn 6.68 and the solution is expressed by eqn 6.69. To find the probability that the system is in state 2, we need to form $P_2 = |a_2(t)|^2$ for a general value of k and then to plot P_2 against t. For example plots, set $\lambda = k/\omega_{21}$ and plot $P_2/(|V|/\omega_{21})^2$, with $|V| = H_{12}^{(1)}/\hbar$, for $\lambda = 0.01$, 0.1, and 1, which corresponds to switching rates increasing in 10-fold steps.

Answer. From eqn 6.69 with $\lambda = k/\omega_{21}$ and $x = \omega_{21}t$,

$$\frac{P}{|V|^2/\omega_{21}^2} = \left(\frac{1}{1+\lambda^2}\right)\left\{1+2\lambda^2 - \left[2\lambda^2\cos x + \left(2-\mathrm{e}^{-\lambda x}\right)\mathrm{e}^{-\lambda x} + 2\lambda\left(1-\mathrm{e}^{-\lambda x}\right)\sin x\right]\right\}$$

This function is plotted for $\lambda = 0.01, 0.1, 1$ in Fig. 6.13.

Comment. Notice how slow switching ($\lambda = 0.01$) generates hardly any transients, whereas rapid switching ($\lambda = 1$) is like an impulsive shock to the system, and causes the population to oscillate violently between the two states. For very rapid switching ($\lambda \gg 1$), $P_2/(|V|^2/\omega_{21}^2)$ varies as $2(1 - \cos x)$, and so it oscillates between 0 and 4 with an average value of 2: such rapid switching is like a hammer blow.

Exercise 6.9. Suppose the constant perturbation was switched on as $\lambda\hbar Vt$ for $0 \leq \lambda t < 1$ and remained at $\hbar V$ for $\lambda t \geq 1$. Investigate how the transients behave.

6.15 The effect of an oscillating perturbation

We now consider a system that is exposed to an oscillating perturbation, such as an atom may experience when it is exposed to electromagnetic radiation in a spectrometer or in sunlight. Once we can deal with oscillating perturbations, we can deal with all perturbations, for a general time-dependent perturbation can be expressed as a superposition of harmonically oscillating functions. In the first stage of the discussion we shall consider transitions between discrete states $|\mathrm{i}\rangle$ and $|\mathrm{f}\rangle$. In the next section we shall allow the final state to be embedded in a continuum.

A perturbation oscillating with an angular frequency $\omega = 2\pi\nu$ and turned on at $t = 0$ has the form

$$H^{(1)}(t) = 2H^{(1)}\cos\omega t = H^{(1)}\left(\mathrm{e}^{\mathrm{i}\omega t} + \mathrm{e}^{-\mathrm{i}\omega t}\right) \tag{71}$$

for $t \geq 0$. If this perturbation is inserted into eqn 6.67 we obtain

$$\begin{aligned} a_{\mathrm{f}}(t) &= \frac{H_{\mathrm{fi}}^{(1)}}{\mathrm{i}\hbar}\int_0^t \left(\mathrm{e}^{\mathrm{i}\omega t} + \mathrm{e}^{-\mathrm{i}\omega t}\right)\mathrm{e}^{\mathrm{i}\omega_{\mathrm{fi}}t}\,\mathrm{d}t \\ &= \frac{H_{\mathrm{fi}}^{(1)}}{\mathrm{i}\hbar}\left\{\frac{\mathrm{e}^{\mathrm{i}(\omega_{\mathrm{fi}}+\omega)t} - 1}{\mathrm{i}(\omega_{\mathrm{fi}}+\omega)} + \frac{\mathrm{e}^{\mathrm{i}(\omega_{\mathrm{fi}}-\omega)t} - 1}{\mathrm{i}(\omega_{\mathrm{fi}}-\omega)}\right\} \end{aligned} \tag{72}$$

As it stands, the last equation is quite obscure (but it is quite easy to compute). However, it can be simplified to bring out its principal content by taking note of the conditions under which it is normally used. In applications in electronic spectroscopy, the frequencies ω_{fi} and ω are of the order of $10^{15}\,\mathrm{s}^{-1}$; in NMR, the lowest frequency form of spectroscopy generally encountered, the frequencies are still higher than $10^6\,\mathrm{s}^{-1}$. The exponential functions in the numerator of the term in braces are of the order of 1 regardless of the frequencies in its argument (because $\mathrm{e}^{\mathrm{i}x} = \cos x + \mathrm{i}\sin x$, and neither harmonic function can exceed 1). However, the denominator in the first term is of the order of the frequencies, and so the first term is unlikely to be larger than about 10^{-6} and may be of the order of 10^{-15} in electronic spectroscopy. In contrast, the denominator in the second term can come arbitrarily close to 0 as the external perturbation approaches a transition frequency of the system. Therefore, the second term is normally larger than the first for absorption, and overwhelms it completely as the frequencies approach one another. Consequently, we can be confident about ignoring the first term. When that is

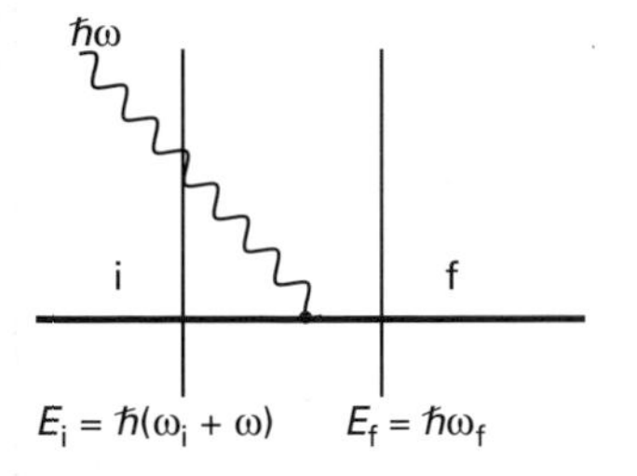

Fig. 6.14 The use of an oscillating perturbation effectively modifies the energy separation between the initial and final states, and at resonance the overall system is effectively degenerate and hence highly responsive.

done, it is easy to conclude that the probability of finding the system in the discrete state $|f\rangle$ after a time t if initially it was in state $|i\rangle$ at $t = 0$ is

$$P_f(t) = \frac{4|H_{fi}^{(1)}|^2}{\hbar^2(\omega_{fi} - \omega)^2} \sin^2 \tfrac{1}{2}(\omega_{fi} - \omega)t \tag{73}$$

Once again we write $|H_{fi}^{(1)}|^2 = \hbar^2|V_{fi}|^2$, in which case we obtain

$$P_f(t) = \frac{4|V_{fi}|^2}{(\omega_{fi} - \omega)^2} \sin^2 \tfrac{1}{2}(\omega_{fi} - \omega)t \tag{74}$$

The last expression should be familiar. Apart from a small but significant modification, it is exactly the same as eqn 6.62, the expression for a static perturbation applied to a two-level system. The one significant difference is that instead of the actual frequency difference ω_{fi} appearing in the expression, it is replaced throughout by $\omega_{fi} - \omega$. This replacement can be interpreted as an effective shift in the energy differences involved in exciting the system as a result of the presence of a photon in the electromagnetic field. As depicted in Fig. 6.14, where the wavy line now represents an oscillating perturbation, the energy difference $E_f - E_i$ would actually be thought of as

$$\begin{aligned} E_f - E_i &= E(\text{excited molecule, no photon}) \\ &\qquad - E(\text{ground state molecule, photon of energy } \hbar\omega) \\ &= \hbar(\omega_{fi} - \omega) \end{aligned}$$

The time-dependence of the probability of being found in state $|f\rangle$ depends on the **frequency offset**, $\omega_{fi} - \omega$ (Fig. 6.15). When the frequency offset is zero, the field and the system are said to be in **resonance**, and the transition probability increases most rapidly with time. To obtain the quantitative form of the time-dependence at resonance, we take the limit of eqn 6.74 as $\omega \to \omega_{fi}$ by using

$$\lim_{x\to 0} \frac{\sin x}{x} = \lim_{x\to 0} \frac{x - \frac{1}{6}x^3 + \ldots}{x} = 1$$

Then, with $x = \frac{1}{2}(\omega_{fi} - \omega)t$,

$$\lim_{\omega\to\omega_{fi}} P_f(t) = |V_{fi}|^2 t^2$$

and the probability increases quadratically with time. This conclusion is valid so long as $|V_{fi}|^2 t^2 \ll 1$, because that is the underlying assumption of first-order perturbation theory. It follows that the transition probability may approach (and, indeed, in this approximation, unphysically exceed) 1 as the applied frequency approaches a transition frequency. This behaviour can be interpreted in terms of the system then becoming, in effect, a loose, degenerate system which can be nudged fully from state to state even by gentle perturbations.

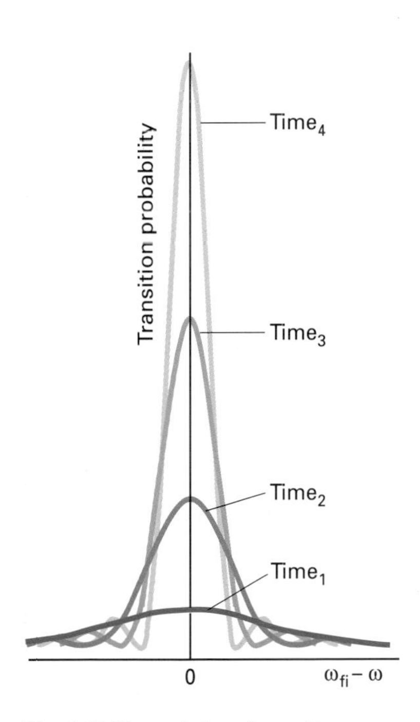

Fig. 6.15 The variation of transition probability with offset frequency and time. Note that the central portion of the curve becomes taller but narrower with time.

6.16 Transition rates to continuum states

We now turn to the case in which the final state is a part of a continuum of states. Although we can still use eqn 6.74 to calculate the transition probability to one member of the continuum, the observed transition rate is an integral over all the transition probabilities to which the perturbation can drive the system.

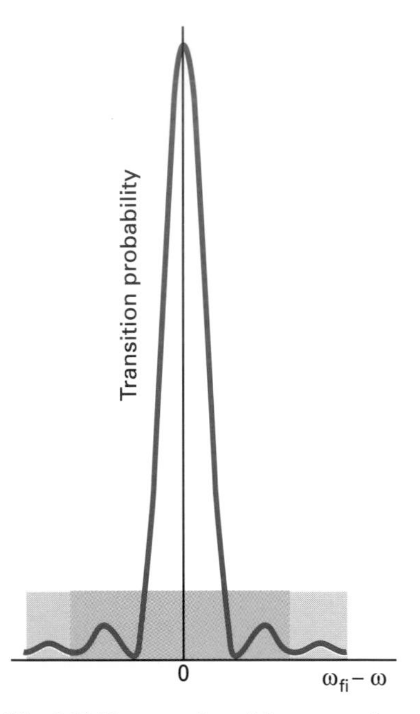

Fig. 6.16 The extension of the range of integration from the actual range (blue shading) to infinity (grey shading) barely affects the value of the integral.

Specifically, if the **density of states** is written $\rho(E)$, where $\rho(E)\mathrm{d}E$ is the number of states in the range E to $E + \mathrm{d}E$, then the total transition probability, $P(t)$, is

$$P(t) = \int_{\text{range}} P_{\text{f}}(t)\rho(E)\,\mathrm{d}E \tag{75}$$

In this expression 'range' means that the integration is over all final states accessible under the influence of the perturbation.

To evaluate the integral, we first express the transition frequency ω_{fi} in terms of the energy E by writing $\omega_{\text{fi}} = E/\hbar$

$$P(t) = \int_{\text{range}} 4|V_{\text{fi}}|^2 \frac{\sin^2 \frac{1}{2}(E/\hbar - \omega)t}{(E/\hbar - \omega)^2} \rho(E)\,\mathrm{d}E$$

The integral can be simplified by noting that the factor $\sin^2 x/x^2$ is sharply peaked close to $E/\hbar = \omega$, the frequency of the radiation. However, for an appreciable transition probability, the frequency of the incident radiation must be close to the transition frequency ω_{fi}, so we can set $E/\hbar \approx \omega_{\text{fi}}$ wherever E occurs. In other words, we can evaluate the density of states at $E_{\text{fi}} = \hbar\omega_{\text{fi}}$, and treat it as a constant. Moreover, although the matrix elements $|V_{\text{fi}}|$ vary with E, such a narrow range of energies contributes to the integral that it is permissible to treat $|V_{\text{fi}}|$ as a constant. The integral then simplifies to

$$P(t) = |V_{\text{fi}}|^2 \rho(E_{\text{fi}}) \int_{\text{range}} \frac{4\sin^2 \frac{1}{2}(E/\hbar - \omega)t}{(E/\hbar - \omega)^2}\,\mathrm{d}E$$

An additional approximation that stems from the narrowness of the function remaining in the integrand is to extend the limits from the actual range to infinity: the integrand is so small outside the actual range that this extension introduces no significant error (Fig. 6.16). At this point it is also convenient to set $x = \frac{1}{2}(E/\hbar - \omega)t$, which implies that $\mathrm{d}E = (2\hbar/t)\mathrm{d}x$. Consequently, the integral becomes

$$P(t) = \left(\frac{2\hbar}{t}\right)|V_{\text{fi}}|^2 \rho(E_{\text{fi}})t^2 \int_{-\infty}^{\infty} \frac{\sin^2 x}{x^2}\,\mathrm{d}x$$

The integral is standard:

$$\int_{-\infty}^{\infty} \left(\frac{\sin x}{x}\right)^2 \mathrm{d}x = \pi \tag{76}$$

Therefore, we conclude that

$$P(t) = 2\pi\hbar t|V_{\text{fi}}|^2 \rho(E_{\text{fi}}) \tag{77}$$

The **transition rate**, W, is the rate of change of probability of being in an initially empty state:

$$W = \frac{\mathrm{d}P}{\mathrm{d}t}$$

and the intensities of spectral lines are proportional to these transition rates because they depend on the rate of transfer of energy between the system and

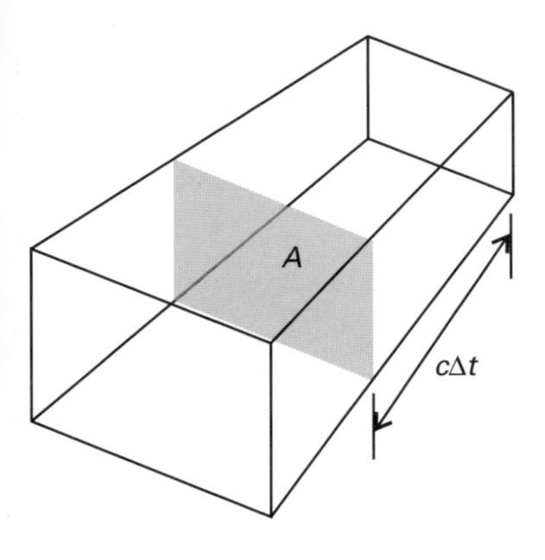

Fig. 6.17 All photons within a distance $c\Delta t$ can reach the right-hand wall in an interval Δt.

the electromagnetic field. It follows that, at resonance,

$$W = 2\pi\hbar|V_{\rm fi}|^2\rho(E_{\rm fi}) \tag{78}$$

This succinct expression is called **Fermi's golden rule**. It asserts that to calculate a transition rate, all we need do is to multiply the square modulus of the transition matrix element between the two states by the density of states at the transition frequency.

6.17 The Einstein transition probabilities

Einstein considered the problem of the transfer of energy between the electromagnetic field and matter and arrived at the conclusion that although eqn 6.78 correctly accounts for the absorption of radiation, it fails to take into account all contribution to the emission of radiation from an excited state. He considered a collection of atoms that were in thermal equilibrium with the electromagnetic field at a temperature T.

First, we note that the quantity $|V_{\rm fi}|^2$ is proportional to the square of the electric field strength of the incident radiation (for a perturbation of the form $-\mu\mathcal{E}$), and hence is proportional to the intensity, I, of the radiation at the frequency of the transition. The intensity is defined so that the energy of radiation in the frequency range ν to $\nu + {\rm d}\nu$ that passes through an area A in an interval Δt is

$${\rm d}E - I(\nu)A\Delta t{\rm d}\nu \tag{79}$$

Because all the radiation within a distance $c\Delta t$ can pass through the area in that time interval (Fig. 6.17), the energy density, $\rho_{\rm rad}(\nu){\rm d}\nu$, in that frequency range is

$$\rho_{\rm rad}(\nu){\rm d}\nu = \frac{{\rm d}E}{Ac\Delta t} = \frac{I(\nu)}{c}{\rm d}\nu$$

Therefore, the energy density of radiation, the energy per unit volume per unit frequency range, is

$$\rho_{\rm rad} = \frac{I(\nu)}{c} \tag{80}$$

Consequently, $|V_{\rm fi}|^2$ is proportional to $\rho_{\rm rad}$ evaluated at the transition frequency, or equivalently through the relation $E_{\rm fi} = h\nu_{\rm fi}$, at the transition energy. It follows that we can write

$$W_{\rm f\leftarrow i} = B_{\rm if}\rho_{\rm rad}(E_{\rm fi}) \tag{81}$$

where $B_{\rm if}$ is the **Einstein coefficient of stimulated absorption**.

Einstein also recognized that the rate at which an excited state $|{\rm f}\rangle$ is induced to make transitions down to the ground state $|{\rm i}\rangle$ is also proportional to the intensity of radiation at the transition frequency:

$$W_{\rm f\rightarrow i} = B_{\rm fi}\rho_{\rm rad}(E_{\rm fi}) \tag{82}$$

The coefficient $B_{\rm fi}$ is the **Einstein coefficient of stimulated emission**. We can infer from eqns 6.78, 6.81, and 6.82 that $B_{\rm if} = B_{\rm fi}$. The argument is based on the hermiticity of the perturbation hamiltonian, which lets us write the following:

$$B_{\rm if} \propto V_{\rm if}V_{\rm if}^* = V_{\rm fi}^*V_{\rm fi} \propto B_{\rm fi}$$

Einstein, however, was able to infer this equality in a different way, as we shall now see. Specifically, for electric-dipole allowed transitions, we show in *Further information 16* that

$$B_{\mathrm{if}} = \frac{|\mu_{\mathrm{fi}}|^2}{6\varepsilon_0\hbar^2} \tag{83}$$

where $\boldsymbol{\mu}_{\mathrm{fi}}$ is the **transition dipole moment**:

$$\boldsymbol{\mu}_{\mathrm{fi}} = \int \psi_{\mathrm{f}}^{*}\boldsymbol{\mu}\psi_{\mathrm{i}}\,d\tau \tag{84}$$

with $\boldsymbol{\mu}$ the electric dipole moment operator.

The transition probabilities we have derived refer to individual atoms. If there are N_{i} atoms in the state $|\mathrm{i}\rangle$ and N_{f} in the state $|\mathrm{f}\rangle$, then at thermal equilibrium, when there is no net transfer of energy between the system and the field, $N_{\mathrm{i}}W_{\mathrm{f}\leftarrow\mathrm{i}} = N_{\mathrm{f}}W_{\mathrm{f}\rightarrow\mathrm{i}}$. However, because the two transition rates are equal, the populations are implied to be equal, which is in conflict with the Boltzmann distribution, which requires from very general principles that

$$\frac{N_{\mathrm{f}}}{N_{\mathrm{i}}} = \mathrm{e}^{-E_{\mathrm{fi}}/kT}$$

Einstein therefore proposed that there was an additional contribution to the emission process that is independent of the presence of radiation of the transition frequency. This additional contribution he wrote

$$W_{\mathrm{f}\rightarrow\mathrm{i}}^{\mathrm{spont}} = A_{\mathrm{fi}} \tag{85}$$

where A_{fi} is the **Einstein coefficient of spontaneous emission**. The total rate of emission is therefore

$$W_{\mathrm{f}\rightarrow\mathrm{i}} = A_{\mathrm{fi}} + B_{\mathrm{fi}}\rho_{\mathrm{rad}}(E_{\mathrm{fi}}) \tag{86}$$

and the condition for thermal equilibrium is now

$$N_{\mathrm{i}}B_{\mathrm{if}}\rho_{\mathrm{rad}}(E_{\mathrm{fi}}) = N_{\mathrm{f}}\{A_{\mathrm{fi}} + B_{\mathrm{fi}}\rho_{\mathrm{rad}}(E_{\mathrm{fi}})\}$$

This expression is no longer in conflict with the Boltzmann distribution. Indeed, if we accept the Boltzmann distribution, it can be rearranged into

$$\rho_{\mathrm{rad}}(E_{\mathrm{fi}}) = \frac{A_{\mathrm{fi}}/B_{\mathrm{fi}}}{(B_{\mathrm{if}}/B_{\mathrm{fi}})\mathrm{e}^{E_{\mathrm{fi}}/kT} - 1}$$

However, it is also known from very general considerations that at equilibrium, the density of states of the electromagnetic field is given by the Planck distribution (see the Introduction):

$$\rho_{\mathrm{rad}}(E_{\mathrm{fi}}) = \frac{8\pi h\nu_{\mathrm{fi}}^3/c^3}{\mathrm{e}^{E_{\mathrm{fi}}/kT} - 1} \tag{87}$$

Comparison of the last two expressions confirms that $B_{\mathrm{if}} = B_{\mathrm{fi}}$ and, moreover, gives a relation between the coefficients of stimulated and spontaneous emission:

$$A_{\mathrm{fi}} = \frac{8\pi h\nu_{\mathrm{fi}}^3}{c^3}B_{\mathrm{fi}} \tag{88}$$

The important point about eqn 6.88 is that it shows that the relative importance of spontaneous emission increases as the cube of the transition frequency, and that it is therefore potentially of great importance at very high frequencies. That is one reason why X-ray lasers are so difficult to make: highly excited populations are difficult to maintain and discard their energy at random instead of cooperating in a stimulated emission process.

Example 6.10 The calculation of transition probabilities

Calculate the rates of stimulated and spontaneous emission for the $3p \rightarrow 2s$ transition in hydrogen: this is the H_α line in its spectrum, the line responsible for the red glow. Suppose that the atoms are exposed to radiation from a source at 1000 K.

Method. We use eqn 6.82 for the rate of stimulated emission, taking the value of B from eqn 6.83 and the density of states of the radiation field from the Planck distribution in eqn 6.87. The transition moment is calculated by using the hydrogen orbitals listed in Tables 3.1 and 3.2 and the transition energy is taken from eqn 3.44. For the rate of spontaneous emission, we use the relation between A and B in eqn 6.88.

Answer. The transition dipole moment for the $3p_z \rightarrow 2s$ transition is

$$\mu_{z,\mathrm{fi}} = -e\int \psi^*_{3p_z} z \psi_{2s}\,\mathrm{d}\tau = -\frac{3^3 \times 2^{10}}{5^6} e a_0$$
$$= -1.769 e a_0 = -1.500 \times 10^{-29}\ \mathrm{C\ m}$$

The square of the transition moment for all three transitions (from $3p_x$, $3p_y$, and $3p_z$) is therefore

$$|\mu|^2 = |\mu_x|^2 + |\mu_y|^2 + |\mu_z|^2 = 3 \times 3.131 e^2 a_0^2 = 6.752 \times 10^{-58}\ \mathrm{C^2\ m^2}$$

All three contributions are the same because the lower state of the atom is spherically symmetrical. The Einstein coefficient of stimulated emission is therefore

$$B = \frac{|\mu|^2}{6\varepsilon_0 \hbar^2} = 1.143 \times 10^{21}\ \mathrm{J^{-1}\ m^3\ s^{-2}}$$

The frequency of the transition is

$$\nu = \left(\frac{1}{2^2} - \frac{1}{3^2}\right) c\mathcal{R} = 4.567 \times 10^{14}\ \mathrm{Hz}$$

and so it follows that

$$A = \frac{8\pi h \nu^3}{c^3} B = 6.728 \times 10^7\ \mathrm{s^{-1}}$$

At 1000 K and for the transition frequency,

$$\rho_{\mathrm{rad}} = \frac{8\pi h\nu^3/c^3}{e^{h\nu/kT} - 1} = 1.782 \times 10^{-23}\ \mathrm{J\ Hz^{-1}\ m^{-3}}$$

It follows that the rate of stimulated emission is $B\rho_{\text{rad}} = 2.036 \times 10^{-2}\ \text{s}^{-1}$ whereas that of spontaneous emission is $A = 6.728 \times 10^{7}\ \text{s}^{-1}$.

Comment. The lifetime of the upper state due to spontaneous emission is $1/A$, or 14.9 ns, whereas the lifetime due to stimulated emission is 49 s under the conditions prevailing.

Exercise 6.10. Repeat the question for the $2p \rightarrow 1s$ transition.

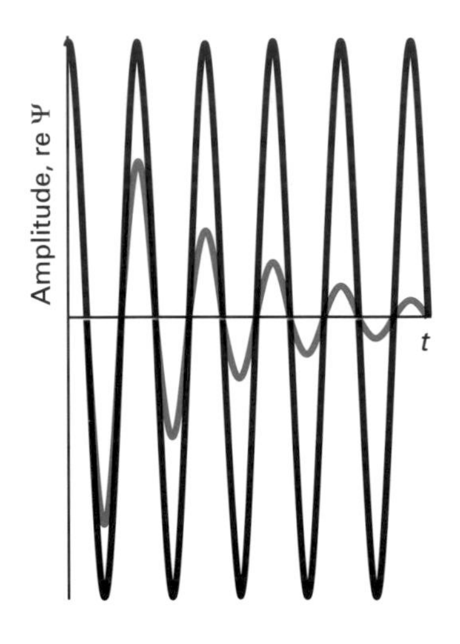

Fig. 6.18 A wavefunction corresponding to a precise energy has a constant maximum amplitude; if the wavefunction decays, then it no longer corresponds to a precise energy.

The presence of the spontaneous emission process can be viewed as the outcome of the presence of zero-point fluctuations of the electromagnetic field. As indicated in footnote 1 in Section 7.3, the electromagnetic field has zero-point oscillations even though there are no photons present, and these fluctuations perturb the excited state and induce the transition to a lower state. 'Spontaneous' transitions are actually caused by these zero-point fluctuations of the electromagnetic vacuum.

6.18 Lifetime and energy uncertainty

We are now in a position to establish the relation between the lifetime of a state and the range of energies that it may possess. We have seen that if a state has a precise energy, then its time-dependent wavefunction has the form $\Psi = \psi e^{-iEt/\hbar}$; such states are steady states in the sense that $|\Psi|^2 = |\psi|^2$, a time-independent probability density. However, if the wavefunction decays with time, perhaps because the system is making transitions to other states, then its energy is imprecise.

We suppose that the probability of finding the system in a particular excited state decays exponentially with time with a time-constant τ:

$$|\Psi|^2 = |\psi|^2 e^{-t/\tau} \tag{89}$$

The justification of this assumption can be found in the references in *Further reading* and Section 14.13. The amplitude therefore has the form

$$\Psi = \psi e^{-iEt/\hbar - t/2\tau} \tag{90}$$

This wavefunction decays as it oscillates (Fig. 6.18), and its energy is not immediately obvious. However, such a function can be modelled as a superposition of oscillating functions by using the techniques of Fourier transform theory, and we write

$$e^{-iEt/\hbar - t/2\tau} = \int g(E')e^{-iE't/\hbar}\,dt$$

where

$$g(E') = \frac{\hbar/\tau}{(E - E')^2 + (\hbar/2\tau)^2} \tag{91}$$

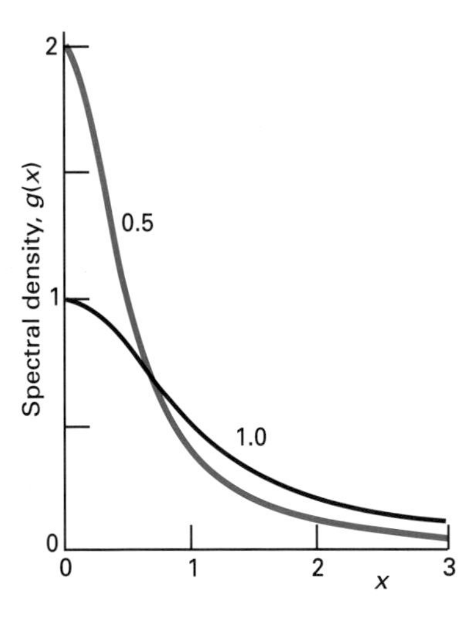

Fig. 6.19 The spectral density function for two wavefunctions that decay at different rates. The labels of the lines are the values of $\hbar/2\tau$, with $x = E - E'$ (in the same units).

This expression shows that the decaying function corresponds to a range of energies (all the values that appear in the superposition), and therefore it implies that any state that has a finite lifetime must be regarded as having an imprecise energy.

We can arrive at the quantitative relation between lifetime and energy by considering the shape of the **spectral density function**, g (Fig. 6.19). The width at half-height is readily shown to be equal to $\hbar/2\tau$, and this quantity can be taken as an indication of the range of energies δE present in the state. It follows that

$$\tau\delta E \approx \tfrac{1}{2}\hbar \tag{92}$$

This **lifetime broadening relation** is reminiscent of the uncertainty principle (Sections 1.16 and 1.18). It shows that the shorter the lifetime of the state (the shorter the time-constant τ for its decay), then the less precise its energy. When a state has zero lifetime, we can say nothing about its energy. Only when the lifetime of a state is infinite can the energy be specified exactly.

Problems

6.1 One excited state of the sodium atom lies at $25\,739.86\,\mathrm{cm^{-1}}$ above the ground state, another lies at $50\,266.88\,\mathrm{cm^{-1}}$. Suppose they are connected by a perturbation equivalent in energy to (a) $100\,\mathrm{cm^{-1}}$, (b) $1000\ \mathrm{cm^{-1}}$, (c) $5000\ \mathrm{cm^{-1}}$. Calculate the energies and composition of the states of the perturbed system. *Hint.* Use eqn 6.6 for the energies and eqn 6.8 for the states, and express the composition as the contribution of the unperturbed states.

6.2 A simple calculation of the energy of the helium atom supposes that each electron occupies the same hydrogenic $1s$-orbital (but with $Z=2$). The electron–electron interaction is regarded as a perturbation, and calculation gives

$$\int \psi_{1s}^{2}(r_1)\left(\frac{e^2}{4\pi\varepsilon_0 r_{12}}\right)\psi_{1s}^{2}(r_2)\,\mathrm{d}\tau = \tfrac{5}{4}\left(\frac{e^2}{4\pi\varepsilon_0}\right)$$

Estimate (a) the binding energy of helium, (b) its first ionization energy. *Hint.* Use eqn 6.6 with $E_1 = E_2 = E_{1s}$. Be careful not to count the electron–electron interaction energy twice.

6.3 Show that the energy of the perturbed levels is related to the mean energy of the unperturbed levels $\bar{E} = \frac{1}{2}(E_1 + E_2)$ by $E_{\pm} - \bar{E} = \pm\frac{1}{2}(E_1 - E_2)\sec 2\zeta$, where ζ is the parameter in eqn 6.9. Devise a diagrammatic method of showing how $E_{\pm} - \bar{E}$ depends on $E_1 - E_2$ and ζ.

6.4 We normally think of the one-dimensional well as being horizontal. Suppose it is vertical; then the potential energy of the particle depends on x because of the presence of the gravitational field. Calculate the first-order correction to the zero-point energy, and evaluate it for an electron in a box on the surface of the Earth. Account for the result. *Hint.* The energy of the particle depends on its height as mgx where $g = 9.81\,\mathrm{m\,s^{-2}}$. Because g is so small, the energy correction is tiny; but it would be significant if the box were on the surface of a neutron star.

6.5 Calculate the second-order correction to the energy for the system described in Problem 6.4 and calculate the ground-state wavefunction. Account for the shape of the distortion caused by the perturbation. *Hint.* Use eqn 6.24 for the energy and eqn 6.22 for the wavefunction. The integral involved is of the form

$$\int x \sin ax \sin bx\,\mathrm{d}x = -\frac{\mathrm{d}}{\mathrm{d}a}\int \cos ax \sin bx\,\mathrm{d}x$$

$$\int \cos ax \sin bx\,\mathrm{d}x = \frac{\cos(a-b)x}{2(a-b)} - \frac{\cos(a+b)x}{2(a+b)}$$

Evaluate the sum over n numerically.

6.6 In the free-electron molecular orbital model (Problem 2.19) the potential energy may be made slightly more realistic by supposing that it varies sinusoidally along the polyene chain. Select a potential energy with suitable periodicity, and calculate the first-order correction to the wavelength of the lowest energy transition.

6.7 Show group-theoretically that when a perturbation of the form $H^{(1)} = az$ is applied to a hydrogen atom, the 1*s*-orbital is contaminated by the admixture of np_z-orbitals. Deduce which orbitals mix into (a) $2p_x$-orbitals, (b) $2p_z$-orbitals, (c) $3d_{xy}$-orbitals.

6.8 The symmetry of the ground electronic state of the water molecule is A_1. (a) An electric field, (b) a magnetic field is applied perpendicular to the molecular plane. What symmetry species of excited states may be mixed into the ground state by the perturbations? *Hint*. The electric interaction has the form $H^{(1)} = ax$; the magnetic interaction has the form $H^{(1)} = bl_x$.

6.9 Repeat Problem 6.5, but estimate the second-order energy correction using the closure approximation. Compare the two calculations and deduce the appropriate value of ΔE.

6.10 Calculate the second-order energy correction to the ground state of a particle in a box for a perturbation of the form $H^{(1)} = -\varepsilon \sin(\pi x/L)$ by using the closure approximation. Infer a value of ΔE by comparison with the numerical calculation in Example 6.2. This and the previous problem show that the parameter ΔE depends on the perturbation and is not simply a characteristic of the system itself.

6.11 The potential energy of a particle on a ring depended on the angle ϕ as $H^{(1)} = \varepsilon \sin^2 \phi$. Calculate the first-order corrections to the energy of the degenerate $m_l = \pm 1$ states, and find the correct linear combinations for the perturbation calculation. Go on to find the second-order correction to the energy. *Hint*. This is an example of degenerate-state perturbation theory, and so find the correct linear combinations by solving eqn 6.35 after deducing the energies from the roots of the secular determinant. For the matrix elements, express $\sin\phi$ as $(1/2\mathrm{i})(\mathrm{e}^{\mathrm{i}\phi} - \mathrm{e}^{-\mathrm{i}\phi})$. Do not forget the $m_l = 0$ state lying beneath the degenerate pair when evaluating eqn 6.35. Energies go as $m_l^2\hbar^2/2mr^2$; use $\psi_{m_l} = (1/2\pi)^{1/2}\mathrm{e}^{\mathrm{i}m_l\phi}$ for the unperturbed states.

6.12 A particle of mass m is confined to a one-dimensional square well of the type treated in Chapter 2. Choose trial functions of the form (a) $\sin kx$, (b) $(x - x^2/L) + k(x - x^2/L)^2$, (c) $\mathrm{e}^{-k(x-\frac{1}{2}L)} - \mathrm{e}^{-\frac{1}{2}kL}$ for $x \geq \frac{1}{2}L$, and $\mathrm{e}^{k(x-\frac{1}{2}L)} - \mathrm{e}^{-\frac{1}{2}kL}$ for $x \leq \frac{1}{2}L$. Find the optimum values of k and the corresponding energies.

6.13 Consider the hypothetical linear H_3 molecule. The wavefunctions may be modelled by expressing them as $\psi = c_A s_A + c_B s_B + c_C s_C$, the s_i denoting hydrogen 1*s*-orbitals of the relevant atom. Use the Rayleigh–Ritz method to find the optimum values of the coefficients and the energies of the orbital. Make the approximations $H_{ss} = \alpha$, $H_{ss'} = \beta$ for neighbours but 0 for non-neighbours, $S_{ss} = 1$, and $S_{ss'} = 0$. *Hint*. Although the basis can be used as it stands, it leads to a 3×3 determinant, and hence to a cubic equation for the energies. A better procedure is to set up symmetry-adapted combinations, and then to use the vanishing of H_{ij} unless $\Gamma_i = \Gamma_j$.

6.14 Repeat the last problem but set $H_{s_A s_C} = \gamma$ and $S_{ss'} \neq 0$. Evaluate the overlap integrals between 1*s*-orbitals on centres separated by R; use

$$S = \left\{1 + \frac{R}{a_0} + \frac{1}{3}\left(\frac{R}{a_0}\right)^2\right\}\mathrm{e}^{-R/a_0}$$

Suppose that $\beta/\gamma = S_{s_A s_B}/S_{s_A s_C}$. For a numerical result, take $R = 80\,\mathrm{pm}$, $a_0 = 53\,\mathrm{pm}$.

6.15 A hydrogen atom in a $2s^1$ configuration passes into a region where it experiences an electric field in the *z*-direction for a time τ. What is its electric dipole moment during its exposure and after it emerges? *Hint*. Use eqn 6.59 with $\omega_{21} = 0$; the dipole moment is the expectation value of ez. The integral $\int \psi_{2s} z \psi_{2p_z}\,\mathrm{d}\tau$ is equal to $3a_0$.

6.16 A biradical is prepared with its two electrons in a singlet state. A magnetic field is present, and because the two electrons are in different environments their interaction with the field is $(\mu_B/\hbar)B(g_1 s_{1z} + g_2 s_{2z})$ with $g_1 \neq g_2$. Evaluate the time-dependence of the probability that the electron spins will take on a triplet configuration (i.e. that the $S = 1$, $M_S = 0$ state will be populated). Examine the role of the energy separation hJ of the singlet state and the $M_S = 0$ state of the triplet. Suppose $g_1 - g_2 \approx 1 \times 10^{-3}$ and

$J \approx 0$; how long does it take for the triplet state to emerge in a field of 1.0 T? *Hint*. Use eqn 6.60; take $|0,0\rangle = (1/\sqrt{2})(\alpha\beta - \beta\alpha)$ and $|1,0\rangle = (1/\sqrt{2})(\alpha\beta + \beta\alpha)$. See Problem 4.21 for a note on the significance of μ_B and g.

6.17 An electric field in the z-direction is increased linearly from zero. What is the probability that a hydrogen atom, initially in the ground state, will be found with its electron in a $2p_z$-orbital at a time t? *Hint.* Use eqn 6.60 with $H^{(1)}_{\mathrm{fi}} \propto t$.

6.18 At $t = \frac{1}{2}T$ the strength of the field used in Problem 6.17 begins to decrease linearly. What is the probability that the electron is in the $2p_z$-orbital at $t = T$? What would the probability be if initially the electron was in a $2s$-orbital?

6.19 Instead of the perturbation being switched linearly it was switched on and off exponentially and slowly, the switching off commencing long after the switching on was complete. Calculate the probabilities, long after the perturbation has been extinguished, of the $2p_z$-orbital being occupied, the initial states being as in Problem 6.17. *Hint.* Take $H^{(1)} \propto 1 - \mathrm{e}^{-kt}$ for $0 \le t \le T$ and $H^{(1)} \propto \mathrm{e}^{-k(t-T)}$ for $t \ge T$. Interpret 'slow' as $k \ll \omega$ and 'long after' as both $kT \gg 1$ (for 'long after switching on') and $k(t-T) \gg 1$ (for 'long after switching off').

6.20 Find the complete atomic-number dependence of the A and B coefficients for the $2p \rightarrow 1s$ transitions of hydrogenic atoms. Calculate how the stimulated emission rate depends on Z when the atom is exposed to black-body radiation at 1000 K. *Hint*. The relevant density of states also depends on Z.

6.21 Examine how the A and B coefficients depend on the length of a one-dimensional square well for the transition $n + 1 \rightarrow n$.

Further reading

Recent developments in perturbation theory. J.O. Hirschfelder, W. Byers Brown, and S.T. Epstein; *Adv. Quantum Chem.*, **1**, 255 (1964).
Perturbation theory and its applications in quantum mechanics. C.H. Wilcox (ed.); Wiley, New York (1966).
Quantum mechanics, Vols 1 and 2. C. Cohen-Tannoudji, B. Diu, and F. Laloë; Wiley, New York (1977).
Intermediate quantum mechanics. H.A. Bethe and R.W. Jackiw; Benjamin/Cummings, Menlo Park (1986).
Quantum mechanics. F. Mandl; Wiley, New York (1992).
Algebraic approach to simple quantum systems: with applications to perturbation theory. B.G. Adams; Springer, New York (1994).
Intermolecular forces and their evaluation by perturbation theory. P. Arrighini; Springer, New York (1981).
Large order perturbation theory and summation methods in quantum mechanics. G.A. Arteca, F.M. Fernandez, and E.A. Castro; Springer, New York (1990).
Quantum theory of finite systems. J.-P. Blaizot and G. Ripka; MIT Press, Cambridge, Mass. (1986).
The mathematics of physics and chemistry. H. Margenau and G.M. Murphy; van Nostrand, Princeton (1956).

7 Atomic spectra and atomic structure

A great deal of chemically interesting information can be obtained by interpreting the **line spectra** of atoms, the frequencies of the electromagnetic radiation that atoms emit when they are excited. We can use the information to establish the electronic structures of the atoms, and then use that information as a basis for discussing the periodicity of the elements and the structures of the bonds they form. Atomic spectra were also of considerable historical importance, because their study led to the formulation of the Pauli principle, without which it would be impossible to understand atomic structure, chemical periodicity, and molecular structure. The information provided by atoms is of considerable importance for the discussion of molecular structure. For example, we need values of ionization energies and spin–orbit coupling parameters if we are to understand the structures of molecules and their properties, particularly their photochemical reactions.

As in the preceding chapters, we begin by describing a system that can be solved exactly: the hydrogen atom. Then we build on our knowledge of that atom's structure and spectra to discuss the properties and structures of many-electron atoms.

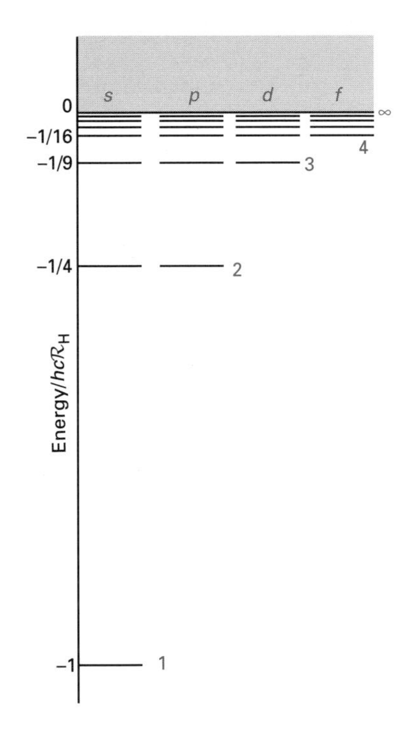

Fig. 7.1 The energy levels of the hydrogen atom. Hydrogenic atoms in general have the same spectrum, but with the energy scale magnified by a factor of Z^2.

The spectrum of atomic hydrogen

So long as we ignore electron spin, the state of an electron in a hydrogen atom is specified by three quantum numbers, n, l, and m_l (Section 3.13) and its energy is given by

$$E = -\frac{\mu e^4}{32\pi^2\varepsilon_0^2\hbar^2 n^2} \qquad n = 1, 2, \ldots \tag{1}$$

This expression is normally written

$$E = -\frac{hc\mathcal{R}_H}{n^2} \qquad \mathcal{R}_H = \frac{\mu e^4}{8\varepsilon_0^2 h^3 c} \tag{2}$$

where $\mathcal{R}_H$ is the **Rydberg constant** for hydrogen. The origin of this expression was explained in Chapter 3 and there is no need to repeat the arguments here, but for convenience the array of energy levels is shown in Fig. 7.1.

7.1 The energies of the transitions

The spectrum of atomic hydrogen arises from transitions between its permitted states, and the difference in energy, ΔE, is discarded as a photon of energy $h\nu$ and wavenumber $\tilde{\nu}$, where $\tilde{\nu} = \nu/c$. For the transition $n_2 \rightarrow n_1$, the wavenumber of the emitted radiation is

$$\tilde{\nu} = \left(\frac{1}{n_1^2} - \frac{1}{n_2^2}\right)\mathcal{R}_{\mathrm{H}} \qquad (3)$$

For a given value of n_1, the set of transitions from $n_2 = n_1 + 1, n_1 + 2, \ldots$ constitutes a **series** of lines, and these series bear the names of their discoverers or principal investigators:

$n_1 = 1, n_2 = 2, 3, \ldots$	Lyman series, ultraviolet
$n_1 = 2, n_2 = 3, 4, \ldots$	Balmer series, visible
$n_1 = 3, n_2 = 4, 5, \ldots$	Paschen series, infrared
$n_1 = 4, n_2 = 5, 6, \ldots$	Brackett series, far infrared
$n_1 = 5, n_2 = 6, 7, \ldots$	Pfund series, far infrared
$n_1 = 6, n_2 = 7, 8, \ldots$	Humphreys series, far infrared

The **limit** of each series is the wavenumber of the transition that just succeeds in ionizing the atom. This limit corresponds to $n_2 = \infty$ in each case, and therefore

$$\tilde{\nu}_\infty = \frac{\mathcal{R}_{\mathrm{H}}}{n_1^2} \qquad (4)$$

The **ionization energy**, I, of the atom is the minimum energy required to ionize it from its ground state, the state with $n_1 = 1$. Hence,

$$I = hc\mathcal{R}_{\mathrm{H}} \qquad (5)$$

The numerical value of the ionization energy is 2.179 aJ, which corresponds to $1312\,\mathrm{kJ\,mol^{-1}}$ and 13.60 eV.

7.2 Selection rules

Not all transitions between states are allowed. The **selection rules** for electric-dipole transitions, the rules that specify the specific transitions that may occur, are based on an examination of the transition dipole moment (Section 6.17) between the two states of interest. They are established by identifying the conditions under which the transition dipole moment is nonzero, corresponding to an **allowed transition**, or zero, for a **forbidden transition**. The transition dipole moment for a transition between states $|\mathrm{i}\rangle$ and $|\mathrm{f}\rangle$ is defined as

$$\boldsymbol{\mu}_{\mathrm{if}} = \langle \mathrm{i}|\boldsymbol{\mu}|\mathrm{f}\rangle \qquad (6)$$

where $\boldsymbol{\mu} = -e\mathbf{r}$ is the electric dipole operator. The transition dipole moment can be regarded as a measure of the size of the electromagnetic jolt that the electron delivers to the electromagnetic field when it makes a transition between orbitals. Large shifts of charge through large distances can deliver strong impulses provided they have a dipolar character (as in the transition between *s*- and *p*-orbitals but not between *s*-orbitals where the shift of charge is spherically symmetric), and such transitions give rise to intense lines.

Group theory (Section 5.16) tells us that a transition dipole moment must be zero unless the integrand is totally symmetric under the symmetry operations of the system, which for atoms is the full rotation group, R_3. The easiest operation to consider is inversion, under which $\mathbf{r} \to -\mathbf{r}$. Under inversion, an atomic orbital with quantum number l has parity $(-1)^l$, as can be appreciated by noting that orbitals with even l (*s*- and *d*-orbitals) do not change sign

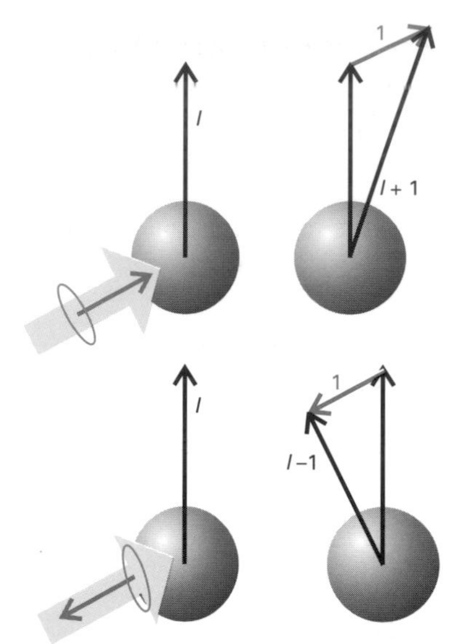

Fig. 7.2 The basis of the $\Delta l = \pm 1$ selection rule is the conservation of angular momentum and the fact that a photon has a helicity (the projection of its spin on its direction of propagation) of ± 1. Note that the absorption of a photon (as depicted in both instances here) can result in either an increase or a decrease of l.

whereas those with odd l (*p*- and *f*-orbitals) do change sign. This behaviour is also apparent from the mathematical form of the spherical harmonics (see Table 3.1). The parity of the integrand is therefore $(-1)^{l_i}(-1)(-1)^{l_f}$, which is even if the two orbitals have opposite parity (one odd, the other even). This argument is the basis of the **Laporte selection rule**:

> The only allowed electric-dipole transitions are those involving a change in parity.

Next, consider the rotational characteristics of the components of the integrand. The atomic orbitals are angular momentum wavefunctions and span the irreducible representations $\Gamma^{(l_i)}$ and $\Gamma^{(l_f)}$ of the full rotation group. The electric dipole moment operator behaves like a translation, and so it spans the irreducible representation $\Gamma^{(1)}$ of the group. The product of $\Gamma^{(l_i)}$ and $\Gamma^{(1)}$ therefore spans

$$\Gamma^{(l_i)} \times \Gamma^{(1)} = \Gamma^{(l_i+1)} + \Gamma^{(l_i)} + \Gamma^{(l_i-1)}$$

as explained in Section 5.20. Therefore, for the product of all three factors in the integrand to span the totally symmetric irreducible representation $\Gamma^{(0)}$, we require $\Gamma^{(l_f)}$ to be equal to $\Gamma^{(l_i+1)}$, $\Gamma^{(l_i)}$, or $\Gamma^{(l_i-1)}$. In other words, $l_f = l_i - 1, l_i$, or $l_i + 1$. However, we have already ruled out transitions that do not change parity, so the only allowed transitions are those to the states with $l_f = l_i \pm 1$. That is:

$$\Delta l = \pm 1 \tag{7}$$

The origin of this selection rule can be put on a more physical basis by noting that the intrinsic spin angular momentum of a photon is 1. Therefore, when it is absorbed or emitted, to conserve total angular momentum, the orbital angular momentum of the electron in the atom must change by ± 1. An increase in orbital angular momentum ($\Delta l = +1$) can accompany either an absorption or an emission of a photon, depending on the orientation of the angular momentum of the photon relative to the angular momentum of the electron in the atom (Fig. 7.2).

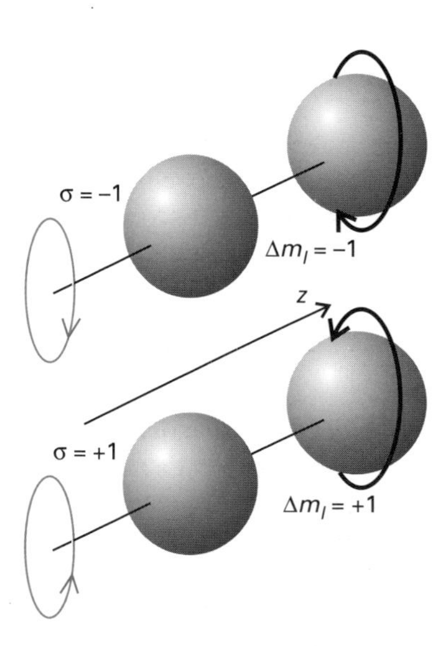

Fig. 7.3 The change in m_l that accompanies the absorption of a photon; the *z*-axis is taken to be the line of flight of the photon.

It is quite easy to extend these pictures to obtain the selection rules for m_l, the magnetic orbital quantum number. Now we need to know that a photon has an intrinsic **helicity**, σ, the spin angular momentum relative to its line of flight, of $\sigma = \pm 1$ (Fig. 7.3). We shall suppose that m_l labels the component of orbital angular momentum on the axis defined by the line of flight of the photon. Then, absorption of a left-circularly polarized photon (with helicity $\sigma = +1$) results in $\Delta m_l = +1$ to preserve overall angular momentum, and its emission results in $\Delta m_l = -1$. The opposite holds for a right-circularly polarized photon. The maximum change in m_l is therefore ± 1. It follows that for an atom that has its electron with a definite value of m_l for the component of angular momentum relative to an *arbitrary* axis, not necessarily the line of flight of the photon, the maximum change in m_l is still ± 1 but an allowed intermediate value may also occur if the photon is travelling in an intermediate direction. Therefore, the general selection rule is

$$\Delta m_l = 0, \pm 1 \tag{8}$$

The selection rule on m_l can also be deduced algebraically. Suppose the radiation is plane-polarized with the electric field in the *z*-direction, then only the z-component of the dipole moment is relevant, and we can write $\mu_z = -er\cos\theta$. The ϕ integral in the transition moment is then proportional to

$$\int_0^{2\pi} \mathrm{e}^{-\mathrm{i}m_{lf}\phi}(-er\cos\theta)\mathrm{e}^{\mathrm{i}m_{li}\phi}\,\mathrm{d}\phi \propto \int_0^{2\pi} \mathrm{e}^{\mathrm{i}(m_{li}-m_{lf})\phi}\,\mathrm{d}\phi$$

The integral over ϕ is zero unless $m_{li} = m_{lf}$. Therefore, for z-polarized radiation, $\Delta m_l = 0$. The selection rules $\Delta m_l = \pm 1$ arise similarly for radiation polarized in the xy-plane.

Example 7.1 The calculation of transition moments

Calculate the electric dipole transition moment for the transition $2p_z \rightarrow 2s$ in a hydrogenic atom.

Method. We use the wavefunctions set out in Tables 3.1 and 3.2 to evaluate the integral $\langle 2p_z|\mu_z|2s\rangle$ with $\mu_z = -er\cos\theta$.

Answer. The wavefunctions for the orbitals are

$$\psi_{2p_z} = \left(\frac{Z^5}{32\pi a_0^5}\right)^{1/2} r\cos\theta\,\mathrm{e}^{-Zr/2a_0}$$

$$\psi_{2s} = \left(\frac{Z^3}{32\pi a_0^5}\right)^{1/2} (2a_0 - Zr)\mathrm{e}^{-Zr/2a_0}$$

The integral we require is therefore

$$\langle 2p_z|\mu_z|2s\rangle = -e\left(\frac{Z^4}{32\pi a_0^5}\right)\int_0^\infty (2a_0 - Zr)r^4\mathrm{e}^{-Zr/a_0}\,\mathrm{d}r$$
$$\times \int_0^\pi \cos^2\theta\sin\theta\,\mathrm{d}\theta \int_0^{2\pi}\mathrm{d}\phi = -\frac{3ea_0}{Z}$$

For a hydrogen atom itself, $\langle 2p_z|\mu_z|2s\rangle = -3ea_0$.

Comment. There is no physical significance in the sign of the transition dipole moment because the relative signs of the wavefunctions used to calculate it are arbitrary. The physical observable, the transition intensity, depends on the square modulus of the transition dipole moment.

Exercise 7.1. Repeat the calculation for the transition $2p_z \rightarrow 1s$.

Electric dipole transitions are not the only types of transition that may occur. Light is an electro*magnetic* phenomenon, and the perturbation arising from the effect of the magnetic component of the field can induce **magnetic dipole transitions**. Such transitions have intensities that are proportional to the squares of matrix elements like $\langle \mathrm{i}|l_z|\mathrm{f}\rangle$. Such transitions are typically about 10^5 times weaker than allowed electric dipole transitions, but because they obey different selection rules, they may give rise to spectral lines where the electric dipole transition is forbidden. Another type of transition is an **electric quadrupole transition** in which the spatial variation of the electric field interacts with the electric quadrupole moment operator. Such transitions have intensities that are proportional to the squares of matrix elements like $\langle \mathrm{i}|xy|\mathrm{f}\rangle$. These transitions are about 10^8 times weaker than electric dipole transitions. Their selection rule is $\Delta l = 0, \pm 2$. The large change in angular momentum that accompanies the transition arises from the fact that the quadrupole transition imparts an orbital angular momentum to the photon (that is, generates it with a non-spherically symmetric wavefunction) in addition to its intrinsic spin. The weakness of magnetic dipole and electric quadrupole transitions stems from the fact that both depend on the variation of the electromagnetic wave over the extent of the atom. As atomic diameters are much smaller than typical wavelengths of radiation, this variation is typically very small and the intensity is correspondingly weak.

In some systems, a transition can result in the generation of two photons by an electric dipole mechanism more efficiently than a single photon is generated by a magnetic dipole transition. An example of this **multiple-quantum dipole transition** is provided by the excited $1s^1 2s^1\ {}^1S$ state of helium: the two-photon process governs the lifetime of the state because the magnetic dipole transition probability is so low.

7.3 Orbital and spin magnetic moments

So far, we have ignored the spin of the electron. Now we consider its effect on the structure and spectra of hydrogenic atoms. Its effect is not very pronounced on the energy levels of hydrogen itself, but it can be of great importance for atoms of high atomic number. We note that an electron has spin quantum number $s = \frac{1}{2}$ and that the spin magnetic quantum number can have the two values $m_s = \pm\frac{1}{2}$.

An electron is a charged particle, and with its angular momentum there is associated a magnetic moment. Because the electron in an atom may have two types of angular momentum, its spin and its orbital angular momentum, there are two sources of magnetic moment. These two magnetic moments can interact and give rise to shifts in the energies of the states of the atom, and these shifts in energy will affect the appearance of the spectrum of the atom. The resulting shifts and splitting of lines is called the **fine structure** of the spectrum.

First, consider the magnetic moment arising from the orbital angular momentum of the electron. The quantum mechanical derivation of its orbital magnetic moment is described in Section 13.6; here we shall use the following classical argument. If a charge $-e$ circulates in an orbit of radius r in the xy-plane at a speed v, the current is

$$I = -\frac{ev}{2\pi r}$$

This current gives rise to a magnetic dipole moment with z-component $m_z = IA$, where A is the area enclosed by the orbit, $A = \pi r^2$. It follows that

$$m_z = IA = -\frac{ev\pi r^2}{2\pi r} = -\tfrac{1}{2}evr$$

The z-component of the orbital angular momentum of the electron is $l_z = m_e vr$, so

$$m_z = -\frac{e}{2m_e} l_z$$

The same argument applies to orbital motion in other planes, and we can therefore write

$$\mathbf{m} = \gamma_e \mathbf{l} \qquad \text{where } \gamma_e = -\frac{e}{2m_e} \tag{9}$$

The constant γ_e is called the **magnetogyric ratio** of the electron.

The properties of the orbital magnetic moment follow from those of the angular momentum itself. In particular, its z-component is quantized and restricted to the values

$$m_z = \gamma_e m_l \hbar \qquad m_l = l, l-1, \ldots, -l \tag{10}$$

The positive quantity

$$\mu_B = -\gamma_e \hbar = \frac{e\hbar}{2m_e} \tag{11}$$

is called the **Bohr magneton**, and is often regarded as the elementary unit of magnetic moment. Its value is $9.274 \times 10^{-24}\,\mathrm{J\,T^{-1}}$. In terms of the Bohr magneton, the z-component of orbital magnetic moment is

$$m_z = -\mu_B m_l$$

Now we consider the magnetic moment that arises from the spin of the electron. By analogy with the orbital magnetic moment, we might expect the spin magnetic moment to be related to the spin angular momentum by $\mathbf{m} = \gamma_e \mathbf{s}$, but this turns out not to be the case. This should not be too surprising, because spin has no classical analogue, yet here we are trying to argue by analogy with orbital angular momentum, which does have a classical analogue. The relation between the spin and its magnetic moment can be derived from the relativistic Dirac equation, which gives $\mathbf{m} = 2\gamma_e \mathbf{s}$, with an additional factor of 2: the magnetic moment due to spin is *twice* the value expected on the basis of a classical analogy. The experimental value of the magnetic moment can be determined by observing the effect of a magnetic field on the motion of an electron beam, and it is found that

$$\mathbf{m} = g_e \gamma_e \mathbf{s} \qquad \text{where } g_e = 2.002\,319\,314 \tag{12}$$

The factor g_e is called the ***g*-factor** of the electron. The small discrepancy between the experimental value and the Dirac value of exactly 2 is accounted for by the more sophisticated theory of **quantum electrodynamics**, in which charged particles are allowed to interact with the quantized electromagnetic field.[1] As for the orbital magnetic moment, the spin magnetic moment has quantized components on the z-axis, and we write

$$m_z = -g_e \mu_B m_s \qquad m_s = \pm\tfrac{1}{2} \tag{13}$$

[1] The following classical picture might be helpful. Quantum electrodynamics expresses the electromagnetic field as a collection of harmonic oscillators. We have seen that a harmonic oscillator has a zero-point energy, and so the electromagnetic vacuum has fluctuating electric and magnetic fields even if no photons are present. These vacuum fluctuations interact with the electron, and instead of moving smoothly it jitterbugs (technically, this motion is called *Zitterbewegung*). It also wobbles as it spins (in so far as spin has any such significance), for the same reason, and the wobble increases its magnetic moment above the value that would be expected for a smoothly spinning object.

7.4 Spin–orbit coupling

We now turn to the energy of interaction between the two magnetic moments of an electron. In fact, we shall use this opportunity to emphasize the danger of arguing by classical analogy, particularly when spin is involved.

The classical calculation of the energy of interaction runs as follows. A particle of mass m_e and charge $-e$ moving at a velocity $\mathbf{v}$ in an electric field $\mathbf{E}$ experiences a magnetic field

$$\mathbf{B} = \frac{\mathbf{E} \times \mathbf{v}}{c^2}$$

If the field is due to an isotropic electric potential ϕ, we can write

$$\mathbf{E} = -\frac{\mathbf{r}}{r}\frac{\mathrm{d}\phi}{\mathrm{d}r}$$

It follows that

$$\mathbf{B} = -\frac{1}{rc^2}\frac{\mathrm{d}\phi}{\mathrm{d}r}\mathbf{r} \times \mathbf{v}$$

The orbital angular momentum of the particle is $\mathbf{l} = \mathbf{r} \times \mathbf{p} = m_e\mathbf{r} \times \mathbf{v}$, and so

$$\mathbf{B} = -\frac{1}{m_e rc^2}\frac{\mathrm{d}\phi}{\mathrm{d}r}\mathbf{l}$$

The energy of interaction between a field $\mathbf{B}$ and a magnetic dipole $\mathbf{m}$ is $-\mathbf{m} \cdot \mathbf{B}$, so we can anticipate that the spin–orbit coupling hamiltonian should be

$$H_{\mathrm{so}} = -\mathbf{m} \cdot \mathbf{B} = \frac{1}{m_e rc^2}\frac{\mathrm{d}\phi}{\mathrm{d}r}\mathbf{m} \cdot \mathbf{l} = -\frac{e}{m_e^2 c^2 r}\frac{\mathrm{d}\phi}{\mathrm{d}r}\mathbf{s} \cdot \mathbf{l} \qquad (14)$$

where we have used eqns 7.9 and 7.12 with $g_e = 2$. It turns out that this is exactly twice the result obtained by solving the Dirac equation. The error is the implicit assumption that one can step from the stationary nucleus to the moving electron without treating the change of viewpoint relativistically.[2] The correct calculation gives

$$H_{\mathrm{so}} = \xi(r)\mathbf{l} \cdot \mathbf{s} \qquad \text{where } \xi(r) = -\frac{e}{2m_e^2 c^2 r}\frac{\mathrm{d}\phi}{\mathrm{d}r} \qquad (15)$$

The radial average of the function $\xi(r)\hbar^2$ is written $hc\zeta$, where ζ is called the **spin–orbit coupling constant**; specifically

$$hc\zeta_{nl} = \langle nlm_l|\xi|nlm_l\rangle\hbar^2 \qquad (16)$$

The same value is obtained regardless of the value of m_l because the potential is isotropic. Defined in this way, ζ is a wavenumber and $hc\zeta$ is an energy. For an electron in a hydrogenic atom, the potential arising from a nucleus of charge Ze

[2] The phenomenon that gives rise to the factor $\frac{1}{2}$ is called 'Thomas precession'. The electron moves in its orbital with speeds that approach the speed of light. To an observer on the nucleus, the coordinate system seems to rotate in the plane of motion, and the electron moves in such a way that its coordinate system appears to rotate by 180° when it has completed one circuit of the nucleus. It is spinning (in a classical sense) within its own frame with only one-half the rate if the frame were stationary, and this virtual slowing of its apparent motion reduces its magnetic moment by a factor of $\frac{1}{2}$.

is Coulombic, and $\phi = Ze/4\pi\varepsilon_0 r$. Consequently

$$\xi(r) = \frac{Ze^2}{8\pi\varepsilon_0 m_e^2 c^2 r^3} \tag{17}$$

The expectation value of r^{-3} for hydrogenic orbitals is

$$\langle nlm_l|r^{-3}|nlm_l\rangle = \frac{Z^3}{n^3 a_0^3 l(l+\frac{1}{2})(l+1)} \tag{18}$$

where a_0 is the Bohr radius ($a_0 = 4\pi\varepsilon_0\hbar^2/m_e e^2$). Therefore, the spin–orbit coupling constant for a hydrogenic atom is

$$hc\zeta_{nl} = \frac{Z^4 e^2\hbar^2}{8\pi\varepsilon_0 m_e^2 c^2 a_0^3 n^3 l(l+\frac{1}{2})(l+1)} \tag{19}$$

It proves useful to express this ungainly formula in terms of the **fine-structure constant**, α, which is defined as

$$\alpha = \frac{e^2}{4\pi\varepsilon_0\hbar c} \tag{20}$$

This dimensionless collection of fundamental constants has a value close to 1/137 (more precisely, $\alpha = 7.297\,35 \times 10^{-3}$) and is of extraordinarily broad significance because it is a fundamental constant for the strength of the coupling of a charge to the electromagnetic field. In the present context, we can use it to write

$$\zeta_{nl} = \frac{\alpha^2 \mathcal{R} Z^4}{n^3 l(l+\frac{1}{2})(l+1)} \tag{21}$$

where $\mathcal{R}$ is the Rydberg constant obtained by replacing μ in eqn 7.2 by m_e.

For hydrogen itself, $Z = 1$, and for a $2p$-electron $\zeta = \alpha^2\mathcal{R}/24$, which is about $2.22 \times 10^{-6} \times \mathcal{R}$. Energy level separations and the wavenumbers of transitions are of the order of $\mathcal{R}$ itself, so the fine structure of the spectrum of atomic hydrogen is a factor of about 2×10^{-6} times smaller, or of the order of 0.2 cm^{-1}, as observed. In passing, note that as $\zeta \propto Z^4$, spin–orbit coupling effects are very much larger in heavy atoms than in light atoms. What may be seen as a niggling problem in hydrogen can be of dominating importance in heavy elements, and the work we are doing here will prepare us for them.

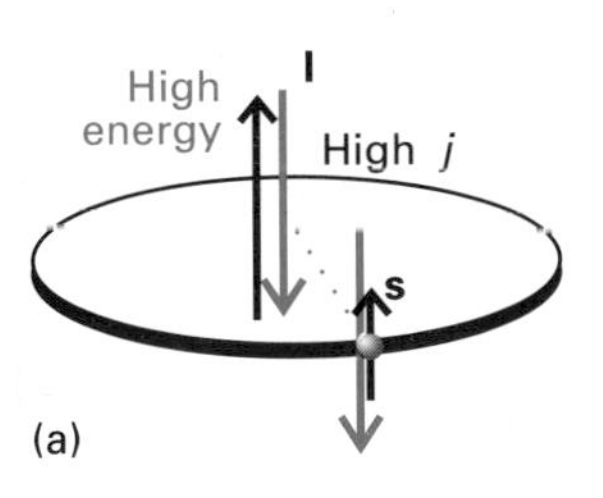

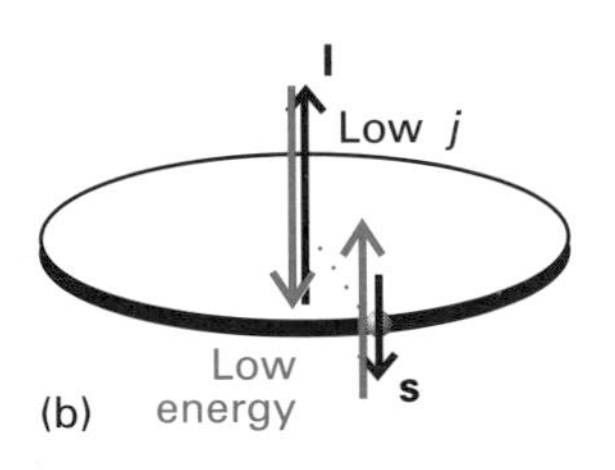

Fig. 7.4 (a) High and (b) low energy relative orientations of spin and orbital angular momenta of an electron as a result of the interaction of the corresponding angular momenta. The black arrows denote angular momenta and the blue arrows denote magnetic moments.

7.5 The fine-structure of spectra

We can now explore how the spin–orbit coupling affects the appearance of spectra. Consider Fig. 7.4. When the spin and orbital angular momenta are parallel, the total angular momentum quantum number, j, takes its highest value ($j = \frac{3}{2}$ for $l = 1$ and $s = \frac{1}{2}$, and $l + \frac{1}{2}$ in general). The corresponding magnetic moments are also parallel, which is a high-energy arrangement. When the two angular momenta are antiparallel, j has its minimum value ($j = \frac{1}{2}$ when $l = 1$, and $j = l - \frac{1}{2}$ in general). The corresponding magnetic moments are now antiparallel, which is a low-energy arrangement. We conclude that the energy of the level with $j = l + \frac{1}{2}$ should lie above the level with $j = l - \frac{1}{2}$, and that the separation should be of the order of the spin–orbit coupling constant since that is a measure of the strength of the magnetic interaction between momenta. Note that the high energy of a state with high j does not stem *directly* from the fact that the total angular momentum is high, but from the fact that a high j indicates that two magnetic moments are parallel and hence interacting adversely. Without that interaction, high j and low j would have the same energy.

Because the spin–orbit interaction is so weak in comparison with the energy-level separations of the atom, we can use first-order perturbation theory to assess its effect. The first-order correction to the energy of the state $|lsjm_j\rangle$ is

$$E_{\mathrm{so}} = \langle lsjm_j|H_{\mathrm{so}}|lsjm_j\rangle = \langle lsjm_j|\xi(r)\mathbf{l}\cdot\mathbf{s}|lsjm_j\rangle \tag{22}$$

(In the language of Section 4.9, note that we are using the coupled representation of the state, which is the natural one to use for the problem.) The matrix elements of a scalar product can be evaluated very simply by noting that

$$j^2 = |\mathbf{l}+\mathbf{s}|^2 = l^2 + s^2 + 2\mathbf{l}\cdot\mathbf{s} \tag{23}$$

Therefore,

$$\begin{aligned}\mathbf{l}\cdot\mathbf{s}|lsjm_j\rangle &= \tfrac{1}{2}(j^2 - l^2 - s^2)|lsjm_j\rangle \\ &= \tfrac{1}{2}\hbar^2\{j(j+1) - l(l+1) - s(s+1)\}|lsjm_j\rangle\end{aligned} \tag{24}$$

Consequently, the interaction energy is

$$\begin{aligned}E_{\mathrm{so}} &= \tfrac{1}{2}\hbar^2\{j(j+1) - l(l+1) - s(s+1)\}\langle lsjm_j|\xi(r)|lsjm_j\rangle \\ &= \tfrac{1}{2}hc\zeta_{nl}\{j(j+1) - l(l+1) - s(s+1)\} \\ &= Z^4\alpha^2 hc\mathcal{R}\left\{\frac{j(j+1) - l(l+1) - s(s+1)}{2n^3 l(l+\frac{1}{2})(l+1)}\right\}\end{aligned} \tag{25}$$

Note that the energy is independent of m_j, the orientation of the total angular momentum in space, as is physically plausible, so each level is $(2j+1)$-fold degenerate. The matrix element $\langle lsjm_j|\xi|lsjm_j\rangle$ is independent of s, j, and m_j because ξ depends only on the radius r; as a result, the matrix element may be identified with $hc\zeta_{nl}/\hbar^2$.

For an s-electron, the spin–orbit interaction is zero because there it has no orbital angular momentum. Specifically, because $j = s$ when $l = 0$,

$$\langle 0ssm_s|\mathbf{l}\cdot\mathbf{s}|0ssm_s\rangle = \tfrac{1}{2}\hbar^2\{s(s+1) - 0 - s(s+1)\} = 0$$

For a p-electron, the separation between levels with $j = \frac{3}{2}$ and $j = \frac{1}{2}$ is $Z^4\alpha^2 hc\mathcal{R}/2n^3$, and so it rapidly becomes negligible as n increases. For a hydrogen $2p$-electron the splitting is $\alpha^2\mathcal{R}/16 \approx 0.365\,\mathrm{cm}^{-1}$. It should be noted that the centre of gravity of the split levels, with each one weighted by its degeneracy, is at the same energy as the unsplit term (Fig. 7.5).

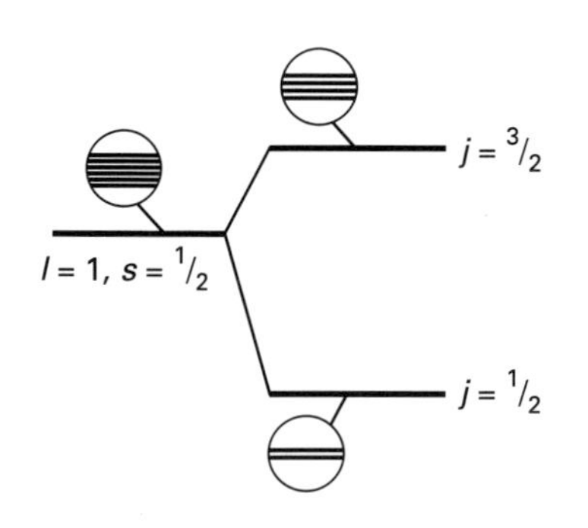

Fig. 7.5 The splitting of the states of a p-electron by spin–orbit coupling. Note that the centre of gravity of the levels is unshifted.

7.6 Term symbols and spectral details

To simplify the discussion of the spectrum that arises from these energy levels we need to introduce some more notation. Spectral lines arise from transitions between **terms**, which is another name for energy levels. The wavenumber, $\tilde{\nu}$, of a transition is the difference between the energies of two terms expressed as

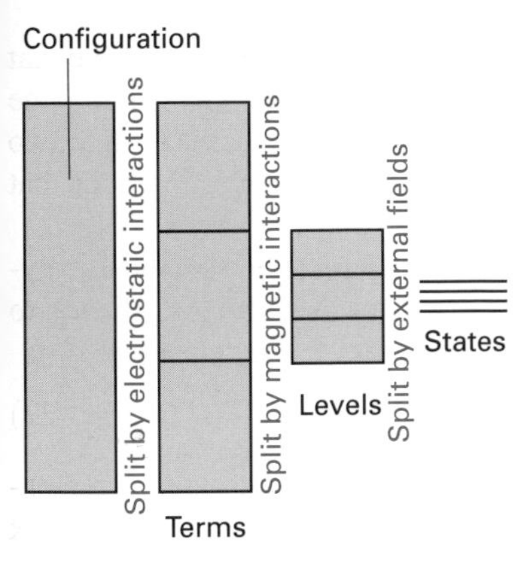

Fig. 7.6 The hierarchy of names and the origin of the splittings that occur in atoms.

wavenumbers:

$$\tilde{\nu} = T' - T \tag{26}$$

A transition is denoted $T' \rightarrow T$ for emission and $T' \leftarrow T$ for absorption, with the term T' higher in energy than the term T.

The **configuration** of an atom is the specification of the orbitals that the electrons occupy. There is only one electron in hydrogen, so we speak of the configuration $1s^1$ if the electron occupies a $1s$-orbital, $2s^1$ if it occupies a $2s$-orbital, and so on. A single configuration (such as $2p^1$) may give rise to several terms. For hydrogen, each configuration with $l > 0$ gives rise to a **doublet** term in the sense that each term splits into two **levels** with different values of j, namely $j = l + \frac{1}{2}$ and $j = l - \frac{1}{2}$. For example, the configuration $2p^1$ gives rise to a doublet term with the levels $j = \frac{3}{2}$ and $j = \frac{1}{2}$, the configuration $3d^1$ gives rise to a doublet term with the levels $j = \frac{5}{2}$ and $j = \frac{3}{2}$, and so on. Each level labelled by the quantum number j consists of $2j + 1$ individual **states** distinguished by the quantum number m_j. The hierarchy of concepts is summarized in Fig. 7.6.

The level of each term arising from a particular configuration is summarized by a **term symbol**:

$$\text{multiplicity}\rightarrow{}^{2S+1}\{L\}_{J\leftarrow\text{level}} \longleftarrow \text{orbital angular momentum}$$

where $\{L\}$ is a letter (S, P, D, F, etc.) corresponding to the value of L (0,1,2,3, etc.). For a hydrogen atom, $L = l$, so a configuration ns^1 gives rise to an S term, a configuration np^1 gives rise to a P term, and so on. The **multiplicity** of a term is the value of $2S + 1$, and provided that $L \geq S$, is the number of levels of the term. For hydrogen, $S = s = \frac{1}{2}$, so $2S + 1 = 2$, and all terms are doublets and are denoted ^{2}S, ^{2}P, etc. As we saw earlier, all terms other than ^{2}S have two levels distinguished by the value of J, and for hydrogen $J = j$. A ^{2}S term has only a single level, with $j = s = \frac{1}{2}$. The precise level of a term is specified by the right subscript of the term symbol, as in $^2S_{1/2}$ and $^2P_{3/2}$. Each of these levels consists of $2J + 1$ states, but these are rarely specified in a term symbol as they are degenerate in the absence of external electric and magnetic fields.

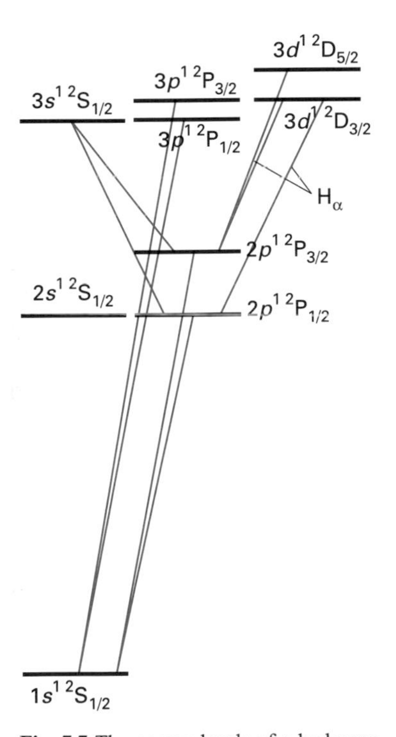

Fig. 7.7 The energy levels of a hydrogen atom showing the fine structure and the transitions that give rise to certain features in the spectrum. Note that in this approximation some degeneracies remain (for states of the same j).

7.7 The detailed spectrum of hydrogen

The transitions responsible for the spectrum of hydrogen can be expressed using term symbols (Fig. 7.7). Consider, for instance, the transitions responsible for the H_α line in the Balmer series (the line responsible for the red glow of excited hydrogen atoms). The upper terms have $n = 3$ and the lower have $n = 2$. The configuration $3s^1$ gives rise to the $^2S_{1/2}$ term with a single level. The configuration $3p^1$ gives rise to $^2P_{3/2}$ and $^2P_{1/2}$, with a very small spin–orbit splitting between the two levels. The $3d^1$ configuration gives rise to the levels $^2D_{5/2}$ and $^2D_{3/2}$. In each case, the level with the lower value of J lies lower in energy. The configuration $2s^1$ similarly gives rise to the term $^2S_{1/2}$ and the configuration $2p^1$ gives rise to $^2P_{3/2}$ and $^2P_{1/2}$ with a splitting of about 0.36 cm^{-1}, as explained before.

One somewhat confusing point is that, according to the Dirac theory of the hydrogen atom, the energy of the $ns^1\,{}^2S_{1/2}$ term is the same as that of the $np^1\,{}^2P_{1/2}$ term (see Fig. 7.7). One way to view this degeneracy is that the

Schrödinger equation ignores the relativistic increase in mass of the electron; when this increase is taken into account (as it is by the Dirac equation), it gives rise to a contribution to the energy which is of the same order of magnitude as the spin–orbit interaction (which is also a relativistic phenomenon), with the result that levels of the same value of j but different values of l are degenerate. Nevertheless, although the Dirac equation predicts exact degeneracy, there is experimentally a small splitting between $^2S_{1/2}$ and $^2P_{1/2}$, which is known as the **Lamb shift**. As in the case of other discrepancies between experiment and the Dirac equation, we have to look for an explanation in the role of the electromagnetic vacuum in which the atom is immersed, and quantum electrodynamics accounts fully for the Lamb shift. The pictorial explanation appeals to the role of the zero-point fluctuations of the oscillations of the electromagnetic field, and their influence on the motion of the electron. This jitterbugging motion of the electron tends to smear its location over a region of space. The effect of this smearing on the energy is most pronounced for s-electrons, for they spend a high proportion of their time close to the nucleus. The smearing tends to reduce the probability that the electron will be found at the nucleus itself, and so the energy of the orbital is raised slightly. There is less effect on the energy of a p-electron because it spends less time close to the nucleus and its interaction with the nucleus is less sensitive to the smearing.

The allowed transitions between terms arising from the configurations with $n = 3, 2$, and 1 are shown in Fig. 7.7 (the selection rules on which this illustration is based are discussed later). Because the only appreciable spin–orbit splitting occurs in the $2p^1$ configuration, the transitions contributing to the H_α line fall into two groups separated by 0.36 cm^{-1}. The doublet structure in the spectrum is therefore a **compound doublet** arising from two almost coincident groups of transitions.

The structure of helium

We now move towards a discussion of many-electron atoms by setting up an approximate description of the simplest example: a helium atom. We shall then use the features that this atom introduces to discuss more complex atoms.

7.8 The helium atom

The hamiltonian for the helium atom ($Z = 2$) is

$$H = -\frac{\hbar^2}{2m_e}(\nabla_1^2 + \nabla_2^2) - \frac{2e^2}{4\pi\varepsilon_0 r_1} - \frac{2e^2}{4\pi\varepsilon_0 r_2} + \frac{e^2}{4\pi\varepsilon_0 r_{12}} \tag{27}$$

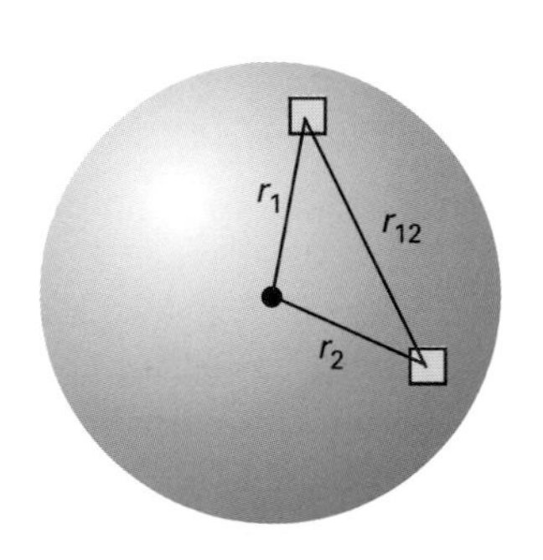

Fig. 7.8 The distances involved in the potential energy of a two-electron atom.

with the distances defined in Fig. 7.8. The first two terms are the kinetic energy operators for the two electrons, the following two are the potential energies of the two electrons in the field of the nucleus of charge $2e$, and the final term is the potential energy arising from the repulsion of the two electrons when they are separated by a distance r_{12}. In a very precise calculation we should use the reduced mass of the electron, but the calculation will be so crude that this refinement is unnecessary.

The Schrödinger equation has the form

$$H\psi(\mathbf{r}_1, \mathbf{r}_2) = E\psi(\mathbf{r}_1, \mathbf{r}_2) \tag{28}$$

and the wavefunction depends on the coordinates of both electrons. It appears to be impossible to find analytical solutions of such a complicated partial differential equation in six variables, and almost all work has been directed towards finding increasingly refined numerical solutions. The simplest version of these approximate solutions is based on a perturbation approach, and this is the line we shall initially take here. The obvious candidate to use as the perturbation is the electron–electron interaction, but as it is not particularly small compared with the other terms in the hamiltonian we should not expect very good agreement with experiment, and will need to make further refinements.

The unperturbed system is described by a hamiltonian that is the sum of two hydrogenic hamiltonians:

$$H^{(0)} = H_1 + H_2, \qquad H_i = -\frac{\hbar^2}{2m_e}\nabla_i^2 - \frac{2e^2}{4\pi\varepsilon_0 r_i} \tag{29}$$

Whenever a hamiltonian is expressed as the sum of two independent terms, the eigenfunction is the product of two factors. This assertion can be proved as follows by writing $\psi = \psi_1\psi_2$; then

$$\begin{aligned}(H_1 + H_2)\psi &= (H_1 + H_2)\psi_1\psi_2 = (H_1\psi_1)\psi_2 + \psi_1(H_2\psi_2)\\ &= E_1\psi_1\psi_2 + E_2\psi_1\psi_2 = (E_1 + E_2)\psi\end{aligned}$$

That is, the product $\psi_1\psi_2$ is an eigenfunction of $H_1 + H_2$. It follows that for helium the wavefunction of the two electrons (with their repulsion disregarded) is the product of two hydrogenic wavefunctions:

$$\psi(\mathbf{r}_1, \mathbf{r}_2) = \psi_{n_1 l_1 m_{l1}}(\mathbf{r}_1)\psi_{n_2 l_2 m_{l2}}(\mathbf{r}_2) \tag{30}$$

and that, from eqn 3.44, the energies are

$$E = -4hc\mathcal{R}\left(\frac{1}{n_1^2} + \frac{1}{n_2^2}\right) \tag{31}$$

We are using $\mathcal{R}$ because we are using the electron mass, not the true reduced mass.

Now consider the influence of the electron–electron repulsion term. The first-order correction to the energy is

$$E^{(1)} = \langle n_1 l_1 m_{l1}; n_2 l_2 m_{l2}|\frac{e^2}{4\pi\varepsilon_0 r_{12}}|n_1 l_1 m_{l1}; n_2 l_2 m_{l2}\rangle = J \tag{32}$$

The term J is called the **Coulomb integral**:

$$J = \frac{e^2}{4\pi\varepsilon_0}\int |\psi_{n_1 l_1 m_{l1}}(\mathbf{r}_1)|^2\left(\frac{1}{r_{12}}\right)|\psi_{n_2 l_2 m_{l2}}(\mathbf{r}_2)|^2\,\mathrm{d}\tau_1\mathrm{d}\tau_2 \tag{33}$$

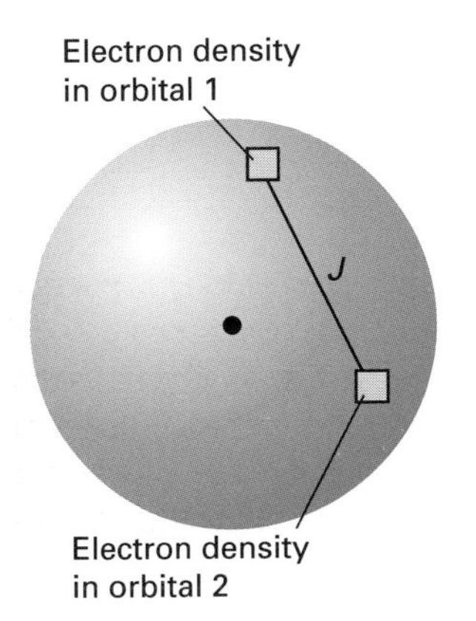

Fig. 7.9 The physical interpretation of the Coulomb integral, J.

This integral (which is positive) has a very simple interpretation (Fig. 7.9). The term $|\psi_{n_1 l_1 m_{l1}}(\mathbf{r}_1)|^2\mathrm{d}\tau_1$ is the probability of finding the electron in the volume element $\mathrm{d}\tau_1$, and so when multiplied by $-e$ it is the charge associated with that region. Likewise, $-e|\psi_{n_2 l_2 m_{l2}}(\mathbf{r}_2)|^2\,\mathrm{d}\tau_2$ is the charge associated with the volume element $\mathrm{d}\tau_2$. The integrand is therefore the Coulombic potential energy of interaction between the charges in these two elements and J is the total contribution to the potential energy arising from electrons in the two orbitals.

Example 7.2 Evaluation of a Coulomb integral

Evaluate the Coulomb integral for the configuration $1s^2$ of a hydrogenic atom given the following expansion:

$$\frac{1}{r_{12}} = \frac{1}{r_1}\sum_{l,m_l}\left(\frac{4\pi}{2l+1}\right)\left(\frac{r_2}{r_1}\right)^l Y_{lm_l}(\theta_1,\phi_1)Y_{lm_l}(\theta_2,\phi_2)$$

when $r_2 > r_1$, and with r_1 and r_2 interchanged when $r_2 < r_1$.

Method. The integral should be evaluated using $\psi = (Z^3/\pi a_0^3)^{1/2}\mathrm{e}^{-Zr/a_0}$ for each electron. Because the wavefunctions are independent of angle, the integration over the angles is straightforward: the integration over Y gives zero except when $l = 0$ and $m_l = 0$. Hence, the sum given above reduces to a single term inside the integral, namely $1/r_{12} = 1/r_1$ when $r_1 > r_2$ and $1/r_{12} = 1/r_2$ when $r_2 > r_1$. The radial integrations should be divided into two parts, one with $r_1 > r_2$ and the other with $r_2 > r_1$.

Answer. The integration is as follows:

$$\begin{aligned}
J &= \left(\frac{e^2}{4\pi\varepsilon_0}\right)\left(\frac{Z^3}{\pi a_0^3}\right)^2 \int_0^{2\pi}\mathrm{d}\phi_1\int_0^{2\pi}\mathrm{d}\phi_2\int_0^{\pi}\sin\theta_1\,\mathrm{d}\theta_1\int_0^{\pi}\sin\theta_2\,\mathrm{d}\theta_2 \\
&\quad\times\int_0^{\infty}\int_0^{\infty}\frac{\mathrm{e}^{-2Z(r_1+r_2)/a_0}}{r_{12}}r_1^2r_2^2\,\mathrm{d}r_1\mathrm{d}r_2 \\
&= \left(\frac{e^2}{4\pi\varepsilon_0}\right)\left(\frac{Z^3}{\pi a_0^3}\right)^2(4\pi)^2 \\
&\quad\times\int_0^{\infty}\left\{\int_0^{r_2}\frac{r_1^2\mathrm{e}^{-2Zr_1/a_0}}{r_2}\mathrm{d}r_1 + \int_{r_2}^{\infty}\frac{r_1^2\mathrm{e}^{-2Zr_1/a_0}}{r_1}\mathrm{d}r_1\right\} \\
&\quad\times r_2^2\mathrm{e}^{-2Zr_2/a_0}\,\mathrm{d}r_2 \\
&= \left(\frac{e^2}{4\pi\varepsilon_0}\right)\left(\frac{Z^3}{\pi a_0^3}\right)^2(4\pi)^2\times\frac{5}{2^7}\left(\frac{a_0}{Z}\right)^5 = \tfrac{5}{8}\left(\frac{e^2}{4\pi\varepsilon_0}\right)\left(\frac{Z}{a_0}\right)
\end{aligned}$$

For helium, $Z = 2$, and so

$$J = \tfrac{5}{4}\left(\frac{e^2}{4\pi\varepsilon_0 a_0}\right) \approx 5.45\ \mathrm{aJ}$$

(where $1\ \mathrm{aJ} = 10^{-18}$ J).

Comment. Take care with the expansion when orbitals other than s-orbitals are involved, because additional terms then survive.

Exercise 7.2. Evaluate J for the configuration $1s^12s^1$.

It is shown in the example that $J \approx 5.45$ aJ, which corresponds to 34 eV or $2.50hc\mathcal{R}$. The total energy of the ground state of the atom in this approximation is therefore

$$E = (-4 - 4 + 2.50)hc\mathcal{R} = -5.50hc\mathcal{R}$$

This value corresponds to -12.0 aJ, or $-7220\,\mathrm{kJ\,mol^{-1}}$. The experimental value, which is equal to the sum of the first and second ionization energies of the atom, is $-7619\,\mathrm{kJ\,mol^{-1}}$ (-12.65 aJ, $-5.804hc\mathcal{R}$). The agreement is not brilliant, but the calculation is obviously on the right track. One of the reasons for the disagreement is that the perturbation is not small, and so first-order perturbation theory cannot be expected to lead to a reliable result.

7.9 Excited states of helium

A new feature comes into play when we consider the excited states of the atom. When the two electrons occupy different orbitals (as in the configuration $1s^1 2s^1$), the wavefunctions are either $\psi_{n_1 l_1 m_{l1}}(\mathbf{r}_1)\psi_{n_2 l_2 m_{l2}}(\mathbf{r}_2)$ or $\psi_{n_2 l_2 m_{l2}}(\mathbf{r}_1)\psi_{n_1 l_1 m_{l1}}(\mathbf{r}_2)$, which we shall denote $a(1)b(2)$ and $b(1)a(2)$, respectively. Both wavefunctions have the same energy and their unperturbed energies are $E_a + E_b$. To calculate the perturbed energy, we use the form of perturbation theory appropriate to degenerate states (Section 6.8), and therefore set up the secular determinant. To do so, we need the following matrix elements, in which we identify state 1 with $a(1)b(2)$ and state 2 with $b(1)a(2)$.

$$H_{11} = \langle a(1)b(2)|H_1 + H_2 + \frac{e^2}{4\pi\varepsilon_0 r_{12}}|a(1)b(2)\rangle = E_a + E_b + J$$

$$H_{22} = E_a + E_b + J$$

$$H_{12} = \langle a(1)b(2)|H_1 + H_2 + \frac{e^2}{4\pi\varepsilon_0 r_{12}}|a(2)b(1)\rangle$$

$$= (E_a + E_b)\langle a(1)b(2)|a(2)b(1)\rangle + \langle a(1)b(2)|\frac{e^2}{4\pi\varepsilon_0 r_{12}}|a(2)b(1)\rangle = H_{21}$$

The first of the integrals in H_{12} is zero because the orbitals a and b are orthogonal:

$$\langle a(1)b(2)|a(2)b(1)\rangle = \langle a(1)|b(1)\rangle\langle b(2)|a(2)\rangle = 0$$

The remaining integral is called the **exchange integral**, K:

$$K = \frac{e^2}{4\pi\varepsilon_0}\langle a(1)b(2)|\frac{1}{r_{12}}|a(2)b(1)\rangle \tag{34}$$

Like J, this integral is positive. The secular determinant is therefore

$$\begin{vmatrix} H_{11} - ES_{11} & H_{12} - ES_{12} \\ H_{21} - ES_{21} & H_{22} - ES_{22} \end{vmatrix} = \begin{vmatrix} H_{11} - E & H_{12} \\ H_{21} & H_{22} - E \end{vmatrix}$$
$$= \begin{vmatrix} E_a + E_b + J - E & K \\ K & E_a + E_b + J - E \end{vmatrix} \tag{35}$$
$$= 0$$

The solutions are

$$E = E_a + E_b + J \pm K \tag{36}$$

and the corresponding wavefunctions are

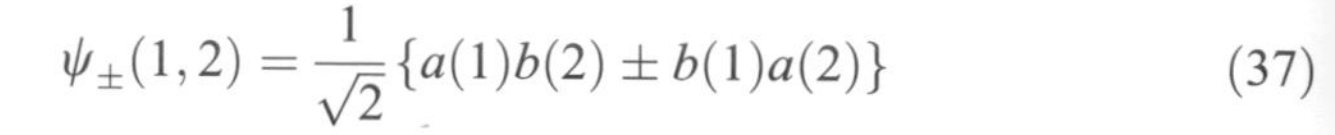

$$\psi_{\pm}(1,2) = \frac{1}{\sqrt{2}}\{a(1)b(2) \pm b(1)a(2)\} \tag{37}$$

or, in more detail,

$$\psi_{\pm}(\mathbf{r}_1,\mathbf{r}_2) = \frac{1}{\sqrt{2}}\{\psi_{n_1l_1m_{l1}}(\mathbf{r}_1)\psi_{n_2l_2m_{l2}}(\mathbf{r}_2) \pm \psi_{n_2l_2m_{l2}}(\mathbf{r}_1)\psi_{n_1l_1m_{l1}}(\mathbf{r}_2)\}$$

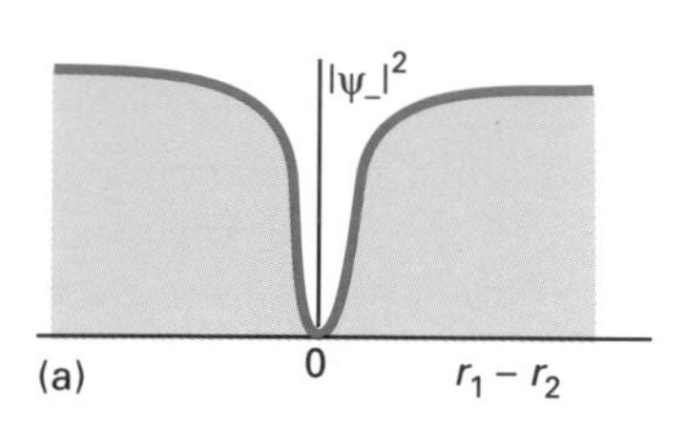

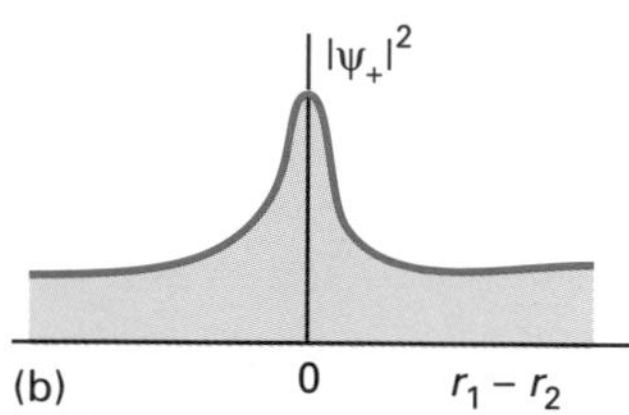

Fig. 7.10 (a) The formation of a Fermi hole by spin-correlation and (b) the formation of a Fermi heap when the spins are paired.

where the individual functions are hydrogenic atomic orbitals with $Z = 2$.

The striking feature of this result is that the degeneracy of the two product functions (*ab* and *ba*) is removed by the electron repulsion, and their two linear combinations $\psi_{\pm}$ differ in energy by $2K$. The exchange integral has no classical counterpart, and should be regarded as a quantum mechanical correction to the Coulomb integral. However, despite its quantum mechanical origin, it is possible to discern the origin of this correction by considering the amplitudes $\psi_{\pm}$ as one electron approaches the other. The crucial point is that $\psi_{-} = 0$ when $\mathbf{r}_1 = \mathbf{r}_2$ whereas ψ_{+} does not necessarily vanish. The corresponding differences in the probability densities are illustrated in Fig. 7.10. We see that there is *zero* probability of finding the two electrons in the same infinitesimal region of space if they are described by the wavefunction ψ_{-}, but there is no such restriction if their wavefunction is ψ_{+} (indeed, there is a small enhancement in the probability that they will both be found together). The dip in the probability density $|\psi_{-}|^2$ wherever $\mathbf{r}_1 \approx \mathbf{r}_2$ is called a **Fermi hole**. It is a purely quantum mechanical phenomenon, and has nothing to do with the charge of the electrons, and even 'uncharged electrons' would show the phenomenon.

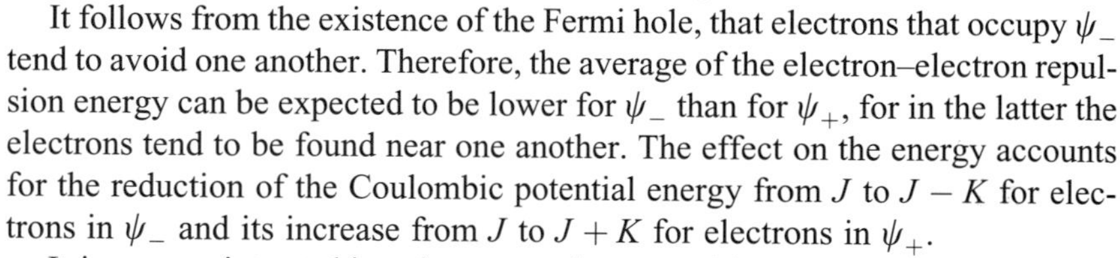

It follows from the existence of the Fermi hole, that electrons that occupy ψ_{-} tend to avoid one another. Therefore, the average of the electron–electron repulsion energy can be expected to be lower for ψ_{-} than for ψ_{+}, for in the latter the electrons tend to be found near one another. The effect on the energy accounts for the reduction of the Coulombic potential energy from J to $J - K$ for electrons in ψ_{-} and its increase from J to $J + K$ for electrons in ψ_{+}.

It is appropriate at this point to mention something that will prove to be of crucial importance shortly. The wavefunction ψ_{-} is *antisymmetric* under the interchange of the names of the electrons:

$$\begin{aligned}\psi_{-}(2,1) &= \frac{1}{\sqrt{2}}\{a(2)b(1) - b(2)a(1)\} \\ &= -\frac{1}{\sqrt{2}}\{a(1)b(2) - b(1)a(2)\} = -\psi_{-}(1,2)\end{aligned}$$

whereas ψ_{+} is *symmetric* under particle interchange:

$$\begin{aligned}\psi_{+}(2,1) &= \frac{1}{\sqrt{2}}\{a(2)b(1) + b(2)a(1)\} \\ &= \frac{1}{\sqrt{2}}\{a(1)b(2) + b(1)a(2)\} = \psi_{+}(1,2)\end{aligned}$$

7.10 The spectrum of helium

At this stage we have seen that when both electrons are in the same orbital (as in $1s^2$, the ground state), the configuration gives rise to a single term with energy $2E_a + J$, with both E_a and J depending on the orbital that is occupied. When the two electrons occupy different orbitals (as in $1s^1 2s^1$), then the configuration gives rise to two terms, one with energy $E_a + E_b + J - K$ and the other with energy $E_a + E_b + J + K$. The separation of the terms by $2K$ should be detectable in the spectrum, and so we shall now consider the transitions in more detail.

The ground configuration is $1s^2$. Its total orbital angular momentum is zero (because $l_1 = l_2 = 0$), so $L = 0$ and it gives rise to an S term. The only excited configurations that we need consider in practice are those involving the excitation of a single electron, and therefore having the form $1s^1 nl^1$, because the excitation of two electrons exceeds the ionization energy of the atom. The configuration $1s^1 nl^1$ gives rise to terms with $L = l$ because only one of the electrons has a nonzero orbital angular momentum. Therefore, the terms we have to consider are $1s^1 2s^1$ S, $1s^1 2p^1$ P, and so on. The selection rule $\Delta l = \pm 1$ implies that transitions may occur between S and P terms, between P and D, etc., but not between S and D.

We need to consider the selection rules governing transitions between states of the form ψ_+ and ψ_- described above. It turns out (as we demonstrate below) that the selection rules are

$$\text{symmetrical} \longleftrightarrow \text{symmetrical} \qquad \text{antisymmetrical} \longleftrightarrow \text{antisymmetrical}$$

but transitions between symmetrical and antisymmetrical combinations are not allowed. The basis of this selection rule is the vanishing of the transition dipole moment for states with different permutation symmetry. The electric dipole moment operator for a two-electron system is equal to $-e\mathbf{r}_1 - e\mathbf{r}_2$, which is symmetric under the permutation of the labels 1 and 2. The dipole moment for the transition between states of different permutation symmetry is

$$\mu_{+-} = -e \int \psi_+^*(\mathbf{r}_1, \mathbf{r}_2)(\mathbf{r}_1 + \mathbf{r}_2)\psi_-(\mathbf{r}_1, \mathbf{r}_2)\, d\tau_1 d\tau_2$$

However, under the permutation of labels, the integrand changes sign. As the value of an integral cannot depend on the names that we give to the electrons, it follows that the only possible value for the integral is zero. Hence, there can be no transitions between symmetric and antisymmetric combinations.

Finally, we need to consider the multiplicities of the terms. Because each electron has $s = \frac{1}{2}$, we expect $S = 0$ and 1, corresponding to singlet and triplet terms, respectively. For the singlet terms, $J = L$; for the triplet terms, the Clebsch–Gordan series gives $J = L + 1, L, L - 1$ provided that $L > 0$. Thus, we can expect the levels such as $^1\text{P}_1$, $^3\text{P}_2$, $^3\text{P}_1$, and $^3\text{P}_0$ to stem from each $1s^1 np^1$ configuration, and these levels are expected to be split by the spin–orbit coupling. At this stage (a phrase intended to strike a note of warning), we expect each of these terms to exist as the symmetric and antisymmetric combinations. So we can expect *eight* terms to stem from a $1s^1 np^1$ configuration, with a symmetric and antisymmetric combination for each of $^1\text{P}_1$, $^3\text{P}_2$, $^3\text{P}_1$, and $^3\text{P}_0$. Similarly we can expect (but see below) four terms from a $1s^1 ns^1$ configuration, corresponding to the symmetric and antisymmetric

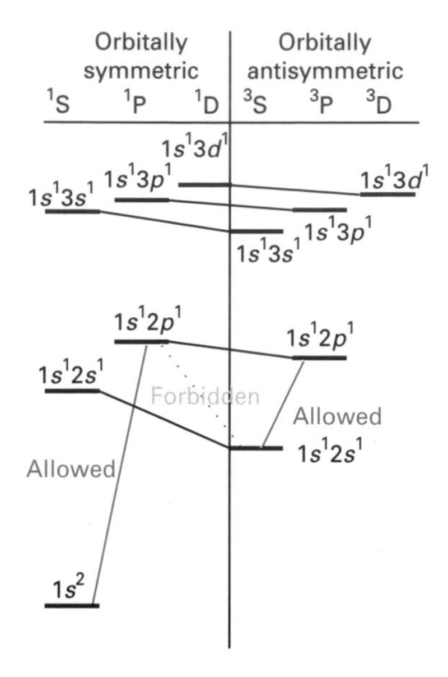

Fig. 7.11 The energy levels of a helium atom, their classification as singlets and triplets, and some of the allowed and forbidden transitions.

combinations for each of ^{1}S and ^{3}S.

The observed spectrum of helium is, to some extent, consistent with these remarks. Each $1s^1nl^1$ configuration gives rise to two types of term (Fig. 7.11), one symmetric and the other antisymmetric. We know which is which, because the ground configuration must be symmetric (both electrons occupy the same orbital), and only symmetric states have appreciable transition intensity to the ground state. Furthermore, wherever both types of term can be identified, the antisymmetrical combination (the one that does not make transitions to the ground state) lies lower in energy than the symmetrical combination, in accord with the discussion in Section 7.9.[3] There is, however, an extraordinary feature. An analysis of the spectrum shows that *all the symmetric states are singlets and all the antisymmetric states are triplets*. There are no symmetric triplets and no antisymmetric singlets. Moreover, there are only four terms from each $1s^1np^1$ configuration, not eight. In fact, half of all possible terms appear to be excluded.

7.11 The Pauli principle

The explanation of the omission of half the expected terms requires the introduction of an entirely new fundamental feature of nature. This was recognized by Wolfgang Pauli, who proposed the following solution.

Consider the state of the system when the spins of the electrons are taken into account. In Section 4.12 we saw that the spin state of two electrons corresponding to $S = 0$ is

$$\sigma_-(1,2) = \frac{1}{\sqrt{2}}\{\alpha(1)\beta(2) - \beta(1)\alpha(2)\}$$

where, as usual, α denotes the state with $m_s = +\frac{1}{2}$ and β denotes the state with $m_s = -\frac{1}{2}$. The state σ_- is antisymmetric under particle exchange:

$$\sigma_-(2,1) = -\sigma_-(1,2)$$

On the other hand, the three states that correspond to $S = 1$ are all symmetric under particle interchange:

$$\sigma_+^{(+1)}(1,2) = \alpha(1)\alpha(2)$$
$$\sigma_+^{(0)}(1,2) = \frac{1}{\sqrt{2}}\{\alpha(1)\beta(2) + \beta(1)\alpha(2)\}$$
$$\sigma_+^{(-1)}(1,2) = \beta(1)\beta(2)$$

[3] Not too much should be made of this point. Although the analysis has shown that it is plausible that an antisymmetric combination, with its Fermi hole, should lie lower in energy, the conclusion was based on first-order perturbation theory and therefore ignored the distortion of the wavefunction that may occur. It turns out that this distortion, which corresponds to the shrinkage of the antisymmetric combination wavefunction so that the electrons lie closer to the nucleus than they do in the symmetric combination wavefunction, is of dominating importance for determining the order of energy levels. It remains true that the antisymmetric combination has a lower energy, but the reason is more complicated than the first-order argument suggests.

(The superscript is the value of M_S.) We can now list all combinations of orbital and spin states that might occur:

$\psi_-\sigma_-$	$\psi_+\sigma_-$	$\psi_-\sigma_+^{(+1)}$	$\psi_+\sigma_+^{(+1)}$
$\psi_-\sigma_+^{(0)}$	$\psi_+\sigma_+^{(0)}$	$\psi_-\sigma_+^{(-1)}$	$\psi_+\sigma_+^{(-1)}$

The experimentally observed states have been printed with a tinted background. It is clear that there is a common feature: *the allowed states are all antisymmetrical overall under particle interchange*. This observation has been elevated to a general law of nature:

> **The Pauli principle**: The total wavefunction (including spin) must be antisymmetric with respect to the interchange of any pair of electrons.

In fact, the Pauli principle can be expressed more broadly by recognising that elementary particles can be classified as fermions or bosons. A **fermion** is a particle with half-integral spin; examples are electrons and protons. A **boson** is a particle with integral spin, including 0. Examples of bosons are photons (spin 1) and α-particles (helium–4 nuclei, spin 0). The more general form of the Pauli principle is then as follows:

> The total wavefunction must be antisymmetric under the interchange of any pair of identical fermions and symmetrical under the interchange of any pair of identical bosons.

We shall consider only the restricted 'electron' form of the principle here, but use the full principle later (in Section 10.7). The principle should be regarded as one more fundamental postulate of quantum mechanics in addition to those presented in Chapter 1. However, it does have a deeper basis, for it can be rationalized to some extent by using relativistic arguments and the requirement that the total energy of the universe be positive. For us, it is a succinct, subtle, summary of experience (the spectrum of helium) which, as we shall see, has wide and never transgressed implications for the structure and properties of matter.

It is a direct consequence of the Pauli principle that there is a restriction on the number of electrons that can occupy the same state. This implication of the Pauli principle is called the **Pauli exclusion principle**:

> No two electrons can occupy the same state.

In its simplest form, the derivation of the exclusion principle from the Pauli principle runs as follows. Suppose the spin states of two electrons are the same. We can always choose the z-direction such that their joint spin state is $\alpha_1\alpha_2$, which is a symmetric state under particle interchange. According to the Pauli principle, the orbital part of the overall wavefunction must be antisymmetric, and hence of the form $a(1)b(2) - b(1)a(2)$. But if a and b are the same wavefunctions, then this combination is identically zero for all locations of the two electrons. Therefore, such a state does not exist, and we cannot have two electrons with the same spins in the same orbital. If the two electrons do not have the same spin, then there does not exist a direction where their joint spin state is $\alpha_1\alpha_2$, so the argument fails. It follows that if two electrons do occupy the same spatial orbital, then they must **pair**; that is, have opposed spins.

Overall wavefunctions that satisfy the Pauli principle are often written as a **Slater determinant**. To see how such a determinant is constructed, consider another way of expressing the (overall antisymmetric) wavefunction of the ground state of helium:

$$\begin{aligned}\psi(1,2) &= \psi_{1s}(\mathbf{r}_1)\psi_{1s}(\mathbf{r}_2)\sigma_-(1,2) \\ &= \frac{1}{\sqrt{2}}\psi_{1s}(\mathbf{r}_1)\psi_{1s}(\mathbf{r}_2)\{\alpha(1)\beta(2) - \beta(1)\alpha(2)\} \\ &= \frac{1}{\sqrt{2}}\begin{vmatrix}\psi_{1s}(\mathbf{r}_1)\alpha(1) & \psi_{1s}(\mathbf{r}_1)\beta(1) \\ \psi_{1s}(\mathbf{r}_2)\alpha(2) & \psi_{1s}(\mathbf{r}_2)\beta(2)\end{vmatrix}\end{aligned}$$

It is easy to show that the expansion of the determinant generates the preceding line. We now introduce the concept of a **spinorbital**, a joint spin-space state of the electron:

$$\psi^{\alpha}_{1s}(1) = \psi_{1s}(\mathbf{r}_1)\alpha(1) \qquad \psi^{\beta}_{1s}(1) = \psi_{1s}(\mathbf{r}_1)\beta(1)$$

Then the ground state can be expressed as the following determinant:

$$\psi(1,2) = \frac{1}{\sqrt{2}}\begin{vmatrix}\psi^{\alpha}_{1s}(1) & \psi^{\beta}_{1s}(1) \\ \psi^{\alpha}_{1s}(2) & \psi^{\beta}_{1s}(2)\end{vmatrix}$$

This is an example of a **Slater determinant**. The determinant displays the overall antisymmetry of the wavefunction very neatly, because if the labels 1 and 2 are interchanged, then the rows of the determinant are interchanged, and it is a general property of determinants that the interchange of two rows results in a change of sign.

Now suppose that the electrons have the same spin and occupy the same orbitals. The Slater determinant for such a state would be

$$\psi(1,2) = \frac{1}{\sqrt{2}}\begin{vmatrix}\psi^{\alpha}_{1s}(1) & \psi^{\alpha}_{1s}(1) \\ \psi^{\alpha}_{1s}(2) & \psi^{\alpha}_{1s}(2)\end{vmatrix}$$

Because a determinant with two identical columns has the value 0, this Slater determinant is identically zero. Such a state, therefore, does not exist.

The general form of a Slater determinant composed of the spinorbitals $\phi_a, \phi_b, \ldots$ and containing N electrons is

$$\psi(1,2,\ldots,N) = \left(\frac{1}{N!}\right)^{1/2}\begin{vmatrix}\phi_a(1) & \phi_b(1) & \cdots & \phi_z(1) \\ \phi_a(2) & \phi_b(2) & \cdots & \phi_z(2) \\ \vdots & \vdots & \cdots & \vdots \\ \phi_a(N) & \phi_b(N) & \cdots & \phi_z(N)\end{vmatrix} \tag{38}$$

A Slater determinant has N rows and N columns because there is one spinorbital for each of the N electrons present. The state is fully antisymmetric under the interchange of any pair of electrons, because that operation corresponds to the interchange of a pair of rows in the determinant. Furthermore, if any two spinorbitals are the same, then the determinant vanishes because it has two columns in common. Instead of writing out the determinant in full, which is tiresome, it is normally summarized by its principal diagonal:

$$\psi(1,2,\ldots,N) = \left(\frac{1}{N!}\right)^{1/2} \det|\phi_a(1)\phi_b(2)\ldots\phi_z(N)| \tag{39}$$

We are now in a position to return to the helium spectrum. We have seen that two electrons tend to avoid each other if they are described by an antisymmetric *spatial* wavefunction. However, if the two electrons are described by such a wavefunction, it follows that their spins must be in a symmetrical state, and hence have $S = 1$. Therefore, we can summarize the effect by saying that *parallel spins tend to avoid one another*. This effect is called **spin correlation**. However, the preceding discussion has shown that spin correlation is only an indirect consequence of spin working through the Pauli principle. That is, if the spins are parallel, then the Pauli principle requires them to have an antisymmetric wavefunction, which implies that they cannot be found at the same point simultaneously.

A consequence of spin correlation is, as we have seen, that the triplet term arising from a configuration lies lower in energy than the singlet term of the same configuration. The point should be noted, however, that the difference in energy is a similar *indirect* consequence of the relative spin orientations of the electrons and does not imply a direct interaction between spins. The difference in energy of terms of different multiplicity is a purely Coulombic effect that reflects the influence of spin correlation on the relative spatial distribution of the electrons.

Many-electron atoms

We have seen that a crude description of the ground state of the helium atom is $1s^2$ with both electrons in hydrogenic $1s$-orbitals with $Z = 2$. An improved description takes into account the repulsion between the electrons and the consequent swelling of the atom to minimize this disadvantageous contribution to the energy. It turns out that the effect of this repulsion on the orbitals occupied can be simulated to some extent by replacing the true nuclear charge, Ze, by an **effective nuclear charge**, $Z_{\text{eff}}e$. The optimum value for helium, in the sense of corresponding to the lowest energy (recall the variation principle, Section 6.9), is $Z_{\text{eff}} \approx 1.3$. This approach to the description of atomic structure can be extended to other many-electron atoms, and we shall give a brief description of what is involved. Some of the principles will be familiar from elementary chemistry and we shall not dwell on them unduly.

7.12 Penetration and shielding

Most descriptions of atomic structure are based on the **orbital approximation**, where it is supposed that each electron occupies its own atomic orbital, and that orbital bears a close resemblance to one of the hydrogenic orbitals. This is the justification of expressing the structure of an atom in terms of a configuration, such as $1s^22s^22p^6$ for neon. Thus, we write the wavefunction for this atom as the approximation

$$\psi = \left(\frac{1}{10!}\right)^{1/2} \det\left|1s^{\alpha}(1)1s^{\beta}(2)\ldots 2p^{\beta}(10)\right|$$

It must clearly be understood that this expression is an approximation, because the actual many-electron wavefunction is not a simple product but is a more general function of $3N$ variables and two spin states for each electron. To reproduce the exact wavefunction, we would have to take a superposition of an infinite number of antisymmetric products, as discussed in Chapter 9.

According to the Pauli exclusion principle, a maximum of two electrons can occupy any one atomic orbital. As a result, the electronic structure of an atom consists of a series of concentric **shells** of electron density, where a shell consists of all the orbitals of a given value of n. We refer to the ***K*-shell** for $n = 1$, the ***L*-shell** for $n = 2$, the ***M*-shell** for $n = 3$, and so on. The Li atom ($Z = 3$), for instance, consists of a complete K-shell and one electron in one of the orbitals of the L-shell. Each shell consists of n **subshells**, which are the orbitals with a common value of l. There are $2l + 1$ individual orbitals in a subshell. In a hydrogenic atom, all subshells of a given shell are degenerate, but the presence of electron–electron interactions in many-electron atoms removes this degeneracy, and although the members of a given subshell remain degenerate (so the three $2p$-orbitals are degenerate in all atoms), the subshells correspond to different energies. It is typically found, for valence (outermost) electrons at least, that subshells lie in the order $s < p < d < f$, but there are deviations from this simple rule.

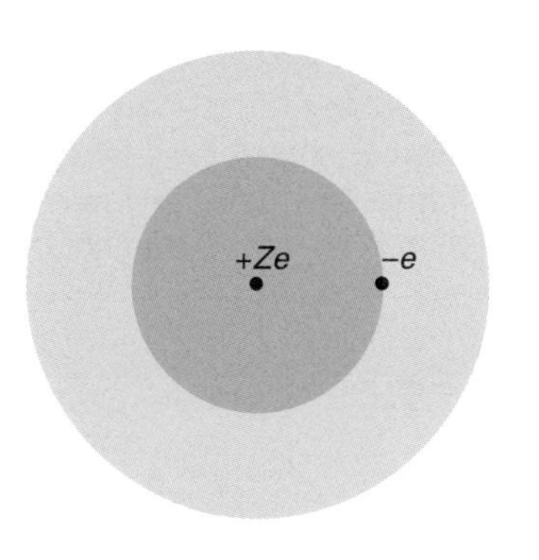

Fig. 7.12 According to classical electrostatics, the charge of a spherically symmetrical distribution can be represented by a point charge equal in value to the total charge of the region and placed at its centre.

The explanation of the order of subshells is based on the **central-field approximation**, in which the highly complicated inter-electronic contribution to the energy, which for electron 1 is

$$V = \sum_i \frac{e^2}{4\pi\varepsilon_0 r_{1i}} \tag{40}$$

with the term $i = 1$ excluded from the sum, is replaced by a single point negative charge on the nucleus, so

$$V \approx \frac{\sigma e^2}{4\pi\varepsilon_0 r_1} \tag{41}$$

where $-\sigma e$ is an effective charge that repels the charge $-e$ of the electron of interest. As a result of this approximation, the nuclear charge Ze is reduced to $(Z - \sigma)e$, and hence we can write

$$Z_{\text{eff}} = Z - \sigma \tag{42}$$

The quantity σ is called the **nuclear screening constant** and is characteristic of the orbital that the electron (which we are calling 1) occupies. Thus, σ is different for $2s$- and $2p$-orbitals. It also depends on the configuration of the atom, and σ for a given orbital has different values in the ground and excited states.

The partial justification for this seemingly (and actually) drastic approximation comes from classical electrostatics. According to classical electrostatics, when an electron is outside a spherical region of electric charge, the potential it experiences is the same as that generated by a single point charge at the centre

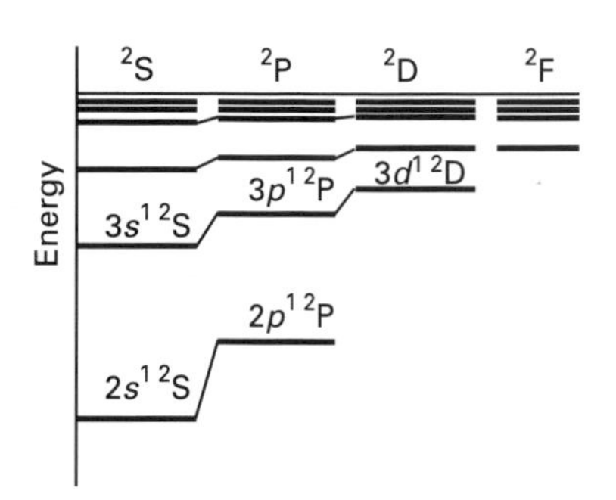

Fig. 7.13 A schematic indication of the orbital energy levels of a many-electron atom (lithium, in fact) showing the removal of the degeneracy characteristic of hydrogenic atoms.

of the region with a magnitude equal to the total charge within a sphere that cuts through the position of the electron (Fig. 7.12). Thus, if the K-shell is full and very compact, the effect of its two electrons can be simulated by placing a point charge $-2e$ on the nucleus provided that the electron of interest stays wholly outside the core region of the atom. If the electron of interest wanders into the core, then its interaction increases the closer it is to the nucleus, and when it is at the nucleus, it experiences the full nuclear charge.

The reduction of the nuclear charge due to the presence of the other electrons in an atom is called **shielding**, and its magnitude is determined by the extent of **penetration** of core regions of the atom, the extent to which the electron of interest will be found close to the nucleus and inside spherical shells of charge. Strictly speaking, the shielding constant varies with distance, and an electron does not have a single value of σ. However, in the next approximation we replace the varying value of σ by its average value, and hence treat Z_{eff} as a constant typical of the atom and of the orbital occupied by the electron of interest. This is the basis of replacing $Z = 2$ by $Z = 1.3$ for each electron in a He atom, for we are ascribing the average value $\sigma = 0.7$ to each electron.

It follows from the discussion of the radial distribution functions for electrons in atoms (Section 3.12) that an *ns*-electron penetrates closer to the nucleus than does an *np*-electron. Hence we can expect an *ns*-electron to be *less* shielded by the core electrons than an *np*-electron, and hence to have a lower energy. There is a similar difference between *np*- and *nd*-electrons, for the wavefunctions of the latter are proportional to r^2 whereas those of the former are proportional to r close to the nucleus; hence, *nd*-electrons are excluded more strongly from the nucleus than *np*-electrons. These effects can be seen in the atomic energy-level diagram for Li (Fig. 7.13), which has been inferred from an analysis of its emission spectrum.

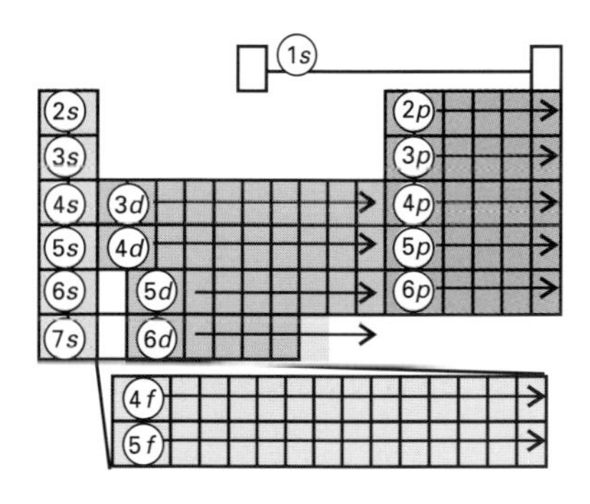

Fig. 7.14 The order of occupation of energy levels as envisaged in the building-up principle. At the end of each period, revert to the start of the next period.

7.13 Periodicity

The ground-state electron configurations are determined experimentally by an analysis of their spectra or, in some cases, by magnetic measurements. These configurations show a periodicity that mirrors the block, group, and period structure of the periodic table. The rationalization of the observed configurations is normally expressed in terms of the **building-up principle**. According to this principle, electrons are allowed to occupy atomic orbitals in an order that mirrors the structure of the periodic table (Fig. 7.14) and subject to the Pauli exclusion principle that no more than two electrons can occupy any one orbital, and if two do occupy an orbital, then their spins must be paired. The order of occupation largely follows the order of energy levels as determined by penetration and shielding, with *ns*-orbitals being occupied before *np*-orbitals. The lowering of energy of *ns*-orbitals is so great that in certain regions of the table they lie below the $(n-1)d$-orbitals of an inner shell: the occupation of 4*s*-orbitals before 3*d*-orbitals is a well-known example of this phenomenon, and it accounts for the intrusion of the *d*-block into the structure of the periodic table.

It is too much to expect such a simple procedure based on the energies of one-electron orbitals to account for all the subtleties of the periodic table. What matters is the attainment of the lowest *total* energy of the atom, not the lowest sum of one-electron energies, for that largely ignores electron–electron interactions (except implicitly). Thus, it is found in some cases that the lowest total energy of the atom is attained by shipping electrons around: the favouring of d^5 and d^{10} configurations is an example of a manner in which the atom can relocate electrons to minimize the total energy, perhaps at the expense of having to occupy an orbital of higher energy. There are various regions of the periodic table where it is necessary to adjust the configuration suggested by the building-up principle, but it is a remarkably simple and generally reliable principle for accounting for the subtleties of the properties of atoms.

There are two features of the building-up principle that should be kept in mind. One is that when more than one orbital is available for occupation, then electrons occupy separate orbitals before entering an already half-occupied orbital. This gives them a greater spatial separation, and hence minimizes the total energy of the atom. Secondly, when electrons occupy separate orbitals, they do so with parallel spins. This rule is often called **Hund's rule** of maximum multiplicity, and it can be traced to the effects of spin correlation, as we have already seen for helium.

Example 7.3 The ground-state electron configurations of atoms

What is the ground-state electron configuration of carbon?

Method. This example is a recapitulation of material normally encountered in introductory chemistry courses, but is included to illustrate the foregoing material. Decide on the number of electrons in the atom, then let them occupy the available orbitals in accord with the scheme in Fig. 7.14 and the restrictions of the Pauli principle. When dealing with the outermost electrons, allow electrons to occupy separate degenerate orbitals, and take note of Hund's rule.

Answer. Carbon has $Z = 6$, so its six electrons give rise to the configuration $1s^2 2s^2 2p^2$. However, in more detail, we expect the configuration $1s^2 2s^2 2p_x^1 2p_y^1$, with the two $2p$-electrons having parallel spins. A triplet is therefore expected for the ground term of the atom.

Comment. It should always be remembered that an electron configuration only has meaning within the orbital approximation.

Exercise 7.3. What is the ground-state electron configuration of an O atom?

[$1s^2 2s^2 2p_x^2 2p_y^1 2p_z^1$, a triplet term]

Once the ground-state electron configuration of an atom is known, it is possible to go on to rationalize a number of the properties. For example, the **ionization energy**, I, of an element, the minimum energy needed to remove an electron from a gas-phase atom,

$$\mathrm{E(g)} \rightarrow \mathrm{E^+(g)} + \mathrm{e^-(g)}$$

generally increases across a period because the effect of nuclear attraction on the outermost electron increases more rapidly than the repulsion from the additional electrons that are present. However, the variation is not uniform

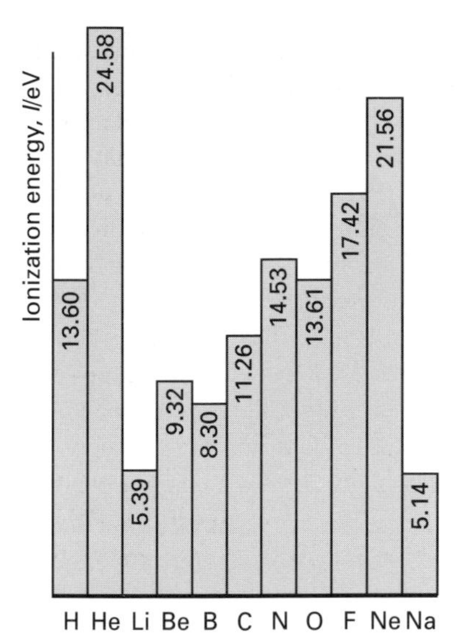

Fig. 7.15 The variation of the first ionization energy through Period 2 of the periodic table.

(Fig. 7.15), because account must be taken of the identity of the orbitals from which the outermost electron is removed and the energy of the ion remaining after the loss of the electron. The dip between Be and B, for instance, can be explained on the grounds that their electron configurations are $1s^22s^2$ and $1s^22s^22p^1$, respectively; so ionization takes place from a $2p$ orbital in B but a $2s$ orbital in Be, and the latter orbital has a lower energy on account of its shielding and penetration. The decrease between N and O reflects the fact that in N ($1s^22s^22p_x^12p_y^12p_z^1$) the electron is removed from a half-filled $2p$-orbital, whereas in O ($1s^22s^22p_x^22p_y^12p_z^1$) the electron is helped on its way by another electron that is present in the $2p_x$-orbital and the fact that the resulting $2p^3$ configuration has a low energy. The steep fall in ionization energy between He and Li, and between Ne and Na, reflects the fact that the electron is being removed from a new shell and so is more distant from the nucleus.

7.14 Slater atomic orbitals

No definitive analytical form can be given for the atomic orbitals of many-electron atoms because the orbital approximation is very primitive. Nevertheless, it is often helpful to have available a set of approximate atomic orbitals which model the actual wavefunctions found by using the more sophisticated numerical techniques that we describe in Chapter 9. These **Slater type orbitals** (STO) are constructed as follows:

1. An orbital with quantum numbers n, l, and m_l belonging to a nucleus of an atom of atomic number Z is written
$$\psi_{nlm_l}(r,\theta,\phi) = Nr^{n_{\text{eff}}-1}\text{e}^{-Z_{\text{eff}}\rho/n_{\text{eff}}}Y_{lm_l}(\theta,\phi)$$
where N is a normalization constant, Y_{lm_l} is a spherical harmonic (Table 3.1), and $\rho = r/a_0$.
2. The effective principal quantum number, n_{eff}, is related to the true principal quantum number, n, by the following mapping:
$$n \to n_{\text{eff}} : 1 \to 1 \quad 2 \to 2 \quad 3 \to 3 \quad 4 \to 3.7 \quad 5 \to 4.0 \quad 6 \to 4.2$$
3. The effective atomic number, Z_{eff}, is taken from Table 7.1.

The values in Table 7.1 have been constructed by fitting STOs to numerically computed wavefunctions, and they supersede the values that were originally given by Slater in terms of a set of rules. Care should be taken when using STOs because orbitals with different values of n but the same values of l and m_l are not orthogonal to one another. Another deficiency of STOs is that ns-orbitals with $n > 1$ have zero amplitude at the nucleus.

7.15 Self-consistent fields

The best atomic orbitals are found by numerical solution of the Schrödinger equation. The original procedure was introduced by D.R. Hartree and is known as the method of **self-consistent fields** (SCF). The procedure was improved by V. Fock and J.C. Slater to include the effects of electron exchange, and the orbitals obtained by their methods are called **Hartree–Fock orbitals**.[4]

[4] See Chapter 9 for a more detailed account of the Hartree–Fock procedure.

Table 7.1 Values of $Z_{eff} = Z - \sigma$ for neutral ground-state atoms

	H							He
1s	1							1.6875
	Li	Be	B	C	N	O	F	Ne
1s	2.6906	3.6848	4.6795	5.6727	6.6651	7.6579	8.6501	9.6421
2s	1.2792	1.9120	2.5762	3.2166	3.8474	4.4916	5.1276	5.7584
2p			2.4214	3.1358	3.8340	4.4532	5.1000	5.7584
	Na	Mg	Al	Si	P	S	Cl	Ar
1s	10.6259	11.6089	12.5910	13.5754	14.5578	15.5409	16.5239	17.5075
2s	6.5714	7.3920	8.2136	9.0200	9.8250	10.6288	11.4304	12.2304
2p	6.8018	7.8258	8.9634	9.9450	10.9612	11.9770	12.9932	14.0082
3s	2.5074	3.3075	4.1172	4.9032	5.6418	6.3669	7.0683	7.7568
3p			4.0656	4.2852	4.8864	5.4819	6.1161	6.7641

Values are from E. Clementi and D.L. Raimondi, Atomic screening constants from SCF functions. IBM Res. Note NJ-27 (1963).

The assumption behind the technique is that any one electron moves in a potential which is a spherical average of the potential due to all the other electrons and the nucleus, and which can be expressed as a single charge centred on the nucleus (this is the central-field approximation; but it is not assumed that the charge has a fixed value). Then the Schrödinger equation is integrated numerically for that electron and that spherically averaged potential, taking into account the fact that the total charge inside the sphere defined by the position of the electron varies as the distance of the electron from the nucleus varies (recall Fig. 7.12). This approach supposes that the wavefunctions of all the other electrons are already known so that the spherically averaged potential can be calculated. That is not in general true, so the calculation starts out from some approximate form of the wavefunctions, such as approximating them by STOs. The Schrödinger equation for the electron is then solved, and the procedure is repeated for all the electrons in the atom. At the end of this first round of calculation, we have a set of improved wavefunctions for all the electrons. These improved wavefunctions are then used to calculate the spherically averaged potential, and the cycle of computation is performed again. The cycle is repeated until the improved set of wavefunctions does not differ significantly from the wavefunctions at the start of the cycle. The wavefunctions are then self-consistent, and are accepted as good approximations to the true many-electron wavefunction.

The Hartree–Fock equations on which the procedure is based are slightly tricky to derive (see *Further information 11*) but they are reasonably easy to interpret. The hamiltonian that we need to consider is

$$H = \sum_i h_i + \tfrac{1}{2}\sum_{i,j}{}' \frac{e^2}{4\pi\varepsilon_0 r_{ij}} \tag{43}$$

where h_i is a hydrogenic hamiltonian for electron i in the field of a bare nucleus of charge Ze. This operator is called the **core hamiltonian**. The factor of $\frac{1}{2}$ in the double sum is to prevent the double-counting of interactions. The prime on the summation excludes terms for which $i = j$ as electrons do not interact with themselves. The Hartree–Fock equation for a space orbital ψ_s occupied by electron 1 is

$$\left\{h_1 + \sum_r (2J_r - K_r)\right\}\psi_s(1) = \varepsilon_s\psi_s(1) \tag{44}$$

The sum is over all occupied spatial wavefunctions. The terms J_r and K_r are *operators* that have the following effects. The **Coulomb operator**, J_r, is defined as follows:

$$J_r\psi_s(1) = \left\{\int \psi_r^*(2)\left(\frac{e^2}{4\pi\varepsilon_0 r_{12}}\right)\psi_r(2)\,\mathrm{d}\tau_2\right\}\psi_s(1) \tag{45}$$

This operator represents the Coulombic interaction of electron 1 with electron 2 in the orbital ψ_r. The **exchange operator**, K_r, is defined similarly:

$$K_r\psi_s(1) = \left\{\int \psi_r^*(2)\left(\frac{e^2}{4\pi\varepsilon_0 r_{12}}\right)\psi_s(2)\,\mathrm{d}\tau_2\right\}\psi_r(1) \tag{46}$$

This operator takes into account the effects of spin correlation. The quantity ε_s in eqn 7.44 is the **one-electron orbital energy**. Equations 7.45 and 7.46 show that it is necessary to know all the other wavefunctions in order to set up the operators J and K and hence to find the form of each wavefunction.

Once the final, self-consistent form of the orbitals has been established, we can find the orbital energies by multiplying both sides of eqn 7.44 by $\psi_s^*(1)$ and integrating over all space. The right-hand side is simply ε_s, and so

$$\varepsilon_s = \int \psi_s^*(1)h_1\psi_s\,\mathrm{d}\tau_1 + \sum_r (2J_{sr} - K_{sr}) \tag{47}$$

where

$$\begin{aligned} J_{sr} &= \int \psi_s^*(1)J_r\psi_s(1)\,\mathrm{d}\tau_1 \\ &= \frac{e^2}{4\pi\varepsilon_0}\int \psi_s^*(1)\psi_r(2)\left(\frac{1}{r_{12}}\right)\psi_r^*(2)\psi_s(1)\,\mathrm{d}\tau_1\mathrm{d}\tau_2 \end{aligned} \tag{48}$$

which, after reorganizing the integrand a little, is seen to be the Coulomb integral introduced in connection with the structure of helium (eqn 7.33). It is the average potential energy of interaction between an electron in ψ_s and an electron in ψ_r. Similarly,

$$\begin{aligned} K_{sr} &= \int \psi_s^*(1)K_r\psi_s(1)\,\mathrm{d}\tau_1 \\ &= \frac{e^2}{4\pi\varepsilon_0}\int \psi_s^*(1)\psi_r^*(2)\left(\frac{1}{r_{12}}\right)\psi_s(2)\psi_r(1)\,\mathrm{d}\tau_1\mathrm{d}\tau_2 \end{aligned} \tag{49}$$

This integral is recognizable, after some reorganization, as the exchange integral (eqn 7.34). In passing, note that $K_{rr} = J_{rr}$.

The sum of the orbital energies is not the total energy of the atom, for such a sum counts all electron–electron interactions twice. So, to obtain the total energy we need to eliminate the effects of double counting:

$$E = 2\sum_{s} \varepsilon_s - \sum_{r,s} (2J_{rs} - K_{rs}) \tag{50}$$

where the sum is over the occupied orbitals (each of which is doubly occupied in a closed-shell species). We can verify that this procedure gives the correct result for helium in the configuration $1s^2$. The one-electron energy is

$$\varepsilon_{1s} = E_{1s} + (2J_{1s,1s} - K_{1s,1s}) = E_{1s} + J_{1s,1s}$$

and the total energy is

$$\begin{aligned} E &= 2\varepsilon_{1s} - (2J_{1s,1s} - K_{1s,1s}) = 2(E_{1s} + J_{1s,1s}) - J_{1s,1s} \\ &= 2E_{1s} + J_{1s,1s} \end{aligned}$$

exactly as before.

The energy required to remove an electron from an orbital ψ_r, on the assumption that the remaining electrons do not adjust their distributions, is the one-electron energy ε_r. Therefore, we may equate the one-electron orbital energy with the ionization energy of the electron from that orbital. This identification is the content of **Koopmans' theorem**:

$$I_r \approx -\varepsilon_r \tag{51}$$

The theorem is only an approximation, because in reality the remaining electrons do relax into a new distribution when one electron is removed.

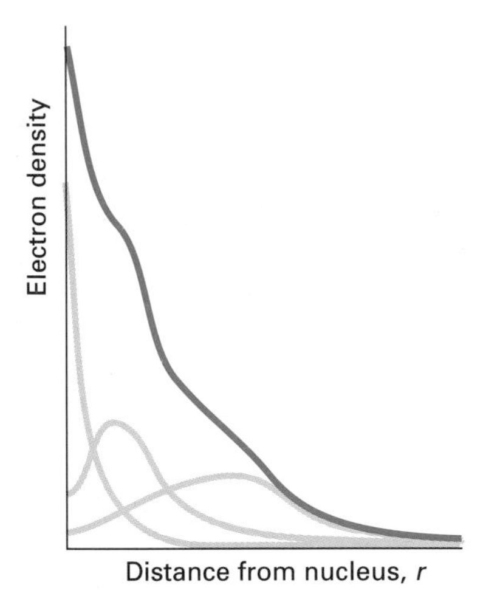

Fig. 7.16 A representation of the electron density calculated for a many-electron atom. Note that the shell structure is apparent, but that the total electron density falls to zero monotonically. The graph is a plot of the electron density along a radius, not the radial distribution function.

Solutions of the Hartree–Fock equations are generally given either as numerical tables or fitted to sets of simple functions. Once they are available, the total electron density in an atom may be calculated very simply by summing the squares of the wavefunctions for each electron. As Fig. 7.16 shows, the calculated value exhibits the shell structure of the atom that more primitive theories have led us to expect. Note that the total electron density shows the shell structure as a series of inflections: it decreases monotonically without intermediate maxima and minima.

Hartree–Fock SCF atomic orbitals are by no means the most refined orbitals that can be obtained. They are rooted in the orbital approximation and therefore to an approximate central-field form of the potential. The true wavefunction for an atom, whatever that may be, depends explicitly on the separations of the electrons, not merely their distances from the nucleus. The incorporation of the separations r_{ij} explicitly into the wavefunction is the background of the **correlation problem**, which is at the centre of much modern work (Chapter 9). Another route to improvement is to use the Dirac equation for the calculation rather than the non-relativistic Schrödinger equation. Relativistic effects are of considerable importance for heavy atoms, and are needed to account for various properties of the elements, including the colour of gold, the lanthanide contraction, the inert-pair effect, and even the liquid character of mercury.

7.16 Term symbols and transitions of many-electron atoms

The state of a many-electron atom is expressed by a term symbol of exactly the same kind as we have already described. To construct the symbol, we need to know the total spin, S, the total orbital angular momentum, L, and the total angular momentum, J, of the atom. These quantities are constructed by an appropriate application of the Clebsch–Gordan series. For instance, in the **Russell–Saunders coupling scheme**, the total angular momenta of the valence electrons are constructed as follows:

$$S = s_1 + s_2, s_1 + s_2 - 1, \ldots, |s_1 - s_2|$$
$$L = l_1 + l_2, l_1 + l_2 - 1, \ldots, |l_1 - l_2|$$
$$J = L + S, L + S - 1, \ldots, |L - S|$$

Each of these series may need to be applied several times if there are more than two electrons in the valence shell. The core electrons can be neglected because the angular momentum of a closed shell is zero.

Example 7.4 The construction of term symbols

Construct the term symbols that can arise from the configurations (a) $2p^1 3p^1$ and (b) $2p^5$.

Method. First, construct the possible values of L by using the Clebsch–Gordan series and identify the corresponding letters. Then construct the possible values of S similarly, and work out the multiplicities. Finally, construct the values of J from the values of L and S for each term by using the Clebsch–Gordan series again. A useful trick for shells that are more than half full is to consider the holes in the shell as particles, and to construct the term symbol for the holes. That is equivalent to treating the electrons, because a closed shell has zero angular momentum, and the angular momentum of the electrons must be equal to (in the sense of cancelling) the angular momentum of the holes.

Answer. (a) For this configuration $l_1 = 1$ and $l_2 = 1$, so $L = 2, 1, 0$, and the configuration gives rise to D, P, and S terms. Two electrons have $S = 1, 0$, giving rise to triplet and singlet terms, respectively, so the complete set of terms is ^{3}D, ^{1}D, ^{3}P, ^{1}P, ^{3}S, and ^{1}S. The values of J that can arise are formed from $J = L + S, L + S - 1, \ldots, |L - S|$, and so the complete list of term symbols is

$$^3\mathrm{D}_3, {}^3\mathrm{D}_2, {}^3\mathrm{D}_1, {}^1\mathrm{D}_2, {}^3\mathrm{P}_2, {}^3\mathrm{P}_1, {}^3\mathrm{P}_0, {}^1\mathrm{P}_1, {}^3\mathrm{S}_1, {}^1\mathrm{S}_0$$

(b) The configuration $2p^5$ is equivalent to a single hole in a shell, so $L = l = 1$, corresponding to a P term. Because $S = s = \frac{1}{2}$ for the hole, the term symbol is ^{2}P. The two levels of this terms are $^2\mathrm{P}_{3/2}$ and $^2\mathrm{P}_{1/2}$.

Comment. The configuration $2p^2$ does not give rise to all the terms that $2p^1 3p^1$ generates because the Pauli principle forbids the occurrence of certain combinations of spin and orbital angular momentum. This point is taken up below.

Exercise 7.4. Construct the term symbols that can arise from the configurations (a) $3d^1 4p^1$ and (b) $3d^9$.

As indicated in the Comment in the example, some care is needed when deriving the term symbols arising from configurations of equivalent electrons, as in the $2p^2$ configuration of carbon. For instance, although the configuration gives rise to D, P, and S terms, and $S = 0, 1$, it is easy to see that ^{3}D is excluded. For this term to occur, we need to obtain a state with $M_L = +2$, as well as the other states that belong to $L = 2$. To obtain $M_L = +2$ means that both electrons must occupy orbitals with $m_l = +1$. However, because the two electrons are in the same orbital, they cannot have the same spins, so the $S = 1$ state is excluded.

A quick way to decide which combinations of L and S are allowed is to use group theory and to identify the antisymmetrized direct product (Section 5.14). For the $2p^2$ configuration we need to form

$$\Gamma^{(1)} \times \Gamma^{(1)} = \Gamma^{(2)} + [\Gamma^{(1)}] + \Gamma^{(0)} \qquad \Gamma^{(1/2)} \times \Gamma^{(1/2)} = \Gamma^{(1)} + [\Gamma^{(0)}]$$

where we have used the notation introduced in Sections 5.14 and 5.20. To ensure that the overall state is antisymmetric, we need to associate symmetric with antisymmetric combinations. In this case the terms are ^{1}D, ^{3}P, and ^{1}S.

A more pedestrian procedure is to draw up a table of **microstates**, or combinations of orbital and spin angular momenta of each electron, and then to identify the values of L and S to which they belong. We shall denote the spin-orbital as m_l if the spin is α and $\overline{m}_l$ if the spin is β. Then one typical microstate of two electrons would be $(1, \overline{1})$ if one electron occupies an orbital with $m_l = +1$ with α spin and the second electron occupies the same orbital with β spin. This microstate has $M_L = +2$ and $M_S = 0$, and is put into the appropriate cell in Table 7.2. The complete set of microstates can be compiled in this way, and ascribed to the appropriate cells in the table. Note that microstates such as $(1, 1)$, which correspond to the two α-spins in an orbital with $m_l = +1$, are excluded by the Pauli principle and have been omitted.

Table 7.2 The microstates of p^2

M_L, M_S:	+1	0	-1
+2		$(1, \overline{1})$	
+1	$(1, 0)$	$(1, \overline{0}), (\overline{1}, 0)$	$(\overline{1}, \overline{0})$
0	$(1, -1)$	$(1, -\overline{1}), (\overline{1}, -1), (0, \overline{0})$	$(\overline{1}, -\overline{1})$
-1	$(-1, 0)$	$(-1, \overline{0}), (-\overline{1}, 0)$	$(-\overline{1}, \overline{0})$
-2		$(-1, -\overline{1})$	

Now we analyse the microstates to see to which values of L and S they belong. The microstate $(1, \overline{1})$ must belong to $L = 2, S = 0$, which identifies it as a state of a ^{1}D term. There are five states with $L = 2$, so we can strike out one microstate in each row of the column headed $M_S = 0$; which one we strike out in each case is immaterial as this is only a book-keeping exercise, and striking out one state is equivalent to striking out one possible linear combination. The next row shows that there is a microstate with $M_L = +1$ and $M_S = +1$. This state must belong to $L = 1$ and $S = 1$ and hence to the term ^{3}P. The nine states of this term span $M_L = +1, 0, -1$ and $M_S = +1, 0, -1$, and so we can strike out nine microstates in this block of nine. Only one microstate remains: it has $M_L = 0$ and $M_S = 0$, and hence belongs to $L = 0$ and $S = 0$, and is therefore a ^{1}S term. Now we have accounted for all the microstates, and have identified the terms as ^{1}D, ^{3}P, and ^{1}S, as we had anticipated.

The transitions that are allowed by the selection rules for a many-electron atom are

$$\Delta J = 0, \pm 1 \text{ but } J = 0 \nleftrightarrow J = 0$$

$$\Delta L = 0, \pm 1 \qquad \Delta l = \pm 1 \qquad \Delta S = 0$$

The rules about ΔJ and ΔL express the general point about the conservation of angular momentum. The rule about Δl is based on the conservation of angular momentum for the actual electron that is excited in the transition and its acquisition of the angular momentum of the photon; it is relevant when using a single Slater determinant to represent a state. The rule about ΔS reflects the fact that the electric component of the electromagnetic field can have no effect on the spin angular momentum of the electron, and in particular that it cannot induce transitions between wavefunctions that have different permutation symmetry (see Section 7.10). The selection rules on J are exact: those concerning l, L, and S presume that these individual angular momenta are well-defined.

7.17 Hund's rules and the relative energies of terms

Friedrich Hund devised a set of rules for identifying the lowest energy term of a configuration with the minimum of calculation.

1. The term with the maximum multiplicity lies lowest in energy.

For the configuration $2p^2$, we expect the ^{3}P term to lie lowest in energy. The explanation of the rule can be traced to the effects of spin correlation. On account of the existence of a Fermi hole, orbitals containing electrons with the same spin can contract towards the nucleus without an undue increase in electron–electron repulsion. The Fermi hole acts as a kind of protective halo around the electrons.

2. For a given multiplicity, the term with the highest value of L lies lowest in energy.

For example, if we had to choose between ^{3}P and ^{3}F in a particular configuration, then we would select the latter as the lower energy term. The classical basis of this rule is essentially that if electrons are orbiting in the same direction (and so have a high value of L), then they will meet less often than when they are orbiting in opposite directions (and so have a low value of L). Because they meet less often, their repulsion is less.

3. For atoms with less than half-filled shells, the level with the lowest value of J lies lowest in energy.

For the $2p^2$ configuration, which corresponds to a shell that is less than half-full, the ground term is ^{3}P. It has three levels, with $J = 2, 1, 0$. We therefore predict that the lowest energy level is $^3\text{P}_0$. When a shell is more than half full, the opposite rule applies (highest J lies lowest). The origin of this rule, in both its forms, is the spin–orbit coupling, and was discussed in Section 7.5.

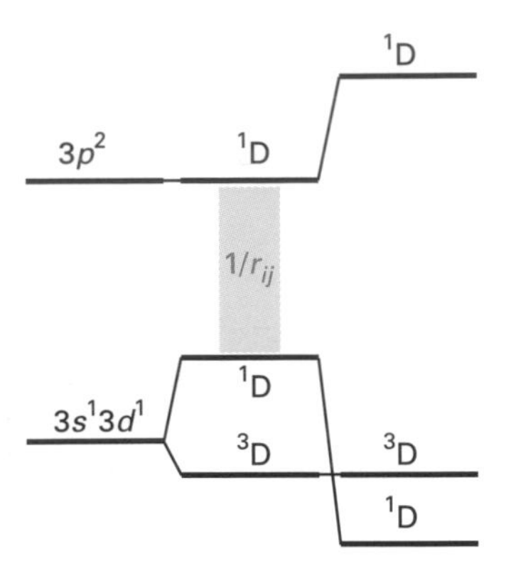

Fig. 7.17 The effect of configuration interaction between two D terms of the same multiplicity and the consequent reversal of the order of the terms of a configuration predicted by Hund's rules.

The rules are reasonably reliable for predicting the term of lowest energy, but are not particularly reliable for ranking all the terms according to their energy. One reason for their failure may be that the structure of an atom is inaccurately described by a single configuration, and a better description is in terms of **configuration interaction**, a superposition of several configurations. An example is found among the excited states of magnesium, and in particular the configuration $3s^1 3d^1$, which is expected to have $^3D < {}^1D$ whereas the opposite is found to be the case. An explanation is that the ^{1}D term is actually a mixture of about 75 per cent $3s^1 3d^1$ and 25 per cent $3p^2$ (which can also give rise to a ^{1}D term). If the two configurations have a similar energy, the electron–electron repulsion term perturbs them and, as for any two-level system, they move apart in energy (Fig. 7.17). The lower combination is pressed down in energy, and as a result may fall below the ^{3}D term, which is unchanged because there is no $3p^2\ {}^3$D term. Configuration interaction is discussed in more detail in Chapter 9.

7.18 Alternative coupling schemes

We have just seen that a configuration should not be taken too literally; the same is true of term symbols too. The statement of a term symbol implies that L and S have definite values, but that may not be true when spin–orbit coupling is appreciable, particularly in heavy atoms.

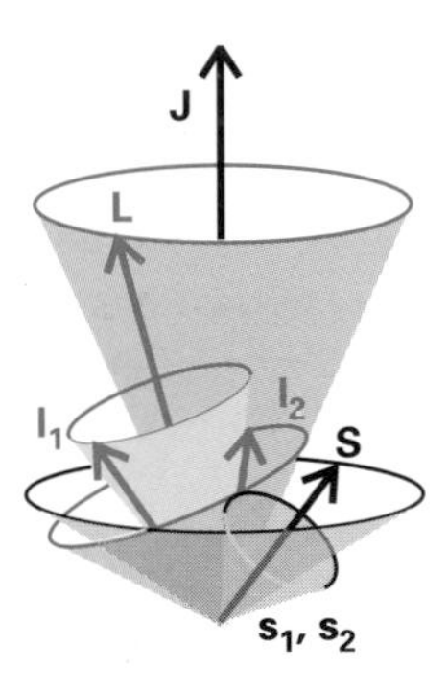

Fig. 7.18 A vector representation of Russell–Saunders (*LS*) coupling in a two-electron atom.

The term symbols we have introduced are based on Russell–Saunders (*LS*) coupling, which is applicable when spin–orbit coupling is weak in comparison with Coulombic interactions between electrons. When the latter are dominant, they result in the coupling of orbital angular momenta into a resultant with quantum number L and the spin angular momenta into a resultant with quantum number S. The weak spin–orbit interaction finally couples these composite angular momenta together into J (Fig. 7.18). To represent the relative strengths of the coupling of the angular momenta, we imagine the component vectors as **precessing**, or migrating around their cones, at a rate proportional to the strength of the coupling. When Russell–Saunders coupling is appropriate, the individual orbital momenta and the spin momenta precess rapidly around their resultants, but the two resultants **L** and **S** precess only slowly around their resultant, **J**.

When spin–orbit coupling is strong, as it is in heavy atoms, we use **jj-coupling**. In this scheme, the orbital and spin angular momenta of individual electrons couple to give a combined angular momentum j, and then these combined angular momenta couple to give a total angular momentum J. Now **l** and **s** each precess rapidly around their resultant **j**, and the various **j**s precess slowly around their resultant **J** (Fig. 7.19). In this scheme, L and S are not specified and so the term symbol loses its significance.

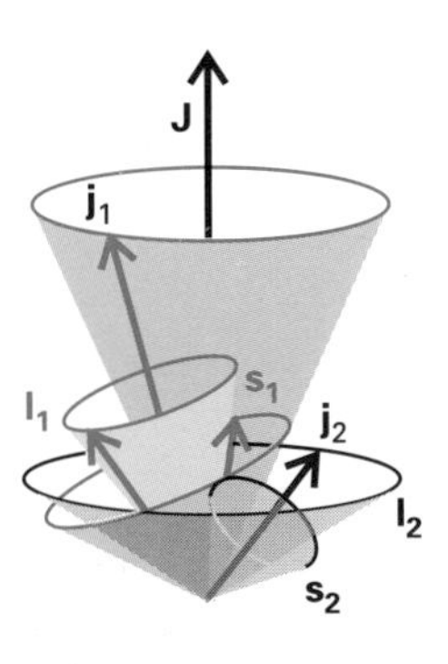

Fig. 7.19 A vector representation of *jj*-coupling in a two-electron atom.

Although the significance of the term symbol is lost when *jj*-coupling is relevant, they can still be used to label the terms because there is a correlation between Russell–Saunders and *jj*-coupled terms. To see that this is so, consider the np^2 configurations of the Group 14 elements carbon to lead. In the Russell–Saunders scheme we expect ^{1}S, ^{3}P, and ^{1}D terms, and the levels $^3P_2$, $^3P_1$, and $^3P_0$ of the ^{3}P term. The energies of these terms are indicated on the left of Fig. 7.20. On the other hand, in *jj*-coupling, each p-electron can have either

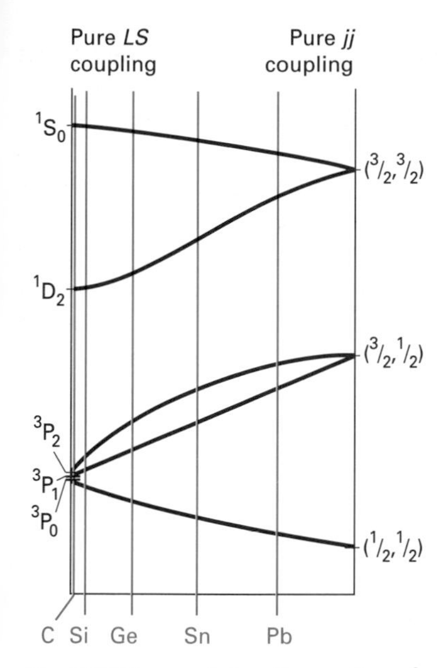

Fig. 7.20 The correlation diagram for a p^2 configuration and the approximate location of Group 14 atoms. Note that Russell–Saunders terms can be used to label the atoms regardless of the extent of *jj*-coupling.

$j = \frac{1}{2}$ or $j = \frac{3}{2}$. The resulting total angular momenta will be

$$\Gamma^{(1/2)} \times \Gamma^{(1/2)} = \Gamma^{(1)} + \Gamma^{(0)}$$
$$\Gamma^{(1/2)} \times \Gamma^{(3/2)} = \Gamma^{(2)} + \Gamma^{(1)}$$
$$\Gamma^{(3/2)} \times \Gamma^{(3/2)} = \Gamma^{(3)} + \Gamma^{(2)} + \Gamma^{(1)} + \Gamma^{(0)}$$

Because an electron with $j = \frac{1}{2}$ can be expected to have a lower energy than one with $j = \frac{3}{2}$ on the basis of spin–orbit coupling in a less than half-filled shell, we expect the order of energies indicated on the right of the illustration. Note that the Pauli principle excludes $J = 3$, because to achieve it, both electrons would need to occupy the same orbital with the same spin.

The states on the two sides can be correlated because J is well-defined in both coupling schemes and we know (Section 6.1) that states of the same symmetry (in this case, the same J) do not cross when perturbations are present. The resulting correlation of states is shown in the illustration, which is called a **correlation diagram**. As can be seen, even though Russell–Saunders coupling is inappropriate for the heavier members of the group, it can still be used to construct labels for the terms.

Atoms in external fields

In this final section, we shall consider how the application of electric and magnetic fields can affect the energy levels and hence the spectra of atoms. We shall describe two effects: the **Zeeman effect** is the response to a magnetic field; the **Stark effect** is the response to an electric field.

7.19 The normal Zeeman effect

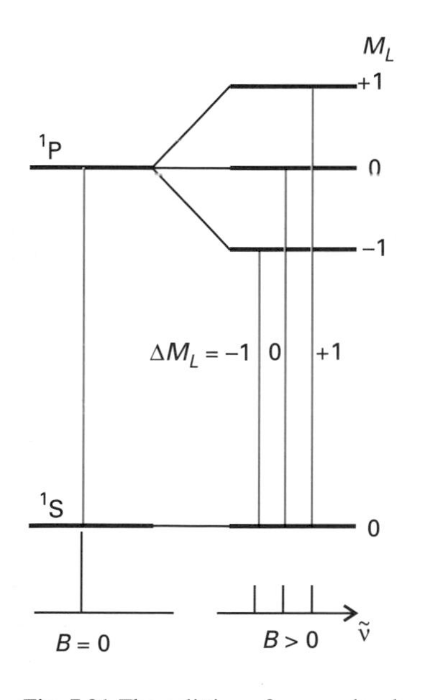

Fig. 7.21 The splitting of energy levels of an atom in the normal Zeeman effect, and the splitting of the transitions into three groups of coincident lines.

Electrons possess magnetic moments as a result of their orbital and spin angular momenta. These moments will interact with an externally applied magnetic field, and the resulting shifts in energy should be apparent in the spectrum of the atom.

Consider first the effect of a magnetic field on a singlet term, such as ^{1}P. Because $S = 0$, the magnetic moment of the atom arises solely from the orbital angular momentum. For a field of magnitude $\mathcal{B}$ in the z-direction the hamiltonian is

$$H^{(1)} = -m_z\mathcal{B} = -\gamma_e l_z \mathcal{B} \tag{52}$$

If several electrons are present,

$$H^{(1)} = -m_z\mathcal{B} = -\gamma_e(l_{z1} + l_{z2} + \ldots)\mathcal{B} = -\gamma_e L_z \mathcal{B} \tag{53}$$

The first-order correction to the energy of the P term is therefore

$$\begin{aligned} E^{(1)} &= \langle {}^1\mathrm{P}^{M_L}|H^{(1)}|{}^1\mathrm{P}^{M_L}\rangle \\ &= -\gamma_e M_L \hbar \mathcal{B} = \mu_B M_L \mathcal{B} \end{aligned} \tag{54}$$

where μ_B is the Bohr magneton. A term ^{1}S has neither orbital nor spin angular momentum, so it is unaffected by a magnetic field. It follows that the transition $^1\mathrm{P} \rightarrow {}^1\mathrm{S}$ should be split into three lines (Fig. 7.21), with a splitting of magnitude $\mu_B\mathcal{B}$. A 1 T magnetic field splits lines by only 0.5 cm^{-1}, so the effect is very

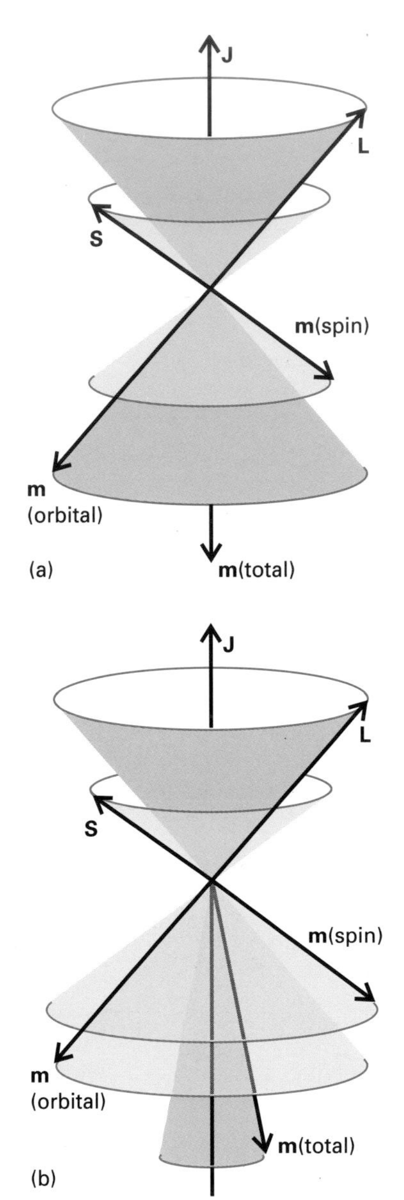

Fig. 7.22 (a) If the spin magnetic moment of an electron bore the same relation to the spin as the orbital moment bears to the orbital angular momentum, the total magnetic moment would be collinear with the total angular momentum. (b) However, because the spin has an anomalous magnetic moment, the total moment is not collinear with the total angular momentum. The surviving component, after allowing for precession, is determined by the Landé g-factor.

small. This splitting of a spectral line into three components is an example of the **normal Zeeman effect**.

The three transitions that make up $^1\text{P} \rightarrow {}^1\text{S}$ correspond to different values of ΔM_L. We have already seen that transitions with different values of ΔM_L correspond to different polarization of electromagnetic radiation. In the present case, an observer perpendicular to the magnetic field sees that the outer lines of the trio (those corresponding to $\Delta M_L = \pm 1$) are circularly polarized in opposite senses. These lines are called the **σ-lines**. The central line (which is due to $\Delta M_L = 0$) is linearly polarized parallel to the applied field. It is called the **π-line**.

The normal Zeeman effect is observed wherever spin is not present. It occurs even for transitions such as $^1\text{D} \rightarrow {}^1\text{P}$, in which the upper term is split into five states and the lower is split into three. In this case, the splittings are the same in the two terms, and the selection rules $\Delta M_L = 0, \pm 1$ limit the transitions to three groups of coincident lines.

7.20 The anomalous Zeeman effect

The **anomalous Zeeman effect**, in which a more elaborate pattern of lines is observed, is in fact more common than the normal Zeeman effect. It is observed when the spin angular momentum is nonzero and stems from the unequal splitting of the energy levels in the two terms involved in the transition. That unequal splitting stems in turn from the anomalous magnetic moment of the electron (Section 7.3).

If the g-value of an electron were 1 and not 2, then the total magnetic moment of the electron would be collinear with its total angular momentum (Fig. 7.22). But in fact, because of the anomaly, the two are not collinear. The spin and orbital momenta precess about their resultant (as a result of spin–orbit coupling), and as a result, the magnetic moment is swept around too. This motion has the effect of averaging to zero all except the component collinear with the direction of J, but the magnitude of this surviving magnetic moment depends on the values of L, S, and J because vectors of different lengths will lie at different angles to one another and give rise to different nonvanishing components of the angular momentum.

The calculation of the surviving component of the magnetic moment runs as follows. The hamiltonian for the interaction of a magnetic field **B** with orbital and spin angular momenta is

$$H^{(1)} = -\mathbf{m}_{\text{orbital}} \cdot \mathbf{B} - \mathbf{m}_{\text{spin}} \cdot \mathbf{B} = -\gamma_{\text{e}}(\mathbf{L} + 2\mathbf{S}) \cdot \mathbf{B} \tag{55}$$

where we have used 2 in place of g_{e}. At this point, we look for a way of expressing the hamiltonian as proportional to **J** by writing

$$H^{(1)} = -g_J \gamma_{\text{e}} \mathbf{J} \cdot \mathbf{B} \tag{56}$$

where g_J is a constant. The two hamiltonians are not equivalent in general, but for a first-order calculation we need only ensure that they have the same diagonal elements.

Consider Fig. 7.23. There are three precessional motions: **S** about **J**, **L** about **J**, and **J** about **B**. The effective magnetic moment can be found by projecting **L** on to **J** and then **J** on to **B**, and then doing the same for **S**. The precession

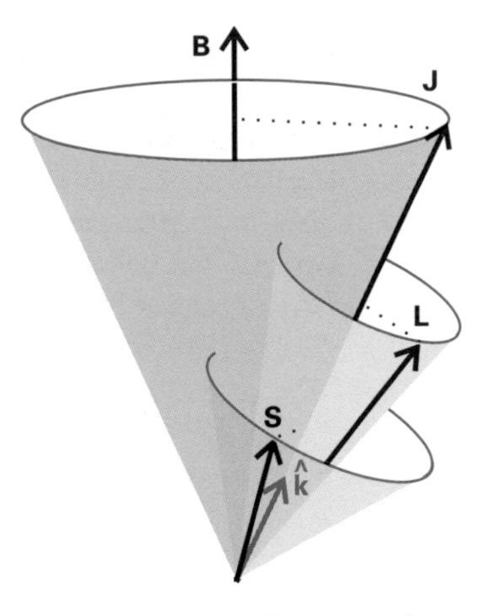

Fig. 7.23 The vector diagram used to calculate the Landé g-factor.

averages to zero all the components perpendicular to this motion (this classical averaging is equivalent to ignoring all off-diagonal components in a quantum mechanical calculation). If $\hat{\mathbf{k}}$ is a unit vector along **J**, it follows that the only surviving terms are

$$\mathbf{L}\cdot\mathbf{B} \rightarrow (\mathbf{L}\cdot\hat{\mathbf{k}})(\hat{\mathbf{k}}\cdot\mathbf{B}) = \frac{(\mathbf{L}\cdot\mathbf{J})(\mathbf{J}\cdot\mathbf{B})}{|J|^2}$$

$$\mathbf{S}\cdot\mathbf{B} \rightarrow (\mathbf{S}\cdot\hat{\mathbf{k}})(\hat{\mathbf{k}}\cdot\mathbf{B}) = \frac{(\mathbf{S}\cdot\mathbf{J})(\mathbf{J}\cdot\mathbf{B})}{|J|^2}$$

Because $\mathbf{J} = \mathbf{L} + \mathbf{S}$, it follows that

$$2\mathbf{L}\cdot\mathbf{J} = J^2 + L^2 - S^2 \qquad 2\mathbf{S}\cdot\mathbf{J} = J^2 + S^2 - L^2$$

If these quantities are now inserted into eqn 7.55 and the quantum mechanical expressions for magnitudes replace the classical values (so that J^2 is replaced by $J(J+1)\hbar^2$, etc), we find

$$\begin{aligned} H^{(1)} &= -\gamma_e(\mathbf{L} + 2\mathbf{S})\cdot\mathbf{B} \\ &= -\gamma_e\left\{1 + \frac{J(J+1) + S(S+1) - L(L+1)}{2J(J+1)}\right\}\mathbf{J}\cdot\mathbf{B} \end{aligned}$$

This is the form we sought. It enables us to identify the **Landé *g*-factor** as

$$g_J(L,S) = 1 + \frac{J(J+1) + S(S+1) - L(L+1)}{2J(J+1)} \qquad (57)$$

When $S = 0$, $g_J = 1$ because then J must equal L. In this case, the magnetic moment is independent of L, and so all singlet terms are split to the same extent. This uniform splitting results in the normal Zeeman effect. When $S \neq 0$, the value of g_J depends on the values of L and S, and so different terms are split to different extents (Fig. 7.24). The selection rule $\Delta M_J = 0, \pm 1$ continues to limit the transitions, but the lines no longer coincide and form three neat groups.

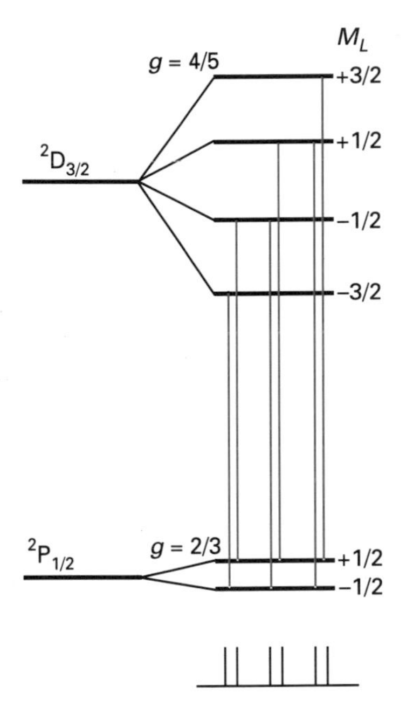

Fig. 7.24 The anomalous Zeeman effect. The splitting of energy levels with different g-values leads to a more complex pattern of lines than in the normal Zeeman effect.

Example 7.5 How to analyse the anomalous Zeeman effect

Account for the form of the Zeeman effect when a magnetic field is applied to the transition $^2D_{3/2} \rightarrow {}^2P_{1/2}$.

Method. Begin by calculating the Landé g-factor for each level, and then split the states by an energy that is proportional to its g-value. Then apply the selection rule $\Delta M_J = 0, \pm 1$ to decide which transitions are allowed.

Answer. For the level $^2D_{3/2}$ we have $L = 2, S = \frac{1}{2}$, and $J = \frac{3}{2}$. It follows that $g_{3/2}(2, \frac{1}{2}) = \frac{4}{5}$. For the lower level, $^2P_{1/2}$, we have $g_{1/2}(1, \frac{1}{2}) = \frac{2}{3}$. The splittings are therefore of magnitude $\frac{4}{5}\mu_B\mathcal{B}$ in the $^2D_{3/2}$ term and $\frac{2}{3}\mu_B\mathcal{B}$ in the $^2P_{1/2}$ term. The six allowed transitions are summarized in Fig. 7.24, where it is seen that they form three doublets.

Exercise 7.5. Construct a diagram showing the form of the Zeeman effect when a field is applied to a $^3D_2 \rightarrow {}^3P_1$ transition.

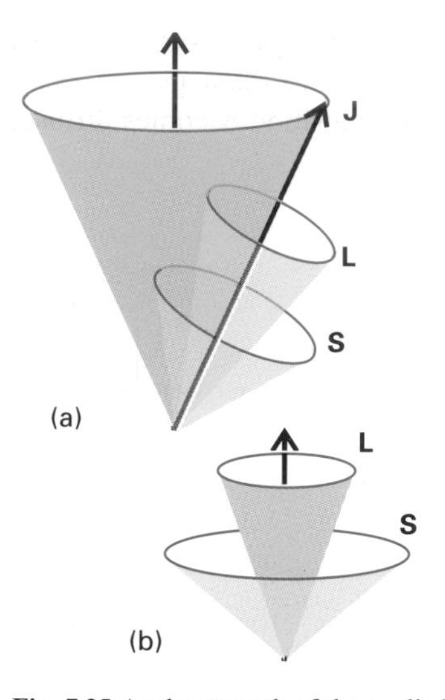

Fig. 7.25 As the strength of the applied field is increased, the precession of angular momenta about their resultant (as in (a)) gives way to precession about the magnetic field (as in (b)).

When the applied field is very strong, the coupling between L and S may be broken in favour of their direct coupling to the magnetic field.[5] The individual momenta, and therefore their magnetic moments, now precess independently about the field direction (Fig. 7.25). As the electromagnetic field couples to the spatial distribution of the electrons, not to the magnetic moment due to the spin, the presence of the spin now makes no difference to the energies of the transitions. As a result, the anomalous Zeeman effect gives way to the normal Zeeman effect. This switch from the anomalous effect to the normal effect is called the **Paschen–Back effect.**

7.21 The Stark effect

The hamiltonian for the interaction with an electric field of strength $\mathcal{E}$ in the z-direction is

$$H^{(1)} = -\mu_z\mathcal{E} = ez\mathcal{E} \tag{58}$$

where μ_z is the z-component of the electric dipole moment operator, $\mu_z = -ez$. This operator has matrix elements between orbitals that differ in l by 1 but which have the same value of m_l (recall Sections 5.16 and 7.2).

The **linear Stark effect** is a modification of the spectrum that is proportional to the strength of the applied electric field. It arises when there is a degeneracy between the two wavefunctions that the perturbation mixes, as for the $2s$ and $2p_z$ orbitals of hydrogen. The matrix element of the perturbation is

$$\langle 2p_z|H^{(1)}|2s\rangle = 3ea_0\mathcal{E} \tag{59}$$

and from Fig. 6.2 (or more formally from eqn 6.6) we know that the two degenerate orbitals mix and give rise to a splitting of magnitude $6ea_0\mathcal{E}$. The two functions that diagonalize the hamiltonian are $N(2s \pm 2p_z)$, with $N = 1/\sqrt{2}$ (Fig. 7.26). It is easy to see that they correspond to a shift of charge density into and out of the direction of the field, and this difference in distribution accounts for their difference in energy. The splitting is very small: even for fields of 1.0 MV m^{-1}, the splitting corresponds to only 2.6 cm^{-1}.

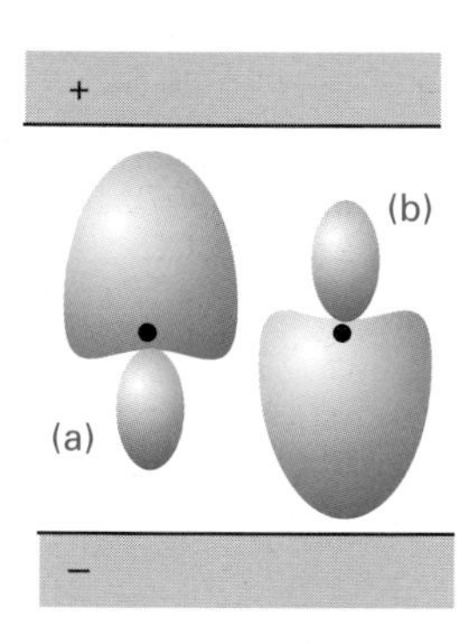

Fig. 7.26 The origin of the first-order Stark effect. The two mixed states give rise to two electron distributions that differ in energy.

The linear Stark effect depends on the peculiar degeneracy characteristic of hydrogenic atoms, and is not observed for many-electron atoms where that degeneracy is absent. In these atoms, it is replaced by the **quadratic Stark effect**, which is weaker still. The origin of the effect is the same, but now the distortion of the charge distribution occurs only as a perturbation and the resulting shifts in energy are proportional to $\mathcal{E}^2$. The field has to distort a non-degenerate and hence 'tight' system, and then interact with the dipole produced by that distortion.

At very high field strengths the H_α line is seen to broaden and its intensity to decrease. These effects are traced to the tunnelling of the electron. In high fields the potential experienced by the electron has the form shown in Fig. 7.27. The tails of the atomic orbitals seep through the region of high potential and penetrate into the external region, where the potential can strip the electron away

[5] This feature is examined in *Further information 15*, where the full significance of the recoupling is seen to be the search for the representation that gave matrices with the smallest off-diagonal elements: the vector recoupling diagram is a pictorial representation of that effect.

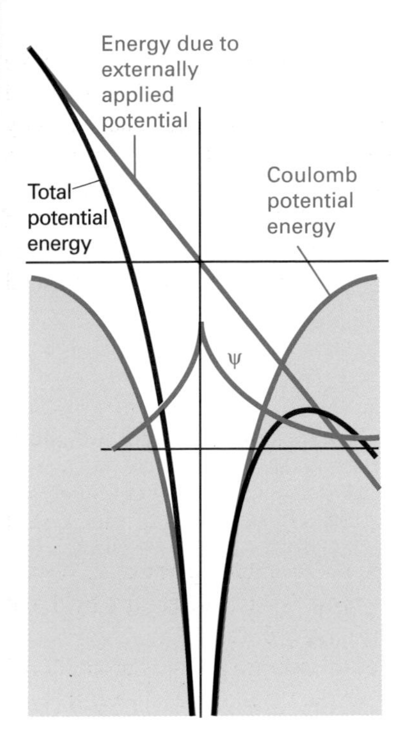

Fig. 7.27 When the applied field is very strong, its contribution to the total potential energy is such as to provide a tunnelling escape route for the originally bound state of an electron.

from the atom. This ionization results in fewer atoms being able to participate in emission, and so the intensity is decreased. Moreover, as the upper state has a shorter lifetime, its energy is less precise and the transition becomes diffuse.

Problems

7.1 Calculate the wavenumbers of the transitions of He^+ for the analogue of the Balmer series of hydrogen. *Hint.* Use eqn 7.2 with the Rydberg constant modified to account for the mass and charge differences.

7.2 Predict the form of the spectrum of the muonic atom formed from a proton in association with a μ-meson ($m_\mu = 207m_e$, charge $-e$).

7.3 Which of the following transitions are electric-dipole allowed: (a) $1s \to 2s$, (b) $1s \to 2p$, (c) $2p \to 3d$, (d) $3s \to 5d$, (e) $3s \to 5p$?

7.4 The spectrum of a one-electron ion of an element showed that its ns-orbitals were at 0, 2 057 972 cm^{-1}, 2 439 156 cm^{-1}, and 2 572 563 cm^{-1} for $n = 1, 2, 3, 4$, respectively. Identify the species and predict its ionization energy.

7.5 Demonstrate that for one-electron atoms the selection rules are $\Delta l = \pm 1$, $\Delta m_l = 0, \pm 1$, and Δn unlimited. *Hint.* Evaluate the electric-dipole transition moment $\langle n'l'm_l'|\boldsymbol{\mu}|nlm_l\rangle$ using $\mu_x = -er\sin\theta\cos\phi$, $\mu_y = -er\sin\theta\sin\phi$, and $\mu_z = -er\cos\theta$. The easiest way of evaluating the angular integrals is to recognize that the components just listed are proportional to Y_{lm_l} with $l = 1$, and to analyse the resulting integral group-theoretically.

7.6 Confirm that in hydrogenic atoms, the spin–orbit coupling constant depends on n and l as in eqn 7.21.

7.7 Calculate the spin–orbit coupling constant for a $2p$-electron in a Slater type atomic orbital, and evaluate it for the neutral atoms of Period 2 of the periodic table (from B to F).

7.8 Deduce the *Landé interval rule*, which states that for a given l and s, the energy difference between two levels differing in j by unity is proportional to j. *Hint.* Evaluate E_{so} in eqn 7.25 for j and $j - 1$; use the second line in the equation (in terms of ζ).

7.9 The ground configuration of an iron atom is $3d^6 4s^2$, and the ^{5}D term has five levels ($J = 4, 3, \ldots, 0$) at relative wavenumbers 0, 415.9, 704.0, 888.1, and 978.1 cm^{-1}. Investigate how well the Landé interval rule is obeyed. Deduce a value of ζ.

7.10 Calculate the energy difference between the levels with the greatest and smallest values of j for given l and s. Each term of a level is $(2j + 1)$-fold degenerate. Demonstrate that the centre of gravity of the energy of a term is the same as the energy in the absence of spin–orbit coupling. *Hint.* Weight each level with $2j + 1$ and sum the energies given in eqn 7.25 from $j = |l - s|$ to $j = l + s$. Use the relations

$$\sum_{s=0}^{n} s = \tfrac{1}{2}n(n+1) \qquad \sum_{s=0}^{n} s^2 = \tfrac{1}{6}n(n+1)(2n+1) \qquad \sum_{s=0}^{n} s^3 = \tfrac{1}{4}n^2(n+1)^2$$

7.11 Deduce what terms may arise from the ground configurations of the atoms of elements of Period 2, and suggest the order of their energies. *Hint.* Construct the term symbols as explained in Section 7.6 and use Hund's rules to arrive at their relative orders. Recall the hole–particle rule explained in Example 7.4.

7.12 Find the first-order corrections to the energies of the hydrogen atom that result from the relativistic mass increase of the electron. *Hint.* The energy is related to the momentum by $E = (p^2c^2 + m^2c^4)^{1/2} + V$. When $p^2c^2 \ll m^2c^4$, $E \approx \mu c^2 + p^2/2\mu + V - p^4/8\mu^3c^2$, where the reduced mass μ has replaced m. Ignore the rest energy μc^2, which simply fixes the zero. The term $-p^4/8\mu^3c^2$ is a perturbation;

hence calculate $\langle nlm_l|H^{(1)}|nlm_l\rangle = -(1/2\mu c^2)\langle nlm_l|(p^2/2\mu)^2|nlm_l\rangle = -(1/2\mu c^2)$ $\langle nlm_l|(E_{nlm_l} - V)^2|nlm_l\rangle$. We know E_{nlm_l}; therefore calculate the matrix elements of $V = -e^2/4\pi\varepsilon_0 r$ and V^2.

7.13 Write the hamiltonian for the lithium atom ($Z = 3$) and confirm that when electron–electron repulsions are neglected the wavefunction can be written as a product $\psi(1)\psi(2)\psi(3)$ of hydrogenic orbitals and the energy is a sum of the corresponding energies.

7.14 The Slater atomic orbitals are normalized but not mutually orthogonal. In the *Schmidt orthogonalization procedure* one orbital ψ is made orthogonal to another orbital ψ' by forming $\psi'' = \psi - c\psi'$, with $c = \int \psi^*\psi'\, d\tau$. Confirm that ψ'' and ψ' are orthogonal and construct a 2*s*-orbital that is orthogonal to a 1*s*-orbital from an STO basis.

7.15 Take a trial function for the helium atom as $\psi = \psi(1)\psi(2)$, with $\psi(1) = (\zeta^3/\pi)^{1/2}e^{-\zeta r_1}$ and $\psi(2) = (\zeta^3/\pi)^{1/2}e^{-\zeta r_2}$, ζ being a parameter, and find the best ground-state energy for a function of this form, and the corresponding value of ζ. Calculate the first and second ionization energies. *Hint*. Use the variation theorem. All the integrals are standard; the electron repulsion term is calculated in Example 7.2. Interpret Z in terms of a shielding constant. The experimental ionization energies are 24.850 eV and 54.40 eV.

7.16 On the basis of the same kind of calculation as in Problem 7.15, but for general Z, account for the first ionization energies of the ions Li^+, Be^{2+}, B^{3+}, C^{4+}. The experimental values are 73.5, 153, 258, and 389 eV, respectively.

7.17 Consider a one-dimensional square well containing two electrons. One electron has $n = 1$ and the other has $n = 2$. Plot a two-dimensional contour diagram of the probability distribution of the electrons when their spins are (a) parallel, (b) antiparallel. Devise a measure of the radius of the Fermi hole. *Hint*. When the spins are parallel (for example, $\alpha\alpha$) the antisymmetric space combination, $\psi_1(1)\psi_2(2) - \psi_2(1)\psi_1(2)$, must be taken, and when antiparallel, the symmetric combination. In each case plot ψ^2 against axes labelled x_1 and x_2. Computer graphics may be used to obtain striking diagrams, but a sketch is sufficient.

7.18 The first few S terms of helium lie at the following wavenumbers: $1s^2\,{}^1S: 0$; $1s^12s^1\,{}^1S: 166\,272\ cm^{-1}$; $1s^12s^1\,{}^3S: 159\,850\ cm^{-1}$; $1s^13s^1\,{}^1S: 184\,859\ cm^{-1}$; $1s^13s^1\,{}^3S: 183\,231\ cm^{-1}$. What are the values of K in the $1s^12s^1$ and $1s^13s^1$ configurations?

7.19 What levels may arise from the following terms: ${}^1S$, ${}^2P$, ${}^3P$, ${}^3D$, ${}^2D$, ${}^1D$, ${}^4D$? Arrange in order of increasing energy the terms that may arise from the following configurations: $1s^12p^1$, $2p^13p^1$, $3p^13d^1$. What terms may arise from (a) a d^2 configuration, (b) an f^2 configuration?

7.20 Write down the Slater determinant for the ground term of the beryllium atom, and find an expression for its energy in terms of Coulomb and exchange integrals. Find expressions for the energy in terms of the Hartree–Fock expression, eqn 7.50. *Hint*. Use eqn 7.50 for the configuration $1s^22s^2$; evaluate the expectation value $\langle\psi|H|\psi\rangle$.

7.21 Calculate the magnetic field required to produce a splitting of $1\ cm^{-1}$ between the states of a ${}^1P$ term. Calculate the Landé *g*-factor for (a) a term in which J has its maximum value for a given L and S; (b) a term in which J has its minimum value.

7.22 Transitions are observed and ascribed to ${}^1F \rightarrow {}^1D$. How many lines will be observed in a magnetic field of 4 T? Calculate the form of the spectrum for the Zeeman effect on a ${}^3P \rightarrow {}^3S$ transition.

Further reading

Atomic spectra. H.G. Kuhn; Longman, London (1969).
Atomic structure. E.U. Condon and H. Odabaşi; Cambridge University Press, Cambridge (1980).
The theory of atomic structure and spectra. R.D. Cowan; University of California Press, Berkeley (1981).
Orbitals, terms, and states. M. Gerloch; Wiley, Chichester (1986).
The theory of atomic spectra. E.U. Condon and G. Shortley; Cambridge University Press, Cambridge (1964).
Atomic spectra and atomic structure. G. Herzberg; Dover, New York (1944).
Data:
American Institute of Physics Handbook. D.E. Gray (ed.) McGraw-Hill, New York (1972).
Atomic energy levels. Vols 1–3. C.E. Moore; NBS-Circ. 467, Washington, DC (1949–58).
Tables of spectral lines of neutral and ionized atoms. A.R. Striganov and N.S. Sventitskii; Plenum, New York (1968).
Atomic energy levels and Grotrian diagrams. S. Bashkin and J.O. Stoner Jr; North-Holland, Amsterdam (1978 *et seq.*).

8 An introduction to molecular structure

Now we come to the heart of chemistry. If we can understand the forces that hold atoms together in molecules, we may also be able to understand why, under certain conditions, initial arrangements of atoms change into new ones in the course of the events we call chemical reactions. The aim of this chapter is to introduce some of the features of **valence theory**, the theory of the formation of chemical bonds. The description of bonding has been greatly enriched by numerical techniques, and these more quantitative aspects of the subject are described in the following chapter.

There are two principal models of molecular structure: molecular orbital theory and valence bond theory. Both models contribute concepts to the everyday language of chemistry and so it is worthwhile to examine them both. However, molecular orbital theory has undergone much more development than valence bond theory, and we shall largely concentrate on it; it will be the most sophisticated of the various approaches outlined in Chapter 9. Valence bond theory (which we shall not consider) has also undergone development in recent years, but even in its newer, more sophisticated form it has not yet displaced molecular orbital theory as the principal mode of molecular structure calculation.

The Born–Oppenheimer approximation

It is an unfortunate fact that, having arrived in sight of the promised land, we are forced to make an approximation at the outset. Even the simplest molecule, H_2^+, consists of three particles, and its Schrödinger equation cannot be solved analytically. To overcome this difficulty, we adopt the **Born–Oppenheimer approximation**, which takes note of the great difference in masses of electrons and nuclei. Because of this difference, the electrons can respond almost instantaneously to displacement of the nuclei. Therefore, instead of trying to solve the Schrödinger equation for all the particles simultaneously, it is possible to regard the nuclei as fixed in position and to solve the Schrödinger equation for the electrons in the static electric potential arising from the nuclei in that particular arrangement. Different arrangements of nuclei may then be adopted and the calculation repeated. The set of solutions so obtained allows us to construct the **molecular potential energy curve** of a diatomic molecule (Fig. 8.1), and in general a potential energy *surface* of a polyatomic species, and to identify the equilibrium conformation of the molecule with the lowest point on this curve (or surface). The Born–Oppenheimer approximation is very reliable for ground electronic states, but it is less reliable for excited states.

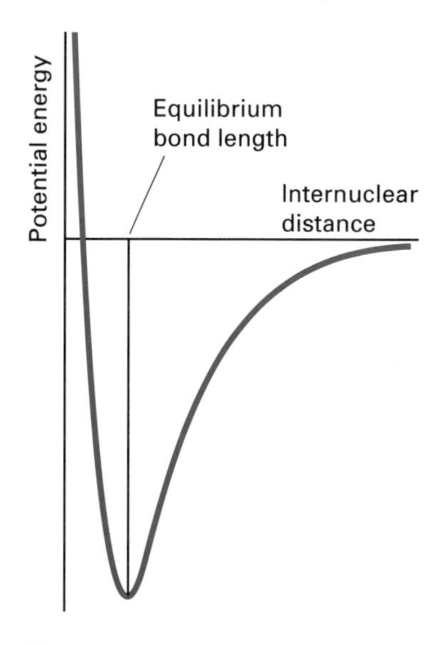

Fig. 8.1 A typical molecular potential energy curve for a diatomic species.

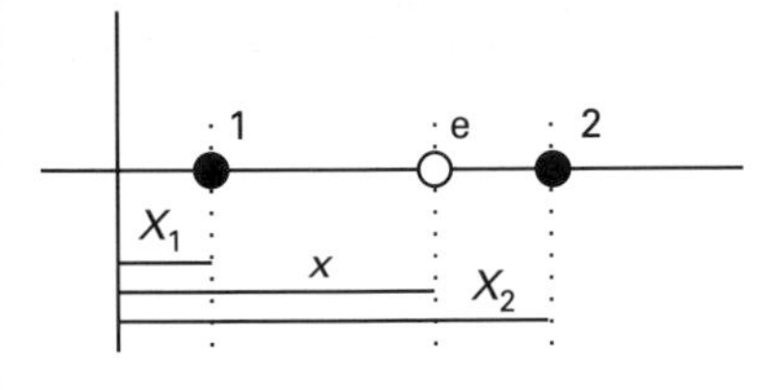

Fig. 8.2 The coordinates used in the discussion of the Born–Oppenheimer approximation.

8.1 The formulation of the approximation

The approximations involved in the Born–Oppenheimer procedure can be seen more sharply by formulating the problem quantitatively. The simplest approach is to consider a one-dimensional analogue of the hydrogen molecule-ion (Fig. 8.2), in which all motion is confined to the x-axis. The full hamiltonian, $\mathcal{H}$, for the problem is

$$\mathcal{H} = -\frac{\hbar^2}{2m_e}\frac{d^2}{dx^2} - \sum_j \frac{\hbar^2}{2m_j}\frac{d^2}{dX_j^2} + V(x, X_1, X_2) \tag{1}$$

where x is the location of the electron and X_j, with $j = 1, 2$, the locations of the two nuclei. The Schrödinger equation is

$$\mathcal{H}\Psi(x, X_1, X_2) = \mathcal{E}\Psi(x, X_1, X_2) \tag{2}$$

We attempt a solution of the form

$$\Psi(x, X_1, X_2) = \psi(x; X_1, X_2)\chi(X_1, X_2) \tag{3}$$

where the notation $\psi(x; X_1, X_2)$ means that the wavefunction for the electron depends parametrically on the coordinates of the two nuclei in the sense that we get a different wavefunction $\psi(x)$ for each arrangement of the nuclei. When this trial solution is substituted into eqn 8.2 we obtain

$$\mathcal{H}\psi\chi = -\frac{\hbar^2}{2m_e}\frac{\partial^2\psi}{\partial x^2}\chi - \sum_j \frac{\hbar^2}{2m_j}\frac{\partial^2\chi}{\partial X_j^2}\psi - \sum_j \frac{\hbar^2}{2m_j}\left(2\frac{\partial\psi}{\partial X_j}\frac{\partial\chi}{\partial X_j} + \frac{\partial^2\psi}{\partial X_j^2}\chi\right) + V(x, X_1, X_2)\psi\chi$$

which can be written as

$$\mathcal{H}\psi\chi = \mathcal{E}'\psi\chi - \sum_j \frac{\hbar^2}{2m_j}\left(2\frac{\partial\psi}{\partial X_j}\frac{\partial\chi}{\partial X_j} + \frac{\partial^2\psi}{\partial X_j^2}\chi\right) \tag{4}$$

where $\mathcal{E}'$ is the eigenvalue of thc cquation

$$-\sum_j \frac{\hbar^2}{2m_j}\frac{\partial^2\chi}{\partial X_j^2} + E(X_1, X_2)\chi = \mathcal{E}'\chi \tag{5}$$

and E is the eigenvalue obtained by solving

$$-\frac{\hbar^2}{2m_e}\frac{\partial^2\psi}{\partial x^2} + V(x, X_1, X_2)\psi = E(X_1, X_2)\psi \tag{6}$$

The Schrödinger equation in eqn 8.4 has almost the form we require, and would have the correct form if the final term

$$-\sum_j \frac{\hbar^2}{2m_j}\left(2\frac{\partial\psi}{\partial X_j}\frac{\partial\chi}{\partial X_j} + \frac{\partial^2\psi}{\partial X_j^2}\chi\right) \tag{7}$$

were zero. It is not zero, but it is very small compared with $\mathcal{E}'\psi\chi$ on account of the appearance of the nuclear masses in the denominator. So, the essence of the Born–Oppenheimer approximation is to set this term equal to zero, in which case eqns 8.5 and 8.6 constitute a solution of the problem. We can interpret these equations as follows.

Equation 8.6 is the Schrödinger equation for the electron in a potential $V(x, X_1, X_2)$ that depends on the fixed locations of the two nuclei. The solution is the electronic wavefunction ψ, and the eigenvalue $E(X_1, X_2)$ is the electronic contribution to the total energy of the molecule plus the potential energy of internuclear repulsion. It is this function that when plotted against the nuclear position (against $X_2 - X_1$ in this case) gives the molecular potential energy curve. Equation 8.5 is the Schrödinger equation for the wavefunction χ of the nuclei when the potential energy has the form of the molecular potential energy curve. Its eigenvalue is the total energy of the molecule within the Born–Oppenheimer approximation.

From now on (in this chapter) we shall concentrate on eqn 8.6, but write it more simply and generally as

$$H\psi = E\psi \qquad H = -\frac{\hbar^2}{2m_e}\nabla^2 + V \tag{8}$$

where V is the potential energy of the electron in the field of the stationary nuclei plus the nuclear interaction contribution and E is the total electronic and nucleus–nucleus repulsion energy for a stationary nuclear conformation. If we needed to investigate departures from the Born–Oppenheimer approximation, we would need to investigate the effect of the neglected part of the hamiltonian, which (in three dimensions) has the following structure:

$$H^{(1)}\psi\chi = -\sum_j \frac{\hbar^2}{2m_j}\{2(\nabla_j\psi)\cdot(\nabla_j\chi) + (\nabla_j^2\psi)\chi\} \tag{9}$$

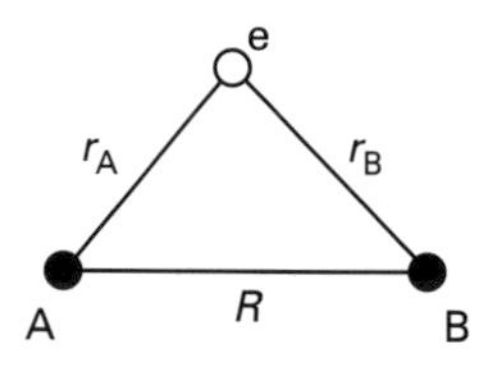

Fig. 8.3 The coordinates used to specify the hamiltonian for the hydrogen molecule-ion.

8.2 An application: the hydrogen molecule-ion

Even within the Born–Oppenheimer approximation there is only one molecular species for which the Schrödinger equation can be solved exactly, which is the hydrogen molecule-ion, H_2^+. The (Born–Oppenheimer) hamiltonian for this species is

$$H = -\frac{\hbar^2}{2m_e}\nabla^2 - \frac{e^2}{4\pi\varepsilon_0 r_A} - \frac{e^2}{4\pi\varepsilon_0 r_B} + \frac{e^2}{4\pi\varepsilon_0 R} \tag{10}$$

where the distances are defined in Fig. 8.3. The final term represents the repulsive interaction between the two nuclei, and within the Born–Oppenheimer approximation it is a constant for a given relative location of the nuclei.

As H_2^+ has only one electron, it has a status in valence theory analogous to the hydrogen atom in the theory of atomic structure. Indeed, just as the Schrödinger equation for the hydrogen atom is separable and solvable if it is expressed in spherical polar coordinates, so the equation for H_2^+ is separable and solvable if it is expressed in **ellipsoidal coordinates** in which the two nuclei lie at the foci of ellipses (Fig. 8.4). The resulting solutions are called **molecular orbitals** and resemble atomic orbitals but spread over both nuclei.

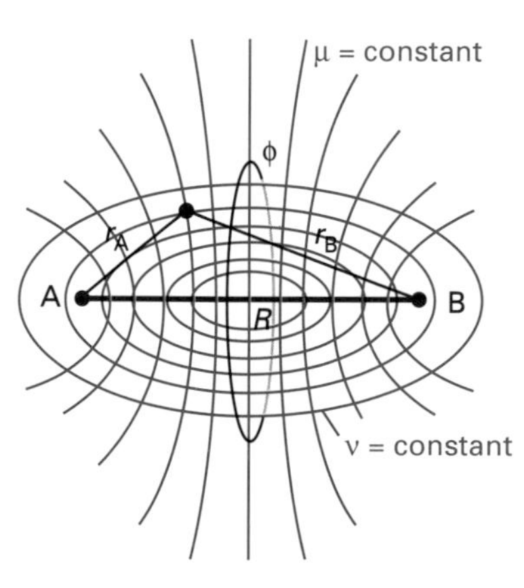

Fig. 8.4 The elliptical coordinates μ, ν, and ϕ used for the separation of variables in the exact treatment (within the Born–Oppenheimer approximation) of the hydrogen molecule-ion.

The 'exact' molecular orbitals of H_2^+ are mathematically much more complicated than the atomic orbitals of hydrogen, and as we shall shortly make yet another approximation, there is little point in giving their detailed form.[1]

[1] A reference to their form is provided in the *Further reading* section.

However, some of their features are very important and will occur in other contexts.

The molecular potential energy curves vary with internuclear distance, R, as shown in Fig. 8.5. The two lowest curves are of the greatest interest, and we concentrate on them. The steep rise in energy as $R \to 0$ is largely due to the increase in the nucleus–nucleus potential energy as the two nuclei are brought close together. At large distances, as $R \to \infty$, the curves tend towards the values typical of a hydrogen atom (with a proton a long way away). The lowest curve passes through a minimum close to $R = 2a_0$, and its energy then lies about 0.20 $hc\mathcal{R}$ (2.7 eV) below the energy of a separated hydrogen atom and proton. This result suggests that H_2^+ is a stable species (in the sense of having a lower energy than its dissociation products, but not in a chemical sense of being non-reactive), and that its bond length will be close to $2a_0$ (106 pm). The species is known spectroscopically: its minimum lies at 2.648 eV and its bond length is 106 pm, in very good agreement.

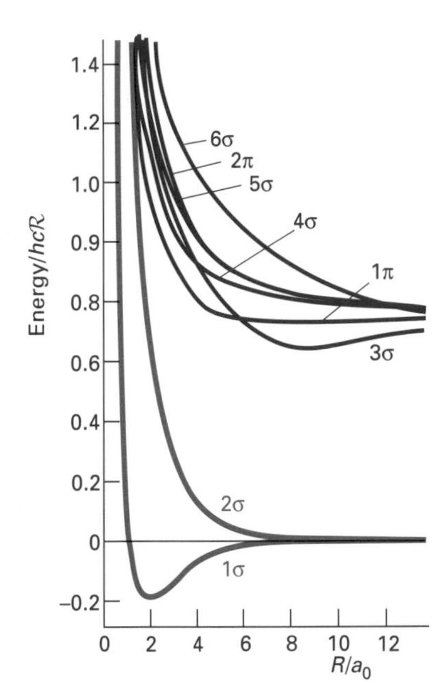

Fig. 8.5 The molecular potential energy curves for the hydrogen molecule-ion.

The origin of the lowering of energy can, to some extent, be discovered by examining the form of the wavefunctions. The two molecular orbitals of lowest energy are drawn as contour diagrams in Fig. 8.6 for various values of R. The striking difference between them is that the higher energy orbital (2σ) has an internuclear node whereas the lower energy orbital (1σ) does not. There is therefore a much greater probability of finding the electron in the internuclear region if it is described by the wavefunction 1σ than if it is described by 2σ. The conventional argument then runs that because the electron can interact with both

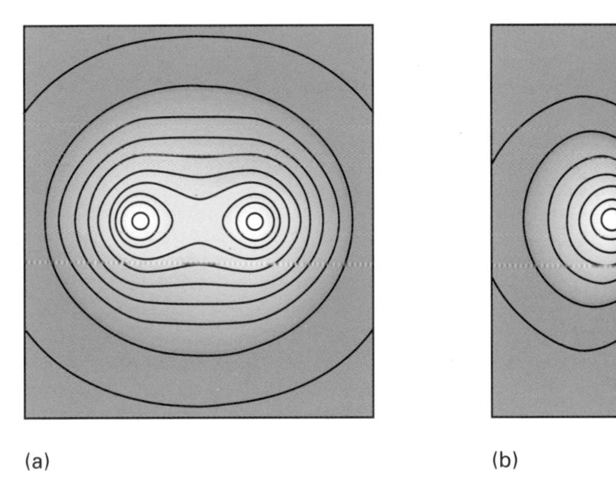

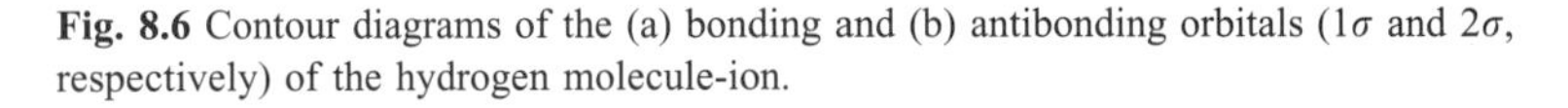

Fig. 8.6 Contour diagrams of the (a) bonding and (b) antibonding orbitals (1σ and 2σ, respectively) of the hydrogen molecule-ion.

nuclei if its wavefunction is 1σ, then it is in a favourable electrostatic environment and will have a lower energy than that of a separated hydrogen atom and proton. It is on the basis of such a simplistic argument that chemical bond formation is generally associated with the accumulation of electron density in an internuclear region.

The actual interpretation of the wavefunctions is, however, a much more delicate problem. The total energy of a molecule has contributions from several sources, including the kinetic energy of the electron. What appears to happen on bond formation (in H_2^+ at least), is that as R is reduced from a large value, the lowest energy wavefunction shrinks on to the nuclei slightly as well as accumulating in the internuclear region. The transfer of electron density into the internuclear region is *dis*advantageous, because it is removed from close to the nuclei. However, the shrinkage of the orbitals overcomes this disadvantage, for although a slight increase in kinetic energy accompanies the shrinking (because the wavefunction becomes more sharply curved), a significant reduction in potential energy overcomes all these unwanted effects, and the net outcome is a lowering of energy. The formation of 2σ, on the other hand, results in a small expansion of the electron distribution around the nuclei, and that has a net energy-raising effect. In other words, it is not the shift of electron density into the internuclear region that lowers the energy of the molecule but the freedom that this redistribution gives for the wavefunction to shrink in the vicinity of the two nuclei.

In what follows, we shall anticipate the formation of a bond whenever there is an enhanced probability density in the internuclear region, but accept that this might be no more than a correlation rather than a direct effect on the energy of the molecule. A detailed analysis has been performed only for H_2^+, and the argument might be quite different in other molecules.[2]

Molecular orbital theory

A difficulty will already have become apparent: the solution of the Schrödinger equation for H_2^+ is so complicated (even after making the Born–Oppenheimer approximation) that there can be little hope that solutions can be found for more complicated molecules. Therefore, we must resort to another approximation, but using the exact solutions for H_2^+ as a guide. Another reason why making a further approximation is quite sensible is that we already have available quite good atomic orbitals for many-electron atoms, and it seems appropriate to try to use them as a starting point for the description of many-electron molecules built from those atoms.

8.3 Linear combinations of atomic orbitals

Inspection of the form of the wavefunctions for H_2^+ shown in Fig. 8.6 suggests that they can be simulated by forming linear combinations of hydrogen atomic orbitals:

$$1\sigma \approx \phi_a + \phi_b \qquad 2\sigma \approx \phi_a - \phi_b \tag{11}$$

where ϕ_a is a H1s-orbital on nucleus A and ϕ_b its analogue on nucleus B.[3] In the first case, the accumulation of electron density in the internuclear region is simulated by the **constructive interference**, or addition of amplitudes, that

[2] See M.J. Feinberg, K. Ruedenberg, and E.L. Mehler, The origin of binding and antibinding in the hydrogen molecule-ion. *Adv. Quantum Chem.*, 27, **5** (1970).

[3] In this chapter, we use ϕ to denote an atomic orbital and ψ to denote a molecular orbital.

takes place between the two waves centred on neighbouring atoms. The nodal plane in the true 2σ wavefunction is recreated by the **destructive interference**, or subtraction of amplitudes, that takes place when two waves are superimposed with opposite signs.

The partial justification for simulating molecular orbitals as a linear combination of atomic orbitals can be appreciated by examining the hamiltonian for the problem, which is

$$H = -\frac{\hbar^2}{2m_e}\nabla^2 + \frac{e^2}{4\pi\varepsilon_0}\left(-\frac{1}{r_A} - \frac{1}{r_B} + \frac{1}{R}\right) \qquad (12)$$

where the distances are defined in Fig. 8.3. When the electron is close to nucleus A, $r_A \ll r_B$, and the hamiltonian is then approximately

$$H = -\frac{\hbar^2}{2m_e}\nabla^2 - \frac{e^2}{4\pi\varepsilon_0 r_A} + \frac{e^2}{4\pi\varepsilon_0 R}$$

Apart from the final, constant term, this hamiltonian is the same as that for a hydrogen atom. So, close to nucleus A, the wavefunction of the electron will resemble a hydrogen atomic orbital. The same is true close to B, and this form of the solution is captured by the two linear combinations constructed above.

The same conclusions can be reached in a more formal way, and which is more readily extended to other species, by writing the molecular orbitals as the following **linear combinations of atomic orbitals** (LCAO):

$$\psi = \sum_r c_r \phi_r \qquad (13)$$

The atomic orbitals used in this expansion constitute the **basis set** for the calculation. In principle, we should use an infinite basis set for a precise recreation of the molecular orbital, but in practice only a finite basis set is used. Throughout this chapter we shall assume that the members of the basis set are real and that each one is normalized to 1. The optimum values of the coefficients are found by applying the variation principle, which means (Section 6.10) that we have to solve the secular equations

$$\sum_r c_r(H_{rs} - ES_{rs}) = 0 \qquad (14)$$

where H_{rs} is a matrix element of the hamiltonian and S_{rs} is an overlap matrix element. These secular equations have non-trivial solutions if

$$\det|H_{rs} - ES_{rs}| = 0 \qquad (15)$$

To make progress with finding the roots of the secular determinant, we need to evaluate the relevant matrix elements. We shall use the following notation and values:

$$S_{AA} = S_{BB} = 1 \qquad S_{AB} = S_{BA} = S$$

where S is the **overlap integral**, and

$$H_{AA} = H_{BB} = \alpha \qquad H_{AB} = H_{BA} = \beta$$

where α is the molecular **Coulomb integral** and β is the **resonance integral**. The secular determinant then becomes

$$\begin{vmatrix} \alpha - E & \beta - ES \\ \beta - ES & \alpha - E \end{vmatrix} = 0$$

The roots of this equation are

$$E_{\pm} = \frac{\alpha \pm \beta}{1 \pm S} \tag{16}$$

and the corresponding values for the real coefficients of the normalized wavefunctions are

$$\begin{aligned} c_{\mathrm{A}} = c_{\mathrm{B}} \qquad c_{\mathrm{A}} &= \frac{1}{\{2(1+S)\}^{1/2}} \quad \text{for } E_{+} = \frac{\alpha+\beta}{1+S} \\ c_{\mathrm{A}} = -c_{\mathrm{B}} \qquad c_{\mathrm{A}} &= \frac{1}{\{2(1-S)\}^{1/2}} \quad \text{for } E_{-} = \frac{\alpha-\beta}{1-S} \end{aligned} \tag{17}$$

The detailed form of the Coulomb and resonance integrals can be established as follows. On inserting the explicit form of the hamiltonian into their definitions, we find that

$$\alpha = \langle \phi_{\mathrm{A}} | H | \phi_{\mathrm{A}} \rangle = E_{1s} - \frac{e^2}{4\pi\varepsilon_0} \langle \phi_{\mathrm{A}} | \frac{1}{r_{\mathrm{B}}} | \phi_{\mathrm{A}} \rangle + \frac{e^2}{4\pi\varepsilon_0 R} \tag{18}$$

The first term follows because ϕ_{A} is an eigenfunction of the atomic hamiltonian. The second term corresponds to the total Coulombic energy of interaction between an electron density ϕ_{A}^2 and the second nucleus B (Fig. 8.7). We shall call this contribution j':

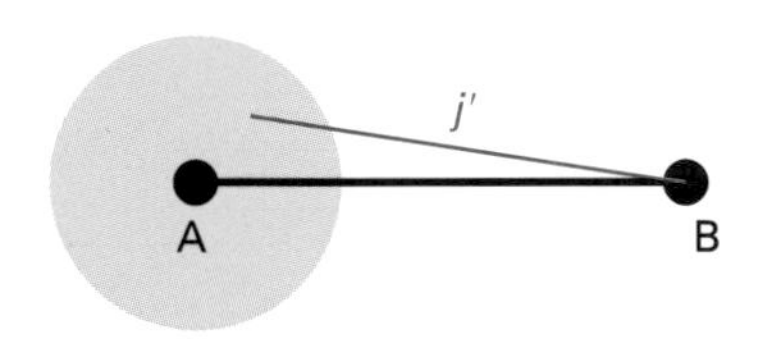

Fig. 8.7 The interpretation of the integral j' as the total Coulombic potential energy arising from a charge distribution on A with nucleus B.

$$j' = \frac{e^2}{4\pi\varepsilon_0} \int \frac{\phi_{\mathrm{A}}^2}{r_{\mathrm{B}}} \, \mathrm{d}\tau \tag{19}$$

This integral is positive, and the total Coulomb integral is

$$\alpha = E_{1s} - j' + \frac{e^2}{4\pi\varepsilon_0 R} \tag{20}$$

Example 8.1 The evaluation of overlap and Coulomb integrals

Evaluate (a) the overlap integral S and (b) the integral j' for an electron in 1σ composed of two hydrogenic $1s$-orbitals.

Method. To evaluate integrals of this kind, it is natural to use ellipsoidal coordinates (μ, ν, ϕ) (Fig. 8.4), where

$$\mu = \frac{r_{\mathrm{A}} + r_{\mathrm{B}}}{R} \qquad \nu = \frac{r_{\mathrm{A}} - r_{\mathrm{B}}}{R}$$

and the volume element is

$$\mathrm{d}\tau = \tfrac{1}{8} R^3 (\mu^2 - \nu^2) \, \mathrm{d}\mu \mathrm{d}\nu \mathrm{d}\phi$$

with $1 \leq \mu \leq \infty$, $-1 \leq \nu \leq 1$, and $0 \leq \phi \leq 2\pi$. The atomic wavefunctions we use are $\phi(r) = (Z^3/\pi a_0^3)^{1/2} \mathrm{e}^{-Zr/a_0}$ centred on either nucleus. All the integrations are straightforward in ellipsoidal coordinates; we write

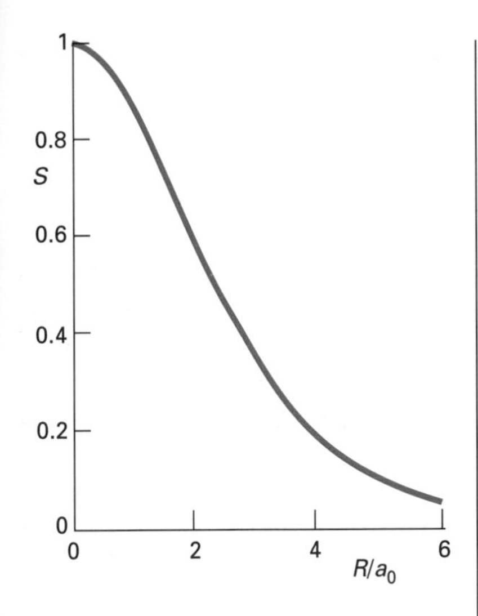

Fig. 8.8 The variation of the overlap integral of two H1s-orbitals with internuclear distance in the hydrogen molecule-ion.

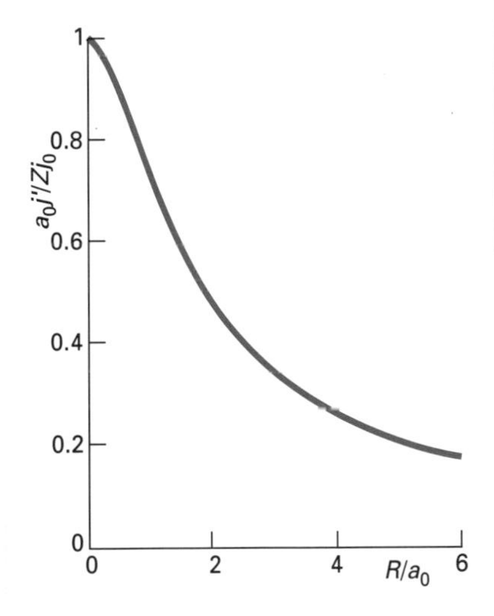

Fig. 8.9 The variation of the integral j' with internuclear distance in the hydrogen molecule-ion.

$j_0 = e^2/4\pi\varepsilon_0$.

Answer. (a) The overlap integral is

$$S = \langle\phi_A|\phi_B\rangle = \frac{Z^3R^3}{8\pi a_0^3}\int_0^{2\pi}\int_1^{\infty}\int_{-1}^{1}(\mu^2-\nu^2)e^{-Z\mu R/a_0}\,d\mu d\nu d\phi$$
$$= \left\{1+\frac{ZR}{a_0}+\tfrac{1}{3}\left(\frac{ZR}{a_0}\right)^2\right\}e^{-ZR/a_0}$$

(b) The contribution j' is similarly

$$j'/j_0 = \frac{Z^3}{\pi a_0^3}\int_0^{2\pi}\int_1^{\infty}\int_{-1}^{1}\frac{\tfrac{1}{8}R^3(\mu^2-\nu^2)e^{-Z(\mu+\nu)R/2a_0}}{\tfrac{1}{2}R(\mu-\nu)}\,d\mu d\nu d\phi$$
$$= \tfrac{1}{2}\left(\frac{Z}{a_0}\right)^3R^2\int_1^{\infty}\int_{-1}^{1}(\mu+\nu)e^{-Z(\mu+\nu)R/2a_0}\,d\mu d\nu$$
$$= \frac{1}{R}\left\{1-\left(1+\frac{ZR}{a_0}\right)e^{-2ZR/a_0}\right\}$$

Comment. These two functions are plotted in Fig. 8.8 and Fig. 8.9. Both S and j' decrease as Z increases because the higher nuclear charge shrinks the orbitals down on to their respective nuclei. A more detailed account of the calculation of molecular integrals is given in S.P. McGlynn, L.G. Vanquickenborne, M. Kinoshita, and D.G. Carroll, *Introduction to applied quantum chemistry*, Holt, Rinehart, and Winston, New York (1972).

Exercise 8.1. Evaluate the overlap integral between two Slater 2s-orbitals on different atoms.

For the resonance integral β, we use the fact that ϕ_B is an eigenfunction of the hamiltonian for hydrogen atom B with eigenvalue E_{1s}, and write

$$\beta = \langle\phi_A|H|\phi_B\rangle$$
$$= E_{1s}\langle\phi_A|\phi_B\rangle - \frac{e^2}{4\pi\varepsilon_0}\langle\phi_A|\frac{1}{r_A}|\phi_B\rangle + \frac{e^2}{4\pi\varepsilon_0 R}\langle\phi_A|\phi_B\rangle \tag{21}$$
$$= \left(E_{1s}+\frac{e^2}{4\pi\varepsilon_0 R}\right)S - k'$$

where

$$k' = \frac{e^2}{4\pi\varepsilon_0}\int\frac{\phi_A\phi_B}{r_A}\,d\tau \tag{22}$$

The integral k', which in this case is positive, has no classical analogue. However, an indication of its significance is that we can think of it as representing the interaction of the **overlap charge density**, $-e\phi_A\phi_B$, with nucleus A (Fig. 8.10). By symmetry, the interaction with nucleus B has the same value. The analytical expression for k' for two H1s-orbitals is

$$k' = \frac{e^2}{4\pi\varepsilon_0 a_0}\left(1+\frac{R}{a_0}\right)e^{-R/a_0} \tag{23}$$

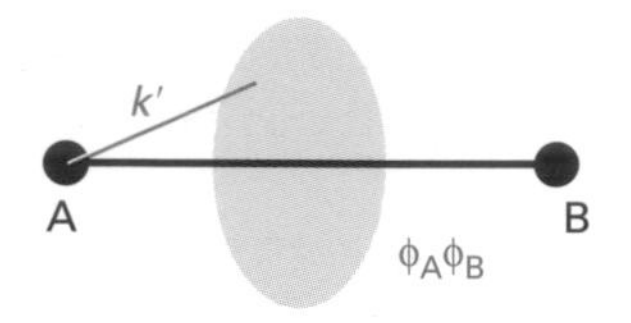

Fig. 8.10 The interpretation of the integral k' as the interaction of an overlap charge distribution with one of the nuclei.

It follows that the energies of the two LCAO-MOs are

$$E_+ = E_{1s} + \frac{e^2}{4\pi\varepsilon_0 R} - \frac{j' + k'}{1+S}$$
$$E_- = E_{1s} + \frac{e^2}{4\pi\varepsilon_0 R} + \frac{j' - k'}{1-S} \qquad (24)$$

The integrals j' and k' are both positive, with $j' > k'$, and so the lower of the two energies is E_+ (Fig. 8.11). The ladder of energy levels is called a **molecular orbital energy level diagram**. The lower-energy orbital, which has the form $1\sigma = \phi_A + \phi_B$, is called a **bonding orbital**. The higher-energy orbital, which has the form $2\sigma = \phi_A - \phi_B$, is called an **antibonding orbital**. Occupation of a bonding orbital lowers the energy of a molecule and helps to draw the two nuclei together; when an antibonding orbital is occupied, the energy of the molecule is raised and the two nuclei tend to be forced apart. One feature that should be noticed is that the diagram is not quite symmetrical: the antibonding orbital lies further above the energy of a hydrogen atom than the bonding orbital lies below it. This asymmetry is largely due to the repulsion between the two nuclei, which pushes both orbitals up in energy. In other words, *an antibonding orbital is more antibonding than a bonding orbital is bonding*.

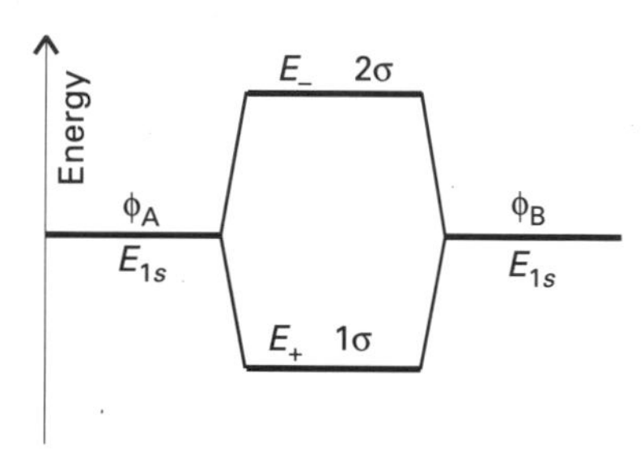

Fig. 8.11 The molecular orbital energy level diagram of the hydrogen molecule-ion in the LCAO approximation. Note that the 2σ-orbital is slightly more antibonding than the 1σ-orbital is bonding.

The analytical expressions for the energies are plotted in Fig. 8.12. As can be seen, the molecular potential energy curve has a minimum close to $R = 2.5a_0$ (130 pm) at a depth of $0.13hc\mathcal{R}$ (170 kJ mol^{-1}). The experimental values are $2.0a_0$ (106 pm) and $0.195hc\mathcal{R}$ (255 kJ mol^{-1}), and so the agreement is not spectacularly good. The principal source of error is that the basis is insufficiently flexible: we saw in the discussion of the exact solutions that a major contribution to the bonding comes from a shrinkage of the orbitals on to their respective nuclei, but this feature cannot be captured by the model.

One final detail of the molecular orbitals can usefully be introduced at this stage. The two molecular orbitals we have constructed can be classified according to their **parity**, their symmetry properties under inversion of the electron coordinates. As indicated in Fig. 8.13, under inversion the orbital 1σ remains indistinguishable from itself, and hence it is classified as having *gerade* symmetry, denoted g, where *gerade* is the German word meaning 'even'. In contrast, 2σ becomes the negative of itself under inversion, so it is classified as *ungerade*, the German word for 'odd', and denoted u. The full-dress versions of the orbitals are therefore $1\sigma_g$ and $2\sigma_u$. We already know that the symmetry classification is important for the discussion of selection rules (Section 7.2); shortly we shall see that the same classification also helps us to understand the electronic structures of molecules.

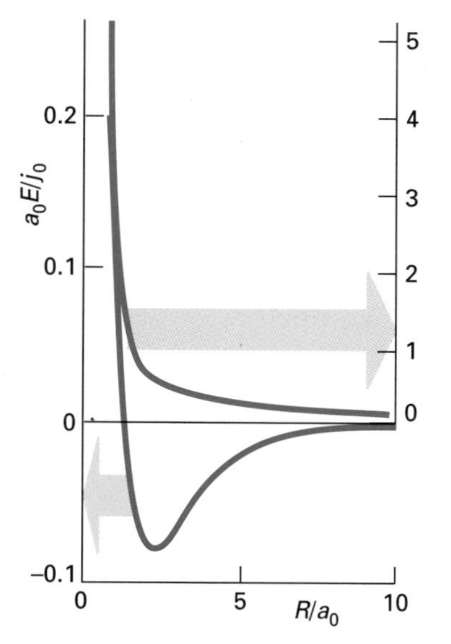

Fig. 8.12 The calculated molecular potential energy curves of the two lowest energy molecular orbitals of the hydrogen molecule-ion within the LCAO approximation. Note the change in scale between the bonding and antibonding curves.

8.4 The hydrogen molecule

The electronic structure of the hydrogen molecule, H_2, is modelled by the addition of a second electron to the 1σ orbital, to give the configuration $1\sigma^2$. The orbital description is therefore $1\sigma(1)1\sigma(2)$, where the 1 and 2 in parentheses are short for $\mathbf{r}_1$ and $\mathbf{r}_2$, respectively. This spatial wavefunction is symmetric under particle interchange, so the spin component must be proportional to $\alpha(1)\beta(2) - \beta(1)\alpha(2)$ to guarantee that the overall wavefunction is antisymmetrical. Therefore, when the two electrons enter a molecular orbital, they do so

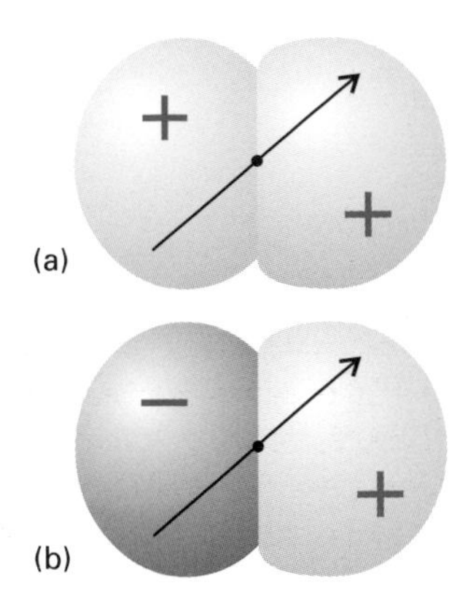

Fig. 8.13 The parity classification of orbitals in a homonuclear diatomic molecule: (a) g, (b) u.

with paired spins (↑↓). Spin-pairing is thus seen not to be an end in itself, but the way that electrons must arrange themselves in order to pack into the lowest energy orbital.

The ground-state configuration of H_2 is classified as $^1\Sigma_g$ in an echo of the term symbols used for atoms. The superscript 1 is the multiplicity of the state, which in this instance corresponds to $S = 0$ because the two electrons are paired. The Σ (the Greek letter Σ is the analogue of the letter S used to denote full spherical rotational symmetry) indicates that the total orbital angular momentum around the internuclear axis is zero because both electrons occupy σ-orbitals, and so neither has orbital angular momentum about the axis. More formally, we denote the component of orbital angular momentum about the axis as λ for each electron, and the total as Λ:

$$\Lambda = \lambda_1 + \lambda_2 \tag{25}$$

and in this case have $\lambda_1 = \lambda_2 = 0$, so $\Lambda = 0$, corresponding to a Σ term. The subscript g indicates that the overall parity of the state is g. To calculate it from the individual values for each electron we use

$$\mathrm{g} \times \mathrm{g} = \mathrm{g} \qquad \mathrm{g} \times \mathrm{u} = \mathrm{u} \qquad \mathrm{u} \times \mathrm{u} = \mathrm{g} \tag{26}$$

which follow from the mathematical properties of the products of odd and even functions, and use the first of these results for this two-electron system in which both electrons occupy g orbitals. Had one electron occupied a $2\sigma_u$ orbital, then the term would have been of overall u parity.

The full form of the H_2 wavefunction is

$$\psi(1,2) = 1\sigma(1)1\sigma(2)\sigma_-(1,2) \tag{27}$$

where the factor σ_- is the spin contribution $(1/\sqrt{2})\{\alpha(1)\beta(2) - \alpha(2)\beta(1)\}$. The energy of the molecule is found by evaluating the expectation value of the hamiltonian. The resulting expression is

$$E = 2E_{1s} + \frac{e^2}{4\pi\varepsilon_0 R} - \frac{2j' + 2k'}{1+S} + \frac{j + 2k + m + 4l}{2(1+S)^2} \tag{28}$$

where, in addition to the integrals already defined,

$$\begin{aligned}
j/j_0 &= \int \frac{\phi_A(1)^2\phi_B(2)^2}{r_{12}}\,\mathrm{d}\tau_1\mathrm{d}\tau_2 = (\mathrm{AA}|\mathrm{BB}) \\
k/j_0 &= \int \frac{\phi_A(1)\phi_B(1)\phi_A(2)\phi_B(2)}{r_{12}}\,\mathrm{d}\tau_1\mathrm{d}\tau_2 = (\mathrm{AB}|\mathrm{AB}) \\
l/j_0 &= \int \frac{\phi_A(1)^2\phi_A(2)\phi_B(2)}{r_{12}}\,\mathrm{d}\tau_1\mathrm{d}\tau_2 = (\mathrm{AA}|\mathrm{AB}) \\
m/j_0 &= \int \frac{\phi_A(1)^2\phi_A(2)^2}{r_{12}}\,\mathrm{d}\tau_1\mathrm{d}\tau_2 = (\mathrm{AA}|\mathrm{AA})
\end{aligned} \tag{29}$$

The values for these integrals are given in *Further information 10*; the notation on the right will be used again in Chapter 9. The molecular potential energy curve calculated from this expression is shown in Fig. 8.14. It has a minimum at $R = 1.4a_0$ (74 pm), and the minimum lies at $0.27hc\mathcal{R}$ (350 kJ mol^{-1}) below $2E_{1s}$, the energy of two separated hydrogen atoms. The experimental values are $1.40a_0$ (74.1 pm) and $0.33hc\mathcal{R}$ (430 kJ mol^{-1}), respectively, and although there is a fair measure of agreement, there is room for improvement. The kind of improvement that can be made includes the use of

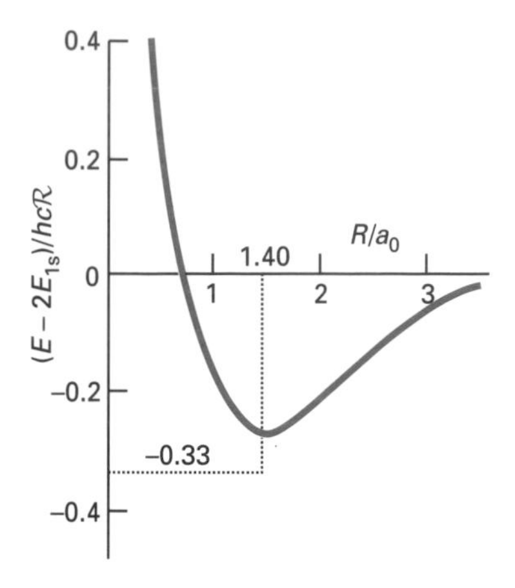

Fig. 8.14 The calculated molecular potential energy curve for the lowest energy orbital of a hydrogen molecule in the LCAO approximation.

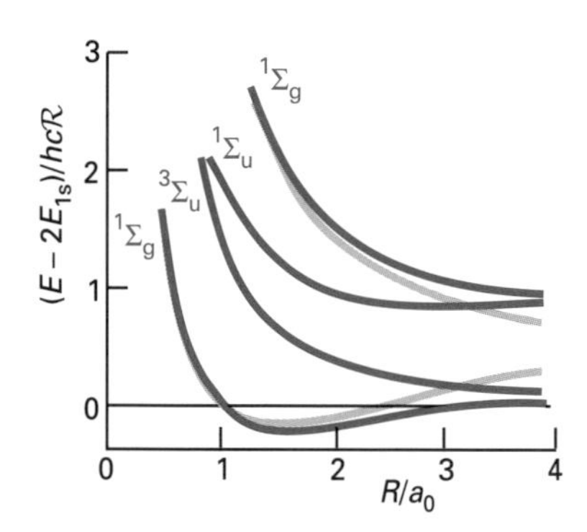

Fig. 8.15 The variation of the energies of four states of the hydrogen molecule with changing internuclear distance and the effect of configuration interaction which pushes the two pale curves apart.

a more flexible basis set, the use of SCF procedures, and the incorporation of electron correlation (see Chapter 9).

8.5 Configuration interaction

One procedure that is widely used and which can be illustrated here is the method of configuration interaction, first mentioned in connection with atoms in Section 7.17. If we continue to stick with the two atomic orbital basis set, then there are two molecular orbitals for the two electrons of H_2 to occupy. The following four configurations are possible:

$$1\sigma(1)1\sigma(2) \qquad 1\sigma(1)2\sigma(2) \qquad 2\sigma(1)1\sigma(2) \qquad 2\sigma(1)2\sigma(2)$$

To ensure that the states are antisymmetric overall, we need to form symmetric and antisymmetric products of the two central terms (the first and last are symmetric), and then multiply each term by the appropriate spin state. There are four possible overall antisymmetric states:

$$\begin{aligned}
\Psi_1(1,2;\,{}^1\Sigma_g) &= 1\sigma(1)1\sigma(2)\sigma_-(1,2) \\
\Psi_2(1,2;\,{}^1\Sigma_u) &= \frac{1}{\sqrt{2}}\{1\sigma(1)2\sigma(2) + 2\sigma(1)1\sigma(2)\}\sigma_-(1,2) \\
\Psi_3(1,2;\,{}^1\Sigma_g) &= 2\sigma(1)2\sigma(2)\sigma_-(1,2) \\
\Psi_4(1,2;\,{}^3\Sigma_u) &= \frac{1}{\sqrt{2}}\{1\sigma(1)2\sigma(2) - 2\sigma(1)1\sigma(2)\}\sigma_+(1,2)
\end{aligned}$$

Each state has been classified according to the procedure indicated in Section 8.4.

The MO description given above considered only $\Psi_1(1,2)$. When the energies of all four terms are calculated, we obtain the molecular potential energy curves shown in Fig. 8.15. One important feature is that two of them, $\Psi_1(1,2)$ and $\Psi_3(1,2)$, converge on the same energy as $R \to \infty$. Moreover, they are both ${}^1\Sigma_g$ terms. We have already seen that states of the same symmetry never cross because the hamiltonian always has nonzero matrix elements between them. As a result, **configuration interaction**, CI, the mixing of configurations of the same overall symmetry, occurs, and instead of crossing the two terms move apart as shown in Fig. 8.15. Configuration interaction lowers the energy of the lower term because their interaction in effect pushes the two states apart (as in the by-now-famous Fig. 6.1), and hence leads to an improved description of the ground state and a lowering of its energy.

The wavefunction of the lower state becomes

$$\psi = c_1\Psi_1 + c_3\Psi_3$$

as a result of CI, and the orbital structure of this function is

$$\begin{aligned}
\psi \approx{}& \tfrac{1}{2}(c_1 + c_3)\{\phi_A(1)\phi_A(2) + \phi_B(2)\phi_B(1)\} \\
&+ \tfrac{1}{2}(c_1 - c_3)\{\phi_A(1)\phi_B(2) + \phi_B(1)\phi_A(2)\}
\end{aligned}$$

It is revealing to compare this wavefunction with the form it has in the absence of CI:

$$\psi \approx \tfrac{1}{2}\phi_A(1)\phi_A(2) + \tfrac{1}{2}\phi_B(2)\phi_B(1) + \tfrac{1}{2}\phi_A(1)\phi_B(2) + \tfrac{1}{2}\phi_B(1)\phi_A(2)$$

The key point is that the former wavefunction is more flexible because the coefficients $c_1 + c_3$ and $c_1 - c_3$ are variable; there is no such flexibility in the latter wavefunction. This relaxation of constraint is an improvement and is reflected in the lowering of energy.

8.6 Diatomic molecules

It is not a long step, at least at the current level of exposition, from H_2 to the molecular-orbital description of other homonuclear diatomic molecules. The basic principle for the construction of molecular orbitals is that linear combinations are formed that have the same symmetry with respect to rotations about the internuclear axis. More formally, we form linear combinations of atomic orbitals that have the same symmetry species (that is, span the same irreducible representation) within the molecular point group. As we established in Section 5.16, only orbitals of the same symmetry species may have nonzero overlap ($S \neq 0$) and hence contribute to bonding. Thus, with the internuclear axis taken as the z-axis, s-, p_z-, and d_{z^2}-orbitals all have symmetry species Σ in $C_{\infty v}$ and may contribute to σ-orbitals. Similarly, p_x- and p_y-orbitals jointly span Π in $C_{\infty v}$, and hence may contribute to π-orbitals (Fig. 8.16); d_{yz}- and d_{zx}-orbitals also span irreducible representations of symmetry species Π, and they too may contribute to π-orbitals. It is rarely necessary to consider δ-orbitals, but the same principles can be applied: we select atomic orbitals of symmetry species Δ, and form linear combinations of them.

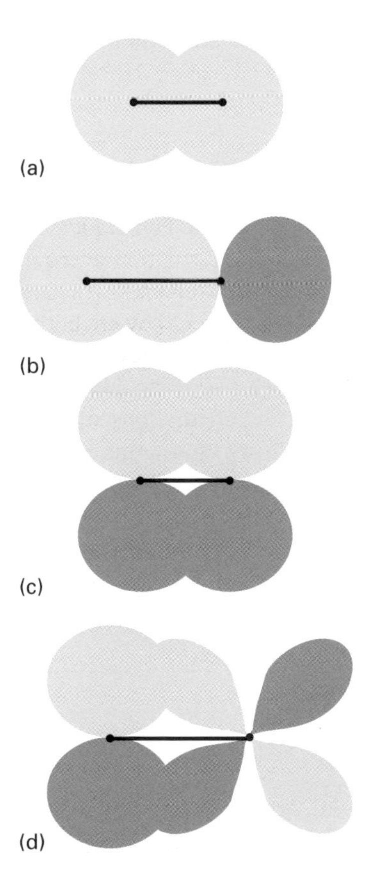

Fig. 8.16 Examples of varieties of molecular orbitals: (a) and (b) σ-orbitals, (c) and (d) π-orbitals.

We have stressed that group theory provides techniques for selecting atomic orbitals that *may* contribute to bonding, but other types of arguments must be used to decide whether these orbitals do in fact contribute, and to what extent. There are essentially two criteria to keep in mind in this connection.

First, to participate significantly to bond formation, *atomic orbitals must be neither too diffuse nor too compact*. In either case, there would be only feeble constructive or destructive overlap between neighbours, and only feeble bonds would result. It follows that in Period 2, $(1s, 1s)$-overlap can be largely neglected in comparison with $(2s, 2s)$-overlap. Indeed, it is generally safe, for qualitative discussions at least, to consider only overlap between orbitals of the valence shell, for only these orbitals are neither too compact nor too diffuse to have significant overlap.

Second, *the energies of the orbitals should be similar*. To see why this is so, consider the following secular determinant for the bond formed between two different atoms:

$$\begin{vmatrix} \alpha_A - E & \beta - ES \\ \beta - ES & \alpha_B - E \end{vmatrix} = 0 \tag{30}$$

The roots may be found by solving the quadratic equation for the energy, and when $\alpha_A - \alpha_B \gg \beta$ and $S = 0$ they are

$$E_+ \approx \alpha_A - \frac{\beta^2}{\alpha_B - \alpha_A} \qquad E_- \approx \alpha_B + \frac{\beta^2}{\alpha_B - \alpha_A} \tag{31}$$

These results (which are illustrated in Fig. 8.17) show that the molecular orbital energies are shifted from the atomic orbital energies (α_A and α_B) by only a small

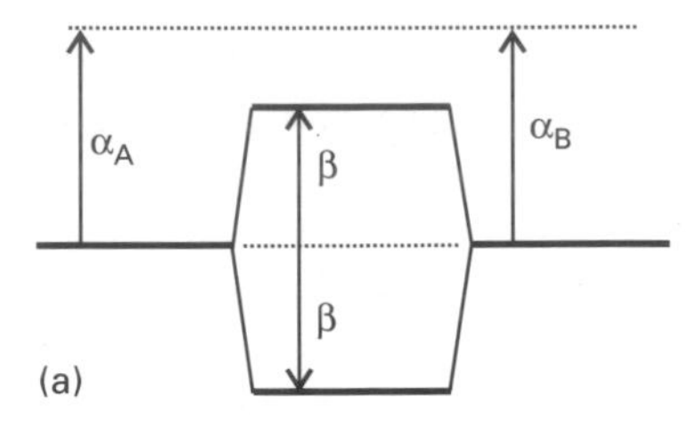

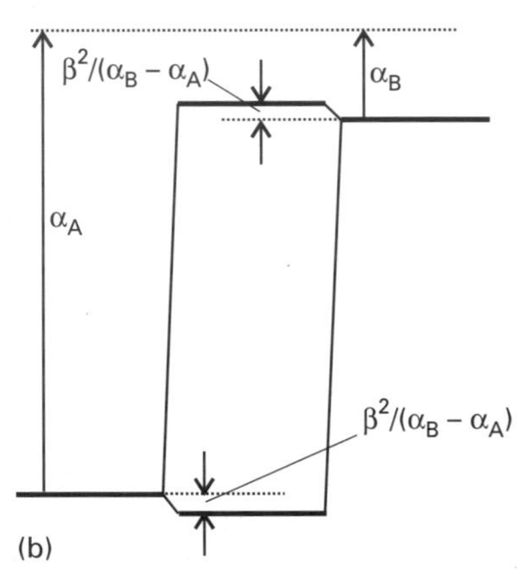

Fig. 8.17 The molecular orbital energy levels stemming from atomic orbitals of (a) the same energy, (b) different energy.

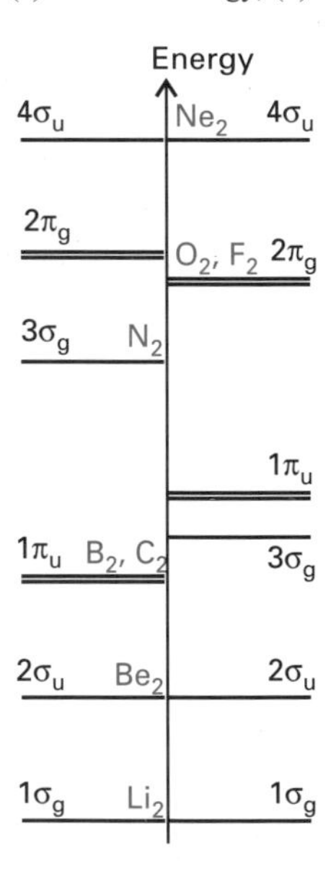

Fig. 8.18 The molecular orbital energy levels of diatomic molecules of the Period 2 elements. The labels indicate the highest occupied level of the specified species. Note the change in the order that appears between dinitrogen and dioxygen.

amount when α_A and α_B are very different. The implication is that in homonuclear diatomic molecules, the atomic orbitals of identical energy dominate the bonding. The strongest bonds will therefore have compositions such as $(2s, 2s)$ and $(2p, 2p)$, and there is no need (for qualitative discussions, at least) to consider $(2s, 1s)$ and $(2p, 1s)$ contributions. There is normally insufficient energy difference between $2s$- and $2p$-orbitals for it to be safe to ignore $(2s, 2p)$ contributions, although in elementary accounts that is often adopted as an initial approximation.

With these rules in mind, it is quite easy to set up a plausible molecular orbital energy level diagram for the Period 2 homonuclear diatomic molecules (Fig. 8.18). We consider only the valence orbitals (and, in due course, the electrons they contain). From the four atomic orbitals of Σ symmetry (the $2s$- and $2p_z$-orbitals on each atom), we can form four linear combinations; these are the four σ-orbitals marked on the diagram. To a first approximation we can think of the $2s$-orbitals as forming bonding and antibonding combinations and the $2p_z$-orbitals as doing the same. However, it is better to think of four combinations emerging from the four atomic orbitals, with increasing energy from the most bonding combination ($1\sigma_g$) to the most antibonding combination ($4\sigma_u$). All four σ-orbitals have mixed $2s$- and $2p_z$-orbital character, with the lowest energy combination predominantly $2s$-orbital in character and the highest energy combination predominantly $2p_z$. The four orbitals with Π symmetry likewise form four combinations, but because they span the two-dimensional irreducible representation, they fall into two doubly degenerate sets, which we call $1\pi_u$ and $2\pi_g$. It is hard to predict the order of energy levels, particularly the relative ordering of the σ and π sets, but it is found experimentally and confirmed by more detailed calculations that the order shown on the left of Fig. 8.18 applies from Li_2 to N_2, whereas the order shown on the right applies to O_2 and F_2.

To arrive at the electron configuration of the neutral molecule, we add the appropriate number of valence electrons to each set of energy levels. The procedure mirrors the building-up principle for atoms in that the electrons are added to the lowest energy available orbital subject to the requirement of the Pauli exclusion principle. If more than one orbital is available (as is the case when electrons occupy the π-orbitals), then electrons first occupy separate orbitals so as to minimize electron–electron repulsions; moreover, to benefit from spin correlation (Section 7.11), they do so with parallel spins. For instance, for N_2 we need to accommodate 10 valence electrons, and the ground-state configuration is

$$N_2 \qquad 1\sigma_g^2 2\sigma_u^2 1\pi_u^4 3\sigma_g^2 \quad {}^1\Sigma_g$$

For O_2, there are 12 valence electrons to accommodate, and the expected configuration is

$$O_2 \qquad 1\sigma_g^2 2\sigma_u^2 3\sigma_g^2 1\pi_u^4 2\pi_g^2 \quad {}^3\Sigma_g$$

Note that because only two electrons occupy the $2\pi_g$ orbital, they will be in separate orbitals and have parallel spins. Hence the ground state is predicted to be a triplet ($S = 1$). The possession of nonzero spin is consistent with the paramagnetic character of oxygen gas. The terms **HOMO** and **LUMO** are used to refer to the highest occupied and lowest unoccupied molecular orbitals,

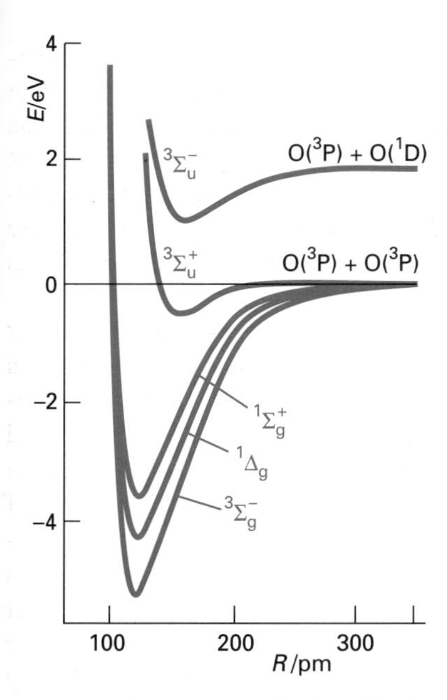

Fig. 8.19 The experimentally determined molecular potential energy curves of some of the lower energy states of dioxygen.

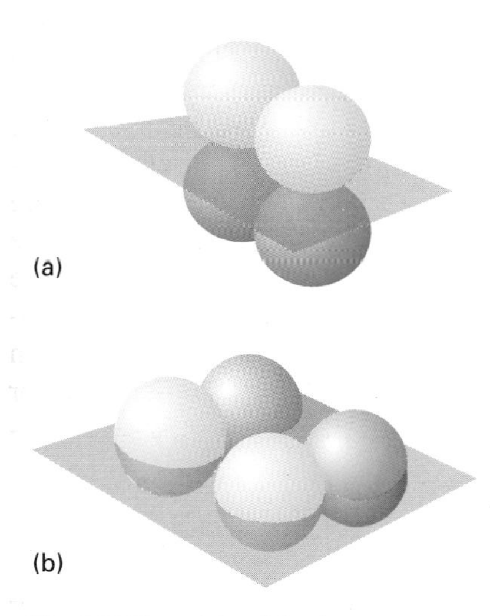

Fig. 8.20 The origin of the ± symmetry classification: (a) a π^--orbital, (b) a π^+-orbital.

respectively. The HOMO and LUMO are referred to jointly as the **frontier orbitals**: the frontier is the site of much of the reactive and spectroscopic activity of the species.

The term symbols that have been attached to the configurations listed above have been worked out in the manner already sketched for H_2. However, the symbol for O_2 is quite instructive (and incomplete!). To determine the value of Λ, the total orbital angular momentum around the internuclear axis, we add together all the individual λs. For σ-orbitals (and each electron they contain), $\lambda = 0$. For π-orbitals, $\lambda = \pm 1$ because each orbital corresponds to a different sense of rotation about the axis. It follows that a π^4 configuration necessarily contributes 0 to Λ, because it has equal numbers of electrons orbiting clockwise and counter-clockwise. A π^2 configuration, as in O_2, however, can contribute 0 or ± 2 because the two electrons can be in different orbitals or the same orbital, respectively. Hence, the configuration can give rise to a Σ and a Δ term, respectively. Because we expect the two electrons to occupy *different* orbitals (to minimize their mutual repulsion), it follows that we expect the ground term to be Σ, with Δ higher in energy. Moreover, because the electrons are in different π-orbitals, they can have either $S = 0$ or $S = 1$, so we expect $^1\Sigma$ and $^3\Sigma$ terms, with the latter lower in energy. On the other hand, the Δ term must have paired spins because both electrons occupy the same π-orbital, and so it must have $S = 0$, so giving a $^1\Delta$ term. The experimental molecular potential energy curves for O_2 are illustrated in Fig. 8.19, and these terms can be identified.

We remarked that the term symbol for O_2 given above is incomplete. Terms designated Σ also require a label to distinguish their behaviour under reflection in a plane that contains the internuclear axis (see the $C_{\infty v}$ character table in Appendix 1). Each σ-orbital has the character $+1$ under this operation. A π-orbital, however, may have the character $+1$ or -1 (Fig. 8.20). If the two electrons of interest occupy different π-orbitals, then one of them will be $+1$ and the other will be -1, and overall the character of the configuration will be $(+1) \times (-1) = -1$. This symmetry is denoted by a right superscript, so the full term symbol for the ground state of O_2 is $^3\Sigma_g^-$.

The case of C_2 is equally instructive. The straightforward application of the building-up principle suggests the ground-state configuration

$$C_2 \qquad 1\sigma_g^2 2\sigma_u^2 1\pi_u^4 \quad {}^1\Sigma_g^+$$

However, we have to be circumspect, because we are dealing with a many-electron molecule, and the occupation of the lowest energy orbitals does not necessarily lead to the lowest energy. We need to allow for the possibility that excitation of an electron to a nearby orbital, as illustrated in Fig. 8.21, might lower the electron–electron repulsion and result in a lower overall energy despite the occupation of a higher energy orbital. The resulting configuration $\ldots 2\sigma_u^2 1\pi_u^3 3\sigma_g^1$ would result in a $^3\Pi_u$ term, with the lowering of energy aided by the presence of spin correlation. Provided that the $1\pi_u$ and $3\sigma_g$ orbitals are quite close in energy, there is no unambiguous way of predicting which is the lower state. Indeed, even the experimental situation was unclear for many years, but it has now been resolved in favour of a $^1\Sigma_g^+$ ground state.

It must be understood that this qualitative approach to the electronic structure of diatomic molecules is only a first stage in reaching an understanding. Modern

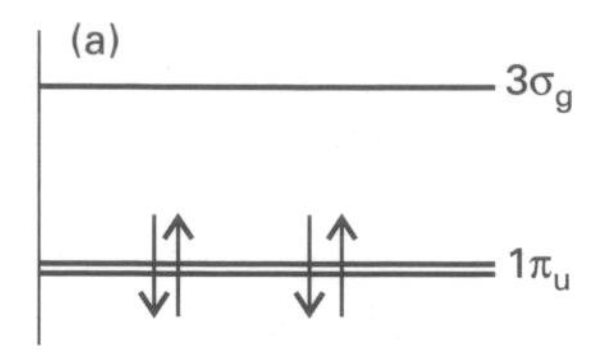

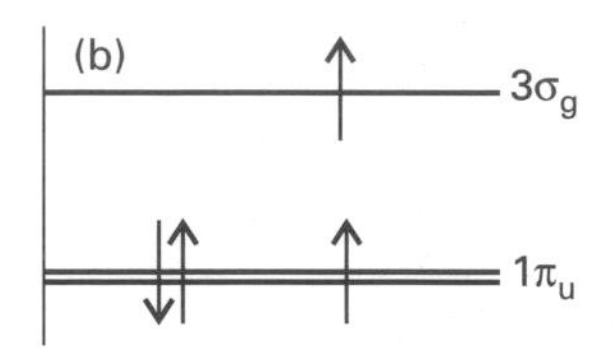

Fig. 8.21 When orbitals have similar energies, there may be a competition to determine whether (a) the lowest energy orbitals are occupied or (b) a higher energy orbital is occupied, with the advantage of the effects of spin correlation.

quantitative theories of structure are based on detailed numerical (SCF) calculations, and are described in Chapter 9.

8.7 Heteronuclear diatomic molecules

The qualitative effect of the presence of two different atoms in a molecule is for there to be a nonuniform distribution of electron density. Specifically, for orbitals of the form

$$\psi = c_A\phi_A + c_B\phi_B \tag{32}$$

it will no longer be true that $|c_A|^2 = |c_B|^2$. A useful rule of thumb is that if A is the more electronegative atom of the two, then the bonding combination will have $|c_A|^2 > |c_B|^2$, as it is a contribution to the lowering of energy for the electron to be found predominantly on A. On the other hand, for the antibonding combination, $|c_A|^2 < |c_B|^2$, and an electron in this orbital will be found predominantly on B. Its occupation of an orbital on B is a contribution to the raising of the energy of this molecular orbital.

A second feature of heteronuclear bonding is that because, except by accident, the energies of the orbitals of one atom do not coincide with those of the second atom, the extent to which the molecular orbitals are shifted in energy from the atomic orbitals is less than for homonuclear species. To borrow a term from classical physics, in homonuclear molecules the orbitals of the same designation 'resonate' with one another and hence couple strongly, whereas the resonance is imperfect in heteronuclear species and the coupling is weaker.

These features suggest that a heteronuclear bonding system can be generated from the homonuclear system by reducing the shifts in energy represented by the molecular orbital energy levels, and moving bonding orbitals towards the lower energy contributing atomic orbitals to represent their greater contribution; antibonding orbitals are similarly shifted towards the higher energy orbitals. The resulting scheme (for CO) is illustrated in Fig. 8.22. Note also that, because a heteronuclear diatomic molecule lacks a centre of inversion, the parity designation (g and u) is no longer relevant.[4] The electron configuration of CO can now be deduced by adding the 10 valence electrons to the lowest five orbitals:

$$\text{CO} \qquad 1\sigma^2 2\sigma^2 1\pi^4 3\sigma^2 \quad {}^1\Sigma^+$$

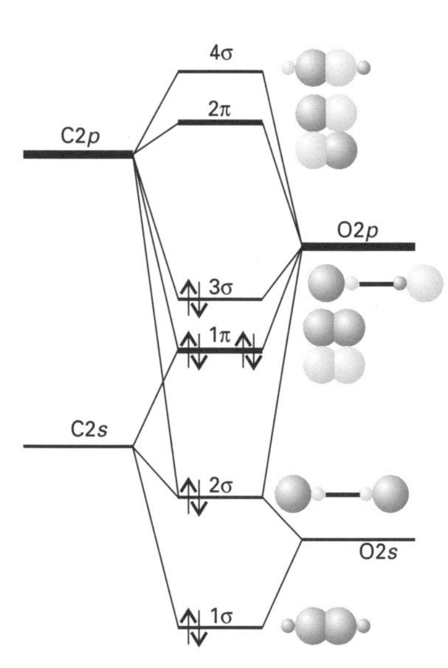

Fig. 8.22 A schematic depiction of the molecular orbital energy levels of the carbon monoxide molecule.

The HOMO is 3σ, which is in fact a largely nonbonding orbital on the C atom, so $3\sigma^2$ corresponds to a lone pair on C. The LUMO is 2π, which is a doubly degenerate pair of orbitals of largely C $2p$-orbital character. This combination of a HOMO that can provide two electrons and a LUMO that can accept them is potent, and accounts for the widespread occurrence of metal carbonyl complexes (and for the ability of carbon monoxide to act as a poison).

Once again, it must be stressed that arguments such as these are little more than a qualitative rule of thumb for rationalizing certain features of the electronic structures of diatomic molecules. For accurate energies and electron

[4] The advantage of numbering the homonuclear molecular orbitals regardless of their parity designation, rather than numbering the g set separately from the u set, is now apparent: the correlation between the homonuclear and heteronuclear orbitals is immediate. However, many authors adopt the alternative numbering scheme.

distributions, and from these wavefunctions reliable molecular properties, it is necessary to use the numerical techniques described in Chapter 9.

Molecular orbital theory of polyatomic molecules

The molecular orbitals of polyatomic species are linear combinations of atomic orbitals:

$$\psi = \sum_i c_i \phi_i \tag{33}$$

The main difference is now that the molecular orbital extends, in principle, over all the atoms of the molecule. However, as for diatomic molecules, only atomic orbitals that have the appropriate symmetry make a contribution, because only they have net overlap with one another. When a molecule lacks any symmetry elements (other than the identity), there is no way of avoiding assembling each molecular orbital from the entire basis set. However, when the molecule has elements of symmetry, group theory can be particularly helpful in deciding which orbitals can contribute to each molecular orbital, and in classifying the resulting orbitals according to their symmetry species. Indeed, the symmetry designations for polyatomic species are the analogues of the σ and π labels for diatomic molecules.

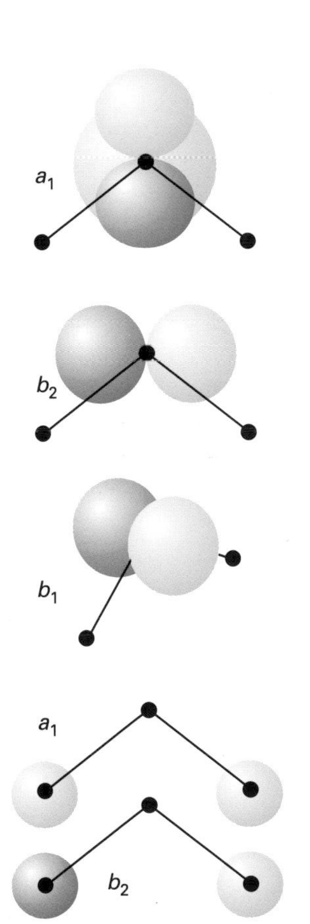

Fig. 8.23 The symmetry classification of the oxygen atomic orbitals in H_2O, a C_{2v} molecule, and the two symmetry-adapted linear combinations of the H1s orbitals.

8.8 Symmetry-adapted linear combinations

The concept behind the construction of a **symmetry-adapted linear combination** (SALC) is to identify two or more equivalent atoms in a molecule, such as the two H atoms in H_2O, and to form combinations of the atomic orbitals they provide that belong to specific symmetry species. Then molecular orbitals are constructed by forming linear combinations of each SALC with an atomic orbital of the same symmetry species on the central atom (the O atom in H_2O). We can be confident that only the SALC with a given symmetry species will have a net overlap with an atomic orbital of the same symmetry species. The effect of using SALCs instead of the raw basis is to factorize the secular determinant into block-diagonal form, because all elements H_{ij} and S_{ij} are zero except between orbitals of the same symmetry species. The secular determinant is thereby factorized into a product of smaller determinants, and we need find the roots of these determinants, which is in general a much simpler task.

An example should make this clear. Consider the H_2O molecule, which belongs to the C_{2v} point group. If we take the basis (H1s_A, H1s_B, O2s, O2p_x, O2p_y, O2p_z), then we should expect a 6×6 determinant and a sixth-order equation to solve for E. However, it should be clear from Fig. 8.23, that the two linear combinations

$$\phi(A_1) = H1s_A + H1s_B \qquad \phi(B_2) = H1s_A - H1s_B$$

can have net overlap with O2s and O2p_z (for $\phi(A_1)$) and with O2p_y (for $\phi(B_2)$), but not with O2p_x. This observation suggests that molecular orbitals in H_2O will fall into the following groups:

$$\psi(\mathrm{A}_1) = c_1\mathrm{O}2s + c_2\mathrm{O}2p_z + c_3\phi(\mathrm{A}_1)$$
$$\psi(\mathrm{B}_1) = \mathrm{O}2p_x$$
$$\psi(\mathrm{B}_2) = c_4\mathrm{O}2p_y + c_5\phi(\mathrm{B}_2)$$

and that when the secular determinant is constructed it will consist of three blocks, one being three-dimensional (involving the solution of a cubic equation for E), one being one-dimensional (involving only a trivial statement of the energy), and one being two-dimensional (and requiring the solution of a quadratic equation). In each case we have identified the symmetry species of the SALC by reference to the character table and have combined it with atomic orbitals of the same symmetry species to form a molecular orbital of the specified symmetry species. The molecular orbitals (not the SALCs) are labelled by lower case italic letters corresponding to the symmetry species, so in H_2O we can expect the orbitals a_1, b_1, and b_2. Each orbital of a particular symmetry species is then numbered sequentially in order of increasing energy, to give a notation such as $1a_1$, $2a_1$, and so on.

The formal procedure for the construction of SALCs was explained in Section 5.12, where we saw that a character table is used to formulate a projection operator, and then that projection operator is applied to a member of the basis. The procedure is illustrated in the following example.

Example 8.2 The construction of symmetry-adapted linear combinations

Construct the SALCs for the basis set given above for H_2O.

Method. Follow the method set out in Example 5.8. The point group is C_{2v} and $h = 4$.

Answer. The effect of the operations of the group on the basis is set out in the following table:

	O2s	O2p_x	O2p_y	O2p_z	H1s_A	H1s_B
E	O2s	O2p_x	O2p_y	O2p_z	H1s_A	H1s_B
C_2	O2s	$-$O2p_x	$-$O2p_y	O2p_z	H1s_B	H1s_A
σ_v	O2s	$-$O2p_x	O2p_y	O2p_z	H1s_A	H1s_B
σ'_v	O2s	O2p_x	$-$O2p_y	O2p_z	H1s_B	H1s_A

For A_1, $d = 1$ and all $\chi(R) = +1$. Hence, column 1 gives O2s, column 2 and column 3 give 0, column 4 gives O2p_z and column 5 gives $\frac{1}{2}(\mathrm{H}1s_\mathrm{A} + \mathrm{H}1s_\mathrm{B})$. This set of orbitals combine to give the molecular orbital $\psi(\mathrm{A}_1)$ listed in the text. For A_2, with characters $(1, 1, -1, -1)$, no column survives. For B_1 with characters $(1, -1, -1, 1)$, column 2 gives O2p_x, and all other columns give 0. The B_1 orbital is therefore a nonbonding orbital confined to the O atom, as no other orbitals present have net overlap with it (see Fig. 8.23). For B_2, with characters $(1, -1, 1, -1)$, column 3 gives O2p_y and columns 5 or 6 give $\frac{1}{2}(\mathrm{H}1s_\mathrm{A} - \mathrm{H}1s_\mathrm{B})$. Hence, the B_2 orbital has the form given in the text.

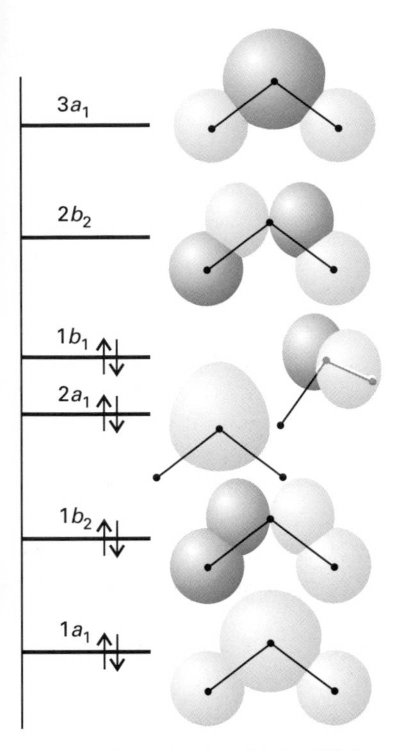

Fig. 8.24 The molecular orbitals of H_2O at its equilibrium bond angle of 104°.

Comment. If there were d-orbitals available (as in H_2S), the d_{z^2}- and $d_{x^2-y^2}$-orbitals would contribute to A_1 orbitals, d_{yz} would contribute to B_2, and d_{zx} and d_{xy} would be nonbonding.

Exercise 8.2. Construct SALCs from the H1s orbitals of NH_3.

The energies of the orbitals and the values of the coefficients are found by solving the secular equations in the normal way. However, there is the added complication that both the bond lengths and the bond angle are variable. These parameters are varied systematically, until the total energy of the molecule is a minimum, and that lowest energy arrangement of the atoms is accepted as the normal state of the molecule. Alternatively, if the geometry of the molecule is known, then a single calculation may be carried out for that arrangement of nuclei. For H_2O, for instance, the bond angle is 104°, and the molecular orbital energy level diagram is as shown in Fig. 8.24. As there are eight valence electrons to accommodate, the ground-state electron configuration of H_2O is expected to be

$$\mathrm{H_2O}\cdot \quad 1a_1^2 1b_2^2 2a_1^2 1b_1^2 \quad {}^1\mathrm{A}_1$$

The overall term symbol is calculated by multiplying together the characters of the occupied orbitals, and then identifying the overall symmetry species of the molecule from the character table. As all the orbitals are doubly occupied, and their characters are ± 1, the outcome is the set $(1, 1, 1, 1)$, which corresponds to A_1. All electrons are paired, so $S = 0$ and the multiplicity is 1.

The same technique may be applied to ammonia, NH_3, which belongs to the point group C_{3v}. Now the **minimum basis set**, the basis set employing only the valence orbitals, consists of N2s, N2p, and three H1s, giving seven members in all. Without adopting symmetry arguments, we would expect to have to solve a 7×7 secular determinant. With symmetry taken into account, we would expect the problem to be reduced to a series of bite-sized determinants. Intuitively, we should expect the N2s- and N2p_z-orbitals to belong to one symmetry species and N2p_x and N2p_y to belong to another. This separation can indeed be seen at a glance by looking at the C_{3v} character table in Appendix 1, because an s-orbital and a p_z-orbital on the central atom both span A_1 whereas (p_x, p_y) jointly span E. The symmetry species of the three H atoms were established in Example 5.8, and we know that the SALCs, which there were called s_1, s_2, and s_3, span $A_1 + E$. These points can be verified by reference to Fig. 8.25 or by reviewing the work that was done in Example 5.8.

The complete 7×7 secular determinant for NH_3 factorizes into a 3×3 determinant (for the A_1 orbitals) and a 4×4 determinant (for the E orbitals). The molecular orbitals are therefore of the form

$$\psi(\mathrm{A}_1) = c_1 s_1 + c_2\phi(\mathrm{N}2s) + c_3\phi(\mathrm{N}2p_z)$$

$$\psi(\mathrm{E}) = \begin{cases} c_1' s_2 + c_2'\phi(\mathrm{N}2p_x) \\ c_1'' s_3 + c_2''\phi(\mathrm{N}2p_y) \end{cases}$$

(The e-orbitals are distinguished by their reflection symmetry.) The solution of the secular determinant for the observed bond angle of 107° gives a set of

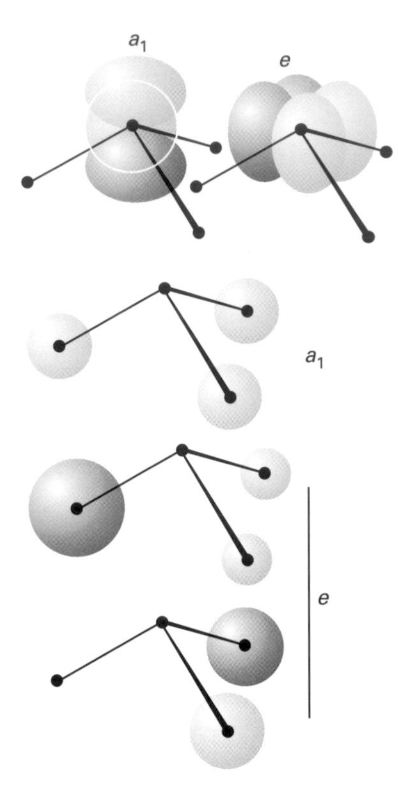

Fig. 8.25 The symmetry classification of the nitrogen atomic orbitals in NH_3, a C_{3v} molecule, and the three symmetry-adapted linear combinations of the H1s orbitals.

energy levels shown in Fig. 8.26. There are eight electrons to accommodate, and so the configuration of the ground state is expected to be

$$\mathrm{NH_3} \qquad 1a_1^2 1e^4 2a_1^2 \quad {}^1\mathrm{A}_1$$

The HOMO is the $2a_1$-orbital, which is largely a nonbonding orbital composed of N2s- and N2p_z-orbitals: the electrons that occupy it therefore constitute a lone pair on the N atom.

8.9 Conjugated π-systems

A special class of polyatomic molecules consists of those containing π-bonded atoms, particularly conjugated polyenes and arenes. They fall into a unique class because the orbitals with local σ and π symmetry can be discussed separately. By 'local' symmetry we mean symmetry with respect to one internuclear axis rather than the global symmetry of the molecule. For global symmetry we have to classify orbitals according to the overall point group of the molecule, and the σ, π designation is relevant only for linear species. However, if we focus on an individual A—B fragment of the molecule, then the orbitals do have a characteristic rotational symmetry about that axis, and they can be classified as locally σ or π.

One reason for the separate treatment of locally σ- and π-orbitals is that the electrons in π-orbitals are typically less strongly bound than those in σ-orbitals, so there is little interaction between the two types of orbital (recall the principles set out in Section 8.6). Another reason for the separation is that as π-orbitals are typically found in planar molecules, they have global symmetry properties that distinguish them from σ-orbitals, and therefore span different irreducible representations of the molecular point group. As a consequence, they factorize and can be discussed separately.

The simplest organic π-system is the ethene molecule, $CH_2{=}CH_2$. The σ-orbitals in ethene are molecular orbitals composed of various symmetry-adapted linear combinations of C2s, C2p_x, C2p_y, and H1s orbitals; the π-orbitals are formed by overlap between C2p_z orbitals (Fig. 8.27). This model immediately accounts for the torsional rigidity of the molecule, because (C2p_z, C2p_z)-overlap is greatest when the molecule is planar. The π-orbital energies are found by solving a 2×2 secular determinant, and the solutions given in eqn 8.17 may be employed as the carbon–carbon fragment is homonuclear.

When the π-system is conjugated, which means that the π-system extends over several neighbouring atoms, the simplest description of the bonding is in terms of the **Hückel approximation**. This drastic approximation makes the following assumptions in the formulation of the secular determinant det $|\mathbf{H} - E\mathbf{S}|$:

1. All overlap integrals are set equal to zero: $S_{ij} = \delta_{ij}$.

This is in fact a poor approximation, because actual overlap integrals are typically close to 0.2. Nevertheless, when the rule is relaxed, the energies are shifted in a simple way and their relative order is not greatly disturbed.

2. All diagonal matrix elements of the hamiltonian are ascribed the same value: $H_{ii} = \alpha$.

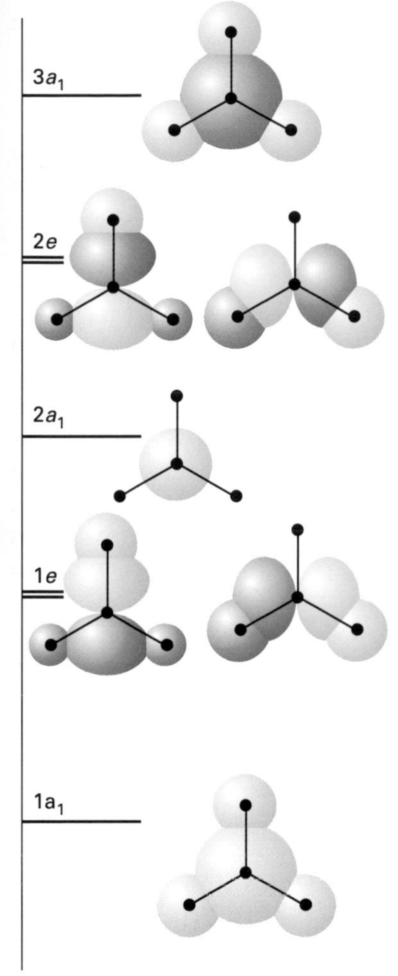

Fig. 8.26 The molecular orbitals of NH_3 at its equilibrium bond angle of 107°.

The parameter α is negative. This approximation is reasonable for species that do not contain heteroatoms because all the conjugated atoms are electronically similar. Some justification comes from the **Coulson–Rushbrooke theorem**, which states that in alternant hydrocarbons[5] the charge density on all the atoms is the same.

3. All off-diagonal elements of the hamiltonian are set equal to zero except for those between neighbouring atoms, all of which are set equal to β.

The parameter β is negative. It is the important parameter characteristic of Hückel theory, in so far as it governs the spacing of the molecular orbital energy levels.

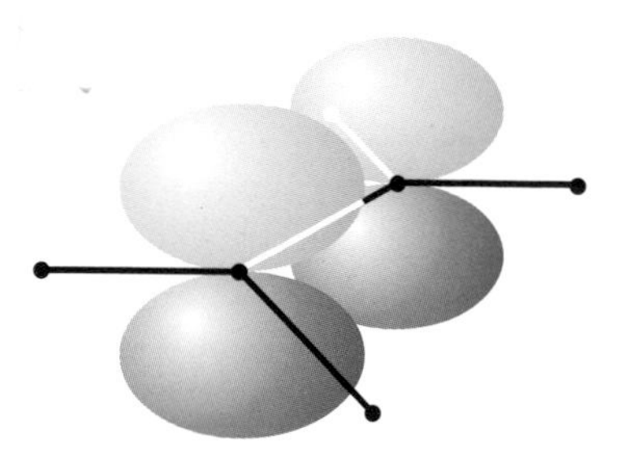

Figure 8.27 The structure of the π-orbital in ethene.

Example 8.3 The implementation of the Hückel approximation

Set up and solve the secular determinant for π-orbitals of the butadiene molecule in the Hückel approximation.

Method. Construct the secular determinant by setting all diagonal elements equal to $\alpha - E$ and off-diagonal elements between neighbouring atoms equal to β; all other elements are zero. Set the secular determinant equal to zero, and solve the resulting quartic equation in $x = \alpha - E$ for x and hence E.

Answer. The equation to solve is

$$\begin{vmatrix} \alpha - E & \beta & 0 & 0 \\ \beta & \alpha - E & \beta & 0 \\ 0 & \beta & \alpha - E & \beta \\ 0 & 0 & \beta & \alpha - E \end{vmatrix} = 0$$

On setting $x = \alpha - E$ and expanding the determinant, we obtain

$$x^4 - 3\beta^2 x^2 + \beta^4 = 0$$

This quartic in x is in fact a quadratic equation in $y = x^2$, so its roots can be found by elementary methods:

$$x = \pm\left(\frac{3 \pm \sqrt{5}}{2}\right)^{1/2}\beta \qquad x = \pm\tfrac{1}{2}(\sqrt{5} \pm 1)\beta$$

We conclude that the energy levels are

$$E = \alpha \pm 1.618\beta \qquad \alpha \pm 0.618\beta$$

as shown in Fig. 8.28.

Comment. The secular determinant for butadiene is an example of a so-called 'tridiagonal determinant', in which the nonzero elements all lie along three neighbouring diagonal lines. From the theory of determinants, an $N \times N$ tridiagonal determinant has the following roots:

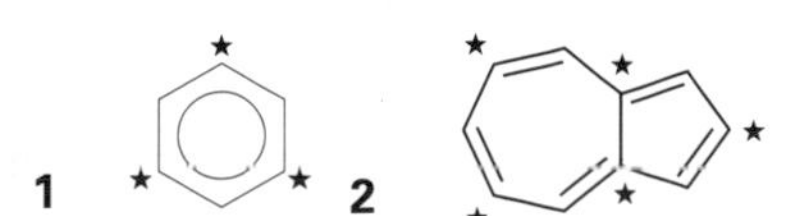

[5] An alternant hydrocarbon is one in which the atoms can be divided into two groups by putting a star on alternate atoms and not having any neighbouring stars when the numbering is complete. Benzene (**1**) is alternant, azulene (**2**) is nonalternant.

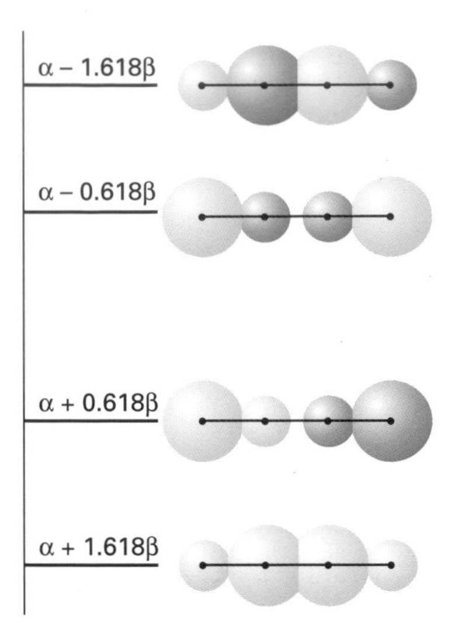

Fig. 8.28 The Hückel molecular orbitals and their energies in butadiene (as viewed down the axis of the p-orbitals).

$$E_k = \alpha + 2\beta\cos\left(\frac{k\pi}{N+1}\right) \qquad k = 1, 2, \ldots, N$$

Exercise 8.3. Find the roots of the secular determinant for the π-orbitals of cyclobutadiene. [$\alpha \pm 2\beta, \alpha \pm \beta$]

The worked example has shown how to calculate the molecular orbital energy levels in a simple case. The coefficients of the orbitals can be found by substituting these energies into the secular equations. However, in practice it is much easier to employ a computer: the roots we have found are the eigenvalues of the secular matrix and the corresponding eigenfunctions of the matrix are the coefficients of the atomic orbitals that contribute to each molecular orbital. For example, the four molecular orbitals of butadiene are found in this way to be

$$\begin{aligned}
1\pi &= 0.372\phi_A + 0.602\phi_B + 0.602\phi_C + 0.372\phi_D \\
2\pi &= -0.602\phi_A - 0.372\phi_B + 0.372\phi_C + 0.602\phi_D \\
3\pi &= -0.602\phi_A + 0.372\phi_B + 0.372\phi_C - 0.602\phi_D \\
4\pi &= 0.372\phi_A - 0.602\phi_B + 0.602\phi_C - 0.372\phi_D
\end{aligned}$$

where the ϕ_J is a $2p_z$-orbital on atom J. The composition of these molecular orbitals is independent of the values of α and β. You should notice that the energy of the orbital increases with the number of nodes, and that the amplitude of each coefficient follows a sine wave fitted to the length of the molecule (Fig. 8.29).

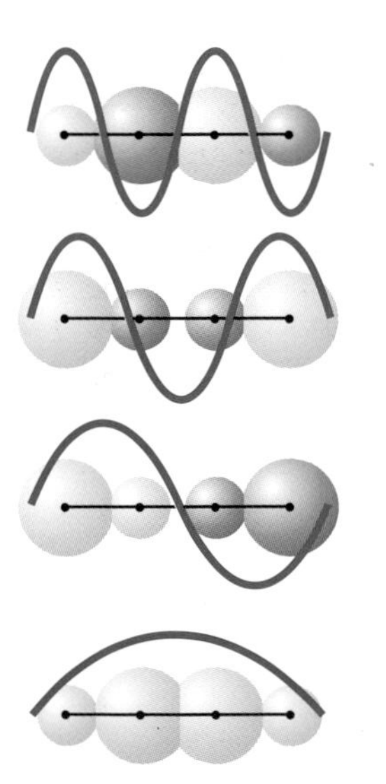

Fig. 8.29 The contributions of the p-orbitals to each π-orbital matches the amplitude of a sine wave (the wavefunction for a particle in a box) at the parent carbon atom.

The ground state configuration of the molecule is $1\pi^2 2\pi^2$, which corresponds to a total π-electron energy of $4\alpha + 2\sqrt{5}\beta$. The energy of a single unconjugated π-orbital is $\alpha + \beta$, and so if the molecule were described as having two unconjugated π-bonds, its total π-electron energy would have been $4\alpha + 4\beta$. The difference, which in this case is $2(\sqrt{5} - 2)\beta = 0.472\beta$, is called the **delocalization energy** of the molecule. The delocalization energy is independent of α within the Hückel approximation largely because all atoms are equivalent and the total electron density on them is the same regardless of the extent of delocalization of the orbitals.

The Hückel procedure leads to secular determinants of large dimension. However, they may often be factorized into more manageable dimensions by making use of the symmetry of the system beyond the simple mirror plane that enables the π-system to be distinguished from the σ-system. This additional factorization follows from the usual arguments about the hamiltonian having no nonzero elements between linear combinations of orbitals that belong to different symmetry species of the molecular point group. For benzene, for instance, the 6×6 determinant can be simplified very considerably by making use of the D_{6h} symmetry of the molecule. In fact, because every $2p_z$-orbital changes sign under reflection in the molecular plane, we lose no information by using the C_{6v} subgroup of the molecule. The procedure involves treating the C atoms as the peripheral atoms of a molecule, and setting up SALCs of their $2p_z$-orbitals; however as there is no 'central' atom, these SALCs are in this instance the actual π molecular orbitals of the molecule. The projection operator technique described in Section 5.12 leads to the fol-

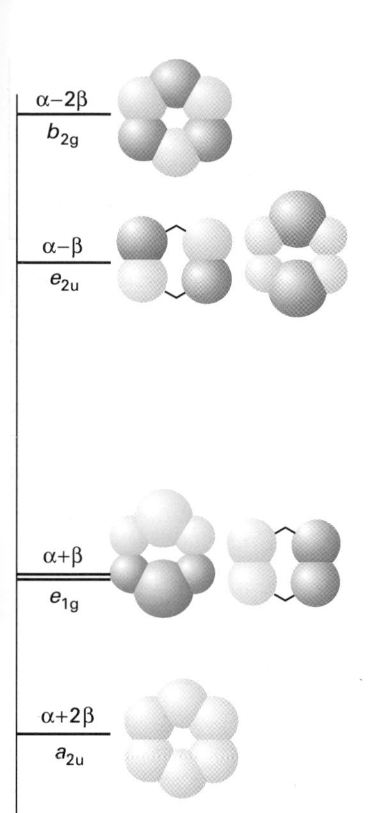

Fig. 8.30 The Hückel molecular orbitals and their energies in benzene.

lowing linear combinations (labelled according to the symmetry species of the group D_{6h}):

$$a_{2u} = \frac{1}{\sqrt{6}}(p_A + p_B + p_C + p_D + p_E + p_F)$$

$$e_{1g} = \begin{cases} (a) \ \dfrac{1}{\sqrt{12}}(2p_A + p_B - p_C - 2p_D - p_E + p_F) \\ (b) \ \frac{1}{2}(p_B + p_C - p_E - p_F) \end{cases}$$

$$e_{2u} = \begin{cases} (a) \ \dfrac{1}{\sqrt{12}}(2p_A - p_B - p_C + 2p_D - p_E - p_F) \\ (b) \ \frac{1}{2}(p_B - p_C + p_E - p_F) \end{cases}$$

$$b_{2g} = \frac{1}{\sqrt{6}}(p_A - p_B + p_C - p_D + p_E - p_F)$$

These orbitals are sketched in Fig. 8.30. Note that the form of the orbitals is determined solely by the symmetry of the molecule and makes no reference to the values of α or β. As we show in the following example, the energy levels are

$$E(a_{2u}) = \alpha + 2\beta \qquad E(e_{1g}) = \alpha + \beta \qquad E(e_{2u}) = \alpha - \beta \qquad E(b_{2g}) = \alpha - 2\beta$$

As we have already remarked, β is negative, so the orbitals lie in the order shown in the illustration.

Example 8.4 The energy levels of the benzene molecule

Determine the π-electron energy levels of the benzene molecule within the Hückel approximation.

Method. The molecular orbitals are specified above. We need form secular determinants for each orbital species separately as the hamiltonian has no off-diagonal elements between orbitals of different symmetry species. Use the Hückel rules for writing the matrix elements after expanding the H_{ij} in terms of the linear combinations of $2p_z$-orbitals. The orbitals that span one-dimensional irreducible representations will give simple 1×1 determinants, which are trivial to solve. The orbitals that span two-dimensional irreducible representations will give 2×2 determinants, which will lead to quadratic equations. However, because the e-orbitals of each set have different reflection symmetry, they too give diagonal determinants, so the roots can be found trivially.

Answer. The matrix elements we require are as follows:

$$\langle a_{2u}|H|a_{2u}\rangle = \tfrac{1}{6}\langle p_A + \ldots + p_F|H|p_A + \ldots + p_F\rangle = \alpha + 2\beta$$

$$\langle b_{2g}|H|b_{2g}\rangle = \tfrac{1}{6}\langle p_A - \ldots - p_F|H|p_A - \ldots - p_F\rangle = \alpha - 2\beta$$

$$\langle e_{1g}(a)|H|e_{1g}(a)\rangle = \alpha + \beta$$

$$\langle e_{1g}(b)|H|e_{1g}(b)\rangle = \alpha + \beta$$

$$\langle e_{2u}(a)|H|e_{2u}(a)\rangle = \alpha - \beta$$

$$\langle e_{2u}(b)|H|e_{2u}(b)\rangle = \alpha - \beta$$

$$\langle e(a)|H|e(b)\rangle = 0 \text{ for both types of } e \text{ orbital}$$

The resulting energies are those quoted in the text and displayed in Fig. 8.30.

Exercise 8.4. Use the C_{2v} subgroup of naphthalene to find the π-electron molecular orbital energy levels within the Hückel approximation.

The ground-state electron configuration of benzene is

$$C_6H_6 \qquad a_{2u}^2 e_{1g}^4 \quad {}^1A_{1g}$$

and the delocalization energy is

$$E_{\text{deloc}} = (6\alpha + 8\beta) - 6(\alpha + \beta) = 2\beta$$

You should notice that the six electrons just complete the molecular orbitals with net bonding effect, leaving unfilled the orbitals with net antibonding character. To some extent this configuration echoes the configuration of N_2, and both molecules have a pronounced chemical inactivity. Another feature of the energy levels of benzene is that the array of levels is symmetrical: to every bonding level there corresponds an antibonding level. This symmetry is a characteristic feature of alternant hydrocarbons and can be traced to the topological character of the molecules. Indeed, many of the results of Hückel theory can be established on the basis of **graph theory**, the branch of topology concerned with the properties of networks. One particular result of this kind of analysis is the justification of the '$(4n + 2)$-rule' for the anticipation of aromatic character, where n is the number of π-electrons.

As we have stressed, Hückel theory, which virtually hijacks the disagreeable integrals that appear in a full treatment, is only the most primitive stage of discussing π-electron molecules. The modern, far more reliable numerical approaches are described in Chapter 9.

8.10 Ligand field theory

The success of Hückel theory is rooted in the fact that the orbitals themselves are determined by the symmetry of the system. These symmetry-determined orbitals are then put into an order of energies, essentially by counting the number and noting the importance of their nodes. The energy differences between the orbitals are typically so large that the coarseness of this procedure does not unduly misrepresent their order. A similar situation occurs in the complexes of d-metal ions. These complexes consist of a central metal ion surrounded by a three-dimensional array of ligands. The compositions of the orbitals of the complex are largely determined by the symmetry of the environment, and a single parameter can be used to give a rough indication of the order of the energies of the molecular orbitals of the complex. **Ligand field theory** is a kind of three-dimensional version of Hückel theory, in which symmetry plays a central role, and in which structural, spectroscopic, magnetic, and thermodynamic properties are parametrized in terms of the **ligand field splitting parameter**, Δ.

We shall denote the central metal ion by M and assume that it has the configuration d^n. The ligands will be denoted L, and we shall confine attention to ML_6 octahedral complexes with O_h symmetry. The orbitals of the ligands will be denoted λ. In particular, we shall suppose that each ligand i supplies an orbital $\lambda_i^{(\sigma)}$ that has local σ symmetry with respect to the M—L bond. Thus, in ligand field theory, each Lewis-base ligand is simulated by a single orbital that supplies two electrons. Later we shall allow for the possibility that the ligands can supply electrons from or accept electrons into their π-orbitals, and will denote the latter by $\lambda_i^{(\pi)}$.

The first step in ligand field theory is to set up symmetry-adapted linear combinations and to identify the symmetry species of the d-orbitals on the central metal ion. A glance at the O_h character table in Appendix 1 shows that in an octahedral environment, two of the d-orbitals (namely d_{z^2} and $d_{x^2-y^2}$) span E_g and the remaining three (d_{xy}, d_{yz}, and d_{zx}) transform as T_{2g}. The standard techniques of group theory show that the ligand σ-orbitals span $A_{1g} + E_g + T_{1u}$ in O_h, and projection operator techniques give the following explicit forms of the corresponding SALCs:

$$\psi(A_{1g}) = \frac{1}{\sqrt{6}}\left(\lambda_1^{(\sigma)} + \lambda_2^{(\sigma)} + \lambda_3^{(\sigma)} + \lambda_4^{(\sigma)} + \lambda_5^{(\sigma)} + \lambda_6^{(\sigma)}\right)$$

$$\psi(E_g) = \begin{cases} (a)\ \frac{1}{\sqrt{12}}\left(2\lambda_5^{(\sigma)} + 2\lambda_6^{(\sigma)} - \lambda_1^{(\sigma)} - \lambda_2^{(\sigma)} - \lambda_3^{(\sigma)} - \lambda_4^{(\sigma)}\right) \\ (b)\ \frac{1}{2}\left(\lambda_1^{(\sigma)} + \lambda_2^{(\sigma)} - \lambda_3^{(\sigma)} - \lambda_4^{(\sigma)}\right) \end{cases}$$

$$\psi(T_{1u}) = \begin{cases} (a)\ \frac{1}{\sqrt{2}}\left(\lambda_1^{(\sigma)} - \lambda_2^{(\sigma)}\right) \\ (b)\ \frac{1}{\sqrt{2}}\left(\lambda_3^{(\sigma)} - \lambda_4^{(\sigma)}\right) \\ (c)\ \frac{1}{\sqrt{2}}\left(\lambda_5^{(\sigma)} - \lambda_6^{(\sigma)}\right) \end{cases}$$

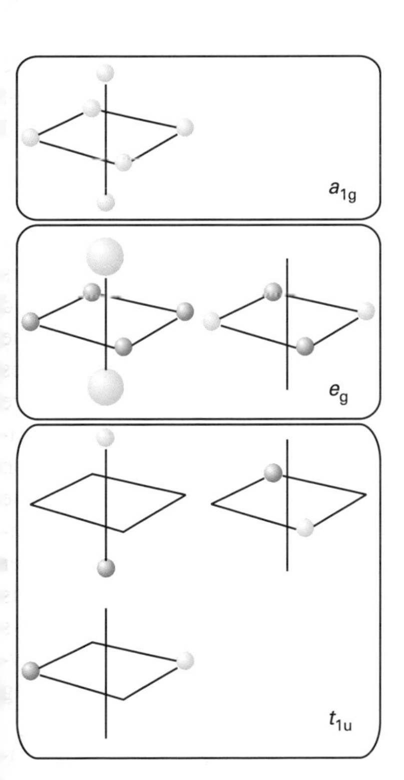

Fig. 8.31 A depiction of the symmetry-adapted linear combination of ligand atomic orbitals in an octahedral complex.

These SALCs are illustrated in Fig. 8.31. Note that there is no T_{2g} combination. These combinations differ in energy slightly when we take into account overlap between ligand orbitals (as distinct from the (M,L) overlap that is predominantly responsible for bonding).

Now we form molecular orbitals as linear combinations of the SALCs and the d-orbitals of the same symmetry species, for only these combinations have nonzero net overlap:

$$a_{1g} = \psi(A_{1g})$$

$$e_g = \begin{cases} (a)\ c_1\phi(d_{z^2}) + c_2\psi(E_g, a) \\ (b)\ c_1'\phi(d_{x^2-y^2}) + c_2'\psi(E_g, b) \end{cases}$$

$$t_{1u} = \psi(T_{1u})$$

$$t_{2g} = \begin{cases} (a)\ d_{xy} \\ (b)\ d_{yz} \\ (c)\ d_{zx} \end{cases}$$

We have not included overlap with s- and p-orbitals: they transform as A_{1g} and T_{1u}, respectively, and so would combine with the SALCs of those symmetry species. Within the d-orbital-only approximation, we see that we can expect an array of energy levels like that shown in Fig. 8.32. The bonding e_g combination

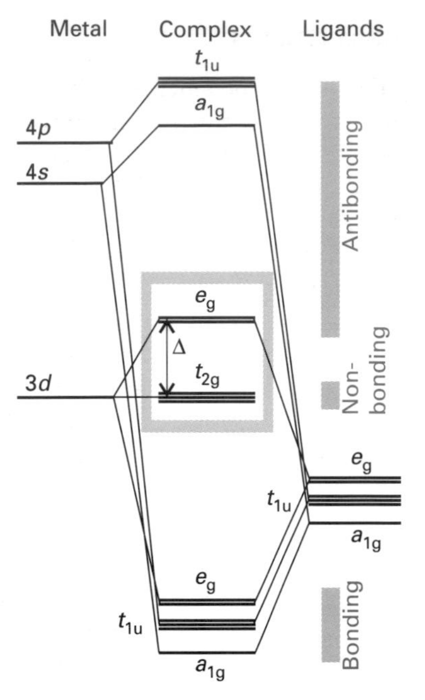

Fig. 8.32 The molecular orbital energy level diagram for an octahedral complex. For accounting purposes, the ligand electrons occupy all the bonding orbitals; the electrons supplied by the metal atom occupy the orbitals in the box.

is largely confined to the ligands (the lower energy orbitals) and the antibonding combination is largely confined to the metal ion. The a_{1g} and t_{1u} combinations are nonbonding, and confined entirely to the ligands. The t_{2g} orbitals are nonbonding atomic orbitals on the metal ion.

There are 12 electrons to accommodate that are supplied by the ligands (two from each Lewis base), and n electrons supplied by the metal ion. Of these $12 + n$ electrons, 12 fill the two bonding e_g and four nonbonding a_{1g} and t_{1u} orbitals: these electrons are largely confined to the ligands. Up to six of the remaining n electrons are free to occupy the three t_{2g} orbitals on the metal ion and the remainder will occupy the antibonding e_g combination, which is largely confined to the metal ion too.

However, at this point there is a complication. The ground-state electron configuration of the complex is the configuration that corresponds to the lowest *total* energy. When the separation between t_{2g} and the antibonding e_g orbitals is small, it may be advantageous to occupy the latter orbital before completely filling the former, because then the electrons occupy spatially distinct regions and may do so with parallel spins and so benefit from spin correlation. The crucial quantity is the ligand field splitting parameter, Δ, the energy separation between e_g and t_{2g}. If this splitting is large, then a d^4 complex, for instance, will adopt the configuration t_{2g}^4 with one orbital doubly occupied. However, if the splitting is small, then it may be energetically advantageous for the complex to adopt the configuration $t_{2g}^3e_g^1$ with all four electrons in separate orbitals with parallel spins.

It follows from this discussion that we should distinguish between the following two cases:

1. The **strong-field case**, in which the ligand field splitting parameter is large and it is energetically favourable to occupy the t_{2g} orbitals first.

2. The **weak-field case**, in which the ligand field splitting parameter is small and it is energetically favourable to occupy the e_g orbitals before the t_{2g} orbitals are completed.

The second category is sometimes further divided into 'weak field' itself and **very weak field**, according to whether the ligand field splitting parameter is stronger or weaker, respectively, than the spin–orbit interaction. The very weak field case is applicable to the f-block elements in which the f-electrons are embedded deeply in the atom and experience the surrounding ligands only very weakly. We shall not consider it further.

The ambiguity in ground-state configuration is found for d^4, d^5, d^6, and d^7 complexes. When the ligand field is so large that a t_{2g}^4, t_{2g}^5, t_{2g}^6, $t_{2g}^6e_g^1$ configuration is adopted, the spins need to pair. As a result, such complexes are classified as **low-spin complexes**. As may be verified from Fig. 8.33, they have 2, 1, 0, and 1 unpaired spins, respectively. When the ligand field is weak, the complexes can be expected to be $t_{2g}^3e_g^1$, $t_{2g}^3e_g^2$, $t_{2g}^4e_g^2$, and $t_{2g}^5e_g^2$, with 4, 5, 4, and 3 unpaired spins. Such complexes are classified as **high-spin complexes**. Because the number of unpaired electrons is responsible for the magnetic properties of complexes, we see that modification of the ligands and consequently the size of Δ may influence the magnetic properties of the species.

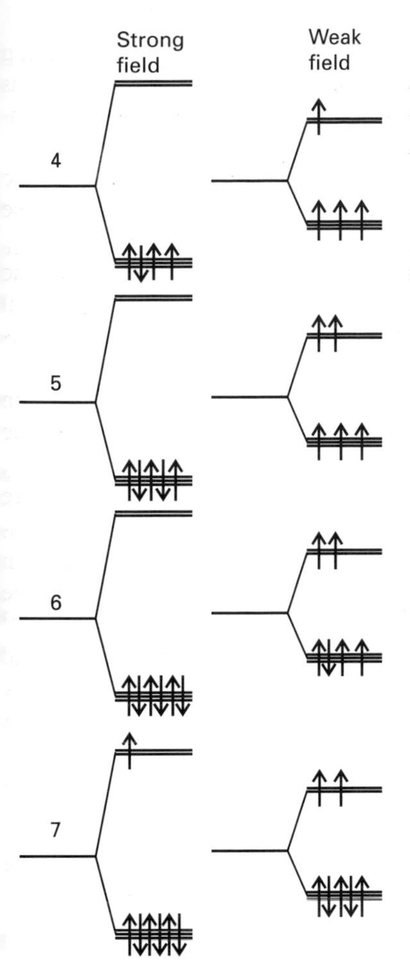

Fig. 8.33 The high- and low-spin arrangements that arise from weak and strong ligand fields for complexes with four to seven d-electrons.

8.11 Further aspects of ligand field theory

There are three aspects of ligand field theory that need to be touched on here. In the first place, we need to be aware that the weak-field case can be quite tricky to handle because the $t_{2g}^{n}e_{g}^{n'}$ configurations are so strongly perturbed by electron–electron interactions. We shall illustrate the difference between high-field and low-field cases by considering a d^2 configuration.

In a free ion, a d^2 configuration can give rise to the terms ^{1}G, ^{3}F, ^{1}D, ^{3}P, and ^{1}S. From Hund's rules, we can expect the ^{3}F term to lie lowest in energy, with perhaps the ^{3}P next above it. When the ligand field is weak, we can think of the formation of molecular orbitals as a small perturbation on the free-ion levels. To determine the effect of this perturbation, we consider the effect of the reduction in symmetry from R_3 to O_h and identify the symmetry species that ^{3}F and ^{3}P become in the octahedral environment. The technique required was described in Section 5.19 and illustrated in Example 5.13:

$$^3\mathrm{P} \rightarrow {}^3\mathrm{T}_{1g} \qquad {}^3\mathrm{F} \rightarrow {}^3\mathrm{A}_{2g} + {}^3\mathrm{T}_{1g} + {}^3\mathrm{T}_{2g}$$

The separation of the terms stemming from ^{3}F increases as the perturbation becomes stronger (Fig. 8.34). In the strong-field case we can discuss the configurations in terms of occupation of the t_{2g} and e_g orbitals, and we can have

$$t_{2g}^2\ {}^3\mathrm{T}_{1g} \qquad t_{2g}^1e_g^1\ {}^3\mathrm{T}_{1g} + {}^3\mathrm{T}_{2g} \qquad e_g^2\ {}^3\mathrm{A}_{2g}$$

(we have retained only the triplet terms). The order of energies can be anticipated by referring to Fig. 8.32, and they are shown on the right of Fig. 8.34.

At this point, we can construct the correlation diagram by connecting states of the same symmetry but allowing for the noncrossing rule. The diagram in Fig. 8.34 is in fact a part of a **Tanabe–Sugano diagram** for the correlation of strong and weak field states of a complex. The actual state of a complex corresponds to an intermediate stage of the diagram, and the location can be determined by fitting the observed spectroscopic transitions to the energy levels. In practice, a Tanabe–Sugano diagram is expressed in terms of quantities that parametrize the strengths of the electron–electron repulsion and the ligand field splitting parameter, so these quantities can be determined. More information on the procedures will be found in the books referred to in the *Further reading* section for this chapter.

The second feature we need to mention concerns deviations from octahedral symmetry that arise spontaneously in certain complexes. Their occurrence is summarized by the **Jahn–Teller theorem**:

> In any nonlinear system, there exists a vibrational mode that removes the degeneracy of an orbitally degenerate state.

The theorem can be illustrated by considering a d^9 octahedral complex. The ground-state configuration is expected to be $t_{2g}^6e_g^3$; it is orbitally degenerate because the 'hole' in the d^{10} configuration can occupy either d_{z^2} or $d_{x^2-y^2}$. If the complex were to distort so that it lengthened along a C_4 axis, then the degeneracy of the antibonding e_g orbital would be removed by the change in overlap. We would expect the antibonding character of the orbital formed by overlap with the d_{z^2}-orbital to be reduced as the M—L length increases, so this

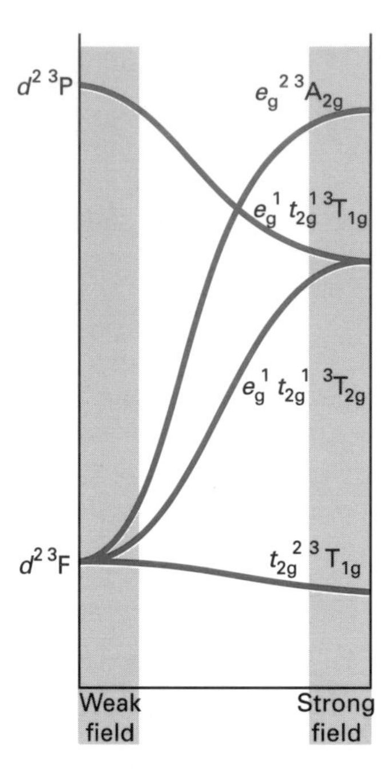

Fig. 8.34 A Tanabe–Sugano type of correlation diagram for the states of an octahedral two-electron complex.

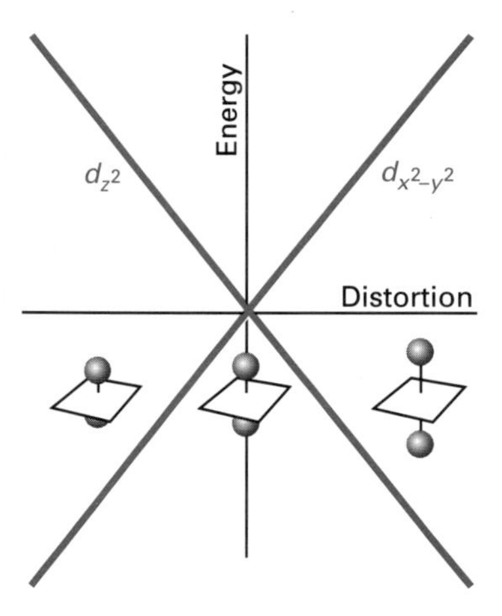

Fig. 8.35 The effect of the distortions envisaged in the Jahn–Teller effect.

molecular orbital will become lower in energy (Fig. 8.35). Alternatively, if the M—L length were to shorten, the same e_g orbital would become higher in energy. In either case, the complex will remain in the distorted shape, because it then has a lower energy than in the undistorted, regular octahedral shape.

A final detail concerns the role of π-bonding between the metal ion and the ligands. We shall suppose that on each ligand there are two orbitals with local π symmetry with respect to the M—L axis. They span $T_{1u} + T_{2u} + T_{1g} + T_{2g}$ (Fig. 8.36), and only the last can have net overlap with the t_{2g} orbitals of the ion. The explicit structures of the SALCs of these orbitals are as follows:

$$t_{1u} = \begin{cases} \frac{1}{2}(p_{3x} + p_{4x} + p_{5x} + p_{6x}) \\ \frac{1}{2}(p_{1y} + p_{2y} + p_{5y} + p_{6y}) \\ \frac{1}{2}(p_{1z} + p_{2z} + p_{3z} + p_{4z}) \end{cases}$$

$$t_{2g} = \begin{cases} \frac{1}{2}(p_{5x} - p_{6x} + p_{1z} - p_{2z}) \\ \frac{1}{2}(p_{5y} - p_{6y} - p_{3z} + p_{4z}) \\ \frac{1}{2}(p_{1y} - p_{3x} - p_{2y} + p_{4x}) \end{cases}$$

$$t_{1g} = \begin{cases} \frac{1}{2}(p_{6x} - p_{5x} + p_{1z} - p_{2z}) \\ \frac{1}{2}(p_{5y} - p_{6y} + p_{3z} - p_{4z}) \\ \frac{1}{2}(p_{1y} - p_{2y} + p_{3x} - p_{4x}) \end{cases}$$

$$t_{2u} = \begin{cases} \frac{1}{2}(p_{5y} + p_{6y} - p_{1y} - p_{2y}) \\ \frac{1}{2}(p_{1z} + p_{2z} - p_{3z} - p_{4z}) \\ \frac{1}{2}(p_{5x} + p_{6x} - p_{3x} - p_{4x}) \end{cases}$$

Two cases may be distinguished, and are illustrated in Fig. 8.37. In the first, the ligand π-orbitals are full (the ligands act as π-donors). In this case, the ligand field splitting parameter is reduced by the formation of (M,L) π-orbitals because the original nonbonding t_{2g} orbitals become slightly antibonding. In the second case, in which the π-orbitals of the ligands are initially empty (so the ligands act as π-acceptors), the t_{2g} orbitals are made slightly bonding, with the result that the ligand field splitting parameter is increased. The correlation of the value of Δ with the identity of the ligand and the metal ion depends critically on the ability of the species to form π-orbitals. Moreover, the stability of complexes such as those formed by CO (a π-acceptor ligand, recall the discussion in Section 8.7) can also be traced to the involvement of π-orbitals.

The band theory of solids

The electronic structures of solids can be regarded as an extension of molecular orbital theory to aggregates consisting of virtually infinite numbers of atoms. However, there are certain features that are unique to solids, particularly the formation of continuous bands of energy levels instead of discrete levels, and the role of the translational symmetry of the lattice. There are in fact two starting points for the discussion of solids. One is the particle-in-a-box wavefunctions described in Chapter 2. The other is the discussion of conjugated molecules presented earlier in this chapter. We shall give a brief introduction to both and see how one mirrors the other. We shall confine our attention to one-dimensional solids because they are so much simpler to treat.

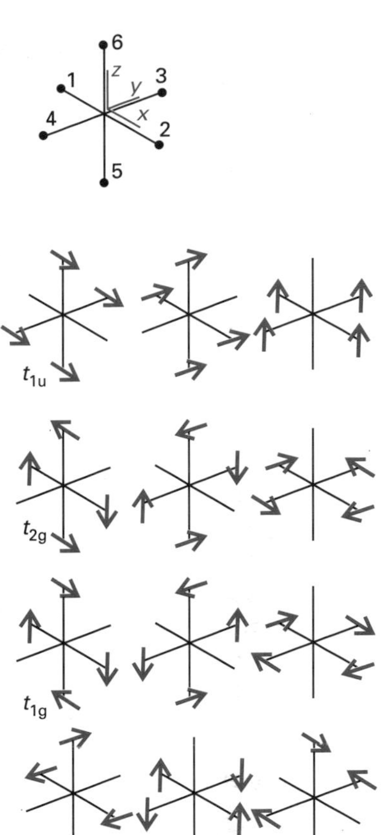

Fig. 8.36 A representation of the symmetry-adapted linear combinations of ligand π-orbitals. Each arrow can be regarded as indicating a p-orbital, the head indicating the positive lobe.

However, such solids do not show all the properties of a true solid, and this material must be regarded as no more than introductory.

8.12 The tight-binding approximation

The **tight-binding approximation** treats a solid as an extended molecule, and takes as its starting point orbitals that are confined to individual atoms (hence the name of the approach). Then molecular orbitals are formed that spread throughout the solid.

The simplest approach of all is to adopt the Hückel approximation and to consider a line of N atoms, each of which has one valence s-orbital that can overlap only its two immediate neighbours (Fig. 8.38). The wavefunctions will be

$$\psi = \sum_i c_i \phi_i \tag{34}$$

where the index i runs over all the atoms in the line. The $N \times N$ secular determinant has the form

$$\begin{vmatrix} \alpha - E & \beta & 0 & 0 & \cdots \\ \beta & \alpha - E & \beta & 0 & \cdots \\ 0 & \beta & \alpha - E & \beta & \cdots \\ \vdots & \vdots & \vdots & \vdots & \cdots \end{vmatrix} = 0$$

This determinant is tridiagonal (see Example 8.3), so we can write down the roots immediately:

$$E_k = \alpha + 2\beta \cos\left(\frac{k\pi}{N+1}\right) \qquad k = 1, 2, \ldots, N \tag{35}$$

Noticc that as $N \to \infty$ the energy separation between neighbouring levels approaches zero but the width of the band remains finite (Fig. 8.39):

$$\lim_{N \to \infty} E_N - E_1 = 4\beta \tag{36}$$

The lowest energy corresponds to a fully bonding linear combination of atomic orbitals and the highest energy corresponds to a molecular orbital that has a node between each neighbour. The molecular orbitals of intermediate energy have $k - 1$ nodes distributed along the chain of atoms.

Example 8.5 The density of states of a linear solid

Inspection of the diagram in Fig. 8.39 indicates that the density of states, $\rho(E)$, the number of states in an energy range divided by the width of the range, increases towards the edges of the bands. Confirm this conclusion analytically.

Method. First, we need to define the density of states analytically. When the states are packed together so closely that to a good approximation they form a continuum, we can write

$$\mathrm{d}k = \rho(E)\mathrm{d}E$$

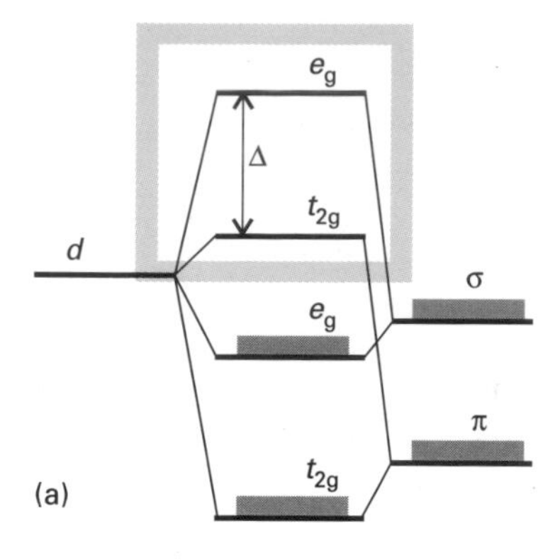

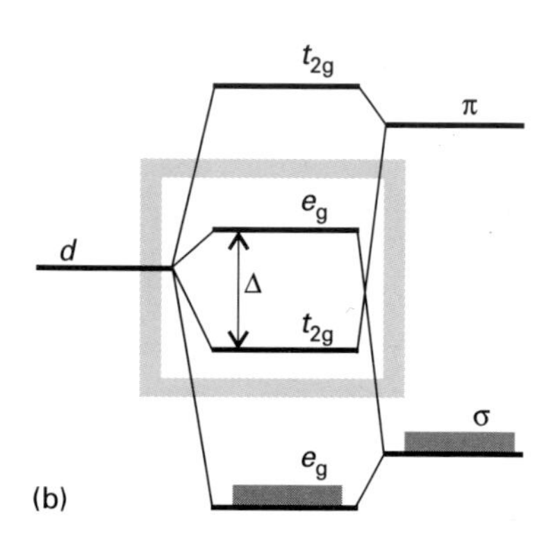

Fig. 8.37 The effect of π-bonding on the ligand field splitting in an octahedral complex: (a) occupied ligand π-orbitals (a π-donor ligand) and (b) unoccupied ligand π-orbitals (a π-acceptor ligand).

because an increase in energy by $\mathrm{d}E$ results in an increase in the quantum number k by $\mathrm{d}k$. Therefore, calculate $\rho(E) = \mathrm{d}k/\mathrm{d}E$. It is simpler to express the energy as $\varepsilon = E - \alpha$.

Answer. We write the energy expression as

$$\varepsilon = 2\beta \cos\left(\frac{k\pi}{N+1}\right)$$

from which it follows that

$$k = \left(\frac{N+1}{\pi}\right) \arccos\left(\frac{\varepsilon}{2\beta}\right)$$

The density of states is therefore

$$\rho(\varepsilon) = \frac{\mathrm{d}k}{\mathrm{d}\varepsilon} = -\frac{(N+1)/2\pi\beta}{\{1 - (\varepsilon/2\beta)^2\}^{1/2}}$$

This function (noting that $\beta < 0$) becomes infinite at the edges of the band, where $\varepsilon = \pm 2\beta$ (Fig. 8.40).

Comment. The increase of the density of states towards the edge of the band is uncharacteristic of higher-dimensional solids. In them, the density is highest towards the centre of the band, and is least at the edges. This difference arises from the possibility of degeneracies in dimensions greater than 1.

Exercise 8.5. Derive an expression for the mean energy of a band that is half full.

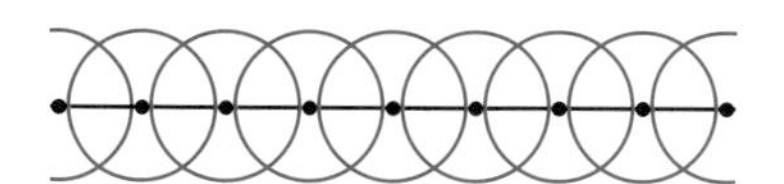

Fig. 8.38 The string of s-orbitals used to discuss the formation of bands in a one-dimensional solid.

The band of orbitals we have constructed is called an **s-band** because it is formed by the linear combination of s-orbitals. If the atoms have valence p-orbitals too, then a similar superposition can take place, with the formation of a **p-band**. In a typical solid, the energy separation of the s- and p-orbitals of the free atoms will be quite large, and as a result the two bands will not overlap. The orbital structure of the solid will therefore consist of two (or more) bands separated by a **band gap**, a region of energy to which no orbitals belong.

If each atom provides one electron, the s-band will be half full. The band is then known as a **conduction band** because the electrons in the highest filled orbitals can travel through the solid in response to the application of electric fields. If, however, each atom provides two electrons, then the s-band will be full. It is then called a **valence band**. Its uppermost electrons are separated by a substantial energy gap from the p-band, and so they are not mobile. Such a material is classified as a **semiconductor**. If the band gap is large compared with kT so that the conductivity is very low, then a material is termed an **insulator**. Semiconductors are characterized by an increase in their electrical conductivity with temperature, because electrons are excited thermally across the band gap, and hence become sufficiently mobile to act as carriers. The artificial manipulation of the properties of conduction and valence bands by the insertion of foreign atoms is the basis of the semiconductor industry, and further information can be found in the references given in *Further reading*.

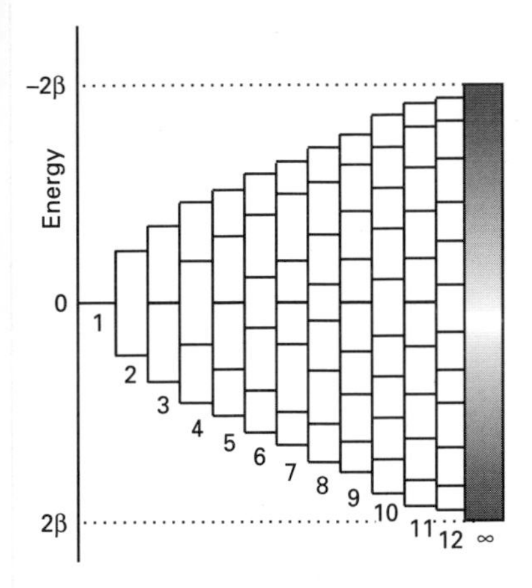

Fig. 8.39 The formation of molecular orbitals from a chain of N atomic orbitals. Note that the separation of the most bonding and most antibonding orbitals remains finite and that the density of orbitals is greatest at the edges of the band.

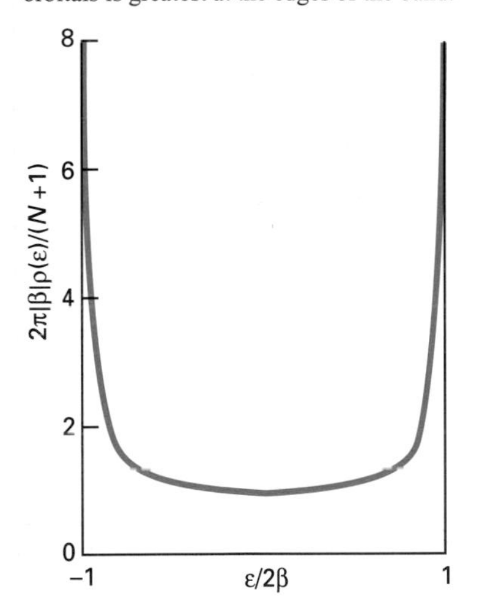

Fig. 8.40 The density of states in a band formed from an infinite chain of atomic orbitals.

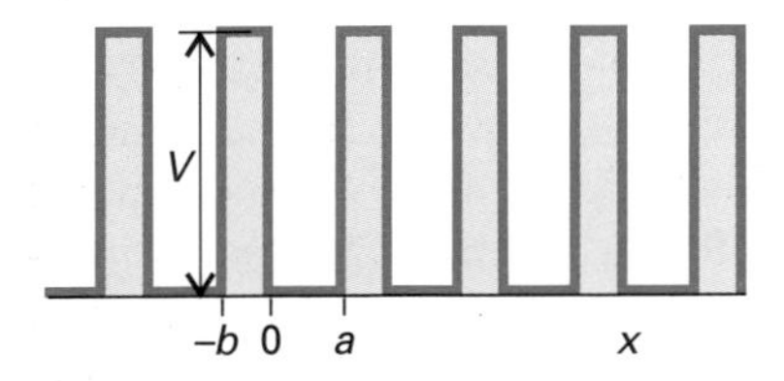

Fig. 8.41 The potential energy of an electron in the Kronig–Penney model.

8.13 The Kronig–Penney model

We now turn to a seemingly entirely different attack on the same problem. This approach will echo the material in Chapter 2, and—most surprisingly—results in a description of solids which, despite the entirely different starting point, mirrors the molecular-orbital approach. The origin of that similarity of conclusion is another manifestation of the power of symmetry, in this case translational symmetry, in determining the general structure of energy levels regardless of the details of physical interactions.

If we were to disregard the variation in the potential energy of an electron as it travels through the lattice, the solutions of the Schrödinger equation would be those of a free particle, and we would write

$$\psi = \mathrm{e}^{\pm ikx} \qquad E_k = \frac{k^2\hbar^2}{2m_\mathrm{e}} \tag{37}$$

Notice how the energy varies quadratically with the wavevector k. In an actual solid, the potential energy varies periodically, and the Schrödinger equation is

$$-\frac{\hbar^2}{2m_\mathrm{e}}\frac{\mathrm{d}^2\psi}{\mathrm{d}x^2} + V(x)\psi = E\psi \tag{38}$$

with $V(x+a) = V(x)$, where a is the spacing of the lattice points. According to the **Bloch theorem**, the solutions of the Schrödinger equation for a periodic potential of this kind have the form

$$\psi = u_k(x)\mathrm{e}^{ikx} \qquad u_k(x+a) = u_k(x) \tag{39}$$

The periodic functions $u_k(x)$ are called **Bloch functions**. Substitution of the function $u_k(x)\mathrm{e}^{ikx}$ into the Schrödinger equation leads to the following equation for the Bloch function:

$$u_k'' + 2iku_k' - \left\{\frac{2m_\mathrm{e}}{\hbar^2}\left(V(x) - E\right) + k^2\right\}u_k = 0 \tag{40}$$

where $u' = \mathrm{d}u/\mathrm{d}x$ and $u'' = \mathrm{d}^2u/\mathrm{d}x^2$. We shall establish the form of the Bloch functions in the particular case of a periodic potential energy like that shown in Fig. 8.41, which is called the **Kronig–Penney model**. It is plainly a great simplification of the true potential energy, but it establishes certain important features that are found in practice.

There are two types of region, one in which the potential is zero and the other in which it has the constant value V. We shall consider solutions for which $E < V$ and shortly simplify the problem still further by letting $V \to \infty$ and $b \to 0$ in such a way that Vb (the area of the rectangular region of nonzero potential energy) remains constant. It will be convenient to introduce the two real parameters

$$\alpha^2 = \frac{2m_\mathrm{e}E}{\hbar^2} \qquad \beta^2 = \frac{2m_\mathrm{e}}{\hbar^2}(V - E) \tag{41}$$

and then to write the equations for the two regions as

$$(a)\ V = 0: \qquad u_k'' + 2iku_k' + (\alpha^2 - k^2)u_k = 0$$
$$(b)\ V \neq 0: \qquad u_k'' + 2iku_k' - (\beta^2 + k^2)u_k = 0$$

subject to the requirement that the wavefunctions and their first derivatives are continuous at the interfaces between the regions.

The solutions of the two differential equations have the form

$$(a)\ u_k = A\mathrm{e}^{\mathrm{i}(\alpha-k)x} + B\mathrm{e}^{-\mathrm{i}(\alpha+k)x}$$
$$(b)\ u_k = C\mathrm{e}^{(\beta-\mathrm{i}k)x} + D\mathrm{e}^{-(\beta+\mathrm{i}k)x}$$

The conditions of continuity of u and u' at the two boundaries of each zone, namely $u_k(a) = u_k(-b)$ and $u'_k(a) = u'_k(-b)$, lead to the following four equations:

$$A + B - C - D = 0$$
$$A\mathrm{e}^{\mathrm{i}(\alpha-k)a} + B\mathrm{e}^{-\mathrm{i}(\alpha+k)a} - C\mathrm{e}^{-(\beta-\mathrm{i}k)b} - D\mathrm{e}^{(\beta+\mathrm{i}k)b} = 0$$
$$\mathrm{i}(\alpha-k)A - \mathrm{i}(\alpha+k)B - (\beta-\mathrm{i}k)C + (\beta+\mathrm{i}k)D = 0$$
$$\mathrm{i}(\alpha-k)A\mathrm{e}^{\mathrm{i}(\alpha-k)a} - \mathrm{i}(\alpha+k)B\mathrm{e}^{-\mathrm{i}(\alpha+k)a} - (\beta-\mathrm{i}k)C\mathrm{e}^{-(\beta-\mathrm{i}k)b} + (\beta+\mathrm{i}k)D\mathrm{e}^{(\beta+\mathrm{i}k)b} = 0$$

For these four simultaneous equations to have a solution, the determinant of the coefficients must be zero, and the condition

$$\begin{vmatrix} 1 & 1 & -1 & -1 \\ \mathrm{e}^{\mathrm{i}(\alpha-k)a} & \mathrm{e}^{-\mathrm{i}(\alpha+k)a} & -\mathrm{e}^{-(\beta-\mathrm{i}k)b} & -\mathrm{e}^{(\beta+\mathrm{i}k)b} \\ \mathrm{i}(\alpha-k) & -\mathrm{i}(\alpha+k) & -(\beta-\mathrm{i}k) & (\beta+\mathrm{i}k) \\ \mathrm{i}(\alpha-k)\mathrm{e}^{\mathrm{i}(\alpha-k)a} & -\mathrm{i}(\alpha+k)\mathrm{e}^{-\mathrm{i}(\alpha+k)a} & -(\beta-\mathrm{i}k)\mathrm{e}^{-(\beta-\mathrm{i}k)b} & (\beta+\mathrm{i}k)\mathrm{e}^{(\beta+\mathrm{i}k)b} \end{vmatrix} = 0 \tag{42}$$

This rather horrendous determinant reduces (as can best be shown by use of a symbolic algebra program, but patience and a pencil also work) to the condition

$$\left(\frac{\beta^2-\alpha^2}{2\alpha\beta}\right)\sinh\beta b\sin\alpha a + \cosh\beta b\cos\alpha a = \cos k(a+b) \tag{43}$$

We promised to use the simplifying condition of $V \to \infty$, $b \to 0$, Vb = constant. Equation 8.43 then simplifies to

$$\gamma\frac{\sin\alpha a}{\alpha a} + \cos\alpha a = \cos ka \qquad \gamma = \frac{m_\mathrm{e}Vba}{\alpha\hbar^2} \tag{44}$$

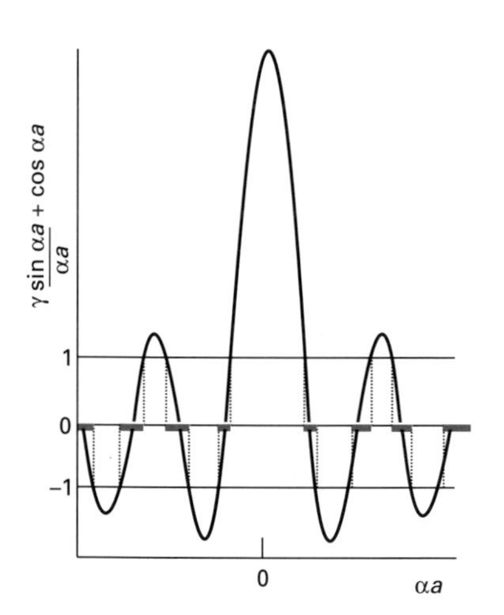

Fig. 8.42 The solution of the equation for the energy levels of the Kronig–Penney model. The only permitted solutions are those that correspond to the regions in which the curve lies between $+1$ and -1.

This is still a transcendental equation, but we can understand its implications by plotting the left-hand side against αa. The left-hand side depends on the value of γ, which is a measure of the height and width of the barrier between neighbouring wells, and one such graph is shown in Fig. 8.42, where we have used $\gamma = \frac{3}{2}\pi$.

The essential point can now be made clear. Because the right-hand side of eqn 8.44 lies between -1 and $+1$, only certain values of αa (that is, only certain values of E, because $\alpha \propto \sqrt{E}$) give rise to solutions. It follows that the solutions of the Schrödinger equation for a periodic potential correspond to a series of allowed bands separated by gaps in which there are no solutions. Moreover, as can be seen from Fig. 8.42 and Fig. 8.43, the widths of the allowed bands increase with increasing energy. A final important point can be seen by comparing the diagrams in Fig. 8.43, which show the effect of changing γ, the depth

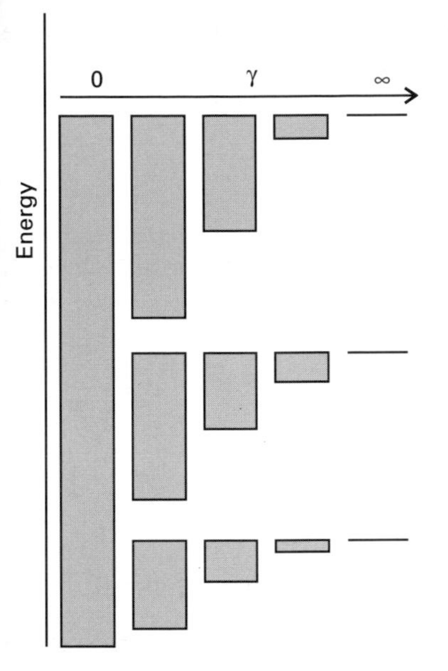

Fig. 8.43 The formation of bands as a function of the parameter γ in the Kronig–Penney model. The limit $\gamma = 0$ corresponds to the absence of barriers, and there is no discrete band structure (the levels are those of a free particle, so there is one infinitely wide band). The other limit, $\gamma = \infty$, corresponds to a series of independent infinitely deep square wells, and each energy level corresponds to those of a particle in a box of width a.

of the potential wells. As can be seen, as the depth increases, the allowed regions become narrower and converge on those of a particle in a square well.

Example 8.6 The asymptotic behaviour of a periodic solid

Show that for infinitely deep wells, the energy spectrum of a periodic solid becomes that of a collection of independent wells.

Method. When examining an equation for its asymptotic solutions, identify the terms that dominate the others as the selected parameter becomes infinite, and retain only them. Find the solution of the remaining terms.

Answer. As $\gamma \to \infty$, eqn 8.44 becomes

$$\frac{\gamma \sin \alpha a}{\alpha a} \simeq 0$$

which has solutions only for $\alpha a = \pm n\pi$ with $n = 1, 2, \ldots$. It follows that the allowed energies are

$$E_n = \frac{\hbar^2 \alpha^2}{2m_e} = \frac{n^2 h^2}{8m_e a^2}$$

exactly as for a particle in a single box.

Comment. The electron cannot tunnel between neighbouring boxes when the wells are infinitely deep.

Exercise 8.6. Derive an expression for the width of the allowed band. [$2 \arctan(\gamma/a\alpha)$]

Equation 8.44 can be solved numerically for α as a function of k, and hence the variation of the energy with k can be determined. The variation for $\gamma = \frac{3}{2}\pi$ is shown in Fig. 8.44. The discontinuities occur at

$$k = \frac{n\pi}{a} \qquad n = \pm 1, \pm 2, \ldots \tag{45}$$

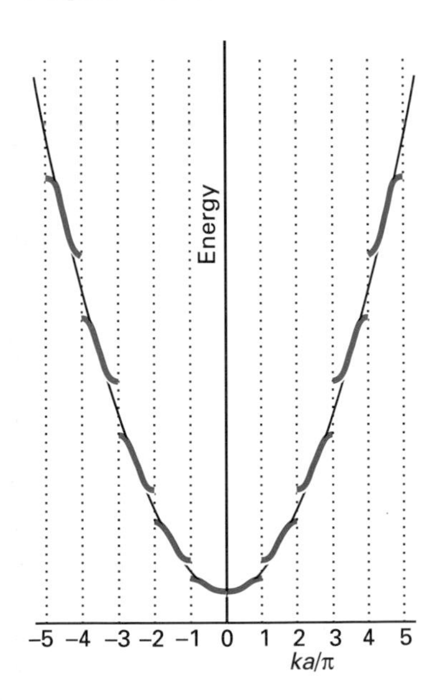

Fig. 8.44 The band structure represented as a plot of energy against the parameter k.

8.14 Brillouin zones

Each region between the discontinuities in eqn 8.45 is called a **Brillouin zone**. The discontinuities occur at the edges of the Brillouin zones, and towards the centres of the zones the variation of E with k is parabolic, exactly as in a free particle model. This observation, together with eqn 8.45, is a clue to the origin of the existence of band gaps, because they occur where the periodicity of the lattice matches the periodicity of the wavefunctions. In the centres of the zones, there is no match between the two, and the combinations $\cos kx$ and $\sin kx$ of the complex wave e^{ikx} are degenerate because, averaged over the lattice, each combination samples favourable regions of the potential equally. However, when the periods match, the cosine function (for instance) has maximum probability in the wells throughout the solid and the sine function has nodes in the wells everywhere. The two combinations are now no longer degenerate, and the perturbation caused by the lattice has driven them apart (just as in famous Fig 6.1).

It should be noted that k in the right-hand side of eqn 8.44 can be changed to $k \pm 2\pi n/a$ without changing the value of the right-hand side. So we can adjust the value of k by this amount at will, yet still obtain the same energies. In the **reduced wavevector representation**, k is modified by a different amount in each Brillouin zone to bring its value into the range

$$-\pi/a \leq k \leq \pi/a$$

This reduction has the effect of compressing Fig. 8.44 into the form shown in Fig. 8.45, where all the values of k lie in a range of width $2\pi/a$.

Finally, we impose a further constraint on the wavefunctions. When there are many atoms (wells) in the lattice, there is little error introduced if we assume that the ends of the lattice can be brought round into a circle and joined. This preserves the translational symmetry of the system throughout its length rather than introducing awkward end effects. (If we were interested in the surface states of metals, such a procedure would be invalid, of course.) The circularity of the system implies that the wavefunctions must satisfy cyclic boundary conditions (Section 3.1), and that

$$\psi(x+L) = \psi(x)$$

In terms of the Bloch functions, this condition is

$$u_k(x+L)\mathrm{e}^{\mathrm{i}k(x+L)} = u_k(x)\mathrm{e}^{\mathrm{i}kx}$$

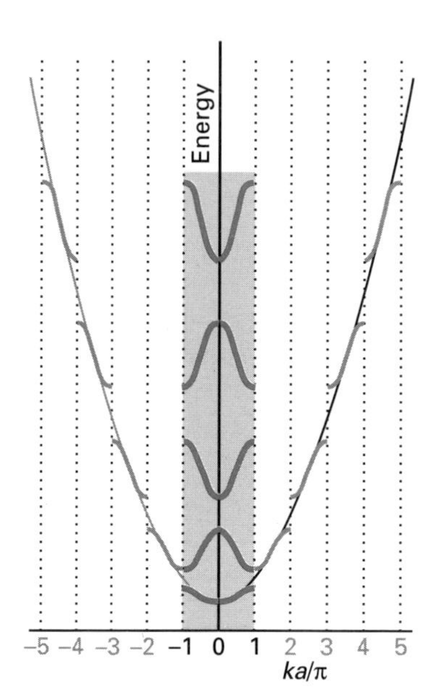

Fig. 8.45 In this depiction of the band structure, the curves illustrated in the previous diagram have all been transferred into the central zone to give a more compact representation.

However, because $u_k(x+L) = u_k(x)$ (as one location is an integral number of lattice periods a away from the other), this condition is equivalent to

$$k = \frac{2\pi n}{L} = \frac{2\pi n}{Na} \qquad n = 0, \pm 1, \pm 2, \ldots$$

where N is the number of atoms in the ring ($L = Na$).

We have seen, however, that an entire zone is expressed by values of k that lie within a length $2\pi/a$. It follows that $|n|$ cannot exceed $\frac{1}{2}N$, for otherwise k would lie outside the range. Therefore the number of spatial states in any band of the system is N. Another way of accepting the validity of this result is to consider the limit of very deep wells, when we have seen that the solid is then equivalent to N independent wells. Each band then consists of an infinitely narrow band of N levels of the same energy (recall Fig. 8.43), and this number is preserved when interactions are allowed between the wells.

The conclusion we have just drawn concerning the number of levels in a band is of the greatest importance for understanding the electronic structure of solids, for it implies that each Brillouin zone can accommodate up to $2N$ electrons. When each atom provides one electron, the zone is only half full and the solid is a metallic conductor. When each atom provides two electrons, the lowest zone is full and there is an energy gap before the next zone becomes available; such a material is a semiconductor (and an effective insulator if the gap is large). This description mirrors exactly the conclusions of the molecular orbital, tight-binding description of solids.

Problems

8.1 The R-dependences of the molecular integrals j', k', and S for the hydrogen molecule-ion are specified in Section 8.3. (a) Plot the R-dependences of the integrals α and β. (b) Plot E_+ and E_- against R and identify the equilibrium bond length and the dissociation energy of the molecule-ion.

8.2 Confirm that $\frac{1}{2}(E_- + E_+) - E_{1s}$ is a positive quantity, and hence that the effect of an antibonding orbital outweighs the effect of a bonding orbital. *Hint*. Set up expressions for the quantity by using eqn 8.17, and then plot it against R.

8.3 Evaluate the probability density of the electron in H_2^+ at the mid-point of the bond, and plot it as a function of R. Evaluate the difference densities $\rho_\pm = \psi_\pm^2 - \frac{1}{2}(\psi_a^2 + \psi_b^2)$ at points along the line joining the two nuclei (including the regions outside the nuclei) for $R = 130\,\text{pm}$. The difference density shows the modification to the electron distribution brought about by constructive (or destructive) overlap. *Hint*. Use the $\psi_\pm$ in eqn 8.17. The overlap integral S is given in Example 8.1. Repeat the calculation for several values of R.

8.4 We shall see in Chapter 10 that the vibrational frequency of a chemical bond is $\omega = \sqrt{k/\mu}$, where $k = (\mathrm{d}^2E/\mathrm{d}R^2)_0$ is the force constant and μ is the effective mass; for a homonuclear diatomic molecule of atoms of mass m, $\mu = \frac{1}{2}m$. Estimate the vibrational frequency of the hydrogen molecule-ion.

8.5 Take the hydrogen molecule wavefunction in eqn 8.27 and find an expression for the expectation value of the hamiltonian in terms of molecular integrals. *Hint*. The outcome of this calculation is eqn 8.28.

8.6 All the integrals involved in the H_2 molecular orbital calculation are listed in eqn 8.29 and *Further information 10*. Write and run a program to calculate $E - 2E_{1s}$ as a function of R, and identify the equilibrium bond length and the dissociation energy.

8.7 Evaluate the probability density for a single electron at a point on a line running between the two nuclei in H_2, and plot the difference density $\rho_1 - \frac{1}{2}(\psi_a^2 + \psi_b^2)$ for $R = 74\,\text{pm}$. *Hint*. Use eqn 8.27. The probability density of electron 1, ρ_1, is obtained from $\psi^2(1,2)$ by integrating over all locations of electron 2, because the latter's position is irrelevant. Therefore, begin by forming $\rho_1 = \int \psi^2(1,2)\,\mathrm{d}\tau_2$.

8.8 Confirm that the CI wavefunction $\psi = c_1\Psi_1 + c_2\Psi_2$ in Section 8.5 can be expressed as shown there, in terms of the sums and differences of the coefficients c_i.

8.9 Predict the ground configuration of (a) C_2, (b) C_2^+, (c) C_2^-, (d) N_2^+, (e) N_2^-, (f) F_2^+, and (g) Ne_2^+. Decide which terms can arise in each case, and suggest which lies lowest in energy.

8.10 Predict the ground configuration of (a) CO, (b) NO. Decide which terms can arise in each case, and suggest which lies lowest in energy.

8.11 Use a minimal basis set for the MO description of the molecule H_2O to show that the secular determinant factorizes into $(1 \times 1) + (2 \times 2) + (3 \times 3)$ determinants. Set up the secular determinant, denoting the Coulomb integrals α_H, α'_O, and α_O for H1s, O2s, and O2p, respectively, and writing the (O2p, H1s) and (O2s, H1s) resonance integrals as β and β'. Neglect overlap. Neglect the 2s-orbital, and find expressions for the energies of the molecular orbitals for a bond angle of 90°.

8.12 Now develop the calculation in Problem 8.11 by taking into account the O2s-orbital. Set up the secular determinant with the bond angle Θ as a parameter. Find expressions for the energies of the molecular orbitals and of the entire molecule. As a first step in analysing the expressions, set $\alpha_H \approx \alpha_O \approx \alpha_{O'}$ and $\beta \approx \beta'$. Can you devise improvements to the values of the Coulomb integrals on the basis of atomic spectral data?

8.13 Set up and solve the secular determinants for (a) hexatriene, (b) the cyclopentadienyl radical in the Hückel π-electron scheme; find the energy levels and

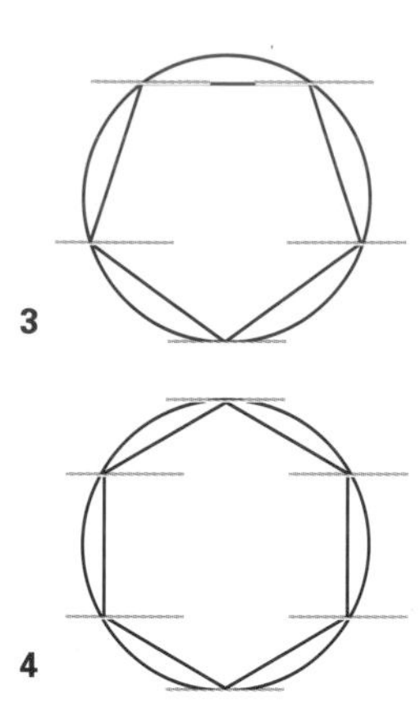
3

4

molecular orbitals, and estimate the delocalization energy. *Hint*. Use the groups C_2 or C_{5v} to factorize the determinants.

8.14 (a) Confirm that the symmetry-adapted linear combinations of $2p$-orbitals for benzene are those set out above Example 8.4. (b) Find the corresponding combinations for naphthalene.

8.15 Confirm that the roots of a tridiagonal determinant are those given in Example 8.3.

8.16 Show that the roots of the secular determinant for a cyclic polyene of N atoms can be constructed by inscribing a regular N-gon in a circle and noting the locations of the corners of the polygon, as in (**3**) and (**4**). *Hint*. See A.A. Frost and B. Musulin, *J. Chem. Phys.*, **21**, 572 (1953).

8.17 Heterocyclic molecules may be incorporated into the Hückel scheme by modifying the Coulomb integral of the atom concerned and the resonance integrals to which it contributes. Consider pyridine, its symmetry group being C_{2v}. Construct and solve the Hückel secular determinant with $\beta_{CC} \approx \beta_{CN} \approx \beta$ and $\alpha_N = \alpha_C + \frac{1}{2}\beta$. Estimate the electron energy and the delocalization energy. *Hint*. The roots of the determinants are best found on a computer.

8.18 Explore the role of p-orbital overlap in π-electron calculations. Take the cyclobutadiene secular determinant, but construct it without neglect of overlap between neighbouring atoms. Show that in place of $x = (\alpha - E)/\beta$ and 1 the elements of the determinant become $w = (\alpha - E)/(\beta - ES)$ and 1 respectively. Hence the roots in terms of w are the same as the roots in terms of x. Solve for E. Typically $S = 0.25$.

8.19 Find the effect of including neighbouring atom overlap on the π-electron energy levels of benzene. Use a computer to explore how the energies depend on the bond lengths, using $\beta \propto S$ and

$$S(2p\pi, 2p\pi) = \left\{1 + s + \tfrac{2}{5}s^2 + \tfrac{1}{15}s^3\right\}e^{-s} \qquad s = \frac{Z^*R}{na_0}$$

Consider the difference in delocalization energy between the cases where the molecule has six equivalent C—C bond lengths of 140 pm (the experimental value) and where it has alternating lengths of 133 pm and 153 pm (typical C=C and C—C lengths, respectively).

8.20 Determine which symmetry species are spanned by d-orbitals in a tetrahedral complex.

8.21 An ion with the configuration f^2 enters an environment of octahedral symmetry. What terms arise in the free ion, and which terms do they correlate with in the complex? *Hint*. Follow the discussion in Section 8.11.

8.22 In the strong field case, the d^2-configuration gives rise to e_g^2, $t_{2g}^1e_g^1$, and t_{2g}^2. (a) What terms may arise? (b) How do the singlet terms of the complex correlate with the singlet terms of the free ion? (c) What configurations arise in a tetrahedral complex, and what are the correlations?

8.23 Find the symmetry-adapted linear combinations of (a) σ-orbitals, (b) π-orbitals on the ligands of an octahedral complex. *Hint*. Set Cartesian axes on each ligand site, with z pointing towards the central ion, determine how the orbitals are transformed under the operations of the group O, and use the procedures for establishing symmetry-adapted orbitals as described in Chapter 5.

8.24 Repeat Problem 8.23 for a tetrahedral complex. What is the role of π-bonding in such complexes?

8.25 Verify that the Kronig–Penney model results in eqn 8.42, and show that this condition can be expressed as eqn 8.43.

8.26 Explore the effect of changing the depth of the potential well by finding the solutions of eqn 8.44 for different values of γ. Solve the equations numerically for (a) $\gamma = \pi$ and (b) $\gamma = 2\pi$.

Further reading

Coulson's valence. R. McWeeny; Oxford University Press, Oxford (1979).
Chemical structure and bonding. R.L. DeKock and H.B. Gray; Benjamin/Cummings, Menlo Park (1980).
The chemical bond. J.N. Murrell, S.F.A. Kettle, and J.M. Tedder; Wiley, Chichester (1985).
Valence theory. J.N. Murrell, S.F.A. Kettle, and J.M. Tedder; Wiley, Chichester (1975).
Chemical bonding theory. B. Webster; Blackwell Scientific, Oxford (1990).
Introduction to applied quantum chemistry. S.P. McGlynn, L.G. Vanquickenborne, M. Kinoshita, and D.G. Carroll; Holt, Rinehart, and Winston, New York (1972).
Chemical applications of group theory. F.A. Cotton; Wiley, New York (1990).
The theory of transition metal ions. J.S. Griffith; Cambridge University Press, Cambridge (1964).
The Jahn–Teller effect in molecules and crystals. R. Englman; Wiley, New York (1972).
Electronic structure and the properties of solids: The physics of the chemical bond. W.A. Harrison; W.H. Freeman and Co., New York (1980).
An introduction to molecular orbitals. Y. Jean and F. Volatron; Oxford University Press, New York (1993).
Orbital interactions in chemistry. T.A. Albright, J.K. Burdett, and M.-H. Whangbo; Wiley, New York (1985).
The electronic structure and chemistry of solids. P.A. Cox; Oxford University Press, Oxford (1987).
Bonding, energy levels, and bands in inorganic solids. J.A. Duffy; Longman, London (1990).
The origin of binding and antibinding in the hydrogen molecule-ion. M.J. Feinberg, K. Ruedenberg, and E.L. Mehler; *Adv. Quantum Chem.*, **5**, 27 (1970).
Aromaticity. P.J. Garratt; Wiley, Chichester (1986).
Hückel molecular orbital theory. K. Yates; Academic Press, New York (1978).
Problems in molecular orbital theory. T.A. Albright and J.K. Burdett; Oxford University Press, New York (1992).
The chemical bond: structure and dynamics. A. Zewail (ed.); Academic Press, Boston (1992).
The concept of the chemical bond. Z.B. Maksic (ed.); Springer, New York (1990).
Electronic structure of materials. A.P. Sutton; Oxford University Press, New York (1993).
Symmetry and structure: readable group theory for chemists. S.F.A. Kettle; Wiley, New York (1995).

9 The calculation of electronic structure

A primary objective of molecular quantum mechanics is the solution of the nonrelativistic, time-independent Schrödinger equation, and in particular the calculation of the electronic structures of atoms and molecules. Chapter 8 established the qualitative features of molecular structure calculations in terms of visualizable concepts. In this chapter, we introduce some of the techniques that are used to solve the Schrödinger equation for electrons in molecules. All such techniques make heavy use of computers, and it would be out of place to go into the technical details of the computations. Instead, we establish the equations that are the basis of such computations and describe some of the approximations that are used to make the computations feasible.

All the material in this chapter is based on the Born–Oppenheimer approximation (Section 8.1) and focuses on the solution of the electronic Schrödinger equation

$$H\psi(\mathbf{r};\mathbf{R}) = E(\mathbf{R})\psi(\mathbf{r};\mathbf{R}) \tag{1a}$$

for a fixed set of locations $\mathbf{R}$ of the nuclei. The hamiltonian is

$$H = -\frac{\hbar^2}{2m_e}\sum_i^n \nabla_i^2 - \sum_i^n\sum_I^N \frac{Z_I e^2}{4\pi\varepsilon_0 r_{Ii}} + \tfrac{1}{2}\sum_{i,j}^n \frac{e^2}{4\pi\varepsilon_0 r_{ij}} \tag{1b}$$

In molecular structure calculations it is conventional not to include the nucleus–nucleus repulsion energy in H, but to add it as a classical term at the end of the calculation. There are two main approaches to the solution of this equation. In ***ab initio*** **calculations**,[1] a model is chosen for the electronic wavefunction and eqn 9.1 is solved using as input only the values of the fundamental constants and the atomic numbers of the nuclei. The accuracy of this approach is determined principally by the model chosen for the wavefunction. For large molecules, accurate *ab initio* calculations are computationally expensive and semiempirical methods have been developed in an attempt to treat a wider variety of chemical species. A **semiempirical method** makes use of a simplified form for the hamiltonian as well as adjustable parameters with values obtained from experimental data. In both cases it is a challenging task to compute 'chemically accurate' energies, that is, energies calculated within about 0.01 eV (about 1 kJ mol^{-1}) of the exact values.

This chapter concentrates on the calculation of the electronic wavefunction and the electronic energy $E(\mathbf{R})$. However, it should be clear that once those quantities are known, a wide range of chemically and physically important properties can be determined. For example, by finding the minimum of the potential energy surface of a stable molecule, it is possible to characterize the

[1] The term *ab initio* comes from the Latin words for 'from the beginning'.

equilibrium structure of the species in terms of its bond lengths and bond angles. The modern trend is to broaden the range of properties that are calculated to include the location of the stationary points of the surface describing a reactive system, and hence to characterize activated complexes and transition states.

The Hartree–Fock self-consistent field method

The self-consistent field method was described in Section 7.15. However, because it is the starting point of many of the *ab initio* methods, we consider it again here in more detail and generalize some of the previous discussion.

9.1 The formulation of the approach

The crucial complication in all electronic structure calculations is the presence of the electron–electron potential energy, which depends on the electron–electron separations r_{ij} as given by the third term in eqn 9.1(b). As a first step, we suppose that the true electronic wavefunction, ψ, is similar in form to the wavefunction ψ° that would be obtained if this complicating feature were neglected. That is, ψ° is a solution of

$$H^\circ\psi^\circ = E^\circ\psi^\circ \qquad H^\circ = \sum_{i=1}^{n} h_i \tag{2}$$

where h_i is the **core hamiltonian** for electron i (see Section 7.15). This n-electron equation can be separated into n one-electron equations, so we can immediately write ψ° as a product of n one-electron wavefunctions (orbitals) of the form $\psi_a^\circ(\mathbf{r}_i; \mathbf{R})$. To simplify the notation, we shall denote the orbital occupied by electron i with coordinate $\mathbf{r}_i$ and parametrically depending on the nuclear arrangement $\mathbf{R}$ as $\psi_a^\circ(i)$. It is a solution of

$$h_i\psi_a^\circ(i) = E_a^\circ\psi_a^\circ(i) \tag{3}$$

where E_a° is the energy of an electron in orbital a in this independent electron model. The overall wavefunction ψ° is the following product of one-electron wavefunctions:

$$\psi^\circ = \psi_a^\circ(1)\psi_b^\circ(2)\ldots\psi_z^\circ(n) \tag{4}$$

The function ψ° depends on all the electron coordinates and, parametrically, on the nuclear locations.

At this stage, we have taken into account neither the spin of the electron nor the requirement that the electronic wavefunction must obey the Pauli principle. To do so, we re-introduce the concept of the **spinorbital**, $\phi_a(i)$, first encountered in Section 7.11. A spinorbital is a product of an orbital wavefunction and a spin function, and in a more elaborate notation would be denoted $\phi_a(\mathbf{x}_i; \mathbf{R})$, where $\mathbf{x}_i$ represents the joint space and spin coordinates of electron i. The overall wavefunction is then written as the following Slater determinant (Section 7.11):

$$\psi^\circ(\mathbf{x}; \mathbf{R}) = (n!)^{-1/2}\det|\phi_a(1)\phi_b(2)\ldots\phi_z(n)| \tag{5}$$

The ϕ_u, with $u = a, b, \ldots, z$, are orthonormal and the label u now incorporates the spin state as well as the spatial state.

9.2 The Hartree–Fock approach

Electron–electron repulsions are significant and must be included in any accurate treatment. In the **Hartree–Fock method** (HF),[2] a product wavefunction of the form of eqn 9.5 is sought, with the electron–electron repulsions treated in an average way. Each electron is considered to be moving in the field of the nuclei and the average field of the other $n-1$ electrons. The spinorbitals that give the 'best' n-electron determinantal wavefunction are found by using variation theory (Section 6.9), which involves minimizing the Rayleigh ratio

$$\mathcal{E} = \frac{\int \psi^*(\mathbf{x};\mathbf{R})H\psi(\mathbf{x};\mathbf{R})\,\mathrm{d}\mathbf{x}}{\int \psi^*(\mathbf{x};\mathbf{R})\psi(\mathbf{x};\mathbf{R})\,\mathrm{d}\mathbf{x}} \tag{6}$$

subject to the constraint that the spinorbitals are orthonormal. The lowest value of $\mathcal{E}$ is identified with the electronic energy for the selected nuclear configuration.

The application of this procedure leads to the **Hartree–Fock equations** for the individual spinorbitals (see *Further information 11*). The Hartree–Fock equation for spinorbital $\phi_a(1)$, where we have arbitrarily assigned electron 1 to spinorbital ϕ_a, is

$$f_1\phi_a(1) = \varepsilon_a\phi_a(1) \tag{7}$$

where ε_a is the orbital energy of the spinorbital and f_1 is the **Fock operator**:

$$f_1 = h_1 + \sum_u \{J_u(1) - K_u(1)\} \tag{8}$$

In this expression, h_1 is the core hamiltonian for electron 1, the sum is over all spinorbitals $u = a, b, \ldots, z$, and the **Coulomb operator**, J_u, and **exchange operator**, K_u, are defined as follows:

$$J_u(1)\phi_a(1) = \left\{\int \phi_u^*(2)\left(\frac{e^2}{4\pi\varepsilon_0 r_{12}}\right)\phi_u(2)\,\mathrm{d}\mathbf{x}_2\right\}\phi_a(1) \tag{9}$$

$$K_u(1)\phi_a(1) = \left\{\int \phi_u^*(2)\left(\frac{e^2}{4\pi\varepsilon_0 r_{12}}\right)\phi_a(2)\,\mathrm{d}\mathbf{x}_2\right\}\phi_u(1) \tag{10}$$

The Coulomb and exchange operators are defined in terms of spinorbitals rather than in terms of spatial wavefunctions, as in Section 7.15,[3] but their meaning is essentially the same: the Coulomb operator takes into account the Coulombic repulsion between electrons, and the exchange operator represents the modification of this energy that can be ascribed to the effects of spin correlation. It follows that the sum in eqn 9.8 represents the average potential energy of

[2] Electronic structure calculations are littered with acronyms: the terms we use in this chapter are collected together in Box 9.1 at the end of the chapter.

[3] To compare eqn 9.8 with eqn 7.44, we need to note that the factor of 2 accompanying J arises when the Hartree–Fock equations for the spinorbitals are converted to equations for spatial orbitals, with each spatial orbital doubly occupied.

electron 1 due to the presence of the other $n-1$ electrons. Note that because

$$J_a(1)\phi_a(1) = K_a(1)\phi_a(1)$$

the summation in eqn 9.8 includes contributions from all spinorbitals ϕ_u except the ϕ_a being computed.

Each spinorbital must be obtained by solving an equation of the form of eqn 9.7 with the corresponding Fock operator f_i. However, because f_i depends on the spinorbitals of all the other $n-1$ electrons, it appears that to set up the HF equations, one must already know the solutions beforehand! This is a common dilemma in electronic structure calculations, and it is commonly attacked by adopting an iterative style of solution, and stopping when the solutions are self-consistent, hence the name **self-consistent field** (SCF) for this approach. In a self-consistent procedure, a trial set of spinorbitals is formulated and used to formulate the Fock operator, then the HF equations are solved to obtain a new set of spinorbitals which are used to construct a revised Fock operator, and so on. The cycle of calculation and reformulation is repeated until a convergence criterion is satisfied.[4]

The Fock operator defined in eqn 9.8 depends on the n occupied spinorbitals. However, once these spinorbitals have been determined, the Fock operator can be treated as a well-defined hermitian operator. As for other hermitian operators (for example the hamiltonian operator), there is an infinite number of eigenfunctions of the Fock operator. In other words, there is an infinite number of spinorbitals ϕ_u, each having an energy ε_u. In practice, of course, we have to be content with solving for a finite number m of spinorbitals with $m \geq n$.

The m optimized spinorbitals obtained on completion of the Hartee–Fock SCF procedure are arranged in order of increasing orbital energy, and the n lowest energy spinorbitals are called the **occupied orbitals**. The remaining unoccupied $m-n$ spinorbitals are called **virtual orbitals**. The Slater determinant (of the form given in eqn 9.5) composed of the occupied spinorbitals is the Hartree–Fock ground-state wavefunction for the molecule; we shall denote it Φ_0.[5] By ordering the orbital energies and analysing the radial and angular nodal patterns of the spatial parts of the spinorbitals, a spinorbital can be identified as a 1s-spinorbital, a 2s-spinorbital, and so on.

9.3 Restricted and unrestricted Hartree–Fock calculations

It is customary in SCF calculations on closed-shell states of atoms (for which the number of electrons, n, is always even) to suppose that the spatial components of the spinorbitals are identical for each member of a pair of electrons. There are then $\frac{1}{2}n$ spatial orbitals of the form $\psi_a(\mathbf{r}_1)$ and the HF wavefunction is

$$\Phi_0 = (n!)^{-1/2} \det |\psi_a^\alpha(1)\psi_a^\beta(2)\psi_b^\alpha(3)\ldots\psi_z^\beta(n)| \quad (11)$$

[4] Convergence problems are sometimes encountered but they usually are not a major problem for many calculations. Several methods have been developed to improve convergence; these include the 'level shifter method' of V.R. Saunders and I.H. Hillier, *Int. J. Quantum Chem.*, 699, **7** (1973), and the direct inversion of the iterative subspace, P. Pulay, *Chem. Phys. Lett.*, 393, **73** (1980).

[5] The HF ground-state wavefunction, Φ_0, is either a single determinant or a linear combination of a small number of Slater determinants chosen to give the correct symmetry of the electronic state.

where we have used the same notation as in Section 7.11. Such a function is called a **restricted Hartree–Fock** (RHF) wavefunction. The HF equations for the spinorbitals are converted to the set of spatial eigenvalue equations given in eqns 7.44–46 by integration over the spin functions and using the orthonormality of α and β.[6]

Two procedures are commonly used for open-shell states of atoms. In the **restricted open-shell formalism**, all the electrons except those occupying open-shell orbitals are forced to occupy doubly occupied spatial orbitals. For example, the restricted open-shell wavefunction for atomic lithium would be of the form

$$\Phi_0 = (6)^{-1/2} \det |\psi_{1s}^{\alpha}(1)\psi_{1s}^{\beta}(2)\psi_{2s}^{\alpha}(3)|$$

in which the first two spinorbitals in the Slater determinant (which we identify as $1s$-spinorbitals) have the same spatial wavefunction. However, the restricted wavefunction imposes a severe constraint on the solution. Whereas the $1s\alpha$ electron has an exchange interaction with the $2s\alpha$ electron, the $1s\beta$ electron does not. In the **unrestricted open-shell Hartree–Fock** (UHF) formalism the two $1s$-electrons are not constrained to the same spatial wavefunction. For instance, the UHF wavefunction for Li would be of the form

$$\Phi_0 = (6)^{-1/2} \det |\psi_a^{\alpha}(1)\psi_b^{\beta}(2)\psi_c^{\alpha}(3)|$$

in which all three spatial orbitals are different. By relaxing the constraint of occupying orbitals in pairs, the open-shell UHF formalism gives a lower variational energy than the open-shell RHF formalism. One disadvantage of the UHF approach is that whereas the RHF wavefunction is an eigenfunction of S^2, the UHF function is not; that is, the total spin angular momentum is not a well-defined quantity for a UHF wavefunction.

Example 9.1 Showing that the RHF wavefunction is an eigenfunction of S^2

Consider the restricted Hartree–Fock wavefunction for the helium atom

$$\Phi_0 = (2)^{-1/2} \det |\psi_a^{\alpha}(1)\psi_a^{\beta}(2)|$$

Show that this Slater determinant is an eigenfunction of S^2 and evaluate its eigenvalue.

Method. We need to expand the Slater determinant and consider the effect of the spin operator, which acts only on the spin states α and β and not on the spatial function ψ_a. Because we are dealing with a two-electron system, $\mathbf{S} = \mathbf{s}_1 + \mathbf{s}_2$, where $\mathbf{s}_i$ acts only on electron i. We use the relation given in Example 4.4

$$S^2 = s_1^2 + s_2^2 + 2s_{1z}s_{2z} + s_{1+}s_{2-} + s_{1-}s_{2+}$$

with the results of the operations of s^2, s_z, s_+, and s_- on α and β given in Section 4.8.

[6] The details of the conversion are given in Section 3.4.1 of the excellent book A. Szabo and N.S. Ostlund, *Modern quantum chemistry: introduction to advanced electronic structure*, Macmillan, New York (1982). This text should be consulted for many of the details of the discussions in this chapter.

Answer. First, we expand the determinant:

$$\Phi_0 = (2)^{-1/2}\{\psi_a(1)\alpha(1)\psi_a(2)\beta(2) - \psi_a(2)\alpha(2)\psi_a(1)\beta(1)\}$$

The effect of S^2 on the first term in Φ_0 is

$$\begin{aligned} S^2\psi_a(1)\alpha(1)\psi_a(2)\beta(2) &= \psi_a(1)s_1^2\alpha(1)\psi_a(2)\beta(2) + \psi_a(1)\alpha(1)\psi_a(2)s_2^2\beta(2) \\ &\quad + 2\psi_a(1)s_{1z}\alpha(1)\psi_a(2)s_{2z}\beta(2) + \psi_a(1)s_{1+}\alpha(1)\psi_a(2)s_{2-}\beta(2) \\ &\quad + \psi_a(1)s_{1-}\alpha(1)\psi_a(2)s_{2+}\beta(2) \\ &= \tfrac{3}{4}\hbar^2\psi_a(1)\alpha(1)\psi_a(2)\beta(2) + \tfrac{3}{4}\hbar^2\psi_a(1)\alpha(1)\psi_a(2)\beta(2) \\ &\quad - \tfrac{1}{2}\hbar^2\psi_a(1)\alpha(1)\psi_a(2)\beta(2) + 0 + \hbar^2\psi_a(1)\beta(1)\psi_a(2)\alpha(2) \\ &= \hbar^2\psi_a(1)\alpha(1)\psi_a(2)\beta(2) + \hbar^2\psi_a(1)\beta(1)\psi_a(2)\alpha(2) \end{aligned}$$

A similar analysis of the effect of S^2 on the second term in Φ_0 yields

$$\begin{aligned} S^2\psi_a(2)\alpha(2)\psi_a(1)\beta(1) \\ = \hbar^2\psi_a(2)\alpha(2)\psi_a(1)\beta(1) + \hbar^2\psi_a(2)\beta(2)\psi_a(1)\alpha(1) \end{aligned}$$

Collecting terms, we obtain

$$\begin{aligned} S^2\Phi_0 = (2)^{-1/2}\{(\hbar^2 - \hbar^2)\psi_a(1)\alpha(1)\psi_a(2)\beta(2) \\ - (\hbar^2 - \hbar^2)\psi_a(2)\alpha(2)\psi_a(1)\beta(1)\} = 0 \times \Phi_0 \end{aligned}$$

Therefore, Φ_0 is an eigenfunction of S^2 with an eigenvalue of zero, as is to be expected because the ground state of the closed-shell helium atom is a singlet.

Exercise 9.1. Confirm that the unrestricted Hartree–Fock wavefunction for helium of the form $\Phi_0 = (2)^{-1/2}\det|\psi_a(1)\alpha(1)\psi_b(2)\beta(2)|$ where $\psi_1 \neq \psi_2$ is not an eigenfunction of S^2.

In practice, the expectation value of S^2 for the unrestricted wavefunction is computed and compared with the true value $S(S+1)\hbar^2$ for the ground state. If the discrepancy is not significant, the UHF method has given a reasonable molecular wavefunction. The UHF wavefunction is often used as a first approximation to the true wavefunction even if the discrepancy is significant.

9.4 The Roothaan equations

We have concealed a difficulty up to this point. The HF procedure is relatively simple to implement for atoms, for their spherical symmetry means that the HF equations can be solved numerically for the spinorbitals. However, such numerical solution is still not computationally feasible for molecules, and a modification of the technique must be used. As long ago as 1951, C.C.J. Roothaan and G.G. Hall independently suggested using a known set of basis functions with which to expand the spinorbitals (or more properly, the spatial parts of the spinorbitals). In this section, which is limited to a discussion of the restricted closed-shell Hartree–Fock formalism, we show how this suggestion transforms the coupled HF equations into a matrix problem which can be solved by using matrix manipulations.

We begin with eqn 7.44 for the spatial function $\psi_a(1)$ occupied by electron 1 and write it in the notation of this chapter as

$$f_1\psi_a(1) = \varepsilon_a\psi_a(1) \tag{12}$$

where f_1 is the Fock operator expressed in terms of the *spatial* wavefunctions:

$$f_1 = h_1 + \sum_u \{2J_u(1) - K_u(1)\} \tag{13}$$

and the Coulomb and exchange operators are defined in eqns 7.45–46, solely in terms of spatial coordinates.

Next, we introduce a set of M basis functions, θ_j (the form of which we justify in Section 9.5 and which are usually taken to be real), and express each spatial wavefunction ψ_i as a linear combination of these functions:

$$\psi_i = \sum_{j=1}^{M} c_{ji}\theta_j \tag{14}$$

where c_{ji} are as yet unknown coefficients. From a set of M basis functions, we can obtain M linearly independent spatial wavefunctions, and the problem of calculating the wavefunctions has been transformed to one of computing the coefficients c_{ji}.

When the expansion in eqn 9.14 is substituted into eqn 9.12, we obtain

$$f_1 \sum_{j=1}^{M} c_{ja}\theta_j(1) = \varepsilon_a \sum_{j=1}^{M} c_{ja}\theta_j(1) \tag{15}$$

Multiplication of both sides of this equation by the basis function $\theta_i^*(1)$ and integration over $d\mathbf{r}_1$ yields

$$\sum_{j=1}^{M} c_{ja} \int \theta_i^*(1) f_1 \theta_j(1)\, d\mathbf{r}_1 = \varepsilon_a \sum_{j=1}^{M} c_{ja} \int \theta_i^*(1)\theta_j(1)\, d\mathbf{r}_1 \tag{16}$$

As so often in quantum chemistry, the structure of a set of equations becomes clear if we introduce a more compact notation. In this case, it proves sensible to introduce the **overlap matrix**, **S**, with elements

$$S_{ij} = \int \theta_i^*(1)\theta_j(1)\, d\mathbf{r}_1 \tag{17}$$

(this matrix is not in general the unit matrix because the basis functions are not necessarily orthogonal) and the **Fock matrix**, **F**, with elements

$$F_{ij} = \int \theta_i^*(1) f_1 \theta_j(1)\, d\mathbf{r}_1 \tag{18}$$

Then eqn 9.16 becomes

$$\sum_{j=1}^{M} F_{ij}c_{ja} = \varepsilon_a \sum_{j=1}^{M} S_{ij}c_{ja} \tag{19}$$

This expression is one in a set of M simultaneous equations (one for each value of i) that are known as the **Roothaan equations**. The entire set of equations can be written as the single matrix equation

$$\mathbf{Fc} = \mathbf{Sc}\boldsymbol{\varepsilon} \tag{20}$$

where $\mathbf{c}$ is an $M \times M$ matrix composed of elements c_{ja} and $\boldsymbol{\varepsilon}$ is an $M \times M$ diagonal matrix of the orbital energies ε_a.

At this stage we can make progress by drawing on some of the properties of matrix equations (see *Further information 23*). The Roothaan equations have a non-trivial solution only if the following secular equation is satisfied:

$$\det|\mathbf{F} - \varepsilon_a \mathbf{S}| = 0 \tag{21}$$

This equation cannot be solved directly because the matrix elements F_{ij} involve Coulomb and exchange integrals which themselves depend on the spatial wavefunctions. Therefore, as before, we must adopt a self-consistent field approach, obtaining with each iteration a new set of coefficients c_{ja} and continuing until a convergence criterion has been reached (Fig. 9.1).

It is instructive to examine the matrix elements of the Fock operator, for in that way we can begin to appreciate some of the computational difficulties of obtaining Hartree–Fock SCF wavefunctions. The explicit form of the matrix

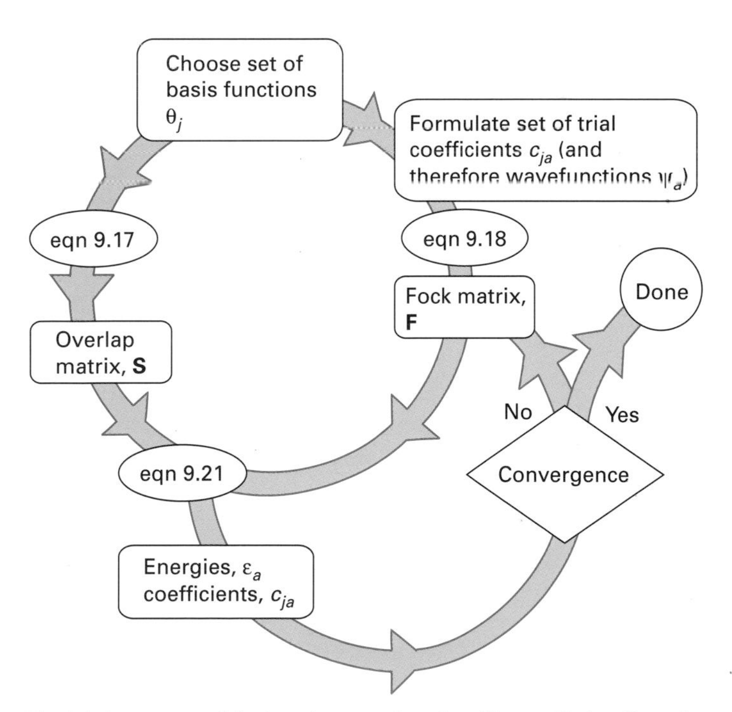

Fig. 9.1 A summary of the iteration procedure for a Hartree–Fock self consistent field calculation.

element F_{ij} is obtained from eqns 9.13, 7.45, and 7.46, and is

$$\begin{aligned}F_{ij} = &\int \theta_i^*(1)h_1\theta_j(1)\,\mathbf{dr}_1 \\ &+ 2\sum_u \int \theta_i^*(1)\psi_u^*(2)\left(\frac{e^2}{4\pi\varepsilon_0 r_{12}}\right)\psi_u(2)\theta_j(1)\,\mathbf{dr}_1\mathbf{dr}_2 \\ &- \sum_u \int \theta_i^*(1)\psi_u^*(2)\left(\frac{e^2}{4\pi\varepsilon_0 r_{12}}\right)\theta_j(2)\psi_u(1)\,\mathbf{dr}_1\mathbf{dr}_2\end{aligned} \tag{22}$$

The first term on the right is a one-electron integral that we shall denote h_{ij}. Insertion of the expansion in eqn 9.14 results in the following expression for F_{ij} solely in terms of integrals over the known basis functions:

$$\begin{aligned}F_{ij} = &h_{ij} \\ &+ 2\sum_{u,l,m} c_{lu}^* c_{mu} \int \theta_i^*(1)\theta_l^*(2)\left(\frac{e^2}{4\pi\varepsilon_0 r_{12}}\right)\theta_m(2)\theta_j(1)\,\mathbf{dr}_1\mathbf{dr}_2 \\ &- \sum_{u,l,m} c_{lu}^* c_{mu} \int \theta_i^*(1)\theta_l^*(2)\left(\frac{e^2}{4\pi\varepsilon_0 r_{12}}\right)\theta_j(2)\theta_m(1)\,\mathbf{dr}_1\mathbf{dr}_2\end{aligned} \tag{23}$$

The appearance of this rather horrendous expression can be greatly simplified by introducing the following notation for the two-electron integrals over the basis functions:

$$(ab|cd) = \int \theta_a^*(1)\theta_b(1)\left(\frac{e^2}{4\pi\varepsilon_0 r_{12}}\right)\theta_c^*(2)\theta_d(2)\,\mathbf{dr}_1\mathbf{dr}_2 \tag{24}$$

for it then becomes

$$F_{ij} = h_{ij} + \sum_{u,l,m} c_{lu}^* c_{mu}\{2(ij|lm) - (im|lj)\} \tag{25}$$

This expression is usually written as

$$F_{ij} = h_{ij} + \sum_{l,m} P_{lm}\{(ij|lm) - \tfrac{1}{2}(im|lj)\} \tag{26}$$

where the P_{lm} are defined as follows:

$$P_{lm} = 2\sum_u c_{lu}^* c_{mu} \tag{27}$$

The P_{lm} are referred to as **density matrix elements**, and are interpreted as the total electron density in the overlap region of θ_l and θ_m. The one-electron matrix elements h_{ij} need to be evaluated only once because they remain unchanged during each iteration. However, the P_{lm}, which depend on the expansion coefficients c_{lu} and c_{mu}, do need to be re-evaluated at each iteration. Because there are of the order of M^4 two-electron integrals to evaluate—so even small basis sets for moderately sized molecules can rapidly approach millions—their efficient calculation poses the greatest challenge in an Hartree–Fock SCF calculation. The problem is alleviated somewhat by the possibility that a number of integrals may be identically zero due to symmetry, some of the nonzero integrals may be equal by symmetry, and some of the integrals may be negligibly small because the basis functions may be centred on atomic nuclei separated by a large distance. Nevertheless, in general, there will be many more two-electron integrals than can be stored in the core memory of the computer, and a large body of work has been done in trying to develop efficient approaches to the calculation of two-electron integrals.[7]

9.5 The selection of basis sets

In principle, a complete set of basis functions must be used to represent spinorbitals exactly, and the use of an infinite number of functions would result in a Hartree–Fock energy equal to that given by the variational expression, eqn 9.6. This limiting energy is called the **Hartree–Fock limit**. The HF limit is not the exact ground-state energy of the molecule because it still ignores effects of electron correlation (a point discussed below). Because an infinite basis set is not computationally feasible, a finite basis set is always used, and the error due to the incompleteness of the basis set is called the **basis-set truncation error**. The difference between the Hartree–Fock limit and the computed lowest energy in an Hartree–Fock SCF calculation is a measure of the basis-set truncation error. A key computational consideration therefore will be to keep the number of basis functions low (to minimize the number of two-electron integrals to evaluate), to choose them cleverly (to minimize the computational effort for the evaluation of each integral), but nevertheless achieve a small basis-set truncation error.

One choice of basis functions for use in eqn 9.14 are the Slater type orbitals (STO) introduced in Section 7.14. The set of STOs with all permitted integral values of n, l, and m_l and all positive values of the **orbital exponents**, ζ, the parameter that occurs in the radial part of the STO ($\psi \propto e^{-\zeta r}$) forms a complete set. In practice, only a small number of all possible functions are used. The best values of ζ are determined by fitting STOs to the numerically computed atomic wavefunctions. For atomic SCF calculations, STO basis functions are centred on the atomic nucleus. For diatomic and polyatomic species, STOs are centred on each of the atoms. However, for Hartree–Fock SCF calculations on molecules with three or more atoms, the evaluation of the many two-electron integrals $(ab|cd)$ is impractical. Indeed, this 'two-electron integral problem' was once considered to be one of the greatest problems in quantum chemistry.

The introduction of **Gaussian-type orbitals** (GTO) by S.F. Boys[8] played a major role in making *ab initio* calculations computationally feasible. Cartesian Gaussians are functions of the form

$$\theta_{ijk}(\mathbf{r}_1 - \mathbf{r}_c) = (x_1 - x_c)^i (y_1 - y_c)^j (z_1 - z_c)^k e^{-\alpha|\mathbf{r}_1 - \mathbf{r}_c|^2} \tag{28}$$

where (x_c, y_c, z_c) are the Cartesian coordinates of the centre of the Gaussian at $\mathbf{r}_c$; (x_1, y_1, z_1) are the Cartesian coordinates of an electron at $\mathbf{r}_1$; i, j, and k are non-negative integers; and α is a positive exponent. When $i = j = k = 0$, the Cartesian Gaussian is an ***s*-type Gaussian**; when $i + j + k = 1$, it is a ***p*-type Gaussian**; when $i + j + k = 2$, it is a ***d*-type Gaussian**, and so on (Fig. 9.2). There are six d-type Gaussians. If preferred, one can use instead six linear

[7] For a discussion of some of these approaches, see Section 2.3 of D.M. Hirst, *A computational approach to chemistry*, Blackwell Scientific Publications, Oxford (1990) and references therein.

[8] S.F. Boys, *Proc. R. Soc. (London)*, 542, **A200** (1950).

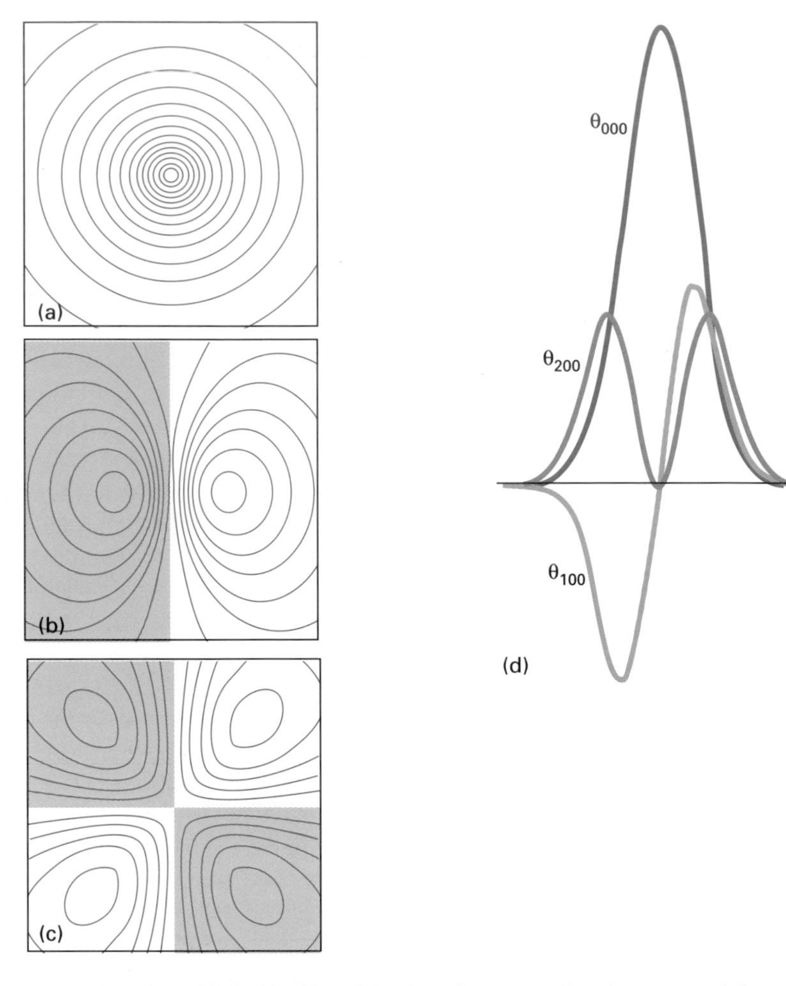

Fig. 9.2 Gaussian orbitals. (a), (b), and (c) show the contour plots for *s*-, *p*-, and *d*-type Gaussians, respectively, with the form e^{-r^2}, xe^{-r^2}, and xye^{-r^2}. (d) Cross-sections through the three wavefunctions.

combinations of these Gaussians, five of them having the angular behaviour of the five real 3*d*-hydrogenic orbitals and the sixth being spherically symmetrical like an *s*-function. This sixth linear combination is sometimes eliminated from the basis set, but its elimination is not essential because it will be recalled that we have not assumed that the basis is orthogonal. Spherical Gaussians, in which factors like $x_1 - x_c$ are replaced by spherical harmonics, are also used.

The central advantage of GTOs is that the product of two Gaussians at different centres is equivalent to a single Gaussian function centred at a point between the two centres (Fig. 9.3). Therefore, two-electron integrals on three and four different atomic centres can be reduced to integrals over two different centres, which are much easier to compute. However, there is also a disadvantage to using GTOs which in part negates the computational advantage. A 1*s* hydrogenic atomic orbital has a cusp at the atomic nucleus; an $n = 1$ STO also has a cusp there, but a GTO does not (Fig. 9.4). Because a GTO gives a poorer representation of the orbitals at the atomic nuclei, a larger basis must be used to

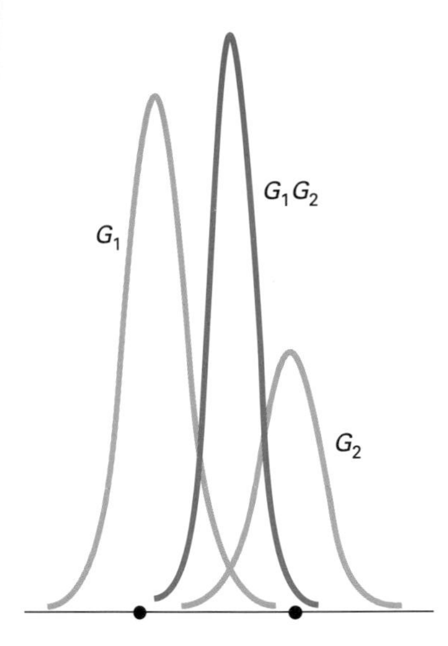

Fig. 9.3 The product of two Gaussians is itself a Gaussian lying between the two original functions. In this illustration, the amplitude of the product has been multiplied by 100.

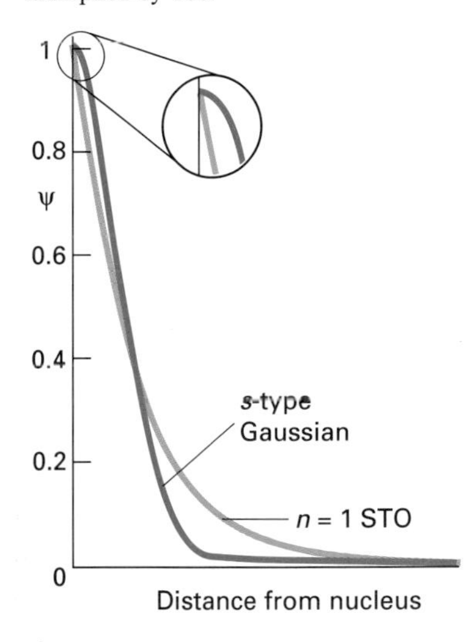

Fig. 9.4 A hydrogenic 1*s*-orbital is an exponential function, so there is a cusp at the nucleus. A Gaussian does not have a cusp at the nucleus.

achieve an accuracy comparable to that obtained from STOs.

To alleviate the latter problem, several GTOs are often grouped together to form what are known as **contracted Gaussian functions**. In particular, each contracted Gaussian, χ, is taken to be a fixed linear combination of the original or **primitive Gaussian functions**, g, centred on the same atomic nucleus:

$$\chi_j = \sum_i d_{ji} g_i \tag{29}$$

with the contraction coefficients d_{ji} and the parameters characterizing g held fixed during the calculation. The spatial orbitals are then expanded in terms of the contracted Gaussians:

$$\psi_i = \sum_j c_{ji} \chi_j \tag{30}$$

The use of contracted rather than primitive Gaussians reduces the number of unknown coefficients c_{ji} to be determined in the HF calculation. For example, if each contracted Gaussian is composed of three primitives from a set of 30 primitive basis functions, then whereas the expansion in eqn 9.14 involves 30 unknown c_{ji} coefficients, the corresponding expansion in eqn 9.30 has only 10 unknown coefficients. This decrease in the number of coefficients leads to potentially large savings in computer time with little loss of accuracy if the contracted Gaussians are well-chosen.

We now need to see how the primitives and the contracted Gaussians are constructed. In most applications, a set of basis functions is chosen and an atomic SCF calculation is performed, resulting in an optimized set of exponents for the basis functions, which can then be used in molecular structure calculations. The simplest type of basis set is a **minimal basis set** in which one function is used to represent each of the orbitals of elementary valence theory. A minimal basis set would include one function each for H and He (for the 1*s* orbital); five basis functions each for Li to Ne (for the 1*s*, 2*s*, and three 2*p* orbitals); nine functions each for Na to Ar, and so on. For instance, a minimal basis set for H_2O consists of seven functions, and includes two basis functions to represent the two H1*s* orbitals, and one basis function each for the 1*s*, 2*s*, $2p_x$, $2p_y$, and $2p_z$ orbitals of oxygen. It is found that a calculation with such a minimal basis set results in wavefunctions and energies that are not very close to the Hartree–Fock limits. Accurate calculations need more extensive basis sets.

A significant improvement is made by adopting a **double-zeta basis set** (DZ), in which each basis function in the minimal basis set is replaced by two basis functions. Compared to a minimal basis set, the number of basis functions has doubled and with it the number of variationally determined expansion coefficients c_{ji}. A DZ basis set for H_2O, for instance, would involve 14 functions. In a **triple-zeta basis set** (TZ), three basis functions are used to represent each of the orbitals encountered in elementary valence theory.

A **split-valence basis set** (SV) is a compromise between the inadequacy of a minimal basis set and the computational demands of DZ and TZ basis sets. Each valence atomic orbital is represented by two basis functions while each inner-shell atomic orbital is represented by a single basis function. For example, for an atomic SCF calculation on C using contracted Gaussians, there is one contracted function representing the 1*s*-orbital, two representing the 2*s*-orbital, and two each for the three 2*p*-orbitals.

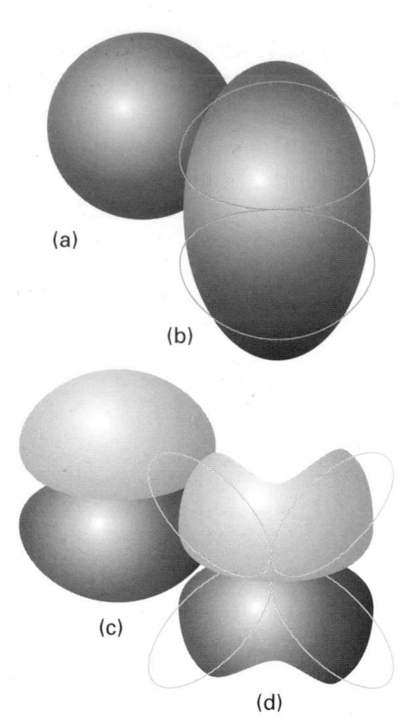

Fig. 9.5 Distortions are introduced into Gaussians to emulate the true angular dependence of p- and d-orbitals (shown by the outline shapes).

The basis sets we have described have ignored possible contributions from functions that represent orbitals for which the value of the quantum number l is larger than the maximum value considered in elementary valence theory. However, when bonds form in molecules, atomic orbitals are distorted (or polarized) by adjacent atoms. This distortion can be taken into account by including basis functions representing orbitals with high values of l. For example, the inclusion of p-type basis functions can model reasonably well the distortion of a $1s$-orbital, and d-type functions are used to describe distortion of p-orbitals (Fig. 9.5). The addition of these polarization functions to a DZ basis set results in what is called a **double-zeta plus polarization basis** (DZP). For example, in a DZP basis for methane, a set of three $2p$-functions is added to each hydrogen atom and a set of six $3d$-functions is added to the carbon atom.

One approach to the construction of a contracted Gaussian set is to make a least-squares fit of N primitive Gaussians to a set of STOs that have been optimized in an atomic SCF calculation. For example, an atomic SCF calculation is performed on carbon using STOs to find the contracted Gaussians best representing the $1s$, $2s$, and $2p$ STOs, and then these contracted Gaussians are used in a subsequent SCF calculation on methane. The expansion of an STO in terms of N primitive Gaussians is designated **STO-NG**. A common choice is $N = 3$, giving a set of contracted Gaussians referred to as **STO-3G**. Alternatively, an atomic SCF calculation can be carried out using a relatively large basis of Gaussian primitives. This procedure results in a set of optimized Gaussian exponents (α) as well as a set of variationally determined SCF coefficients (c_{ji}) for the primitives of each spatial orbital ψ_i. The optimized exponents and coefficients of the primitive Gaussians can then be used to obtain contracted Gaussian basis sets for use in molecular calculations. In the **(4s)/[2s] contraction scheme**,[9] four primitive s-type Gaussians are used to construct two basis set functions for atomic hydrogen. As in many contraction schemes, the most diffuse primitive (the one with the smallest value of the exponent α) is left uncontracted, and each of the remaining primitives appears in only one contracted Gaussian. That is, in the $(4s)/[2s]$ scheme, three of the primitives are used to form a contracted Gaussian basis set function.

In the **(9s5p)/[3s2p] contraction scheme**,[10] nine s-type and five p-type primitive Gaussians (which have been optimized in an atomic SCF calculation on a Period 2 element) are contracted into three and two basis functions, respectively. This contraction scheme usually results in a split-valence basis set containing one basis function representing the inner shell $1s$-orbital, two basis functions for the valence $2s$-orbital, and two for each of the three $2p$-orbitals. Therefore, it reduces the total number of basis functions from 24 (five p-type primitives for each of $2p_x$, $2p_y$, and $2p_z$, and nine s-type primitives) to nine. This reduction achieves substantial decrease in computer time because the number of two-electron integrals to be evaluated is proportional to the fourth power of the number of basis functions.

[9] S. Huzinaga, *J. Chem. Phys.*, 1293, **42** (1965).
[10] T.H. Dunning, *J. Chem. Phys.*, 2823, **53** (1970).

Other contraction schemes have also given valuable savings. In the **3-21G basis set**,[11] one contracted Gaussian composed of three primitives is used to represent each inner shell atomic orbital. Each valence shell orbital is represented by two functions, one a contracted Gaussian of two primitives and one a single (and usually diffuse) primitive. The primitives are first optimized in a prior SCF calculation on the atoms, and the contracted sets are then used in the molecular calculation. The **6-31G*** basis set adds polarization functions in the form of six *d*-type functions for each atom other than H to the split-valence 6-31G basis. Another star, an additional polarization function: **6-31G**** indicates the addition to 6-31G* of a set of three *p*-type polarization functions for each H atom.

Example 9.2 Determining the number of basis set functions in a molecular structure calculation

Determine the total number of Gaussian basis set functions in an *ab initio* calculation of C_2H_2 using a 6-31G* basis set.

Method. Start with the 6-31G basis, which, for each atom in the molecule, consists of (a) one contracted Gaussian composed of six primitives for each inner shell orbital and (b) two functions for each valence shell orbital, a contracted Gaussian of three primitives and a single uncontracted primitive. Then add six *d*-type polarization functions for each atom other than hydrogen.

Answer. Each 1*s*-orbital of H is represented by two basis set functions, using a total of four primitive Gaussians. Each 1*s*-orbital of C is represented by one contracted Gaussian of six primitives. The 2*s*-, $2p_x$-, $2p_y$-, and $2p_z$-orbitals of each C atom are each represented by two basis set functions, one a contraction of three primitives and one a single uncontracted primitive. In addition, each C atom will also have six *d*-type polarization functions. Therefore, the total number of 6-31G* basis set functions for C_2H_2 is

$$2(1+4\times 2+6)+2\times 2=34$$

and the total number of primitives used is

$$2\{6+4\times(3+1)+6\}+2\times 4=64$$

Exercise 9.2. How many functions in a 6-31G** basis set would there be in a molecular structure calculation on H_2O? [25 basis functions composed of 42 primitives]

An additional contribution to the inaccuracy of calculations that stems from the use of a finite basis set is the **basis-set superposition error** which may arise in the calculation of the interaction energy of two weakly bound systems. For example, suppose we were interested in characterizing the energetics of the

[11] W.J. Hehre, L. Radom, P.v.R. Schleyer, and J.A. Pople, *Ab initio molecular orbital theory*, Wiley, New York (1986).

dimerization of hydrogen fluoride and defined the interaction energy as the energy of the dimer minus the energies of the two infinitely separated monomers. If one used, for example, a 6-31G basis set for each of the atoms in hydrogen fluoride, it might seem that the obvious choice would be to use a 6-31G basis set on each of the four atoms of the dimer. However, when the energy of an individual hydrogen fluoride molecule is computed, only the basis set functions on two atoms are used to describe each electronic spatial orbital ψ. On the other hand, the electrons in the dimer have associated orbitals composed of linear combinations of the basis set functions on all four atoms. In other words, the basis set for the dimer is larger than that of either monomer, and this enlargement of the basis results in a nonphysical lowering of the energy of the dissociated dimer relative to the separated monomers.

A common method used to correct the basis-set superposition error is the **counterpoise correction**,[12] in which the energies of the monomer systems are computed by using the full basis set used for the dimer. For example, in the case of the hydrogen fluoride dimer, when computing the energy of an individual molecule, one would use a basis set for each nucleus of the monomer as well as the same basis set functions centered at the two points in space that would correspond to the equilibrium positions of the other two nuclei in the dimer.

9.6 Calculational accuracy and the basis set

Table 9.1 presents results of *ab initio* Hartree–Fock SCF calculations on the ground states of several closed-shell molecules and shows how the SCF energy varies with the basis set used in the calculation. The reported SCF energies correspond to geometries at or nearly at equilibrium. The energies represent the sum of the electronic energy (the result of the *ab initio* calculation) and the nucleus–nucleus repulsion energy for the selected geometry. We see clearly from the table that as the basis set becomes more complete, the energy approaches the Hartree–Fock limit.

We have already mentioned that the electronic potential energy surface (or curve for a diatomic molecule) can be used to predict the equilibrium geometries of molecules. This prediction can then be compared directly with the best experimental values. Table 9.2 shows a number of calculated equilibrium bond lengths using the basis sets used in Table 9.1. A good *ab initio* SCF calculation

Table 9.1 Self-consistent field energies with a variety of basis sets*

Basis set	H_2	N_2	CH_4	NH_3	H_2O
STO-3G	−1.117	−107.496	−39.727	−55.454	−74.963
4-31G	−1.127	−108.754	−40.140	−56.102	−75.907
6-31G*	−1.127	−108.942	−40.195	−56.184	−76.011
6-31G**	−1.131	−108.942	−40.202	−56.195	−76.023
HF limit	−1.134	−108.997	−40.225	−56.225	−76.065

* The energies are expressed as multiples of the hartree ($E_h = 2hc\mathcal{R}$), where 1 $E_h = 4.359\,75$ aJ.

[12] S.F. Boys and F. Bernardi, *Mol. Phys.*, 553, **19** (1970).

Table 9.2 Self-consistent field equilibrium bond lengths with a variety of basis sets*

Basis set	H_2	N_2	CH_4	NH_3	H_2O
STO-3G	1.346	2.143	2.047	1.952	1.871
4-31G	1.380	2.050	2.043	1.873	1.797
6-31G*	1.380	2.039	2.048	1.897	1.791
6-31G**	1.385	2.039	2.048	1.897	1.782
Observed	1.401	2.074	2.050	1.912	1.809

* The bond lengths are expressed as multiples of the Bohr radius ($a_0 = 52.91772$ pm).

typically is in error by 0.02–$0.04a_0$ and bond lengths shorten as more basis functions are added.

Electron correlation

However good Φ_0 may appear to be, it is not the 'exact' wavefunction. The Hartree–Fock method relies on averages: it does not consider the *instantaneous* electrostatic interactions between electrons; nor does it take into account the quantum mechanical effects on electron distributions because the effect of $n-1$ electrons on an electron of interest is treated in an average way. We summarize these deficiencies by saying that the HF method ignores **electron correlation**. A great deal of modern work in the field of electronic structure calculation is aimed at taking electron correlation into account.

9.7 Configuration state functions

The HF method yields a finite set of spinorbitals when a finite basis set expansion is used. In general, a basis with M members results in $2M$ different spinorbitals. As we have already mentioned, by ordering the spinorbitals energetically and taking the n lowest in energy (to be occupied by the n electrons), we form the Hartree–Fock wavefunction Φ_0. However, there remain $2M-n$ virtual orbitals. Clearly, many Slater determinants can be formed from the $2M$ spinorbitals; Φ_0 is just one of them. By using the single determinantal wavefunction Φ_0 as a convenient reference, it is possible to classify all other determinants according to how many electrons have been promoted from occupied orbitals to virtual orbitals. To simplify the appearance of the following expressions, we omit the normalization factors from the Slater determinants, and hence denote Φ_0 as

$$\Phi_0 = |\phi_1\phi_2\ldots\phi_a\phi_b\ldots\phi_n|$$

where ϕ_a and ϕ_b are among the n occupied spinorbitals for the HF ground state.

A **singly excited determinant** corresponds to one for which a single electron in occupied spinorbital ϕ_a has been promoted to a virtual spinorbital ϕ_p (Fig. 9.6):

$$\Phi_a^p = |\phi_1\phi_2\ldots\phi_p\phi_b\ldots\phi_n|$$

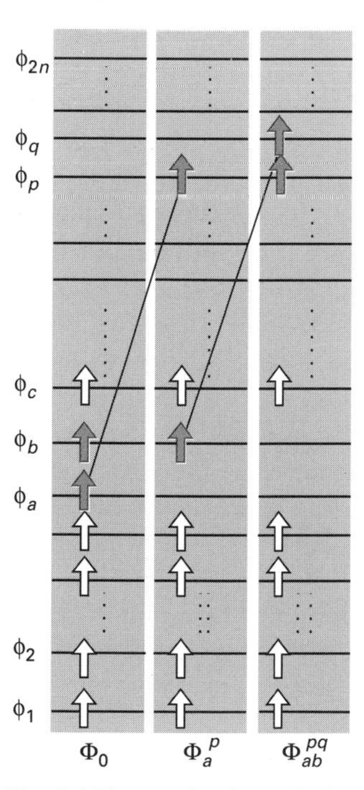

Fig. 9.6 The notation for excited determinants.

A **doubly excited determinant** is one in which two electrons have been promoted, one from ϕ_a to ϕ_p and one from ϕ_b to ϕ_q

$$\Phi_{ab}^{pq} = |\phi_1\phi_2 \ldots \phi_p\phi_q \ldots \phi_n|$$

In a similar manner, we can form other multiply excited determinants. Each of the determinants, or a linear combination of a small number of them constructed so as to have the correct electronic symmetry (for example, to be an eigenfunction of S^2), is called a **configuration state function** (CSF). More precisely, a CSF is an eigenfunction of all the operators that commute with H. These excited CSFs can be taken to approximate excited-state wavefunctions or, as we now see, they can be used in linear combination with Φ_0 to improve the representation of the ground-state (or any excited-state) wavefunction.

9.8 Configuration interaction

The exact ground-state and excited-state wavefunctions can be expressed as a linear combination of all possible n-electron Slater determinants arising from a complete set of spinorbitals.[13] Therefore, we can write the exact electronic wavefunction ψ for any state of the system in the form

$$\psi = C_0\Phi_0 + \sum_{a,p} C_a^p\Phi_a^p + \sum_{\substack{a<b \\ p<q}} C_{ab}^{pq}\Phi_{ab}^{pq} + \sum_{\substack{a<b<c \\ p<q<r}} C_{abc}^{pqr}\Phi_{abc}^{pqr} + \ldots \qquad (31)$$

where the Cs are expansion coefficients and where the limits in the summations ensure that we sum over all unique pairs of spinorbitals in doubly excited determinants, over all unique triplets of spinorbitals in triply excited determinants, and so on. In other words, a given excited determinant appears only once in the summation. An *ab initio* method in which the wavefunction is expressed as a linear combination of determinants is called **configuration interaction** (CI). A primitive example of CI was described in Section 8.5 in connection with the structure of H_2.

The energy associated with the exact ground-state wavefunction of the form of eqn 9.31 is the exact nonrelativistic ground-state energy (within the Born–Oppenheimer approximation.) The difference between this exact energy and the HF limit is called the **correlation energy**. Configuration interaction accounts for the electron correlation neglected in the Hartree–Fock method.

At this point the familiar refrain is inevitable: in practice, it is not computationally possible to handle an infinite basis set of n-electron determinants with each determinant constructed from an infinite set of spinorbitals. Furthermore, it becomes computationally very demanding (both in computer time and storage) to handle extremely large numbers of determinants. The latter problem is slightly alleviated by the observation that a number of the determinants in eqn 9.31 can be eliminated on the basis of symmetry. For example, if we are interested in computing an accurate wavefunction for the ${}^1\Sigma_g^+$ ground state of H_2, we do not need to include CSFs that do not correspond to the required ${}^1\Sigma_g^+$ symmetry. For instance, we can neglect CSFs that have u parity or which have nonzero eigenvalues of S_z.

[13] A proof of this statement can be found in P.-O. Löwdin, *Adv. Chem. Phys.*, 207, **2** (1959), a classic review on electron correlation.

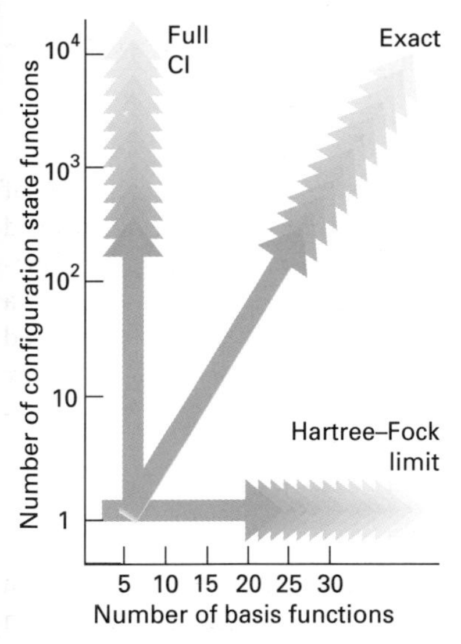

Fig. 9.7 The approach to the exact energy of a system as the numbers of basis functions and configuration state functions increase.

There is another point that emphasizes—if further emphasis is required—the difficulty of carrying out molecular structure calculations reliably. Even if we could include all CSFs of the desired symmetry in eqn 9.31, we must also remember that the CSFs themselves are constructed from a finite set of one-electron spinorbitals. A calculation is classified as **full CI** if all CSFs of the appropriate symmetry are used *for a given finite basis set*. For a given basis, full CI is the best CI calculation we can do. The difference between the ground-state energies obtained from a Hartree–Fock SCF calculation and a full CI calculation using the same basis set is called the **basis-set correlation energy** (Fig. 9.7). As the number of one-electron spinorbitals computed from the HF equations gets larger and larger, the basis-set correlation energy gets closer and closer to the exact correlation energy.

Unfortunately, even with a small number (n) of electrons and a relatively small number (M) of basis set functions θ (and subsequently a small number $2M$ of spinorbitals), the total number of determinants can be extremely large. For example, with 10 electrons and 20 basis set functions, the number of determinants to consider is

$$\binom{2M}{n} = 8.477\ldots \times 10^8$$

In practice, therefore, the expansion in eqn 9.31 must almost always be truncated. Nonetheless, although the calculation is limited to a finite set of spinorbitals and only a fraction of all possible determinants, CI is a popular method for the calculation of accurate molecular wavefunctions and potential energy surfaces. Even with a small number of CSFs it can correct for one of the deficiencies that stem from the use of only doubly occupied orbitals in the restricted HF method, the incorrect behaviour for the dissociation of a molecule. This point was illustrated in Section 8.5. A calculation in which the incorrect behaviour of the HF wavefunction upon dissociation is corrected accounts for an important part of the correlation energy called **nondynamic correlation** or **structural correlation**. On the other hand, **dynamic correlation** accounts for the incorrect HF wavefunction at short interatomic distances.

9.9 CI calculations

In CI calculations, the ground- or excited-state wavefunction, ψ, for state s (which we will denote for the remainder of this chapter by Ψ_s to avoid confusion with the spatial function ψ_i) is represented as a linear combination of n-electron Slater determinants. Equation 9.31 can be written in a notationally simpler form as

$$\Psi_s = \sum_{J=1}^{L} C_{Js}\Phi_J \qquad (32)$$

where the sum is over a finite number L of determinants Φ_J with expansion coefficients C_{Js} for the state s. The expansion coefficients C_{Js} are determined variationally by minimizing the Rayleigh ratio as given by eqn 9.6 but using Ψ_s as the trial function. As in all applications of variation theory (see Sections 6.9 and 6.10), this minimization is equivalent to solving the following set of simultaneous equations for the coefficients C_{Js} for each state s

$$\sum_{J=1}^{L} H_{IJ}C_{Js} = E_s \sum_{J=1}^{L} S_{IJ}C_{Js} \tag{33}$$

where

$$H_{IJ} = \int \Phi_I^* H \Phi_J \, \mathrm{d}\mathbf{x}_1 \mathrm{d}\mathbf{x}_2 \ldots \mathrm{d}\mathbf{x}_n \tag{34}$$

and

$$S_{IJ} = \int \Phi_I^* \Phi_J \, \mathrm{d}\mathbf{x}_1 \mathrm{d}\mathbf{x}_2 \ldots \mathrm{d}\mathbf{x}_n \tag{35}$$

The set of equations can be written in matrix notation as

$$\mathbf{HC} = \mathbf{ESC} \tag{36}$$

where the elements of the $L \times L$ square matrices $\mathbf{H}$ and $\mathbf{S}$ are H_{IJ} and S_{IJ}, respectively; $\mathbf{E}$ is the diagonal matrix of energies E_s; and $\mathbf{C}$ is an $L \times L$ matrix of coefficients. Because the Slater determinants form an orthonormal set ($S_{IJ} = \delta_{IJ}$), eqn 9.36 becomes

$$\mathbf{HC} = \mathbf{EC} \tag{37}$$

This matrix equation can be solved by diagonalizing $\mathbf{H}$, and yields a total of L wavefunctions (eigenfunctions) Ψ_s with energies (eigenvalues) E_s. The lowest energy eigenvalue represents an upper bound to the ground-state energy of the molecule; it would be the exact ground-state energy in a full CI calculation with an infinite number of spinorbitals. Provided the excited-state wavefunction is orthogonal to the ground-state wavefunction, the next lowest energy eigenvalue represents an upper bound for the first excited-state energy of the molecule, and so on.

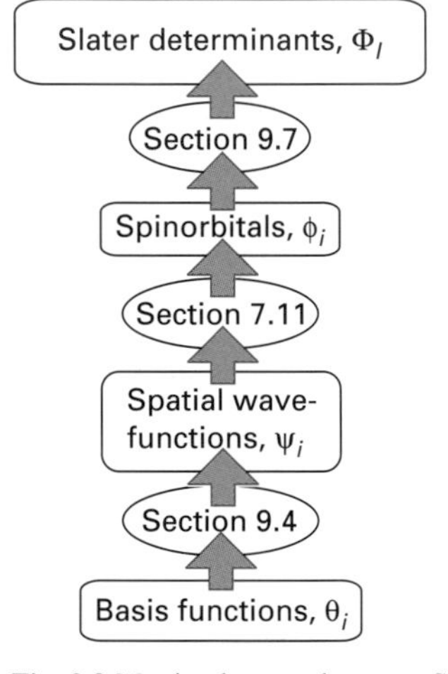

Fig. 9.8 Matrix elements between Slater determinants can ultimately be expressed in terms of matrix elements between basis functions.

The evaluation of the matrix elements H_{IJ} is critically important in CI calculations. These matrix elements can ultimately be expressed in terms of the basis functions θ, because the determinants are composed of spinorbitals expressed in terms of the basis functions (Fig. 9.8). For a large number of determinants, there may be too many one- and two-electron integrals to hold in the memory of the computer simultaneously, so their computation may have to be done in groups. In any event, conventional CI calculations are usually limited to a number of CSFs of the order of 10^4, and because full CI usually results in a list far exceeding this number, it is necessary to employ a truncation scheme so that the list of CSFs is kept at a manageable size.

The use of a truncated CSF list is referred to as **limited CI**. A systematic approach to the selection of determinants for use in eqn 9.32 is to include all those differing from the HF wavefunction Φ_0 by no more than some predetermined number of spinorbitals. Because the spinorbitals cannot be improved once the SCF calculation has been completed, the best we can do is systematically to include more and more excited determinants in the expansion in eqn 9.31. Hamiltonian matrix elements H_{0J} between the HF wavefunction Φ_0 and determinants Φ_J that are more than doubly excited are zero. In addition, by **Brillouin's theorem**,[14] hamiltonian matrix elements between Φ_0 and all singly

[14] See Problem 9.17.

excited determinants also vanish. Thus, a first approach (which can be expected to be reasonably accurate when Φ_0 is an approximation to the exact wavefunction) is to limit the list of excited determinants to those that are singly and doubly excited.[15] Limitation of the list of determinants to Φ_0 and determinants that are singly and doubly excited with respect to Φ_0 is denoted **SDCI**. If only Φ_0 and doubly excited determinants are used, then the technique is denoted **DCI**.

Table 9.3 presents results for CI calculations on H_2 using the various basis sets of Table 9.1. In particular, we compare results from DCI and SDCI. As dihydrogen is a two-electron species, SDCI on H_2 is the same as full CI. The entries for the energies in Table 9.3 represent the differences between the SCF and CI ground-state energies, both computed using the same basis set. There are several things to be noted from the table. First, the single excitations make an insignificant contribution to the energy. Secondly, the contribution to the energy from the double excitations very much depends on the basis set. As the basis set gets larger, so we recover a larger fraction of the exact correlation energy of $-0.0409(2hc\mathcal{R})$ from the doubly excited configurations. Even the largest basis set of Table 9.3, 6-31G**, gives an energy that is significantly different from the correlation energy, primarily because $l \geq 2$ functions have not been included in the basis.

Table 9.3 also presents the equilibrium bond length of H_2 obtained by finding the minimum in the calculated potential energy curve as a function of the bond length. We see that as the basis set is improved, the computed bond length is closer to the experimental result of $1.401a_0$. By comparing Tables 9.2 and 9.3, one can see that for a given basis set, the full CI calculation is superior to the Hartree–Fock SCF result.

Table 9.3 Calculated properties of dihydrogen*

Basis set	$\{E(\mathrm{DCI}) - E(\mathrm{SCF})\}$ $/(10^{-3}\,E_\mathrm{h})$	$\{E(\mathrm{SDCI}) - E(\mathrm{SCF})\}$ $/(10^{-3}\,E_\mathrm{h})$	$R_\mathrm{e}(\mathrm{SDCI})$ $/a_0$
STO-3G	−20.56	−20.56	1.389
4-31G	−24.87	−24.94	1.410
6-31G**	−33.73	−33.87	1.396

* The energy difference (in hartrees, Table 9.1) are calculated at $1.4\,a_0$.

Configuration interaction calculations that include single, double, triple, and quadruple excitations are designated **SDTQCI**. However, for basis sets large enough to recover most of the correlation energy, SDTQCI involves too many determinants to be computationally practicable. As the quadruply excited determinants can be important in computing the correlation energy, a simple formula known as the **Davidson correction** has been proposed for estimating the contribution ΔE_Q of quadruply excited determinants to the

[15] Singly excited determinants will have a small but nonzero effect on the calculation of the ground-state energy because they have nonzero matrix elements with doubly excited determinants, which themselves mix with Φ_0. Moreover, single excitations do affect the electronic charge distribution and therefore properties such as the dipole moment.

correlation energy:[16]

$$\Delta E_Q = (1 - C_0^2)(E_{DCI} - E_{SCF}) \tag{38}$$

where E_{DCI} is the ground-state energy computed in a CI calculation using Φ_0 and all its double excitations; C_0 is the coefficient of Φ_0 for the normalized wavefunction of eqn 9.32 obtained in the DCI calculation; E_{SCF} is the ground-state energy associated with Φ_0 obtained in a Hartree–Fock SCF calculation. By using eqn 9.38 in a CI study of the ground state of N_2, Langhoff and Davidson estimated that the contribution of quadruple excitations to the correlation energy was 7.6 per cent or $0.048hc\mathcal{R}$.

We conclude this section by mentioning one serious deficiency that plagues limited CI calculations, that of the lack of **size-consistency**. A method is size-consistent if the energy of a many-electron system is proportional to the number of electrons n in the limit of $n \to \infty$; in particular, the energy of a system AB computed when subsystems A and B are infinitely far apart should be equal to the sum of the energies of A and B separately computed using the same method. That the physical requirement of size-consistency is not satisfied, for example, by a SDCI wavefunction is demonstrated in the following example.

Example 9.3 Demonstrating the lack of size-consistency

Demonstrate that the SDCI calculation of the energy of the dimer He_2 is not size-consistent.

Method. To show that the calculation is not size-consistent, we need to show that the energy of two infinitely separated He atoms is not equal to the energy of the dimer He_2 when the internuclear distance in the dimer is infinite. Recalling that a full CI is size-consistent but limited CI is not, we should compare SDCI calculations to full CI calculations on both He + He and He_2.

Answer. First, consider the energies of the two helium atoms separately computed via SDCI. The infinitely separated helium atoms each have a (restricted) HF wavefunction given by

$$\Phi_0 = (1/\sqrt{2}) \det |\psi_{1s}^{\alpha}(1)\psi_{1s}^{\beta}(2)|$$

Because He has only two electrons, a calculation involving all single and double excitations would involve all possible determinants: it would be a full CI calculation for each individual helium atom. Therefore, the SDCI four-electron calculation on the two independent two-electron He systems includes contributions from *quadruply excited* determinants in which both electrons on each independent He atom are excited. Now consider the SDCI calculation on the composite four-electron dimer with an infinite internuclear distance. A SDCI calculation involving only singly and doubly excited determinants will not be the same as full CI; in particular, it will not include contributions from quadruply excited determinants. Therefore, the SDCI calculation on the composite four-electron system

[16] S.R. Langhoff and E.R. Davidson, *Int. J. Quantum Chem.*, 61, **8** (1974).

He_2 will result in both a different wavefunction and a different energy than the SDCI treatment of the two independent two-electron He systems. Thus, the limited CI calculation is not size-consistent.

Exercise 9.3. What level of CI calculation is necessary to ensure that the He_2 dimer CI calculation is size-consistent? [SDTQCI]

It should be noted that the magnitude of the size-consistency error increases as the size of the molecule increases. However, using the Davidson correction can reduce the error significantly.

9.10 Multiconfiguration and multireference methods

In the **multiconfiguration self-consistent field method** (MCSCF), in addition to optimizing the coefficients C_{Js} in eqn 9.32, the expansion coefficients c_{ji} in eqn 9.14 are also optimized.[17] This simultaneous optimization of both sets of expansion coefficients makes MCSCF computationally demanding, but by optimizing c_{ji}, accurate results can be obtained with the inclusion of even a relatively small number of CSFs.

The development of efficient MCSCF methods is actively being pursued and is particularly important for excited states. One such scheme is the **complete active-space self-consistent field method** (CASSCF).[18] In this approach, the spinorbitals (which are themselves optimized during the calculation by determining the optimal values of the c_{ji}) are divided into three classes:

1. A set of **inactive orbitals** composed of the lowest energy spinorbitals which are doubly occupied in all determinants.
2. A set of **virtual orbitals** of very high energy spinorbitals which are unoccupied in all determinants.
3. A set of **active orbitals** which are energetically intermediate between the inactive and virtual orbitals.

The **active electrons** are the electrons that are not in the doubly occupied inactive orbital set. The CSFs included in the CASSCF calculation are configurations (of the appropriate symmetry and spin) that arise from all possible ways of distributing the active electrons over the active orbitals.

There is a choice as to which orbitals to include as active orbitals. One approach is to take the active orbitals as those (bonding, nonbonding, and antibonding) orbitals that arise in qualitative MO theory from the valence atomic orbitals forming the molecule. For example, in a CASSCF calculation of the ground-state wavefunction and energy of the homonuclear diatomic B_2, one might take the inactive orbitals to be the σ-orbitals formed from the B$1s$ atomic orbitals and the active orbitals to be the σ- and π-orbitals that are formed from the $2s$ and $2p$ atomic orbitals. The choice of active orbitals is critical because the number of CSFs rises very quickly as the number of active orbitals increases.

[17] In the CI method described previously, the coefficients c_{ji} are held fixed during the calculation once they have been determined by an initial Hartree–Fock SCF calculation.

[18] B.O. Roos, *Int. J. Quantum Chem. Symp.*, 175, **14** (1980); L.M. Cheung, K.R. Sundberg, and K. Ruedenberg, *Int. J. Quantum Chem.*, 1103, **16** (1979).

In the conventional CI methods described above, the Hartree–Fock SCF wavefunction Φ_0 is used as a reference configuration and configuration state functions are formed by moving electrons out of the occupied spinorbitals of Φ_0 into unoccupied spinorbitals. In **multireference configuration interaction** (MRCI), a set of reference configurations is created, from which excited determinants are formed for use in a CI calculation. For example, an MCSCF calculation could be carried out and a set of reference configurations composed of those determinants having a coefficient C_{Js} larger than some threshold value (such as 0.05 or 0.1) in the final normalized MCSCF wavefunction. For each of the reference determinants, electrons are moved from occupied spinorbitals to unoccupied spinorbitals to create more determinants for inclusion in the CI expansion in eqn 9.32. Then CI is performed, optimizing all the coefficients C_{Js} of the determinants that have been included. The reference determinants will themselves often be singly and doubly excited determinants with respect to Φ_0 and single and double excitations from the reference determinants are included. Therefore, the final MRCI wavefunction will include determinants that are triply and quadruply excited from Φ_0. Because MRCI calculations often include the most important quadruply excited determinants, they usually significantly reduce the size-consistency error encountered in SDCI calculations. In addition, it is often the case that a large fraction of the exact correlation energy can be recovered from MRCI calculations with a much smaller number of determinants.

9.11 Møller–Plesset many-body perturbation theory

Configuration interaction calculations provide a systematic approach for going beyond the Hartree–Fock level, by including determinants that are successively singly excited, doubly excited, triply excited, and so on, from a reference configuration. One important feature of the method is that it is variational, but one disadvantage is its lack of size-consistency (with the exception of full CI). Perturbation theory (PT, Section 6.2) provides an alternative systematic approach to finding the correlation energy: whereas its calculations are size-consistent, they are not variational in the sense that it does not in general give energies that are upper bounds to the exact energy.

The application of PT to a system composed of many interacting particles is generally called **many-body perturbation theory** (MBPT). Because we want to find the correlation energy for the ground state, we take the zero-order hamiltonian from the Fock operators of the Hartree–Fock SCF method. This choice of $H^{(0)}$ was made in the early days of quantum mechanics (in 1934) by C. Møller and M.S. Plesset, and the procedure is called **Møller–Plesset perturbation theory** (MPPT). Applications of MPPT to molecular systems did not actually begin till some 40 years later.[19]

In MPPT, the zero-order hamiltonian $H^{(0)}$ (in this context denoted H_{HF}) is given by the sum of one-electron Fock operators defined in eqn 9.8:

$$H_{\mathrm{HF}} = \sum_{i=1}^{n} f_i \tag{39}$$

[19] See J.A. Pople, J.S. Binkley, and R. Seeger, *Int. J. Quant. Chem. Symp.*, 1, **10** (1976) for an early reference.

The HF ground-state wavefunction Φ_0 is an eigenfunction of H_{HF} with an eigenvalue $E^{(0)}$ given by the sum of the orbital energies of all the occupied spinorbitals, as demonstrated in the next worked example.

Example 9.4 Showing that Φ_0 is an eigenfunction of H_{HF} and determining its eigenvalue

Show that the HF ground-state wavefunction is an eigenfunction of the zero-order MPPT hamiltonian and that its eigenvalue equals the sum of the occupied spinorbital energies.

Method. Consider the hamiltonian H_{HF} given in eqn 9.39 and the effect of the one-electron Fock operator f_i on the spinorbital $\phi_a(1)$ as given in eqn 9.7. Analyse the effect of H_{HF} on each term in the expansion of Φ_0. Use the fact that a linear combination of eigenfunctions of an hermitian operator all having an identical eigenvalue is itself an eigenfunction with the same eigenvalue.

Answer. The HF ground-state wavefunction is denoted

$$\Phi_0 = |\phi_a(1)\phi_b(2)\ldots\phi_z(n)|$$

An expansion of the Slater determinant yields a sum of terms, each of which involves a product of the n spinorbitals $\phi_a\phi_b\ldots\phi_z$ with the electrons $1, 2, \ldots n$ distributed differently in each term in the summation. Consider the effect of H_{HF} on one of these terms, the principal diagonal of the determinant, using eqns 9.39 and 9.7:

$$\begin{aligned} &H_{HF}\phi_a(1)\phi_b(2)\ldots\phi_z(n) \\ &\quad = \sum_{i=1}^{n} f_i\{\phi_a(1)\phi_b(2)\ldots\phi_z(n)\} \\ &\quad = \{f_1\phi_a(1)\}\phi_b(2)\ldots\phi_z(n) + \phi_a(1)\{f_2\phi_b(2)\}\ldots\phi_z(n) + \ldots \\ &\qquad + \phi_a(1)\phi_b(2)\ldots\{f_n\phi_z(n)\} \\ &\quad = (\varepsilon_a + \varepsilon_b + \ldots + \varepsilon_z)\phi_a(1)\phi_b(2)\ldots\phi_z(n) \end{aligned}$$

Because each term in the expansion of Φ_0 has each occupied spinorbital appearing once, each term in the expansion is an eigenfunction of H_{HF} with the same eigenvalue $\varepsilon_a + \varepsilon_b + \ldots + \varepsilon_z$. Therefore, we can immediately conclude that Φ_0 is an eigenfunction of the MPPT zero-order hamiltonian with an eigenvalue given by the sum of the orbital energies of all occupied spinorbitals.

Exercise 9.4. Show that all singly and multiply excited determinants Φ_J are also eigenfunctions of H_{HF} with eigenvalues E_J equal to the sums of the orbital energies of the spinorbitals occupied in that particular determinant.

The perturbation $H^{(1)}$ is given by

$$H^{(1)} = H - \sum_{i=1}^{n} f_i \tag{40}$$

where, as before, H is the electronic hamiltonian. The HF energy E_{HF} associated with the (normalized) ground-state HF wavefunction Φ_0 is the expectation value

$$E_{\mathrm{HF}} = \langle \Phi_0 | H | \Phi_0 \rangle \tag{41}$$

or, equivalently,

$$E_{\mathrm{HF}} = \langle \Phi_0 | H_{\mathrm{HF}} + H^{(1)} | \Phi_0 \rangle \tag{42}$$

It is easy to show that E_{HF} is equal to the sum of the zero-order energy $E^{(0)}$ and the first-order energy correction $E^{(1)}$. From eqns 6.12 and 6.20 and the fact that Φ_0 is an eigenfunction of H_{HF}, we have

$$E^{(0)} = \langle \Phi_0 | H_{\mathrm{HF}} | \Phi_0 \rangle \tag{43}$$

$$E^{(1)} = \langle \Phi_0 | H^{(1)} | \Phi_0 \rangle \tag{44}$$

From eqns 9.42–44, we conclude that

$$E_{\mathrm{HF}} = E^{(0)} + E^{(1)}$$

Therefore, the first correction to the ground-state energy is given by second-order perturbation theory as (see eqn 6.24)

$$E^{(2)} = \sum_{J \neq 0} \frac{\langle \Phi_J | H^{(1)} | \Phi_0 \rangle \langle \Phi_0 | H^{(1)} | \Phi_J \rangle}{E^{(0)} - E_J} \tag{45}$$

To evaluate eqn 9.45, we need to be able to evaluate the off-diagonal matrix elements $\langle \Phi_J | H^{(1)} | \Phi_0 \rangle$. First, note that the matrix element

$$\langle \Phi_J | H_{\mathrm{HF}} | \Phi_0 \rangle = 0$$

because Φ_0 is an eigenfunction of H_{HF} and the spinorbitals, and hence the determinants, are orthogonal. Therefore,

$$\text{if} \quad \langle \Phi_J | H | \Phi_0 \rangle = 0, \quad \text{then} \quad \langle \Phi_J | H^{(1)} | \Phi_0 \rangle = 0$$

From Brillouin's theorem and the discussion in Section 9.9, we conclude that only the doubly excited determinants have nonzero $H^{(1)}$ matrix elements with Φ_0 and therefore only double excitations contribute to $E^{(2)}$. An analysis of these nonvanishing matrix elements[20] yields the following expression:

$$E^{(2)} = \tfrac{1}{4} \sum_{a,b}^{\mathrm{occ}} \sum_{p,q}^{\mathrm{vir}} \frac{(ab\|pq)(pq\|ab)}{\varepsilon_a + \varepsilon_b - \varepsilon_p - \varepsilon_q} \tag{46}$$

where

$$\begin{aligned}(ab\|pq) = {} & \int \phi_a^*(1)\phi_b^*(2)\left(\frac{e^2}{4\pi\varepsilon_0 r_{12}}\right)\phi_p(1)\phi_q(2)\,\mathrm{d}\mathbf{x}_1\mathrm{d}\mathbf{x}_2 \\ & - \int \phi_a^*(1)\phi_b^*(2)\left(\frac{e^2}{4\pi\varepsilon_0 r_{12}}\right)\phi_q(1)\phi_p(2)\,\mathrm{d}\mathbf{x}_1\mathrm{d}\mathbf{x}_2\end{aligned} \tag{47}$$

[20] See, for example, Section 2.3.4.3 of D.M. Hirst, *A computational approach to chemistry*, Blackwell Scientific Publications, Oxford (1990).

with ϕ_a, ϕ_b occupied spinorbitals and ϕ_p, ϕ_q virtual spinorbitals. The inclusion of the second-order energy correction is designated **MP2**.

In general, bond lengths based on MP2 are in excellent agreement with experiment for bonds involving hydrogen. However, the same cannot be said in general for multiple bonds. For example, the bond lengths for N_2 from MP2 are $2.322a_0$ (STO-3G basis), $2.171a_0$ (4-31G basis), and $2.133a_0$ (6-31G** basis) compared to the experimental result of $2.074a_0$.

It is possible to extend the MPPT to include third- and fourth-order energy corrections, and the procedures are then referred to as MP3 and MP4.[21] As one moves to higher orders of perturbation theory, the algebra involved becomes more and more complicated and it is common to use diagrammatic techniques to classify and represent the various terms that appear in the perturbation series expressions. These diagrammatic representations can be used to prove that MPPT is size-consistent in all orders.

Density functional theory

The *ab initio* methods described above all start with the Hartree–Fock approximation in that the HF equations are first solved to obtain spinorbitals, which can then be used to construct configuration state functions. These methods are widely used by quantum chemists today. However, they do have limitations, in particular the computational difficulty of performing accurate calculations with large basis sets on molecules containing many atoms.

An alternative to the HF methods that has been growing in popularity over the past decade is **density functional theory** (DFT). In contrast to the methods described above, which use CSFs, DFT begins with the concept of the electron probability density. One reason for the growing popularity is that DFT, which takes into account electron correlation, is less demanding computationally than, for example, CI and MP2. It can be used to do calculations on molecules of 100 or more atoms in significantly less time than these HF methods. Furthermore, for systems involving *d*-block metals, DFT yields results that very frequently agree more closely with experiment than HF calculations do.

The basic idea behind DFT is that the energy of an electronic system can be written in terms of the electron probability density, ρ.[22] For a system of n electrons, $\rho(\mathbf{r})$ denotes the total electron density at a particular point in space $\mathbf{r}$. The electronic energy E is said to be a **functional** of the electron density, denoted $E[\rho]$, in the sense that for a given *function* $\rho(\mathbf{r})$, there is a single corresponding energy.[23]

The concept of a density functional for the energy was the basis of some early but useful approximate models such as the Thomas–Fermi method (which emerged from work in the late 1920s by E. Fermi and L.H. Thomas)

[21] For a detailed discussion, see S. Wilson, *Electron correlation in molecules*, Clarendon Press, Oxford (1984).

[22] For more extensive treatments, the reader is encouraged to see the qualitative discussion in S. Borman, *Chem. Eng. News*, April 9, 1990, p22 and a more detailed quantitative discussion in the review by T. Ziegler, *Chem. Rev.*, 651, **91** (1991).

[23] We have encountered the idea of a functional before, but we did not use this name. An expectation value of a hamiltonian is the energy as a functional of the wavefunction, ψ: each well-behaved function ψ is associated with a single expectation value of the energy.

and the Hartree–Fock–Slater or Xα method, which emerged from the work of J.C. Slater in the 1950s. However, it was not until 1964 that a formal proof was given[24] that the ground-state energy and all other ground-state electronic properties are uniquely determined by the electron density. Unfortunately, the **Hohenberg–Kohn theorem** does not tell us the *form* of the functional dependence of energy on the density: it confirms only that such a functional exists. The next major step in the development of DFT came with the derivation of a set of one-electron equations from which in theory the electron density ρ could be obtained[25].

We shall focus exclusively on systems in which paired electrons have the same spatial one-electron orbitals (as in restricted Hartree–Fock theory). As shown by Kohn and Sham, the exact ground-state electronic energy E of an n-electron system can be written as

$$E[\rho] = -\frac{\hbar^2}{2m_e}\sum_{i=1}^{n}\int \psi_i^*(\mathbf{r}_1)\nabla_1^2\psi_i(\mathbf{r}_1)\,d\mathbf{r}_1 - \sum_{I=1}^{N}\int \frac{Z_I e^2}{4\pi\varepsilon_0 r_{I1}}\rho(\mathbf{r}_1)\,d\mathbf{r}_1 + \frac{1}{2}\int \frac{\rho(\mathbf{r}_1)\rho(\mathbf{r}_2)e^2}{4\pi\varepsilon_0 r_{12}}\,d\mathbf{r}_1 d\mathbf{r}_2 + E_{XC}[\rho] \tag{48}$$

where the one-electron spatial orbitals ψ_i $(i = 1, 2, \ldots n)$ are the **Kohn–Sham orbitals**, the solutions of the equations given below. The exact ground-state charge density ρ at a location $\mathbf{r}$ is given by

$$\rho(\mathbf{r}) = \sum_{i=1}^{n} |\psi_i(\mathbf{r})|^2 \tag{49}$$

where the sum is over all the occupied Kohn–Sham orbitals and is known once these orbitals have been computed. The first term in eqn 9.48 represents the kinetic energy of the electrons; the second term represents the electron–nucleus attraction with the sum over all N nuclei with index I and atomic number Z_I; the third term represents the Coulomb interaction between the total charge distribution (summed over all orbitals) at $\mathbf{r}_1$ and $\mathbf{r}_2$; the last term is the **exchange-correlation energy** of the system which is also a functional of the density and takes into account all nonclassical electron–electron interactions. Of the four terms, E_{XC} is the one we do not know how to obtain exactly. Although the Hohenberg–Kohn theorem tells us that E and therefore E_{XC} must be functionals of electron density, we do not know the latter's exact analytical form, and so are forced to use approximate forms for it.

The Kohn–Sham (KS) orbitals are found by solving the **Kohn–Sham equations**, which can be derived by applying a variational principle to the electronic energy $E[\rho]$ with the charge density given by eqn 9.49. The KS equations for the one-electron orbitals $\psi_i(\mathbf{r}_1)$ have the form

$$\left\{-\frac{\hbar^2}{2m_e}\nabla_1^2 - \sum_{I=1}^{N}\frac{Z_I e^2}{4\pi\varepsilon_0 r_{I1}} + \int \frac{\rho(\mathbf{r}_2)e^2}{4\pi\varepsilon_0 r_{12}}\,d\mathbf{r}_2 + V_{XC}(\mathbf{r}_1)\right\}\psi_i(\mathbf{r}_1) = \varepsilon_i\psi_i(\mathbf{r}_1) \tag{50}$$

[24] P. Hohenberg and W. Kohn, *Phys. Rev.*, 864, **B136** (1964).
[25] W. Kohn and L.J. Sham, *Phys. Rev.*, 1133, **A140** (1965).

where ε_i are the KS orbital energies and the **exchange-correlation potential**, V_{XC}, is the functional derivative of the exchange-correlation energy:[26]

$$V_{XC}[\rho] = \frac{\delta E_{XC}[\rho]}{\delta \rho} \tag{51}$$

If E_{XC} is known, then V_{XC} is readily obtained. The significance of the KS orbitals is that they allow the density ρ to be computed from eqn 9.49.

The KS equations are solved in a self-consistent fashion. Initially, we guess the charge density ρ (to do so, a superposition of atomic densities for molecular systems is often used). By using some approximate form (which is fixed during all iterations) for the functional dependence of E_{XC} on density, we next compute V_{XC} as a function of $\mathbf{r}$. The set of KS equations is then solved to obtain an initial set of KS orbitals. This set of orbitals is then used to compute an improved density from eqn 9.49, and the process is repeated until the density and exchange-correlation energy have converged to within some tolerance. The electronic energy is then computed from eqn 9.48.

The KS orbitals on each iteration can be computed numerically or they can be expressed in terms of a set of basis functions; in the case of the latter, solving the KS equations amounts to finding the coefficients in the basis set expansion. As in the HF methods, a variety of basis set functions can be used and the wealth of experience gained in HF calculations can prove to be useful in the choice of DFT basis sets. The computation time required for a DFT calculation formally scales as the third power of the number of basis functions.

Several different schemes have been developed for obtaining approximate forms for the functional for the exchange-correlation energy. The main source of error in DFT usually arises from the approximate nature of E_{XC}. In the **local density approximation** (LDA), it is

$$E_{XC} = \int \rho(\mathbf{r})\varepsilon_{XC}[\rho(\mathbf{r})]\,\mathbf{dr} \tag{52}$$

where $\varepsilon_{XC}[\rho(\mathbf{r})]$ is the exchange-correlation energy per electron in a homogeneous electron gas of constant density. In a hypothetical homogeneous electron gas, an infinite number of electrons travel throughout a space of infinite volume in which there is a uniform and continuous distribution of positive charge to retain electroneutrality.[27]

This expression for the exchange-correlation energy is clearly an approximation because neither positive charge nor electronic charge are uniformly distributed in actual molecules. To account for the inhomogeneity of the electron density, a nonlocal correction involving the gradient of ρ is often added to the exchange-correlation energy of eqn 9.52. The LDA with nonlocal

[26] Consider a functional $G[f]$ which depends on the function $f(\mathbf{r})$. When $\mathbf{r}$ undergoes an arbitrarily small change $\delta\mathbf{r}$ and the function changes to $f + \delta f$, the functional undergoes a corresponding change to $G[f + \delta f]$. We then define the functional derivative $\delta G/\delta f$ as

$$\frac{\delta G}{\delta f} = \lim_{|\delta f| \to 0} \frac{G[f + \delta f] - G[f]}{|\delta f|}$$

where the manner in which $|\delta f| \to 0$ must be specified explicitly.

[27] An accurate expression for $\varepsilon_{XC}[\rho]$ will be found in R.G. Parr and W. Yang, *Density functional theory of atoms and molecules*, Oxford University Press, Oxford (1989).

corrections (LDA-NL) appears to be one of the most accurate and efficient methods for calculations involving d-metal complexes within DFT. Table 9.4 compares calculated and experimental values of M—CO bond strengths for several d-block metals. The calculated metal–ligand mean bond energies are almost of chemical accuracy ($\pm 20\,\mathrm{kJ\,mol^{-1}}$).

Table 9.4 Calculated and experimental metal–ligand mean bond energies*

	Calculated	Observed
$Cr(CO)_6$	107	110
$Mo(CO)_6$	126	151
$W(CO)_6$	156	179

* Energies are in kilojoules per mole of M—L bonds ($\mathrm{kJ\,mol^{-1}}$).

Gradient methods and molecular properties

Although this chapter has been concerned primarily with the solution of the electronic Schrödinger equation within the Born–Oppenheimer approximation, the resulting potential energy surface (or curve for a diatomic) is invaluable for computing a number of molecular properties, perhaps the most important being the equilibrium molecular geometry. The calculation of molecular structures can be a valuable supplement to experimental data in areas of structural chemistry, including X-ray crystallography, electron diffraction, and microwave spectroscopy. Calculation of derivatives of the potential energy with respect to nuclear coordinates is crucial to the efficient determination of equilibrium structures. The derivatives can be computed numerically by calculating the potential energy at many geometries and determining the change in energy as each nuclear coordinate is varied. However, **gradient methods**, which determine energy derivatives *analytically*,[28] are computationally faster and more accurate than numerical differentiation. Since 1969, when P. Pulay wrote the first computer program for determining first derivatives of SCF energies analytically, gradient methods have developed into one of the most vigorously studied areas of modern quantum chemistry. First applied to closed-shell SCF calculations, gradient methods were then generalized to open-shell RHF and UHF calculations. In addition to the development of gradient methods for *ab initio* techniques based on Slater determinants, analytical expressions have also been derived for DFT. In general, analytical first and second energy derivatives are now available for a number of levels of *ab initio* calculations.

For a diatomic molecule, the molecular potential energy, E, depends only on the internuclear distance, R; therefore, to find the potential minimum (more generally, any stationary point) we need to locate a zero in $\mathrm{d}E/\mathrm{d}R$.[29] The search is more complicated for polyatomic molecules because the potential energy is a function of many nuclear coordinates, q_i. At the equilibrium

[28] The gradient of a function of several variables is defined in *Further information 22*.
[29] In this section, nucleus–nucleus repulsion is included in the potential energy.

geometry, each of the forces f_i exerted on a nucleus by electrons and other nuclei must vanish:

$$f_i = -\frac{\partial E}{\partial q_i} = 0$$

Therefore, in principle, the equilibrium geometry can be found by computing all the forces at a given molecular geometry and seeing if they vanish. If they do not vanish, we vary the geometry until it does correspond to zero forces (a gradient vector of zero length). Computationally, the forces will not vanish identically, but by computing their magnitudes we can stop the iterative search for the equilibrium geometry when the magnitudes are close to zero. Typically, to obtain (bond and torsional) angles and bond distances within 1° and $0.002a_0$, respectively, of their optimal computed equilibrium values, all forces should be below about 10 pN. Such tolerances typically require between N and $2N$ cycles.

Before one optimizes the equilibrium geometry, a coordinate system must be chosen to represent the potential surface and molecular structure. This choice is important because it affects the ease of optimization, and the internal coordinates (the bond lengths, bond angles, and torsional angles) are often used. An additional point is that it is important to recognize that experimental data usually refer to averaged molecular quantities. Therefore, the comparison should use vibrationally averaged bond lengths and bond angles. The differences between experimental and computed equilibrium geometries for a series of related compounds are often found to be very consistent. For instance, in a Hartree–Fock SCF study of 30 organic compounds using a 4-21G basis, the optimized CH bond distance tended to be consistently about $0.07a_0$ smaller than the experimental value.[30] This consistency makes it possible in many cases to construct a set of empirical corrections for *ab initio* geometries yielding absolute accuracies of about $0.02a_0$.

A zero gradient characterizes a stationary point on the surface but does not differentiate among minima, maxima, or saddle points.[31] Therefore, the searching procedure lets us identify not only an equilibrium geometry but also the transition state of a chemical reaction, the latter corresponding to a saddle point on the surface. To distinguish the types of stationary points, it is necessary to consider the second derivatives of the energy with respect to the nuclear coordinates. The quantities $\partial^2 E/\partial q_i \partial q_j$ comprise the **Hessian matrix**. Whereas a minimum (maximum) of a curve corresponds to a positive (negative) second derivative, a minimum (maximum) of a multidimensional potential energy surface is characterized by the *eigenvalues* of the Hessian matrix all being positive (negative). A transition state (a first-order saddle point) corresponds to one negative eigenvalue and all the rest positive.

There are a number of algorithms available for finding stationary points of a potential surface. In general, the stability, reliability, and computational cost of the algorithm as well as its speed of convergence should be considered. The

[30] An extensive comparison of theory and experiment is made in W.J. Hehre, L. Radom, P.v.R. Schleyer, and J.A. Pople, *Ab initio molecular orbital theory*, Wiley, New York (1986).

[31] A (first-order) saddle point is a potential maximum along one nuclear coordinate and a potential minimum along all others.

algorithms can be broadly classified into three groups. Those using only the energy are the slowest to converge but are useful if analytical derivatives are unavailable. Those using both the energy and its analytical first derivatives are significantly more efficient (by almost an order of magnitude). Furthermore, their rate of convergence can be improved if one has a good initial estimate of the Hessian matrix, perhaps obtained from lower level *ab initio* calculations (for example, one that uses smaller basis sets). Algorithms that use the energy together with its analytical first and second derivatives are the most accurate and efficient methods. Whichever algorithm is used, all nuclear coordinates should be optimized; this optimization is especially important for transition states where optimizing a subset of all the nuclear coordinates might locate a saddle point that changes significantly when all coordinates are optimized.

Energy derivatives are also useful for determining molecular properties other than equilibrium and transition state geometries. The second derivatives of the energy with respect to nuclear coordinates (the Hessian matrix elements) are the force constants for normal mode frequencies within the harmonic approximation (Section 10.13). The third, fourth, and higher derivatives give anharmonic corrections to vibrational frequencies (Section 10.16). Energy derivatives are not limited to nuclear coordinates; for example, it is sometimes useful to consider derivatives with respect to electric field components. Mixed second derivatives with respect to one nuclear coordinate and one electric field component yield dipole moment derivatives which are used to determine infrared intensities within the harmonic approximation (see Section 10.10).

To calculate the analytical derivatives of the energy with respect to nuclear coordinates it is necessary to compute derivatives of one- and two-electron integrals over basis functions. Because the basis functions are centred on atomic nuclei, when derivatives of the integrals are determined, we need the derivatives of the basis set functions with respect to nuclear coordinates. Whether or not derivatives of various expansion coefficients are also required depends on if they were determined variationally and on the order of the energy derivative under consideration.

As an example, consider derivatives of an expansion coefficient which in general we denote c_j. The contribution to the first derivative of the energy with respect to q_i from the derivative of c_j will include (by the chain rule) a contribution of the form $(\partial E/\partial c_j)(\partial c_j/\partial q_i)$. However, for variationally determined c_j, $(\partial E/\partial c_j)$ is zero; therefore, we have the important result that to evaluate the gradient (first derivative) of the energy, we do not need derivatives of variationally determined coefficients. As a result, in HF and MCSCF, the analytical determination of the energy gradient requires derivatives of only the one- and two-electron integrals. Similarly, evaluation of the second derivative of the energy does not require the second derivative of the variationally determined coefficients. However, it does require their first derivatives, $(\partial c_j/\partial q_i)$, which are computed by solving the **coupled perturbed Hartree–Fock equations** (CPHF) for those *ab initio* methods that use a single reference CSF (for instance, SDCI and MP2) or by solving the **coupled perturbed MCSCF equations** (CPMCSCF)[32] for those using multiple reference CSFs

[32] When the expansion coefficients are constrained, for example by orthonormality, appropriate Lagrangian multipliers must be introduced into the coupled perturbed equations.

(for instance, MCSCF and MRCI). In general, to compute energy derivatives through order $(2n+1)$ requires derivatives of variationally determined coefficients through order n.[33] Therefore, the third derivative of the energy requires first derivatives of variational coefficients. An efficient method for solving the CPHF equations was first developed by J.A. Pople and coworkers in 1979 that made energy second derivative calculations practicable for RHF and UHF.

Analytical first derivatives of the energy do require determination of the first derivatives of non-variationally determined coefficients. The latter derivatives are obtained by solving the appropriate coupled perturbed equations. However, a significant simplification in the solution of the coupled perturbed equations was discovered by N.C. Handy and H.F. Schaefer in 1984, who showed that rather than solving all the CPHF or CPMCSCF equations, it is in fact necessary to solve only a much smaller set of equations. This simplification is also applicable to higher energy derivatives; for example, when determining second derivatives of the energy, the full set of coupled perturbed equations for the second derivatives of the non-variational coefficients can also be reduced to a smaller set.

A critical step in calculating analytical derivatives of the energy is the evaluation of the derivatives of the one- and two-electron integrals. These derivatives require evaluation of derivatives of the basis functions with respect to the nuclear coordinates of their centres. For Gaussian-type orbitals (eqn 9.28), the derivatives of the basis functions may be computed analytically and result in other GTOs. For example, the first derivative of an s-type Gaussian with respect to x_c yields a p-type GTO; the first derivative of the p-type GTO θ_{100} yields an s- and a d-type GTO. Much work has gone into the efficient evaluation of integral derivatives (see *Further reading*). The efficiency can be increased by using translational and rotational invariance properties and by using molecular symmetry.

Semiempirical methods

It should be apparent by now that there are computational limitations in the accurate treatment of molecular systems with large numbers of electrons. Even with increases in computer speed and memory and the development of efficient algorithms, *ab initio* methods are not applied routinely to molecules with several dozen atoms. On the other hand, semiempirical methods are both accurate and fast enough to be applied routinely to larger systems. Thus, semiempirical methods make electronic structure calculations available for a wider range of molecules. Nevertheless, *ab initio* methods represent a more theoretically 'pure' approach, and one of the limitations to the accuracy of the semiempirical methods in addition to the approximations inherent in their formulation is the accuracy of experimental data used to obtain the parameters. However, in large part because adjustable parameters are optimized to reproduce a number of important chemical properties, semiempirical methods have become widely popular.

[33] The analogy to perturbation theory (Section 6.6) should be noted. In fact, the connection between PT and gradient methods runs deeper than may be apparent; early investigations of SCF derivatives were often described in the language of PT.

The optimization of parameters is, in general, a difficult task for several reasons. For one, accurate experimental data are often not available. Also, the simultaneous optimization of several parameters for a large number of molecules is very time-consuming. The parameters are interconnected in the sense that significantly varying the value of one parameter in a nearly optimal parameter set needs to be accompanied by variations in several other parameters too. The successive optimization of each parameter is not feasible. Semiempirical methods were first developed for conjugated π-electron systems. We shall therefore begin our discussion with them and later describe more general methods.

9.12 Conjugated π-electron systems

We consider the case of a conjugated π-system with a total of n_π π-electrons. The π-electrons are dealt with separately from the σ-electrons partly because their energies are so different and partly on account of the different symmetries of their orbitals. The effective π-electron hamiltonian H_π is given by

$$H_\pi = -\frac{\hbar^2}{2m_e}\sum_{i=1}^{n_\pi}\nabla_i^2 + \sum_{i=1}^{n_\pi} V_i^{\pi,\text{eff}} + \tfrac{1}{2}\sum_{i,j}^{n_\pi}\frac{e^2}{4\pi\varepsilon_0 r_{ij}} \tag{53}$$

where the first term is the kinetic energy operator for the π-electrons, $V_i^{\pi,\text{eff}}$ is the effective potential energy for π-electron i resulting from the potential field of the nuclei and all σ-electrons, and the final term represents the repulsive potential energy due to π-interelectronic interactions. The core hamiltonian h_i^π for π-electron i is defined by

$$h_i^\pi = -\frac{\hbar^2}{2m_e}\nabla_i^2 + V_i^{\pi,\text{eff}} \tag{54}$$

so we can write

$$H_\pi = \sum_{i=1}^{n_\pi} h_i^\pi + \tfrac{1}{2}\sum_{i,j}^{n_\pi}\frac{e^2}{4\pi\varepsilon_0 r_{ij}} \tag{55}$$

The analogy with eqn 7.43 is apparent. The hamiltonian H_π is approximate because the π- and σ-electrons have been treated separately and the effect of the latter in H_π appears only in the effective potential $V_i^{\pi,\text{eff}}$. The use of an approximate form for the hamiltonian in eqn 9.1 is characteristic of semiempirical methods.

The most famous semiempirical π-electron theory is **Hückel molecular orbital theory** (HMO). As this method has already been described in some detail in Section 8.9, here we shall only point out some of the features of this method that characterize it as semiempirical. In the HMO method, H_π is approximated as a sum of one-electron terms:

$$H_\pi = \sum_{i=1}^{n_\pi} h_i^{\pi,\text{eff}} \tag{56}$$

where $h_i^{\pi,\text{eff}}$ is an effective hamiltonian for π-electron i. The form of $h_i^{\pi,\text{eff}}$ is left unspecified; only its matrix elements appear in HMO. Because H_π is a sum of one-electron terms, the wavefunction Ψ_π can be written in terms of a product of one-electron (molecular) orbitals ψ_i, each of which is a solution of the eigenvalue equation

$$h_i^{\pi,\text{eff}}\psi_i = E_i\psi_i \tag{57}$$

where E_i is the energy associated with the molecular orbital labelled i. Each molecular orbital is written as a linear combination of atomic orbitals (LCAO). For example, in an HMO treatment of a conjugated hydrocarbon (such as benzene), the atomic orbitals are usually composed of the set of C2p_z atomic orbitals. The variation principle is then applied, and gives rise to a set of secular equations, which have non-trivial solutions only if

$$\det|\mathbf{h}^{\pi,\mathrm{eff}} - E\mathbf{S}| = 0 \tag{58}$$

where $h_{rs}^{\pi,\mathrm{eff}}$ is the matrix element of $\mathbf{h}^{\pi,\mathrm{eff}}$ between the atomic orbitals of the conjugated atoms r and s and S_{rs} is their overlap integral. This expression is the analogue of the Roothaan equations, eqn 9.21, but differs in the restriction of the hamiltonian to a sum of one-electron terms and the orbitals to π-orbitals. Solution of the secular determinant then yields the set of molecular orbital energies E_i as well as the expansion coefficients of the LCAO. As described in Section 8.9, HMO makes some assumptions about the values of the matrix elements $h_{rs}^{\pi,\mathrm{eff}}$ and S_{rs}:

1. For all overlap integrals, $S_{rs} = \delta_{rs}$.
2. Diagonal elements $h_{rr}^{\pi,\mathrm{eff}} = \alpha$.
3. Off-diagonal elements $h_{rs}^{\pi,\mathrm{eff}} = \beta$ if atoms r and s are neighbours and 0 otherwise.

The setting of selected matrix elements to zero and the parametrizing of nonzero matrix elements are also common features in semiempirical methods.

Because H_π is written as a sum of one-electron terms with explicit forms left unspecified, the HMO method treats repulsions between the π-electrons very poorly (if at all!). As a result, it is only useful for qualitative discussions of π-conjugated systems.

The **Pariser–Parr–Pople method** (PPP) is a much more substantial procedure, but nevertheless quite primitive when compared with current semiempirical procedures. It starts with the hamiltonian H_π of eqn 9.55, which includes π-interelectronic repulsions, and writes the π-electron wavefunction Ψ_π as a Slater determinant of π-electron spinorbitals ϕ_i^π:

$$\Psi_\pi = (n_\pi!)^{-1/2} \det|\phi_a^\pi(1)\phi_b^\pi(2)\ldots\phi_z^\pi(n_\pi)| \tag{59}$$

The optimal spinorbitals are determined by using the variation principle, and satisfy

$$f_1^\pi \phi_a^\pi(1) = \varepsilon_a^\pi \phi_a^\pi(1) \tag{60}$$

where ε_a^π is the orbital energy of spinorbital ϕ_a^π, and where

$$f_1^\pi = h_1^\pi + \sum_u \{J_u(1) - K_u(1)\} \tag{61}$$

The Coulomb (J_u) and exchange (K_u) operators are defined as in eqns 9.9 and 9.10.

At this stage, the calculation is following the *ab initio* route described in Sections 9.1 and 9.2. Indeed, proceeding as in Section 9.4 for the closed-shell case, we can write the π-electron spinorbital as a product of a spin function and a space function; the space function is then expanded in a basis of known functions. The space functions are the π molecular orbitals and the known basis functions are atomic orbitals θ_i centred on each π-conjugated atom i. For example, in a conjugated hydrocarbon, a basis set consisting of C2p_z atomic orbitals is typically employed. The use of the basis set results in a set of equations analogous to the Roothaan equations, eqn 9.19, with F_{ij} in the latter replaced by the matrix elements F_{ij}^{π}:

$$F_{ij}^{\pi} = \int \theta_i^*(1) f_1^{\pi} \theta_j(1)\,\mathrm{d}\mathbf{r}_1 \tag{62}$$

and with ε_a replaced by ε_a^{π}. Next, we simplify the notation by defining the one-electron integral

$$h_{ij}^{\pi} = \int \theta_i^*(1) h_1^{\pi} \theta_j(1)\,\mathrm{d}\mathbf{r}_1 \tag{63}$$

and using the symbol $(ab|cd)$ to represent two-electron integrals over the atomic orbitals, we can write the matrix element F_{ij}^{π} as

$$F_{ij}^{\pi} = h_{ij}^{\pi} + \sum_{l,m} P_{lm}\{(ij|lm) - \tfrac{1}{2}(im|lj)\} \tag{64}$$

where P_{lm} is defined in eqn 9.27.

At this point the PPP method makes some approximations beyond the separation of π and σ orbitals. First, we set $S_{ij} = \delta_{ij}$, as in HMO theory. Then we set some of the two-electron integrals

$$(ab|cd) = \int \theta_a^*(1)\theta_b(1)\left(\frac{e^2}{4\pi\varepsilon_0 r_{12}}\right)\theta_c^*(2)\theta_d(2)\,\mathrm{d}\mathbf{r}_1\mathrm{d}\mathbf{r}_2$$

to zero too, but in a more subtle way than in HMO. The product $\theta_i^*(1)\theta_j(1)$ in which $i \neq j$ is designated a **differential overlap** term (it is the integrand of an overlap integral, so can formally be obtained from an overlap integral by differentiation; hence the name). In the **zero differential overlap approximation** (ZDO), the two-electron integral vanishes unless $a = b$ and $c = d$. In other words, we set the product of atomic orbitals

$$\theta_a^*(1)\theta_b(1) = 0 \qquad \text{if } a \neq b \tag{65}$$

As a result, the two-electron integrals are given by

$$(ab|cd) = \delta_{ab}\delta_{cd}(aa|cc) \tag{66}$$

and the integral $(aa|cc)$, which could be computed theoretically, is often treated as an empirical parameter. In the ZDO approximation, all three-centre and four-centre two-electron integrals are neglected.

In addition, the PPP method usually does not calculate the integrals h_{ij}^{π} theoretically but instead takes some to be empirical parameters and sets the remainder to zero. In particular, for atomic orbitals θ_i and θ_j centred on atoms i and j which are *not* bonded together, h_{ij}^{π} is set to zero, and for atomic orbitals centred on atoms which *are* bonded together, the matrix element is taken to be an empirical parameter β_{ij} which varies with the nature of the atoms i and j. The diagonal elements h_{ii}^{π} are usually set to an empirical parameter α_i. (Note the resemblance to HMO theory at this point.)

If all two-electron integrals $(ab|cd)$ are set to zero and the matrix elements h_{ij}^{π} replaced by the matrix elements $h_{ij}^{\pi,\mathrm{eff}}$, then the PPP method (an SCF treatment) is 'reduced' to the HMO method (which is not an SCF treatment).

9.13 Neglect of differential overlap

The development of semiempirical methods to treat general molecular systems (and not just π-conjugated systems) has made significant progress due, in large part, to the efforts of J.A. Pople and M.J.S. Dewar and their coworkers. These methods explicitly treat valence electrons and the names of the various methods are suggestive of which two-electron integrals are set to zero in the treatment. We shall first set up the general equations for the treatment of the valence electrons and then describe some of these semiempirical methods without going into too much detail.[34] One point to keep in mind in the following discussion is that (except for hydrogen) there will be several basis functions (atomic orbitals) on each atom; this was not the case for conjugated systems and it makes the bookkeeping of neglected and non-neglected two-electron integrals more complicated.

Consider a closed-shell molecule with n_V valence electrons. The valence-electron hamiltonian H_V is given by

$$H_\mathrm{V} = \sum_{i=1}^{n_\mathrm{V}} h_i^\mathrm{V} + \tfrac{1}{2}\sum_{i,j}^{n_\mathrm{V}} \frac{e^2}{4\pi\varepsilon_0 r_{ij}} \tag{67}$$

where h_i^V is the core hamiltonian for valence electron i given by

$$h_i^\mathrm{V} = -\frac{\hbar^2}{2m_\mathrm{e}}\nabla_i^2 + V_i^{\mathrm{V,eff}} \tag{68}$$

and $V_i^{\mathrm{V,eff}}$ is the effective potential energy for valence electron i resulting from the potential field of the nuclei and all of the inner-shell electrons. We then proceed exactly as we did in Section 9.12 for the PPP method, and so obtain a set of equations identical to those of eqns 9.59–64 but with all the quantities previously labelled π now labelled V. This procedure results in a set of equations analogous to the Roothaan equations (eqn 9.19) to be solved in a self-consistent fashion with F_{ij} in eqn 9.26 replaced by

$$F_{ij}^\mathrm{V} = h_{ij}^\mathrm{V} + \sum_{l,m} P_{lm}\{(ij|lm) - \tfrac{1}{2}(im|lj)\} \tag{69}$$

The most primitive approach is to use the zero differential overlap approximation, as in the PPP method. In the level of approximation known as the **complete neglect of differential overlap** (CNDO) we write

$$(ij|lm) = \delta_{ij}\delta_{lm}(ii|ll) \tag{70}$$

The two-electron integral is taken to be zero even when different atomic orbitals θ_i and θ_j belong to the same atom. The surviving integrals are often taken to be parameters with values that are adjusted until the results of the CNDO calculations resemble those of Hartree–Fock SCF minimal basis set calculations.

To discuss the next level of approximation, which is not as drastic as CNDO, we need to introduce some terminology. The 'exchange integral' was defined in eqn 7.34 in terms of the spatial parts of the spinorbitals; here we shall refer to the two-electron integrals over basis functions θ_i as **exchange integrals** if they

[34] For details about the following procedures, see J.A. Pople and D.L. Beveridge, *Approximate molecular orbital theory*, McGraw-Hill, New York (1970).

are of the form $(ij|ij)$ (which, we note, is equal to $(ij|ji)$ if the basis functions are real). In the level of approximation known as **intermediate neglect of differential overlap** (INDO) we retain exchange integrals $(ij|ij)$ for which atomic orbitals θ_i and θ_j belong to the same atom. These one-centre exchange integrals turn out to be important for explaining the splitting between electronic states that come from the same electronic configuration; thus INDO will give vastly improved results over CNDO when spectroscopic terms are of interest.

We shall now show in a little detail which two-electron integrals are retained in INDO. Consider a diagonal element F^{V}_{ii}. The first integral in the summation $(ij|lm)$ becomes $(ii|lm)$ and the only contribution comes from the integral with $l = m$. This integral $(ii|ll)$ would also be retained in CNDO. The second integral in the summation $(im|lj)$ becomes $(im|li)$, and there are two contributions to it. One contribution comes from $i = l = m$, giving the integral $(ii|ii)$; the other contribution comes from $m = l$ with the stipulation that atomic orbital m belongs to the same atom as atomic orbital i. This one-centre exchange integral $(im|mi)$ would not be retained in CNDO.

Example 9.5 Identifying nonzero two-electron integrals in INDO

What two-electron integrals should be retained in the off-diagonal elements of F^{V}_{ij} if atomic orbitals i and j are centred on the same atom?

Method. We need to examine eqn 9.69 and consider each term in the summation separately. We retain one-centre exchange integrals $(ij|ij)$ in addition to those two-electron integrals $(ii|ll)$ retained in CNDO.

Answer. The two-electron integral $(ij|lm)$ (the first term in the summation of eqn 9.69) contributes when $i = l$ and $j = m$, or $i = m$ and $j = l$, giving one-centre exchange integrals. The second term in the summation $(im|lj)$ contributes (1) when $i = m$ and $j = l$, giving an integral $(ii|ll)$ that is also retained in CNDO or (2) when $i = l$ and $m = j$, giving the one-centre two-electron integral $(ij|ij)$.

Exercise 9.5. What two-electron integrals contribute to the off-diagonal elements F^{V}_{ij} when atomic orbitals i and j belong to different atoms?

[First term $(ij|lm)$ never contributes; second term $(im|lj)$ contributes only when $i = m$ and $l = j$.]

As in CNDO, parameters are chosen in INDO to give as close agreement as possible to the results of minimal basis set Hartree–Fock SCF calculations. Thus, although CNDO and INDO calculations give reasonable equilibrium geometries when compared to experiment, they give poor results (as do Hartree–Fock SCF methods) when compared with experimental quantities such as standard enthalpies of formation. During the past three decades, Dewar and his coworkers have developed a variety of semiempirical methods with the aim of reproducing not Hartree–Fock SCF wavefunctions but rather four gas-phase molecular properties, namely molecular geometries, enthalpies of formation, dipole moments, and ionization energies. The hope was to achieve this goal by careful selection (that is, optimization) of the values of the parameters.

Dewar first used the INDO approach and produced several versions of a semiempirical method he termed **modified intermediate neglect of differential overlap** (MINDO). The first two versions were called MINDO/1 and MINDO/2; a much improved version is MINDO/3.[35]

A much less severe approximation than INDO is the **neglect of diatomic differential overlap** (NDDO) in which only *diatomic* differential overlap is not retained: that is, differential overlap $\theta_i^*(1)\theta_j(1)$ is neglected only when the basis functions belong to different atoms. Therefore, in the NDDO formalism, we retain all one-centre two-electron integrals involving differential overlap and not just the one-centre exchange integrals; for example, we retain $(im|li)$ where i, l, m are different basis functions on the same atom. Also retained, for example, are two-centre integrals of the form $(ab|cd)$ where a and b are different orbitals on one atom and c and d are different orbitals on a different atom.

The NDDO method was proposed by Pople in the mid 1960s; however, it was not until Dewar developed the **modified neglect of differential overlap method** (MNDO) based on the NDDO formalism that the latter became more widely used as a predictive semiempirical method. In general, the MNDO method gives substantially better agreement with experiment than the MINDO/3 method. For example, in a study of 138 closed-shell molecules,[36] the mean absolute error in enthalpies of formation decreased from about $50\,\text{kJ mol}^{-1}$ in MINDO/3 to $30\,\text{kJ mol}^{-1}$ in MNDO. It should be mentioned that molecules were deliberately included in the study that had presented difficulties in earlier calculations and therefore the reported errors are larger than they would have been for a randomly selected set of molecules. Similarly, in a treatment of 80 different molecules and a total of 228 bonds,[36] the mean absolute error in equilibrium bond length decreased from 2.2 pm in MINDO/3 to 1.4 pm in MNDO.

Although MNDO was a significant improvement over MINDO/3, there remained deficiencies in the MNDO method, such as an inadequate description of systems with hydrogen bonds. In 1985, Dewar developed an improved version of MNDO called **Austin model 1** (AM1; the procedure is named after the University of Texas at Austin) which overcomes the major weaknesses of MNDO, in particular the failure to reproduce hydrogen bonds, without any significant increase in computing time. The method also provides more accurate enthalpies of formation and (through the application of Koopmans' theorem, Section 7.15) ionization energies than MNDO.

Semiempirical methods continue to be developed because there is always room for improvement of the parametrization scheme and the use of experimental data. For instance, a third parametrization, designated **PM3**, of the MNDO method (MNDO and AM1 being versions 1 and 2) has been developed,[37] and in general has been found to give better bond lengths, ionization energies, and enthalpies of formation than the two other MNDO schemes.

[35] For a discussion of the parametrization of MINDO/3, see R.C. Bingham, M.J.S. Dewar and D.H. Lo, *J. Am. Chem. Soc.*, 1285, **97** (1975) and Section 2.4.1 of D.M. Hirst, *A computational approach to chemistry*, Blackwell Scientific Publications, Oxford (1990).

[36] M.J.S. Dewar and W. Thiel, *J. Am. Chem. Soc.*, 4907, **99** (1977).

[37] J.J.P. Stewart, *J. Comput. Chem.*, 209, **10** (1989); *idem.*, 221, **10** (1989).

Nevertheless, despite this progress, it is not uniform: there are cases where PM3 is much worse than MNDO, and AM1 can result in the prediction of some very peculiar geometries for hydrogen bonds. It is essential to compare the outcome of several methods.

Software packages for electronic structure calculations

Sophisticated software packages have been developed over the past three decades to perform electronic structure calculations using the *ab initio* and semiempirical methods we have described. These packages are widely available and are becoming increasingly easy to use. They are often accompanied by sophisticated graphical user interfaces that allow visualization of results of calculations. Many packages can be used to compute both potential energies and analytic derivatives, and so are useful for determining equilibrium geometries, transition state geometries, and vibrational frequencies. We mention here only some of the available packages and emphasize methods we have discussed in this chapter. See the *Further reading* section for a more complete discussion, including sources for the software packages.

Packages capable of self-consistent field, Møller–Plesset, and configuration interaction calculations include Gaussian, GAMESS (General Atomic and Molecular Electronic Structure System), CADPAC (Cambridge Analytical Derivatives Package), and HONDO. Several other *ab initio* software programs with strengths in particular types of approach that we have described are MOLCAS (for CASSCF), COLUMBUS (for MRCI), ACES II (for MBPT), and MOLPRO (for MRCI). Density functional theory software includes DGauss, deMon, and DMol. Gaussian is also capable of DFT.

Two widely-used semiempirical packages are MOPAC and AMPAC. Both packages include MINDO/3, MNDO, AM1, and PM3, as does MNDO94, a very sophisticated semiempirical package. The semiempirical package AMSOL includes solvation models for water and alkane solvents. Semiempirical capabilities are also available in both Gaussian and GAMESS. Be aware, however, that because the semiempirical methodologies are parametrization-dependent, AM1 in one software package, for example, need not be the same as AM1 in another software package.

The availability of all of these software packages, together with a variety of commercial versions (such as SPARTAN, HyperChem, and Mulliken), have made electronic structure calculations accessible to a wide range of scientists.

Box 9.1 Acronyms for electronic structure calculations

Acronym	**Name/Description**
AM1	Austin model 1 (version 2 of MNDO)
CASSCF	complete active-space self-consistent field
CI	configuration interaction
CNDO	complete neglect of differential overlap
CPHF	coupled perturbed Hartree–Fock
CPMCSCF	coupled perturbed MCSCF
CSF	configuration state function
DCI	CI including doubly excited Slater determinants
DFT	density functional theory
DZ	double-zeta basis
DZP	double-zeta plus polarization basis
GTO	Gaussian-type orbital
HF	Hartree–Fock
HMO	Hückel molecular orbital
INDO	intermediate neglect of differential overlap
KS	Kohn–Sham
LCAO	linear combination of atomic orbitals
LDA	local density approximation
LDA-NL	local density approximation with nonlocal corrections
MBPT	many-body perturbation theory
MCSCF	multiconfiguration self-consistent field
MINDO	modified intermediate neglect of differential overlap
MNDO	modified neglect of differential overlap
MPn	Møller–Plesset perturbation theory including nth-order energy correction
MPPT	Møller–Plesset perturbation theory
MRCI	multireference configuration interaction
NDDO	neglect of diatomic differential overlap
PM3	parametrization model 3 (version 3 of MNDO)
PPP	Pariser–Parr–Pople
PT	perturbation theory
RHF	restricted Hartree–Fock
SCF	self consistent field
SDCI	CI including singly and doubly excited Slater determinants
SDTQCI	CI including singly, doubly, triply, and quadruply excited Slater determinants
STO	Slater type orbital
STO-NG	representation of STO as linear combination of N primitive Gaussians
SV	split-valence basis
TZ	triple-zeta basis
UHF	unrestricted Hartree–Fock
ZDO	zero differential overlap
m-npG	one contracted Gaussian composed of m primitives for each inner shell atomic orbital; two contracted Gaussians of n and p primitives, respectively, for each valence shell atomic orbital
m-npG*	m-npG basis plus d-type polarization functions for non-hydrogen atoms
m-npG**	m-npG* basis plus p-type polarization functions for hydrogen atoms

Problems

9.1 Confirm that the product in eqn 9.4 of one-electron wavefunctions is an eigenfunction of the hamiltonian $H°$ of eqn 9.2 and determine its corresponding eigenvalue.

9.2 Show that $(n!)^{-1/2}$ is the correct normalization factor for a single Slater determinant consisting of n orthonormal spinorbitals.

9.3 Show that the Slater determinant $\Phi = (6)^{-1/2}\det|\psi_{1s}^{\alpha}(1)\psi_{1s}^{\beta}(2)\psi_{1s}^{\alpha}(3)|$ for the He^- ion is identically zero.

9.4 Show that in the closed-shell restricted Hartree–Fock case the general spinorbital Hartree–Fock equation (eqns 9.7 and 9.8) can be converted to the HF equation in eqn 7.44 for the spatial wavefunction ψ. *Hint*. To convert from spinorbitals to spatial orbitals, you will need to integrate out the spin functions. Begin with eqn 9.7 and let $\phi_a(1) = \psi_a^{\alpha}(1)$; an identical result will be obtained if you assume that $\phi_a(1) = \psi_a^{\beta}(1)$.

9.5 Give an example of a restricted Hartree–Fock wavefunction and an unrestricted Hartree–Fock wavefunction for an aluminium atom.

9.6 In a Hartree–Fock SCF calculation on the chlorine atom using 20 (spatial) basis functions, how many virtual orbitals are determined?

9.7 Consider the two-electron integrals over the basis functions specified in eqn 9.24. If the basis functions are taken to be real, a number of the integrals are equivalent; for example, $(ab|cd) = (ba|cd)$. Find the other integrals that are equal to $(ab|cd)$.

9.8 Show that the product of an s-type Gaussian centred at $\mathbf{R}_A$ with exponent α_A and an s-type Gaussian centred at $\mathbf{R}_B$ with exponent α_B can be written in terms of a single s-type Gaussian centred between $\mathbf{R}_A$ and $\mathbf{R}_B$.

9.9 In an electronic structure calculation on chloromethane, CH_3Cl, describe briefly what would be meant by (i) a minimal basis set, (ii) a split-valence basis set, (iii) a DZP basis set. How many basis functions are needed in each basis case?

9.10 In a Hartree–Fock calculation on atomic hydrogen using four primitive s-type Gaussian functions, S. Huzinaga (*J. Chem. Phys.*,1293, **42** (1965)) found that optimized results were obtained with a linear combination of Gaussians with coefficients c_{ji} and exponents α of 0.50907, 0.123317; 0.47449, 0.453757; 0.13424, 2.01330; and 0.01906, 13.3615. Describe how these primitives would be utilized in a $(4s)/[2s]$ contraction scheme.

9.11 Determine the number of basis set functions in a molecular electronic structure calculation on ethanol, CH_3CH_2OH, using (i) a 6-31G; (ii) a 6-31G *; (iii) a 6-31G ** basis.

9.12 Determine the total number of different Slater determinants for an electronic structure calculation on ethanol, CH_3CH_2OH, that can be formed from a 6-31G ** basis set.

9.13 A single Slater determinant is not necessarily an eigenfunction of the total electron spin operator. Therefore, even within the Hartree–Fock approximation, for the wavefunction Φ_0 to be an eigenfunction of S^2, it might have to be expressed as a linear combination of Slater determinants. The linear combination is referred to as a *spin-adapted configuration*. As a simple example, consider a two-electron system with the four possible Slater determinants $\Phi_1 = (2)^{-1/2}\det|\psi_1^{\alpha}(1)\psi_2^{\alpha}(2)|$, $\Phi_2 = (2)^{-1/2}\det|\psi_1^{\alpha}(1)\psi_2^{\beta}(2)|$, $\Phi_3 = (2)^{-1/2}\det|\psi_1^{\beta}(1)\psi_2^{\alpha}(2)|$, and $\Phi_4 = (2)^{-1/2}\det|\psi_1^{\beta}(1)\psi_2^{\beta}(2)|$. First, show that the Slater determinants Φ_1 and Φ_4 are themselves eigenfunctions of S^2 with eigenvalue $2\hbar^2$ (corresponding to $S = 1$). Then, from Φ_2 and Φ_3, determine two linear combinations, one of which corresponds to $S = 1, M_S = 0$ and the other of which corresponds to $S = 0, M_S = 0$.

9.14 In a CI calculation on the ^{2}S ground state of lithium, which of the following Slater determinants can contribute to the ground-state wavefunction? (a) $|\psi_{1s}^{\alpha}\psi_{1s}^{\beta}\psi_{2s}^{\alpha}|$; (b) $|\psi_{1s}^{\alpha}\psi_{1s}^{\beta}\psi_{2s}^{\beta}|$; (c) $|\psi_{1s}^{\alpha}\psi_{1s}^{\beta}\psi_{2p}^{\alpha}|$; (d) $|\psi_{1s}^{\alpha}\psi_{2p}^{\alpha}\psi_{2p}^{\beta}|$; (e) $|\psi_{1s}^{\alpha}\psi_{3d}^{\alpha}\psi_{3d}^{\beta}|$; (f) $|\psi_{1s}^{\alpha}\psi_{2s}^{\alpha}\psi_{3s}^{\alpha}|$.

9.15 In a CI calculation on the excited $^3\Sigma_{\mathrm{u}}$ electronic state of H_2, which of the following Slater determinants can contribute to the excited-state wavefunction? (a) $|1\sigma_{\mathrm{g}}^{\alpha}1\sigma_{\mathrm{u}}^{\alpha}|$; (b) $|1\sigma_{\mathrm{g}}^{\alpha}1\pi_{\mathrm{u}}^{\alpha}|$; (c) $|1\sigma_{\mathrm{u}}^{\alpha}1\pi_{\mathrm{g}}^{\beta}|$; (d) $|1\sigma_{\mathrm{g}}^{\beta}3\sigma_{\mathrm{u}}^{\beta}|$; (e) $|1\pi_{\mathrm{u}}^{\alpha}1\pi_{\mathrm{g}}^{\alpha}|$; (f) $|1\pi_{\mathrm{u}}^{\beta}2\pi_{\mathrm{u}}^{\beta}|$.

9.16 Consider a configuration interaction calculation which employs three orthonormal n-electron Slater determinants Φ_1, Φ_2, and Φ_3. Write out the secular determinant from which the three lowest energies would be found.

9.17 Prove Brillouin's theorem; that is, show that hamiltonian matrix elements between the HF wavefunction Φ_0 and singly excited determinants are identically zero.

9.18 Hamiltonian matrix elements between two n-electron Slater determinants can be conveniently expressed in terms of integrals over the orthonormal spinorbitals of which the determinants are comprised. This was first done by E.U. Condon and J.C. Slater and the resulting expressions are sometimes referred to as the *Slater–Condon rules.* Consider two Slater determinants Φ_1 and Φ_2 that differ by only one spinorbital; that is,

$$\Phi_1 = (n!)^{-1/2}\det|\ldots\phi_m\phi_i\ldots| \qquad \Phi_2 = (n!)^{-1/2}\det|\ldots\phi_p\phi_i\ldots|$$

Derive the following Slater–Condon rule.

$$\langle\Phi_1|H|\Phi_2\rangle = \langle\phi_m(1)|h_1|\phi_p(1)\rangle + \sum_i\{[\phi_m\phi_p|\phi_i\phi_i] - [\phi_m\phi_i|\phi_i\phi_p]\}$$

where we have used the notation (see *Further information 11*)

$$[\phi_a\phi_b|\phi_c\phi_d] = \int\phi_a^*(1)\phi_b(1)\left(\frac{e^2}{4\pi\varepsilon_0 r_{12}}\right)\phi_c^*(2)\phi_d(2)\,\mathrm{d}\mathbf{x}_1\,\mathrm{d}\mathbf{x}_2$$

9.19 Using the notation $[\phi_a\phi_b|\phi_c\phi_d]$ given in Problem 9.18 for a two-electron integral over the spinorbitals, show that $[\phi_a\phi_b|\phi_c\phi_d] = [\phi_c\phi_d|\phi_a\phi_b]$ and $[\phi_a\phi_b|\phi_c\phi_d] = [\phi_b\phi_a|\phi_d\phi_c]^*$.

9.20 For a CASSCF calculation of the ground-state wavefunction of the diatomic molecule C_2, describe a reasonable choice for the distribution of σ and π molecular orbitals into active, inactive, and virtual orbitals. How many inactive and active electrons are there in the calculation?

9.21 Show that the Møller–Plesset perturbation $H^{(1)}$ can be written in terms of the Coulomb and exchange operators as

$$H^{(1)} = \sum_{i,j=1}^{n}\left(\frac{e^2}{8\pi\varepsilon_0 r_{ij}} - J_j(i) + K_j(i)\right)$$

9.22 Use Møller–Plesset perturbation theory to obtain an expression for the ground-state wavefunction corrected to first order in the perturbation.

9.23 (a) Which of the following methods are capable of yielding an energy below the exact ground-state energy? (b) Which of the following methods are not assured of being size-consistent? (i) Hartree–Fock SCF; (ii) full CI; (iii) SDCI; (iv) MP2; (v) MRCI; (vi) MP4.

9.24 In a SDCI calculation using gradient methods to compute the force constants of NH_3, which analytical derivatives are needed to calculate the required energy derivatives?

9.25 Show that the derivative of an s-type GTO with respect to the nuclear coordinate x_{c} yields a p-type GTO and that the derivative of the p-type Gaussian θ_{100} yields a sum of s- and d-type Gaussians. (The GTO is given in eqn 9.28.)

9.26 Demonstrate explicitly the relation between the PPP and the HMO methods described in the last paragraph of Section 9.12.

9.27 Which of the following two-electron integrals (over real basis functions) are *not* neglected in (i) CNDO; (ii) INDO; (iii) MNDO? (a) $(ii|jj)$ with θ_i and θ_j belonging to different atoms; (b) $(ij|ji)$ with θ_i and θ_j belonging to the same atom; (c) $(ij|ji)$ with θ_i and θ_j belonging to different atoms; (d) $(ij|ki)$ with θ_i, θ_j, and θ_k belonging to the same atom; (e) $(ij|kl)$ with θ_i and θ_j belonging to one atom and θ_k and θ_l belonging to another; (f) $(ii|ii)$.

Further reading

E.R. Davidson (ed.), *Chem. Rev.*, **91**, 649–1108 (1991), a special issue on theoretical chemistry.
A computational approach to chemistry. D.M. Hirst; Blackwell Scientific, Oxford (1990).
A handbook of computational chemistry: A practical guide to chemical structure and energy calculations. T. Clark; Wiley, New York (1985).
Acronyms used in theoretical chemistry. R.D. Brown, J.E. Boggs, R. Hilderbrandt, K.F. Lim, I.M. Mills, E. Nikitin, and M.H. Palmer, *Pure and Applied Chemistry*, **68** (2), 387–456 (1966).

Ab initio methods
Modern electronic structure theory. D.R. Yarkony (ed.); World Scientific, London (1995).
Ab initio calculation of the structures and properties of molecules. C.E. Dykstra; Elsevier, Amsterdam (1988).
Ab initio methods in quantum chemistry Parts I and II. K.P. Lawley (ed.); Wiley, Chichester (1987).
Electron correlation in molecules. S. Wilson; Clarendon Press, Oxford (1984).
Modern quantum chemistry: introduction to advanced electronic structure theory. A. Szabo and N.S. Ostlund; Macmillan, New York (1982).
The historical development of the electron correlation problem. P.-O. Löwdin, *Int. J. Quant. Chem.*, **55**, 77 (1995).
Handbook of Gaussian basis sets: A compendium for ab initio molecular orbital calculations. R. Poirier, R. Kari, and I.G. Csizmadia; Elsevier, Amsterdam (1985).
A new dimension to quantum chemistry. Analytical derivative methods in ab initio molecular electonic structure theory. Y. Yamaguchi, Y. Osamura, J. Goddard, and H.F. Schaefer III; Oxford University Press, Oxford (1994).

Semiempirical methods
AM1: A new general purpose quantum mechanical molecular model. M.J.S. Dewar, E.G. Zoebisch, E.F. Healy, and J.J.P. Stewart; *J. Am. Chem. Soc.*, **107**, 3902 (1985).
Approximate molecular orbital theory. J.A. Pople and D.L. Beveridge; McGraw-Hill, New York (1970).
Semi-empirical methods of quantum chemistry. J. Sadlej; Halstead Press, New York (1985).

Density functional theory
Modern density functional theory: a tool for chemistry. J.M. Seminario and P. Politzer (ed.); Elsevier, Amsterdam (1995).
Density-functional theory of atoms and molecules. R.G. Parr and W. Yang; Oxford University Press, New York (1989).

Software packages
Compendium of software for molecular modeling. D.B. Boyd; *Rev. Comp. Chem.*, **6**, 383 (1995).
The following are some software packages currently available and their sources: ACES II (Quantum Theory Project, University of Florida, USA); AMPAC (Semichem, Shawnee, Kansas, USA); AMSOL (Quantum Chemistry Program Exchange (QCPE),

Indiana University, USA); CADPAC (Lynxvale WCIU Programs, Cambridge, England; it is available from Cray Research, Inc., Eagen, Minnesota, USA, as part of the UniChem software package and from Oxford Molecular Ltd., Oxford, UK); COLUMBUS (Ohio State University, USA); deMon (University of Montreal, Canada); DGauss (part of the UniChem package); DMol (BIOSYM/Molecular Simulations, Inc., San Diego, California, USA); GAMESS (Iowa State University, USA); GAMESS-UK (Engineering and Physical Sciences Research Council, Daresbury Laboratory, Warrington, UK); Gaussian (Gaussian, Inc., Pittsburgh, Pennsylvania, USA); HONDO (QCPE); HyperChem (Hypercube, Inc., Waterloo, Ontario, Canada); MNDO94 (part of the UniChem package); MOLCAS (University of Lund, Sweden); MOLPRO (University of Sussex, UK and University of Bielefeld, Germany); MOPAC (QCPE); Mulliken (IBM Almaden Research Center, San Jose, California, USA); SPARTAN (Wavefunction, Inc., Irvine, California, USA).

10 Molecular rotations and vibrations

Molecular spectra are both more complex and richer in information than atomic spectra. Their greater complexity arises from the more complicated structures of molecules, for whereas the spectra of atoms are due only to their electronic transitions, the spectra of molecules arise from electronic, vibrational, and rotational transitions. These modes are not independent of one another, and the complexity of the spectra is enriched by the interactions between them. We shall see that an interpretation of molecular spectra yields a great deal of information about the shapes and sizes of molecules, the strengths and stiffnesses of their bonds, and other information that is needed to account for chemical reactions.

The energy associated with rotational transitions is usually less than that involved in vibrational transitions, and the energy of vibrational transitions is usually less than that of electronic transitions. As a consequence of this hierarchy, although it is possible to observe pure rotational transitions, a vibrational transition is normally accompanied by rotational transitions. Electronic transitions are accompanied by both vibrational and rotational transitions, and are correspondingly complicated. Because of this hierarchy, we shall deal with transitions in order of increasing size of the quanta involved.

Spectroscopic transitions

There are certain features that are common to all forms of spectroscopy, particularly relating to the intensities of lines. The background to this material was presented in Chapter 6. We shall need to distinguish between the **gross selection rules**, which are statements about the properties that a molecule must possess in order for it to be capable of showing a particular type of transition, and the **specific selection rules**, the changes in quantum numbers that may occur during such a transition.

10.1 Absorption and emission

We saw in Chapters 6 and 7 that the most intense transitions are induced by the interaction of the electric component of the electromagnetic field with the electric dipole associated with the transition. We also saw that the intensity of the transition is proportional to the square of the transition dipole moment, $\boldsymbol{\mu}_{\mathrm{fi}}$, where

$$\boldsymbol{\mu}_{\mathrm{fi}} = \langle \mathrm{f}|\boldsymbol{\mu}|\mathrm{i}\rangle \tag{1}$$

in which $\boldsymbol{\mu}$ is the electric dipole moment operator (a vector). The selection rules for absorption and emission of radiation are based on the criteria for this integral being nonzero, as explained in Section 5.16. The physical interpretation of the transition dipole moment is that it is a measure of the magnitude of the dipolar

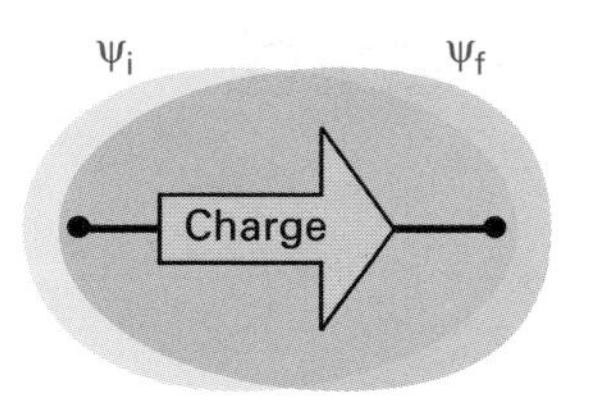

Fig. 10.1 In order for a transition to be electric-dipole allowed, it must possess a degree of dipolar character. A purely spherically symmetrical (or some other non-dipolar) redistribution of charge cannot interact with the electric field vector of the electromagnetic field.

migration of charge that accompanies the transition (Fig. 10.1). Molecular collisions obey different selection rules, and may induce a wide variety of transitions. Their effect is usually to establish thermal equilibrium populations of rotational, vibrational, and electronic states. Collisions affect the appearance of spectra, because spectral intensities depend on the populations of the states involved in the transition, and lifetime broadening (Section 6.18) affects their widths.

Once a transition dipole moment has been calculated it can be used in the expressions derived in Section 6.17 for the rates of transitions:

$$\text{Stimulated: } W = B\rho_{\text{rad}}(E) \qquad \text{Spontaneous: } W = A \tag{2}$$

with

$$A = \frac{8\pi h\nu_{\text{fi}}^3}{c^3}B \qquad B = \frac{|\mu_{\text{fi}}|^2}{6\varepsilon_0\hbar^2} \tag{3}$$

If it is safe to ignore spontaneous emission (for transition frequencies of less than about 1 THz, or when considering systems in which only the ground state is significantly populated), the net rate of absorption of energy is the difference between the rate of absorption and the rate of stimulated emission multiplied by the energy change that accompanies each transition ($h\nu_{\text{fi}} = E$):

$$\begin{aligned}\frac{\mathrm{d}E}{\mathrm{d}t} &= N_{\text{l}}h\nu W_{\text{u}\leftarrow\text{l}} - N_{\text{u}}h\nu W_{\text{u}\rightarrow\text{l}} \\ &= (N_{\text{l}} - N_{\text{u}})h\nu B\rho_{\text{rad}}(E)\end{aligned} \tag{4}$$

where N_{u} is the population of the upper state and N_{l} is the population of the lower state. That is, the net rate of energy extraction from the incident radiation is proportional to the population difference between the two states. For electronic transitions and most vibrational transitions the upper state is virtually unpopulated at normal temperatures, and so only absorption processes are significant; then

$$\frac{\mathrm{d}E}{\mathrm{d}t} = Nh\nu B\rho_{\text{rad}}(E) \tag{5}$$

where N is the total number of molecules in the sample. If the sample is at thermal equilibrium at a temperature T, and the transitions are such that $h\nu \ll kT$ (as is the case for most rotational transitions), then the population difference can be expressed in terms of a Boltzmann distribution,

$$\frac{N_{\text{u}}}{N_{\text{l}}} = \mathrm{e}^{-(E_{\text{u}}-E_{\text{l}})/kT} = \mathrm{e}^{-h\nu/kT}$$

and the exponential term expanded:

$$N_{\text{l}} - N_{\text{u}} = N_{\text{l}}\left(1 - \mathrm{e}^{-h\nu/kT}\right) \approx \frac{N_{\text{l}}h\nu}{kT}$$

In that case it follows that

$$\frac{\mathrm{d}E}{\mathrm{d}t} \approx N_{\text{l}}\left(\frac{h^2\nu^2}{kT}\right)B\rho_{\text{rad}}(E) \tag{6}$$

and the absorption intensity is lower the higher the temperature because as T increases the two states involved in the transition acquire equal populations.

10.2 Raman processes

A special class of transitions gives rise to **Raman spectra**. The process involves the inelastic scattering of a photon when it is incident on a molecule. The photon loses some of its energy to the molecule or gains some from it, and so leaves the molecule with a lower or a higher frequency, respectively. The lower frequency components of the scattered radiation are called the **Stokes lines** and the higher frequency components are called the **anti-Stokes lines**.

The selection rules for Raman transitions are based on aspects of the polarizability of the molecule, the measure of its response to an electric field (see Section 12.1). Their origin can be appreciated by a classical argument in which we consider the time-variation of the magnitude of the dipole moment induced in a molecule by an electromagnetic field $\mathcal{E}(t)$:

$$\mu(t) = \alpha(t)\mathcal{E}(t) \tag{7}$$

where $\alpha(t)$ is the polarizability. If the incident radiation has frequency ω and the polarizability of the molecule changes between α_{min} and α_{max} at a frequency ω_{int} as a result of its rotation or vibration, we can write

$$\mu(t) = (\alpha + \tfrac{1}{2}\Delta\alpha \cos \omega_{\text{int}} t)\mathcal{E}_0 \cos \omega t$$

where α is the mean polarizability and $\Delta\alpha = \alpha_{\text{max}} - \alpha_{\text{min}}$ is its range of variation. This product expands to

$$\mu(t) = \alpha\mathcal{E}_0 \cos \omega t + \tfrac{1}{4}\Delta\alpha\mathcal{E}_0\{\cos(\omega + \omega_{\text{int}})t + \cos(\omega - \omega_{\text{int}})t\} \tag{8}$$

We see that the induced dipole moment has three components. One has the incident frequency and gives rise to the unshifted **Rayleigh line** in the spectrum. The other two are shifted in frequency by the frequency at which the molecular motion causes the polarizability to oscillate and give rise to the Stokes and anti-Stokes lines. It is clear that these Raman frequencies will be observed only if $\Delta\alpha \neq 0$, so rotational Raman transitions require the molecule to have an anisotropic polarizability. Vibrational Raman transitions require the polarizability to change as the molecule vibrates.

Molecular rotation

The strategy for each section of this chapter will be to establish the energy levels of molecules for each mode of motion, and then to apply the selection rules to determine the appearance of the relevant spectrum. We begin here with the rotations of molecules. The treatment is considerably simplified by drawing on the properties of angular momentum obtained in Chapter 4.

We shall need the concept of the **moment of inertia**, I, of a body, a property first introduced in connection with rotational motion in Chapter 3. The moment of inertia about an axis q set in the molecule is defined as

$$I_{qq} = \sum_i m_i x_i^2(q) \tag{9}$$

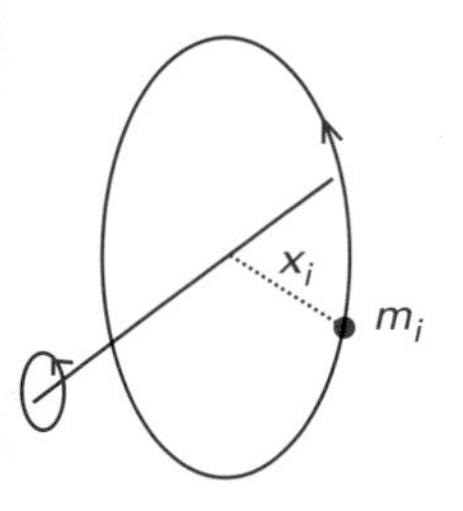

Fig. 10.2 The basis of the definition of the moment of inertia about a selected axis in terms of the mass of a particle and its vertical distance from the axis.

where $x_i(q)$ is the perpendicular distance of the atom i of mass m_i from the axis q (Fig. 10.2). The double subscript is used on I for technical reasons, but broadly speaking it echoes the presence of the distances $x_i(q)$ as their squares. The moment of inertia of a diatomic molecule with bond length R and atomic masses m_A and m_B is particularly simple and will be useful later. For rotation about an axis perpendicular to the bond it is:

$$I = \mu R^2 \qquad \frac{1}{\mu} = \frac{1}{m_A} + \frac{1}{m_B} \tag{10}$$

A molecule with heavy atoms well away from its centre of mass has a large moment of inertia and, in classical physics, accelerates only slowly when subjected to a torque (a turning force), $\mathcal{T}$:

$$\frac{d\omega}{dt} = \frac{\mathcal{T}}{I}$$

where ω is the angular velocity (the rate of change of orientation). In this respect, the moment of inertia plays in rotational motion the same role as inertial mass plays in linear motion. The expressions for the moments of inertia of other molecules are more complex (Table 10.1).

10.3 Rotational energy levels

According to classical physics, the kinetic energy of rotation of a body of moment of inertia I_{qq} about an axis q is

$$T = \tfrac{1}{2}\sum_q I_{qq}\omega_q^2 = \sum_q \frac{J_q^2}{2I_{qq}} \tag{11}$$

where ω_q is the angular frequency about the axis and we have used the classical expression for the component of angular momentum around each axis, $J_q = I_{qq}\omega_q$. There is no contribution from the potential energy, so the hamiltonian for the problem is

$$H = \frac{J_x^2}{2I_{xx}} + \frac{J_y^2}{2I_{yy}} + \frac{J_z^2}{2I_{zz}}$$

with each J_q to be interpreted as an operator for the q component of angular momentum.

Consider first a **symmetric rotor**, which is a rigid body with one symmetry axis C_n with $n \geq 3$.[1] As a consequence of this symmetry, two of the moments of inertia are the same, and we shall write $I_\parallel = I_{zz}$ and $I_\perp = I_{xx} = I_{yy}$, where z is the figure axis of the molecule (the axis parallel to C_n). It follows that

$$H = \frac{J_x^2 + J_y^2}{2I_\perp} + \frac{J_z^2}{2I_\parallel}$$

[1] Asymmetric rotors, which are rigid bodies with three different moments of inertia, are too difficult to treat by elementary methods, and we shall not consider them. For an account, see the references in *Further reading*.

Table 10.1 Moments of inertia†

1. Diatomics

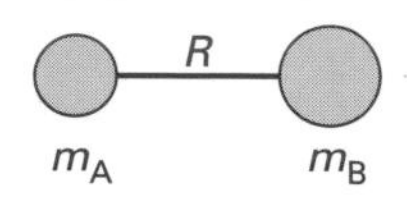

$$I = \frac{m_A m_B}{m} R^2 = \mu R^2$$

2. Linear rotors

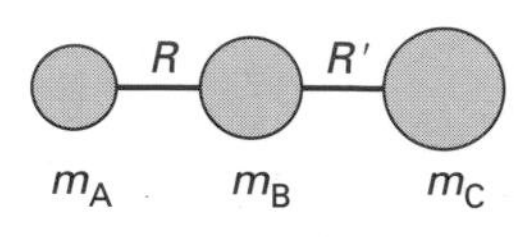

$$I = m_A R^2 + m_C R'^2 - \frac{(m_A R - m_C R')^2}{m}$$

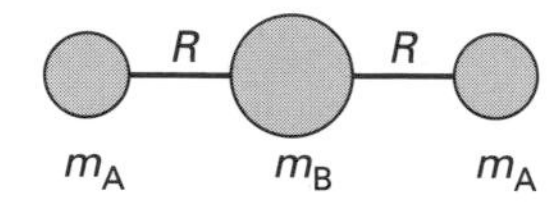

$$I = 2m_A R^2$$

3. Symmetric rotors

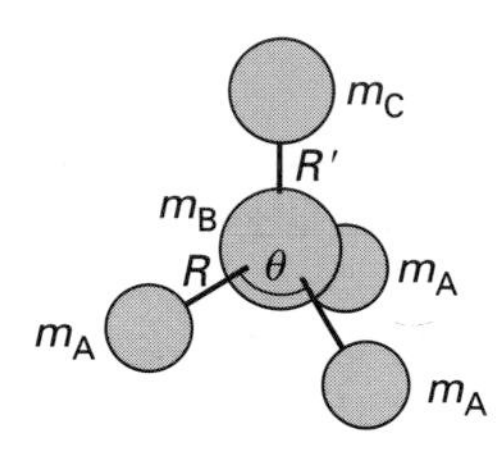

$$I_{\parallel} = 2m_A R^2(1 - \cos\theta)$$

$$I_{\perp} = m_A R^2(1 - \cos\theta) + \frac{m_A}{m}(m_B + m_C)R^2(1 + 2\cos\theta) + \frac{m_C R'}{m}\{(3m_A + m_B)R' + 6m_A R[\tfrac{1}{3}(1 + 2\cos\theta)]^{1/2}\}$$

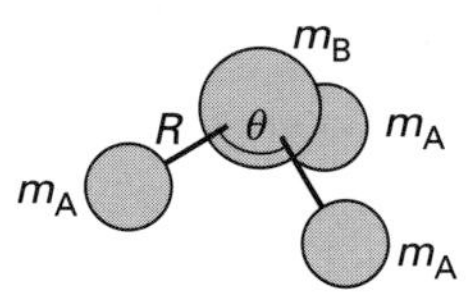

$$I_{\parallel} = 2m_A R^2(1 - \cos\theta)$$

$$I_{\perp} = m_A R^2(1 - \cos\theta) + \frac{m_A m_B}{m} R^2(1 + 2\cos\theta)$$

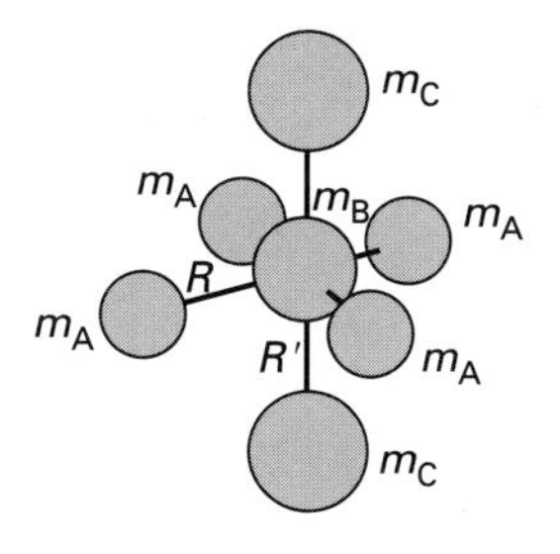

$$I_{\parallel} = 4m_A R^2$$

$$I_{\perp} = 2m_A R^2 + 2m_C R'^2$$

4. Spherical rotors

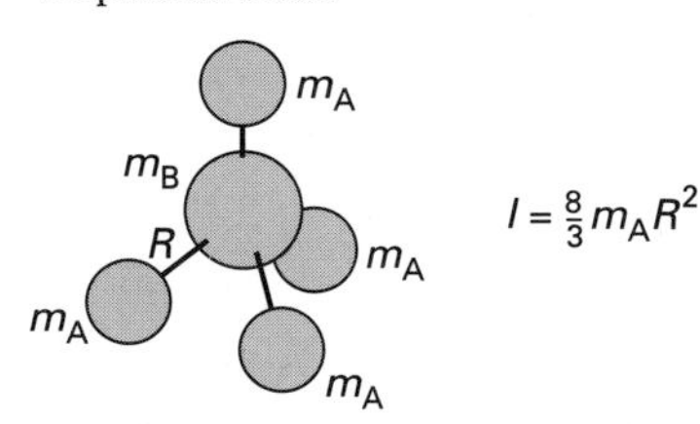

$$I = \tfrac{8}{3} m_A R^2$$

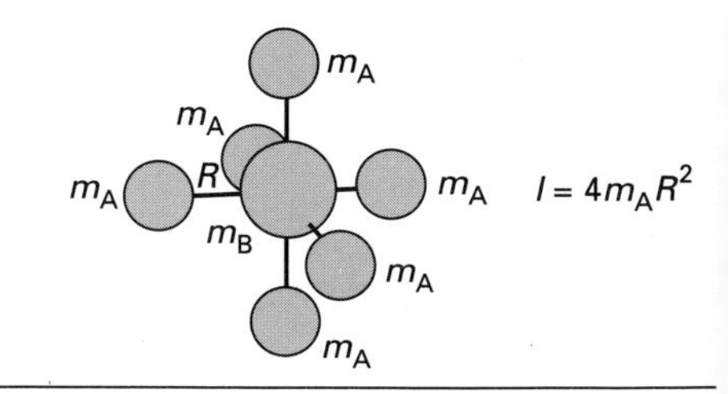

† In each case m is the total mass of the molecule.

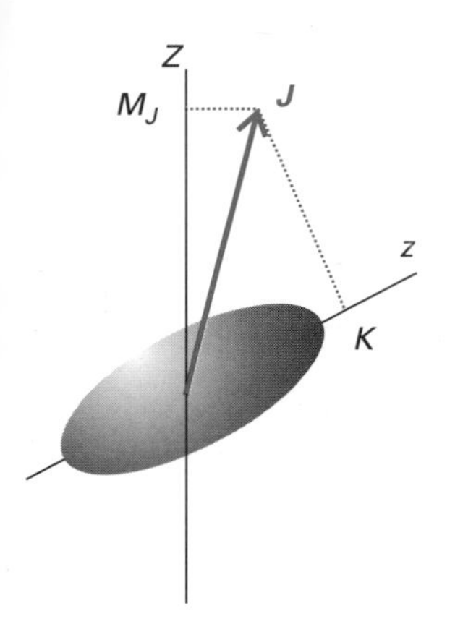

Fig. 10.3 The physical significance of the quantum numbers J, K, and M_J for a rotating nonlinear molecule.

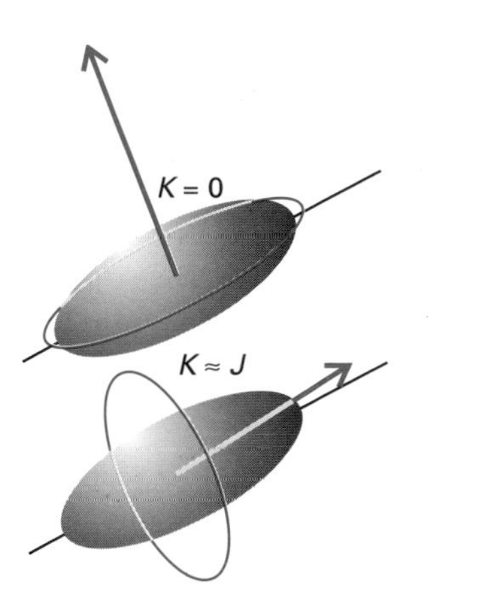

Fig. 10.4 When $K = 0$ the rotation of the molecule is entirely about an axis that is perpendicular to its figure axis. When K has its maximum value (of J), most of the rotational motion is around the figure axis.

This hamiltonian can be expressed in terms of the operator $J^2 = J_x^2 + J_y^2 + J_z^2$ for the magnitude of the angular momentum, when it becomes

$$H = \frac{J^2}{2I_\perp} + \left(\frac{1}{2I_\parallel} - \frac{1}{2I_\perp}\right)J_z^2 \tag{12}$$

It follows that in order to establish the eigenvalues of this hamiltonian, all we need do is to import the eigenvalues of the operators J^2 and J_z, which were established in Chapter 4. It is conventional to use K for the quantum number specifying the component of angular momentum on the internal figure axis of a molecule and to reserve M_J for its component on the laboratory fixed z-axis. Then we can write the eigenvalues of H as

$$E(J,K,M_J) = \frac{J(J+1)\hbar^2}{2I_\perp} + \left(\frac{1}{2I_\parallel} - \frac{1}{2I_\perp}\right)K^2\hbar^2 \tag{13}$$

with

$$J = 0, 1, 2, \ldots \qquad K = J, J-1, \ldots, -J \qquad M_J = J, J-1, \ldots, -J$$

It is important to note that although the component of angular momentum on the laboratory axis does not appear explicitly in the hamiltonian, it is nevertheless required to specify the complete angular momentum state of the molecule (Fig. 10.3). Its absence from the expression for E is consistent with the fact that in the absence of external fields, the energy of the molecule is independent of the orientation of its angular momentum in space. The significance of the quantum number K is that it tells us how the total angular momentum of the molecule is distributed over the molecular axes: when $|K| \approx J$, then almost the whole of the molecule's angular momentum is around its figure axis (Fig. 10.4); if $|K| \approx 0$, then most of its angular momentum is about an axis perpendicular to the figure axis. Note that the energy depends on K^2, so the energy is independent of the direction of rotation about the figure axis, as is physically plausible.

It is a further convention in the discussion of molecular rotation to express the energy in terms of the **rotational constants** A and B:

$$A = \frac{\hbar}{4\pi c I_\parallel} \qquad B = \frac{\hbar}{4\pi c I_\perp} \tag{14}$$

Then, with $E(J,K,M_J) = hcF(J,K,M_J)$, where F is a wavenumber,

$$F(J,K,M_J) = BJ(J+1) + (A-B)K^2 \tag{15}$$

The degeneracy of each level with $K \neq 0$ is $g_J = 2(2J+1)$, because M_J can take $2J+1$ different values for a given value of J, and K can be either positive or negative. If $K = 0$, $g_J = 2J+1$ because K then has only a single value. When $K = 0$, the motion is entirely around an axis perpendicular to the figure axis, and

$$F(J,0,M_J) = BJ(J+1) \tag{16}$$

As expected, the energy of rotation now depends solely on the moment of inertia about that perpendicular axis. When $|K| = J$, its maximum value,

$$F(J,\pm J,M_J) = AJ^2 + BJ \tag{17}$$

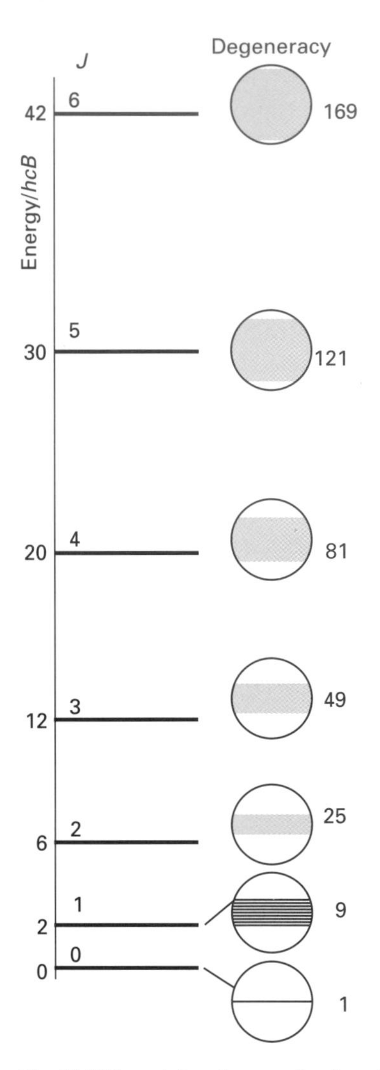

Fig. 10.5 The rotational energy levels and their degeneracies of a spherical rotor. Note the very rapid increase in degeneracy (which at high values of J is proportional to J^2).

Now the main contribution (the term proportional to J^2) comes from the moment of inertia about the figure axis. The perpendicular component continues to contribute because the component of angular momentum about the figure axis is always less than the magnitude of the angular momentum, so even if $|K| = J$, the molecule continues to rotate at least slowly around the perpendicular axis.

There are two special cases that we need to consider. A **spherical rotor** is a rigid molecule that belongs to a cubic (tetrahedral and octahedral) or icosahedral point group. Such molecules have all three moments of inertia equal, so $A = B$. It follows that

$$F(J, K, M_J) = BJ(J+1) \tag{18}$$

and the energy of the molecule is independent of both K and M_J. However, as both quantum numbers are still needed to specify the angular momentum of the molecule, each level is now $(2J+1)^2$-fold degenerate (Fig. 10.5). The K-degeneracy reflects the fact that it is immaterial what component the angular momentum has on the now arbitrary molecular z-axis. A **linear rotor** is a rigid linear molecule, one that belongs to the point group $C_{\infty v}$ or $D_{\infty h}$. In such molecules, the angular momentum vector is necessarily perpendicular to the axis of the molecule, and so $K \equiv 0$ in all states. Substitution of this value in eqn 10.15 gives

$$F(J, M_J) = BJ(J+1) \tag{19}$$

This equation resembles the last one, but note that K does not appear in the specification of the state as it is identically zero. One implication of the absence of K is that the degeneracy of a linear rotor is only $g_J = 2J+1$, for now only M_J can range over a series of values and K is fixed at zero (Fig. 10.6).

10.4 Centrifugal distortion

The treatment of a molecule as a rigid rotor is only an approximation. As the degree of rotational excitation increases, the bonds are put under stress and are stretched. The increase in moment of inertia that accompanies this **centrifugal distortion** results in a lowering of the rotational constants, and so the energy levels are less far apart at high J than expected on the basis of the rigid rotor assumption. We shall now show that a first approximation to the effect of centrifugal distortion on the energy levels of a linear rotor is obtained by writing

$$F(J, M_J) = BJ(J+1) - DJ^2(J+1)^2 \tag{20}$$

where D, which is called the **centrifugal distortion constant**, depends on the stiffness of the bonds.

Consider a diatomic molecule of reduced mass μ (see eqn 10.10). If it is rotating at an angular velocity ω, it will experience a centrifugal force of magnitude $\mu R\omega^2$ that tends to stretch the bond. A bond acts like a spring, and the restoring force obeys Hooke's law to a good approximation, that it is proportional to the displacement from equilibrium, R_0; the magnitude of this restoring force is written $k(R - R_0)$ where k is the force constant (this molecular parameter will figure in the discussion of molecular vibrations later in the chapter). At equilibrium the centrifugal and restoring forces are in balance, and from the

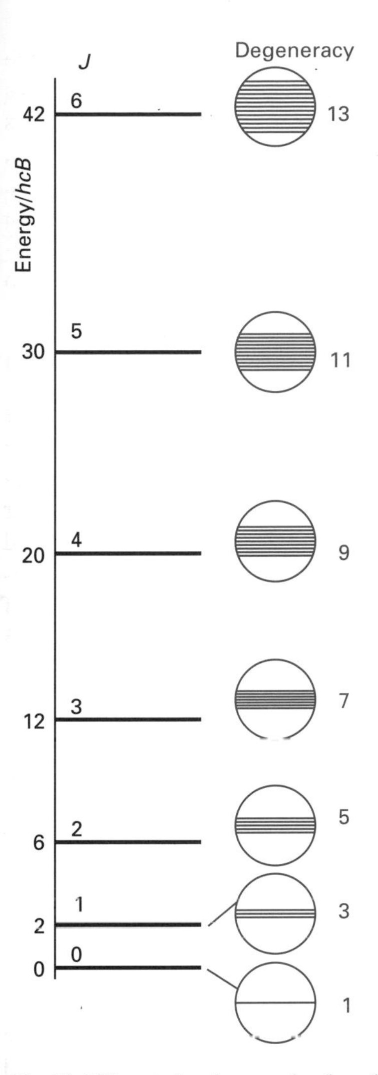

Fig. 10.6 The rotational energy levels and their degeneracies of a linear rotor. Note that the degeneracy increases more slowly (at high values of J the number is proportional to J) than for a spherical rotor, and the rotational states are much more sparse.

condition

$$\mu R\omega^2 = k(R - R_0) \tag{21}$$

we can deduce that

$$R = \frac{kR_0}{k - \mu\omega^2} \approx R_0\left(1 + \frac{\mu\omega^2}{k}\right) \tag{22}$$

This approximation holds for $\mu\omega^2/k \ll 1$, which corresponds to small displacements ($R - R_0 \ll R_0$). The classical hamiltonian for the molecule is

$$H = \frac{J^2}{2\mu R^2} + \tfrac{1}{2}k(R - R_0)^2$$

where the first term is the rotational kinetic energy and the second is the potential energy arising from the stretching of the bond (recall that $F = -\mathrm{d}V/\mathrm{d}R$). It follows from the introduction of eqn 10.21 into this equation and the use of $J = \mu R^2\omega$ that

$$H = \frac{J^2}{2\mu R^2} + \frac{J^4}{2k\mu^2 R^6} \tag{23}$$

Now we confine attention to small displacements and use eqn 10.22 in the form

$$\frac{1}{R^2} = \frac{1}{R_0^2}\left(1 - \frac{\mu\omega^2}{k}\right)^2 \approx \frac{1}{R_0^2}\left(1 - \frac{2\mu\omega^2}{k}\right)$$

which, with $J = \mu R^2\omega$, is equivalent to

$$\frac{1}{R^2} = \frac{1}{R_0^2}\left(1 - \frac{J^2}{k\mu R^4}\right)^2 \approx \frac{1}{R_0^2} - \frac{2J^2}{k\mu R_0^6}$$

With this expression substituted into the first term of eqn 10.23 and R^6 in the second term approximated by R_0^6, we obtain

$$H \approx \frac{J^2}{2\mu R_0^2} - \frac{J^4}{k\mu^2 R_0^6} + \frac{J^4}{2k\mu^2 R_0^6} = \frac{J^2}{2\mu R_0^2} - \frac{J^4}{2k\mu^2 R_0^6}$$

We can now interpret the J^2 and J^4 terms as operators and immediately write down the eigenvalues:

$$E(J, M_J) \approx \frac{J(J+1)\hbar^2}{2\mu R_0^2} - \frac{J^2(J+1)^2\hbar^4}{2k\mu^2 R_0^6}$$

It follows that the wavenumbers of the rotational terms have the form

$$F(J, M_J) \approx BJ(J+1) - DJ^2(J+1)^2 \qquad D = \frac{\hbar^3}{4\pi kc\mu^2 R_0^6} \tag{24}$$

where D is the centrifugal distortion constant.

The centrifugal distortion constant is larger for molecules with bonds that have low force constants, for then the centrifugal distortion caused by a given angular momentum is large. However, because a small force constant is often associated with long bond lengths and high reduced mass, the effect of the latter terms may overcome the effect of changes in k itself.

10.5 Pure rotational selection rules

Consider a linear molecule in the state $|\varepsilon, J, M_J\rangle$, where ε is a label for the electronic (and possibly vibrational state) of the molecule. The electric transition dipole matrix element that we need to consider to establish the rotational selection rules is $\langle \varepsilon, J', M_J'|\boldsymbol{\mu}|\varepsilon, J, M_J\rangle$, where $\boldsymbol{\mu}$ is the electric dipole moment operator. According to the Born–Oppenheimer approximation (Section 8.1), we can separate the rotation of the molecule as a whole from the motion of the electrons, and presume that because the vibrations are so much faster than the rotations, we may also separate them too. Therefore, we write the overall wavefunction of the molecule as the product $|\varepsilon\rangle|J, M_J\rangle$. The transition matrix element then factorizes into

$$\begin{aligned}\langle \varepsilon, J', M_J'|\boldsymbol{\mu}|\varepsilon, J, M_J\rangle &= \langle J', M_J'|\langle \varepsilon|\boldsymbol{\mu}|\varepsilon\rangle|J, M_J\rangle \\ &= \langle J', M_J'|\boldsymbol{\mu}_\varepsilon|J, M_J\rangle\end{aligned} \tag{25}$$

where $\boldsymbol{\mu}_\varepsilon$ is the permanent electric dipole moment of the molecule in the state ε. In other words, the transition matrix element is the matrix element of the permanent electric dipole moment between the two states connected by the transition. We can immediately conclude that only polar molecules (those with a nonzero permanent electric dipole moment) can have a pure rotational spectrum.

To establish the specific selection rules governing rotational transitions, we have to investigate the values of J' and M_J' for which the matrix element $\langle J', M_J'|\boldsymbol{\mu}_\varepsilon|J, M_J\rangle$ is nonzero for given values of J and M_J. For a linear molecule, the rotational wavefunctions are eigenfunctions of the operators J^2 and J_Z (where Z denotes the laboratory axis). As we established in Section 4.7 in connection with orbital angular momenta, these eigenfunctions are the spherical harmonics $Y_{JM_J}(\theta, \phi)$. It follows that for a component Q (where $Q = X$, Y, or Z in the laboratory-fixed axes)

$$\langle J', M_J'|\mu_{\varepsilon Q}|J, M_J\rangle = \int_0^\pi \int_0^{2\pi} Y_{J'M_J'}^* \mu_{\varepsilon Q} Y_{JM_J} \sin\theta \mathrm{d}\theta \mathrm{d}\phi \tag{26}$$

The most efficient way to evaluate this integral is to recognize that the components of the dipole moment operator may be written in terms of spherical harmonics (by using the information in Table 3.1):

$$\begin{aligned}\mu_{\varepsilon X} &= \mu_\varepsilon \sin\theta\cos\phi = -\tfrac{1}{2}\left(\frac{8\pi}{3}\right)^{1/2} \mu_\varepsilon (Y_{1,+1} - Y_{1,-1}) \\ \mu_{\varepsilon Y} &= \mu_\varepsilon \sin\theta\sin\phi = \tfrac{1}{2}\mathrm{i}\left(\frac{8\pi}{3}\right)^{1/2} \mu_\varepsilon (Y_{1,+1} + Y_{1,-1}) \\ \mu_{\varepsilon Z} &= \mu_\varepsilon \cos\theta = \left(\frac{4\pi}{3}\right)^{1/2} \mu_\varepsilon Y_{1,0}\end{aligned} \tag{27}$$

So, to evaluate the matrix elements we need to evaluate integrals of the form

$$I_M = \int_0^\pi \int_0^{2\pi} Y_{J'M_J'}^* Y_{1M} Y_{JM_J} \sin\theta \mathrm{d}\theta \mathrm{d}\phi \tag{28}$$

with $M = 0, \pm 1$.

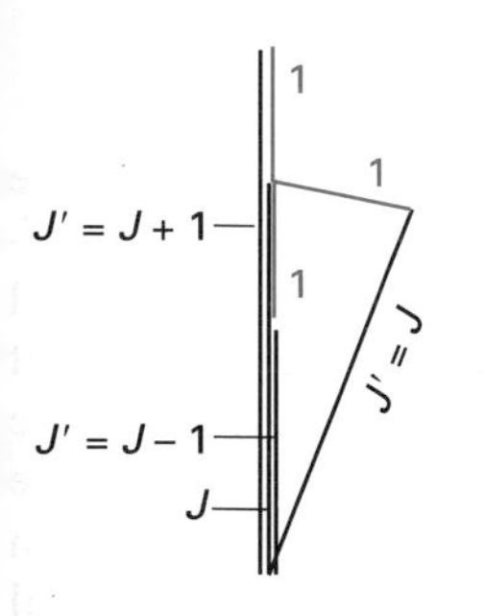

Fig. 10.7 The vector basis of the selection rule for rotational transitions of polar molecules.

The integral I_M is in an ideal form for the application of group theoretical arguments. We saw in Section 5.19 that Y_{JM_J} is a member of the basis that spans the irreducible representation $\Gamma^{(J)}$ of the full rotation group. The integrand therefore spans the completely symmetric irreducible representation only if

$$\Gamma^{(J')} \times \Gamma^{(1)} \times \Gamma^{(J)} = \Gamma^{(0)} \tag{29}$$

which it does only if $J' = J, J \pm 1$ (Fig. 10.7). Therefore, one selection rule is $\Delta J = \pm 1$. (The integral with $J' = J$ does not correspond to an observable transition in pure rotational spectroscopy.) The integral over ϕ has the form

$$I_M \propto \int_0^{2\pi} \mathrm{e}^{\mathrm{i}(M_J+M-M'_J)\phi}\,\mathrm{d}\phi$$

This integrand is completely symmetric (that is, independent of ϕ) only if $M_J + M - M'_J = 0$, so we can conclude that the selection rule for M_J is $\Delta M_J = 0, \pm 1$. The joint selection rules are therefore

$$\Delta J = \pm 1 \qquad \Delta M_J = 0, \pm 1 \tag{30}$$

for a polar linear rotor.

For symmetric rotors we need to consider the possibility of transitions that involve changes in the quantum number K. Because in a symmetric rotor any permanent electric dipole moment must lie parallel to the C_n axis, there is no component perpendicular to the principal axis. Hence, the electromagnetic field cannot couple to transitions that correspond to changes in the component of angular momentum around the principal axis, and hence to changes in K. In a sense, there is no 'handle' perpendicular to the principal axis on which an electric field can exert a torque. The selection rules for polar symmetric rotors are therefore

$$\Delta J = \pm 1 \qquad \Delta M_J = 0, \pm 1 \qquad \Delta K = 0 \tag{31}$$

Spherical rotors do not have permanent dipole moments (by symmetry), so they do not show pure rotational transitions.

When the selection rules are applied to the expressions for the energy levels of linear and symmetric rotors, we find the following expressions for the wavenumbers of the transitions:

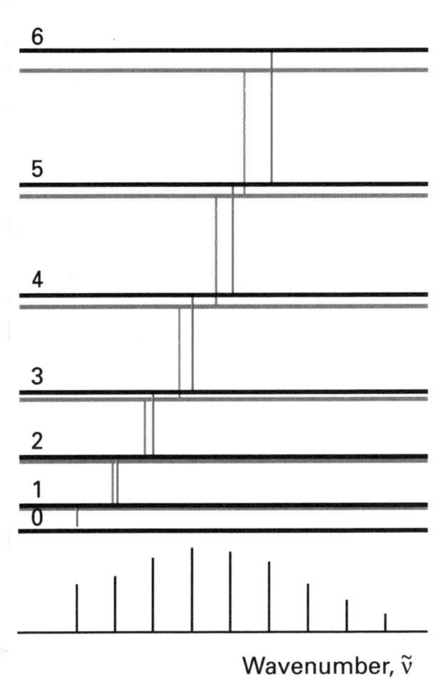

Fig. 10.8 The first few rotational transitions of a linear molecule. The pale lines indicate the effect of centrifugal distortion, which leads to a reduction in the separation of the energy levels at high rotational quantum numbers.

$$\tilde{\nu}_J = F(J+1, K, M_J) - F(J, K, M_J) = 2B(J+1) \tag{32}$$

with $J = 0, 1, 2, \ldots$. For a diatomic linear rotor that displays centrifugal distortion we would use eqn 10.24 to write

$$\tilde{\nu}_J = F(J+1, M_J) - F(J, M_J) \approx 2B(J+1) - 4D(J+1)^3 \tag{33}$$

A pure rotational spectrum therefore consists of a series of lines, which in the absence of centrifugal distortion have uniform spacing $2B$ (Fig. 10.8). Such transitions typically lie in the microwave region of the electromagnetic spectrum.

10.6 Rotational Raman selection rules

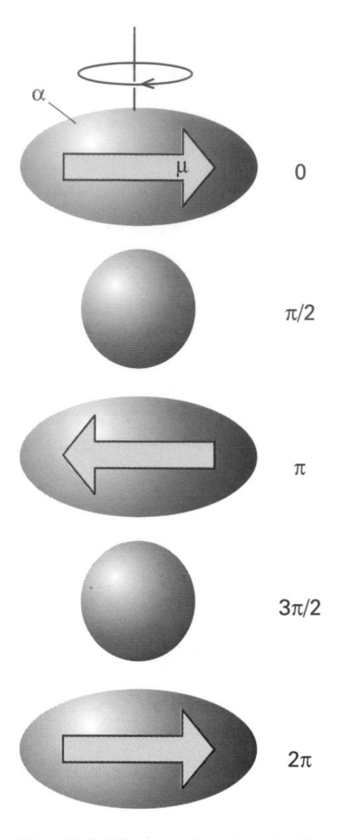

Fig. 10.9 Whereas the electric dipole moment of a molecule requires a rotation through 2π to restore it to its initial value, the polarizability requires a rotation of only π. Thus, the polarizability tensor appears to rotate at twice the rate of the electric dipole.

Molecules with anisotropic electric polarizabilities can show pure rotational Raman lines. The selection rules are now

$$\Delta J = \pm 2, \pm 1 \qquad \Delta K = 0 \text{ but } K = 0 \to 0 \text{ is forbidden for } \Delta J = \pm 1 \quad (34)$$

Note that the restriction on transitions between states with $K = 0$ rules out $\Delta J = \pm 1$ for linear molecules.

There are several ways of understanding the occurrence of 2 in the selection rule for J. In the first place, we saw in Section 10.2 that the classical origin of the Raman effect depends on the polarizability of a molecule changing with time as

$$\alpha(t) = \alpha + \tfrac{1}{2}\Delta\alpha \cos \omega_{\text{int}} t \quad (35)$$

where ω_{int} is some 'internal' frequency of the molecule. For rotation, the polarizability returns to its original value twice per revolution (Fig. 10.9), so we should interpret ω_{int} as $2\omega_{\text{rot}}$. From the point of view of the polarizability, the molecule appears to be rotating twice as fast as its mechanical motion. As a result, lines at $\omega \pm 2\,\omega_{\text{rot}}$ are observed in the scattered radiation. For symmetric tops the possibility of angular momentum around the figure axis complicates the analysis and allows for transitions with $\Delta J = \pm 1$ also.

The more formal procedure for establishing the selection rules is to recognize that the anisotropy of the polarizability has components that vary with angle as $Y_{2M}(\theta, \phi)$. To see that this is so, we consider a diatomic molecule with polarizabilities $\alpha_\parallel$ and $\alpha_\perp$ and an electric field $\mathcal{E}$ applied in the laboratory Z-direction. The induced dipole moment will be parallel to the Z-axis, so we can write $\mu_Z = \alpha_{ZZ}\mathcal{E}$. In the molecular frame, the components of the dipole moment will be μ_x, μ_y, and μ_z, and from Fig. 10.10 we see that

$$\begin{aligned} \mu_Z &= \mu_x \sin\theta\cos\phi + \mu_y \sin\theta\sin\phi + \mu_z\cos\theta \\ \mathcal{E}_x &= \mathcal{E}\sin\theta\cos\phi \qquad \mathcal{E}_y = \mathcal{E}\sin\theta\sin\phi \qquad \mathcal{E}_z = \mathcal{E}\cos\theta \end{aligned} \quad (36)$$

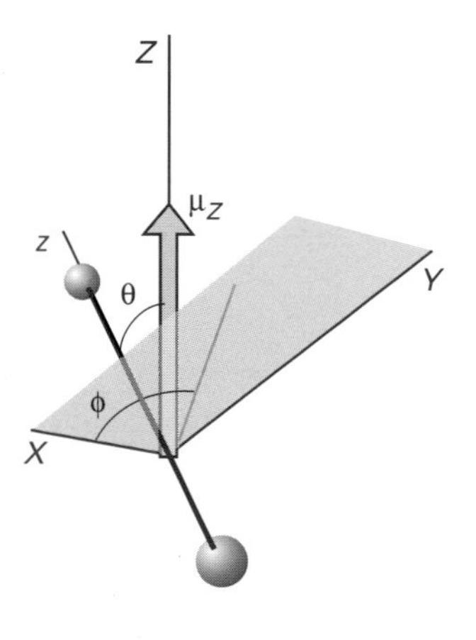

Fig. 10.10 The quantities used to relate the component of electric dipole moment in the laboratory axes to the component in the molecular axes.

Because the molecular component of the induced electric dipole moment is related to the molecular component of the electric field by $\mu_q = \alpha_{qq}\mathcal{E}_q$, it follows that

$$\begin{aligned} \mu_Z &= \alpha_{xx}\mathcal{E}_x \sin\theta\cos\phi + \alpha_{yy}\mathcal{E}_y \sin\theta\sin\phi + \alpha_{zz}\mathcal{E}_z\cos\theta \\ &= \alpha_\perp\mathcal{E}\sin^2\theta\cos^2\phi + \alpha_\perp\mathcal{E}\sin^2\theta\sin^2\phi + \alpha_\parallel\mathcal{E}\cos^2\theta \\ &= \alpha_\perp\mathcal{E}\sin^2\theta + \alpha_\parallel\mathcal{E}\cos^2\theta \end{aligned} \quad (37)$$

where we have identified $\alpha_\perp = \alpha_{xx} = \alpha_{yy}$ and $\alpha_\parallel = \alpha_{zz}$. The mean polarizability is $\alpha = \frac{1}{3}(\alpha_\parallel + 2\alpha_\perp)$; therefore, with $Y_{20} = (5/16\pi)^{1/2}(3\cos^2\theta - 1)$ from Table 3.1 and $\Delta\alpha = \alpha_\parallel - \alpha_\perp$, it follows that

$$\mu_Z = \left\{\alpha + \tfrac{4}{3}\left(\frac{\pi}{5}\right)^{1/2}\Delta\alpha Y_{20}(\theta,\phi)\right\}\mathcal{E} \quad (38)$$

The first term does not contribute any off-diagonal elements, but the second term gives a contribution to the transition dipole moment of the form

$$\langle J', M'_J|\mu_Z|J, M_J\rangle = \tfrac{4}{3}\left(\frac{\pi}{5}\right)^{1/2}\mathcal{E}\Delta\alpha\langle J', M'_J|Y_{20}|J, M_J\rangle \tag{39}$$

The integral that determines whether or not this matrix element vanishes is

$$I = \int_0^{\pi}\int_0^{2\pi} Y^*_{J'M'_J}(\theta,\phi)Y_{20}(\theta,\phi)Y_{JM_J}(\theta,\phi)\ \sin\theta\mathrm{d}\theta\mathrm{d}\phi \tag{40}$$

By the same argument as before, and as illustrated in the following example, the integral is zero unless $J' = J \pm 2$.

Example 10.1 The deduction of the rotational Raman selection rules

Show that the rotational Raman selection rules for a linear rotor are $\Delta J = \pm 2$.

Method. We have to investigate the conditions under which the integral I in eqn 10.40 is nonzero. Group theory is the tool: we need to decide the conditions under which the integrand is a basis for the totally symmetric irreducible representation of the full rotation group. Some care must be taken to take into account the full symmetry of the system.

Answer. The irreducible representation spanned by the integrand is $\Gamma^{(J')} \times \Gamma^{(2)} \times \Gamma^{(J)}$; this direct product includes $\Gamma^{(0)}$ if $J' = J, J \pm 1, J \pm 2$. However, $J' = J \pm 1$ is excluded by the fact that the spherical harmonics change phase by $(-1)^J$ when θ is increased by π, so the overall change in the integrand under this symmetry operation is a change of phase by $(-1)^{J'+2+J}$, which is $+1$ only if $J + J'$ is an even number, which rules out $J' = J \pm 1$. The contribution $J' = J$ is also excluded for rotational Raman spectra because it does not correspond to a change in energy of the system.

Exercise 10.1. Establish the selection rules on M_J for pure rotational Raman transitions.

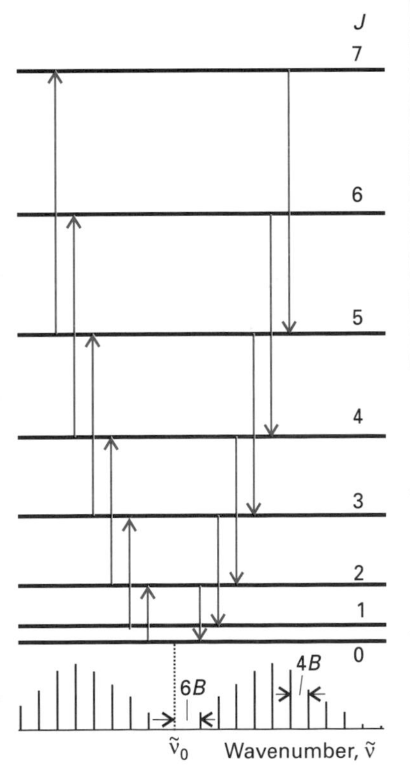

Fig. 10.11 The rotational Raman transitions of a linear molecule.

It follows from the selection rule $\Delta J = \pm 2$ that rotational Raman lines can be expected at the following wavenumbers:

$$\text{Stokes lines } (\Delta J = +2): \quad \tilde{\nu}_J = \tilde{\nu}_0 - 4B(J + \tfrac{3}{2}) \qquad J = 0, 1, 2, \ldots$$

$$\text{Anti-Stokes lines } (\Delta J = -2): \quad \tilde{\nu}_J = \tilde{\nu}_0 + 4B(J - \tfrac{1}{2}) \qquad J = 2, 3, \ldots$$

where $\tilde{\nu}_0$ is the wavenumber of the incident radiation (Fig. 10.11).

10.7 Nuclear statistics

The rotational Raman spectra of certain molecules show a peculiar alternation in intensity. A linear molecule of $C_{\infty v}$ symmetry, such as HCl or OCS, displays the intensity distribution that would be expected on the basis of a Boltzmann

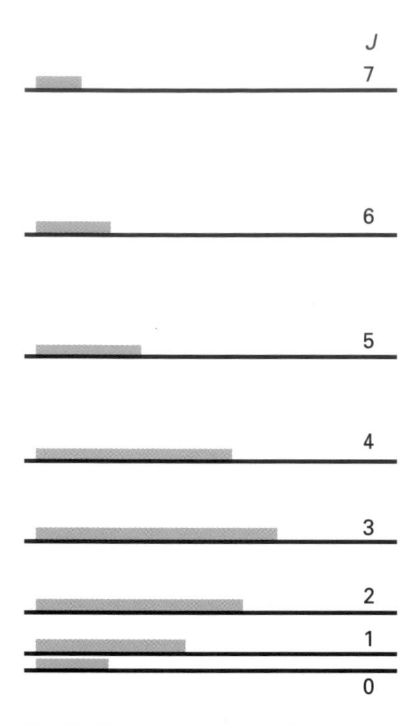

Fig. 10.12 A representation of the Boltzmann distribution of populations in the rotational energy levels of a linear rotor. The populations pass through a maximum on account of the increasing degeneracy of the levels.

distribution of populations over the rotational states:

$$\frac{N_J}{N} = \frac{(2J+1)}{q} \mathrm{e}^{-hcBJ(J+1)/kT} \tag{41}$$

(N_J is the total population of a rotational energy level J, which consists of $2J+1$ individual, degenerate states, and q is the rotational partition function.) Although the transition matrix elements depend on J, the dependence is not very strong and to a good approximation the intensity distribution in the spectrum follows the distribution of populations (Fig. 10.12). The population, and hence the intensity, passes through a maximum at

$$J \approx \tfrac{1}{2}\left\{\left(\frac{2kT}{hcB}\right)^{1/2} - 1\right\} \tag{42}$$

In contrast to this behaviour, a linear molecule of $D_{\infty h}$ symmetry, such as H_2 or CO_2, shows an *alternation* in intensity. Indeed, in CO_2 alternate lines are completely missing, and there are no transitions from states with J odd. We shall now see, in fact, that certain states of symmetrical molecules are not populated and hence make no contribution to spectra.

The key to understanding the absence of certain rotational states is the Pauli principle (Section 7.11) and the fact that the rotation of a molecule may interchange identical nuclei. Nuclei have spin (denoted by the quantum number I, the analogue of s for electrons), which may be integral or half integral depending on the specific nuclide. According to the Pauli principle the interchange of identical fermions (fractional-spin particles, such as protons or carbon-13 nuclei) or bosons (integral-spin particles, such as carbon-12 or oxygen-16 nuclei) must obey the following relation:

$$\Psi(2,1) = \begin{cases} +\Psi(1,2) & \text{for bosons} \\ -\Psi(1,2) & \text{for fermions} \end{cases}$$

and we must ensure that these symmetries are obeyed when a molecule rotates through π (or some other equivalent angle for symmetric rotors such as NH_3).

Consider CO_2 first. The two ^{16}O nuclei are bosons, so the total wavefunction of the molecule must be unchanged when they are interchanged. However, rotation of the molecule by π about a perpendicular axis results in a change in phase of the rotational wavefunction by $(-1)^J$ (Fig. 10.13):

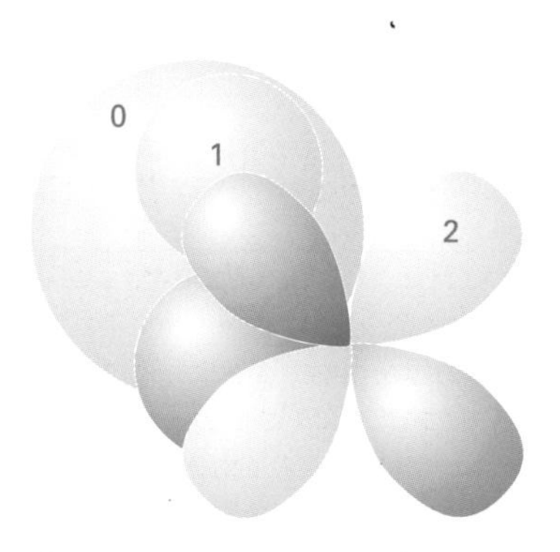

Fig. 10.13 A representation of the (real parts of the) rotational wavefunctions of a rotor for $J = 0, 1, 2$; these wavefunctions are in fact the spherical harmonics encountered in Chapter 3.

$$Y_{JM_J}(\theta+\pi, \phi) = (-1)^J Y_{JM_J}(\theta, \phi) \tag{43}$$

Therefore, to be consistent with the Pauli principle, only even values of J are allowed. This argument accounts for the absence of alternate lines in the rotational Raman spectrum of CO_2.

The discussion of CO_2 that we have just given is in fact somewhat simplistic and does not apply to molecules with nuclei having spin greater than 0; nor does it apply to molecules with incomplete shells or in vibrationally excited states. Consider $^1H_2$, in which the two nuclei are spin-$\frac{1}{2}$ protons. The interchange of the two protons (which are fermions) must result in a change in sign of the overall wavefunction of the molecule. But by 'overall wavefunction' is not meant simply the rotational component: it means the entire wavefunction for all the

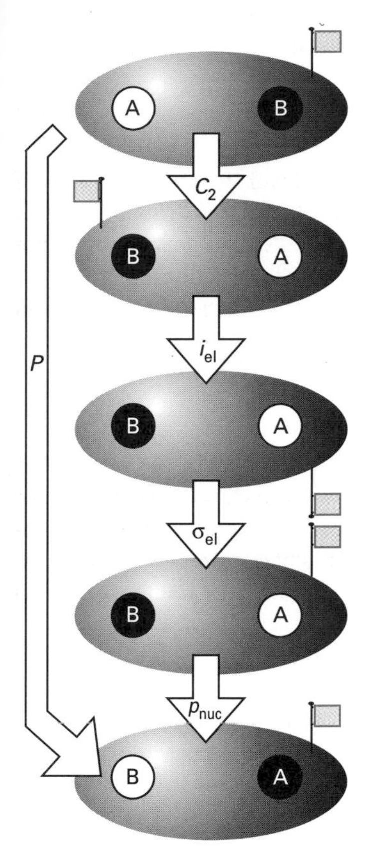

Fig. 10.14 The sequence of transformations involved in the examination of the role of nuclear statistics on the existence of rotational states. The symbol p_{nuc} denotes the permutation of nuclear spin states.

modes of motion:

$$\psi = \psi^{E}\psi^{V}\psi^{R}\psi^{N}$$

where E, V, R, and N denote the electronic, vibrational, rotational, and nuclear spin degrees of freedom. This factorization is valid provided the modes can be separated. When the molecule is rotated by π, the rotational wavefunction is multiplied by $(-1)^J$ and the nuclei are interchanged. However, the electronic wavefunction is also rotated, whereas we want only to interchange the nuclei. The electronic wavefunction can be returned to its original position by an inversion (i^{E}) followed by a reflection (σ_h^{E}) in a plane perpendicular to the rotation (Fig. 10.14). As we saw in Section 8.4, the outcome of the first operation is ± 1 according to whether the molecular state is g or u; similarly, the outcome of the second operation is ± 1 according to whether the state is $\Sigma^{\pm}$ (see Section 8.6). For H_2, which has a ${}^1\Sigma_g^+$ ground state, both operations give a factor of $+1$. The rotation of the molecule also changes the relative displacement coordinate of the atoms into the negative of itself. Oscillator wavefunctions for diatomic molecules depend only on the separation of the nuclei. However, we must distinguish the behaviour of the Hermite polynomials, which have parity $(-1)^v$ under replacement of x by $-x$ (that is, by the replacement of extension of the bond by compression), from the separation itself: a compression is still a compression regardless of the labels on the nuclei. Because ψ^{V} is even for all v, we can ignore it when considering nuclear statistics for diatomics (but not for polyatomics in general).

So far, only factors of $+1$ have occurred other than the factor of $(-1)^J$ for the rotational wavefunction. However, we now need to consider ψ^{N}, the nuclear spin state. There are four possible spin states for two spin-$\frac{1}{2}$ nuclei:

$$\sigma_-(1,2) = \frac{1}{\sqrt{2}}\{\alpha(1)\beta(2) - \beta(1)\alpha(2)\}$$

$$\sigma_+(1,2) = \begin{cases} \alpha(1)\alpha(2) \\ \frac{1}{\sqrt{2}}\{\alpha(1)\beta(2) + \beta(1)\alpha(2)\} \\ \beta(1)\beta(2) \end{cases}$$

The three states labelled σ_+ are symmetric under relabelling whereas the single state σ_- is antisymmetrical.

We can now bring these features together for H_2. Suppose the two nuclei happen to have antiparallel spins, in which case they will be described by the singlet nuclear state σ_-. Relabelling the two nuclei introduces a phase factor of $(-1)^J$ for the rotational wavefunction and -1 for the nuclear state, giving a factor of $(-1)^{J+1}$ overall. For the overall wavefunction to be antisymmetric, it follows that J must be even. On the other hand, if the nuclear spins are parallel, in which case they would be described by one of the three symmetric states σ_+, then the overall phase factor would be $(-1)^J$, and J would be allowed only odd values.

This argument leads to the following remarkable conclusion. Dihydrogen consists of two types of molecule. One, in which the nuclear spins are parallel, is called ***ortho*-hydrogen** and can exist only in rotational states with odd values of J ($J = 1, 3, 5, \ldots$). The other, in which the nuclear spins are antiparallel, is called ***para*-hydrogen** and can exist only in rotational states with even values of J ($J = 0, 2, 4, \ldots$). Because there are three symmetric nuclear spin states and only one antisymmetric state, in a sample at thermal equilibrium at high temperatures we should expect *ortho*-hydrogen to be three times as abundant as *para*-hydrogen. This in turn implies that the Raman lines should show a 3:1 alternation in intensity, with odd J transitions dominant.

At very low temperatures, we would expect only $J = 0$ to be occupied, so the thermal equilibrium sample should consist of pure *para*-hydrogen at low temperatures. However, the conversion of *ortho*-hydrogen to *para*-hydrogen is very slow because it involves the reorientation of one nuclear spin relative to the other, and nuclear magnetic moments are so small that they interact only weakly with external perturbations. Therefore, when a sample of hydrogen gas at room temperature is cooled, the *ortho*-hydrogen component settles into its lowest rotational state ($J = 1$) but cannot readily undergo conversion to *para*-hydrogen. To bring about the conversion more rapidly, a catalyst may be introduced. The gas chemisorbs on the surface of the catalyst as atoms, and the atoms, and their nuclear spins, recombine at random; in due course the equilibrium populations are attained. Interconversion can also be brought about non-dissociatively by bubbling the gas through a solution of a paramagnetic species. The species gives rise to a magnetic field that is inhomogeneous on an atomic scale, and this field can induce the relative reorientation of nuclear spins (as in singlet–triplet transitions between electronic states, Section 11.9).

Example 10.2 The nuclear statistics of linear molecules

What rotational states are occupied in the ground state of dioxygen?

Method. We first decide whether rotation interchanges bosons or fermions. Then we consider the effect of interchanging the labels of the nuclei, considering the electronic, vibrational, rotational, and nuclear components in turn. The electronic ground state of O_2 was deduced in Section 8.6.

Answer. Oxygen–16 nuclei are bosons, so overall the wavefunction must not change sign when the nuclei are relabelled. The electronic ground state is ${}^3\Sigma_g^-$, so the electronic wavefunction changes sign under $\sigma_h^E i^E$. The vibrational wavefunction contributes a factor of $+1$. The nuclear spin state is necessarily symmetric, as both nuclei have zero spin and the only spin state is $|0,0\rangle$. Overall, therefore, a rotation of π results in a phase factor of $(-1)^{J+1}$. For this factor to be even, only odd J states are allowed.

Comment. It follows that in the rotational Raman spectrum of O_2, only transitions between odd J states will occur. This is in contrast to CO_2 in which only even J states contribute.

Exercise 10.2. What rotational states may be occupied by (a) ${}^{12}C_2$ and (b) ${}^{13}C_2$? Carbon-12 and carbon-13 nuclei have spin 0 and $\frac{1}{2}$, respectively.

[(a) Even J only, (b) as for H_2]

Similar arguments can be applied to molecules with nuclei of general spin I, and it is quite easy to derive a general rule for Σ_g^+ linear molecules in their vibrational ground states. If both nuclei that are interchanged have spin I, there are $(2I+1)^2$ product functions of the form $|I_1 m_{I1} I_2 m_{I2}\rangle$. Of these products, $2I+1$ will have $m_{I1} = m_{I2}$ and hence will be symmetric. Of the remaining states, which number

$$(2I+1)^2 - (2I+1) = 2I(2I+1)$$

half will be symmetric (and have the form $|I_1 m_{I1} I_2 m_{I2}\rangle + |I_2 m_{I2} I_1 m_{I1}\rangle$) and half will be antisymmetric ($|I_1 m_{I1} I_2 m_{I2}\rangle - |I_2 m_{I2} I_1 m_{I1}\rangle$). Therefore, the total numbers of each type are

$$N_+ = (2I+1) + I(2I+1) = (I+1)(2I+1) \qquad N_- = I(2I+1)$$

The ratio of the numbers is

$$\frac{N_+}{N_-} = \frac{I+1}{I} \tag{44}$$

Thus, when $I = \frac{1}{2}$, the ratio is 3:1, as we have already seen. For dideuterium ($^2H_2$), for which $I = 1$, the ratio is 2:1. Moreover, because deuterium is a boson, it is the symmetrical states that are associated with even values of J. The rotational Raman spectrum of $^2H_2$ will therefore show an alternation of intensities with even-J lines having about twice the intensity of their neighbouring odd-J lines.

These arguments can be applied to molecules containing more than two identical nuclei (such as NH_3 and CH_4), but the considerations rapidly become quite complicated. Nevertheless, their correct investigation is crucial to a full interpretation of spectra and to the proper implementation of statistical mechanical calculations of thermodynamic properties, which depend on the correct counting of available states.

Molecular vibration

Once again, we pursue the strategy of establishing the energy levels of a molecule, this time of its vibration, and then derive and apply the selection rules for pure vibrational and vibrational Raman transitions. However, there are two main elaborations. One is that we shall need to generalize our conclusions from diatomic molecules, in which there is only one degree of vibrational freedom (the stretching of the bond) to polyatomic molecules, in which there are more. We shall also need to consider the possibility of rotational transitions accompanying vibrational transitions.

10.8 The vibrational energy levels of diatomic molecules

The molecular potential energy of a diatomic molecule increases if the nuclei are displaced from their equilibrium positions. When the displacement $x = R - R_e$ is small, we can express the potential energy as the first few terms of a Taylor series:

$$V(x) = V(0) + \left(\frac{dV}{dx}\right)_0 x + \frac{1}{2}\left(\frac{d^2V}{dx^2}\right)_0 x^2 + \frac{1}{3!}\left(\frac{d^3V}{dx^3}\right)_0 x^3 + \dots \tag{45}$$

where the subscript 0 indicates that the derivatives are to be evaluated at the equilibrium bond length (at $x = 0$). We are not interested in the absolute potential energy of the molecule for the present purposes, and so we can set $V(0) = 0$. The first derivative is zero at the equilibrium separation, for there the molecular potential energy curve goes through a minimum. Provided the displacement from equilibrium is small, the third-order term may be neglected. The only remaining term is the one proportional to x^2, and so we may write

$$V(x) = \tfrac{1}{2}kx^2 \qquad k = \left(\frac{d^2V}{dx^2}\right)_0 \tag{46}$$

and the potential energy close to equilibrium is parabolic. It follows that the hamiltonian for the two atoms of masses m_1 and m_2 is

$$H = -\frac{\hbar^2}{2m_1}\frac{d^2}{dx_1^2} - \frac{\hbar^2}{2m_2}\frac{d^2}{dx_2^2} + \tfrac{1}{2}kx^2$$

We saw in connection with the discussion of the hydrogen atom in Section 3.9 that when the potential energy depends only on the separation of the particles, the hamiltonian can be expressed as a sum, one term referring to the motion of the centre of mass of the system and the other to the relative motion. The former is of no concern here; the latter is

$$H = -\frac{\hbar^2}{2\mu}\frac{d^2}{dx^2} + \tfrac{1}{2}kx^2 \tag{47}$$

where μ is the **effective mass**:[2]

$$\frac{1}{\mu} = \frac{1}{m_1} + \frac{1}{m_2} \tag{48}$$

The appearance of μ in the hamiltonian is physically plausible, because we expect the motion to be dominated by the lighter atom. When $m_1 \gg m_2$, $\mu \approx m_2$, the mass of the lighter particle. Think of a small particle attached by a spring to a brick wall: it is the mass of the particle that determines the vibrational characteristics of the system, not the mass of the wall.

A hamiltonian with a parabolic potential energy is characteristic of a harmonic oscillator, and so we may immediately adopt the solutions found for the harmonic oscillator in Section 2.16:

$$E_v = (v + \tfrac{1}{2})\hbar\omega \qquad \omega = \left(\frac{k}{\mu}\right)^{1/2}$$

with $v = 0, 1, 2, \ldots$. These levels lie in a uniform ladder with separation $\hbar\omega$. The corresponding wavefunctions are bell-shaped gaussian functions multiplied by an Hermite polynomial (Section 2.16 and Fig. 2.27). All the remarks we made about the properties of the solutions of the harmonic oscillator are applicable to small displacements of diatomic molecules from their equilibrium bond lengths.

[2] This quantity is termed the 'reduced mass' in the hydrogen atom, and most people use that name in this connection too. There are, however, advantages in the more general term 'effective mass' as will become apparent when we consider polyatomic molecules.

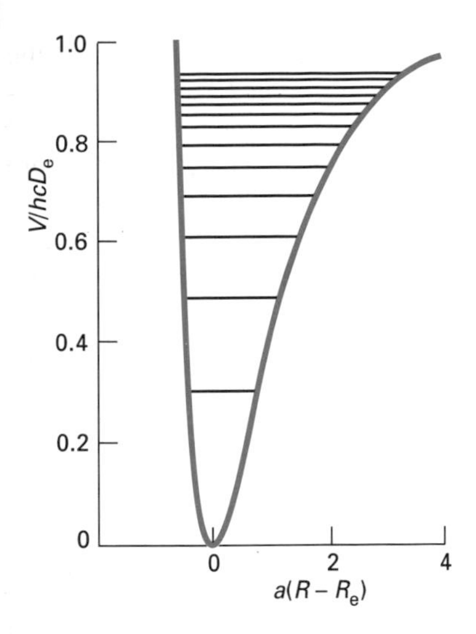

Fig. 10.15 The vibrational energy levels of a molecular oscillator. Note the convergence of levels as the potential becomes less confining. The curve is a plot of the Morse potential.

10.9 Anharmonic oscillation

The truncation of the Taylor expansion of the molecular potential energy after the quadratic term is only an approximation, and in real molecules the neglected terms are important, particularly for large displacements from equilibrium. The typical form of the potential energy is shown in Fig. 10.15, and because at high excitations it is less confining than a parabola, the energy levels converge instead of staying uniformly separated. It follows that **anharmonic vibration**, vibrational behaviour that differs from that of a harmonic oscillator, is increasingly important as the degree of vibrational excitation of a molecule is increased.

One procedure for coping with anharmonicities is to solve the Schrödinger equation with a potential energy term that matches the true potential energy over a wider range better than does a parabola. One of the most useful, but still very approximate, functions is the **Morse potential energy**:

$$V(x) = hcD_e\{1 - e^{-ax}\}^2 \qquad a = \left(\frac{k}{2hcD_e}\right)^{1/2} \tag{49}$$

The parameter D_e is the depth of the minimum of the curve. The Schrödinger equation can be solved analytically with this potential energy (although the techniques required are quite advanced), and the quantized energy levels are

$$E_v = (v + \tfrac{1}{2})\hbar\omega - (v + \tfrac{1}{2})^2\hbar\omega x_e \tag{50}$$

with

$$\omega x_e = \frac{a^2\hbar}{2\mu} \qquad \omega = \left(\frac{k}{\mu}\right)^{1/2} \tag{51}$$

The quantity x_e is called the **anharmonicity constant**. The additional term subtracts from the harmonic expression and becomes more important as v becomes large, which corresponds to the convergence of levels at high excitations. One feature of the Morse potential energy is that the number of bound levels is finite, and $v = 0, 1, 2, \ldots, v_{\max}$ where

$$v_{\max} < \frac{hcD_e}{\hbar\omega/2} - \tfrac{1}{2} \tag{52}$$

(See Problem 10.24.) The zero-point energy of a Morse oscillator is

$$E_0 = \tfrac{1}{2}\hbar\omega(1 - \tfrac{1}{2}x_e)$$

and the **dissociation energy**, hcD_0, is related to the depth of the well by

$$D_0 = D_e - E_0/hc$$

As we have remarked, the Morse oscillator is only an approximation to an actual molecular oscillator. The form of its solution points towards the following polynomial to which experimental data can be fitted:

$$E_v = (v + \tfrac{1}{2})\hbar\omega - (v + \tfrac{1}{2})^2\hbar\omega x_e + (v + \tfrac{1}{2})^3\hbar\omega y_e + \ldots$$

Modern computer methods for the treatment of data and the use of polynomials of this kind are described in the books referred to in *Further reading*.

10.10 Vibrational selection rules

The transition dipole moment for $v' \leftarrow v$ is $\langle \varepsilon, v'|\boldsymbol{\mu}|\varepsilon, v\rangle$ because at this stage we are interested only in transitions within a given electronic state ε. The integration over the electron coordinates can be carried out as before, because we are assuming the validity of the Born–Oppenheimer approximation and the separability of electron and nuclear motion. The transition matrix element is therefore

$$\boldsymbol{\mu}_{v'v} = \langle v'|\boldsymbol{\mu}|v\rangle$$

where now $\boldsymbol{\mu}$ is the dipole moment of the molecule when it is in the electronic state ε and has a bond length R. Because the dipole moment depends on R (because the electronic wavefunction depends parametrically on the internuclear separation) we can express its variation with displacement of the nuclei from equilibrium as

$$\boldsymbol{\mu} = \boldsymbol{\mu}_0 + \left(\frac{\mathrm{d}\boldsymbol{\mu}}{\mathrm{d}x}\right)_0 x + \tfrac{1}{2}\left(\frac{\mathrm{d}^2\boldsymbol{\mu}}{\mathrm{d}x^2}\right)_0 x^2 + \ldots \tag{53}$$

where $\boldsymbol{\mu}_0$ is the dipole moment when the displacement is zero. The transition matrix element is therefore

$$\begin{aligned}\boldsymbol{\mu}_{v'v} &= \boldsymbol{\mu}_0\langle v'|v\rangle + \left(\frac{\mathrm{d}\boldsymbol{\mu}}{\mathrm{d}x}\right)_0 \langle v'|x|v\rangle + \tfrac{1}{2}\left(\frac{\mathrm{d}^2\boldsymbol{\mu}}{\mathrm{d}x^2}\right)_0 \langle v'|x^2|v\rangle + \ldots \\ &= \left(\frac{\mathrm{d}\boldsymbol{\mu}}{\mathrm{d}x}\right)_0 \langle v'|x|v\rangle + \tfrac{1}{2}\left(\frac{\mathrm{d}^2\boldsymbol{\mu}}{\mathrm{d}x^2}\right)_0 \langle v'|x^2|v\rangle + \ldots\end{aligned} \tag{54}$$

The term proportional to $\boldsymbol{\mu}_0$ is zero on account of the orthogonality of the states when $v' \neq v$. The first conclusion we can draw, therefore, is that the transition matrix is nonzero only if the molecular dipole moment varies with displacement for otherwise the derivatives in eqn 10.54 would be zero. The gross selection rule for the vibrational transitions of diatomic molecules is that *they must have a dipole moment that varies with extension*. It follows that homonuclear diatomic molecules do not undergo electric dipole vibrational transitions.

For small displacements, the electric dipole moment of a molecule can be expected to vary linearly with the extension of the bond. This would be the case for a heteronuclear molecule in which the partial charges on the two atoms were independent of the internuclear distance. In such cases, the quadratic and higher terms in the expansion can be ignored and

$$\boldsymbol{\mu}_{v'v} \approx \left(\frac{\mathrm{d}\boldsymbol{\mu}}{\mathrm{d}x}\right)_0 \langle v'|x|v\rangle \tag{55}$$

The specific selection rule can be established by investigating the conditions under which the matrix element in this equation is nonzero. The elementary procedure is to express the matrix element in terms of the harmonic oscillator wavefunctions, and to use the following property of Hermite polynomials:

$$2yH_v(y) = H_{v+1}(y) + 2vH_{v-1}(y) \tag{56}$$

(In this expression, y is proportional to the displacement x: see eqn 2.43.) Even without going into details of the calculation, it can be seen that $x|v\rangle$ can be expected to produce two terms, one proportional to $|v+1\rangle$ and the other to $|v-1\rangle$. That being so, we can anticipate that the only nonzero contributions to $\boldsymbol{\mu}_{v'v}$ will be obtained when $v' = v \pm 1$, and hence conclude that the selection rule for electric dipole transitions within the harmonic approximation is

$$\Delta v = \pm 1 \qquad (57)$$

The detailed calculation is left as an exercise (see Problem 10.16). The alternative procedure for establishing this selection rule makes use of the annihilation and creation operators introduced in *Further information 6*, and is illustrated in the following example.

Example 10.3 The selection rules for a harmonic oscillator

Use the annihilation and creation operators introduced in *Further information 6* to establish the selection rules for electric dipole transitions of a harmonic oscillator, and deduce the explicit forms of the transition dipole moments.

Method. The displacement x can be expressed in terms of annihilation (a) and creation (a^+) operators by using eqn 3 of *Further information 6*. The matrix elements of these operators are given in eqn 10 of the same section, so it is a simple matter to glue the pieces of the calculation together.

Answer. The displacement x is related to the operators by

$$x = \left(\frac{\hbar}{2\mu\omega}\right)^{1/2}(a - a^+)$$

Because $a|v\rangle \propto |v-1\rangle$ and $a^+|v\rangle \propto |v+1\rangle$, we know immediately that $\boldsymbol{\mu}_{v'v} = 0$ unless $v' = v \pm 1$. For the explicit form of the matrix elements we use

$$a|v\rangle = v^{1/2}|v-1\rangle \qquad a^+|v\rangle = -(v+1)^{1/2}|v+1\rangle$$

to write

$$\boldsymbol{\mu}_{v+1,v} = \left(\frac{\mathrm{d}\boldsymbol{\mu}}{\mathrm{d}x}\right)_0\left(\frac{\hbar}{2\mu\omega}\right)^{1/2}\langle v+1|a - a^+|v\rangle = \left(\frac{\hbar}{2\mu\omega}\right)^{1/2}(v+1)^{1/2}\left(\frac{\mathrm{d}\boldsymbol{\mu}}{\mathrm{d}x}\right)_0$$

$$\boldsymbol{\mu}_{v-1,v} = \left(\frac{\mathrm{d}\boldsymbol{\mu}}{\mathrm{d}x}\right)_0\left(\frac{\hbar}{2\mu\omega}\right)^{1/2}\langle v-1|a - a^+|v\rangle = \left(\frac{\hbar}{2\mu\omega}\right)^{1/2}v^{1/2}\left(\frac{\mathrm{d}\boldsymbol{\mu}}{\mathrm{d}x}\right)_0$$

Comment. This procedure is readily extended to the evaluation of matrix elements of higher powers of the displacement, such as we meet in a moment. Note that transition matrix elements are proportional to $v^{1/2}$ or $(v+1)^{1/2}$: this is another example of the population of a state not being the sole determinant of the transition intensity in a spectrum.

Exercise 10.3. Evaluate the value of $\boldsymbol{\mu}_{v\pm 2,v}$ by taking into account the quadratic term in eqn 10.54.

From these considerations it follows that the wavenumbers of the transitions that can be observed by electric dipole transitions in a harmonic oscillator are

$$\tilde{\nu} = \frac{E_{v+1} - E_v}{hc} = \frac{\hbar\omega}{hc} = \frac{\omega}{2\pi c} \tag{58}$$

and that the spectrum should therefore consist of a single line regardless of the initial vibrational state. However, in practice anharmonicities need to be taken into account, so that different transitions occur with different wavenumbers.

A further complication is that it may be necessary to use the quadratic (and higher) terms in the expression for the transition dipole moment. There is no guarantee that the electric dipole moment of the molecule is proportional to the displacement from equilibrium, and for large displacements the partial charges adjust as the internuclear distance changes. As a result, contributions to the matrix element arising from terms in x^2, etc. play a role. These **electrical anharmonicities** permit transitions with $\Delta v = \pm 2$ (for x^2 contributions) and so on. Transitions with $\Delta v = \pm 2$ are the **first overtones** or **second harmonics** of the vibrational spectrum. Even without electrical anharmonicity, overtones can occur if the oscillator is anharmonic, because then the wavefunctions differ from those of a harmonic oscillator, and the selection rules then relax and allow $\Delta v = \pm 2$, etc.

10.11 Vibration–rotation spectra of diatomic molecules

The vibrational transition of a diatomic molecule is accompanied by a simultaneous rotational transition in which $\Delta J = \pm 1$. The total energy change, and hence the frequency of the transition, then depends on the rotational constant, B, of the molecule and the initial value of J. We also need to note that the rotational constant depends on the vibrational state of the molecule, because vibrations modify the average value of R^{-2}, and so we need to attach a label to B and write it B_v.

The energy of a rotating, vibrating molecule is

$$\begin{aligned} E(v,J) = (v+\tfrac{1}{2})\hbar\omega - (v+\tfrac{1}{2})^2\hbar\omega x_e + \ldots \\ + hcB_vJ(J+1) - hcD_vJ^2(J+1)^2 + \ldots \end{aligned} \tag{59}$$

The transitions with $\Delta v = +1$ and $\Delta J = -1$ give rise to the **P-branch** of the vibrational spectrum. The wavenumbers of the transitions are

$$\begin{aligned} \tilde{\nu}^{P}(v,J) = \tilde{\nu} - 2(v+1)\tilde{\nu}x_e + \ldots \\ - (B_{v+1} + B_v)J + (B_{v+1} - B_v)J^2 + \ldots \end{aligned} \tag{60}$$

A series of lines is obtained because many initial rotational states are occupied. Transitions with $\Delta J = 0$ give rise to the **Q-branch** of the vibrational spectrum. This branch is allowed only when the molecule possesses angular momentum parallel to the internuclear axis, and so a diatomic molecule can possess a Q-branch only if $\Lambda \neq 0$ (where Λ is the total orbital angular momentum of the electrons around the internuclear axis). The wavenumbers of this branch,

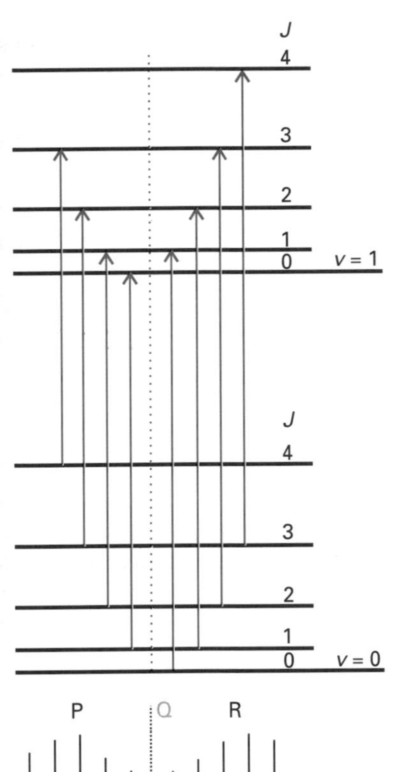

Fig. 10.16 The formation of P and R branches in a linear vibrating rotor and the location of the (usually invisible) Q branch.

when it is allowed, are

$$\tilde{\nu}^{\mathrm{Q}}(v,J) = \tilde{\nu} - 2(v+1)\tilde{\nu}x_{\mathrm{e}} + \ldots + (B_{v+1} - B_v)J + (B_{v+1} - B_v)J^2 + \ldots \tag{61}$$

The transitions with $\Delta J = +1$ give rise to the **R-branch** of the vibrational spectrum. The wavenumbers are

$$\begin{aligned}\tilde{\nu}^{\mathrm{R}}(v,J) &= \{E(v+1,J+1) - E(v,J)\}/hc \\ &= \tilde{\nu} - 2(v+1)\tilde{\nu}x_{\mathrm{e}} + \ldots + 2B_{v+1} + (3B_{v+1} - B_v)J \\ &\quad + (B_{v+1} - B_v)J^2 + \ldots\end{aligned} \tag{62}$$

When the rotational constants are the same in the upper and lower vibrational states ($B_{v+1} = B_v = B$) and we can disregard the effects of anharmonicity and centrifugal distortion, these three expressions simplify to

$$\begin{aligned}\tilde{\nu}^{\mathrm{P}}(v,J) &= \tilde{\nu} - 2BJ \qquad J = 1,2,\ldots \\ \tilde{\nu}^{\mathrm{Q}}(v,J) &= \tilde{\nu} \qquad J = 0,1,2\ldots \\ \tilde{\nu}^{\mathrm{R}}(v,J) &= \tilde{\nu} + 2B(J+1) \qquad J = 0,1,2,\ldots\end{aligned} \tag{63}$$

These expressions show that the P- and R-branches consist of a series of lines separated by $2B$ with an intensity distribution that mirrors the thermal population of the rotational states (Fig. 10.16). The Q-branch, if it is present, consists of a single line at the vibrational wavenumber (and, in practice, of a narrow band of many lines).

When the rotational constants are markedly different in the two vibrational states, the spacing within the branches is no longer regular, and one of the branches may start to converge. If at high values of J the quantity $(B_{v+1} - B_v)J^2$ becomes large enough, it may dominate the term linear in J and the branch may pass through a **head**, a turning point in the spectrum, after which successive lines approach the location of the Q-branch instead of moving away from it. The effect is much more pronounced when transitions are between different electronic states.

10.12 Vibrational Raman transitions of diatomic molecules

The gross selection rule for the observation of vibrational Raman spectra of diatomic molecules is that the molecular polarizability should vary with internuclear separation. That is universally the case with diatomic molecules regardless of their polarity, and so all diatomic molecules, including homonuclear diatomic molecules, are vibrationally Raman active.

The origin of the gross selection rule, and the derivation of the particular selection rules, can be discovered by considering once again the transition dipole moment in much the same way as we did in Section 10.6 but without, at this stage, troubling about the orientation dependence of the interaction between the electromagnetic field and the molecule:

$$\mu_{v'v} = \langle \varepsilon, v'|\mu|\varepsilon, v\rangle = \langle \varepsilon, v'|\alpha|\varepsilon, v\rangle \mathcal{E}$$

Within the Born–Oppenheimer approximation we are free to separate the electronic and vibrational wavefunctions, and hence to evaluate $\alpha(x) = \langle \varepsilon|\alpha|\varepsilon\rangle$ for a

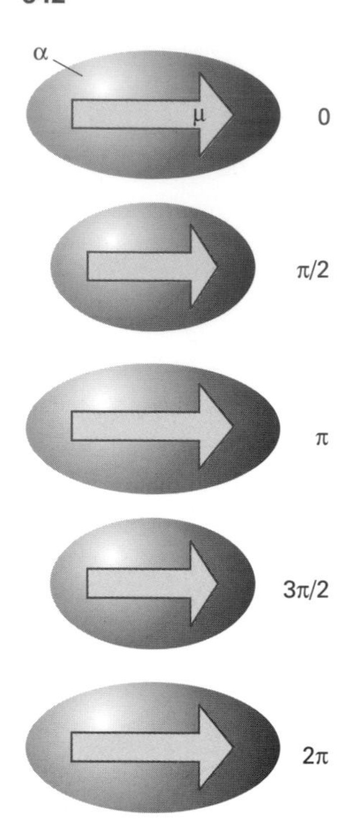

Fig. 10.17 The electric dipole and the polarizability vary with time at the same rate as a result of a molecular vibration.

series of selected displacements, x, from equilibrium. Then, as in the treatment of pure vibrational transitions, we can expand the polarizability as a Taylor series in the displacement:

$$\begin{aligned}\mu_{v'v} &= \langle v'|\alpha(0) + \left(\frac{\mathrm{d}\alpha}{\mathrm{d}x}\right)_0 x + \ldots|v\rangle\mathcal{E} \\ &= \langle v'|v\rangle\alpha(0)\mathcal{E} + \left(\frac{\mathrm{d}\alpha}{\mathrm{d}x}\right)_0 \langle v'|x|v\rangle\mathcal{E} + \ldots\end{aligned}$$

The first matrix element is zero on account of the orthogonality of the vibrational states when $v' \neq v$. Therefore,

$$\mu_{v'v} \approx \left(\frac{\mathrm{d}\alpha}{\mathrm{d}x}\right)_0 \langle v'|x|v\rangle\mathcal{E} \tag{64}$$

This equation shows explicitly that the transition dipole moment is zero unless the polarizability varies with the displacement of the nuclei. Moreover, because the same matrix element occurs on the right as for vibrational transitions, we can also conclude that the selection rule is

$$\Delta v = \pm 1 \tag{65}$$

The selection rule is the same as for vibrational absorption and emission because the polarizability, like the electric dipole moment, returns to its initial value once during each oscillation (Fig. 10.17), not twice. So, in the classical picture presented in Section 10.2, $\omega_{\text{int}} = \omega_{\text{vib}}$. The transitions with $\Delta v = +1$ give rise to the Stokes lines in the spectrum, and those with $\Delta v = -1$ give the anti-Stokes lines. Only the Stokes lines are normally observed, because most molecules have $v = 0$ initially.

In the gas phase, both the Stokes and the anti-Stokes lines of the vibrational Raman spectrum show branch structure. The selection rules for diatomic molecules are

$$\Delta J = 0, \pm 2 \tag{66}$$

and in addition to the Q-branch, there are also O- and S-branches for $\Delta J = -2$ and $\Delta J = +2$, respectively. Note that a Q-branch is observed for all diatomic molecules regardless of their orbital angular momentum.

The vibrations of polyatomic molecules

For a nonlinear molecule consisting of N atoms, there are $3N - 6$ displacements corresponding to vibrations of the molecule. This figure is arrived at as follows. To specify the locations of N atoms we need to specify $3N$ coordinates. These coordinates can be grouped together in a physically sensible manner. Three of them, for instance, can be used to specify the centre of mass of the molecule, leaving $3N - 3$ coordinates for the location of the atoms relative to the centre of mass. The orientation of a nonlinear molecule requires the specification of three angles (Fig. 10.18), so leaving $3N - 6$ coordinates which, when varied, neither change the location of the centre of mass nor the orientation of the molecule. Displacements along these coordinates therefore correspond to vibrations of the molecule. If the molecule is linear, then only two angles are needed to specify its orientation (Fig. 10.19), and so the number of coordinates that correspond to

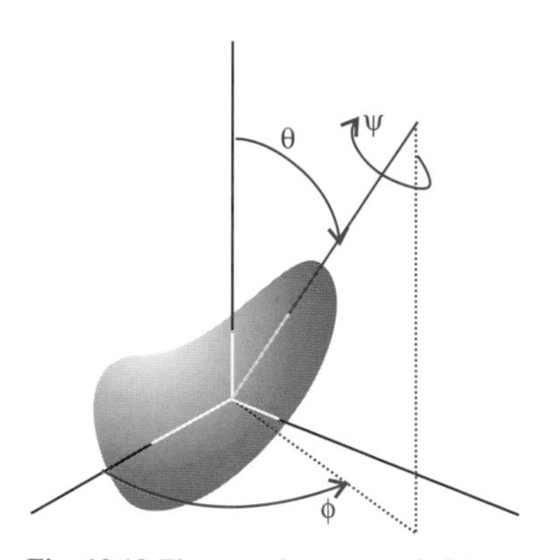

Fig. 10.18 Three angles are needed to specify the orientation of a nonlinear molecule. In other words, a nonlinear molecule has three degrees of rotational freedom.

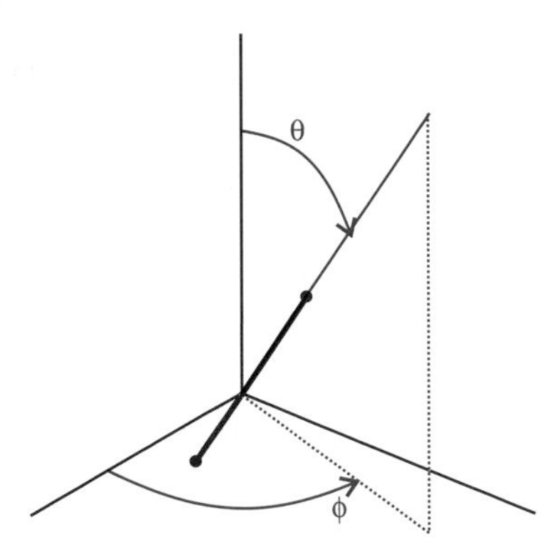

Fig. 10.19 Only two angles are needed to specify the orientation of a linear molecule. In other words, a linear molecule has two degrees of rotational freedom.

vibrational modes of the molecule is $3N - 5$.

The first problem we must tackle is the description of vibrations in molecules. As we shall see, it is possible to express the numerous vibrations of polyatomic species in a manner that brings out clear analogies with the material covered so far. We shall also see that group theory is of the greatest usefulness in deciding which of these vibrational modes are active spectroscopically.

10.13 Normal modes

In principle, all the atoms participate in the vibrations of a polyatomic molecule. Thus, if one bond of a triatomic molecule is vibrationally excited, the energy of vibration will rapidly be transferred to the other bond through the motion of the central atom. Moreover, the potential energy of a nonlinear polyatomic molecule depends on all the displacements of the atoms from their equilibrium positions, and we should write

$$V = V(0) + \sum_i \left(\frac{\partial V}{\partial x_i}\right)_0 x_i + \tfrac{1}{2}\sum_{i,j}\left(\frac{\partial^2 V}{\partial x_i \partial x_j}\right)_0 x_i x_j + \ldots \tag{67}$$

as a generalization of eqn 10.45. The sum is over all $3N$ displacements of the N atoms, so some displacements (those corresponding to translation and rotation of the molecule as a whole) will turn out to have zero force constant. As for diatomic molecules, $V(0)$ may be set equal to 0 and the first derivatives are all zero at $x = 0$. Therefore, for small displacements from equilibrium,

$$V = \tfrac{1}{2}\sum_{i,j} k_{ij} x_i x_j \qquad k_{ij} = \left(\frac{\partial^2 V}{\partial x_i \partial x_j}\right)_0 \tag{68}$$

Here, k_{ij} is a **generalized force constant**. When there is only one vibrational displacement, this expression reduces to eqn 10.46. When there is more than one vibrational displacement, a displacement of one atom may influence the restoring force experienced by another: this possibility is reflected in the occurrence of partial derivatives with respect to two displacements (both x_i and x_j) in the definition of k_{ij}.

As a first step in the simplification of the problem we introduce the **mass-weighted coordinates**, q_i, where

$$q_i = m_i^{1/2} x_i \tag{69}$$

with m_i the mass of the atom being displaced by x_i. The potential energy then becomes

$$V = \tfrac{1}{2}\sum_{i,j} K_{ij} q_i q_j \qquad K_{ij} = \frac{k_{ij}}{(m_i m_j)^{1/2}} = \left(\frac{\partial^2 V}{\partial q_i \partial q_j}\right)_0 \tag{70}$$

The total kinetic energy becomes

$$T = \tfrac{1}{2}\sum_i m_i \dot{x}_i^2 = \tfrac{1}{2}\sum_i \dot{q}_i^2$$

where the dot signifies differentiation with respect to time. The classical expression for the total energy is therefore

$$E = \tfrac{1}{2}\sum_i \dot{q}_i^2 + \tfrac{1}{2}\sum_{i,j} K_{ij} q_i q_j \tag{71}$$

The hamiltonian can now be constructed in the normal way, as prescribed in Section 1.4 and as we illustrate below.

The difficult terms in eqn 10.71 are the cross-terms in the potential (those with $i \neq j$). The question therefore arises as to whether it is possible to find linear combinations Q_i of the mass-weighted coordinates q_i such that the total energy can be expressed in the form

$$E = \tfrac{1}{2}\sum_i \dot{Q}_i^2 + \tfrac{1}{2}\sum_i \kappa_i Q_i^2 \tag{72}$$

in which there are no cross terms. Some combinations Q will also turn out to correspond to translations and rotations, and for them we can expect $\kappa = 0$. The linear combinations that achieve this separation of modes are called **normal coordinates**. We can suspect that they do exist, because an alternative picture of the two stretching modes of a molecule such as CO_2 is as the sum and difference of the two displacements (Fig. 10.20). When the **symmetric stretch**, the mode in which the O atoms move away from or towards the C atom in unison, is excited, the central C atom is buffeted simultaneously from both sides and the **antisymmetric stretch**, the mode in which one bond shortens as the other lengthens, remains unexcited.

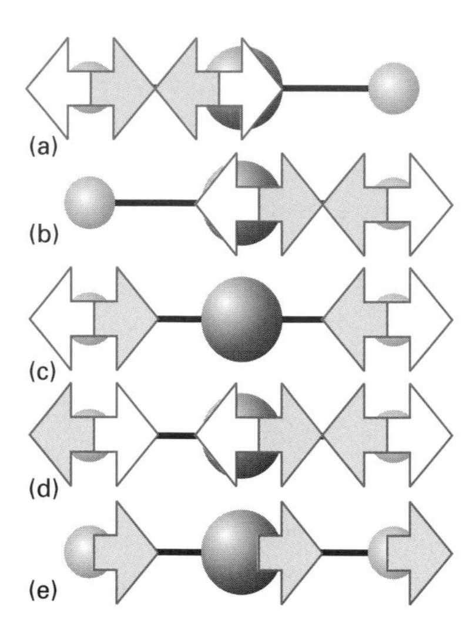

Fig. 10.20 (a) and (b) show two of the vibrations of individual CO bonds in carbon dioxide; (c) and (d) show two linear combinations that preserve the location of the centre of mass of the molecule and which can be excited independently of one another. (e) Another combination of atomic displacements corresponds to the translation of the molecule as a whole.

The formal procedure for determining normal coordinates is described in *Further information 19*. When it is applied to a linear AB_2 triatomic molecule (like CO_2) we find the following expressions for the three normal coordinates corresponding to displacements parallel to the molecular axis:

$$\begin{aligned}
Q_1 &= \frac{1}{\sqrt{m}}(m_B^{1/2} q_1 + m_A^{1/2} q_2 + m_B^{1/2} q_3) \qquad \kappa_1 = 0 \\
Q_2 &= \frac{1}{\sqrt{2}}(q_1 - q_3) \qquad \kappa_2 = \frac{k}{m_B} \\
Q_3 &= \frac{1}{\sqrt{2m}}(m_A^{1/2} q_1 - 2m_B^{1/2} q_2 + m_A^{1/2} q_3) \qquad \kappa_3 = \frac{km}{m_A m_B}
\end{aligned} \tag{73}$$

where $m = m_A + 2m_B$, the total mass of the molecule. These coordinates, and the bending modes, are illustrated in Fig. 10.21. Note that Q_1 has a zero force constant, and corresponds to the translation of the molecule as a whole; Q_2 corresponds to the symmetric stretch and Q_3 corresponds to the antisymmetric stretch. As the mass of the central atom A is increased relative to the outer two atoms, the coordinate Q_2 and its force constant remain unchanged. On the other hand, the coordinate Q_3 approaches $(q_1 + q_3)/\sqrt{2}$ in which the central atom makes no contribution to the vibration and the force constant changes to k/m_B. The same results for Q_2, Q_3, κ_2, and κ_3 would be obtained for two small masses attached by springs on opposite sides of a brick. The important point to note is that the relative masses of the atoms govern both the details of the normal coordinates and, through their influence on the effective force constants, their vibrational frequencies. From now on, we shall discard the normal

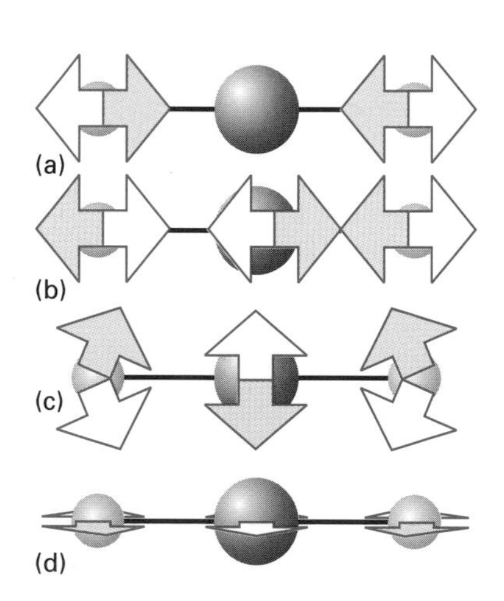

Fig. 10.21 The normal modes of carbon dioxide. (a) Symmetric stretch, (b) antisymmetric stretch, (c) and (d) orthogonal bending modes.

coordinates that correspond to translation and rotation of the entire molecule, and consider only the $3N-6$ (or $3N-5$) vibrational modes. The vibrations that correspond to displacements along these normal coordinates are called the **normal modes** of the molecule.

Because eqn 10.72 is the sum of terms, the hamiltonian operator is also a sum of terms, and in the position representation is

$$H=\sum_i H_i \qquad H_i=-\tfrac{1}{2}\hbar^2\frac{\partial^2}{\partial Q_i^2}+\tfrac{1}{2}\kappa_i Q_i^2 \tag{74}$$

Because the hamiltonian is a sum of terms, the vibrational wavefunction of the molecule is a product of wavefunctions for each mode:

$$\psi=\psi_{v_1}(Q_1)\psi_{v_2}(Q_2)\ldots=\prod_i \psi_{v_i}(Q_i) \tag{75}$$

There are $3N-6$ factors for a nonlinear molecule and $3N-5$ factors for a linear molecule. Each factor satisfies a Schrödinger equation of the form

$$-\tfrac{1}{2}\hbar^2\frac{\partial^2\psi(Q_i)}{\partial Q_i^2}+\tfrac{1}{2}\kappa_i Q_i^2\psi(Q_i)=E\psi(Q_i) \tag{76}$$

which is the Schrödinger equation for a harmonic oscillator of unit mass and force constant κ_i. It follows that the energy levels are

$$E_{v_i}=(v_i+\tfrac{1}{2})\hbar\omega_i \qquad \omega_i=\kappa_i^{1/2} \qquad v_i=0,1,2,\ldots \tag{77}$$

and that the wavefunctions are

$$\psi_{v_i}=N_{v_i}H_{v_i}(y_i)e^{-y_i^2/2} \qquad y_i=\left(\frac{\omega_i}{\hbar}\right)^{1/2}Q_i \tag{78}$$

where N_{v_i} is a normalization constant (Table 2.2). It follows that the total vibrational energy of the molecule in the harmonic approximation is

$$E=\sum_i (v_i+\tfrac{1}{2})\hbar\omega_i \tag{79}$$

and the overall vibrational wavefunction is the product of the factors given in eqn 10.78. A general vibrational state is $|v_1 v_2\ldots\rangle$, with $v_1, v_2\ldots$ the quantum numbers of the modes $1, 2, \ldots$. The ground state is $|0_1 0_2\ldots\rangle$.

The vibrational ground state of a polyatomic molecule, $|0_1 0_2\ldots\rangle$, is of some interest. In the first place, it has a zero-point energy

$$E_0=\tfrac{1}{2}\sum_i \hbar\omega_i \tag{80}$$

For a medium-to-large molecule consisting of 50 atoms, there are 144 modes of vibration, so the total zero-point energy can be substantial. (If the wavenumber of each mode is 300 cm^{-1}, then the total zero-point energy would be close to 260 kJ mol^{-1}, or about 2.7 eV.) The wavefunction of the vibrational ground state is a product of Gaussian functions because $H_0(y)=1$:

$$\psi_0=N\prod_i e^{-y_i^2/2}=Ne^{-y^2/2} \qquad y^2=\sum_i y_i^2 \tag{81}$$

where N is the product of all the normalization constants of the modes. The important feature of this result is that because the normal coordinates appear symmetrically and as their squares,

> In the harmonic approximation, the ground-state vibrational wavefunction of a molecule is totally symmetric under all symmetry operations of the molecule.

The ground-state vibrational wavefunction therefore spans the completely symmetric irreducible representation (A_1, for instance) of the molecular point group. The great significance of this point will become clear when we consider the group theoretical aspects of normal coordinates.

10.14 Vibrational selection rules for polyatomic molecules

We have already seen that the selection rules for harmonic oscillators are $\Delta v = \pm 1$; we shall now see that each normal mode of vibration obeys this selection rule within the harmonic approximation. Moreover, it is easy to establish that electric dipole transitions can occur only for normal modes that correspond to a change in the electric dipole moment of the molecule. The molecular dipole moment depends on an arbitrary displacement as follows:

$$\boldsymbol{\mu} = \boldsymbol{\mu}_0 + \sum_i \left(\frac{\partial \boldsymbol{\mu}}{\partial Q_i}\right)_0 Q_i + \ldots \tag{82}$$

This is a generalization of eqn 10.53. It follows that the transition dipole moment for the individual excitation of a single mode is

$$\langle 00 \ldots v_i' \ldots 0 | \boldsymbol{\mu} | 00 \ldots v_i \ldots 0 \rangle \approx \left(\frac{\partial \boldsymbol{\mu}}{\partial Q_i}\right)_0 \langle v_i' | Q_i | v_i \rangle \tag{83}$$

Consequently, by the same argument as in Section 10.10, $v_i' = v_i \pm 1$. The **fundamental transition** of a single mode is the transition from $v_i = 0$ to $v_i' = 1$. In simple cases it is easy to judge whether $\boldsymbol{\mu}$ varies with displacement along the normal coordinate (which in general involves a composite motion of several atoms). The displacement of the atoms in CO_2 along the normal coordinate corresponding to the symmetric stretch Q_2 leaves the electric dipole moment unchanged, so $(\partial \boldsymbol{\mu}/\partial Q_2)_0 = 0$ and this mode does not couple to the electromagnetic field. On the other hand, displacement of the atoms along Q_3 does result in a change in dipole moment, so $(\partial \boldsymbol{\mu}/\partial Q_3)_0 \neq 0$ and the mode does couple to the electromagnetic field. Normal modes for which $(\partial \boldsymbol{\mu}/\partial Q_i)_0 \neq 0$ are said to be **infrared active** as they can contribute to a vibrational (infrared) absorption or emission spectrum. Group theory greatly aids the determination of which modes are infrared active, as we shall establish shortly.

The corresponding selection rules for vibrational Raman transitions are based, like eqn 10.64, on the expansion of the molecular polarizability in terms of displacements along normal coordinates:

$$\alpha = \alpha_0 + \sum_i \left(\frac{\partial \alpha}{\partial Q_i}\right)_0 Q_i + \ldots \tag{84}$$

It follows that the transition dipole moment for a mode is

$$\mu_{v_i'v_i} \approx \left(\frac{\partial\alpha}{\partial Q_i}\right)_0 \langle v_i'|Q_i|v_i\rangle\mathcal{E} \qquad (85)$$

This equation is a generalization of eqn 10.64. It follows that a transition is Raman active only if the polarizability varies as the atoms are displaced collectively along a normal coordinate (so that $(\partial\alpha/\partial Q_i)_0 \neq 0$), and if that is so, then the particular selection rule for that mode is $\Delta v_i = \pm 1$, as for emission and absorption. Normal modes for which $(\partial\alpha/\partial Q_i)_0 \neq 0$ are classified as **Raman active** as they can contribute to a vibrational Raman spectrum. It is usually much harder to judge whether a mode is Raman active, and group theory becomes almost essential, and certainly much more reliable than intuition.

10.15 Group theory and molecular vibrations

The detailed form of the normal coordinates does not need to be known in order to decide which normal modes are infrared and Raman active. Thus, although the detailed form of the normal coordinates depends on the masses of the atoms, and different species with the same type of molecular formula (such as AB_2 or AB_3, etc.) have different normal coordinates, the *symmetries* of the normal coordinates remain the same regardless of the masses of the atoms.

The first step is to establish the symmetry species of the irreducible representations spanned by the displacement coordinates x_i or (because they are proportional) of the mass-weighted coordinates q_i. The procedure has already been described in Section 5.11 in connection with an arbitrary basis set. Here we need to see how to apply the same procedure to the explicit problem of atomic displacements. We shall illustrate the calculation by means of an example.

Example 10.4 The symmetries of normal modes

Determine the symmetry species of the vibrations of H_2O.

Method. First, identify the point group of the molecule. Then treat the set of mass-weighted coordinates q_i as a basis, and determine the characters of the irreducible representations they span by noting how they transform into one another under the operations of the molecular point group. For a group with only one-dimensional representations the characters of the operations are best found by counting $+1$ whenever a coordinate is left unchanged, -1 when changed into the negative of itself, and 0 if the operation carries it away from its site in the row. That set of characters is then used to determine the symmetry species of the irreducible representations by using eqn 5.29. Three of the symmetry-adapted linear combinations correspond to translations, and their symmetry species (which are the same as those spanned by x, y, and z) can be subtracted. Three more (or two for linear molecules) correspond to rotations and may also be subtracted by reference to the positions occupied by R_x, R_y, and R_z in the character table. The remaining symmetry species are those spanned by the vibrational displacements.

Answer. The $3N = 9$ mass-weighted coordinates are shown in Fig. 10.22. They span a nine-dimensional reducible representation of the group C_{2v}. As an illustration of the determination of the characters, consider the effect of the operation C_2:

$$C_2(q_1, q_2, \ldots, q_9) = (-q_1, -q_2, q_3, -q_7, -q_8, q_9, -q_4, -q_5, q_6)$$

$$= (q_1, q_2, \ldots, q_9)\begin{bmatrix} -1 & 0 & 0 & 0 & 0 & 0 & 0 & 0 & 0 \\ 0 & -1 & 0 & 0 & 0 & 0 & 0 & 0 & 0 \\ 0 & 0 & 1 & 0 & 0 & 0 & 0 & 0 & 0 \\ 0 & 0 & 0 & 0 & 0 & 0 & -1 & 0 & 0 \\ 0 & 0 & 0 & 0 & 0 & 0 & 0 & -1 & 0 \\ 0 & 0 & 0 & 0 & 0 & 0 & 0 & 0 & 1 \\ 0 & 0 & 0 & -1 & 0 & 0 & 0 & 0 & 0 \\ 0 & 0 & 0 & 0 & -1 & 0 & 0 & 0 & 0 \\ 0 & 0 & 0 & 0 & 0 & 1 & 0 & 0 & 0 \end{bmatrix}$$

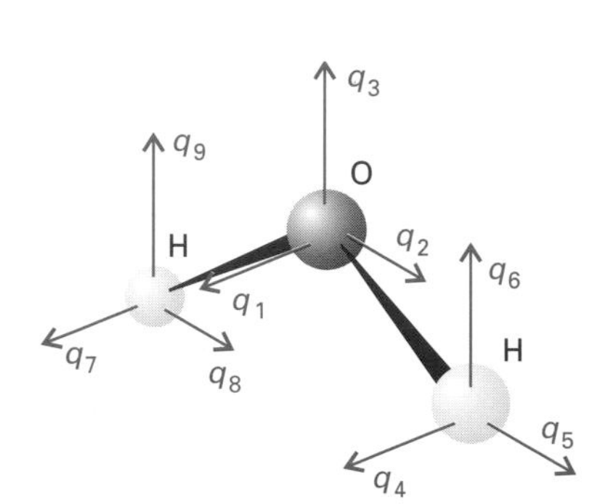

Fig. 10.22 The displacements used for the discussion of the normal modes of a water molecule.

It follows that $\chi(C_2) = -1$. The same result can be obtained much more quickly by inspection of Fig. 10.22. Continuation of this procedure gives the characters $9, -1, 3, 1$ for the four operations of the group, which decomposes (eqn 5.29) into $3A_1 + A_2 + 2B_1 + 3B_2$. In C_{2v}, translations transform as $A_1 + B_1 + B_2$, and rotations transform as $A_2 + B_1 + B_2$. Subtraction of these symmetry species leaves $2A_1 + B_2$.

Comment. As we see, there are three normal modes (the special case of $3N - 6$ with $N = 3$). An example when rotations of the molecule mix coordinates in a more complex manner is illustrated in Example 10.5 later in this section.

Exercise 10.4. Determine the symmetry species of the vibrations of a planar AB_4 (D_{4h}) molecule.

Once the symmetry species of normal modes have been established, we can do a great deal with very little additional calculation. The argument is based on the fact that within the harmonic approximation the ground vibrational state wavefunction is totally symmetric under all the operations of the group. This should be obvious for one-dimensional bases because each operation multiplies Q_i by either $+1$ or -1, and so Q_i^2 remains unchanged; because the wavefunction with $v_i = 0$ is a function of Q_i^2, it follows that the wavefunction is totally symmetric and spans, for instance, A_1. We also need to note that because the Hermite polynomial $H_1(y)$ is proportional to y, and hence to the relevant Q_i, the symmetry species of the first excited vibrational state of a mode is the same as that of the normal coordinate for the mode.

The last point provides a powerful method for determining what transitions are allowed. We consider a fundamental transition in which only one mode is undergoing excitation. The transition dipole moment between the ground state and the first excited state of a normal mode i is $\langle 1_i|\boldsymbol{\mu}|0_i\rangle$. This matrix element is zero unless the direct product of the components of the integrand contains the totally symmetric irreducible representation of the molecular point group. But

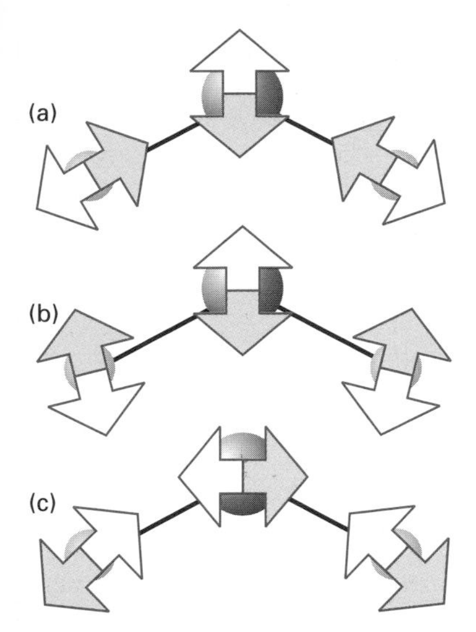

Fig. 10.23 The three normal modes of vibration of a water molecule.

we have seen that ψ_{0i} is a basis for A_1. Therefore, ψ_{1i} and $\boldsymbol{\mu}$ must span the same irreducible representation if their product is to contain A_1. We know that ψ_{1i} is a basis for the same irreducible representation as Q_i; therefore,

> For a fundamental transition to be infrared active, a normal mode must belong to the same symmetry species as the components of the electric dipole moment.

The components of the latter transform as translations, so we know from the character table to which symmetry species it belongs. In C_{2v}, for instance, translations span $B_1(x)$, $B_2(y)$, and $A_1(z)$. The three normal modes of a C_{2v} molecule span $2A_1$ and B_2 (Fig. 10.23). Therefore, all three modes are infrared active. We can go on to say that the A_1 fundamental modes are excited by radiation that is z-polarized and the B_2 fundamental mode is y-polarized.

As a second example, consider CO_2 again. It belongs to the point group $D_{\infty h}$ and has four normal modes of vibration. By using the same techniques as were illustrated in Example 10.4, we can conclude that the normal coordinates span $\Sigma_g^+ + \Sigma_u^+ + \Pi_u$, the last being doubly degenerate (see Fig. 10.21). In $D_{\infty h}$, translations, and hence the components of the dipole moment, span $\Sigma_u^+ + \Pi_u$. It follows that the fundamental transitions of $\Sigma_u^+ + \Pi_u$ modes are active but the Σ_g^+ mode is inactive. A glance at the illustration confirms that this mode is the symmetric stretch, and that it results in no change in the electric dipole moment of the molecule.

The same style of argument may be applied to determine which normal modes are vibrationally Raman active within the harmonic approximation. Instead of the transformation properties of the electric dipole moment, we now have to consider the transformations of the polarizability, α. The electric polarizability transforms in the same way as the quadratic forms x^2, xy, etc. (as will be explained when its origin is established in Section 12.1). The symmetry species of the irreducible representations spanned by these forms are also listed in the character tables (see Appendix 1), and so exactly the same procedure can be followed. Now, though, we use the following rule:

> For a fundamental transition to be Raman active, a normal mode must belong to the same symmetry species as the components of the electric polarizability.

In the group C_{2v}, for instance, the components of the polarizability span all the symmetry species of the group, and so all three normal modes are Raman active. In $D_{\infty h}$, the quadratic forms span $\Sigma_g^+ + \Pi_g + \Delta_g$. It follows that only the Σ_g^+ mode is Raman active. We see that in CO_2 the fundamental modes are either infrared active or Raman active, but not both.

The following **exclusion rule** is a generalization of the last remark:

> In a molecule with a centre of inversion, a normal mode cannot be both infrared and Raman active.

(A mode may be inactive for both.) The justification of this rule is that the components of the electric dipole moment (the translations) have odd parity under inversion whereas the components of the polarizability (the quadratic forms) have even parity. Therefore, because the final state in the matrix element $\langle f|\Omega|i\rangle$ cannot simultaneously have both odd and even parity under inversion, the matrix element cannot be nonzero for both types of transition. The exclusion rule is silent on H_2O because the molecule has no centre of inversion, and the same modes can be both infrared and Raman active, as we have seen.

Example 10.5 The activities of molecular vibrations

Establish the symmetry species of the vibrations of CH_4 and decide which fundamental modes are infrared active and which are Raman active.

Method. We proceed as in Example 10.4, but meet the complication that some of the operations mix the coordinates in a complicated manner. However, all is not lost. First, we note that because operations in the same class have the same character, we need consider only one operation of each class $(E, C_3, C_2, S_4, \sigma_d)$. The only tricky operation is C_3, which partially rotates one coordinate into another. Reference to Section 5.13, though, shows that the character of C_3 in the basis (x, y, z) is 0, so the *net* effect of this rotation on the C and H atoms through which the symmetry axis runs is 0 even though individual coordinates are changed in a more complex manner. With the characters established, subtract the symmetry species of the translations and rotations, and then apply the two rules above to determine the activities of the remaining vibrational modes. The character table for the point group T_d is given in Appendix 1.

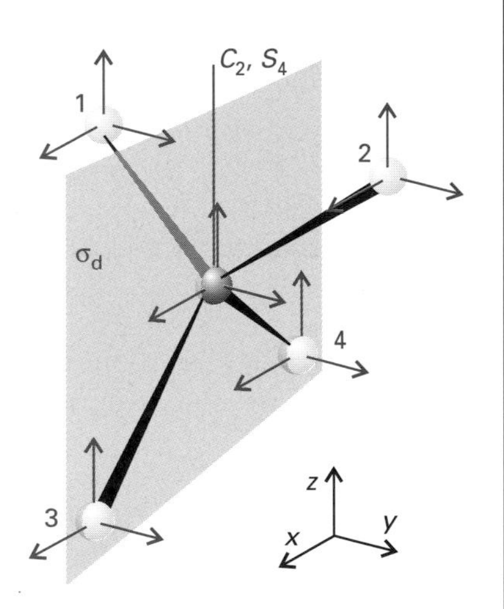

Fig. 10.24 The displacements used in the discussion of the normal modes of a tetrahedral methane molecule.

Answer. There are 15 displacements to consider (Fig. 10.24). Under E, all 15 remain unchanged, so $\chi(E) = 15$. Under C_3, the six displacements on the axial C and H atoms contribute 0 to the character overall, as explained above, and all other displacements are removed completely from their locations in the set $(q_1, q_2, \ldots, q_{15})$, so they too make no contribution to the character, giving $\chi(C_3) = 0$. Under C_2, only the displacements on the central C atom contribute: two displacements become the negative of themselves, and the third remains the same; hence $\chi(C_2) = -1$. Under S_4, the z-displacement on the central atom is reversed, and all others move; so $\chi(S_4) = -1$. Under σ_d, the x- and z-displacements on C, H(3), and H(4) are unchanged, but their y-displacements change sign; all other displacements are moved. Therefore $\chi(\sigma_d) = 3 + 3 - 3 = 3$. The characters $(15, 0, -1, -1, 3)$ span $A_1 + E + T_1 + 3T_2$. Translations span T_2 and rotations span T_1. When these symmetry species are subtracted, we are left with $A_1 + E + 2T_2$ for the vibrations. Infrared active vibrations have T_2 symmetry (the species of translations), and Raman active vibrations have $A_1 + E + T_2$ symmetry (the species of quadratic forms).

Comment. Note that the T_2 modes are both infrared and Raman active (the molecule has no centre of symmetry) and that the T_1 modes are inactive in both. The modes are illustrated in Fig. 10.25 and the physical basis of these conclusions should be apparent.

Exercise 10.5. Repeat the analysis for SF_6, which belongs to the point group O_h.

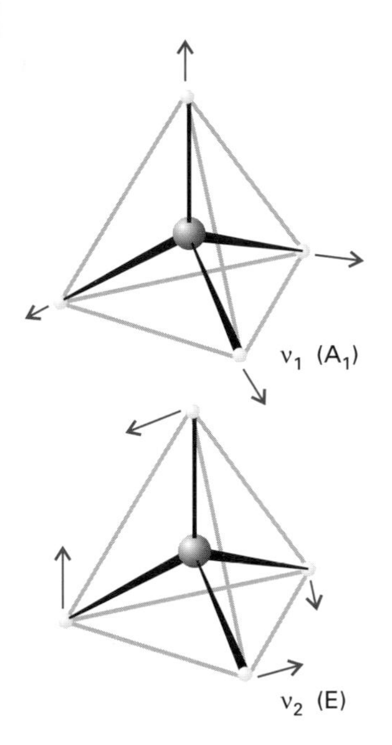

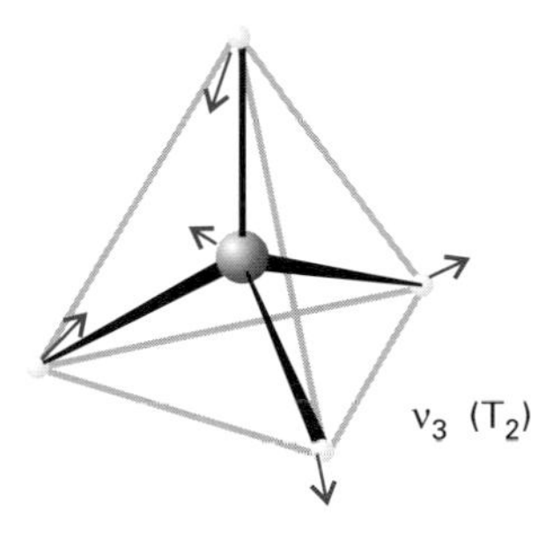

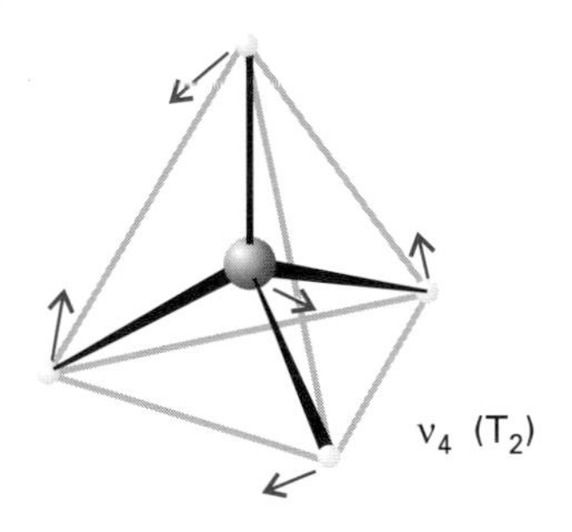

Fig. 10.25 Representative normal modes of a tetrahedral molecule.

10.16 The effects of anharmonicity

We need to distinguish between the effects of electrical and mechanical anharmonicity, and will deal with the former first.

In the first place, symmetry arguments do not yet appear to have ruled out the appearance of **overtones**, transitions for which $\Delta v > 1$ in a single mode. For example, in H_2O, because the Hermite polynomial $H_2(y)$ is symmetrical under all operations of C_{2v}, the $v = 2$ states of all the normal modes are symmetric, and as the z-component of the dipole moment has symmetry species A_1, it looks as though the transition $2 \leftarrow 0$ of a single mode is allowed, because $A_1 \times A_1 \times A_1 = A_1$. It must never be forgotten, however, that group theory asserts when an integral must be zero, but says nothing about the values of integrals that are not necessarily zero. It is often found that there are other reasons why such integrals are in fact zero. This is the case with overtones when there is neither electrical nor mechanical anharmonicity, for then the z-component of the electric dipole moment has the form

$$\mu_z = \mu_{0z} + \sum_i \left(\frac{\partial \mu_z}{\partial Q_i}\right)_0 Q_i \tag{86}$$

with no further terms in the expansion. The transition dipole moment of the first overtone of mode i is then

$$\langle 2_i|\mu_z|0_i\rangle = \left(\frac{\partial \mu_z}{\partial Q_i}\right)_0 \langle 2_i|Q_i|0_i\rangle$$

This matrix element vanishes if the wavefunctions are those of a harmonic oscillator. The overtone becomes allowed for a harmonic oscillator only if there is electrical anharmonicity, because then eqn 10.86 is replaced by

$$\mu_z = \mu_{0z} + \sum_i \left(\frac{\partial \mu_z}{\partial Q_i}\right) Q_i + \tfrac{1}{2}\sum_{i,j} \left(\frac{\partial^2 \mu_z}{\partial Q_i \partial Q_j}\right)_0 Q_i Q_j + \ldots \tag{87}$$

and terms of the form $\langle 2_i|Q_i^2|0_i\rangle$ are not necessarily zero. Therefore, even in the absence of mechanical anharmonicity, this overtone may be visible. Group theory tells us nothing about the stage at which the Taylor series should terminate, but takes a *global* view of the symmetry. We need physical information beyond symmetry to decide whether an *individual* term, even though it has the appropriate symmetry, can actually contribute.

It will have been noticed that eqn 10.87 contains cross terms proportional to Q_iQ_j. These terms can result in **combination bands** in which more than one mode is excited simultaneously. The group theoretical possibility of such an event is quite easy to see. Consider the excitation of an H_2O molecule with y-polarized radiation. The ground vibrational state is A_1. The y-component of the electric dipole moment transforms as B_2 in C_{2v}; therefore, the vibrationally excited state must also be B_2. Such a symmetry can be achieved by the single excitation of the B_2 normal mode, but it can also be achieved by the simultaneous excitation of the B_2 and A_1 modes because their overall symmetry is $B_2 \times A_1 = B_2$. To determine whether the transition can actually occur, we need to consider the following transition dipole moment:

$$\langle 1_a 1_b|\mu_y|0_a 0_b\rangle = \left(\frac{\partial \mu_y}{\partial Q_a}\right)_0 \langle 1_a|Q_a|0_a\rangle\langle 1_b|0_b\rangle + \left(\frac{\partial \mu_y}{\partial Q_b}\right)_0 \langle 1_b|Q_b|0_b\rangle\langle 1_a|0_a\rangle$$
$$+\tfrac{1}{2}\left(\frac{\partial^2 \mu_y}{\partial Q_a \partial Q_b}\right)_0 \langle 1_a 1_b|Q_a Q_b|0_a 0_b\rangle + \ldots$$

The first two terms are zero on account of the orthogonality of the vibrational states. However, the remaining term displayed,

$$\langle 1_a 1_b|\mu_y|0_a 0_b\rangle = \tfrac{1}{2}\left(\frac{\partial^2 \mu_y}{\partial Q_a \partial Q_b}\right)_0 \langle 1_a|Q_a|0_a\rangle\langle 1_b|Q_b|0_b\rangle \tag{88}$$

is not necessarily zero, and so the combination band can occur.

Combination bands are also observed as a result of mechanical anharmonicity. In a polyatomic molecule, the potential energy varies with displacement as

$$V = \tfrac{1}{2}\sum_{i,j}\left(\frac{\partial^2 V}{\partial q_i \partial q_j}\right)_0 q_i q_j + \tfrac{1}{3!}\sum_{i,j,k}\left(\frac{\partial^3 V}{\partial q_i \partial q_j \partial q_k}\right)_0 q_i q_j q_k + \ldots \tag{89}$$

The presence of the cubic terms removes the independence of the normal modes because the transformation that separates the hamiltonian with its quadratic terms does not simultaneously separate the remaining terms in the expansion.

Group theory provides a simplification of the mixing of normal modes by noting that the potential energy, regardless of whether it is harmonic or anharmonic, must be totally symmetric under every symmetry operation of the molecular point group. (The hamiltonian, of which the potential energy is part, always has the full symmetry of the point group: the energy cannot depend on how the molecule is orientated in field-free space.) Therefore, each term in eqn 10.89 must be a basis for the totally symmetric irreducible representation of the group. It follows that the anharmonic contribution to the potential mixes states of the same overall symmetry because only then may its matrix elements be nonzero.

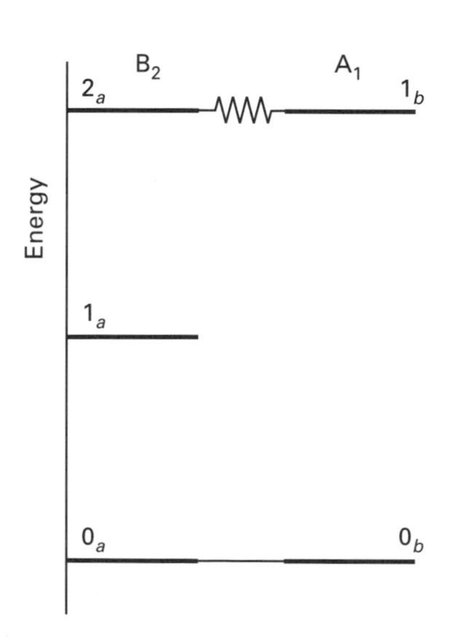

Fig. 10.26 A Fermi resonance between an overtone of B_2 and the A_1 fundamental.

As an example of the interaction caused by anharmonicity, consider the case in which an overtone of one mode (*a*) coincides in energy with the fundamental of another (*b*), as depicted in Fig. 10.26. We need to investigate whether the anharmonic contribution to the potential, V_{an}, has matrix elements of the form $\langle 2_a 0_b|V_{\text{an}}|0_a 1_b\rangle$: if it does, then mixing may occur and it will be possible to excite the molecule from its ground vibrational state to a final state in which mode *a* is doubly excited and mode *b* is not excited, or in which mode *b* is singly excited and mode *a* is not excited. Suppose that mode *b* has symmetry A_1; the overtone of mode *a* will necessarily be A_1 also, because its wavefunctions depend only on Q_a^2 (recall the form of the Hermite polynomials, Table 2.1). Therefore, $\langle 2_a 0_b|V_{\text{an}}|0_a 1_b\rangle$ may be nonzero. Whether it is actually nonzero

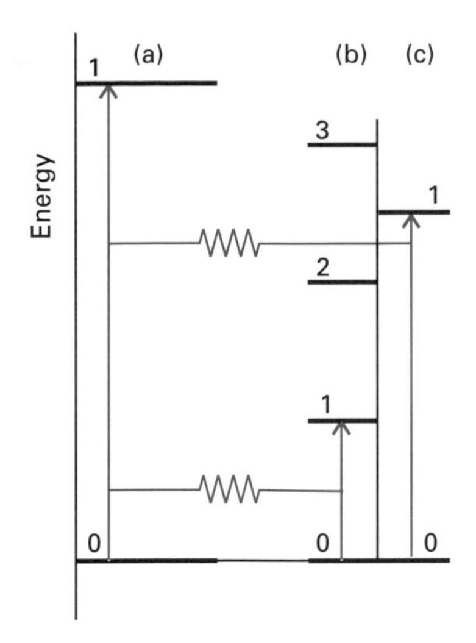

Fig. 10.27 The combination band (b,c) borrows intensity from the allowed (a) fundamental.

depends on the evaluation of the matrix element, which will be of the form

$$\langle 2_a 0_b | V_{\rm an} | 0_a 1_b \rangle = \tfrac{1}{3!}\left(\frac{\partial^3 V}{\partial Q_a^2 \partial Q_b}\right)_0 \langle 2_a | Q_a^2 | 0_a \rangle \langle 0_b | Q_b | 1_b \rangle$$

The matrix elements of Q are nonzero (recall Exercise 10.3), and so long as the third derivative of V is nonzero too, then the modes will mix. This type of mode mixing, in which the interaction is between a fundamental and a combination band or overtone, is called a **Fermi resonance**. Fermi resonance can be viewed as the vibrational analogue of configuration interaction (Section 8.5).

The consequence of the interactions that we have just described is that the energy levels change as a result of their mixing under the influence of a perturbation ($V_{\rm an}$). Furthermore, the transitions take on different intensities because wavefunctions mix and so acquire characteristics of one another. This is most striking in the case of an allowed fundamental and a forbidden combination, for the latter may acquire intensity by virtue of the component of the allowed fundamental that the anharmonicity mixes into it (Fig. 10.27).

10.17 Coriolis forces

Another type of interaction that can affect the appearance of vibrational spectra is the **Coriolis force**, the interaction between vibrational and rotational modes of the molecule.[3]

In classical physics, the Coriolis force is a force that appears to be necessary to an observer in a rotating system in order to account for the motion of particles from their point of view. In particular, it is the tangential component of the force; the radial component is the centrifugal force that we discussed earlier. The source of the tangential effective force can be appreciated by considering the paths taken by balls rolled outwards from the centre of a rotating disk (Fig. 10.28). An external observer sees the ball roll in a straight line toward the edge. An observer stationed at the centre of the disk and rotating with it, misinterprets this straight line as an arc, and therefore concludes that there must be a tangential force in operation. A standard illustration of the Coriolis force is the fact that, because the Earth rotates from west to east, a projectile fired towards the equator from the north pole seems to drift to the west.

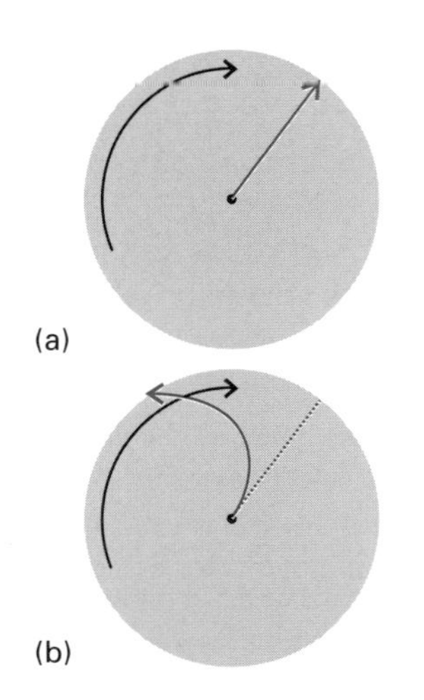

Fig. 10.28 An external observer sees (a) motion in a straight line, but an observer in the rotating frame sees (b) apparently curving motion and concludes that a force must be present.

Consider now the rotation of a mass on a spring (Fig. 10.29). As the mass moves out radially, the rotating observer perceives it as moving in an arc, and concludes that a Coriolis force has retarded its motion. As the particle moves in towards the centre, it appears to accelerate in the direction of travel. Therefore, if it is vibrating, the rotation of the particle is periodically accelerated and decelerated.

Now consider how the Coriolis force affects a rotating molecule when its antisymmetric vibrational mode has been excited (Fig. 10.30). When one of the bonds stretches, it experiences a retarding Coriolis force; at the same time, the bond that is shortening experiences an accelerating Coriolis force. As a result, the molecule tends to bend. As the bonds next contract and lengthen, respectively, the Coriolis force acts in the opposite way, and the molecule is forced to bend in the opposite direction. The effect of the rotation on the anti-

[3] The following discussion is largely qualitative; the *Further reading* section points the way to more quantitative classical and quantum mechanical treatments.

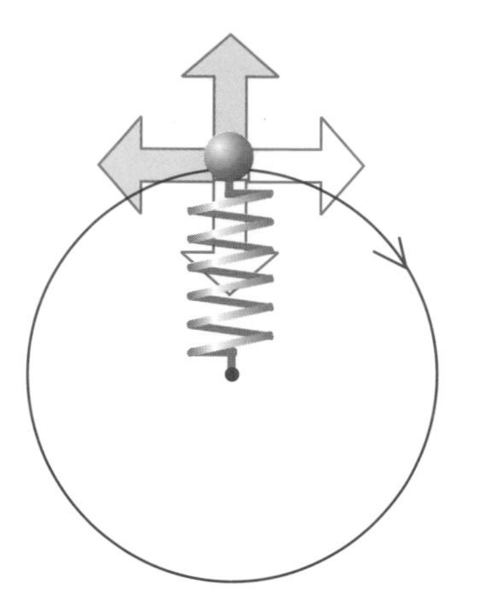

Fig. 10.29 Coriolis forces on a rotating, oscillating mass: the direction of the force (which accelerates or decelerates the particle) is colour-coded to the direction of travel of the oscillator.

symmetric stretch, therefore, is to induce one of the bending modes. Quantum mechanically, we would say that the rotation provides a perturbation that mixes the antisymmetric stretch with one of the components of the doubly-degenerate pair of bending modes. As a result, these two levels move apart in energy, and the bending mode in the plane of rotation is no longer degenerate with the bending mode perpendicular to the plane. Transitions to these two levels no longer fall at the same energy and so the lines are doubled by the rotation. This effect is called ***l*-type doubling.** [4]

10.18 Inversion doubling

Consider a pyramidal (C_{3v}) AB_3 molecule. If we were to plot its potential energy as it is flattened and the pyramid inverted, then we would expect a curve like that shown in Fig. 10.32. Either the barrier is high and the inversion very difficult (as for a well-made umbrella), or the barrier is low and the inversion is easy. In the first case, the molecule vibrates around its AB_3 equilibrium conformation, and does not undergo inversion except perhaps at high excitations. The wavefunctions of these vibrations we denote ψ_L. If the molecule were to invert, then its vibrations would be those of the species B_3A, which we denote ψ_R. The two ladders of vibrational energy levels for the two wells match, and so for a given quantum number ψ_L and ψ_R are degenerate. When the barrier is infinite (in practice, very high), as far as AB_3 is concerned the wavefunctions ψ_R represent states of an inaccessible other world and it is completely oblivious of them. The interesting case, however, is when the barrier is so low that AB_3 can invert and take on the character of B_3A.

For simplicity, suppose that there is only one level on the left and one on the right (Fig. 10.33). The wavefunction of the (almost) harmonic oscillator on the left seeps through the barrier and has nonzero amplitude where ψ_R is also nonzero. The two levels therefore perturb one another and, being degenerate, effect each other strongly. The two wavefunctions mix to form the combinations $\psi_L \pm \psi_R$ and their energies move apart. Where initially there were two degenerate states, there are now two non-degenerate states that are delocalized over both wells. This removal of degeneracy is called **inversion doubling**.

In a more realistic case, there are several levels in each well, but the matching pairs of degenerate states interact with one another most strongly and we can think of the inversion doubling as involving each pair separately (at least to a first approximation). This doubling results in the levels shown in Fig. 10.34. The difference in energy depends on the energies of the states relative to the height of the barrier, and penetration from one well to the other is greatest at high energies, as we saw in Section 2.10. The magnitude of the splitting depends on the state and the identity of the molecule: for the lowest energy states of NH_3 it corresponds to $0.79\ cm^{-1}$ or 24 GHz; the latter figure is known

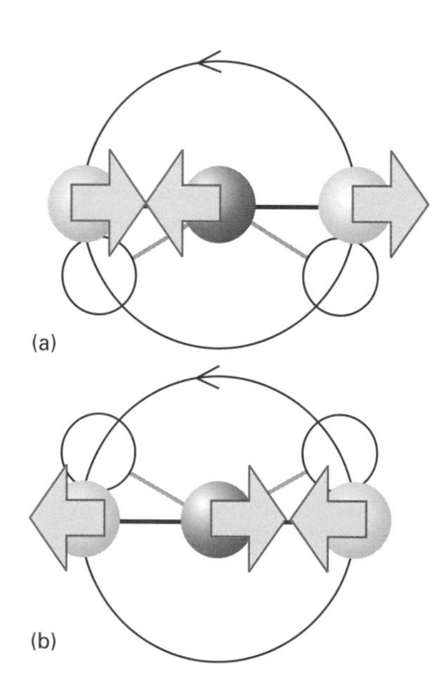

Fig. 10.30 Normal mode coupling in a rotating molecule: (a) and (b) show different stages of the antisymmetric stretch.

[4] Why '*l*-type'? When a linear molecule is not rotating, the two bending modes are degenerate, and we can take any linear combination of them. Two such combinations correspond to rotations of the bent molecule around the previous internuclear axis, in opposite directions. These rotations correspond to an angular momentum of the molecule about its axis (Fig. 10.31), and is described by the quantum number *l*. The Coriolis interaction removes the degeneracy of the bending modes, and so upsets this description. 'Doubling' is a general term signifying the effect on the appearance of the spectrum of the removal of degeneracy.

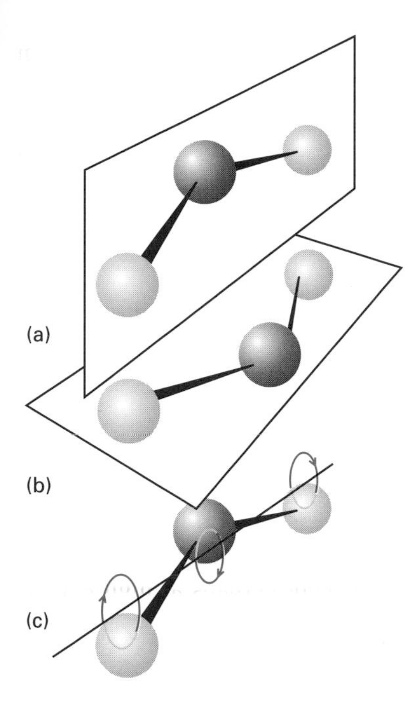

Fig. 10.31 (a,b) Two orthogonal bending modes of a linear triatomic molecule and (c) a linear combination with definite angular momentum about an axis.

as the **inversion frequency**. The origin of this name can be traced back to the discussion of time-dependent behaviour in a 2-level system (Section 6.11). We saw there that if initially the system is in one state, then it periodically visits another degenerate state with a frequency determined by the strength of the perturbation that couples them (Fig. 6.8). In the present case, an NH_3 molecule in its ground vibrational state could be pictured as oscillating between the two inversion-related wells at a frequency of 24 GHz.

The combinations $\psi_L \pm \psi_R$ are respectively even and odd under the inversion of the molecule, and so electric dipole transitions can take place between them. This transition is the basis of 'maser action', the early forerunner of lasers. The ammonia maser operates at $0.79\,\text{cm}^{-1}$ (wavelength 13 mm, frequency 24 GHz), in the microwave region of the spectrum.

Problems

10.1 What is the moment of inertia of (a) a solid disc of mass m, radius R, about its axis, (b) a solid sphere of mass m, radius R?

10.2 Find expressions for the moments of inertia of an AB_3 molecule that is (a) planar, (b) trigonal pyramidal.

10.3 Express the moment of inertia of an octahedral AB_6 molecule in terms of its bond lengths and the masses of the B atoms.

10.4 Show that the moment of inertia I' about an axis parallel to an axis that passes through the centre of mass of a molecule and at a distance R from it is related to the moment of inertia I about the latter axis by $I' = I + mR^2$, where m is the total mass of the body.

10.5 Show that, for a planar lamina (a two-dimensional sheet) in the xy-plane, the moments of inertia parallel and perpendicular to the plane satisfy $I_{xx} + I_{yy} - I_{zz}$.

10.6 Show that the rotational energy levels of a square-planar AB_4 molecule may be expressed solely in terms of the rotational constant B.

10.7 Show that if a time-dependent electric field $\mathcal{E}_0 \cos \omega t$ can induce a nonlinear response, then the scattered light may contain a frequency-doubled (2ω) component. *Hint*. Write $\mu(t) = \alpha\mathcal{E} + \frac{1}{2}\beta\mathcal{E}^2$, and consider an argument like that relating to eqn 10.8.

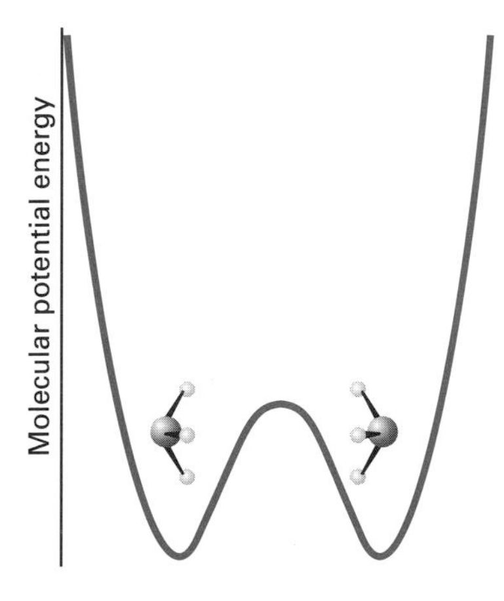

Fig. 10.32 The molecular potential energy curve for a molecule that undergoes inversion.

10.8 Show that the moment of inertia of a diatomic molecule formed from atoms of masses m_A and m_B and bond length R is given by $I = \mu R^2$, where $\mu = m_A m_B/(m_A + m_B)$. Calculate the moments of inertia of (a) $^1H_2$, $R = 75.09$ pm, (b) $^2H_2$, $R = 75.09$ pm, (c) $^1H^{35}Cl$, $R = 127.5$ pm. ($m(^1H) = 1.0078$ u, $m(^2H) = 2.0141$ u, $m(^{35}Cl) = 34.9688$ u.)

10.9 The microwave spectrum of $^1H^{127}I$ consists of a series of lines separated by $12.8\,\text{cm}^{-1}$. Compute its bond length. What would be the separation of $^2H^{127}I$? ($m(^{127}I) = 126.9045$ u.)

10.10 The $J+1 \leftarrow J$ rotational transitions of $^{16}O^{12}C^{32}S$ and $^{16}O^{12}C^{34}S$ occur at the following frequencies (ν/GHz):

J:	1	2	3	4
$^{16}O^{12}C^{32}S$	24.325 92	36.488 82	48.651 64	60.814 08
$^{16}O^{12}C^{34}S$	23.732 23		47.462 40	

Find the rotational constants, the moments of inertia, and the CS and CO bond lengths. *Hint*. Begin by finding expressions for the moment of inertia I through $I = m_A R_A^2 + m_B R_B^2 + m_C R_C^2$, where R_X is the distance of atom X from the centre of mass. The easiest procedure is to use the result established in Problem 10.4, which leads to $I = (m_A m_C/m)(R_{AB} + R_{BC})^2 + (m_B/m)(m_A R_{AB}^2 + m_C R_{BC}^2)$. The lengths R_{AB} and

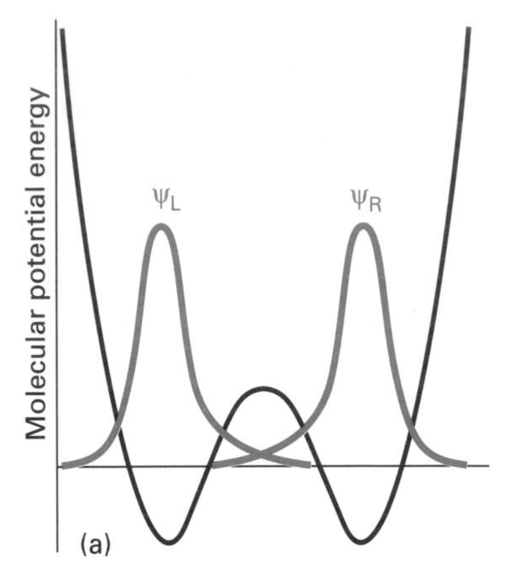

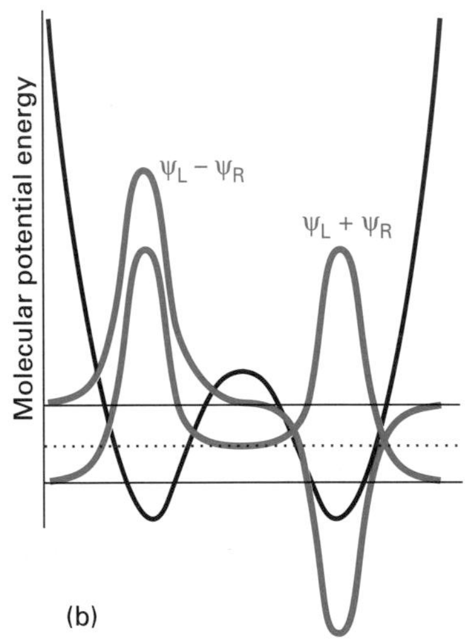

Fig. 10.33 (a) To a first approximation, the molecule oscillates like a harmonic oscillator in either of the two wells: the wavefunctions shown correspond to the ground state of each oscillation. (b) When inversion is allowed, the wavefunctions of the molecule can be modelled as linear combinations of the two independent-well oscillators.

R_{BC} may be found only if two values of I are known. Assume the bond lengths are the same in isotopomeric molecules.

10.11 In PCl_3 the bond length is 204.3 pm and the ClPCl angle is 100.1°. Predict the form of (a) its microwave spectrum, (b) its rotational Raman spectrum, including the general structure of the line intensities. Ignore the effects of nuclear spin statistics. *Hint.* Establish that $I_\perp = m_B R^2(1-\cos\theta) + (m_A m_B/m)R^2(1+2\cos\theta)$ for AB_3, with $m = m_A + 3m_B$, and $I_\parallel = 2m_B R^2(1-\cos\theta)$. Suppose that the intensities are governed predominantly by the Boltzmann distribution.

10.12 The square of the electric transition dipole moment depends on J as $|\mu_{J+1,J}|^2 = \mu^2(J+1)/(2J+1)$. Predict the form of the $^1H^{35}Cl$ spectrum at 300 K (a) without taking account of this dependence, (b) taking this dependence into account. Estimate the values of J in each case corresponding to the most intense transition. *Hint.* Only relative intensities are important. Find the relative populations from the Boltzmann factor and the degeneracies. For (a) examine $(2J+I)e^{-hcBJ(J+1)/kT}$; for (b) examine $(J+1)/(2J+1)$ times this factor.

10.13 Confirm that J_{max}, the value of J corresponding to the maximum in the rotational Boltzmann distribution, is given by eqn 10.42.

10.14 The ethyne molecule (HC≡CH) consists of two fermions (^1H) and two bosons, (^{12}C). What implications are there for the statistical weights of the levels of various J? What are the implications of replacing (a) one ^{12}C by ^{13}C, (b) both ^{12}C by ^{13}C. (The ^{13}C nucleus is a fermion, $I = \frac{1}{2}$.)

10.15 Calculate the effective vibrational masses of (a) $^1H_2$, (b) $^1H^{19}F$, (c) $^1H^{35}Cl$, (d) $^1H^{81}Br$, (e) $^1H^{127}I$. The wavenumbers of the vibrations of these molecules are (a) 4400.39 cm^{-1}, (b) 4138.32 cm^{-1}, (c) 2990.95 cm^{-1}, (d) 2648.98 cm^{-1}, (e) 2308.09 cm^{-1}; calculate the force constants of the bonds. Predict the vibrational wavenumbers of the deuterium halides. ($m(^{19}F) = 18.9984$ u, $m(^{81}Br) = 80.9163$ u; more data are available from Problem 10.8.)

10.16 One way of establishing the harmonic oscillator selection rules is described in Example 10.3. Another way is to use the recursion relation for the Hermite polynomials, eqn 10.56. Calculate the transition moment for transitions commencing in the state with quantum number v. *Hint.* The integral $\int \psi_{v'} x \psi_v \, dx$ can be evaluated very simply by using the orthonormality of the oscillator functions that arise from using the recursion relation.

10.17 The rotational constant of $^1H^{35}Cl$ is 10.4400 cm^{-1} in the ground vibrational state and 10.1366 cm^{-1} in the state $v = 1$. Plot the wavenumbers of the P-, Q-, and R-branches against J as a representation of the structure of the 1–0 transition. (The Q-branch is not observed.)

10.18 The effect of vibrational excitation on the rotational constant can be modelled as follows. First, interpret $B = \hbar/4\pi c\mu R^2$ as the expectation value $(\hbar/4\pi c\mu)\langle 1/R^2\rangle$. Model the vibrational wavefunction by a rectangular probability amplitude, a constant from $R_e - \frac{1}{2}\delta R$ to $R_e + \frac{1}{2}\delta R$, and zero elsewhere. Evaluate $\langle 1/R^2\rangle$, and explore the approximation $\delta R^2 \ll 4R_e^2$. The magnitude of δR^2 can be estimated from $\langle (R-R_e)^2\rangle$ calculated from harmonic oscillator wavefunctions, and expressed in terms of v. Hence arrive at B in terms of v.

10.19 The three fundamental vibrations of CO_2 are observed at 1340 cm^{-1}, 667 cm^{-1}, and 2349 cm^{-1}, the second being the bending mode. Determine the force constant of the CO bond.

10.20 Show that the vibrations of any nonlinear AB_2 molecule span $2A_1 + B_2$ in C_{2v}. Which vibrations are (a) infrared, (b) Raman active?

10.21 Establish the symmetries of the vibrations of the ethene molecule, and classify their activities.

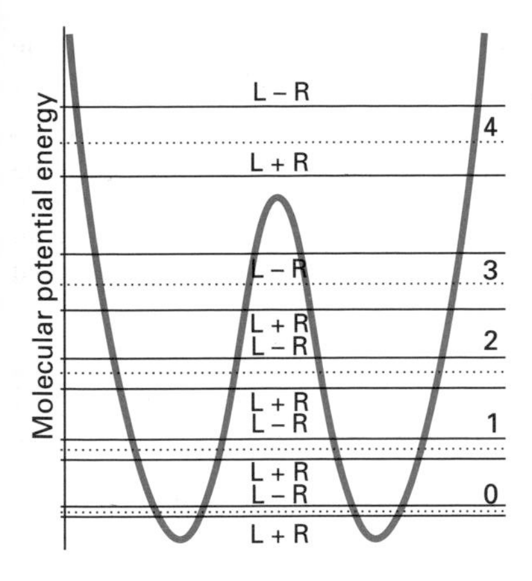

Fig. 10.34 The effect of inversion doubling. The dotted lines are the energy levels of the independent-well oscillators; the full lines are the levels in which inversion through the barrier has removed the degeneracy.

10.22 Consider a two-dimensional harmonic oscillator with displacements in the x- and y-directions, the force constants being the same for each direction (as for the two bending modes of CO_2). Show that the state resulting from the excitation of the oscillator to its first excited state can be regarded as possessing one unit of angular momentum about the z-axis. *Hint*. Show that $\psi(x)\psi(y) \propto e^{i\phi}$.

10.23 Identify the conditions for the existence of and the locations of heads in the P- and R-branches of a diatomic molecule.

10.24 Confirm that a Morse oscillator has a finite number of bound states, and determine the value of v_{max} for the highest bound state.

Further reading

Fundamentals of molecular spectroscopy. C.N. Banwell; McGraw-Hill, New York (1983).
Theory and methods of calculation of molecular spectra. L.A. Gribov and W.J. Orville-Thomas; Wiley, Chichester (1988).
Molecular spectroscopy. J.D. Greybeal; McGraw-Hill, New York (1993).
Fundamentals of molecular spectroscopy. W.S. Struve; Wiley, New York (1989).
Molecular structure and dynamics. W.H. Flygare; Prentice-Hall, Englewood Cliffs, NJ (1978).
Microwave spectroscopy. C.H. Townes and A.L. Schawlow; McGraw-Hill, New York (1955).
Molecular vibrations: the theory of infrared and Raman vibrational spectra. E.B. Wilson, J.C. Decius, and P.C. Cross; McGraw-Hill, New York (1955).
Molecular vib-rotors: the theory and interpretation of high resolution infra-red spectra. H.C. Allen and P.C. Cross; Wiley, New York (1963).
Vibronic coupling: The interaction between the electronic and nuclear motions. G. Fischer; Academic Press, London (1984).
Group theory in chemistry and spectroscopy: a simple guide to advanced usage. B.S. Tsukerblat; Academic Press, London (1994).
Symmetry and spectroscopy: an introduction to vibrational and electronic spectroscopy. D.C. Harris and M.D. Bertolucci; Oxford University Press, New York (1977).
Atomic and molecular spectroscopy: basic aspects and practical applications. S. Svanberg; Springer, New York (1992).
Quantum chemistry and molecular spectroscopy. C.E. Dykstra; Prentice-Hall, Englewood Cliffs, NJ (1992).
Spectra of atoms and molecules. P.F. Bernath; Oxford University Press, New York (1995).
Spectroscopy of molecular rotation in gases and liquids. A.I. Burshtein and S.I. Temkin; Cambridge University Press, Cambridge (1994).
Acronyms and abbreviations in molecular spectroscopy: an encyclopedic dictionary. D.A.W. Wendish; Springer, New York (1990).
Introduction to infrared and Raman spectroscopy. N.B. Colthup, L.H. Daly, and S.E. Wiberley; Academic Press, Boston (1990).
Molecular spectra and molecular structure. II. Infrared and Raman spectra of polyatomic molecules. G. Herzberg; van Nostrand, New York (1945).

11 Molecular electronic transitions

The complexity of electronic spectra of molecules, which occur in the visible and ultraviolet regions, arises in part from the stimulation of simultaneous vibrational and rotational transitions. An electronic transition changes the distribution of the electrons, and the nuclei respond to the new force field by breaking into vibration. In turn, the stimulation of vibration results in rotational transitions, just as in ice-skating skaters change the speed of their rotation by pulling in or throwing out their arms. We shall pick our way through this forest of complication by concentrating initially on diatomic molecules, and then seeing how the concepts generalize to polyatomic molecules.

The states of diatomic molecules

A complication in addition to those already mentioned is that in molecules there are several sources of angular momentum, and to make headway it is necessary to understand how they couple together. The coupling of angular momenta enables us to construct term symbols, and then to use those term symbols to express the selection rules.

11.1 The Hund coupling cases

There are four sources of angular momentum in a diatomic molecule: the spin of the electrons (**S**), their orbital angular momenta (**L**), the rotation of the nuclear framework (**O**), and the nuclear spin (**I**). There are interactions that couple these momenta together to varying extents. For example, the electric field arising from the nuclei couples the orbital angular momentum of the electrons to the internuclear axis, in the sense that only that component is well defined. It is denoted by the quantum number Λ. In highly excited rotational states, however, the nuclear framework may be moving so fast that electrons may be unable to follow the nuclear motions precisely, and the orbital angular momentum is **decoupled** from the internuclear axis. This decoupling is a breakdown of the Born–Oppenheimer approximation (Section 8.1). When the spin–orbit interaction (Section 7.4) is strong, the spin angular momentum of the electrons is coupled to the orbital angular momentum; if the latter is coupled to the internuclear axis, then indirectly the spin is coupled to the axis too, and we speak of the component of electron spin (Σ) on the axis. On the other hand, if the spin–orbit coupling is weak, then the dominant coupling may be between the spin and the magnetic moment arising from the rotation of the molecule as a whole. The nuclear spin may couple to the magnetic field arising from any of the other angular momenta, or it may couple to any of

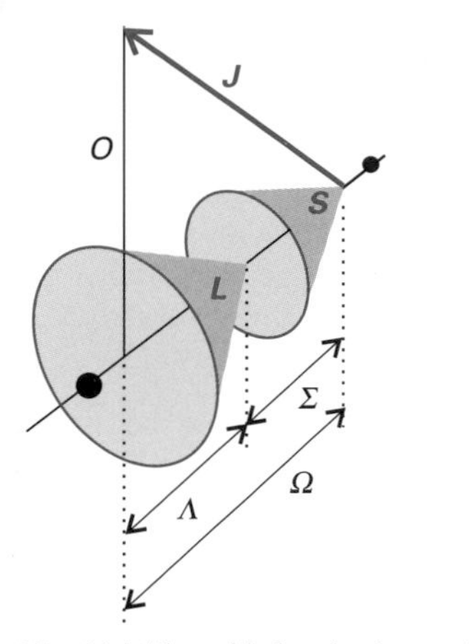

Fig. 11.1 The orbital and spin angular momenta and their projections in Hund's case (a).

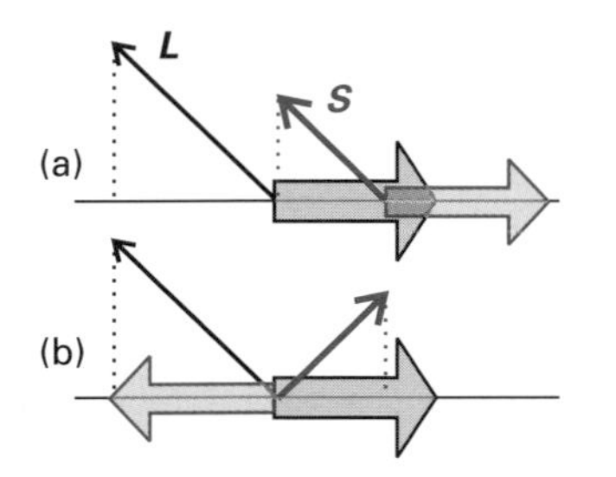

Fig. 11.2 The arrows show the components of magnetic moment that survive after precession: (a) a state of high $|\Omega|$ corresponds to high energy and (b) a state of low $|\Omega|$ corresponds to low energy.

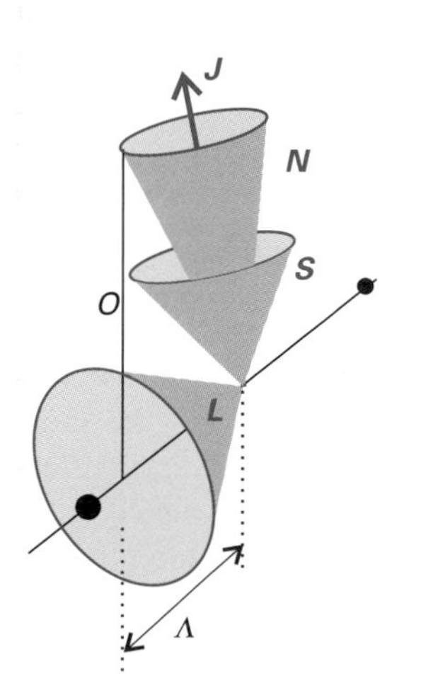

Fig. 11.3 A vector diagram for Hund's case (b).

their resultants: this coupling gives rise to **nuclear hyperfine effects**. Nuclear hyperfine structure of electronic spectra is typically very small and we shall not consider it further.

The spectroscopist F. Hund attempted to impose order on the discussion of all these possibilities by focusing attention on four basic types of coupling. **Hund's case (a)** is depicted in Fig. 11.1; it is appropriate when the orbital angular momentum is coupled strongly to the internuclear axis. The total angular momentum of the molecule is denoted **J** and has magnitude $\{J(J+1)\}^{1/2}\hbar$. It has a component of magnitude $O\hbar$ perpendicular to the internuclear axis, which arises from the rotation of the nuclear framework. It also has a component $\Omega\hbar$ parallel to the internuclear axis which arises from the electronic angular momentum around the axis. This component is related to the components of orbital and spin angular momenta by

$$\Omega = \Lambda + \Sigma \tag{1}$$

As remarked above, the orbital angular momentum is pinned to the axis by the Coulombic field of the nuclei, and the spin angular momentum is brought into line with the orbital angular momentum by the spin–orbit coupling. For the $^2\Pi$ molecule NO, for instance, for which $\Lambda = \pm 1$ and $\Sigma = \pm\frac{1}{2}$, Ω can take the values $\pm\frac{3}{2}$ and $\pm\frac{1}{2}$.

We shall mostly confine our attention to Hund's case (a). In particular, the validity of this coupling scheme means that we can describe the electronic state of a molecule by giving the term symbol constructed on the basis of the point group $C_{\infty v}$ or $D_{\infty h}$, as already explained in Section 8.6. The ground term of NO, for example, is reported as $^2\Pi$, where Π signifies that $\Lambda = \pm 1$. To complete the specification we need to report the value of Ω, and then decide which term, the one with $\Omega = \pm\frac{3}{2}$ or $\Omega = \pm\frac{1}{2}$, lies lower. If the rules described for atoms in Section 7.17 are applicable, we can predict that the two states with $\Omega = \pm\frac{1}{2}$ lie lower because in them the spin and orbital angular momenta, and hence the associated magnetic moments, are opposed (Fig. 11.2). This prediction turns out to be true, and the levels with $|\Omega| = \frac{1}{2}$ lie about 121 cm^{-1} lower than the two levels with $|\Omega| = \frac{3}{2}$.

In **Hund's case (b)** (Fig. 11.3), the spin–orbit coupling is so weak that the spin is not coupled to the orbital angular momentum, but the latter is still coupled to the internuclear axis. Instead, the spin couples to the resultant, **N**, of **O** and $\Lambda\hbar\hat{\mathbf{k}}$, where $\hat{\mathbf{k}}$ is a unit vector parallel to the internuclear axis. The coupling of **S** and **N** gives **J**. Although Λ is a good quantum number, that is no longer true of Ω.

In **Hund's case (c)** (Fig. 11.4), the spin–orbit coupling is so strong that the electron spin and orbital angular momenta couple to give a resultant **E**. This angular momentum has a component $\Omega\hbar$ on the internuclear axis, so Ω is again a good quantum number but Λ and Σ are not.

The final case, **Hund's case (d)** (Fig. 11.5), is rare in practice, and arises when the coupling between the electrons and the molecular axis is so weak that they do not follow the molecular rotation strongly. Now the axial symmetry of the molecule is barely noticed by the electrons, and so the orbital angular momentum, **L**, is well defined. It couples to the angular momentum of the molecule, **O**, to give the resultant **N**. Then the spin couples to that resultant, so forming the overall angular momentum **J**. This coupling is appropriate to the

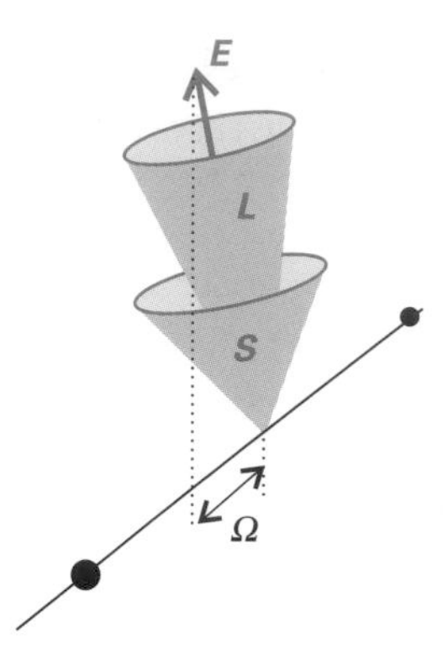

Fig. 11.4 A vector diagram for Hund's case (c), in which the spin–orbit coupling is very strong.

Rydberg levels of diatomic molecules, in which an electron has been excited from the valence shell orbitals of the atoms into orbitals of higher principal quantum number. For instance, an electron in H_2 may be excited from the $1\sigma_g$ orbital into a molecular orbital formed from H2*s*-orbitals. Rydberg orbitals are very diffuse, and the electron is so far from the nuclei of the molecule that it experiences a potential similar to that of a single point charge. As a result, the shape of the molecule is not transmitted to the excited electron and the rotation of the molecule is barely noticed.

11.2 Decoupling and Λ-doubling

In principle, any scheme could be used to describe any molecule. However, the 'correct' scheme is the one for which the hamiltonian of the molecule, with all the interactions included, has the smallest off-diagonal elements. No molecule has a hamiltonian matrix that is exactly diagonal in any one of these schemes, and so if we use one scheme, we can expect it to be contaminated by at least a small admixture of the features of the other schemes. The tendency of one coupling scheme to be contaminated by the other is called the **decoupling** of the angular momenta. Decoupling often increases as J increases because the electrons become increasingly incapable of following the motion of the nuclear framework: this is the phenomenon of **electron slip**.

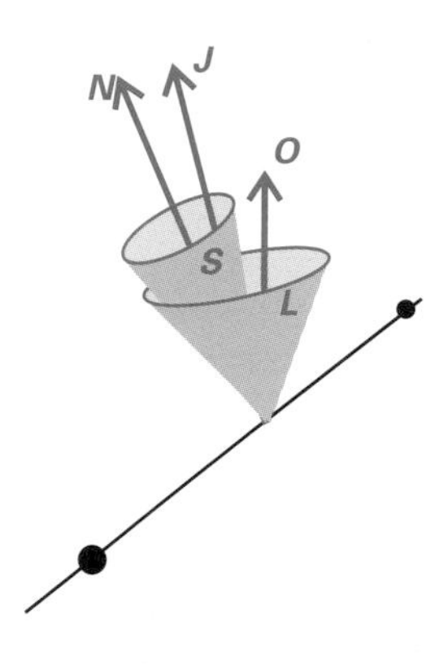

Fig. 11.5 A vector diagram for Hund's case (d).

As an illustration of electron slip, consider a ${}^1\Pi$ term of a diatomic molecule in case (a). In the stationary molecule, the two states $\Lambda = \pm 1$ are degenerate because they differ only in the sense of rotation of the electrons about the axis. The degeneracy is lost when the molecule rotates because the Π_x linear combination of the two orbital angular momentum states mixes with a nearby ${}^1\Sigma^+$ term, but the Π_y term remains unaffected (Fig. 11.6). As a result, the two combinations Π_x and Π_y are no longer degenerate and the energy levels 'double'. A qualitative interpretation is suggested in Fig. 11.7. This effect is called ***Λ*-doubling**. It can be regarded as the outcome of the contamination of case (a) by case (d).

The quantitative treatment of Λ-doubling depends on setting up the appropriate perturbation hamiltonian and then using second-order perturbation theory (Section 6.5). The hamiltonian for the rotation of the nuclear framework is expressed in terms of the angular momentum **O**. This angular momentum has no z-component in the molecular frame (where z is the internuclear axis), and for singlet states is related to the overall angular momentum by $O_x = J_x - L_x$ and $O_y = J_y - L_y$ (Fig. 11.8). It follows that the rotational hamiltonian is

$$\begin{aligned} H &= \frac{1}{2I}(O_x^2 + O_y^2) = \frac{1}{2I}\{(J_x - L_x)^2 + (J_y - L_y)^2\} \\ &= \frac{1}{2I}\{J^2 - J_z^2 + (L_x^2 + L_y^2) - 2(J_xL_x + J_yL_y)\} \end{aligned} \qquad (2)$$

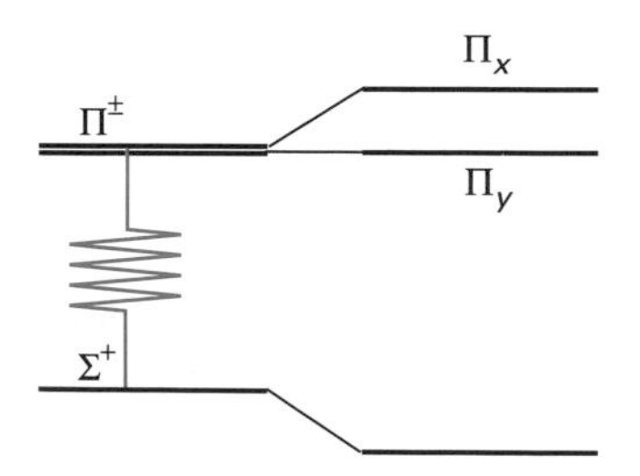

Fig. 11.6 The interaction between states that results in Λ-doubling as a result of the rotation of the molecule.

The term proportional to $L_x^2 + L_y^2$ is independent of the rotational state of the molecule and can be ignored for the present purposes. The term proportional to $J_xL_x + J_yL_y$ can be expressed in terms of raising and lowering operators:

$$J_xL_x + J_yL_y = \tfrac{1}{2}(J_+L_- + J_-L_+)$$

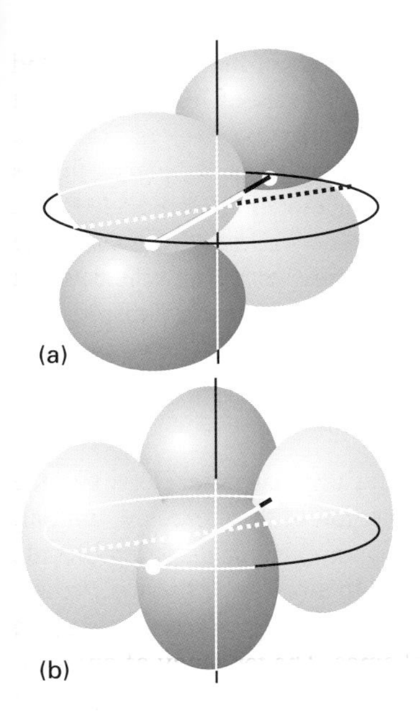

Fig. 11.7 A pictorial interpretation of the effects of molecular rotation and electron 'slip'. In (a), the nuclei slip in the node of the orbital, whereas in (b) they slip into a region of high electron density. The latter corresponds to the partial admixture of Σ-character into the electron wavefunction, as indicated in Fig. 11.6.

and is plainly off-diagonal in Λ on account of the shift operators.[1] These off-diagonal terms, which result in the Σ state removing the degeneracy of the Π states, can be regarded as a perturbation of the major part of the hamiltonian, and we write

$$H^{(0)} = \frac{1}{2I}(J^2 - J_z^2) \qquad H^{(1)} = -\frac{1}{2I}(J_+L_- + J_-L_+) \tag{3}$$

A singlet molecular term with quantum numbers J and Λ has the eigenvalues

$$E(J, \Lambda) = hcB\{J(J+1) - \Lambda^2\} \tag{4}$$

and, at this stage, the states $\pm|\Lambda|$ are degenerate. However, we now allow for the perturbation. The first-order correction is zero because $H^{(1)}$ is off-diagonal in Λ. The second-order contribution to the energy is calculated by using eqn 6.24, and is

$$E^{(2)}(J, \Lambda) = \frac{|\langle J, \Lambda|J_+L_- + J_-L_+|J', \Lambda'\rangle|^2}{(2I)^2\{E(J, \Lambda) - E(J', \Lambda')\}} \tag{5}$$

This expression can be used to evaluate the correction to the energies of the linear combinations Π_x and Π_y, and it turns out that the difference in energy of the two combinations is

$$\Delta E^{(2)} = \frac{2(hcB)^2 L(L+1)J(J+1)}{E(\Pi) - E(\Sigma_g^+)} \tag{6}$$

We see that it is indeed the Σ_g^+ term that is mixed (and, as the details of the calculation show, mixed only with the Π_x component) and that the extent of mixing increases with J.

11.3 Selection rules

Chapters 8 and 9 showed how the electronic energies of diatomic molecules can be calculated. Now that we have some idea of how these energy levels are modified by rotation, we can move on to the prediction of the appearance of electronic spectra by imposing the selection rules. These selection rules have already been introduced in various parts of the text, and may be summarized (and slightly elaborated) as follows:

$\mathrm{g} \to \mathrm{u}$ but not $\mathrm{g} \to \mathrm{g}, \mathrm{u} \to \mathrm{u}$

$\Sigma^+ \to \Sigma^+, \Sigma^- \to \Sigma^-$ but not $\Sigma^+ \to \Sigma^-, \Sigma^- \to \Sigma^+$

$\Delta\Lambda = 0, \pm 1 \qquad \Delta\Omega = 0, \pm 1$

$\Delta S = 0, \Delta\Sigma = 0$ for weak spin–orbit coupling

$\Delta J = 0, \pm 1$ but not $J = 0 \to J = 0$, and for $\Omega = 0 \to \Omega = 0$, $\Delta J \neq 0$

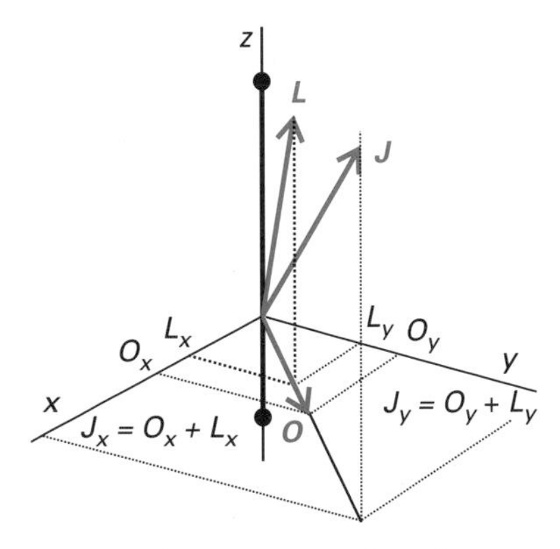

Fig. 11.8 The components of angular momentum that are used to express the hamiltonian for a rotating molecule.

[1] There is a subtlety here. In the rotating molecular framework, J_+ is a *lowering* operator and J_- is a *raising* operator. This reversal of the normal roles follows from the fact that although we know the commutation relations of angular momentum in a laboratory frame, we need to transform them into a rotating frame before we can draw any conclusions from them. When this transformation is carried out, it turns out that $[J_x, J_y] = -i\hbar J_z$. This change of sign compared to the fixed frame interchanges the roles of the shift operators. For more information, see B.R. Judd, *Angular momentum theory for diatomic molecules*, Academic Press, New York (1975).

All these rules are established by detailed consideration of the symmetry properties of the transition dipole moment.

Vibronic transitions

Whenever an electronic transition occurs in a molecule the nuclei are subjected to a change in Coulombic force as a result of the redistribution of electronic charge that accompanies the transition. As a result, the nuclei respond by breaking into more vigorous vibration and the absorption spectrum shows a structure characteristic of the vibrational energy levels of the molecule. Simultaneous electronic and vibrational transitions are known as **vibronic transitions**. We shall begin this section by seeing to what extent the vibrational structure can be predicted and explained.

11.4 The Franck–Condon principle

The analysis of vibronic transitions is based on the **Franck–Condon principle** that, because the mass of an electron is so different from that of nuclei, an electronic transition occurs within a stationary nuclear framework. As a result, the nuclear locations remain unchanged during the actual transition, but then readjust once the electrons have adopted their final distribution.

The qualitative implications of the principle are illustrated in Fig. 11.9. This illustration shows two molecular potential energy curves for two electronic states of a diatomic molecule. The upper curve is typically displaced to the right relative to the lower curve because excitation of electrons generally introduces more antibonding character into the molecular orbitals and the equilibrium bond length increases. The force constants of the two states also differ, for the same reason. We shall confine our attention to the **fundamental progression**, the transitions starting in the ground vibrational state of the lower electronic state. Classically, the transition occurs when the internuclear separation is equal to the equilibrium bond length of the lower electronic state, when the nuclei are stationary, and that internuclear separation and state of motion are preserved during the transition. As a result, the transition terminates where a vertical line cuts through the upper molecular potential energy curve. At the point of intersection, the excited molecule is at a turning point of a vibration, so the nuclei are stationary, and the internuclear separation is the same as it was initially. Such a transition is called **vertical**. Once the electronic transition is complete, the molecule begins to vibrate at an energy corresponding to the intersection.

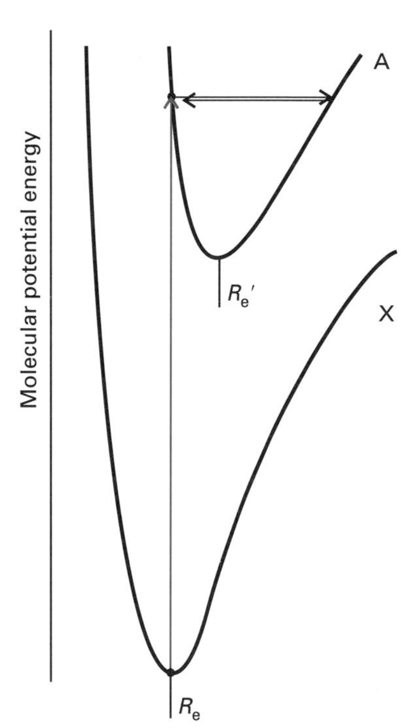

Fig. 11.9 The classical basis of the Franck–Condon principle in which the molecule makes a vertical transition that terminates at the turning point of the excited state. The nuclei neither change their locations nor accelerate while the transition is in progress.

The quantum mechanical description of the process echoes the classical description (Fig. 11.10). Qualitatively, the transition occurs from the ground vibrational state of the lower electronic state to the vibrational state that it most resembles in the upper electronic state. In that way, the vibrational wavefunction undergoes least change, which corresponds to the preservation of the dynamical state of the nuclei as required by the Franck–Condon principle. The vibrational state with a wavefunction that most resembles the original bell-shaped gaussian of the vibrational ground state is one with a peak immediately above the ground state. As can be seen from the illustration, this wavefunction corresponds to an energy level that lies in much the same position as in the

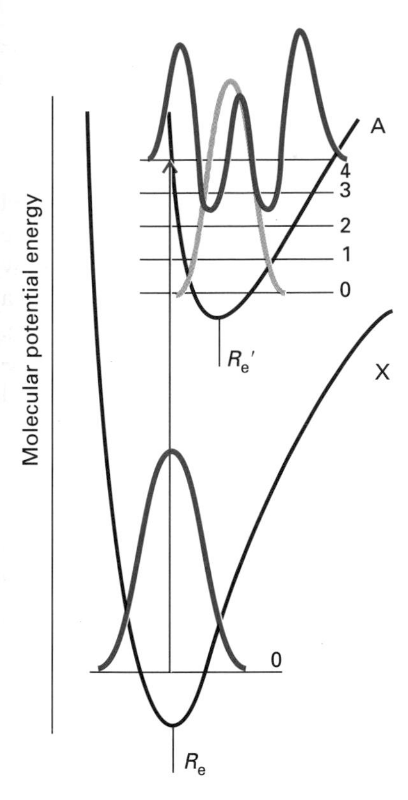

Fig. 11.10 The quantum mechanical version of the Franck–Condon principle. The molecule makes a transition from the ground vibrational state to the state with a vibrational wavefunction that most strongly resembles the initial vibrational wavefunction.

vertical transition of the classical description.

The justification of the quantum mechanical description is based on the evaluation of the transition dipole moment between the ground vibronic state, $|\varepsilon v\rangle$ and the upper vibronic state $|\varepsilon' v'\rangle$. In a molecule, the electric dipole moment operator depends on the locations and charges of the electrons, $\mathbf{r}_i$ and $-e$, and the locations and charges of the nuclei, which we denote $\mathbf{R}_s$ and $Z_s e$, respectively:

$$\boldsymbol{\mu} = -e\sum_i \mathbf{r}_i + e\sum_s Z_s\mathbf{R}_s = \boldsymbol{\mu}_e + \boldsymbol{\mu}_N \qquad (7)$$

Within the Born–Oppenheimer approximation, the vibronic state $|\varepsilon v\rangle$ is described by the wavefunction $\psi_\varepsilon(\mathbf{r};\mathbf{R})\psi_v(\mathbf{R})$, where $\mathbf{r}$ and $\mathbf{R}$ denote the electronic and nuclear coordinates collectively. Note that the electronic wavefunction depends parametrically on the nuclear coordinates (that is, there is a separate electronic wavefunction for each nuclear arrangement). The transition moment is therefore

$$\begin{aligned}\langle\varepsilon' v'|\boldsymbol{\mu}|\varepsilon v\rangle &= \int \psi^*_{\varepsilon'}(\mathbf{r};\mathbf{R})\psi^*_{v'}(\mathbf{R})(\boldsymbol{\mu}_e+\boldsymbol{\mu}_N)\psi_\varepsilon(\mathbf{r};\mathbf{R})\psi_v(\mathbf{R})\,d\tau_e d\tau_N \\ &= \int \psi^*_{v'}(\mathbf{R})\left\{\int \psi^*_{\varepsilon'}(\mathbf{r};\mathbf{R})\boldsymbol{\mu}_e\psi_\varepsilon(\mathbf{r};\mathbf{R})\,d\tau_e\right\}\psi_v(\mathbf{R})\,d\tau_N \\ &\quad + \int \psi^*_{v'}(\mathbf{R})\boldsymbol{\mu}_N\left\{\int \psi^*_{\varepsilon'}(\mathbf{r};\mathbf{R})\psi_\varepsilon(\mathbf{r};\mathbf{R})\,d\tau_e\right\}\psi_v(\mathbf{R})\,d\tau_N\end{aligned}$$

The integral over the electron coordinates in the final term is zero because the electronic states are orthogonal to one another for each selected value of $\mathbf{R}$. The integral over the electron coordinates in the remaining integral is the electric dipole moment for the transition when the nuclei have coordinates $\mathbf{R}$. To a reasonable first approximation, this transition moment is independent of the locations of the nuclei so long as they are not displaced by a large amount from equilibrium, and so the integral may be approximated by a constant $\boldsymbol{\mu}_{\varepsilon'\varepsilon}$. Therefore, the overall transition dipole moment is

$$\langle\varepsilon' v'|\boldsymbol{\mu}|\varepsilon v\rangle = \boldsymbol{\mu}_{\varepsilon'\varepsilon}\int \psi^*_{v'}(\mathbf{R})\psi_v(\mathbf{R})\,d\tau_N = \boldsymbol{\mu}_{\varepsilon'\varepsilon}S(v',v) \qquad (8)$$

where

$$S(v',v) = \int \psi^*_{v'}(\mathbf{R})\psi_v(\mathbf{R})\,d\tau_N \qquad (9)$$

is the overlap integral between the two vibrational states in their respective electronic states. The transition dipole moment is therefore largest between vibrational states that have the greatest overlap. This is the quantitative version of the previous qualitative discussion, where we looked for the upper vibrational state that had a local bell-shaped region above the gaussian function of the ground vibrational state of the lower electronic state.

No one vibrational state has a nonzero value of $S(v',v)$ (unless the two molecular potential energy curves happened to be perfect replications of one another). Indeed, it is generally the case that several vibrational states have similar values of $S(v',v)$, and so transitions occur to all of them. Thus, a **progression** of transitions is stimulated and a series of lines are observed in the electronic spectrum. The relative intensities of the lines are proportional to the square of the transition dipole moments and hence to the **Franck–Condon factors**, $|S(v', v)|^2$.

Example 11.1 The calculation of Franck–Condon factors

Consider a case in which two electronic states have the same force constant but in which the equilibrium bond lengths differ by ΔR. Find an expression for the relative intensity of the 0–0 transition as a function of ΔR.

Method. We need to evaluate the Franck–Condon factor $|S(0,0)|^2$. To do so we calculate the overlap integral $S(0,0)$ using harmonic oscillator wavefunctions (Tables 2.1 and 2.2), one centred on $x = 0$ and the other on $x = \Delta R$. We shall need the following integral:

$$\int_{-\infty}^{\infty} \mathrm{e}^{-ax^2}\,\mathrm{d}x = \left(\frac{\pi}{a}\right)^{1/2}$$

Answer. The wavefunctions for the two states are

$$\psi_0 = \left(\frac{\alpha}{\pi}\right)^{1/4} \mathrm{e}^{-\alpha x^2/2} \qquad \psi_{0'} = \left(\frac{\alpha}{\pi}\right)^{1/4} \mathrm{e}^{-\alpha(x-\Delta R)^2/2}$$

where $\alpha = m\omega/\hbar$ and the wavefunctions are normalized in the sense

$$\int_{-\infty}^{\infty} |\psi_0|^2\,\mathrm{d}x = 1$$

It then follows that

$$S(0,0) = \left(\frac{\alpha}{\pi}\right)^{1/2} \int_{-\infty}^{\infty} \mathrm{e}^{-\alpha x^2/2-\alpha(x-\Delta R)^2/2}\,\mathrm{d}x$$

$$= \left(\frac{\alpha}{\pi}\right)^{1/2} \mathrm{e}^{-\alpha(\Delta R/2)^2} \int_{-\infty}^{\infty} \mathrm{e}^{-\alpha(x-\Delta R/2)^2}\,\mathrm{d}x = \mathrm{e}^{-\alpha(\Delta R/2)^2}$$

The Franck–Condon factor for the transition is therefore

$$|S(0,0)|^2 = \mathrm{e}^{-\alpha(\Delta R)^2/2}$$

This function is plotted in Fig. 11.11, and the strong dependence on ΔR should be noticed.

Exercise 11.1. Show that the sum of all Franck–Condon factors for transitions from a given state v is equal to 1.

$[\sum_{v'} |S(v', v)|^2 = 1]$

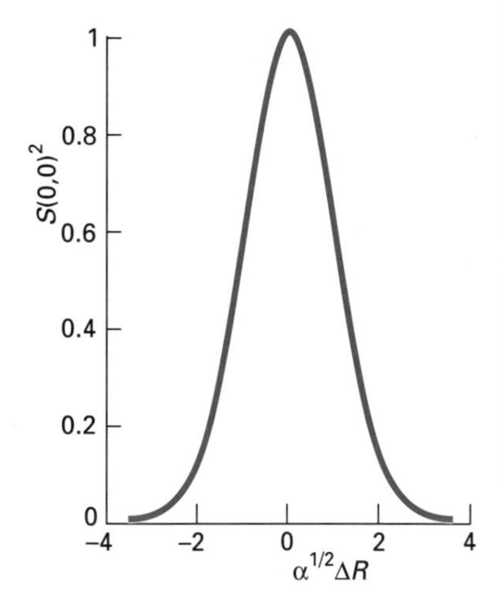

Fig. 11.11 The variation of the Franck–Condon factor with displacement of the minimum of the upper electronic state. Note that the factor is a maximum (1) when the two curves lie exactly over one another.

11.5 The structure of vibronic transitions

Superimposed on the vibronic transitions are rotational transitions that occur according to the selection rules set out in Section 10.5. The $\Delta J = -1, 0, +1$ transitions give rise, respectively, to the P-, Q-, and R-branches of the spectrum, and their appearance (for gas phase species) is similar to the structure of vibration–rotation spectra discussed in Section 10.11. There is, however, one

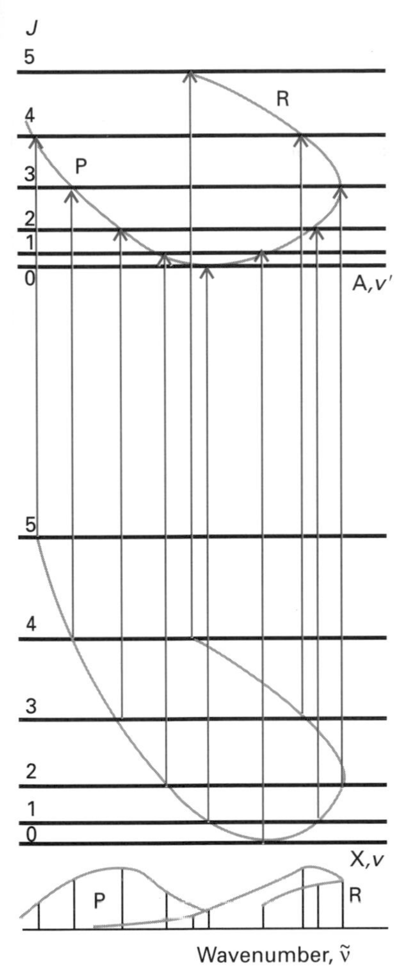

Fig. 11.12 The formation of P- and R-branches for a vibronic transition, showing the formation of a head in the R-branch.

important exception. Because the rotational constants of the upper and lower electronic states are likely to be so different from one another (because the bond lengths are so different in the two states), head formation is likely to occur. It is commonly found, for instance, that the spectrum has the appearance shown in Fig. 11.12, with the R-branch showing a head at high frequencies.

The presence of Λ-doubling affects the spectrum in a subtle way. In a $^1\Pi \leftarrow {}^1\Sigma$ transition, the P- and R-branches arise from the combination of the ground term with one of the components of the Π term whereas the Q-branch arises from a transition to the other component of the Π term. A consequence is that the Q-branch is slightly shifted relative to the other two branches, and the magnitude of the shift gives the magnitude of the Λ-doubling.

Further complications arise when one state is perturbed by another, and perturbations can be very effective in shifting the energy levels. A particular phenomenon that tends to obscure regions of the spectrum is **predissociation**, in which the vibrational structure is blurred in one region of the spectrum, but resumes at higher frequencies before the true dissociation and its associated structureless absorption begin. The mechanism of predissociation is illustrated in Fig. 11.13. As shown there, the upper electronic state A is perturbed by a dissociative state C, and a molecule excited to a vibrational state close to the intersection of the two electronic states may take on dissociative character, and fly apart. This probability of dissociation reduces the lifetime of molecules in energy levels close to the intersection, and they are broadened by the lifetime-broadening effect (Section 6.18). The coupling of the discrete states to the continuum also results in small shifts in the former.

The electronic spectra of polyatomic molecules

We have seen the complexity of diatomic molecule electronic spectra, and can therefore imagine the complexity that sets in when we examine polyatomic molecules. Happily, there is an effective simplification. Many polyatomic molecules are studied in solution, and as a result of the collisions that occur between solvent and solute species, the rotational structure of the bands is blurred. In weakly interacting solvents, such as hydrocarbons, the vibrational structure of bands may still be present, but in interacting solvents even that may be lost. Therefore, we shall mainly be concerned with spectra in which most of the details of the vibrational and rotational structure have disappeared. Another effective simplification, especially in considerations of the spectra of organic molecules, is that often the absorption in a particular region of the spectrum may be ascribed to a transition involving a particular group of atoms in the molecule. Such a group, which is called a **chromophore**, may occur in a number of different types of molecule, and gives rise to an absorption band at about the same wavenumber. Thus, an introductory discussion of the spectra of molecules may be based on their chromophores and the perturbations caused by other groups.

11.6 Symmetry considerations

For small molecules, the transitions are discussed in terms of the entire molecule rather than identifiable chromophores because electronic excitation involves the entire structure. Therefore, the selection rules for the transitions

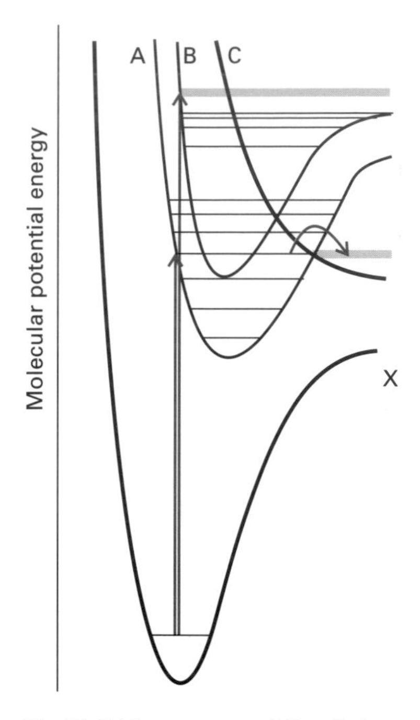

Fig. 11.13 The processes of dissociation (in the B←X transition) and predissociation (in the A←X transition).

must be expressed in terms of the point group of the whole molecule rather than a localized group of atoms. This is in fact a simple task, because if the irreducible representations spanned by the electric dipole moment operator are known (as they are, by a quick reference to the character table), then the selection rules can be expressed in terms of the direct product rule (Section 5.16).

As an example, consider the NO_2 molecule. Its point group is C_{2v} and its ground state configuration is

$$NO_2 \qquad \ldots b_2^2 a_2^2 a_1^1 \quad {}^2A_1$$

The three highest energy orbitals are illustrated in Fig. 11.14. In C_{2v}, the electric dipole moment operator spans $B_1(x) + B_2(y) + A_1(z)$. It follows that transitions may be stimulated from the ground state to excited states of symmetry species B_1, B_2, and A_1 by irradiation with *x*-, *y*-, and *z*-polarized light, respectively. The axes refer to the molecular system, and so the polarizations are relevant only if the NO_2 is trapped in a solid in a well-defined orientation. Because excitation involves a considerable reorganization of the distribution of the electrons, each electronic transition is accompanied by extensive vibrational structure. For example, the transition $\ldots b_2^2 a_2^2 b_2^1\,{}^2B_2 \leftarrow \ldots b_2^2 a_2^2 a_1^1\,{}^2A_1$ redistributes the electron that can be regarded as responsible for holding the molecule in its angular shape in the ground state. As illustrated by this example, it is conventional to write the upper term first and the lower second; then the direction of the arrow indicates emission (A→X) or absorption (A←X). The ground state is usually labelled X (unless its full symmetry designation is given), and the excited states are labelled A, B, C,. . ..

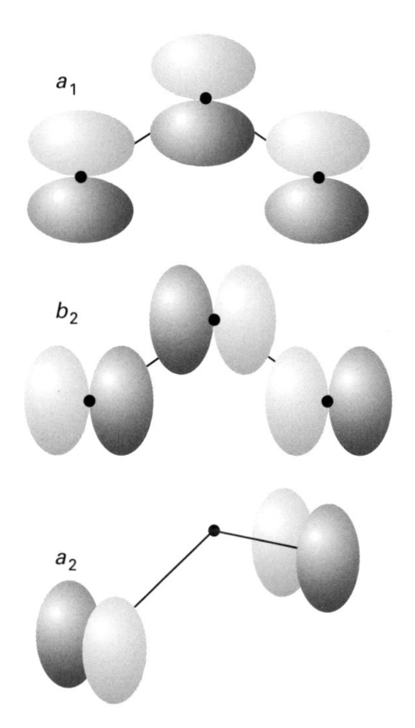

Fig. 11.14 Three of the molecular orbitals of a C_{2v} species.

11.7 Chromophores

The spectra of larger molecules may often be discussed in terms of the chromophores they contain. Among the most common chromophores are the carbonyl and nitro groups and the carbon–carbon double bond. The transitions responsible for their absorptions are typically classified as $\pi^* \leftarrow n$ ('n-to-pi star') and $\pi^* \leftarrow \pi$ ('pi-to-pi star'), where *n* represents a nonbonding orbital (Fig. 11.15). An $\pi^* \leftarrow n$ transition of the carbonyl group, which occurs near 290 nm, involves the partial removal of electron density on the O atom to the C atom, because the *n* orbital is confined to the O atom whereas the antibonding π^*-orbital spreads over both atoms. This migration of charge also helps to explain the shift to higher absorption frequencies that occurs when the chromophore is immersed in a polar or hydrogen-bonding solvent. In such an environment, the ground state of the molecule favours a particular arrangement of solvent molecules. However, the electronic transition occurs too rapidly for the complete reorientation of the solvent molecules to adjust to the new electron distribution, and so whereas the ground state is stabilized, the upper state is stabilized to a lesser extent. Consequently, the energy separation of the two states is larger than in a nonpolar solvent.

There is, however, one difficulty: the $\pi^* \leftarrow n$ transition is forbidden. To see that this is so, we note that the nonbonding orbital is confined to the O atom, and to a good approximation is $O2p_y$; so, if $\psi_{\pi^*} = c'\phi(C2p_x) + c\phi(O2p_x)$, as illustrated in Fig. 11.15(a), we have (for real c)

$$\langle \pi^* | \boldsymbol{\mu} | n \rangle \approx c \langle \mathrm{O2}p_x | \boldsymbol{\mu} | \mathrm{O2}p_y \rangle \tag{10}$$

This matrix element is zero for each component μ_q, as may easily be verified, and so the transition is forbidden. However, as is always the case with forbidden transitions, their 'forbidden' character is an aspect of the adoption of a simpified hamiltonian, and the presence of additional terms in the hamiltonian may relax the constraints on the transitions. In this case, intensity may be acquired by the transition by virtue of the fact that the nonbonding orbital is not strictly localized and is not pure $\mathrm{O2}p_y$. Another source of intensity is the coupling of the electronic and vibrational modes of the molecule. We discuss its details in the following section.

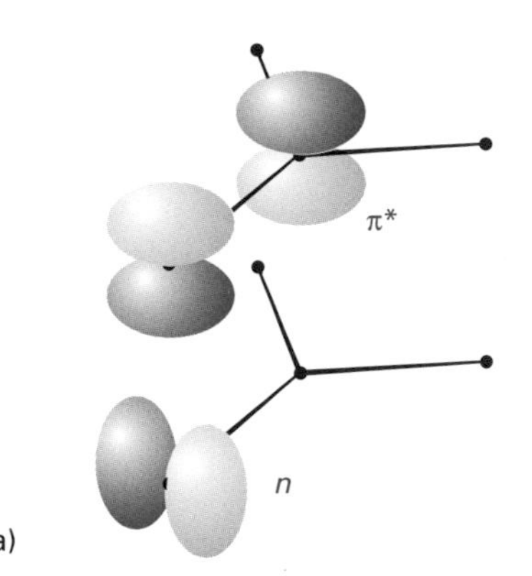

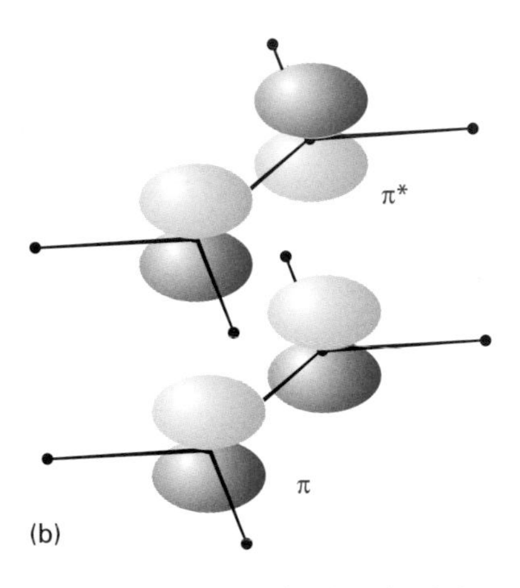

Fig. 11.15 The orbitals involved in (a) the $\pi^* \leftarrow n$ transition of a carbonyl group and (b) the $\pi^* \leftarrow \pi$ transition of a carbon–carbon double bond.

The $\pi^* \leftarrow \pi$ transition is allowed, and the transition dipole moment is directed along the internuclear axis (Fig. 11.15(b)). The transition reduces the strength of the bond because a bonding electron is transferred into an antibonding orbital. This reduction in strength may be so great that the bonded groups twist about the bond direction in order to minimize the antibonding effect. Thus, in ethene itself, the CH_2 groups are perpendicular in the $\sigma^2\pi^1\pi^{*1}$ excited state.

The benzene molecule, C_6H_6, is an interesting but complex example of transitions that involve the π-electrons of a molecule. There are three major bands. The one at about 260 nm, which is called the **benzenoid band**, is weak because it is symmetry-forbidden. A second, at 185 nm, is symmetry-allowed and is reasonably intense. There is also a band at 200 nm. The ground state of the D_{6h} molecule is $^1\mathrm{A}_{1g}$. The electric dipole moment operator spans $\mathrm{A}_{2u}(z) + \mathrm{E}_{1u}(x, y)$ in the group D_{6h}, where the z-axis lies perpendicular to the ring. Therefore, the allowed transitions can be expected to be $^1\mathrm{E}_{1u} \leftarrow {}^1\mathrm{A}_{1g}$ and $^1\mathrm{A}_{2u} \leftarrow {}^1\mathrm{A}_{1g}$. The strong transition at 185 nm has been identified as the former, with the upper term arising from the configuration $a_{2u}^2 e_{1g}^3 e_{2u}^1$, which gives rise to the terms $\mathrm{B}_{1u} + \mathrm{B}_{2u} + \mathrm{E}_{1u}$. This assignment has been confirmed by checking the polarization of the transition moment in a crystalline sample. The band at 200 nm is $^1\mathrm{B}_{1u} \leftarrow {}^1\mathrm{A}_{1g}$, and the benzenoid band at 260 nm has been ascribed to the transition to the other term arising from the $a_{2u}^2 e_{1g}^3 e_{2u}^1$ configuration (see Fig. 8.30), namely $^1\mathrm{B}_{2u} \leftarrow {}^1\mathrm{A}_{1g}$. Here is another problem that we need to address, for these two transitions are forbidden, yet manage to obtain intensity somehow. The same configuration can also give rise to triplet terms, but the **intercombination transitions**, the transitions between terms of different multiplicity, are weak in a molecule built from light atoms in which the spin–orbit coupling is small.

11.8 Vibronically allowed transitions

The forbidden transitions in the carbonyl chromophore and in benzene acquire intensity by coupling to the vibrations of the molecule. They are therefore classified as **vibronic transitions**.

The potential energy of an electron in a molecule depends on the locations of the nuclei. Therefore, the electronic hamiltonian also depends on their locations in a manner that may be expressed in terms of a Taylor expansion with respect to displacement along the normal coordinates:

$$H = H^{(0)} + \sum_i \left(\frac{\partial H}{\partial Q_i}\right)_0 Q_i + \dots \tag{11}$$

The eigenfunctions of $H^{(0)}$ are ψ_ε and their energies are E_ε. The presence of the additional terms in the hamiltonian mixes these eigenstates together, and to first order in the perturbation a particular electronic eigenfunction ψ'_ε becomes

$$\psi = \psi'_\varepsilon + \sum_\varepsilon{}' a_\varepsilon \psi_\varepsilon \qquad a_\varepsilon = \frac{\langle\varepsilon|\sum_i(\partial H/\partial Q_i)_0|\varepsilon'\rangle Q_i}{E_{\varepsilon'} - E_\varepsilon} \tag{12}$$

Within the Born–Oppenheimer approximation, the coefficient a_ε depends parametrically on the nuclear coordinate Q_i. Suppose now that only the upper state of a pair is perturbed, then the transition dipole moment for $\varepsilon' \leftarrow \varepsilon''$ is

$$\boldsymbol{\mu}_{\varepsilon',\varepsilon''} = \langle\varepsilon'|\boldsymbol{\mu}|\varepsilon''\rangle + \sum_\varepsilon{}' a^*_\varepsilon \langle\varepsilon|\boldsymbol{\mu}|\varepsilon''\rangle$$

If the transition between the unperturbed levels ε' and ε'' is forbidden, the first matrix element is zero, and we are left with

$$\boldsymbol{\mu}_{\varepsilon',\varepsilon''} = \sum_\varepsilon{}' a^*_\varepsilon \langle\varepsilon|\boldsymbol{\mu}|\varepsilon''\rangle \tag{13}$$

When transitions between ε'' and ε are allowed, and the perturbation can mix the states ε' and ε, the transition $\varepsilon' \leftarrow \varepsilon''$ can 'borrow' intensity from the allowed transitions.

The next step is to see which states can be mixed. The hamiltonian transforms as A_1 (or its equivalent); therefore, so too must each term in its expansion. In particular, the second term in eqn 11.11 must transform as A_1. However, one factor in that term is the normal coordinate Q_i, which transforms as $\Gamma^{(i)}$; therefore, the term $(\partial H/\partial Q_i)_0$ must also transform as $\Gamma^{(i)}$ if its product with Q_i is to be totally symmetric. This partial derivative term is the part of the hamiltonian that acts as the perturbation and mixes the electronic states (it has terms that depend on the electron coordinates), and so we can conclude that the matrix element in eqn 11.12 is nonzero only if $\Gamma^{(\varepsilon)} \times \Gamma^{(i)} \times \Gamma^{(\varepsilon')}$ contains the totally symmetric irreducible representation (such as A_1).

One final important point can be made before we give an example. The presence of the factor Q_i in the perturbation implies that when the perturbation acts, it leaves its footprint on both the electronic and the vibrational states. Therefore, a more complete form of eqn 11.12 is

$$\psi = \psi'_\varepsilon + \sum_{\varepsilon,v}{}' a_{\varepsilon,v} \psi_{\varepsilon v} \qquad a_{\varepsilon,v} = \frac{\sum_i \langle\varepsilon|(\partial H/\partial Q_i)_0|\varepsilon'\rangle \langle v|Q_i|v'\rangle}{E_{\varepsilon' v'} - E_{\varepsilon v}} \tag{14}$$

and the more complete version of eqn 11.13 is

$$\boldsymbol{\mu}_{\varepsilon' v',\varepsilon'' v''} = \sum_\varepsilon{}' a^*_{\varepsilon,v} \langle\varepsilon v|\boldsymbol{\mu}|\varepsilon'' v''\rangle \tag{15}$$

The implication of this more complete formulation is that, in a vibronically allowed transition, a vibrational excitation accompanies the electronic transition. The interpretation of the borrowing of intensity can now be expressed in a new light: we need to apply symmetry selection rules to entire vibronic states, not simply to electronic states.

It is time for an example. Consider the forbidden $B_{2u} \leftarrow A_{1g}$ band in benzene. Suppose an E_{2g} vibration can be excited at the same time as an electronic transition. Then, because the overall symmetry of the upper vibronic state is $E_{2g} \times B_{2u} = E_{1u}$ and the transition $E_{1u} \leftarrow A_{1g}$ is electric dipole allowed, the vibronic transition is allowed even though the pure electronic $B_{2u} \leftarrow A_{1g}$ transition is forbidden. In other words, the $B_{2u} \leftarrow A_{1g}$ transition acquires intensity through its coupling to the E_{2g} vibrations of the molecule.

Example 11.2 Intensity borrowing in vibronic systems

Account for the intensity of the $\pi^* \leftarrow n$ transition in the carbonyl group in terms of a vibronic process.

Method. First, identify the local point group symmetry of the chromophore and the symmetry species of the transition. Then decide what transitions are in fact allowed by considering the symmetry species of the components of the electric dipole moment operator. Then identify the symmetry species of the vibration which, when mixed with the upper state, leads to an allowed transition.

Answer. We shall treat the CO group as locally C_{2v} and for simplicity regard the nonbonding orbital as O2p_y (Fig. 11.16) and the π^* orbital as built from $2p_x$ orbitals. The $\pi^* \leftarrow n$ transition is $n^1\pi^{*1}$ $A_2 \leftarrow n^2$ A_1 in the coordinate system shown in the illustration (we have used $B_1 \times B_2 = A_2$ to work out the symmetry species of the upper state). Because the electric dipole moment operator transforms as $B_1(x) + B_2(y) + A_1(z)$, the only purely electronic transitions from the A_1 ground state are to $B_1(x) + B_2(y) + A_1(z)$, which does not include A_2. However, if this electronic state couples with a vibration of B_1 symmetry, then its overall symmetry is $B_1 \times A_2 = B_2$, which is an accessible state for y-polarized radiation. Similarly, if it couples with a vibration of B_2 symmetry, then the overall symmetry is $B_2 \times A_2 = B_1$, which is accessible with x-polarized radiation. The vibrations mentioned are illustrated in Fig. 11.17.

Comment. It should not be forgotten that there are other reasons why a transition acquires intensity, including the departure from the assumed point group symmetry as a result of the presence of substituents.

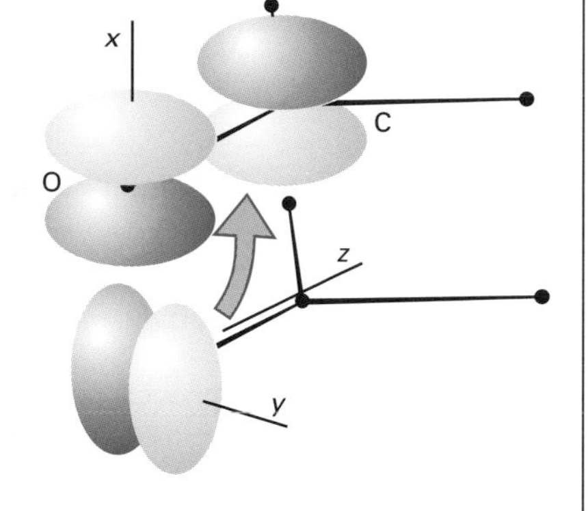

Fig. 11.16 The shift in electron distribution associated with a $\pi^* \leftarrow n$ transition in the carbonyl group.

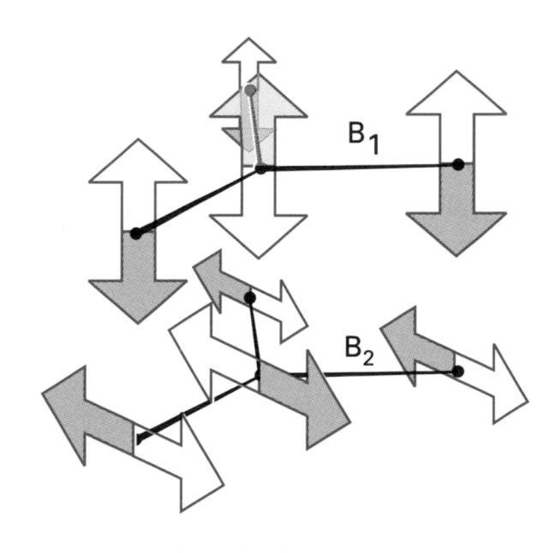

Fig. 11.17 The vibrations of the molecular framework involved in the vibronic transitions of a carbonyl group.

The intensities of d–d transitions in d-metal octahedral complexes also arise from vibronic effects. It is easy to see that some such mechanism is necessary, because an octahedral complex has a centre of inversion, and $g \leftarrow g$ transitions are forbidden by the Laporte selection rule (Section 7.2). However, if there is a coupling of the electronic transition to a vibration that destroys the centre of inversion, then the transition may acquire intensity. One way of interpreting this acquisition of intensity is to imagine the loss of inversion symmetry as permitting the mixing together of d- and p-orbitals, which are g and u, respectively, and $p \leftarrow d$ transitions are allowed.

11.9 Singlet–triplet transitions

Intercombination bands, which include transitions between singlet and triplet terms, are observed when the spin–orbit coupling is significant, such as when a heavy atom is present in the molecule. In this section, we shall see how the spin–orbit coupling term in a molecular hamiltonian can act as a perturbation that mixes states of different multiplicity.

Consider an operator that can be written in the form

$$\Omega = \sum_i R(i)s_z(i) \tag{16}$$

where the sum is over all the electrons in the molecule and R is an operator for the spatial component of the wavefunction. The effect of Ω on a singlet state, $|0,0\rangle$, of two electrons can be demonstrated quite easily as follows:

$$\begin{aligned}\Omega|0,0\rangle &= \{R(1)s_z(1) + R(2)s_z(2)\}N\phi(1)\phi(2)\{\alpha(1)\beta(2) - \beta(1)\alpha(2)\}\\ &= \tfrac{1}{2}\hbar N\big(\{R(1)\phi(1)\}\phi(2)\{\alpha(1)\beta(2) + \beta(1)\alpha(2)\}\\ &\qquad -\phi(1)\{R(2)\phi(2)\}\{\alpha(1)\beta(2) + \beta(1)\alpha(2)\}\big)\\ &= N'\{\phi'(1)\phi(2) - \phi(1)\phi'(2)\}\{\alpha(1)\beta(2) + \beta(1)\alpha(2)\} \propto |1,0\rangle\end{aligned}$$

where $\phi' = R\phi$, various constants have been absorbed into the normalization constants N and N', and $|1,0\rangle$ is a triplet state with $S = 1$ and $M_S = 0$. We see that the effect of Ω is to mix the $M_S = 0$ state of the triplet into the singlet.

The spin–orbit interaction (Section 7.4) is

$$H_{\text{so}} = \sum_i \xi_i \mathbf{l}_i \cdot \mathbf{s}_i \tag{17}$$

This operator has the same form as Ω (as well as having analogous terms in x and y), and so we should expect it to induce singlet–triplet mixing. For two electrons it takes the form

$$\begin{aligned}H_{\text{so}} &= \xi_1\mathbf{l}_1 \cdot \mathbf{s}_1 + \xi_2\mathbf{l}_2 \cdot \mathbf{s}_2\\ &= \tfrac{1}{2}(\xi_1\mathbf{l}_1 + \xi_2\mathbf{l}_2)\cdot(\mathbf{s}_1 + \mathbf{s}_2) + \tfrac{1}{2}(\xi_1\mathbf{l}_1 - \xi_2\mathbf{l}_2)\cdot(\mathbf{s}_1 - \mathbf{s}_2)\end{aligned}$$

The operator $\mathbf{s}_1 + \mathbf{s}_2$ commutes with S^2, the total spin operator, and so it cannot mix states of different multiplicity. However, the operator $\mathbf{s}_1 - \mathbf{s}_2$ does not commute with S^2, and so this component of the operator is the one that is responsible for singlet–triplet mixing:

$$\langle 1, M_S|H_{\text{so}}|0,0\rangle = \tfrac{1}{2}\langle 1, M_S|(\xi_1\mathbf{l}_1 - \xi_2\mathbf{l}_2)\cdot(\mathbf{s}_1 - \mathbf{s}_2)|0,0\rangle$$

For the z-component of the spin–orbit coupling, the spin operator is $s_{1z} - s_{2z}$ and its effect is

$$\begin{aligned}(s_{1z} - s_{2z})|0,0\rangle &= (s_{1z} - s_{2z})\frac{1}{\sqrt{2}}\{\alpha(1)\beta(2) - \beta(1)\alpha(2)\}\\ &= \hbar\frac{1}{\sqrt{2}}\{\alpha(1)\beta(2) + \beta(1)\alpha(2)\} = \hbar|1,0\rangle\end{aligned}$$

Consequently, the remaining orbital operator part of the spin–orbit coupling hamiltonian is

$$\langle 1,0|H_{\text{so}}|0,0\rangle = \tfrac{1}{2}\hbar(\xi_1 l_{1z} - \xi_2 l_{2z}) \tag{18}$$

(The bra and ket on the left simply integrate out the spin operators, leaving an orbital operator.) This operator has components that transform as rotations about the z-axis. The x- and y-components transform analogously. Because the transformation properties of rotations are listed in the character tables (such as those in Appendix 1), it is a simple job to decide which terms the spin–orbit coupling can mix together.

Example 11.3 State mixing by spin–orbit coupling

Show that spin–orbit coupling in a C_{2v} molecule can provide intensity to a ${}^3B_2 \leftarrow {}^1A_1$ transition.

Method. Decide which states can be mixed into the ground and excited states by noting how rotations transform in the group. Then decide whether any transitions between the contributing states of the same multiplicity are electric-dipole allowed.

Answer. In C_{2v}, rotations transform as $B_2(R_x) + B_1(R_y) + A_2(R_z)$. Therefore, the spin–orbit coupling can mix ${}^3B_2$, ${}^3B_1$, and ${}^3A_2$ terms into the ${}^1A_1$ ground state. It can also mix ${}^1A_1$, ${}^1A_2$, and ${}^1B_1$ terms into the ${}^3B_2$ excited state. The electric dipole moment operator transforms as $B_1(x) + B_2(y) + A_1(z)$, and so the transitions shown in Fig. 11.18 are allowed. Thus, the ${}^3B_2 \leftarrow {}^1A_1$ transition acquires intensity from these allowed components.

Exercise 11.3. Identify a mechanism for the ${}^3B_{1u} \leftarrow {}^1A_{1g}$ transition in benzene.
[Spin–orbit coupling mixes ${}^3B_{1u}$ with ${}^1B_{2u}$ and ${}^1E_{2u}$, then vibronic coupling makes these states accessible from ${}^1A_{1g}$.]

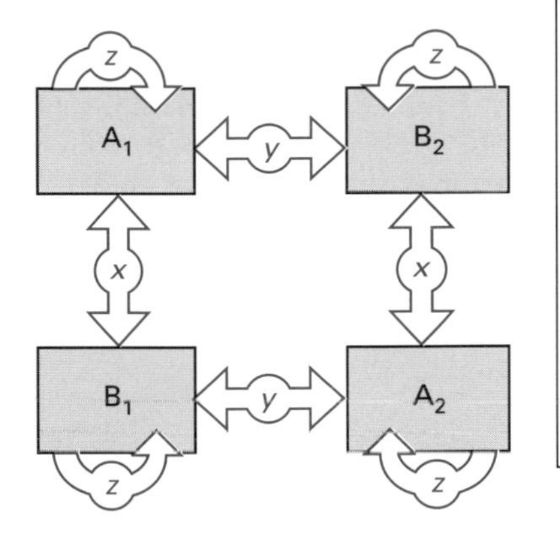

Fig. 11.18 The allowed transitions and their polarizations for electric dipole transitions in a C_{2v} species.

The fates of excited species

Electronically excited states discard or utilize their excess energy in a number of ways. These include its dissipation as heat and the rather more interesting processes of fluorescence and phosphorescence. Chemical reactions also often ensue after an initial electronic transition, and interesting phenomena are often observed. We shall look briefly at each of these processes.

11.10 Non-radiative decay

The most common mode is **thermal decay**, in which the energy is dissipated as thermal motion in the surroundings. The mechanism of this relaxation to equilibrium is a sequence of **radiationless transitions**, in which energy is transferred from the excited species to the molecules in its immediate environment. The initial transfer of energy is typically into the vibrational modes of the surrounding medium, and the efficiency of the transfer, because it involves the perturbation mixing of the states of the two systems, depends on how well the energy separations of the excited molecules match those of the surroundings. As a result, the lifetime of an excited state may be affected quite considerably by varying the solvent. Water has rather high vibrational wavenumbers (1595,

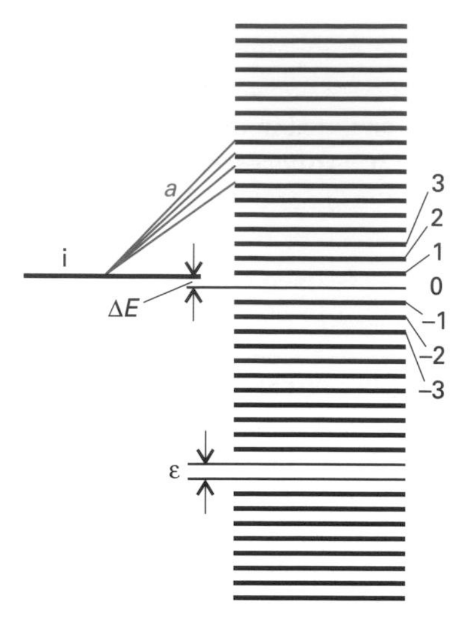

Fig. 11.19 The model used for the discussion of non-radiative energy transfer into a system with a high density of states.

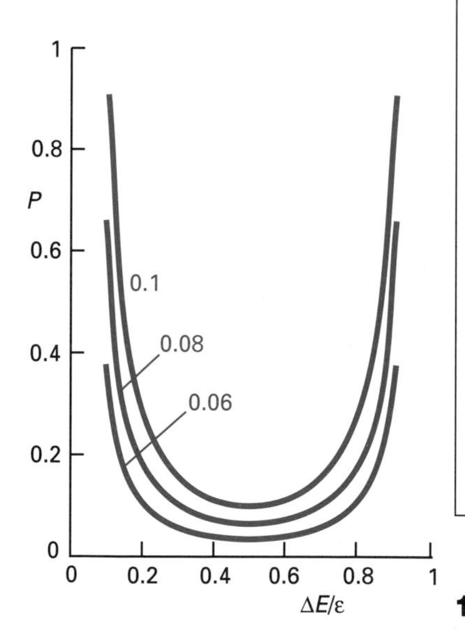

Fig. 11.20 The probability of energy transfer for the model in the previous illustration as a function of the energy separation ΔE. The numbers labelling the curves are the values of a/ε. Perturbation theory fails unless $P \ll 1$.

3652, and 3756 cm^{-1} for its three normal modes), and its higher harmonics coincide with a range of typical electronic excitation energies; hence, lifetimes are often short in water. A solvent such as selenium oxochloride, $SeOCl_2$, on the other hand, for which the wavenumber of the highest fundamental is only 995 cm^{-1}, acts as only a poor receptor for electronic energy transfer.

Example 11.4 Modelling non-radiative energy transfer

Consider the following mode of a non-radiative transition. Let the initial state be $|\text{i}\rangle$, and let there be a uniform ladder of states $|v\rangle$ of spacing ε that acts as a thermal reservoir (Fig. 11.19). Take the matrix elements of the perturbation that mixes the states of the two systems to be real and equal to a for all values of v. Calculate the probability that the original excitation has been transferred to the reservoir.

Method. We use first-order perturbation theory (Section 6.4), which tells us that the probability amplitude for finding the system in the state $|v\rangle$ of the reservoir is $a/(E_\text{i} - E_v)$ (see eqn 6.21). The probability is the square of this amplitude, and the total probability is the sum over all v. Set $E_v = E_\text{f} + v\varepsilon$ with $v = 0, \pm 1, \pm 2, \ldots$ (see the illustration). We shall write $\Delta E = E_\text{i} - E_\text{f}$.

Answer. It follows from eqn 6.21 that the total probability is

$$P = \sum_v \frac{a^2}{(E_\text{i} - E_v)^2} = \left(\frac{a}{\varepsilon}\right)^2 \sum_v \frac{1}{(\Delta E/\varepsilon - v)^2}$$
$$= \left(\frac{a\pi}{\varepsilon}\right)^2 \text{cosec}^2\left(\frac{\pi \Delta E}{\varepsilon}\right)$$

(For the sum, see M. Abramowitz and I.A. Stegun, *Handbook of mathematical functions*, Dover (1965), eqn 4.3.92.) The variation of P with the parameters is illustrated in Fig. 11.20.

Comment. The model is a greatly simplified version of the Bixon–Jortner theory of radiationless transitions.[2] The quantity $\rho = 1/\varepsilon$ is the density of states in the reservoir, so an alternative version of the result is

$$P = (a\pi\rho)^2 \text{cosec}^2\, \pi\rho\Delta E$$

If E_i lies half way between the $v = 0$ and $v = 1$ levels, then $\Delta E = \frac{1}{2}\varepsilon$, and $P = (a\pi\rho)^2$. Although P appears to be infinite for $\Delta E = 0$, that is a limitation of first-order perturbation theory.

11.11 Radiative decay

Decay by a radiative process in which the excess energy is discarded as a photon may also occur. There are two main types of process, fluorescence and phosphorescence. The distinction between the two processes was originally made on the basis of the lifetime of the radiation: in fluorescence, the radiation ceased as

[2] M. Bixon and J. Jortner, *J. Chem. Phys.*, 3284, **50** (1969).

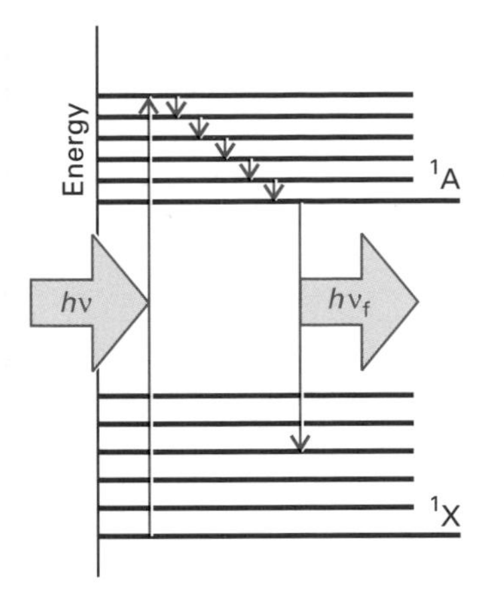

Fig. 11.21 The mechanism of fluorescence. The vibrational relaxation is non-radiative.

soon as the exciting radiation was removed, but in phosphorescence it continues for at least a short time. The distinction is now made on the basis of their mechanisms. In **fluorescence**, the radiation is generated in the course of transitions between states of the same multiplicity. In **phosphorescence** the radiation is generated in a sequence of steps that involve changes in multiplicity.

The steps that give rise to fluorescence are shown in Fig. 11.21. The initial absorption is $^1\text{A} \leftarrow {}^1\text{X}$ (here, A is not a symmetry designation, just a label; X is the ground state). The transitions are governed by the Franck–Condon principle, and so a range of vibrationally excited states of the upper electronic state is populated. Intermolecular collisions result in vibrational de-excitation, but the solvent may be such that the excess electronic excitation energy cannot easily be discarded on account of the mismatch of energy separations. The molecules persist in the lowest vibrational states of the excited singlet, and if their lifetime is long enough, spontaneous emission may occur as the molecule generates a photon. The emission also occurs in accord with the Franck–Condon principle, and the emission spectrum will show vibrational structure characteristic of the electronic ground state. The fluorescence spectrum will also be shifted to longer wavelengths than the absorption spectrum, because some of the initial excitation energy has been discarded into the surroundings. It follows that fluorescence spectra can be used to gather valuable information about the shape of the ground-state molecular potential energy surface, and from the variation of the overall intensity with solvent, to investigate the mechanism of energy transfer between species.

A spontaneous emission process is responsible for conventional fluorescence. In general, the fluorescence radiation has a different frequency from the incident radiation, so the incident photons do not stimulate the fluorescence. However, under certain circumstances, such as when the fluorescence radiation is trapped in a cavity, it can stimulate further emission, and there is a growth in intensity. This stimulated process in conjunction with the maintenance of a population inversion (with a greater number of species in the upper electronic state than in the lower state) is the basis of laser action.

The transitions leading to phosphorescence are illustrated in Fig. 11.22. The first step is the absorption $^1\text{A} \leftarrow {}^1\text{X}$. Thermal degradation within the state ^{1}A then occurs, and if it is not too fast, the spin–orbit coupling in the molecule might succeed in causing an **intersystem crossing**, a radiationless transition involving a change of multiplicity, into a nearby triplet (perhaps arising from the same configuration), which we shall denote ^{3}A. The crossing occurs in accord with the Franck–Condon principle, at the intersection of the molecular potential energy surfaces for the two electronic states, which is where the vibrational wavefunctions of the two electronic states match one another best. (In classical terms, at the intersection the oscillators share the same turning point.) This intersystem crossing will occur most rapidly if spin–orbit coupling is large, and so it is encouraged by the presence of heavy atoms in the molecule.

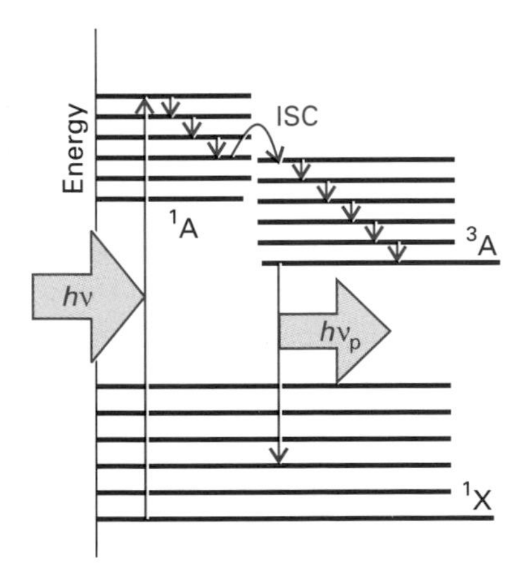

Fig. 11.22 The mechanism of phosphorescence. The vibrational relaxation is non-radiative; ISC stands for intersystem crossing, and is induced by spin–orbit coupling.

If intersystem crossing takes place, thermal degradation will continue, but now the molecule is lowered down the stack of vibrational states of the triplet ^{3}A and becomes trapped in the vibrational ground state. There is now little that the molecule can do. It cannot return to the ground state because singlet–triplet transitions are forbidden. It cannot return to ^{1}A because it has insufficient energy. However, it is not quite true that the molecule can do nothing

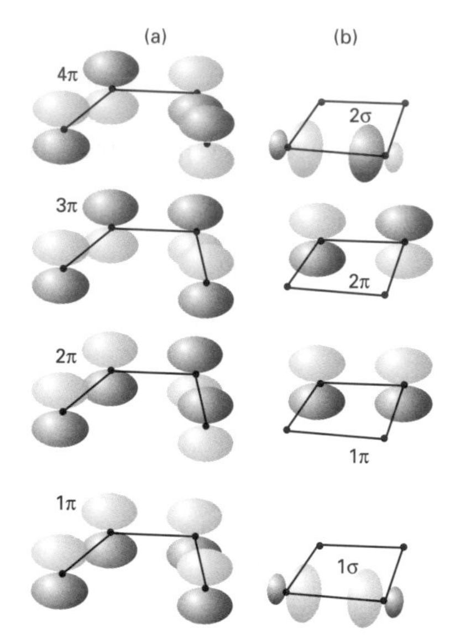

Fig. 11.23 A schematic representation of the molecular orbitals of (a) butadiene and (b) cyclobutene.

because the fact that intersystem crossing has occurred implies that the spin–orbit coupling is strong enough to mix states of different multiplicity, and hence the forbidden $^3A \rightarrow {}^1X$ transition is in fact weakly allowed. It follows that the system can slowly radiate its excess energy as the spin–orbit coupling enables this transition, and the photons produced are the radiation we call phosphorescence.

11.12 The conservation of orbital symmetry

The final fate of energetically excited molecules that we shall consider is their chemical reaction, when they change their identity. A knowledge of the way in which electron distributions are reorganized in the course of reactions is essential for understanding these processes, and we shall see in this section how the interplay of ideas stemming from molecular orbital theory, electron transition processes, and group theory account for a range of organic reactions.

We shall consider a **pericyclic reaction**, which is a concerted process (that is, a reaction in which bond breaking and bond formation occur simultaneously) which takes place by the reorganization of electron pairs within a closed chain of interacting atomic orbitals. We shall concentrate on two types of pericyclic reactions. In an **electrocyclic reaction**, ring closure or opening occurs in a single molecule. In a **cycloaddition reaction**, two or more molecules condense to form a ring and form new σ-bonds at the expense of old π-bonds. An example of an electrocyclic reaction is the ring-opening of cyclobutene (**1**) to form butadiene (**2**), and vice versa. An example of a cycloaddition reaction is the Diels–Alder reaction, which includes the reaction of ethene and butadiene to form cyclohexene (**3**). Each of these types of reaction has interesting features which can be explained very readily on the basis that orbital symmetry is conserved (in a sense we shall explain) as it takes place.

11.13 Electrocyclic reactions

Consider the electrocyclic reaction butadiene $\rightarrow$ cyclobutene. The four butadiene π-orbitals were derived in Section 8.9 and are drawn again on the left of Fig. 11.23. As a result of the formation of a ring, a π-bond turns into a σ-bond, and the orbital scheme for cyclobutene is shown on the right of the illustration. Next, we note that the two molecules have symmetry elements in common. For instance, both have a C_2 axis, and both have mirror planes (Fig. 11.24). Therefore, it should be possible to keep track of the molecular orbitals as they change from one molecule to the other by keeping an eye on their symmetries with respect to the elements they have in common. In other words, we should be able to set up a correlation diagram showing how the orbitals of butadiene change into the orbitals of cyclobutene. When that has been done, we should be in a position to describe the orbitals of the **transition state**, the state through which the molecule must pass as it changes from reac-

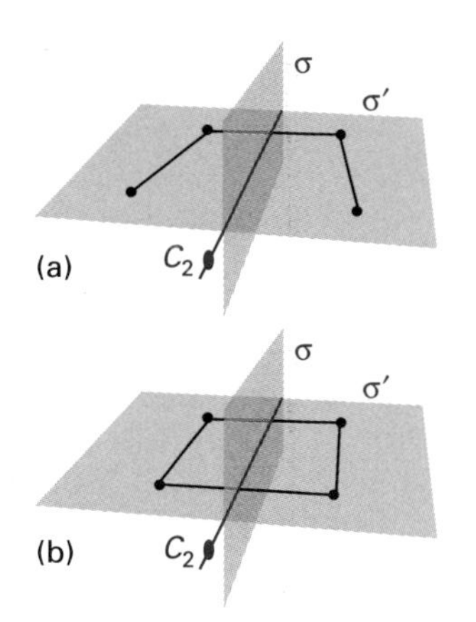

Fig. 11.24 The common symmetry elements of (a) butadiene and (b) cyclobutene.

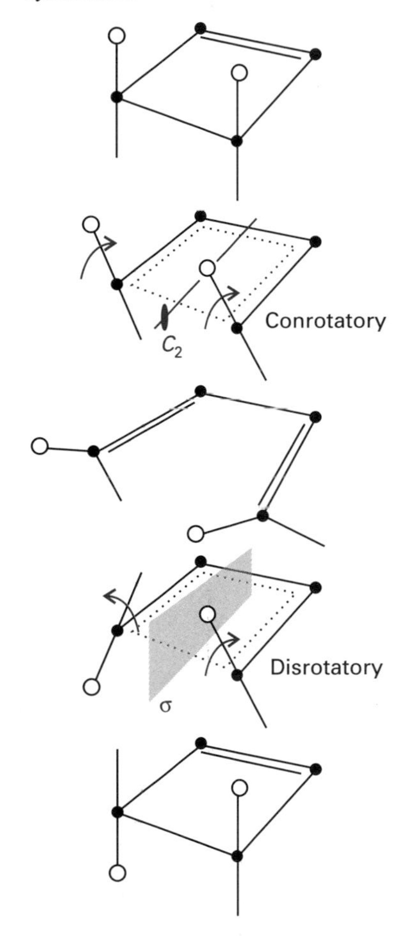

Fig. 11.25 The conrotatory and disrotatory ring closures of butadiene. The small spheres serve merely to identify protons; they do not necessarily correspond to substituents.

tants to products.

There is, however, a crucial complication; but it is this complication that makes pericyclic reactions so interesting. We see from Fig. 11.25 that there are two pathways for the reaction. In one, the **conrotatory path**, the CH_2 groups rotate in the same sense as one another. In the **disrotatory path** they rotate in opposite senses. Neither transition state (for each path) possesses the full common symmetry of the reactants and products. The conrotatory path preserves the C_2 axis throughout the reaction with the mirror planes present only at the beginning and end. The disrotatory path preserves one of the mirror planes but the axis and the other plane are present only at the beginning and end. It follows that, to construct the correlation diagram, we must examine the evolution of the orbitals in these two different *reduced* point groups.

We deal first with the conrotatory path, the path that preserves C_2. The four orbitals $1\pi, \ldots, 4\pi$ of butadiene have characters $-1, 1, -1, 1$ under C_2 (see Fig. 11.23). In the application of group theory to organic reaction mechanisms it is conventional to be less formal with the notation, and orbitals are classified as S (for symmetric, character $+1$) or A (for antisymmetric, character -1). We shall use this notation from now on. The classification of the butadiene molecule in this way is shown on the middle of the correlation diagram in Fig. 11.26 and the classification of the cyclobutene orbitals is shown on the left. Because the C_2 symmetry element is common to the reactant, the transition state, and the product, the symmetry labels S and A are applicable throughout the course of the reaction: they are 'good quantum numbers'. It follows that the S orbitals of the reactants correlate with the S orbitals of the products, and likewise for the A orbitals. The ambiguity about which S orbital correlates with which S orbital, and which A orbital correlates with which A orbital, is resolved by the noncrossing rule (Section 6.1), which forbids the crossing of states of the same symmetry. It follows that the correlation diagram for the conrotatory electrocyclic reaction is as shown on the left of Fig. 11.26.

A similar argument may be applied to the disrotatory path and the preservation of the single mirror plane. The orbital classification of butadiene is shown on the middle of Fig. 11.26, and the classification for cyclobutene is shown on the right. Once again, we can use the noncrossing rule to construct the correlation diagram shown in the right half of the illustration.

It should now be clear that there is a substantial difference between the two pathways. Suppose that there is insufficient energy available for the electrons to be excited out of the ground state of the reactant molecule. That is the case in a thermal reaction pathway, when the reaction is induced by heating. In a conrotatory process, the ground configuration $1\pi^2 2\pi^2$ of butadiene goes smoothly over into the ground configuration $1\sigma^2 1\pi^2$ of cyclobutene and the energy demands of the reaction are minimal. On the other hand, in a disrotatory path, one of the electron pairs ends up in a high energy orbital, and the product is the excited configuration $1\sigma^2 2\pi^2$. There is insufficient energy available for this process to occur, and so we can conclude that in the thermal cyclization of butadiene, only the conrotatory path is taken. Likewise, in the thermal ring-opening reaction of cyclobutene, similar arguments lead to the conclusion that the conrotatory path will be taken because it has low energy demands; the disrotatory path evolves into the excited state $1\pi^2 3\pi^2$. It should also be noticed that the HOMO dominates the conclusions, for it correlates

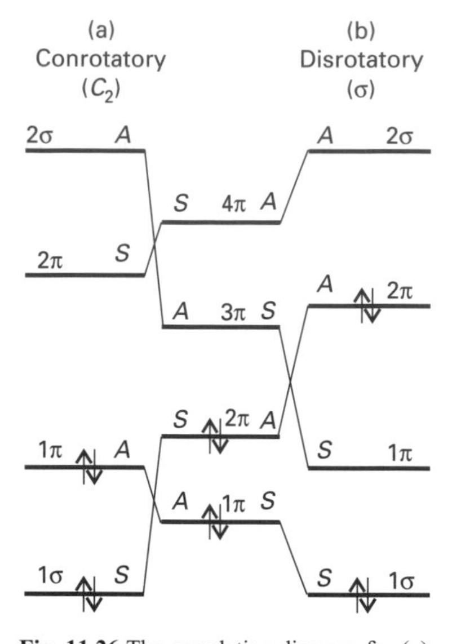

Fig. 11.26 The correlation diagram for (a) the conrotatory and (b) disrotatory butadiene–cyclobutene interconversion.

strongly upwards in energy in the thermally forbidden reaction. This is a general feature, and accounts for the importance of the frontier orbitals, the HOMO and LUMO, in reaction mechanisms.

There are two experimentally verifiable predictions that come from this account. In the first place, we expect the activation energy for ring opening to be quite small because it can occur without the promotion of electrons to excited states. The experimental value is in fact only about 80 kJ mol^{-1}. It can be ascribed largely to changes in the σ-framework of the molecule and changes in orbital composition, which are effects ignored in the correlation scheme. Secondly, the conrotatory path has specific stereochemical implications. Take, for example, the analogous six π-electron reaction shown in Fig. 11.27. Substituents rarely perturb the symmetry of a molecule sufficiently to upset orbital correlation arguments, and so they may be treated as labels. An analysis of the relevant correlation diagram shows that the thermally feasible reaction takes place along the disrotatory path, and gives stereochemically distinct products from the thermally forbidden conrotatory path. This difference is confirmed experimentally. The alternation conrotatory, disrotatory,... for the thermally feasible reaction as the number of electrons in the π-system changes along the series $4, 6, \ldots$ is a general prediction for electrocyclic reactions, and is one of the **Woodward–Hoffmann rules** devised by R. Hoffmann and R.B. Woodward.

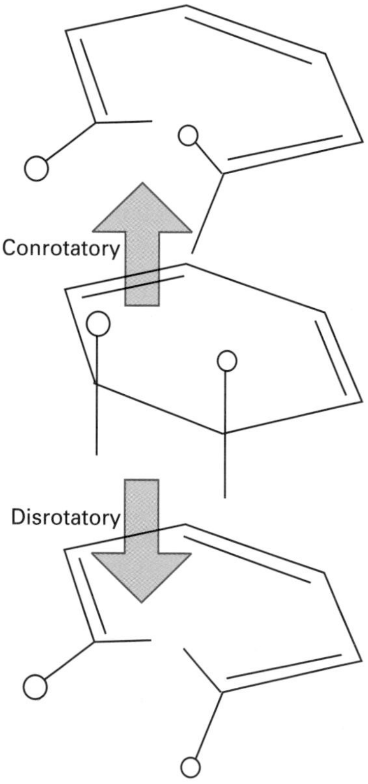

Conrotatory

Disrotatory

Fig. 11.27 The stereochemical consequences of different reaction paths.

11.14 Cycloaddition reactions

The same kind of argument can be used to explain the stereochemical consequences of cycloaddition reactions. We shall investigate the contrast between the negligibly slow thermal dimerization of ethene to cyclobutane and the much faster Diels–Alder addition of ethene to butadiene. We shall see that the difference can be expressed in terms of symmetry arguments. In other words, chemical reactions, like spectroscopic transitions, obey selection rules.

Consider the face-to-face approach of two ethene molecules. In the arrangement shown in Fig. 11.28, the two mirror planes are preserved throughout the reaction: they occur in the initial encounter and in the transition state. They can therefore be used for the symmetry analysis of the orbitals. The bonding and antibonding orbitals of the ethene molecule are A or S with respect to each of the two mirror planes, and their joint classification is shown on the left in Fig. 11.29. The designation SA, for example, signifies a joint molecular orbital that is S with respect to σ and A with respect to σ'. The σ bonds they form may also be classified as A or S with respect to each plane, and their order of energies can be assessed by judging the importance of their nodes. This assessment can often be done intuitively, and by supposing that there is very little interaction between different σ-bonds across a cyclobutane ring. The correlation diagram is then constructed by connecting orbitals of the same symmetry but by avoiding crossings. It is quite clear that the HOMO of the reactants rises steeply and the dimerization leads to a cyclobutane molecule in an excited state if the populations migrate adiabatically (that is, along the connecting lines, without making transitions between them). Therefore, we conclude that the ethene–ethene cycloaddition reaction (and the reverse cycloreversion reaction) with face-to-

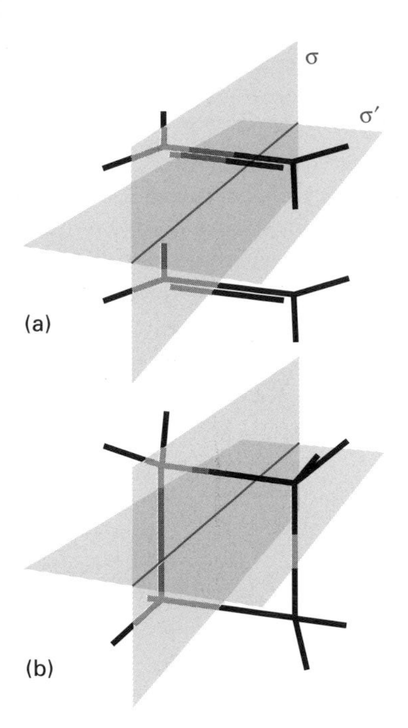

Fig. 11.28 The common symmetry elements of (a) an ethene dimer and (b) cyclobutane.

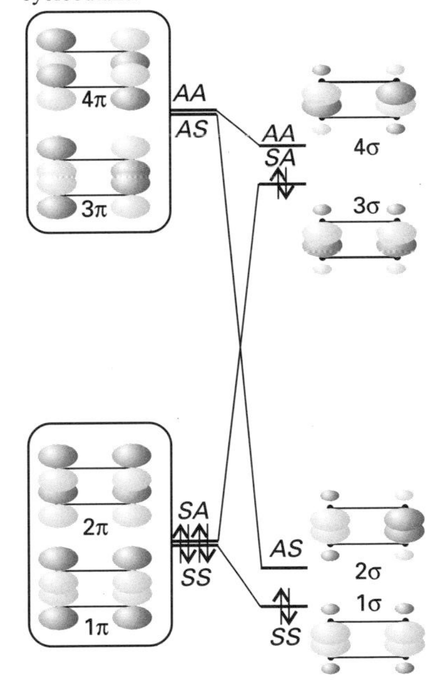

Fig. 11.29 The correlation diagram for the dimerization of ethene to cyclobutane. The (SA, SS) pair is degenerate when the ethene molecules are far apart; the same is true of the (AA, AS) pair. The symmetry classification refers to the elements $\sigma\sigma'$ illustrated in the preceding diagram.

face geometry is thermally forbidden. This conclusion is in accord with observation.

Now we apply the same argument to the ethene–butadiene reaction, which is the prototype of the wide class of Diels–Alder reactions. We continue to consider the face-to-face approach of the molecules, which preserves the mirror plane shown in Fig. 11.30 throughout the course of the reaction. The orbitals of the cluster of molecules is depicted on the left of Fig. 11.31, and are classified with respect to the preserved mirror plane. The left side of the diagram is simply the superposition of the butadiene and ethene energy levels with the disregarded σ-framework indicated throughout. In the course of the reaction, two new σ-bonds are formed at the expense of two π-bonds and one π-bond is relocated. The orbitals and energy levels of the product, cyclohexene, are shown on the right of the illustration, and have been classified with respect to the same mirror plane.

At this point we can construct the correlation diagram by using the non-crossing rule, and then trace the evolution of the bonding electron pairs of the reactants as they change adiabatically into products. The obvious feature is that the ground-state configuration correlates with the ground-state configuration of the products. The activation energy for the reaction can therefore be expected to be sufficiently low for it to be thermally feasible. This is in accord with the readiness with which Diels–Alder reactions are known to take place: they are thermally allowed reactions. Another Woodward–Hoffmann rule is exemplified by the two reactions we have described: a $4+2$ π-electron cycloaddition reaction is thermally allowed, whereas a $2+2$ π-electron reaction is thermally forbidden in the same face-to-face geometry.

11.15 Photochemically induced electrocyclic reactions

Reactions are thermally allowed when there is a transfer of electron pairs from bonding orbitals in the ground state of the reactant molecules to bonding orbitals in the products. A reaction that is thermally forbidden may become photochemically allowed when electrons are excited into higher orbitals. Excitation permits reaction not only because more energy is available to overcome activation barriers but also because the consequences of orbital symmetry are different. In other words, because the initially occupied orbitals are different, the same selection rules permit the exploration of different reaction channels.

We shall illustrate this feature by considering once again the ring closure of butadiene. This time, though, we shall consider a photochemical mechanism in which the absorption of a photon has led to the excitation of a single electron (Fig. 11.32). The disrotatory adiabatic correlation of the excited butadiene configuration leads to an excited cyclobutene configuration of similar energy to the starting point whereas the conrotatory path involves a significant increase in energy. Hence, in contrast to the thermal electrocyclic reaction, the disrotatory path is open to the photochemically induced reaction and the conrotatory path is closed. The reversal of the thermal prediction is another general feature of electrocyclic reactions, and is another of the Woodward–Hoffmann rules. The photochemical ring-closure of butadiene is known, and it does in fact proceed by the predicted disrotatory path. Nevertheless, there are complications

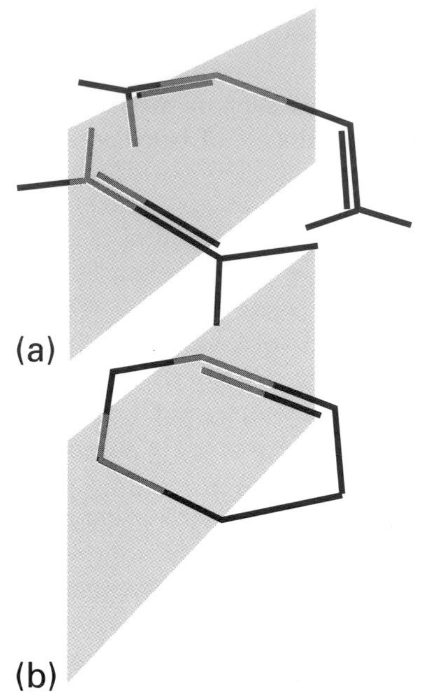

Fig. 11.30 The common symmetry elements of (a) an ethene and butadiene pair and (b) cyclohexene.

(as in most photochemical processes), for the cyclobutene is produced in its electronic ground state, not the excited state the correlation diagram suggests. We need to resolve this discrepancy.

A problem with the correlation diagrams presented so far is that they focus attention on the individual orbitals. We should in fact consider the overall states of molecules, and apply our arguments to them. To illustrate what is involved, we consider the first few excited states of butadiene and cyclobutene. Their symmetry species are obtained in the normal way, by taking the direct product of the symmetry species of the individual, occupied orbitals, all doubly occupied orbitals being totally symmetric. Because the disrotatory path preserves the single mirror plane, the relevant state classification is in terms of S and A with respect to the plane. To work out the direct products, we use

$$S \times S = S \qquad S \times A = A \qquad A \times A = S \tag{19}$$

which follow from the characters $+1$ and -1 for S and A, respectively.

The ground states are S (they are closed-shell species). The first excited configuration of butadiene is $1\pi^2 2\pi^1 3\pi^1$, which has symmetry species $A \times S = A$. Because the two outermost electrons occupy different orbitals this configuration can give rise to both singlet and triplet terms, with the triplet lower than the singlet. The next higher configuration is $1\pi^2 3\pi^2$, which is S overall and necessarily a singlet. The cyclobutene states are set out on the diagram in Fig. 11.33. The correlation diagram in Fig. 11.26 can be used to simplify the construction of the state correlations because since butadiene$(1\pi, 2\pi)$ correlates with cyclobutene$(1\sigma, 2\pi)$, it follows that butadiene$(1\pi^2 2\pi^2, {}^1S)$ correlates with cyclobutene$(1\sigma^2 2\pi^2, {}^1S)$. This connection lets us draw the lines in the illustration. Now we see an important point: overall states of the same symmetry have incorrectly crossed. Such crossing is forbidden by the noncrossing rule, so the light lines in the illustration should be replaced by the heavy lines.[3]

Now consider the disrotatory ring closure in terms of the overall states of the molecule. If the butadiene is in its ground state and we are considering a thermal reaction, then although in principle the ground state of cyclobutene can be reached without electronic excitation, the reaction involves a considerable activation energy and is therefore forbidden. This conclusion modifies the earlier discussion, where we decided that it is because the disrotatory path leads to an excited state of the product that it is forbidden. We now see that the forbidden nature of the reaction stems from the activation barrier, and that that barrier exists for two reasons: the rise in energy is a consequence of orbital correlation (so that remains an important part of the argument), and the existence of the peak is a consequence of the noncrossing of states of the same overall symmetry.

If the butadiene is initially in a triplet excited state, then disrotatory motion moves it to the point P_1 on the correlation diagram in Fig. 11.33. There is sufficient spin–orbit coupling to induce intersystem crossing, and so it switches to the lower ${}^1S$ curve. It cannot go forward to cyclobutene because that would

[3] The interaction of two states of the same overall symmetry is another example of the configuration interaction introduced in Section 8.5.

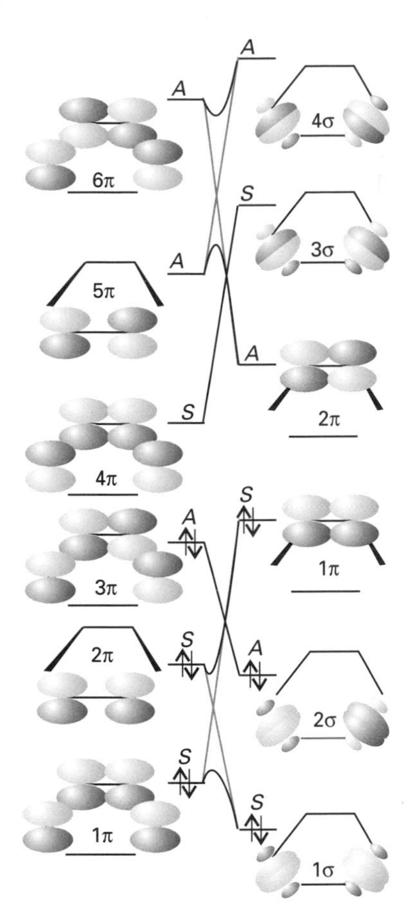

Fig. 11.31 The correlation diagram for the cycloaddition of ethene to butadiene to form cyclohexene.

require a further injection of energy to overcome the barrier at P_4. Therefore, the molecule loses its energy non-radiatively and converts back to ground-state butadiene. This behaviour is actually observed. Now suppose that absorption results in the population of the first singlet excited state of butadiene. Then the simple conclusion would be that it can pass over into the first excited singlet state of cyclobutene, as we concluded from the individual orbital analysis. The crossing at P_2, however, plays a significant role because there may be a strong enough perturbation present (such as rapid nuclear motion and the failure of the Born–Oppenheimer approximation) to induce an internal conversion between curves at P_2. In other words, the ^1A state can convert into the ^1S state when its geometry corresponds to the point P_2. As the reaction proceeds, the state of the molecule moves on to P_3 where it is sufficiently close to the lower curve for nuclear motion to induce a second internal conversion to the lower ^1S curve. This curve-jumping is an example of a **nonadiabatic process**. The second internal conversion leaves the molecule at point P_4. Now it needs no activation energy to go on to ground-state cyclobutene (or back to ground-state butadiene). Hence, ground-state cyclobutene appears in the products of singlet excited butadiene, exactly as observed.

11.16 Photochemically induced cycloaddition reactions

The same kind of analysis accounts for the characteristics of photochemically induced cycloaddition reactions. The strategy is to use the orbital correlation diagrams to construct first approximations to the state correlation diagrams for the lowest few configurations. Then we allow for interaction between states of the same symmetry so that crossings are eliminated from the diagrams. Finally, we recognize that all the intersections and the noncrossings are leaky on account of the presence of ignored perturbations, such as spin–orbit coupling (which mixes states of different multiplicity) and the breakdown of the Born–Oppenheimer approximation (which gives rise to interaction between states of the same multiplicity).

To see the strategy in action, consider the dimerization of ethene once again. The orbital correlation diagram lets us construct the state correlation diagram shown in Fig. 11.34. The ground states of the ethene pair and the cyclobutane are each of *SS* symmetry with respect to the two preserved mirror planes, and the forbidden character of the thermal reaction can be ascribed to the existence of the high activation barrier. On the other hand, the first excited configuration $(1\pi^2 2\pi^1 3\pi^1)$ correlates, with little change of energy, with the first excited state $(1\sigma^2 2\sigma^1 3\sigma^1)$ of the cyclobutane. A simple analysis would lead us to expect the dimerization to be photochemically allowed (which it is) and the products to be excited (which they are not).

To explain the last point we need to consider the conversions that can occur at intersections of the state lines. The intersection at P_1 permits one internal conversion, and the close approach of the two interacting curves near P_2 allows a second conversion to the lower curve to take place. With that accomplished, the molecule can slide down to either reactant or product, each being produced in its ground state, as observed.

The face-to-face dimerization of ethene is thermally forbidden but photochemically allowed. This reversal of cycloaddition behaviour is a general feature of

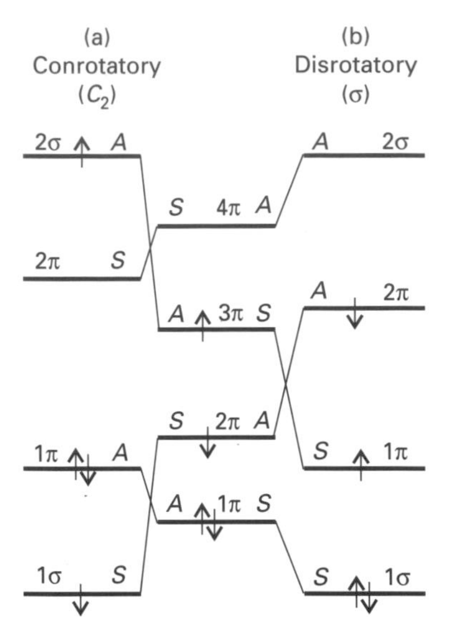

Fig. 11.32 The correlation diagram for the photochemical interconversion of butadiene and cyclobutene.

such reactions, and is yet another of the Woodward–Hoffmann rules. The Diels–Alder ethene–butadiene cycloaddition is thermally allowed. We can see that it is photochemically forbidden by reference to the state correlation diagram (Fig. 11.35) which has been constructed by using the orbital correlation diagram in Fig. 11.31. The most obvious feature is the absence of any energy barrier in the correlation of the two ground-state configurations: the reaction is therefore predicted to be thermally allowed. The first excited configuration ($1\pi^2 2\pi^2 3\pi^1 4\pi^1$) correlates with a highly excited configuration ($1\sigma^2 2\sigma^2 1\pi^1 2\pi^1$) of the addition product, and so on simple grounds we do not expect it to occur. To some extent interaction between configurations alleviates the energy requirements because there is a crossing with a configuration ($1\sigma^2 2\sigma^1 1\pi^2 3\sigma^1$) of the same symmetry, and so the adiabatic evolution of the first excited state ends up in the first excited state of the cyclohexene. Nevertheless, this still leaves a barrier, and so the photochemical process remains forbidden, as observed.

The consequences of orbital correlation diagrams, and of their more sophisticated interpretation in terms of state correlations, has led to a much deeper understanding of some aspects of organic chemistry. Indeed, orbital correlation is a prime example of how much theory can contribute to experimental chemistry.

Problems

11.1 Consider the Rydberg state of H_2^+ that arises from the overlap of two H2*s*-orbitals as resembling a single 2*s*-orbital of He^+ centred on the midpoint of the bond. Determine (a) the mean radius of the orbital and (b) the radius of the 90 per cent boundary surface. For comparison, the bond length of the ground state of the molecule-ion is 106 pm.

11.2 Which of the following transitions are electric-dipole allowed? (a) ${}^2\Pi \rightarrow {}^2\Pi$, (b) ${}^1\Sigma \rightarrow {}^1\Sigma$, (c) $\Sigma \rightarrow \Delta$, (d) $\Sigma^+ \rightarrow \Sigma^-$, (e) $\Sigma^+ \rightarrow \Sigma^+$, (f) ${}^1\Sigma_g^+ \rightarrow {}^1\Sigma_u^+$, (g) ${}^3\Sigma_g^- \rightarrow {}^3\Sigma_u^+$.

11.3 Show that in the carbonyl group the $\pi^* \leftarrow \pi$ transition is allowed, its transition dipole moment lying along the bond.

11.4 Show that the transition ${}^1A_2 \leftarrow {}^1A_1$ is electric-dipole forbidden in H_2O but may become allowed as a vibronic transition involving one of the molecule's vibrational modes.

11.5 Assess the polarization of the ${}^1A_2 \leftarrow {}^1A_1$ transition in H_2CO and of the ${}^1B_{2u} \leftarrow {}^1A_g$ transition in $CH_2{=}CH_2$. *Hint*. Use C_{2v} and D_{2h} respectively; consider the role of vibrational coupling.

11.6 In a diamagnetic octahedral complex of Co^{3+}, two transitions can be assigned to ${}^1T_{1g} \leftarrow {}^1A_{1g}$ and ${}^1T_{2g} \leftarrow {}^1A_{1g}$. Are these transitions forbidden? If they are forbidden, what symmetries of vibrations would provide intensity? Can the intensities be ascribed to the admixture of configurations involving *p*-orbitals?

11.7 Consider the molecular potential energy curves of two electronic states; let their force constants be the same, but the minima offset by a distance ΔR. Find an expression for the Franck–Condon factors $|S_{v0}|^2$ for $v = 0, 1, 2$ as a function of ΔR. What value of ΔR is needed for the transition intensity to $v = 1$ to dominate the other two?

11.8 Deduce the effect of the operator H_{so} in eqn 11.17 on a two-electron singlet state. *Hint*. Proceed as in the discussion following eqn 11.17 but include s_+ and s_-.

11.9 At time $t = 0$ a molecule is known to be in a singlet state. The energy separation of the singlet and triplet states is hJ. Deduce an expression for the time dependence of

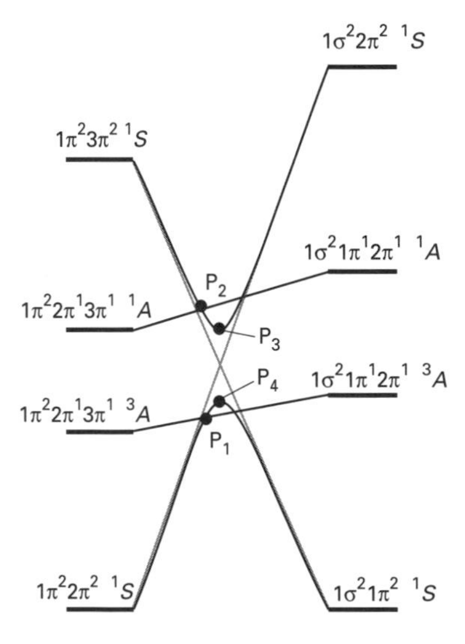

Fig. 11.33 The state correlation diagram for the butadiene–cyclobutene interconversion.

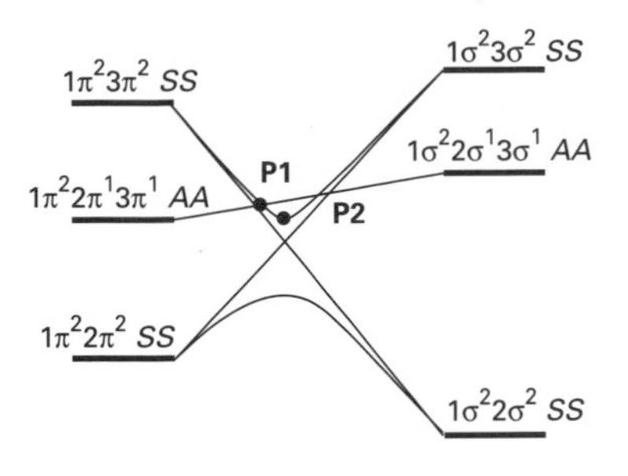

Fig. 11.34 The state correlation diagram for the dimerization of ethene.

the probability that the system is in any of the three states of the triplet at some later time as a result of the spin–orbit interaction. Suppose that the sample consists of a large number of molecules which are excited photochemically to a singlet state over a range of time $0 \le t_0 \le T$ with equal probability. What is the probability that any molecule is in a triplet state at some time later than T? *Hint.* The basic equation to use is eqn 6.60. For the second part, average this equation over a uniform distribution of starting times in the range $0 \le t_0 \le T$ (that is, multiply by dt/T and integrate between the appropriate limits).

11.10 Which states of benzene may be mixed with $^3B_{1u}$ and $^3B_{2u}$ by spin–orbit coupling?

11.11 In an aromatic molecule of D_{2h} symmetry the lowest triplet term was identified as $^3B_{1u}$. What is the polarization of its phosphorescence? *Hint.* Decide which singlet terms can mix with $^3B_{1u}$ and assess the polarization of the light involved in the return of that state to the ^1A_g ground state.

11.12 The Bixon–Jortner approach to radiationless transitions was sketched in a very simplified form in Example 11.4. The following is a slightly more elaborate version. Let ψ, an eigenstate of the system hamiltonian $H(\text{sys})$ with eigenvalue E, be the state populated initially, and let ϕ_n, an eigenstate of the bath hamiltonian $H(\text{bath})$ with eigenvalue E_n, be a state of the bath. Let $\Psi = a\psi + \Sigma_n b_n \phi_n$ be an eigenstate of the true hamiltonian H with energy $\mathcal{E}$. Let $\langle\psi|\phi_n\rangle = 0$ and $H' = H - H(\text{sys}) - H(\text{bath})$ have constant matrix elements $\langle\phi_n|H'|\psi\rangle = V$ for all n. Show that $H\Psi = \mathcal{E}\Psi$ leads to $Va + (E_n - \mathcal{E})b_n = 0$ and $(E - \mathcal{E})a + V\sum_n b_n = 0$. Hence find an expression for a and b_n. Letting $\mathcal{E} - E_n = \gamma\varepsilon - n\varepsilon$ and using $\sum_{n=-\infty}^{\infty}\{1/(\gamma - n)\} = -\pi \cot \pi\gamma$ and $\rho = 1/\varepsilon$, show that $E - \mathcal{E} - \pi\rho V^2 \cot \pi\gamma = 0$, an equation for $\mathcal{E}$. Go on to show on the basis that $a^2 + \sum_n b_n^2 = 1$, that $a^2 = V^2/\{(E - \mathcal{E})^2 + V^2 + (\pi V^2 \rho)^2\}$. *Hint.* See M. Bixon and J. Jortner, *J. Chem. Phys.*, 3284, **50** (1969).

Further reading

Modern spectroscopy. J.M. Hollas; Wiley, Chichester (1992).
Molecules and radiation: An introduction to modern molecular spectroscopy. J.I. Steinfeld; MIT Press, Cambridge, Mass. (1985).
Molecular spectra and molecular structure. I. Spectra of diatomic molecules. G. Herzberg; van Nostrand, Princeton (1950).
Molecular spectra and molecular structure. III. Electronic spectra and electronic structure of polyatomic molecules. G. Herzberg; van Nostrand, Princeton (1967).
The theory of transition metal ions. J.S. Griffith; Cambridge University Press, Cambridge (1964).
Photochemistry. J.G. Calvert and J.N. Pitts; Wiley, New York (1966).
Non-radiative decay of ions and molecules in solids. R. Englman; North-Holland, Amsterdam (1979).
Radiationless transitions. S.H. Lin (ed.); Academic Press, New York (1980).
Perturbations in the spectra of diatomic molecules. H. Lefebvre-Brion and R.W. Field; Academic Press, Orlando, Florida (1986).
The conservation of orbital symmetry. R.B. Woodward and R. Hoffmann; VCH and Academic Press, New York (1970).
Organic reactions and orbital symmetry. T.L. Gilchrist and R.C. Storr; Cambridge University Press, Cambridge (1979).
Advanced organic chemistry. Part A: Structure and mechanisms. F.A. Carey and R.J. Sundberg; Plenum Press, New York (1990).
Frontier orbitals and organic chemical reactions. I. Fleming; Wiley, Chichester (1976).
Frontier orbitals and properties of organic molecules. V.F. Traven; Ellis Horwood, New York (1992).
Orbital interaction theory of organic chemistry. A. Rauk; Wiley, New York (1994).
Radiationless transitions in polyatomic molecules. E.S. Medvedev and V.I. Osherov; Springer, New York (1995).

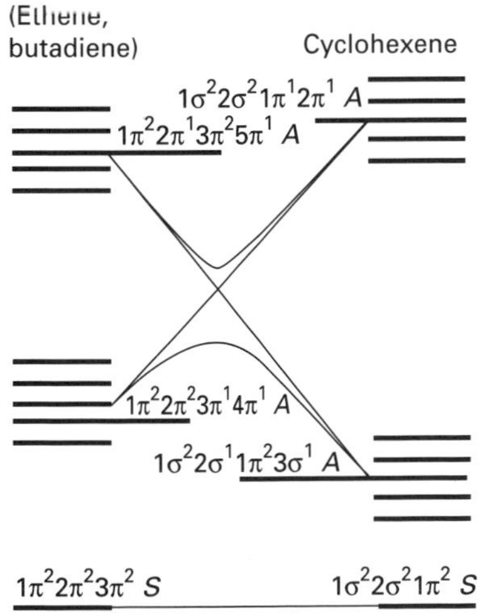

(Ethene, butadiene)
Cyclohexene
$1\sigma^2 2\sigma^2 1\pi^1 2\pi^1$ A
$1\pi^2 2\pi^1 3\pi^2 5\pi^1$ A
$1\pi^2 2\pi^2 3\pi^1 4\pi^1$ A
$1\sigma^2 2\sigma^1 1\pi^2 3\sigma^1$ A
$1\pi^2 2\pi^2 3\pi^2$ S
$1\sigma^2 2\sigma^2 1\pi^2$ S

Fig. 11.35 The state correlation diagram for the cycloaddition of ethene to butadiene.

12 The electric properties of molecules

This chapter explores the properties of molecules that are exposed to electric fields. The sources of the fields may be external electrodes or some other molecule. We shall also allow them to be either constant in time or oscillatory. A knowledge of these properties will enable us to discuss a variety of related molecular properties, which includes the relative permittivities (dielectric constants) of bulk samples, refractive indices, and optical activity. We shall also be able to use the information to discuss intermolecular forces. Throughout the chapter we shall draw on the material on perturbation theory introduced in Chapter 6.

The response to electric fields

The electric **polarizability**, $\boldsymbol{\alpha}$, of a molecule is a measure of its ability to respond to an electric field and acquire an electric dipole moment, $\boldsymbol{\mu}$. The perturbation caused by an electric field $\mathbf{E}$ is

$$H^{(1)} = -\boldsymbol{\mu} \cdot \mathbf{E} \qquad \boldsymbol{\mu} = \sum_i q_i \mathbf{r}_i \tag{1}$$

where q_i is the charge of the particle i at the location $\mathbf{r}_i$. We shall suppose that the field is uniform over the molecule, and so avoid having to deal with its interaction with higher multipoles (the quadrupole moment, for instance, interacts with the field gradient). To keep the notation simple, we shall suppose that the electric field is applied in the z-direction, and write $\mathbf{E} = \mathcal{E}\hat{\mathbf{k}}$, where $\hat{\mathbf{k}}$ is a unit vector in the z-direction. Then

$$H^{(1)} = -\mu_z \mathcal{E} \tag{2}$$

The rest of this chapter explores the consequences of this simple perturbation.

12.1 Molecular response parameters

In Chapter 6, time-independent perturbation theory was set up to provide expressions for the energy in powers of the perturbation. Our first task in this chapter is to adapt those expressions to give formulae for properties other than the energy. There are two approaches. One is to set up an operator for the property of interest and to evaluate its expectation value by using the perturbed wavefunctions. An alternative approach is to set up an expression for the energy in terms of the property of interest and then to compare that expression with the one obtained for the energy by using perturbation theory. We shall illustrate both techniques.

The key to the extraction of the polarizability from the perturbation expression for the energy is the Hellmann–Feynman theorem (eqn 6.42):

$$\frac{\mathrm{d}E}{\mathrm{d}P} = \left\langle \frac{\partial H}{\partial P} \right\rangle \tag{3}$$

In the present case, the parameter P is the electric field strength $\mathcal{E}$, and so we need to use

$$\frac{\mathrm{d}E}{\mathrm{d}\mathcal{E}} = \left\langle \frac{\partial H}{\partial \mathcal{E}} \right\rangle \tag{4}$$

The partial derivative of the hamiltonian is simply

$$\frac{\partial H}{\partial \mathcal{E}} = \frac{\partial H^{(1)}}{\partial \mathcal{E}} = -\mu_z$$

It follows that the variation of the energy with the electric field strength is given by

$$\frac{\mathrm{d}E}{\mathrm{d}\mathcal{E}} = -\langle \mu_z \rangle \tag{5}$$

In the next step, we note that the energy E of the molecule in the presence of the electric field can be developed in terms of a Taylor expansion relative to its energy $E(0)$ in the absence of the field:

$$E = E(0) + \left(\frac{\mathrm{d}E}{\mathrm{d}\mathcal{E}}\right)_0 \mathcal{E} + \tfrac{1}{2}\left(\frac{\mathrm{d}^2E}{\mathrm{d}\mathcal{E}^2}\right)_0 \mathcal{E}^2 + \tfrac{1}{3!}\left(\frac{\mathrm{d}^3E}{\mathrm{d}\mathcal{E}^3}\right)_0 \mathcal{E}^3 + \ldots \tag{6}$$

where the subscript 0 implies that the derivative is evaluated at $\mathcal{E} = 0$. It then follows from eqn 12.5 that

$$\langle \mu_z \rangle = -\left(\frac{\mathrm{d}E}{\mathrm{d}\mathcal{E}}\right)_0 - \left(\frac{\mathrm{d}^2E}{\mathrm{d}\mathcal{E}^2}\right)_0 \mathcal{E} - \tfrac{1}{2}\left(\frac{\mathrm{d}^3E}{\mathrm{d}\mathcal{E}^3}\right)_0 \mathcal{E}^2 - \ldots \tag{7}$$

The expectation value of the electric dipole moment in the presence of the electric field is the sum of a permanent dipole moment and the contribution induced by the field, so we can also write

$$\langle \mu_z \rangle = \mu_{0z} + \alpha_{zz}\mathcal{E} + \tfrac{1}{2}\beta_{zzz}\mathcal{E}^2 + \ldots \tag{8}$$

In this expression, α_{zz} is the polarizability in the z-direction and β_{zzz} is the **first hyperpolarizability** in the z-direction. There are higher-order hyperpolarizabilities too, but we shall not consider them.

Before going further, it is appropriate to explain why there are two subscripts on α_{zz}. The polarizability is properly regarded as a matrix (or, more loosely, as a second rank 'tensor'). When a field is applied along the z-axis, a dipole may be induced with components μ_x, μ_y, and μ_z (Fig. 12.1), where

$$\mu_q = \alpha_{qz}\mathcal{E} \qquad q = x, y, z$$

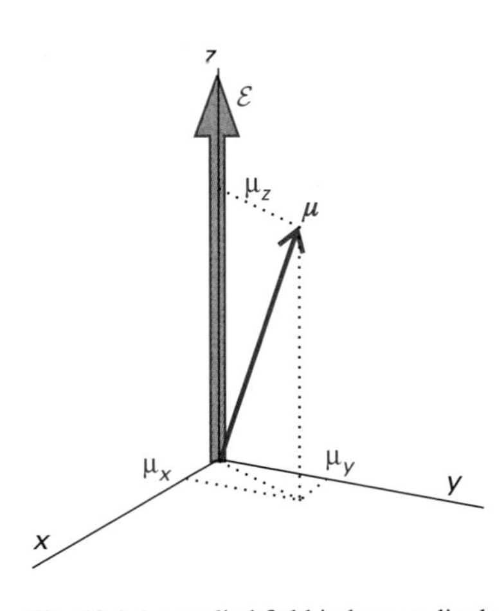

Fig. 12.1 An applied field induces a dipole moment that might not be parallel to the field. The off-diagonal components of the polarizability tensor determine the non-parallel components of the induced dipole moment.

The three components of the matrix therefore relate the magnitude of each component to the strength of the field in the z-direction. Normally, the diagonal element (α_{zz}) dominates the other two, because the induced moment is usually almost parallel to the applied field. There are in general three directions relative to the molecule which, when the field is applied along them in turn, give rise to strictly parallel induced dipole moments. These directions are called the **principal axes** of the polarizability. For similar reasons β is written with three subscripts: β_{qzz} is its contribution to the q-component of the electric dipole when the electric field is applied along the z-axis. A field with both x- and y-components would lead to components of the dipole moment equal to $\beta_{qxy}\mathcal{E}_x\mathcal{E}_y$, etc.

Comparison of eqns 12.7 and 12.8 lets us make the following identifications:

$$\mu_{0z} = -\left(\frac{\mathrm{d}E}{\mathrm{d}\mathcal{E}}\right)_0 \qquad \alpha_{zz} = -\left(\frac{\mathrm{d}^2E}{\mathrm{d}\mathcal{E}^2}\right)_0 \qquad \beta_{zzz} = -\left(\frac{\mathrm{d}^3E}{\mathrm{d}\mathcal{E}^3}\right)_0 \tag{9}$$

and so on. These expressions are the links we need between the properties we want to calculate and the energy of the system, which we can calculate by using perturbation theory. With these relations established, we can write eqn 12.6 in terms of molecular properties:

$$E = E(0) - \mu_{0z}\mathcal{E} - \tfrac{1}{2}\alpha_{zz}\mathcal{E}^2 - \tfrac{1}{3!}\beta_{zzz}\mathcal{E}^3 - \ldots \tag{10}$$

12.2 The static electric polarizability

From now on we shall confine our attention to the calculation of the polarizability of a molecule. To implement eqn 12.9 we need the perturbation expression for the energy, which in Section 6.5 was found to be as follows for the state $|0\rangle$:

$$E_0 = E_0^{(0)} + \langle 0|H^{(1)}|0\rangle + \langle 0|H^{(2)}|0\rangle + {\sum_n}' \frac{\langle 0|H^{(1)}|n\rangle\langle n|H^{(1)}|0\rangle}{E_0 - E_n} + \ldots \tag{11}$$

There is no second-order hamiltonian in the present problem, so the third term on the right makes no contribution. Substitution of $H^{(1)} = -\mu_z\mathcal{E}$ gives

$$E_0 = E_0^{(0)} - \langle 0|\mu_z|0\rangle\mathcal{E} + \left\{{\sum_n}' \frac{\langle 0|\mu_z|n\rangle\langle n|\mu_z|0\rangle}{E_0 - E_n}\right\}\mathcal{E}^2 + \ldots \tag{12}$$

At this point we can use the first relation in eqn 12.9 to write

$$\mu_{0z} = -\left(\frac{\mathrm{d}E_0}{\mathrm{d}\mathcal{E}}\right)_0 = \langle 0|\mu_z|0\rangle \tag{13}$$

because only the second term on the right survives after taking the first derivative with respect to $\mathcal{E}$ and then setting $\mathcal{E} = 0$. This relation tells us nothing new: it states that the permanent electric dipole moment of the molecule is the expectation value of the dipole moment operator in the unperturbed state of the system. Of more interest is the second derivative, which gives the following result after setting $\mathcal{E} = 0$:

$$\alpha_{zz} = -2{\sum_n}' \frac{\langle 0|\mu_z|n\rangle\langle n|\mu_z|0\rangle}{E_0 - E_n} \tag{14}$$

This equation is an explicit expression for the polarizability of the molecule in terms of integrals over its wavefunctions. It is clear from eqn 12.14 that because μ_z transforms as z, α_{zz} transforms as z^2; in general, $\alpha_{qq'}$ transforms as qq'.

To make progress, we shall write $\Delta E_{n0} = E_n - E_0$, which is a positive quantity if 0 denotes the ground state of the molecule. We shall also write the matrix elements $\langle m|\mu_z|n\rangle$ as $\mu_{z,mn}$; then eqn 12.14 becomes

$$\alpha_{zz} = 2\sum_n{}' \frac{\mu_{z,0n}\mu_{z,n0}}{\Delta E_{n0}} \tag{15}$$

Similar expressions for the polarizability when the field is applied along the x- and y-axes can be written down by analogy with this one. The **mean polarizability**, α, is the property observed when a molecule is rotating in a fluid and it presents all orientations to the applied field:

$$\begin{aligned}\alpha &= \tfrac{1}{3}(\alpha_{xx} + \alpha_{yy} + \alpha_{zz}) \\ &= \tfrac{2}{3}\sum_n{}' \frac{\mu_{x,0n}\mu_{x,n0} + \mu_{y,0n}\mu_{y,n0} + \mu_{z,0n}\mu_{z,n0}}{\Delta E_{n0}}\end{aligned}$$

The appearance of this expression can be simplified by writing the numerator as a scalar product of two vectors:

$$\boldsymbol{\mu}_{0n} \cdot \boldsymbol{\mu}_{n0} = \mu_{x,0n}\mu_{x,n0} + \mu_{y,0n}\mu_{y,n0} + \mu_{z,0n}\mu_{z,n0} \tag{16}$$

However, because $\boldsymbol{\mu}$ is an hermitian operator, $\boldsymbol{\mu}_{0n} = \boldsymbol{\mu}_{n0}^*$, it is convenient to express the left-hand side of this expression as $|\mu_{n0}|^2$, and so we obtain

$$\alpha = \tfrac{2}{3}\sum_n{}' \frac{|\mu_{n0}|^2}{\Delta E_{n0}} \tag{17}$$

To use this expression, we interpret the numerator as the sum of the three terms in eqn 12.16.

A final point concerns the units of polarizability. With the dipole moment operators expressed in coulomb-metre (C m) and the energy differences in joule (J), the polarizability is expressed in (coulomb metre)^2 per joule ($C^2\,m^2\,J^{-1}$). These units are disagreeably cumbersome, and it is common to introduce the **polarizability volume**, α', which is defined as

$$\alpha' = \frac{\alpha}{4\pi\varepsilon_0} \tag{18}$$

where ε_0 is the vacuum permittivity. The polarizability volume has the dimensions of volume and its units are metre cubed (m^3). The use of the polarizability volume also simplifies some expressions, and we shall use it when it is convenient to do so.

Example 12.1 The polarizability of a harmonic oscillator

Calculate the polarizability of a one-dimensional system of two charges, e and $-e$, bound together to form a harmonic oscillator by a spring of force constant k, with the electric field applied parallel to the x-axis (the inter-charge direction).

Method. Let the equilibrium distance between the charges be R and the extension x. The dipole moment operator for the system is $\mu = e(R + x)$. When evaluating the sum in eqn 12.17 we should take note of the fact that, as established in Example 10.3, the only matrix elements of x are from $|v\rangle$ to $|v \pm 1\rangle$, so there are only two terms in the sum, which may therefore be written down and evaluated term by term. For the energies, use $E_v = (v + \frac{1}{2})\hbar\omega_0$ with $\omega_0 = (k/m)^{1/2}$, where m is the effective mass of the oscillator.

Answer. The matrix elements we require were evaluated in Example 10.3 and are

$$\langle v+1|\mu_x|v\rangle = e\langle v+1|x|v\rangle = e(v+1)^{1/2}\left(\frac{\hbar}{2m\omega_0}\right)^{1/2}$$

$$\langle v-1|\mu_x|v\rangle = e\langle v-1|x|v\rangle = ev^{1/2}\left(\frac{\hbar}{2m\omega_0}\right)^{1/2}$$

In each case, the matrix elements of eR are zero. The polarizability parallel to x is therefore

$$\alpha_{xx} = 2\sum_{v'}{}' \frac{|\langle v|\mu_x|v'\rangle|^2}{(v'-v)\hbar\omega_0} = \frac{2}{\hbar\omega_0}\left\{|\langle v|\mu_x|v+1\rangle|^2 - |\langle v|\mu_x|v-1\rangle|^2\right\}$$
$$= \frac{e^2}{m\omega_0^2}\{(v+1)-v\} = \frac{e^2}{k}$$

Comment. The polarizability is independent of the state of the oscillator and of its mass. The latter independence arises from the fact that the static (zero-frequency) polarizability is a response to a stationary electric field and does not depend on the inertial properties of the oscillator (the rate at which it responds to a changing force). For comparison, see later (Example 12.3), where the dynamic problem is treated. This calculation models the distortion contribution to the polarizability of a molecule, the contribution to the polarizability of a distortion of the molecular geometry.

12.3 Polarizability and molecular properties

To use the expressions we have derived, it is in principle necessary to know the wavefunctions and energies of all the excited states of the molecule, for only then can the sum in eqn 12.17 be evaluated. Usually this formidable task is impossible, and it is necessary to resort to an approximate procedure. Such additional approximations should not be scorned: they can provide valuable pointers to the variation of molecular properties with a variety of parameters, such as molecular size, and can provide links between observables. The numerical values they suggest, however, must be viewed with great suspicion.

One way forward is to invoke the closure approximation (Section 6.7). If the excitation energies are replaced by a mean value ΔE, we obtain

$$\alpha \approx \frac{2}{3\Delta E}\sum_n{}' \boldsymbol{\mu}_{0n}\cdot\boldsymbol{\mu}_{n0} \approx \frac{2}{3\Delta E}\left\{\sum_n \boldsymbol{\mu}_{0n}\cdot\boldsymbol{\mu}_{n0} - \boldsymbol{\mu}_{00}\cdot\boldsymbol{\mu}_{00}\right\}$$
$$\approx \frac{2}{3\Delta E}\left\{\langle\mu^2\rangle - \langle\mu\rangle^2\right\}$$

If we write $\langle\mu^2\rangle - \langle\mu\rangle^2 = (\Delta\mu)^2$, we obtain

$$\alpha \approx \frac{2(\Delta\mu)^2}{3\Delta E} \tag{19}$$

We shall refer to $\Delta\mu$ as the **fluctuation** in the mean electric dipole moment: it is the root mean square deviation of the dipole moment from its mean value. Even a nonpolar molecule with a zero permanent electric dipole moment ($\langle\mu\rangle = 0$) has a nonzero dipole fluctuation. To some extent, we can guardedly think of the fluctuation as arising from an actual classical fluctuation of the electron density in the molecule about its average value. As we see from eqn 12.19, the polarizability of a molecule is proportional to the square of the magnitude of these fluctuations. This result is consistent with the view that the molecule can be easily distorted by an applied electric field if its electrons are not under the tight control of the nuclei. Indeed, there is a much deeper result lurking beneath this physically plausible remark, for the **fluctuation–dissipation theorem** establishes a proportionality between the response of a system and the square of the magnitude of the fluctuations that occur in the unperturbed system (see *Further reading* for more information).

If we continue with this line of argument, we can expect the polarizability to increase with the radius of the molecule and the number of electrons it contains, because in each case we can expect the nuclei to have less control over their electrons. To illustrate this conclusion, consider a one-electron atom. Because the electric dipole moment operator is then $\boldsymbol{\mu} = -e\mathbf{r}$, and the unperturbed species is nonpolar, we can conclude from eqn 12.19 that

$$\alpha \approx \frac{2e^2\langle r^2\rangle}{3\Delta E} \tag{20}$$

where $\langle r^2\rangle$ is the mean square radius of the electron's orbital. This expression confirms that the polarizability increases as the radius increases. This conclusion is consistent with a progressive loss of control by the nucleus over its electron as the orbital expands. In a many-electron atom, we can expect each electron to contribute a similar term, resulting in a polarizability that is proportional to $N_e\langle r^2\rangle$, where N_e is the number of electrons in the atom and $\langle r^2\rangle$ is the mean square radius of all the occupied orbitals. Because $\langle r^2\rangle \approx R_a^2$, where R_a is the radius of the atom, it follows that

$$\alpha \approx \frac{2e^2N_eR_a^2}{3\Delta E} \approx \frac{e^2N_eR_a^2}{I} \tag{21}$$

In the second step we have made yet another approximation: that the mean excitation energy is approximately the same as the ionization energy, I. This approximation is so questionable that we have also discarded the factor of $\frac{2}{3}$. It follows that as the size of the atom increases, either as a result of an expansion of its orbitals or an increase in the number of electrons, the polarizability will increase. As ionization energies generally follow the opposite trend, the presence of I in the denominator reinforces this trend.

12.4 Polarizabilities and molecular spectroscopy

We can in fact develop another line of argument in a similar way. First, we note that the polarizability depends on the square of transition dipole moments. But we have already met such squares in the context of the intensities of spectroscopic transitions. Specifically (see *Further information 17*), a useful measure of absorption intensity is the oscillator strength, which for the transition $n \leftarrow 0$ is

$$f_{n0} = \left(\frac{4\pi m_e}{3e^2\hbar}\right)\nu_{n0}|\mu_{n0}|^2 \tag{22}$$

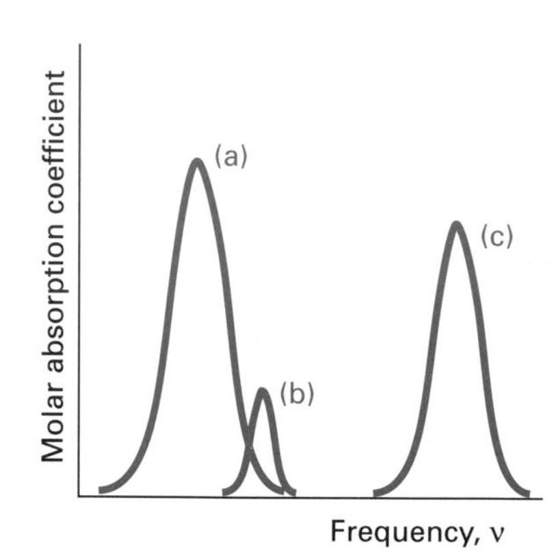

Fig. 12.2 (a) A strong absorption at low energy gives a large contribution to the polarizability of a molecule. (b) A weak absorption at low energy and (c) a strong absorption at high energy each give small contributions to the polarizability.

It follows that

$$\alpha = \frac{\hbar^2 e^2}{m_e}\sum_n{}' \frac{f_{n0}}{\Delta E_{n0}^2} \tag{23}$$

This simple expression provides a link between spectroscopy and the prediction of polarizabilities, because the oscillator strengths of the transitions of a molecule can be determined from band intensities (*Further information 17*) and their energies can be determined from their locations on a frequency scale. The expression indicates that large contributions come from low energy, high intensity transitions; high energy or weak transitions make little contribution (Fig. 12.2). An implication is that if a molecule has intense, low-frequency transitions in its absorption spectrum, then it can be expected to be highly polarizable. Hence, intensely coloured molecules should be highly polarizable. In contrast, molecules that absorb only weakly or at high frequencies (such as the colourless hydrocarbons, which absorb only in the ultraviolet and then only weakly) are expected to be only weakly polarizable.

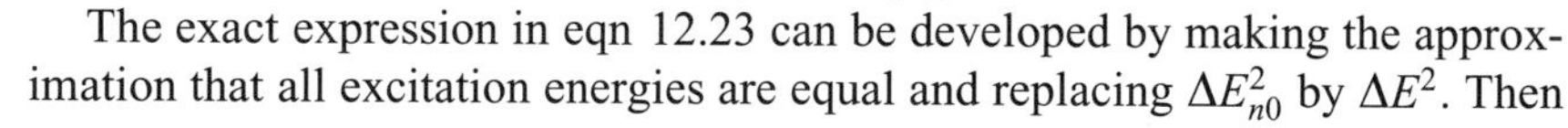

The exact expression in eqn 12.23 can be developed by making the approximation that all excitation energies are equal and replacing ΔE_{n0}^2 by ΔE^2. Then

$$\alpha \approx \frac{\hbar^2 e^2}{m_e \Delta E^2}\sum_n{}' f_{n0}$$

The sum over oscillator strengths is a standard result known as the **Kuhn–Thomas sum rule**:

$$\sum_n{}' f_{n0} = N_e \tag{24}$$

It is proved in *Further information 18*; in practice, interpreting N_e as the number of *valence* electrons, N_V, tends to give better results for the sum of measured oscillator strengths. Therefore,

$$\alpha \approx \frac{\hbar^2 e^2 N_V}{m_e \Delta E^2} \tag{25}$$

This expression shows that the polarizability increases as the number of (valence) electrons increases and as the mean excitation energy decreases. The two effects generally reinforce one another, so we can expect molecules composed of heavy atoms to be strongly polarizable.

12.5 Polarizabilities and dispersion forces

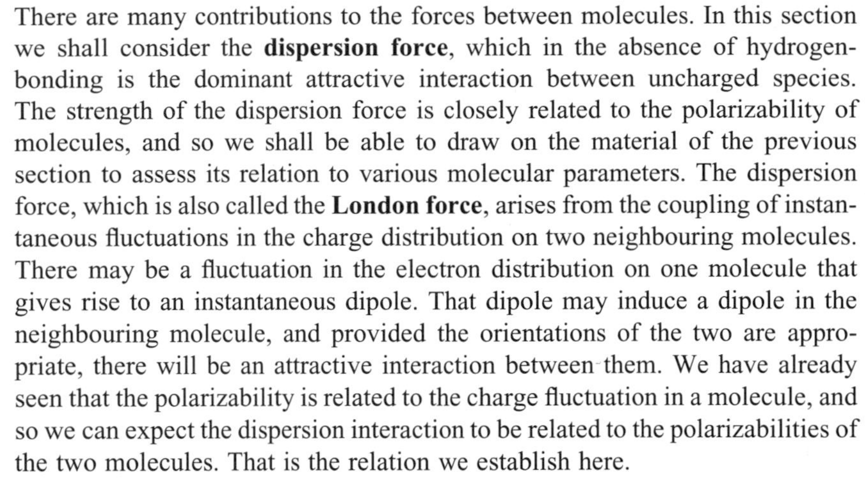

There are many contributions to the forces between molecules. In this section we shall consider the **dispersion force**, which in the absence of hydrogen-bonding is the dominant attractive interaction between uncharged species. The strength of the dispersion force is closely related to the polarizability of molecules, and so we shall be able to draw on the material of the previous section to assess its relation to various molecular parameters. The dispersion force, which is also called the **London force**, arises from the coupling of instantaneous fluctuations in the charge distribution on two neighbouring molecules. There may be a fluctuation in the electron distribution on one molecule that gives rise to an instantaneous dipole. That dipole may induce a dipole in the neighbouring molecule, and provided the orientations of the two are appropriate, there will be an attractive interaction between them. We have already seen that the polarizability is related to the charge fluctuation in a molecule, and so we can expect the dispersion interaction to be related to the polarizabilities of the two molecules. That is the relation we establish here.

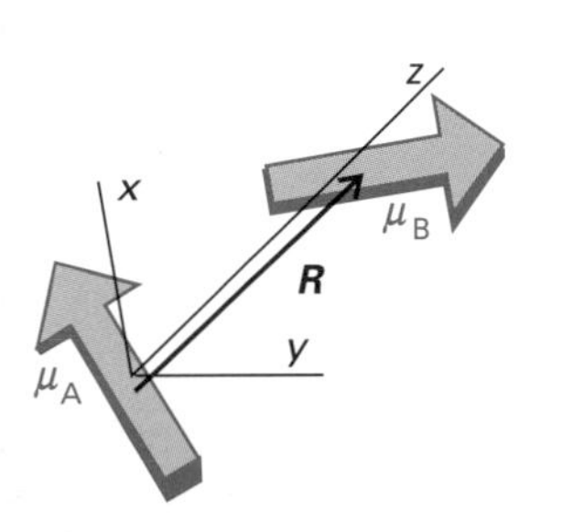

Fig. 12.3 The coordinate system used for setting up the dipole–dipole interaction hamiltonian for the discussion of dispersion forces.

We shall use perturbation theory to calculate the lowering in energy when two closed-shell atoms are brought to a separation R. The perturbation hamiltonian is the interaction of two electric dipole operators based on the two atoms. It follows from classical electrostatics, that such an interaction for the orientations shown in Fig. 12.3 is

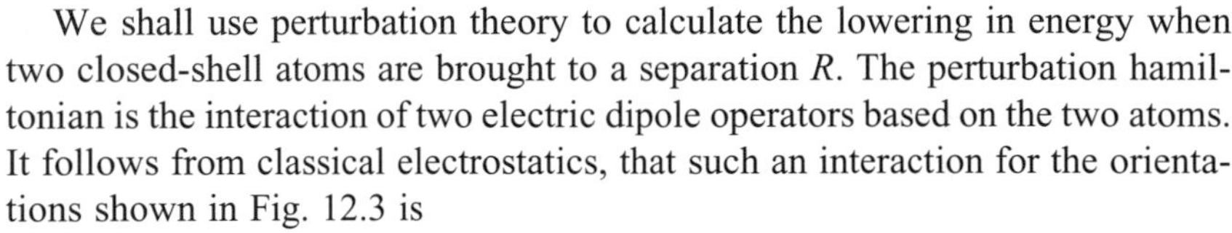

$$H^{(1)} = \left(\frac{1}{4\pi\varepsilon_0 R^3}\right)\left\{\boldsymbol{\mu}_A\cdot\boldsymbol{\mu}_B - \frac{3(\boldsymbol{\mu}_A\cdot\mathbf{R})(\mathbf{R}\cdot\boldsymbol{\mu}_B)}{R^2}\right\} \tag{26}$$

It will be simplest to select as the z-axis the axis that joins the centres of the two atoms, then with the axes arranged as in Fig. 12.3 the perturbation is

$$H^{(1)} = \left(\frac{1}{4\pi\varepsilon_0 R^3}\right)\{\mu_{Ax}\mu_{Bx} + \mu_{Ay}\mu_{By} - 2\mu_{Az}\mu_{Bz}\} \tag{27}$$

The total hamiltonian of the system is

$$H = H^{(0)} + H^{(1)}, \qquad H^{(0)} = H_A + H_B$$

The unperturbed states of the pair of atoms are $|n_A n_B\rangle$, with

$$H^{(0)}|n_A n_B\rangle = (E_{n_A} + E_{n_B})|n_A n_B\rangle$$

We shall write $E_{n_A n_B} = E_{n_A} + E_{n_B}$ and consider interactions between the atoms in their ground states, $|00\rangle$.

It is quite easy to show that the first-order correction to the energy is zero:

$$E^{(1)} = \langle 00|H^{(1)}|00\rangle \propto \langle 00|\mu_{Ax}\mu_{Bx} + \ldots|00\rangle = \langle 0|\mu_{Ax}|0\rangle\langle 0|\mu_{Bx}|0\rangle + \ldots = 0$$

because every matrix element is the ground-state expectation value of the electric dipole moment operator, which is zero for a nonpolar species.

Because the first-order terms are zero, we have to consider the second-order contribution. Physically, this means that we must allow for the distortion of the wavefunction of each atom as a result of the presence of the second atom. That corresponds, in the classical picture, to the correlation of the fluctuating instantaneous dipole moments when one dipole drives the other into existence. The second order energy is

$$E^{(2)} = \sum_{n_A, n_B}{}' \frac{\langle 00|H^{(1)}|n_A n_B\rangle\langle n_A n_B|H^{(1)}|00\rangle}{E_{00} - E_{n_A n_B}} \tag{28}$$

As before, we shall express the denominator in terms of excitation energies, and this time will write

$$\Delta E_{n_A 0} + \Delta E_{n_B 0} = E_{n_A n_B} - E_{00}$$

The perturbation hamiltonian has three terms, so the second-order energy expression has nine terms. Happily, though, most of them vanish. Consider, for instance, one of the cross-terms

$$\langle 00|\mu_{Ax}\mu_{Bx}|n_A n_B\rangle\langle n_A n_B|\mu_{Ay}\mu_{By}|00\rangle$$

This term includes the factor

$$\langle 0|\mu_{Ax}|n_A\rangle\langle n_A|\mu_{Ay}|0\rangle$$

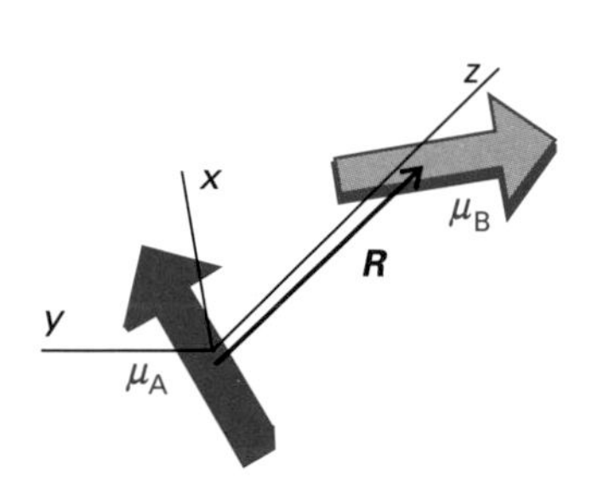

Fig. 12.4 Reversal of the direction of the y-axis must leave the calculated interaction energy unchanged.

To see that this term is zero, we make use of the fact that we are free to choose an alternative coordinate system on A with the y-axis pointing in the opposite direction but with the x-axis unchanged (Fig. 12.4). The product of matrix elements then changes sign. However, a contribution to the energy cannot depend on the choice of axes, and so the contribution must be zero. The same argument applies to all the cross-terms in eqn 12.28, and so only the three terms of the form

$$\langle 00|\mu_{Aq}\mu_{Bq}|n_A n_B\rangle\langle n_A n_B|\mu_{Aq}\mu_{Bq}|00\rangle$$

survive. For atoms, these three terms are all the same (by spherical symmetry of each atom). Moreover, by spherical symmetry,

$$\begin{aligned}\langle 0|\mu_{Ax}|n_A\rangle\langle n_A|\mu_{Ax}|0\rangle &= \langle 0|\mu_{Ay}|n_A\rangle\langle n_A|\mu_{Ay}|0\rangle \\ &= \langle 0|\mu_{Az}|n_A\rangle\langle n_A|\mu_{Az}|0\rangle\end{aligned} \tag{29}$$

from which it follows that any one is one-third the sum of the three, and hence

$$\langle 0|\mu_{Ax}|n_A\rangle\langle n_A|\mu_{Ax}|0\rangle = \tfrac{1}{3}\boldsymbol{\mu}_{A,0n_A}\cdot\boldsymbol{\mu}_{A,n_A 0} \tag{30}$$

and likewise for the other two components for A and for all three components for B. Therefore, the entire expression reduces to

$$E^{(2)} = -\tfrac{2}{3}\left(\frac{1}{4\pi\varepsilon_0 R^3}\right)^2 \sum_{n_A, n_B}{}' \left\{\frac{(\boldsymbol{\mu}_{A,0n_A}\cdot\boldsymbol{\mu}_{A,n_A 0})(\boldsymbol{\mu}_{B,0n_B}\cdot\boldsymbol{\mu}_{B,n_B 0})}{\Delta E_{n_A 0} + \Delta E_{n_B 0}}\right\} \tag{31}$$

This expression confirms that there is a nonzero interaction energy which is attractive ($E^{(2)} < 0$) and inversely proportional to the sixth-power of the separation ($E^{(2)} \propto 1/R^6$).

Example 12.2 Dispersion interactions between oscillators

Calculate the energy of the dispersion interaction between two electrons oscillating harmonically and isotropically in three dimensions about centres separated by a distance R, and express the answer in terms of their polarizabilities. Let the state be characterized by the ket $|v, v\rangle$.

Method. We base the answer on eqn 12.31. For the matrix elements, we use the values in Example 12.1, but we need to distinguish the frequencies and force constants by subscripts A and B for the two 'atoms'. The selection rules result in the restriction of the sum in eqn 12.31 to only four terms, so it may be evaluated explicitly. For the relation to the polarizabilities, use the results obtained in Example 12.1.

Answer. The sum we require has the following four nonzero terms:

$$E^{(2)} = -\tfrac{2}{3}\left(\frac{1}{4\pi\varepsilon_0 R^3}\right)^2 \left\{ \frac{|\mu_{\mathrm{A};v,v+1}|^2 |\mu_{\mathrm{B};v,v+1}|^2}{\hbar(\omega_\mathrm{A}+\omega_\mathrm{B})} + \frac{|\mu_{\mathrm{A};v,v+1}|^2 |\mu_{\mathrm{B};v,v-1}|^2}{\hbar(\omega_\mathrm{A}-\omega_\mathrm{B})} \right.$$
$$\left. + \frac{|\mu_{\mathrm{A};v,v-1}|^2 |\mu_{\mathrm{B};v,v+1}|^2}{\hbar(-\omega_\mathrm{A}+\omega_\mathrm{B})} + \frac{|\mu_{\mathrm{A};v,v-1}|^2 |\mu_{\mathrm{B};v,v-1}|^2}{\hbar(-\omega_\mathrm{A}-\omega_\mathrm{B})} \right\}$$
$$= -\tfrac{2}{3}\left(\frac{1}{4\pi\varepsilon_0 R^3}\right)^2 \left(\frac{3\hbar e^2}{2m_\mathrm{A}\omega_\mathrm{A}}\right)\left(\frac{3\hbar e^2}{2m_\mathrm{B}\omega_\mathrm{B}}\right)$$
$$\times \left\{ \frac{(v_\mathrm{A}+1)(v_\mathrm{B}+1) - v_\mathrm{A}v_\mathrm{B}}{\hbar(\omega_\mathrm{A}+\omega_\mathrm{B})} + \frac{(v_\mathrm{A}+1)v_\mathrm{B} - v_\mathrm{A}(v_\mathrm{B}+1)}{\hbar(\omega_\mathrm{A}-\omega_\mathrm{B})} \right\}$$

We have used the relation

$$|\mu_{v,v+1}|^2 = |\mu_{x;v,v+1}|^2 + |\mu_{y;v,v+1}|^2 + |\mu_{z;v,v+1}|^2 = \left(\frac{3\hbar e^2}{2m\omega_0}\right)(v+1)$$

and its analogues. It then follows that

$$E^{(2)} = -\tfrac{3}{2}\left(\frac{1}{4\pi\varepsilon_0 R^3}\right)^2 \left(\frac{e^4\hbar}{m_\mathrm{A}m_\mathrm{B}\omega_\mathrm{A}\omega_\mathrm{B}}\right)\left\{\frac{(1+2v_\mathrm{B})\omega_\mathrm{A} - (1+2v_\mathrm{A})\omega_\mathrm{B}}{\omega_\mathrm{A}^2 - \omega_\mathrm{B}^2}\right\}$$

When the two oscillators are in their ground states, this expression simplifies to

$$E^{(2)} = -\tfrac{3}{2}\left(\frac{1}{4\pi\varepsilon_0 R^3}\right)^2 \left(\frac{e^4\hbar}{m_\mathrm{A}m_\mathrm{B}\omega_\mathrm{A}\omega_\mathrm{B}(\omega_\mathrm{A}+\omega_\mathrm{B})}\right)$$

At this stage, we introduce the expressions for the polarizabilities (Example 12.1) and convert them to polarizability volumes (to simplify the appearance of the final expression), and obtain

$$E^{(2)} = -\tfrac{3}{2}\left(\frac{\hbar\omega_\mathrm{A}\omega_\mathrm{B}}{\omega_\mathrm{A}+\omega_\mathrm{B}}\right)\left(\frac{\alpha'_\mathrm{A}\alpha'_\mathrm{B}}{R^6}\right)$$

Comment. Keep this exact result in mind and compare it with the approximate London formula that we derive below: the two expressions have identical structures. In this case only a very limited number of transitions are allowed, and the closure approximation on which the London formula is based is exact.

An approximate, revealing, and useful form of eqn 12.31 can be obtained by making use of the closure approximation.[1] Therefore, we write $\Delta E_{n_A 0} \approx \Delta E_A$, and likewise for B, and obtain

$$E^{(2)} \approx -\left(\frac{1}{24\pi^2\varepsilon_0^2 R^6}\right)\left(\frac{1}{\Delta E_A + \Delta E_B}\right) \times \sum_{n_A, n_B}{}' (\boldsymbol{\mu}_{A,0n_A} \cdot \boldsymbol{\mu}_{A,n_A 0})(\boldsymbol{\mu}_{B,0n_B} \cdot \boldsymbol{\mu}_{B,n_B 0}) \approx -\left(\frac{1}{24\pi^2\varepsilon_0^2 R^6}\right)\left(\frac{1}{\Delta E_A + \Delta E_B}\right)\langle\mu_A^2\rangle\langle\mu_B^2\rangle \tag{32}$$

where $\langle\mu_A^2\rangle = \langle 0|\mu_A^2|0\rangle$, and likewise for B. The terms $\langle\mu_A\rangle^2$ and $\langle\mu_B\rangle^2$ are both zero for nonpolar species. This expression can be taken further by using the relation between the mean square dipole moment and the polarizability (eqn 12.19), which for nonpolar species simplifies to $\langle\mu_A^2\rangle \approx \frac{3}{2}\alpha_A \Delta E_A$, and likewise for B. With substitution of this term, we obtain

$$E^{(2)} \approx -\left(\frac{3}{32\pi^2\varepsilon_0^2}\right)\left(\frac{\Delta E_A \Delta E_B}{\Delta E_A + \Delta E_B}\right)\left(\frac{\alpha_A \alpha_B}{R^6}\right) = -\frac{3}{2}\left(\frac{\Delta E_A \Delta E_B}{\Delta E_A + \Delta E_B}\right)\left(\frac{\alpha'_A \alpha'_B}{R^6}\right) \tag{33}$$

A general indication of the magnitudes of the mean excitation energy is the ionization energy of each atom, and if we write $\Delta E_A \approx I_A$, and likewise for B, we arrive at the **London formula**:

$$E^{(2)} \approx -\frac{3}{2}\left(\frac{I_A I_B}{I_A + I_B}\right)\left(\frac{\alpha'_A \alpha'_B}{R^6}\right) \tag{34}$$

The London formula, although only approximate, reveals the essential character of the dispersion energy and may be used to make rough estimates of its magnitude. We see, for instance, that the interaction is greatest between atoms of high polarizability. We have already seen how the polarizability is related to the structures of atoms, and the remarks made in Sections 12.3 and 12.4 may be extended to the interactions between atoms and molecules. Thus, we expect intensely coloured, large, many-electron species to have strong dispersion interactions. One consequence of this dependence of dispersion interactions on polarizability is the high volatility of low molar mass hydrocarbons, which have low polarizabilities.

12.6 Retardation effects

At this point it is appropriate to admit that the starting point of Section 12.5, the hamiltonian in eqn 12.26, is only an approximation. The true description of the interaction between two atoms should be expressed in terms of their joint interaction with the electromagnetic field. Thus, when a fluctuation in electron density occurs on A, it generates a photon that travels through the vacuum at the speed of light. It stimulates a fluctuation on B, and that fluctuation in turn generates a photon that travels back to A. The interaction therefore takes place

[1] Other interesting forms can be obtained by using the oscillator strengths.

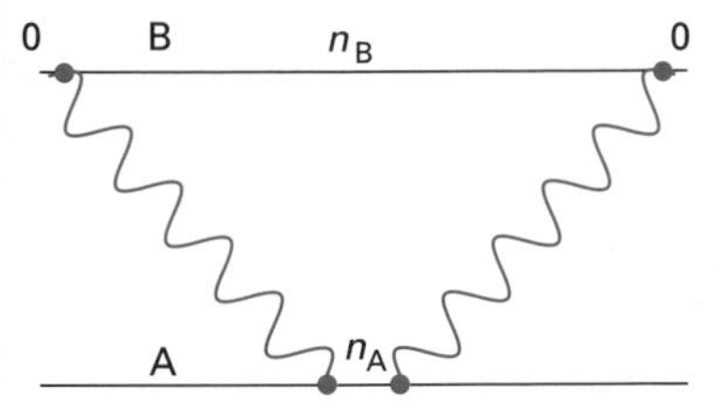

Fig. 12.5 One of the Feynman diagrams that contribute to the dispersion interaction. The interaction is mediated by photons that are generated by transition dipoles on each molecule.

by an exchange of photons between the two atoms.[2] An example of a Feynman diagram that contributes to the dispersion interaction is shown in Fig. 12.5.

It takes a time R/c for the photon from A to arrive at B, and the response takes the same time to return to A. The fluctuations on the atoms occur on a time scale of approximately $\Delta E/h$, where ΔE is a typical excitation energy. If the time it takes for the round trip, $2R/c$, is longer than the fluctuation time, the dipole on A will have migrated to a new position. As a result of this **retardation**, or finite travel time for signals, the dispersion interaction is weakened. Only when the atoms are so close that $2R/c \ll \Delta E/h$ will the correlation of the dipoles be perfect and the interaction have the full strength represented by eqn 12.34. When $2R/c \gg \Delta E/h$ (when R exceeds about 10 nm), the weakening effect of retardation is so great that the $1/R^6$ form of the interaction changes to a more rapidly decaying $1/R^7$ form. Specifically, at such distances

$$E^{(2)} \approx -\left(\frac{23\hbar c}{4\pi}\right)\left(\frac{\alpha'_A \alpha'_B}{R^7}\right) \tag{35}$$

(For a derivation of this expression, see *Further reading*.) The formula is much more complicated when $2R/c \approx \Delta E/h$ because the conventional $1/R^6$ expression is in the middle of turning into a $1/R^7$ expression. Retardation effects are important for colloids and macromolecules.

Bulk electrical properties

Now that we have an expression for the polarizability of an individual molecule, we can move on to a discussion of some of the properties of **dielectric media**, nonconducting bulk substances. These properties include relative permittivity and refractive index. A property related to the refractive index is optical activity; so we shall also investigate its origin.

12.7 The relative permittivity and the electric susceptibility

In a vacuum, the Coulomb potential of a charge q at a distance r is

$$\phi = \frac{q}{4\pi\varepsilon_0 r} \tag{36}$$

where ε_0 is vacuum permittivity ($\varepsilon_0 = 8.854 \times 10^{-12}\,\mathrm{J^{-1}\,C^2\,m^{-1}}$). In a dielectric medium, the same charge gives rise to a potential

$$\phi = \frac{q}{4\pi\varepsilon r} \tag{37}$$

where ε is the **permittivity** of the medium. The dimensionless ratio

$$\varepsilon_r = \frac{\varepsilon}{\varepsilon_0} \tag{38}$$

is called the **relative permittivity** of the medium; its older name is 'dielectric constant'. In practice, the relative permittivity is measured as the ratio of the capacitances of a capacitor with and without the dielectric between the plates.

[2] It is a very general feature of the composition of the universe that interactions between matter (agglomerations of fermions) take place by the exchange of bosons.

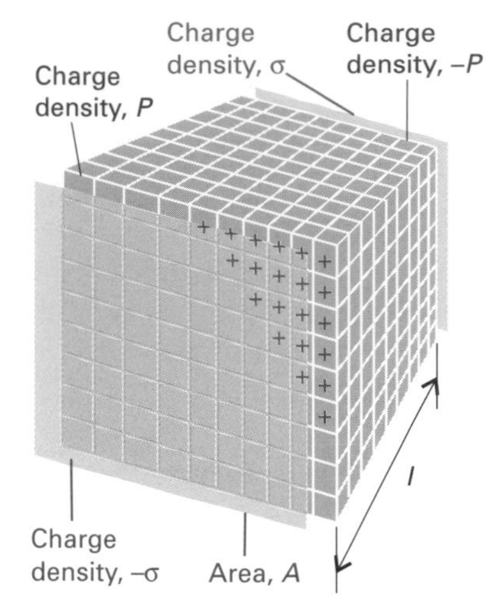

Fig. 12.6 The relation between the polarization of a medium and the mean dipole-moment density.

Now consider the electric field between two plates each of area A, each one with a charge density of magnitude σ, so the total charge on one plate is σA and on the other is $-\sigma A$. A result from electrostatics is that the electric field strength between the plates is σ/ε_0 if the intervening medium is a vacuum but

$$\mathcal{E} = \frac{\sigma}{\varepsilon} \tag{39}$$

if it is a dielectric medium. Instead of thinking of the reduction of the electric field as stemming from the permittivity, we could think of it instead as arising from the surface charge polarizing the medium and inducing on it a surface charge density P (Fig. 12.6). This induced surface charge density is called the **polarization** of the medium. From this point of view, the electric field between the plates would be written

$$\mathcal{E} = \frac{\sigma - P}{\varepsilon_0} \tag{40}$$

Because eqns 12.39 and 12.40 are merely different ways of expressing the same electric field, we can equate them to find an expression for P:

$$P = \left(\frac{\varepsilon - \varepsilon_0}{\varepsilon}\right)\sigma = (\varepsilon_r - 1)\varepsilon_0\mathcal{E} \tag{41}$$

The **electric susceptibility**, χ_e, of a medium is defined through

$$P = \chi_e\varepsilon_0\mathcal{E} \tag{42}$$

so it follows (by comparing the last two equations) that the electric susceptibility is related to the relative permittivity by

$$\chi_e = \varepsilon_r - 1 \tag{43}$$

The next stage in the argument involves relating the polarization of the medium to the polarizability of its molecules. To do so, we need to know that as well as being the induced surface charge density, P is also the dipole-moment density of the medium, the dipole moment divided by the volume of the sample. This interpretation is established by referring again to Fig. 12.6, which shows that the sample can be regarded as having charges PA and $-PA$ separated by a distance l, and hence a dipole moment PAl. However, as the volume of the sample is Al, the dipole moment divided by the volume is $PAl/Al = P$, as we set out to show.

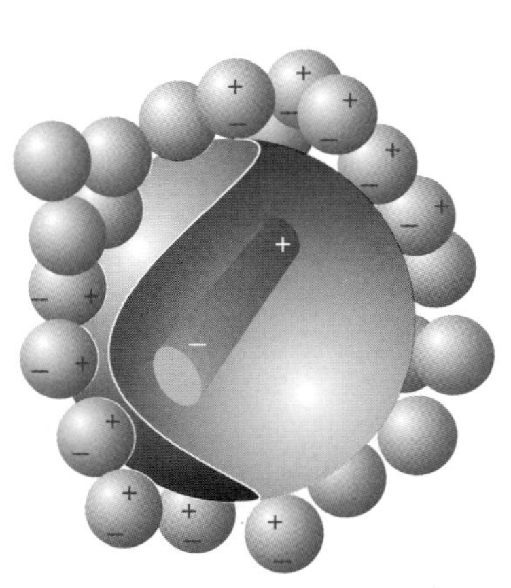

Fig. 12.7 The polarization of the surroundings by the polarized molecule contributes to the total electric field experienced by the molecule.

Now that we know that the polarization is the dipole-moment density, we can relate it to molecular properties, because the dipole-moment density is the mean dipole moment of a molecule in the medium, $\langle\mu\rangle$, multiplied by the number density of molecules, $\mathcal{N}$. If for the time being we suppose that the molecules are nonpolar, then $\langle\mu\rangle$ is the induced dipole moment. At this point, though, we cannot simply write $\langle\mu\rangle = \alpha\mathcal{E}$ because the molecule experiences the **local electric field**, $\mathcal{E}^*$, not the applied field. The local electric field is the total field arising from the applied field and the electric dipoles that that field stimulates in the medium (Fig. 12.7). It follows that

$$P = \alpha\mathcal{N}\mathcal{E}^* \tag{44}$$

The **Lorentz local field** is an approximate relation between $\mathcal{E}^*$ and the applied field $\mathcal{E}$ which is based on the assumption that the medium is a continuous dielectric:

$$\mathcal{E}^* = \mathcal{E} + \frac{P}{3\varepsilon_0} \tag{45}$$

For the derivation of this field, texts on electrostatics should be consulted (see *Further reading*). This expression can be used in eqn 12.44 to give

$$P = \left(\frac{3\alpha\mathcal{N}}{3\varepsilon_0 - \alpha\mathcal{N}}\right)\varepsilon_0\mathcal{E} \tag{46}$$

Comparison of this equation with eqn 12.42 lets us identify the electric susceptibility as

$$\chi_e = \frac{\alpha\mathcal{N}/\varepsilon_0}{1 - \alpha\mathcal{N}/3\varepsilon_0} \tag{47}$$

It immediately follows from eqn 12.43 that the relative permittivity is related to the polarizability of the molecules by

$$\varepsilon_r = \frac{1 + (2\alpha\mathcal{N}/3\varepsilon_0)}{1 - (\alpha\mathcal{N}/3\varepsilon_0)} \tag{48}$$

Before discussing this result, we shall develop equations that are applicable when the molecules have permanent dipole moments too.

12.8 Polar molecules

Although molecules may be tumbling in their fluid environment, the orientating effect of the external field will favour particular orientations, and the net dipole-moment density will differ from zero. The magnitude of the effect can be calculated from the Boltzmann distribution, because the most favoured orientations are the ones with lowest energy. The energy of a dipole in a local electric field $\mathcal{E}^*$ directed along the z-axis is

$$E(\theta) = -\mu_{0z}\mathcal{E}^* = -\mu_0\mathcal{E}^* \cos\theta \tag{49}$$

where θ is the angle the dipole moment of magnitude μ_0 makes to the direction of the local field. At a temperature T, the proportion of N molecules in the orientation range θ to $\theta + \mathrm{d}\theta$ is

$$\frac{\mathrm{d}N(\theta)}{N} = \frac{\mathrm{e}^{-E(\theta)/kT}\sin\theta\,\mathrm{d}\theta}{\int_0^\pi \mathrm{e}^{-E(\theta)/kT}\sin\theta\,\mathrm{d}\theta} = \frac{\mathrm{e}^{\mu_0\mathcal{E}^*\cos\theta/kT}\sin\theta\,\mathrm{d}\theta}{\int_0^\pi \mathrm{e}^{\mu_0\mathcal{E}^*\cos\theta/kT}\sin\theta\,\mathrm{d}\theta}$$

The denominator can be evaluated quite readily if we write $x = \mu_0\mathcal{E}^*/kT$ and note that $\sin\theta\,\mathrm{d}\theta = -\mathrm{d}\cos\theta$:

$$\int_0^\pi \mathrm{e}^{\mu_0\mathcal{E}^*\cos\theta/kT}\sin\theta\,\mathrm{d}\theta = \int_{-1}^1 \mathrm{e}^{x\cos\theta}\,\mathrm{d}\cos\theta = \frac{\mathrm{e}^x - \mathrm{e}^{-x}}{x}$$

Then the Boltzmann distribution is

$$\frac{\mathrm{d}N(\theta)}{N} = \frac{x\mathrm{e}^{x\cos\theta}\sin\theta\,\mathrm{d}\theta}{\mathrm{e}^x - \mathrm{e}^{-x}} \tag{50}$$

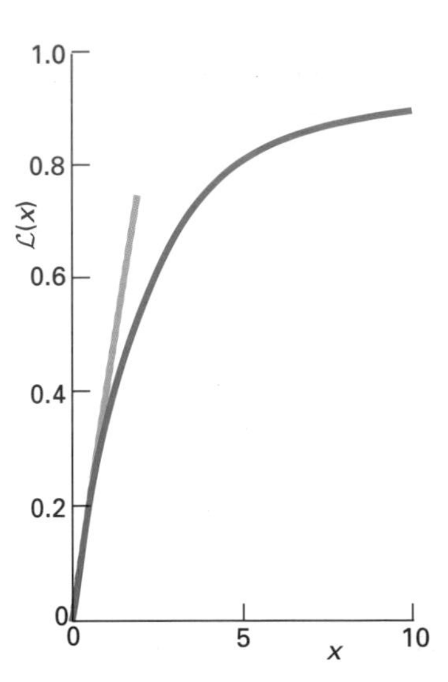

Fig. 12.8 The Langevin function and the linear approximation when $x \ll 1$.

The dipole-moment density is the average of $\mu_0 \cos\theta$ weighted by the Boltzmann factor and divided by the volume, V, of the sample:

$$P = \frac{Nx\mu_0 \int_0^{\pi} \cos\theta \, e^{x\cos\theta} \sin\theta \, d\theta}{V(e^x - e^{-x})} = \mu_0 \mathcal{N} \mathcal{L}(x) \tag{51}$$

where the function $\mathcal{L}$ is the **Langevin function**:

$$\mathcal{L}(x) = \frac{e^x + e^{-x}}{e^x - e^{-x}} - \frac{1}{x} \tag{52}$$

This function is plotted in Fig. 12.8.

When $\mu_0 \mathcal{E}^* \ll kT$, which corresponds to $x \ll 1$, and is the case at all normal temperatures and field strengths,

$$\mathcal{L}(x) \approx \tfrac{1}{3}x = \frac{\mu_0 \mathcal{E}^*}{3kT}$$

It follows that the permanent dipole moments of the molecules contribute

$$P \approx \frac{\mu_0^2 \mathcal{N} \mathcal{E}^*}{3kT} \tag{53}$$

The total polarization of a medium composed of polarizable polar molecules is therefore

$$P = \left(\alpha + \frac{\mu_0^2}{3kT}\right) \mathcal{N} \mathcal{E}^* \tag{54}$$

The development that led to eqns 12.47 and 12.48 can now be repeated, but the simplest procedure is simply to add in the additional terms representing the contribution of the polar molecules. In this way we obtain

$$\begin{aligned} \chi_e &= \frac{\{\alpha + (\mu_0^2/3kT)\}\mathcal{N}/\varepsilon_0}{1 - \{\alpha + (\mu_0^2/3kT)\}\mathcal{N}/3\varepsilon_0} \\ \varepsilon_r &= \frac{1 + 2\{\alpha + (\mu_0^2/3kT)\}\mathcal{N}/3\varepsilon_0}{1 - \{\alpha + (\mu_0^2/3kT)\}\mathcal{N}/3\varepsilon_0} \end{aligned} \tag{55}$$

A practical form of these expressions is obtained by replacing the number density by the mass density, ρ:

$$\mathcal{N} = \frac{N}{V} = \frac{N_A(m/M)}{V} = \frac{N_A \rho}{M}$$

where m is the mass of the sample, M is the molar mass of the molecules, and N_A is the Avogadro constant. Then, converting at the same time to polarizability volume (eqn 12.18), we find

$$\varepsilon_r = \frac{1 + 2\beta}{1 - \beta} \qquad \beta = \left(\frac{4\pi N_A \rho}{3M}\right)\left(\alpha' + \frac{\mu_0^2}{12\pi\varepsilon_0 kT}\right) \tag{56}$$

The dependence of the permittivity of a medium on the characteristics of the molecules of which it is composed can now be discussed in the same way as before, because we know how they determine the polarizability. Thus, we expect a medium to have a high relative permittivity if α is large and, if the molecules are polar, if their permanent dipole moment is large. Hence, media composed of molecules in which the electrons are relatively mobile (atoms with large numbers of electrons with low-lying energy levels) can be expected to have high relative permittivities.

12.9 Refractive index

The **refractive index**, n_r, is the ratio of the speed of light in a vacuum, c, to its speed in a medium, c_{med}:

$$n_r = \frac{c}{c_{med}} \tag{57}$$

It follows from Maxwell's equations (which describe the propagation of electromagnetic radiation, *Further information 20*), that

$$n_r = \varepsilon_r^{1/2} \tag{58}$$

Because we have an expression for the relative permittivity in terms of the molecular polarizability (eqn 12.48), we should now be in a position to calculate n_r and relate it to molecular properties. There is one simplification we can make, and one unavoidable complication.

The simplification is that the permanent electric dipole moment of a molecule is too sluggish to respond to the high-frequency alternation in the direction of the electric field in a light ray. A molecule needs about 10^{-12} s to tumble into a significantly new orientation, but the electric vector changes direction every 10^{-15} s for visible light. It follows that we can ignore the contribution of the permanent electric dipole to the permittivity, and use eqn 12.48 for both polar and nonpolar molecules. We shall suppose that the refractive index does not differ much from 1, and use the relations $(1-x)^{-1} \approx 1+x$ and $(1+x)^{1/2} \approx 1+\frac{1}{2}x$, which are valid if $x \ll 1$. As a result, the combination of eqn 12.48 and eqn 12.58 becomes

$$n_r = \left(\frac{1+(2\alpha\mathcal{N}/3\varepsilon_0)}{1-(\alpha\mathcal{N}/3\varepsilon_0)}\right)^{1/2} \approx 1 + \frac{\alpha\mathcal{N}}{2\varepsilon_0} \tag{59}$$

On making the same substitutions that led to eqn 12.56 we obtain

$$n_r \approx 1 + \frac{2\pi N_A \alpha' \rho}{M} \tag{60}$$

This expression shows that the refractive index increases with the polarizability volume and the density of the medium.

The complication is rather deeper and will take more work to resolve. The refractive index is a property relating to the response of the sample to an *oscillating* electric field. Therefore, we cannot use eqn 12.17 directly, because it was derived by using time-independent perturbation theory. We need to calculate $\alpha(\omega)$, the polarizability of a molecule exposed to an electric field that is oscillating at a frequency ω, and to do so we have to use time-dependent perturbation theory.

It is at this point that we shall use the alternative approach to the calculation of molecular properties mentioned in the introduction to this chapter: we shall calculate the expectation value of the electric dipole moment operator using the first-order perturbed wavefunctions. The calculation runs as follows.

The perturbation due to a field that lies in the z-direction and is oscillating at a frequency ω is

$$H^{(1)}(t) = -2\mu_z\mathcal{E}\cos\omega t \tag{61}$$

The factor of 2 is included by convention with an eye on future convenience. The expectation value of the z-component of the electric dipole moment is

$$\langle\mu_z\rangle = \int \Psi^*(t)\mu_z\Psi(t)\,\mathrm{d}\tau \tag{62}$$

where the time-dependent wavefunction is given by eqn 6.64 as

$$\Psi(t) = \psi_0\mathrm{e}^{-\mathrm{i}E_0t/\hbar} + \sum_n{}' a_n(t)\psi_n\mathrm{e}^{-\mathrm{i}E_nt/\hbar} \tag{63}$$

where, as usual, the prime signifies the omission of the term with $n = 0$ and to first order, $a_0(t) = 1$. It will be convenient to replace the wavefunctions ψ_n by the states $|n\rangle$, and we shall do so in the following. Because we are looking for the field-induced contribution to the electric dipole moment, we need to evaluate $\langle\mu_z\rangle$ to first-order in $\mathcal{E}$, which means that we must evaluate

$$\begin{aligned}\langle\mu_z\rangle &= \langle 0|\mu_z|0\rangle + \sum_n{}'\left\{\langle 0|\mu_z|n\rangle a_n(t)\mathrm{e}^{-\mathrm{i}\omega_{n0}t} + \langle n|\mu_z|0\rangle a_n^*(t)\mathrm{e}^{\mathrm{i}\omega_{n0}t}\right\}\\ &= \mu_{0,z} + \sum_n{}'\left\{\mu_{z;0n}a_n(t)\mathrm{e}^{-\mathrm{i}\omega_{n0}t} + \mu_{z;n0}a_n^*(t)\mathrm{e}^{\mathrm{i}\omega_{n0}t}\right\}\end{aligned} \tag{64}$$

where $\hbar\omega_{n0} = E_n - E_0$. Because we are working only to first order in the perturbation, terms such as $a_na_m^*$ have been ignored.

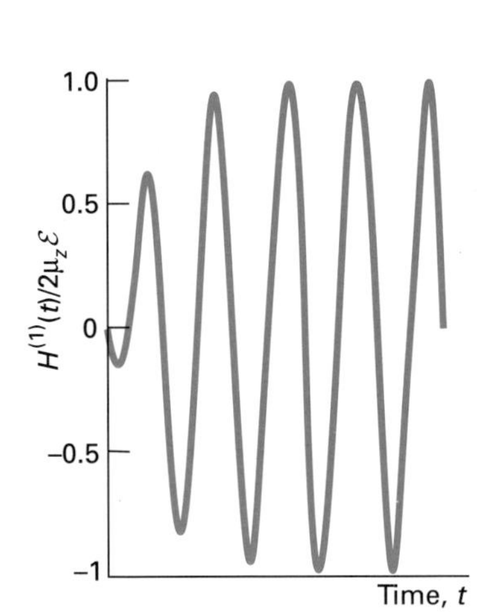

Fig. 12.9 The early stages of an exponentially switched oscillating perturbation.

One problem with this approach, as in all time-dependent perturbation calculations, is that when the perturbation is applied, it may result in the generation of transient oscillations of the electron density, which confuses the analysis. Therefore, we ensure that all transients have died away by starting to switch on the oscillating field long ago and allowing it to rise to full strength very slowly. The same procedure was adopted in Section 6.14, where we switched on a static perturbation; here we modify eqn 12.61 to

$$H^{(1)}(t) = -2\mu_z\mathcal{E}\left(1 - \mathrm{e}^{-t/\tau}\right)\cos\omega t = -\mu_z\mathcal{E}\left(1 - \mathrm{e}^{-t/\tau}\right)\left(\mathrm{e}^{\mathrm{i}\omega t} + \mathrm{e}^{-\mathrm{i}\omega t}\right)$$

where τ is the time-constant for switching on the perturbation. The early moments of this perturbation are illustrated in Fig. 12.9. Because we are interested in times that are very long compared with the switching time τ, we can set $t \gg \tau$ when we evaluate eqn 6.67 for the coefficients $a_n(t)$. We can also suppose that the perturbation is switched on very slowly in the sense that $|\tau(\omega \pm \omega_{n0})| \gg 1$. Then we obtain

$$\begin{aligned}a_n(t) &= \frac{1}{\mathrm{i}\hbar}\int_0^t H_{n0}^{(1)}(t)\mathrm{e}^{\mathrm{i}\omega_{n0}t}\,\mathrm{d}t\\ &= \frac{\mu_{z;n0}\mathcal{E}}{\hbar}\left\{\frac{\mathrm{e}^{\mathrm{i}(\omega+\omega_{n0})t}}{\omega + \omega_{n0}} - \frac{\mathrm{e}^{-\mathrm{i}(\omega-\omega_{n0})t}}{\omega - \omega_{n0}}\right\}\end{aligned}$$

It then follows, after some straightforward algebra, that

$$\langle \mu_z \rangle = \mu_{0z} + \frac{2}{\hbar}\sum_n{}' \left(\frac{\omega_{n0}|\mu_{z;n0}|^2}{\omega_{n0}^2 - \omega^2} \right) \times 2\mathcal{E}\cos\omega t \tag{65}$$

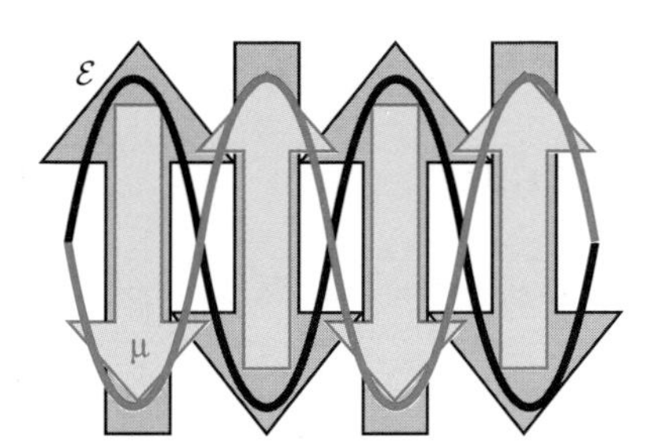

Fig. 12.10 When the frequency of the incident radiation is greater than the transition frequency of the molecule, the induced electric dipole moment is out of phase by 180°.

At this point we can compare this expression with

$$\langle \mu_z \rangle = \mu_{0z} + \alpha_{zz}(\omega) \times 2\mathcal{E}\cos\omega t + \ldots$$

(see eqn 12.8) and so derive an expression for the **dynamic polarizability**:

$$\alpha_{zz}(\omega) = \frac{2}{\hbar}\sum_n{}' \frac{\omega_{n0}|\mu_{z;n0}|^2}{\omega_{n0}^2 - \omega^2} \tag{66}$$

The mean dynamic polarizability, $\alpha(\omega)$, is the average of α_{xx}, α_{yy}, and α_{zz}, and so

$$\alpha(\omega) = \frac{2}{3\hbar}\sum_n{}' \frac{\omega_{n0}|\mu_{n0}|^2}{\omega_{n0}^2 - \omega^2} \tag{67}$$

where $|\mu_{n0}|^2 = \boldsymbol{\mu}_{0n} \cdot \boldsymbol{\mu}_{n0}$. Notice how this expression reduced to the static polarizability (eqn 12.17) when $\omega \to 0$. Furthermore, when the incident radiation has such a high frequency that $\omega^2 \gg \omega_{n0}^2$, we find

$$\begin{aligned} \alpha(\omega) &\simeq \frac{2}{3\hbar}\sum_n{}' \left(\frac{\omega_{n0}}{-\omega^2} \right) |\mu_{n0}|^2 \\ &\simeq -\frac{e^2}{m_e\omega^2}\sum_n{}' f_{n0} = -\frac{e^2 N_e}{m_e\omega^2} \end{aligned} \tag{68}$$

According to this expression, the polarizability goes to zero as $\omega \to \infty$ because the electrons cannot contribute to the induced moment if the field changes direction too quickly for them to follow. At high frequencies the polarizability is negative, which implies that the induced dipole moment is in the opposite direction to the instantaneous electric field (Fig. 12.10). This behaviour is an echo of the classical behaviour of a forced oscillator, which shifts in phase by 180° in advance of the driving force when the latter's frequency exceeds the natural frequency of the driven oscillator (Fig. 12.11).

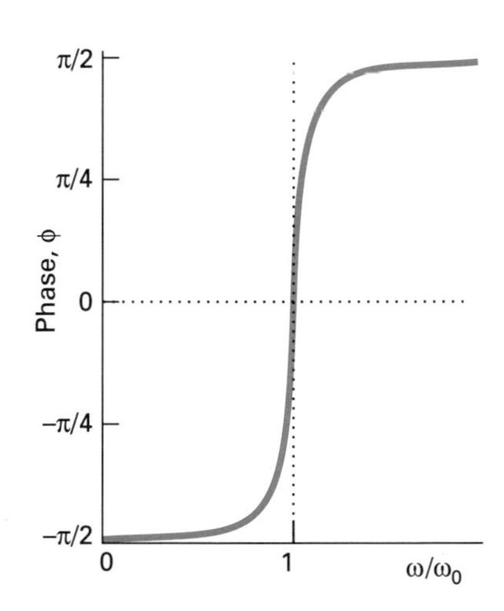

Fig. 12.11 The variation of the phase of a driven, damped harmonic oscillator as the driving frequency passes through resonance at the natural frequency of the oscillator. Note the change in phase by π.

Example 12.3 The dynamic polarizability of an oscillator

Calculate the dynamic polarizability of the oscillator used in Example 12.1 when it is exposed to a field of frequency ω applied along the x-axis of the oscillator.

Method. We need to use eqn 12.66 with z replaced by x. All the matrix elements are the same as in Example 12.1, and there are still only two terms in the sum. As in that example, we develop the equation for a general state of the oscillator; so the label 0 becomes v and n becomes v'. The frequency differences are $\omega_{v'v} = (v' - v)\omega_0$.

Answer. Substitution of the matrix elements into eqn 12.66 gives

$$\alpha_{xx}(\omega) = \frac{2}{\hbar}\sum_{v'}{}' \frac{\omega_{v'v}|\mu_{x;v'v}|^2}{\omega_{v'v}^2 - \omega^2} = \frac{2}{\hbar}\left\{\frac{\omega_0|\mu_{x;v+1,v}|^2}{\omega_0^2-\omega^2} - \frac{\omega_0|\mu_{x;v-1,v}|^2}{\omega_0^2-\omega^2}\right\}$$
$$= \frac{2}{\hbar}\left(\frac{\omega_0}{\omega_0^2-\omega^2}\right)(|\mu_{x;v+1,v}|^2 - |\mu_{x;v-1,v}|^2) = \frac{e^2}{m(\omega_0^2-\omega^2)}$$
$$= \frac{e^2}{k - m\omega^2}$$

Comment. This calculation is exact and reduces to the static polarizability calculated in Example 12.1 when $\omega = 0$. In this case the polarizability depends on the mass of the oscillator because the mass determines how rapidly it responds to the changing direction of the applied field. Note that if the effective mass of the oscillator is infinite, then the dynamic polarizability is zero at all frequencies greater than zero, but it still has a static polarizability. The polarizability is very small for finite-mass oscillators when $\omega \gg \omega_0$.

With the dynamic polarizability available, we can now complete the calculation of the refractive index, because all we need do is substitute eqn 12.67 into eqn 12.60:

$$n_\mathrm{r} \approx 1 + \left(\frac{2\pi N_\mathrm{A}\rho}{M}\right)\alpha'(\omega)$$
$$\approx 1 + \left(\frac{N_\mathrm{A}\rho}{3\hbar\varepsilon_0 M}\right)\sum_n{}' \frac{\omega_{n0}|\mu_{n0}|^2}{\omega_{n0}^2 - \omega^2} \qquad (69)$$

When the term $(2\pi N_\mathrm{A}\rho/M)\alpha(\omega)$ is not small enough for the approximations that led to eqn 12.59 to be used, we should use

$$n_\mathrm{r}^2(\omega) = \frac{1 + 2\alpha(\omega)\mathcal{N}/3\varepsilon_0}{1 - \alpha(\omega)\mathcal{N}/3\varepsilon_0} \qquad (70)$$

This expression is a version of the **Lorenz–Lorentz formula**:

$$\frac{n_\mathrm{r}^2 - 1}{n_\mathrm{r}^2 + 2} = \frac{\mathcal{N}\alpha(\omega)}{3\varepsilon_0} \qquad (71)$$

The right-hand side can be replaced by $4\pi\alpha'(\omega)N_\mathrm{A}\rho/3M$ in practical applications. Indeed, the Lorenz–Lorentz formula is normally expressed as

$$R_\mathrm{m} = \frac{M}{\rho}\left(\frac{n_\mathrm{r}^2 - 1}{n_\mathrm{r}^2 + 2}\right) \qquad (72)$$

where the **molar refractivity**, R_m, is

$$R_\mathrm{m} = \tfrac{4}{3}\pi N_\mathrm{A}\alpha'(\omega) \qquad (73)$$

The dimensions of the molar refractivity are the same as those of molar volume.

The advantage of concentrating on the molar refractivity is that it eliminates the molar mass and mass density dependence of the refractive index itself and focuses attention on the molecular property, the dynamic polarizability volume,

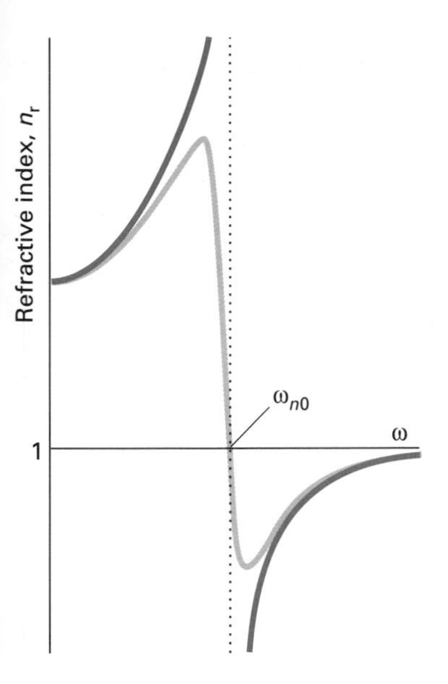

Fig. 12.12 The refractive index of a molecule close to a transition frequency.

$\alpha'(\omega)$. This property is more likely to be additive than the refractive index itself, in the sense that the refractivity of a molecule may be expressed, approximately at least, as the sum of the refractivities of its component atoms or groups. To some extent, this is confirmed, and tables of molecular refractivities have been compiled. The molar refractivity of the molecule as a whole is approximately the sum of its component refractivities, and the refractive index is then obtained by the appropriate manipulation of eqn 12.72.

Now consider the dispersion characteristics of the refractive index. We shall suppose that the density is always small enough for eqn 12.69 to be applicable as this simplifies the discussion. Suppose that ω is so close to one of the electronic transition frequencies of the molecule that its contribution dominates the frequency dependence as a whole (because the terms in the sum in eqn 12.69 are proportional to $1/(\omega_{n0}^2 - \omega^2)$). In this case

$$n_r \approx 1 + A\left(\frac{\omega_{n0}}{\omega_{n0}^2 - \omega^2}\right) \qquad A = \frac{N_A \rho |\mu_{n0}|^2}{3\hbar\varepsilon_0 M} \tag{74}$$

The frequency dependence of this expression is sketched in Fig. 12.12. We see that so long as $\omega < \omega_{n0}$, the refractive index is greater than 1 and increases as ω increases. This behaviour is a reflection of the effective degeneracy brought about by an oscillating perturbation, as described in Section 6.15, in which the *overall* difference in energy of the molecule and the field is close to zero. The increase of refractive index with frequency means that blue light is refracted more than red light. As a result, white light is 'dispersed' into its constituent colours when it passes through a prism. The term **dispersion** is borrowed from this behaviour and generalized to mean the frequency dependence of any property. The underlying cause of dispersion is the effective-degeneracy effect. At resonance, when $\omega = \omega_0$, eqn 12.74 appears to indicate an infinite refractive index. However, perturbation theory breaks down at this point and close to it, and the dispersion curve will be more like that shown as the pale line in Fig. 12.12.

It should be observed that the refractive index is less than 1 at frequencies $\omega > \omega_{n0}$. This conclusion appears to suggest that the radiation propagates at greater than the speed of light. However, a detailed analysis shows that it is the *phase* of the wave that propagates faster than c, and information cannot be propagated by phase alone. Hence, a refractive index $n_r < 1$ is not in conflict with special relativity. The origin of this very speedy propagation of phase of the radiation is related to the phase shift of the induced dipole moment when $\omega > \omega_0$, which was described above. In the present case, the incident radiation drives an induced dipole in a molecule, and that dipole has an advanced phase if $\omega > \omega_0$; that dipole generates a phase-advanced wave, and stimulates its neighbours. As a result, the phase of the incident wave propagates rapidly through the medium.

Optical activity

In **optical activity**, a plane-polarized wave of electromagnetic radiation is rotated as it passes through a medium. This behaviour can be traced to the **circular birefringence** of the medium, its possession of different refractive indices for left- and right-circularly polarized radiation. Circular birefringence is a special case of the property of **optical birefringence**, the possession of different refractive indices for radiation with different polarizations.

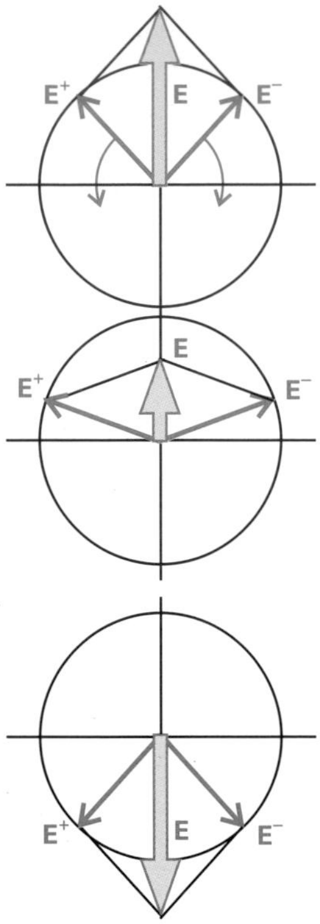

12.10 Circular birefringence and optical rotation

First, we shall deal with the relation between the angle of rotation of the beam as it propagates through the medium and the different refractive indices. Then we shall relate the difference in refractive indices to molecular properties by using perturbation theory in a similar manner to the previous section.

Consider the diagram in Fig. 12.13. It shows how a plane-polarized ray can be expressed as the superposition of two counter-rotating components $\mathbf{E}^+$ (left-circularly polarized light) and $\mathbf{E}^-$ (right-circularly polarized light). The expressions for the components in terms of the time and the location along the propagation direction (z) are

$$\mathbf{E}^{\pm} = \mathcal{E}\hat{\mathbf{i}}\cos\phi_{\pm} \pm \mathcal{E}\hat{\mathbf{j}}\sin\phi_{\pm} \tag{75}$$

with $\hat{\mathbf{i}}$ and $\hat{\mathbf{j}}$ unit vectors perpendicular to the propagation direction and

$$\phi_{\pm} = \omega t - \frac{2\pi z}{\lambda_{\pm}} \qquad \lambda_{\pm} = \frac{v_{\pm}}{\nu} = \frac{c}{n_{\pm}\nu} \tag{76}$$

Fig. 12.13 The resolution of a plane-polarized wave into two counter-rotating circularly polarized components.

The relation between the wavelength and frequency in this expression allows for the possibility that light of different senses of circular polarization propagates through the medium with different speeds, and so has different refractive indices n_+ and n_-. Because $\omega = 2\pi\nu$, we can write

$$\phi_{\pm} = \phi \mp \frac{\omega z \Delta n}{2c} \qquad \begin{cases} \phi = \omega t - n\omega z/c \\ n = \frac{1}{2}(n_+ + n_-) \\ \Delta n = n_+ - n_- \end{cases} \tag{77}$$

When the medium is not circularly birefringent, $\Delta n = 0$; then

$$\mathbf{E}^{\pm} = \mathcal{E}\hat{\mathbf{i}}\cos\phi \pm \mathcal{E}\hat{\mathbf{j}}\sin\phi$$

and the superposition of the two components gives a ray with electric vector

$$\mathbf{E} = \mathbf{E}^+ + \mathbf{E}^- = 2\mathcal{E}\hat{\mathbf{i}}\cos\phi \tag{78}$$

This field oscillates in the plane defined by the direction of propagation and the unit vector $\hat{\mathbf{i}}$. When the ray enters a circularly birefringent medium, one of the components propagates faster than the other and their phases diverge from one another. The superposition is now

$$\begin{aligned} \mathbf{E} = \mathbf{E}^+ + \mathbf{E}^- &= \mathcal{E}\{(\cos\phi_+ + \cos\phi_-)\hat{\mathbf{i}} + (\sin\phi_+ - \sin\phi_-)\hat{\mathbf{j}}\} \\ &= 2\mathcal{E}\left\{\hat{\mathbf{i}}\cos\left(\frac{z\omega\Delta n}{2c}\right) - \hat{\mathbf{j}}\sin\left(\frac{z\omega\Delta n}{2c}\right)\right\}\cos\phi \end{aligned} \tag{79}$$

This is still a plane-polarized ray, but its plane of polarization is rotated by

$$\Delta\theta = \frac{z\omega\Delta n}{2c} \tag{80}$$

to the original direction (Fig. 12.14). The sample is **dextrorotatory**, $\Delta\theta > 0$, if $n_+ > n_-$, and **laevorotatory**, $\Delta\theta < 0$, if $n_+ < n_-$.

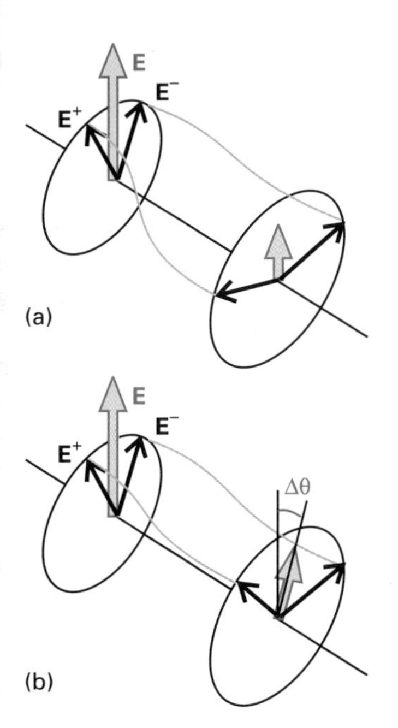

Fig. 12.14 (a) If the two circularly polarized components travel at the same speed through a medium, then their resultant remains plane-polarized in the original direction. (b) However, if one component is faster than the other, then the resultant rotates away from the plane of polarization of the incident ray.

The fundamental reason why the refractive indices are different for left- and right-circularly polarized radiation lies in the spatial variation of the electromagnetic field over the extent of the molecule. Because the molecules sample the electric fields in different ways, they have different polarizabilities, and hence different refractive indices. According to Maxwell's equations (*Further information 20*), the spatial variation of the electric field is proportional to the time variation of the magnetic field ($\partial \mathbf{E}/\partial x \propto \partial \mathbf{B}/\partial t$). Therefore, the part of the electric dipole, or the electric polarization $\mathbf{P}$, induced by the spatial variation of the field ($\mathbf{P} \propto \partial \mathbf{E}/\partial x$) can be expressed as a proportionality to the time variation of the magnetic field ($\mathbf{P} \propto \partial \mathbf{B}/\partial t = \dot{\mathbf{B}}$). It follows that when the spatial variation is taken into account, the total polarization of the medium should be written

$$\mathbf{P} = \mathcal{N}\alpha\mathbf{E} - \mathcal{N}\beta\dot{\mathbf{B}} \tag{81}$$

where β is a molecular characteristic (not the hyperpolarizability). That the polarization does in fact have a term proportional to the rate of change of the magnetic field will be confirmed more formally later. In this formulation, we are ignoring the fact that the effective electric field experienced by the molecules differs from the applied field, for that introduces a considerable complication: in other words, we are dealing with the optical activity of an isolated molecule.

We can see that we are on the right track. In the first place, the magnetic component of an electromagnetic field is perpendicular to the electric component. Therefore, whereas the term $\alpha\mathbf{E}$ corresponds to the induction of an electric dipole moment in the same plane as the electric vector, the term $\beta\dot{\mathbf{B}}$ corresponds to the induction of an electric moment in a plane parallel to $\mathbf{B}$ and hence perpendicular to $\mathbf{E}$. The resultant of these two dipole moments lies in a plane that is rotated from the direction of $\mathbf{E}$, with the result that the plane of polarization of the propagating ray is rotated. This conclusion is confirmed by solving the Maxwell equations for a medium with a polarization given by eqn 12.81 (see *Further information 20*); the calculation shows that in the presence of the β term the refractive indices of the medium are

$$n_{\pm} \approx 1 + \frac{\alpha\mathcal{N}}{2\varepsilon_0} \pm \frac{\omega\beta\mathcal{N}}{2c\varepsilon_0} \tag{82}$$

It follows that the difference in refractive indices is

$$\Delta n = \frac{\omega\beta\mathcal{N}}{c\varepsilon_0} \tag{83}$$

and that the angle of rotation after the radiation has passed through a length l of the medium is

$$\Delta\theta = \frac{\omega^2\beta\mathcal{N}l}{2c^2\varepsilon_0} = \tfrac{1}{2}\beta\mu_0\omega^2 l\mathcal{N} \tag{84}$$

In the second equality we have used $\varepsilon_0\mu_0 = 1/c^2$, where μ_0 is the vacuum permeability.

12.11 Magnetically induced polarization

The calculation of the angle of rotation now boils down to the calculation of β. That is, we must calculate the polarization of a medium in response to the changing magnetic component of the electromagnetic field. The strategy involves adapting the calculation of n_r, which was based on the perturbation $H^{(1)} = -\boldsymbol{\mu}\cdot\mathbf{E}(t)$, to the case in which

$$H^{(1)}(t) = -\boldsymbol{\mu}\cdot\mathbf{E}(t) - \mathbf{m}\cdot\mathbf{B}(t) \tag{85}$$

where $\mathbf{m}$ is the magnetic dipole moment operator for the molecule. For all cases of interest to us, $\mathbf{m} = \gamma_e\mathbf{l}$ (Section 7.3), where γ_e is the magnetogyric ratio of the electron and $\mathbf{l}$ is the orbital angular momentum operator. The precise form of the perturbation depends on which component of circular polarization we are considering, so we write

$$H_{\pm}^{(1)}(t) = -\boldsymbol{\mu}\cdot\mathbf{E}^{\pm}(t) - \mathbf{m}\cdot\mathbf{B}^{\pm}(t) \tag{86}$$

with

$$\mathbf{E}^{\pm}(t) = \mathcal{E}(\hat{\mathbf{i}}\cos\omega t \pm \hat{\mathbf{j}}\sin\omega t)$$
$$\mathbf{B}^{\pm}(t) = \mathcal{B}(\pm\hat{\mathbf{i}}\sin\omega t - \hat{\mathbf{j}}\cos\omega t)$$

The magnetic field vector is in step with the electric vector, but perpendicular to it.

The detailed form of the adiabatically switched hamiltonian is obtained by inserting the expressions for the fields into eqn 12.86 and including a factor of $1 - e^{-t/\tau}$ to represent the switching:

$$\begin{aligned} H_{\pm}^{(1)}(t) = &-\tfrac{1}{2}\mathcal{E}(1 - e^{-t/\tau})\Big\{(e^{i\omega t} + e^{-i\omega t})\mu_x \mp i(e^{i\omega t} - e^{-i\omega t})\mu_y\Big\} \\ &- \tfrac{1}{2}\mathcal{B}(1 - e^{-t/\tau})\{-(e^{i\omega t} + e^{-i\omega t})m_y \mp i(e^{i\omega t} - e^{-i\omega t})m_x\} \end{aligned} \tag{87}$$

From now on, we proceed just like in Section 12.9. The coefficients in the perturbed wavefunctions are

$$a_n^{\pm}(t) = \frac{1}{i\hbar}\int_0^t H_{\pm;n0}^{(1)}(t)e^{i\omega_{n0}t}\,dt$$

and the induced electric dipole moment is the expectation value of the operator using these perturbed wavefunctions. The result of the calculation is

$$\begin{aligned} \langle\boldsymbol{\mu}^{\pm}\rangle &= \boldsymbol{\mu}_0 + \sum_n{}'\{\boldsymbol{\mu}_{0n}a_n^{\pm}(t)e^{-i\omega_{n0}t} + \boldsymbol{\mu}_{n0}a_n^{\pm *}(t)e^{i\omega_{n0}t}\} \\ &= \frac{2}{\hbar}\,\mathrm{re}\sum_n{}'\Bigg\{\boldsymbol{\mu}_{0n}(\mathcal{E}\mu_{x;n0} - \mathcal{B}m_{y;n0})\left(\frac{\omega_{n0}\cos\omega t - i\omega\sin\omega t}{\omega_{n0}^2 - \omega^2}\right) \\ &\qquad \mp i\boldsymbol{\mu}_{0n}(\mathcal{E}\mu_{y;n0} + \mathcal{B}m_{x;n0})\left(\frac{i\omega_{n0}\sin\omega t - \omega\cos\omega t}{\omega_{n0}^2 - \omega^2}\right)\Bigg\} \end{aligned}$$

In the second line, we have supposed that the unperturbed molecule is nonpolar. All the unperturbed wavefunctions may be taken as real; therefore all the matrix elements $\boldsymbol{\mu}_{n0}$ are real ($\boldsymbol{\mu}$ is a real operator) whereas all the $\mathbf{m}_{n0}$ are imaginary (because $\mathbf{l}$ is an imaginary operator). The real part of the last expression is therefore

$$\begin{aligned}
\langle \boldsymbol{\mu}^{\pm} \rangle &= \frac{2}{\hbar}\,\mathrm{re}\sum_n{}' \left(\frac{\mathcal{E}\omega_{n0}}{\omega_{n0}^2 - \omega^2}\right)\boldsymbol{\mu}_{0n}(\mu_{x;n0}\cos\omega t \pm \mu_{y;n0}\sin\omega t) \\
&\quad - \frac{2}{\hbar}\,\mathrm{im}\sum_n{}' \left(\frac{\mathcal{B}\omega}{\omega_{n0}^2 - \omega^2}\right)\boldsymbol{\mu}_{0n}(m_{y;n0}\sin\omega t \pm m_{x;n0}\cos\omega t) \\
&= \frac{2}{\hbar}\,\mathrm{re}\sum_n{}' \left(\frac{\mathcal{E}\omega_{n0}}{\omega_{n0}^2 - \omega^2}\right)\boldsymbol{\mu}_{0n}\boldsymbol{\mu}_{n0}\cdot(\hat{\mathbf{i}}\cos\omega t \pm \hat{\mathbf{j}}\sin\omega t) \\
&\quad - \frac{2}{\hbar}\,\mathrm{im}\sum_n{}' \left(\frac{\mathcal{B}\omega}{\omega_{n0}^2 - \omega^2}\right)\boldsymbol{\mu}_{0n}\mathbf{m}_{n0}\cdot(\hat{\mathbf{j}}\sin\omega t \pm \hat{\mathbf{i}}\cos\omega t) \\
&= \frac{2}{\hbar}\,\mathrm{re}\sum_n{}' \left(\frac{\omega_{n0}}{\omega_{n0}^2 - \omega^2}\right)\boldsymbol{\mu}_{0n}\boldsymbol{\mu}_{n0}\cdot\mathbf{E}^{\pm}(t) \\
&\quad - \frac{2}{\hbar}\,\mathrm{im}\sum_n{}' \left(\frac{1}{\omega_{n0}^2 - \omega^2}\right)\boldsymbol{\mu}_{0n}\mathbf{m}_{n0}\cdot\mathbf{B}^{\pm}(t)
\end{aligned} \tag{88}$$

When this expression is compared with eqn 12.8 (after multiplication by $\mathcal{N}$), we obtain

$$\boldsymbol{\beta} = \frac{2}{\hbar}\,\mathrm{im}\sum_n{}' \frac{\boldsymbol{\mu}_{0n}\mathbf{m}_{n0}}{\omega_{n0}^2 - \omega^2} \tag{89}$$

We can now readily pick out the β_{xx}, β_{yy}, and β_{zz} components of $\boldsymbol{\beta}$, and hence arrive at an expression for its rotational average in solution:

$$\beta = \frac{2}{3\hbar}\,\mathrm{im}\sum_n{}' \frac{\boldsymbol{\mu}_{0n}\cdot\mathbf{m}_{n0}}{\omega_{n0}^2 - \omega^2} \tag{90}$$

We are now at the end of the calculation, because we have seen how to express the angle of optical rotation in terms of β (eqn 12.84). By combining that equation with eqn 12.90 we obtain the **Rosenfeld equation**:

$$\Delta\theta = \left(\frac{\mathcal{N}l\mu_0}{3\hbar}\right)\sum_n{}' \frac{\omega^2 R_{n0}}{\omega_{n0}^2 - \omega^2} \tag{91}$$

where R_{n0} is the **rotational strength** of a transition:

$$R_{n0} = \mathrm{im}\,\boldsymbol{\mu}_{0n}\cdot\mathbf{m}_{n0} \tag{92}$$

It follows, that to discuss the optical activities of molecules, we need to investigate the properties of their rotational strengths.

12.12 Rotational strength

The first property of molecules that we need to know is that their rotational strengths are necessarily zero if they possess an axis of improper rotation (S_n, Section 5.1). The symmetry argument is based on the fact that the electric dipole operator transforms as translations whereas the magnetic moment operator transforms as rotations. In groups that have an S_n symmetry element, no component of translation and rotation belongs to the same symmetry species, so the product of matrix elements in the definition of rotational strength does not transform as the totally symmetric irreducible representation

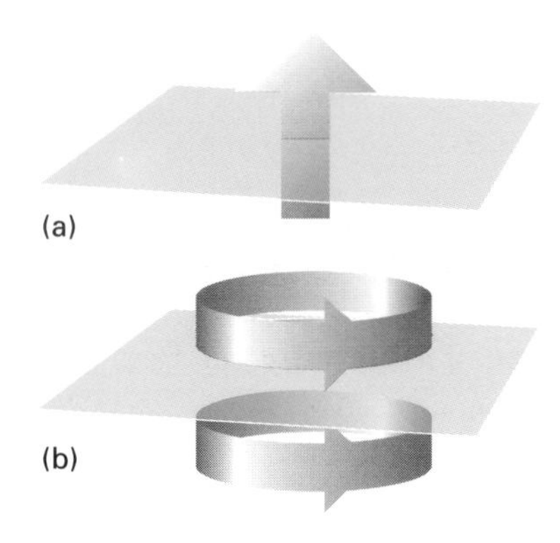

Fig. 12.15 (a) Under reflection, an electric dipole moment changes sign but (b) a magnetic dipole moment (which can be treated as a rotation) does not.

of the group, and hence must be zero. The special cases of improper rotations are S_1, which is equivalent to a mirror plane, and S_2, which is equivalent to an inversion. Under a reflection, μ_q and m_q have different symmetries (Fig. 12.15), and so the rotational strength changes sign. Similarly, under inversion, translations change sign but rotations do not; so in this case too, the rotational strength changes sign. Because the rotational strength cannot change sign under a symmetry transformation of a molecule, it must be equal to zero for molecules with a mirror plane or a centre of inversion.

The second property that stems from symmetry is that if two molecules form an **enantiomeric pair**, in the sense that one is the mirror image of the other, then they will have equal and opposite rotational strengths. As a result, they will rotate light of a given frequency in equal but opposite directions. When a reflection operation is applied to the rotational strength, it changes sign (as we have seen). However, the reflection converts one enantiomer into the other.

A third property stems from the following sum rule:

$$\begin{aligned}\sum_n R_{n0} &= \mathrm{im}\sum_n \boldsymbol{\mu}_{0n}\cdot\mathbf{m}_{n0} = \mathrm{im}\sum_n \langle 0|\boldsymbol{\mu}|n\rangle\cdot\langle n|\mathbf{m}|0\rangle \\ &= \mathrm{im}\,\langle 0|\boldsymbol{\mu}\cdot\mathbf{m}|0\rangle = 0\end{aligned} \tag{93}$$

The last equality stems from the relation

$$\boldsymbol{\mu}\cdot\mathbf{m}\propto\mathbf{r}\cdot\mathbf{l}\propto\mathbf{r}\cdot(\mathbf{r}\times\mathbf{p}) = (\mathbf{r}\times\mathbf{r})\cdot\mathbf{p}\equiv 0$$

This sum rule has the important consequence that the angle of optical rotation tends to zero at both high and low frequencies. At very high frequencies ($\omega^2 \gg \omega_{n0}^2$), the rotation angle is

$$\begin{aligned}\Delta\theta &\simeq \left(\frac{\mathcal{N}l\mu_0}{3\hbar}\right)\sum_n{}'\left(\frac{\omega^2 R_{n0}}{-\omega^2}\right) \\ &\simeq -\left(\frac{\mathcal{N}l\mu_0}{3\hbar}\right)\sum_n{}' R_{n0} = 0\end{aligned}$$

Although the sum omits $n = 0$, the omitted term R_{00} is zero because it is the imaginary part of the product of two expectation values, which are real (a property of hermitian operators, Section 1.15). At the other extreme of frequency, when $\omega^2 \ll \omega_{n0}^2$, we have

$$\Delta\theta \approx \left(\frac{\mathcal{N}l\mu_0}{3\hbar}\right)\sum_n{}'\frac{\omega^2 R_{n0}}{\omega_{n0}^2} \approx 0$$

on account of the ω^2 factor in the numerator.

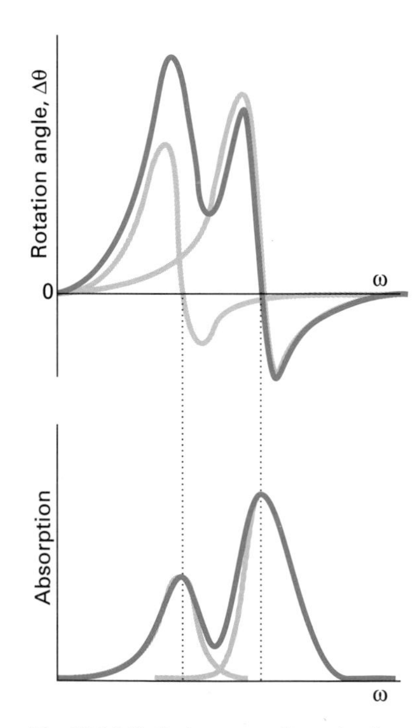

Fig. 12.16 Optical rotatory dispersion in the vicinity of two absorption bands.

The variation of the angle of rotation with frequency is called **optical rotatory dispersion** (ORD). A typical ORD curve is shown in Fig. 12.16. The rotation is close to zero at frequencies far from absorption bands, but may become quite large close to an absorption where $\omega_{n0}^2 - \omega^2$ approaches zero. The rotation does not actually rise to infinity as eqn 12.91 suggests because perturbation theory fails in this region and special techniques have to be used instead. When the incident frequency is close to an absorption frequency (for the $k \leftarrow 0$ transition, for instance), that transition's contribution to the optical rotation dominates and the angle of rotation is given by

$$\Delta\theta \approx \frac{\mathcal{N} l \mu_0 \omega^2 R_{k0}}{3\hbar(\omega_{k0}^2 - \omega^2)} \tag{94}$$

The area under the dispersion curve in this region can then be used to estimate the value of R_{k0} in much the same way as the area under an absorption curve is used to determine the oscillator strength (see *Further information 17*).

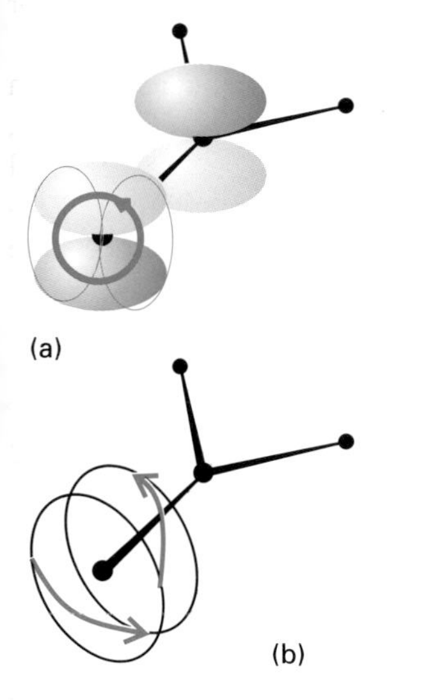

Fig. 12.17 (a) The rotational character of a $\pi^* \leftarrow n$ transition and (b) its helical character when the chromophore is perturbed by the adjacent groups in a chiral molecule.

Much work has been put into the estimation of rotational strengths of molecular transitions and the transitions of chromophores in chiral environments. The carbonyl group has received a lot of attention, and we shall consider it briefly to illustrate the basic ideas and difficulties. We saw in Section 11.7 that the transition in the region of 290 nm in carbonyl compounds can be ascribed to the $\pi^* \leftarrow n$ transition of the carbonyl chromophore. The nonbonding orbital n is almost pure $\mathrm{O2}p_y$ and the π^*-orbital is built from $2p_x$ orbitals on the C and O atoms. The transition is electric-dipole forbidden in a pure C_{2v} environment, but it is magnetic-dipole allowed because the rotation of $\mathrm{O2}p_x$ into $\mathrm{O2}p_y$ can be brought about by the operator $m_z \propto l_z$, which transforms as a rotation about the z-axis. The motion of the electron density during the transition can be thought of as describing a circle around the z-axis. Because it is electric-dipole forbidden, the transition has no rotational strength because $\boldsymbol{\mu}_{0k} \cdot \mathbf{m}_{k0} = 0$. However, we should take into account the possibility that the local environment of the carbonyl group may distort its orbitals. If the environment causes the migration of electron to follow a helical path (Fig. 12.17), then it can acquire a rotational strength.

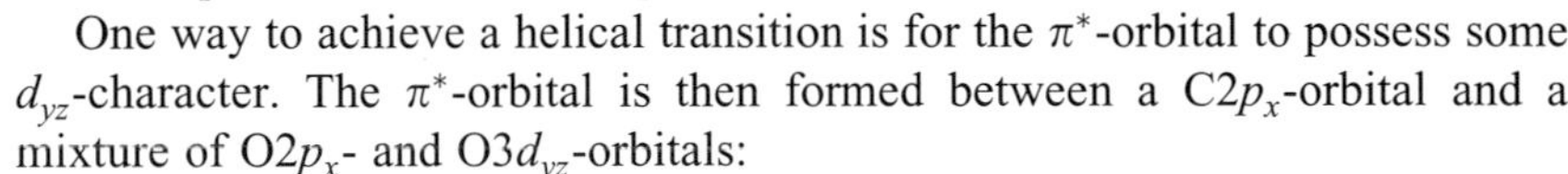

One way to achieve a helical transition is for the π^*-orbital to possess some d_{yz}-character. The π^*-orbital is then formed between a $\mathrm{C2}p_x$-orbital and a mixture of $\mathrm{O2}p_x$- and $\mathrm{O3}d_{yz}$-orbitals:

$$\psi(\pi^*) = c_1\phi(\mathrm{C2}p_x) + c_2\phi(\mathrm{O2}p_x) + c_3\phi(\mathrm{O3}d_{yz})$$

Now the transition has some electric-dipole character parallel to the z-axis as well as some magnetic dipole character around that axis:

$$\begin{aligned}\langle\pi^*|\mu_z|n\rangle &= c_3^*\langle\mathrm{O3}d_{yz}|\mu_z|\mathrm{O2}p_y\rangle \\ \langle\pi^*|m_z|n\rangle &= c_1^*\langle\mathrm{O2}p_x|m_z|\mathrm{O2}p_y\rangle\end{aligned} \tag{95}$$

and both matrix elements may be nonzero. The rotational strength now has a magnitude proportional to $c_1^* c_3^*$, and so the helically distorted carbonyl group is optically active. You should also note that the presence of the d-orbital component removes the plane of symmetry of the group, and so the group becomes chiral and potentially optically active.

Example 12.4 The estimation of rotational strengths

Suppose that there is a single centre to which an electron is confined, and that in the ground state it occupies a pure $2p_y$-orbital but in the upper state the orbital is a mixture of the form $|1\rangle = |2p_x\rangle\cos\zeta + |3d_{yz}\rangle\sin\zeta$, where ζ is a parameter (we encountered this parametrization of normalized two-component superpositions in Section 6.1) and the orbitals are Slater orbitals (Section 7.14). Evaluate the rotational strength of the transition as a function of the parameter ζ.

Method. The expression for the rotational strength is given in eqn 12.92. In this model, only the z-components contribute. For the matrix elements of $m_z = \gamma_e l_z$ we use $l_z \propto \partial/\partial\phi$ and recognize that $p_x \propto \cos\phi$ and $p_y \propto \sin\phi$. For the matrix elements of $\mu_z = -ez$, write $z = r\cos\theta$ and use the form of the STOs specified in Section 7.14.

Answer. For the matrix elements of l_z we use

$$l_z\psi(2p_y) = \frac{\hbar}{\mathrm{i}}\frac{\partial}{\partial\phi}f(r)\sin\theta\sin\phi = -\mathrm{i}\hbar f(r)\sin\theta\cos\phi = -\mathrm{i}\hbar\psi(2p_x)$$

From this relation, it follows that

$$\begin{aligned}\langle 1|m_z|0\rangle &= \gamma_e\{\langle 2p_x|l_z|2p_y\rangle\cos\zeta + \langle 3d_{yz}|l_z|2p_y\rangle\sin\zeta\} \\ &= \gamma_e(-\mathrm{i}\hbar)\langle 2p_x|2p_x\rangle\cos\zeta = \mathrm{i}\mu_B\cos\zeta\end{aligned}$$

(We have introduced the Bohr magneton through $\mu_B = -\gamma_e\hbar$.) For the electric transition dipole we need the explicit form of the orbitals, and will use

$$\psi(2p_y) = \left(\frac{3}{4\pi}\right)^{1/2}\sin\theta\sin\phi\left(\frac{2^5\zeta_p^5}{4!}\right)^{1/2} r\mathrm{e}^{-\zeta_p r} \qquad \zeta_p = \frac{Z_p^*}{n_p a_0}$$

$$\psi(3d_{yz}) = \tfrac{1}{2}\left(\frac{15}{4\pi}\right)^{1/2}\sin 2\theta\sin\phi\left(\frac{2^7\zeta_d^7}{6!}\right)^{1/2} r^2\mathrm{e}^{-\zeta_d r} \qquad \zeta_d = \frac{Z_d^*}{n_d a_0}$$

The matrix element evaluates to

$$\begin{aligned}\langle 0|\mu_z|1\rangle &= -e\langle 2p_y|r\cos\theta|3d_{yz}\rangle\sin\zeta \\ &= -ea_0\left(\frac{2^6(6\zeta_p^5\zeta_d^7)^{1/2}}{(\zeta_p+\zeta_d)^7}\right)\sin\zeta\end{aligned}$$

Therefore, from eqn 12.92 with $2\sin\zeta\cos\zeta = \sin 2\zeta$, we find

$$R_{10} = -ea_0\mu_B\left(\frac{2^5(6\zeta_p^5\zeta_d^7)^{1/2}}{(\zeta_p+\zeta_d)^7}\right)\sin 2\zeta$$

Comment. The rotational constant is greatest when $\zeta = \pi/4$ or $5\pi/4$. For an O atom, $\zeta_p = 2.25/a_0$ and $\zeta_d = 0.33/a_0$, and then

$$R_{10} = (1.30\times 10^{-54}\ \mathrm{C^2\,m^3\,s^{-1}})\times\sin 2\zeta$$

The principal difficulty with this kind of calculation is the estimation of the extent of distortion induced in a chromophore by the asymmetry of its environment. This delicate problem can be explored in the references in *Further reading*.

Problems

12.1 The polarizability volume of tetrachloromethane is $1.05 \times 10^{-29}\,\mathrm{m}^3$. Calculate (a) the magnitude of the dipole moment induced by an electric field of strength $10\,\mathrm{kV\,m^{-1}}$, (b) the change in molar energy.

12.2 Model an atom by an electron in a one-dimensional box of length L (there is assumed to be an 'invisible' positive charge at the centre of the box which provides the positive end of the dipole but does not affect the wavefunctions). Calculate the polarizability of the system parallel to its length. *Hint.* Use eqn 12.14; the wavefunctions are given in eqn 2.33. The procedure and results of Problems 6.4 and 6.5 can be used.

12.3 Repeat the calculation in Problem 12.2, but use the closure approximation. Explore the validity of using the value of ΔE calculated in Problem 6.10.

12.4 Evaluate the polarizability and polarizability volume of a hydrogen atom; for simplicity, confine the perturbation sum to the $2p$-orbitals. Explore the consequences of making the closure approximation, and compare the calculations with the approximate expression, eqn 12.25.

12.5 Devise a variational calculation of the polarizability of the hydrogen atom. *Hint.* A simple procedure would be to take as a trial function the linear combination $c_{1s}\psi_{1s} + c_{2p_z}\psi_{2p_z}$ (the basis could be lengthened in a more sophisticated treatment) or alternatively $(1 + az)\psi_{1s}$ with a the variation parameter. The hamiltonian is $H = H_0 + ez\mathcal{E}$ in each case. Find the best energy and identify α_{zz}. The experimental value of α'_{zz} is $6.6 \times 10^{-31}\,\mathrm{m}^3$.

12.6 Establish a perturbation theory expression for the components of the first hyperpolarizability of a molecule. *Hint.* Refer to eqn 12.9. You need to return to Chapter 6 and to derive the following expression for the third-order correction to the energy:

$$E^{(3)} = \sum_{mn}{}' \frac{H^{(1)}_{0m}H^{(1)}_{mn}H^{(1)}_{n0}}{(E_m - E_0)(E_n - E_0)} - H^{(1)}_{00}\sum_{n}{}' \frac{H^{(1)}_{0n}H^{(1)}_{n0}}{(E_n - E_0)^2}$$

12.7 Show group theoretically that in a tetrahedral molecule (a) the mean hyperpolarizability is zero, (b) the only nonzero components are β_{xyz} and the permutations of its indices. *Hint.* The mean is defined as $\frac{3}{5}(\beta_{xxz} + \beta_{yyz} + \beta_{zzz})$; and so (b) implies (a). For (b) consider the symmetry characteristics of $E = -(1/3!)\sum_{abc}\beta_{abc}\mathcal{E}_a\mathcal{E}_b\mathcal{E}_c$, the generalization of eqn 12.10.

12.8 Evaluate the first hyperpolarizability of a one-dimensional system of two charges $+e$ and $-e$ bound together by a spring of force constant k, the electric field being applied parallel to the x-axis. *Hint.* Use the matrix elements set out in Example 10.3; the result can be obtained by inspection.

12.9 Prove the sum rule

$$\sum_f x_{mf}x_{fm}\omega_{fn} = (\hbar/2m_e)\delta_{mn} + \tfrac{1}{2}\omega_{mn}(x^2)_{mn}$$

Hint. Consider the matrix elements of the commutator $[H, x^2]$.

12.10 Use the closure expressions to estimate the contribution to the polarizability of a carbon atom of one of its $2p$-electrons when the field is applied (a) parallel, (b) perpendicular to the axis. Assess the contributions of the $1s$-electrons and the $2s$-electrons, and estimate the total mean polarizability by adding all the contributions. *Hint.* Use Slater atomic orbitals and eqn 12.20. The $2s, 2p$ energy separation is about $7.5 \times 10^4\,\mathrm{cm}^{-1}$; the first ionization energy corresponds to 11.264 eV. The energies of the $1s$-electrons can be estimated by regarding them as hydrogenic.

12.11 The oscillator strength of a transition at about 160 nm in ethene is about 0.3. Estimate the mean polarizability volume of the molecule. (The experimental value is $4.22 \times 10^{-30}\,\mathrm{m}^3$.)

12.12 Deduce an expression for the refractive index of a gas of free electrons. *Hint.* Take the limit of eqn 12.69 when $\omega_{\mathrm{fi}}^2 \gg \omega_{n0}^2$ and refer to eqn 12.68. This calculation leads to the *Thomson formula* for the refractive index.

12.13 A region of interstellar space contained a diffuse gas of hydrogen atoms at a number density of $1 \times 10^5\ \text{m}^{-3}$. What is the refractive index for visible (590 nm) light in the region?

12.14 Consider two particles, each in a one-dimensional box, with the centres of the boxes separated by a distance R. Each system may be regarded as a model of an atom in the same sense as in Problem 12.2. Calculate the dispersion energy when the boxes are (a) in line, (b) broadside on. *Hint*. Base the calculation on eqn 12.28, noting that the dipole moment operators have only one component in a one-dimensional system. Much of the calculational work has been done in Problem 12.2.

12.15 Investigate the usefulness of the closure approximation in the calculation of the dispersion energy of the system described in Problem 12.14. What values of ΔE_{A} and ΔE_{B} should be used?

12.16 Estimate the dispersion energy between two hydrogen atoms. Use the experimental value of α' given in Problem 12.5.

12.17 Devise a variational calculation of the dispersion interaction between two hydrogen atoms. Start by using the trial functions suggested in Problem 12.5, but note that the dipolar hamiltonian also introduces distortions perpendicular to the line of centres of the atoms; ignore this initially, but include it in an improved trial function. The hamiltonian to use in the evaluation of the Rayleigh ratio (or the secular determinant, depending on the trial function) is $H_{\text{A}} + H_{\text{B}} + H^{(1)}$, where $H^{(1)}$ is given in eqn 12.26.

12.18 Evaluate the rotational strength of a transition of an electron from a $2p_x$-orbital to a $2p_z, 3d_{xy}$-hybrid orbital. Assume the orbitals are on a carbon atom. Estimate the optical rotation angle for 590 nm light. *Hint*. Follow Example 12.4, with changes of detail. For carbon, take $\zeta_p = 1.95a_0$ and $\zeta_d = 0.33a_0$, and use $\lambda_{k0} = 200\,\text{nm}$.

12.19 An electron is bound to a nucleus and undergoes harmonic vibrations in three dimensions, the frequencies being ω_X, ω_Y, and ω_Z. It is subjected to a perturbation of the form $H^{(1)} = Axyz$. Calculate the rotational strength and the optical rotation angle to first order in the parameter A. *Hint*. Base the answer on eqn 12.92, evaluating the matrix elements using the first-order perturbed wavefunctions, eqn 6.22. Express all the operators (including the angular momentum operators) via eqn 4.3 in terms of annihilation and creation operators, and use the matrix elements established in Example 10.3.

Further reading

Classical electrodynamics. J.D. Jackson; Wiley, New York (1975).

Electromagnetic fields. R.K. Wangsness; Wiley, New York (1986).

The theory of the electric and magnetic properties of molecules. D.W. Davies; Wiley, New York (1967).

Theory of electric polarization. Vols 1 and 2. C.J.F. Böttcher, O.C. Van Belle, P. Bordewijk, and A. Rip; Elsevier, Amsterdam (1978).

Intermolecular forces: their origin and determination. G.C. Maitland, M. Rigby, E.B. Smith, and W.A. Wakeham; Oxford University Press, Oxford (1981).

The theory of optical activity. D.J. Caldwell and H. Eyring; Wiley–Interscience, New York (1971).

The molecular basis of optical activity: Optical rotatory dispersion and circular dichroism. E. Charney; Wiley, New York (1979).

Molecular light scattering and optical activity. L.D. Barron; Cambridge University Press, Cambridge (1982).

Molecular optical activity and the chiral discriminations. S.F. Mason; Cambridge University Press, Cambridge (1982).

Optical polarization of molecules. M. Auzinsh and R. Ferber; Cambridge University Press, Cambridge (1995).

13 The magnetic properties of molecules

The difference between electric and magnetic perturbations is that whereas the former stretch a molecule, the latter twist it (this will be demonstrated explicitly in due course). The effect of a twisting perturbation is to induce electronic currents that circulate through the framework of the molecule. These currents give rise to their own magnetic fields. One effect is to modify the magnetic flux density in the material. If the flux density is increased beyond that due to the applied field alone, then the substance is classified as **paramagnetic**. If the flux density is reduced, then the substance is classified as **diamagnetic**. The latter is the much more common property. If there are unpaired electrons present in the molecule, then those spins may interact with the local currents induced by the applied field, and give rise to the ***g*-value** of electron spin resonance (ESR or EPR). Similarly, magnetic nuclei can also interact with the induced electronic currents, and this interaction is responsible for the **chemical shift** of nuclear magnetic resonance (NMR). A nuclear spin can itself give rise to electronic currents in a molecule, and the interaction of this nucleus-induced current with another magnetic nucleus is responsible for the **fine structure** in NMR.

We shall introduce a number of ways of discussing the magnetic properties of materials, and then apply them to the calculation of some of these properties.

The description of magnetic fields

We shall assume that the description of the magnetic field is largely unfamiliar, and introduce some of the concepts involved. One of these concepts, the 'vector potential', is of the greatest importance for this chapter, because it is at the root of the formulation of perturbation hamiltonians.

13.1 The magnetic susceptibility

The electric properties discussed in Chapter 12 have analogues in magnetism. In particular, a molecule may possess a **permanent magnetic dipole moment**, $\mathbf{m}_0$. It may also acquire a contribution to its total magnetic moment by virtue of an applied magnetic field and its **magnetizability**, ξ. A bulk sample subjected to a magnetic field of strength $\mathcal{H}$ (which is usually expressed in ampere per metre, A m^{-1}) acquires a **magnetization**, $\mathcal{M}$, just as a dielectric medium acquires a polarization in an electric field. We write

$$\mathbf{M} = \chi \mathbf{H} \tag{1}$$

where χ is the dimensionless **volume magnetic susceptibility**. There are a number of advantages obtained by expressing the susceptibility as the **molar magnetic susceptibility**, χ_m:

$$\chi_m = \chi V_m \tag{2}$$

where V_m is the molar volume of the sample. The units of molar susceptibility are the same as those of molar volume.

The **magnetic induction**, or 'flux density', $\mathcal{B}$, depends on the magnetic field strength, and we write

$$\mathcal{B} = \mu\mathcal{H} \tag{3}$$

where μ is the **magnetic permeability** of the medium. Equation 13.3 and a number of the equations that follow are normally written as vector equations, but we shall suppose that the magnetic induction is parallel to the inducing field, and so express everything in terms of magnitudes. The magnetic permeability is normally expressed as a multiple of the **vacuum permeability**, μ_0:

$$\mu = \mu_r\mu_0 \tag{4}$$

where, by definition, $\mu_0 = 4\pi \times 10^{-7}\,\mathrm{N\,A^{-2}}$, and μ_r is the **relative permeability** of the medium. Because the flux density can be regarded as arising from the applied field (through a contribution equal to $\mu_0\mathcal{H}$) together with a correction arising from the magnetization that the applied field has induced, we can write

$$\mathcal{B} = \mu_0(\mathcal{H} + \mathcal{M}) = \mu_0(1 + \chi)\mathcal{H} \tag{5}$$

It follows that

$$\mu_r = 1 + \chi \tag{6}$$

which is the magnetic analogue of the relation between the electric susceptibility and the permittivity (eqn 12.43).

The magnetic susceptibility may be either positive or negative. When $\chi < 0$, the magnetization opposes the applied field and the magnetic induction in the medium is lower than it would be in a vacuum; such materials are classified as **diamagnetic**. When $\chi > 0$, the magnetization adds to the applied field and increases the magnetic induction inside the material. Such substances are called **paramagnetic**.[1]

13.2 Paramagnetism

The magnetization of a medium is its magnetic-dipole density (recall the analogous interpretation of the polarization in Section 12.7). Therefore, we can write

$$\mathbf{M} = \mathcal{N}\langle\mathbf{m}\rangle \tag{7}$$

where $\mathcal{N}$ is the number density of molecules and $\langle\mathbf{m}\rangle$ is the mean magnetic dipole. There are two contributions to the latter. One is a contribution from the permanent magnetic dipole moments of the molecules. Their contribution depends on the orientating effect of the applied field as expressed through the Boltzmann distribution. For a field in the z-direction, the energy of interaction of a magnetic dipole is $-m_z\mathcal{B}$. It follows from exactly the same argument as we presented in Section 12.8 that the Boltzmann-weighted average of m_z in a

[1] The names come from the behaviour of a long, thin cylinder of the material which, if supported in the field of a magnet of small cross-section, tends to lie across (*dia* means across in Greek) the field so as to minimize its energy. A paramagnetic substance would tend to lie parallel to the field (*para* means beside or along in Greek).

sample at a temperature T is

$$\langle m_z \rangle = m_0 \mathcal{L}(x) \qquad x = \frac{m_0 \mathcal{B}}{kT} \tag{8}$$

where $\mathcal{L}$ is the Langevin function (eqn 12.52). The magnetic induction $\mathcal{B}$ is playing the role here of the total effective electric field $\mathcal{E}^*$ in the electrical case. The magnetization of the sample is therefore

$$\mathcal{M} = m_0 \mathcal{N} \mathcal{L}(x) \approx \frac{\mathcal{N} m_0^2 \mathcal{B}}{3kT} \tag{9}$$

where we have assumed that $m_0 \mathcal{B} \ll kT$, which implies $x \ll 1$ (which is almost always true). Then, by combining eqns 13.1 and 13.5, we obtain

$$\mathcal{M} = \frac{1}{\mu_0} \left(\frac{\chi}{1+\chi} \right) \mathcal{B} \approx \frac{\chi}{\mu_0} \mathcal{B} \tag{10}$$

provided that $\chi \ll 1$. It follows that the permanent moment contributes

$$\chi = \frac{\mu_0 m_0^2 \mathcal{N}}{3kT} \tag{11}$$

to the magnetic susceptibility. This contribution is positive, so the permanent moments contribute to the paramagnetic susceptibility.

The last expression depends on the number density of the sample. However, if we note that

$$\mathcal{N} V_{\mathrm{m}} = \frac{N V_{\mathrm{m}}}{V} = \frac{n N_{\mathrm{A}} V_{\mathrm{m}}}{n V_{\mathrm{m}}} = N_{\mathrm{A}}$$

where N_{A} is the Avogadro constant and n is the amount of substance, then we see that the molar susceptibility is simply

$$\chi_{\mathrm{m}} = \frac{\mu_0 m_0^2 N_{\mathrm{A}}}{3kT} \tag{12}$$

independent of the number density. This independence is the basic reason for introducing the molar susceptibility. This expression has the form of the **Curie law** for the magnetic susceptibility of paramagnetic substances:

$$\chi_{\mathrm{m}} = \frac{C}{T} \qquad C = \frac{\mu_0 m_0^2 N_{\mathrm{A}}}{3k} \tag{13}$$

All that remains now is to estimate the magnitude of the permanent magnetic moment. That is easy when there is no orbital contribution, for then we have **spin-only paramagnetism**, with the magnetic moment arising solely from the electron spin. If the spin quantum number is S, the spin magnetic moment is given by

$$m_0^2 = S(S+1) g_{\mathrm{e}}^2 \mu_{\mathrm{B}}^2 \tag{14}$$

where μ_{B} is the Bohr magneton (Section 7.3). The spin-only paramagnetic susceptibility is therefore given by eqn 13.13 with

$$C = \frac{S(S+1) g_{\mathrm{e}}^2 \mu_0 \mu_{\mathrm{B}}^2 N_{\mathrm{A}}}{3k} \tag{15}$$

For $S = \frac{1}{2}$, we have $C \approx 4.7 \times 10^{-6}\,\mathrm{m^3\,K\,mol^{-1}}$, so at 300 K, $\chi_\mathrm{m} \approx 1.6 \times 10^{-8}\,\mathrm{m^3\,mol^{-1}}$ (or $16\,\mathrm{mm^3\,mol^{-1}}$).

The spin-only formula is applicable when the orbital angular momentum of the electrons makes no contribution: we say that the momentum is **quenched**. Orbital angular momentum is quenched when the electrons are described by real wavefunctions: if the wavefunctions are real, then by hermiticity

$$\langle 0|l_q|0\rangle = \langle 0|l_q|0\rangle^* = -\langle 0|l_q|0\rangle$$

because $l_q^* = -l_q$.[2] This relation implies that the expectation value of l_q is zero. Because the wavefunctions of electrons in orbitally non-degenerate states may be chosen to be real (Section 2.6), it follows that orbitally nondegenerate systems have quenched orbital angular momentum and display spin-only paramagnetism.

13.3 Vector functions

The prime task in setting up a perturbation procedure for calculating the magnetizability of molecules is to formulate the perturbation hamiltonian. It turns out that we cannot simply argue by analogy with the electric susceptibility and use a perturbation of the form $-m_z\mathcal{B}$. To find the actual hamiltonian, we need to dig deeper into the description of the electromagnetic field.

The electric field $\mathcal{E}$ can be expressed as the gradient of a potential ϕ. Indeed, in the theory of the electromagnetic field, a 'potential' is perhaps so called because it is potentially capable of telling us the magnitude and direction of the electric field, so long as we know how to derive that information from it, that is, evaluate some kind of derivative. The Schrödinger equation for a charged particle in an electric field (such as the electron in a hydrogen atom) is expressed in terms of the potential ϕ that describes the electric field (it is the Coulomb potential for a hydrogen atom). Similarly, we need to identify a potential that describes a magnetic field if we are to formulate the Schrödinger equation for a particle in a magnetic field, and then see how to derive the field from it.

The idea of a **scalar function** should be familiar: it is a function that associates a single number with each point in space. The Coulomb potential is an example of a scalar function. It is for this reason that the electric potential ϕ is called a **scalar potential**. For this chapter, though, we shall also need to consider a **vector function**, a function that attaches three numbers to each point in space. We can think of these numbers as being the three components of a vector, and a vector function associates a vector of a certain magnitude and direction with each point in space. The electric and magnetic vectors of a plane-polarized light ray are examples of vector functions (Fig. 13.1).

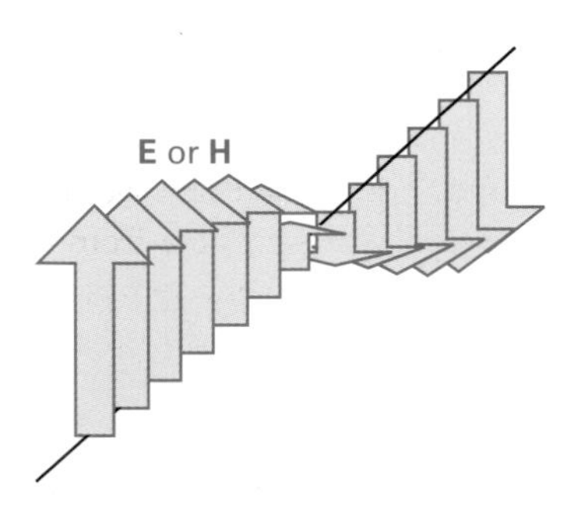

Fig. 13.1 The variation of the electric (or magnetic) field in an electromagnetic wave is an example of a vector field, with a vector associated with each point in space.

Vector functions are more difficult to represent diagrammatically than scalar functions because we have to display direction as well as magnitude at each point. As an illustration, consider the vector function

$$\mathbf{V} = -y\hat{\mathbf{i}} + x\hat{\mathbf{j}} \tag{16}$$

[2] The notation $\Omega^* = -\Omega$ here and in the following is an abbreviation of the relation $(\Omega f)^* = -\Omega f^*$ for all functions f.

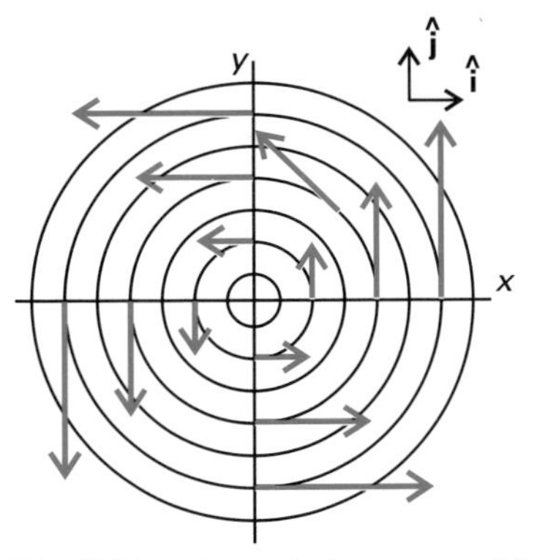

Fig. 13.2 Equal-magnitude contours of the vector function $\mathbf{V} = -y\hat{\mathbf{i}} + x\hat{\mathbf{j}}$; this function has nonzero curl but zero divergence.

where $\hat{\mathbf{i}}$ and $\hat{\mathbf{j}}$ are unit vectors in the (x, y)-plane. This function is drawn in Fig. 13.2. It can be constructed by first concentrating on the values it takes at points along the line $y = 0$, for then $\mathbf{V} = x\hat{\mathbf{j}}$. Along this line, the magnitude of the vector increases in proportion to x and it points in the direction of $\hat{\mathbf{j}}$ for $x > 0$ and along $-\hat{\mathbf{j}}$ for $x < 0$. These values are denoted by the arrows sprouting from the x-axis. Next, take $x = 0$, when $\mathbf{V} = -y\hat{\mathbf{i}}$. The magnitude of the vector increases in proportion to $|y|$, and the function points along $-\hat{\mathbf{i}}$ for $y > 0$ but along $\hat{\mathbf{i}}$ for $y < 0$. The same technique can be used to find the function at any point in the plane, and overall the function can be represented in terms of a series of contours carrying directional arrows. The function $\mathbf{V}$ obviously represents a circulation of some kind around the unit vector $\hat{\mathbf{k}}$ that points parallel to the z-axis. In contrast, the vector function

$$\mathbf{V}' = x\hat{\mathbf{i}} + y\hat{\mathbf{j}} \tag{17}$$

which is illustrated in Fig. 13.3, suggests a radial flow away from a central point.

13.4 Derivatives of vector functions

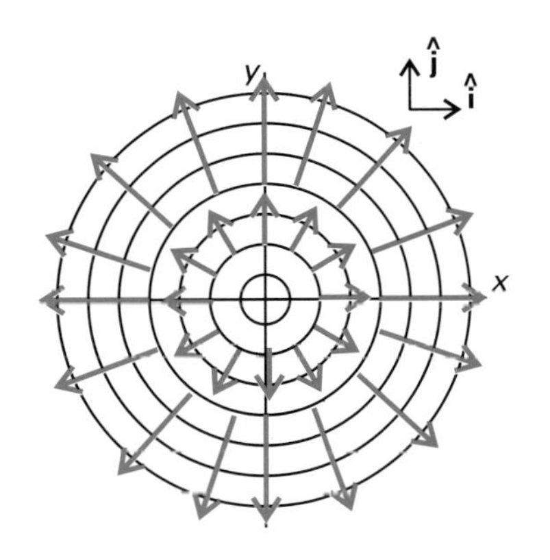

Fig. 13.3 Equal-magnitude contours of the vector function $\mathbf{V}' = x\hat{\mathbf{i}} + y\hat{\mathbf{j}}$; this function has nonzero divergence but zero curl.

We shall need the derivatives of a general vector function

$$\mathbf{F} = f_x\hat{\mathbf{i}} + f_y\hat{\mathbf{j}} + f_z\hat{\mathbf{k}} \tag{18}$$

where each of the f_q is in general a function of x, y, and z. There are two of importance for us. The **divergence** of a vector function is defined as

$$\nabla\cdot\mathbf{F} = \left(\frac{\partial f_x}{\partial x}\right) + \left(\frac{\partial f_y}{\partial y}\right) + \left(\frac{\partial f_z}{\partial z}\right) \tag{19}$$

The origin of the name 'divergence' can be appreciated by evaluating the divergences of the two vector functions $\mathbf{V}$ and $\mathbf{V}'$. We find

$$\nabla\cdot\mathbf{V} = 0 \qquad \nabla\cdot\mathbf{V}' = 2$$

These values reflect the appearances of the functions in the diagrams: $\mathbf{V}$ does not diverge but $\mathbf{V}'$ does. Note that the divergence of a vector function is a scalar function (or a constant).

The other derivative we require is the **curl** of a vector function. A curl is defined as follows:

$$\nabla\times\mathbf{F} = \begin{vmatrix} \hat{\mathbf{i}} & \hat{\mathbf{j}} & \hat{\mathbf{k}} \\ \partial/\partial x & \partial/\partial y & \partial/\partial z \\ f_x & f_y & f_z \end{vmatrix} \tag{20}$$

The origin of the name 'curl' can also be understood by evaluating the curl of the two vector functions in the illustrations. Even before we evaluate the curls, we can anticipate that $\mathbf{V}$ has nonzero curl because it circulates around the z-axis, whereas the curl of $\mathbf{V}'$, which does not circulate, is zero. To verify these intui-

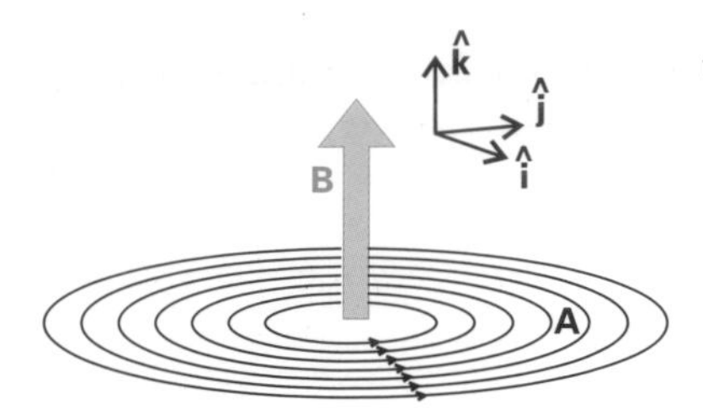

Fig. 13.4 The relation between the vector potential and the magnetic field to which it corresponds. A uniform magnetic field is described by a vector potential like the one illustrated that extends throughout the region of nonzero field. A nonuniform field has a vector potential that is like this one over an infinitesimal region.

tions we perform the following two calculations:

$$\nabla \times \mathbf{V} = \begin{vmatrix} \hat{\mathbf{i}} & \hat{\mathbf{j}} & \hat{\mathbf{k}} \\ \partial/\partial x & \partial/\partial y & \partial/\partial z \\ -y & x & 0 \end{vmatrix} = 2\hat{\mathbf{k}}$$

$$\nabla \times \mathbf{V}' = \begin{vmatrix} \hat{\mathbf{i}} & \hat{\mathbf{j}} & \hat{\mathbf{k}} \\ \partial/\partial x & \partial/\partial y & \partial/\partial z \\ x & y & 0 \end{vmatrix} = 0 \qquad (21)$$

Note that the curl of a vector function is a vector. Moreover, the curl conveys the sense of rotation according to the right-hand screw rule (the same as for angular momentum, Section 3.4).

13.5 The vector potential

We are now at the point where we can introduce the **vector potential**, **A**, the vector function from which the magnetic field may be derived. The vector potential corresponding to a magnetic induction **B** is defined such that

$$\mathbf{B} = \nabla \times \mathbf{A} \qquad (22)$$

For example, suppose that we are given the vector potential

$$\mathbf{A} = \tfrac{1}{2}\mathcal{B}\mathbf{V} = \tfrac{1}{2}\mathcal{B}(-y\hat{\mathbf{i}} + x\hat{\mathbf{j}}) \qquad (23)$$

then the induction to which it corresponds is

$$\mathbf{B} = \tfrac{1}{2}\mathcal{B}\nabla \times \mathbf{V} = \mathcal{B}\hat{\mathbf{k}} \qquad (24)$$

In other words, the vector potential $\frac{1}{2}\mathcal{B}\mathbf{V}$ describes a uniform magnetic field of induction $\mathcal{B}$ pointing in the direction $\hat{\mathbf{k}}$ (Fig. 13.4). It is quite easy to generalize this important result and to show that

$$\mathbf{A} = \tfrac{1}{2}\mathbf{B} \times \mathbf{r} \qquad (25)$$

corresponds to a uniform induction **B**.[3] Therefore, we can always set up the vector potential for a uniform field by forming $\frac{1}{2}\mathbf{B} \times \mathbf{r}$.

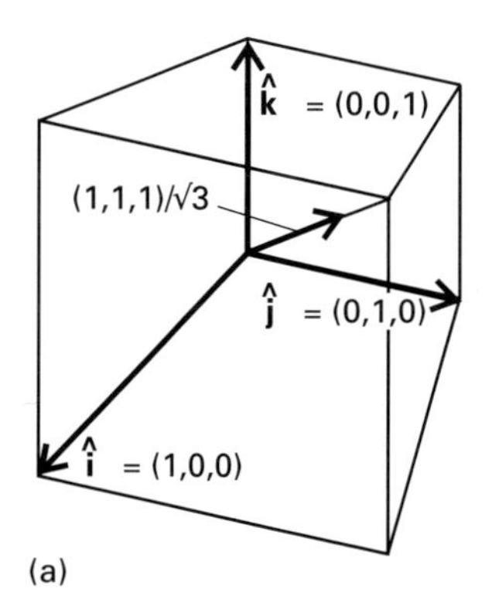

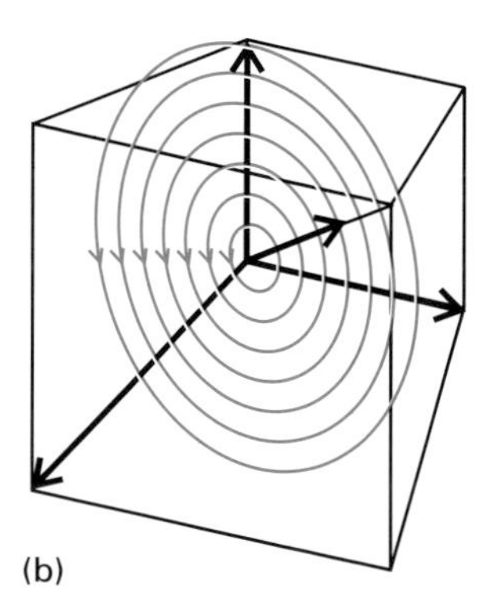

Fig. 13.5 (a) The unit vectors used to describe the field in Example 13.1. (b) The vector potential for the field is like the function **V** but it swirls around the direction of the field, the direction of the vector $(1, 1, 1)$.

Example 13.1 Setting up a vector potential

Construct a vector potential for a uniform magnetic field that points in the direction shown in Fig. 13.5a.

Method. The key to setting up the vector potential is eqn 13.25: all we need do is to form the vector of magnitude $\mathcal{B}$ orientated towards the corner of a unit cube. So, begin by constructing a unit vector in the direction required, and then use eqn 13.25.

[3] The relations needed to evaluate derivatives of vector functions are set out in *Further information 22.*

Answer. The unit vector in the direction shown in the illustration is $(1/\sqrt{3})(1,1,1)$. Therefore, the magnetic induction is $(\mathcal{B}/\sqrt{3})(1,1,1)$ and the vector potential is

$$\mathbf{A} = \left(\frac{\mathcal{B}}{2\sqrt{3}}\right)\begin{vmatrix} \hat{\mathbf{i}} & \hat{\mathbf{j}} & \hat{\mathbf{k}} \\ 1 & 1 & 1 \\ x & y & z \end{vmatrix} = \left(\frac{\mathcal{B}}{2\sqrt{3}}\right)\{(z-y)\hat{\mathbf{i}} + (x-z)\hat{\mathbf{j}} + (y-x)\hat{\mathbf{k}}\}$$

Comment. The vector potential is a function like **V**, but now swirling about the $(1,1,1)$ direction (as in Fig. 13.5b).

Exercise 13.1. Confirm that the vector function has zero divergence, and show that the magnetic induction is indeed that specified.

[$\nabla \cdot \mathbf{A} = 0$; $\nabla \times \mathbf{A} = \mathbf{B}$]

Two points now need to be made. The first is that not all magnetic fields are uniform, and then the vector potential takes on a more complicated form. *Locally*, however, a vector potential can always be imagined as resembling those we have already seen, but the direction of swirl and the closeness of the contour lines change from place to place. We shall see an example in Section 13.8. The second point is that the choice of vector potential corresponding to a given field is not unique. It is always possible to add to a given vector potential a vector function of the form ∇f, where f is an arbitrary scalar (ordinary) function, and leave the field unchanged. This property of **gauge invariance** stems from the vector identity $\nabla \times \nabla f \equiv 0$ and hence for any constant λ,

$$\mathbf{B} = \nabla \times (\mathbf{A} + \lambda \nabla f) = \nabla \times \mathbf{A} \tag{26}$$

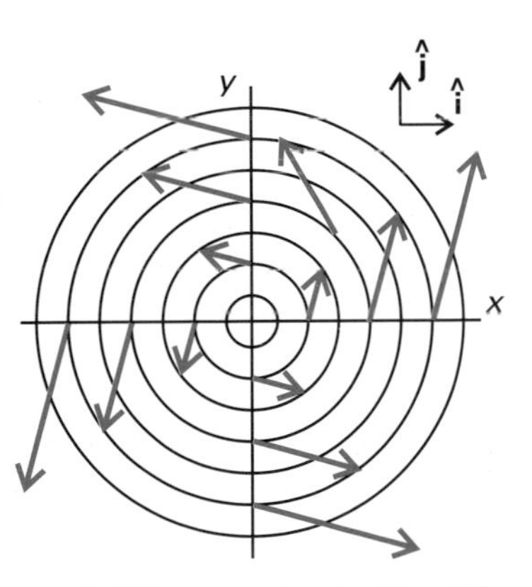

Fig. 13.6 The vector function $\mathbf{V} + \lambda\mathbf{V}'$ with nonzero divergence and curl. The change from Fig. 13.2 to this illustration corresponds to a gauge transformation.

The vector function $\mathbf{V}'$ is a special case of ∇f:

$$\mathbf{V}' = x\hat{\mathbf{i}} + y\hat{\mathbf{j}} = \tfrac{1}{2}\nabla(x^2 + y^2)$$

Therefore, all vector potentials of the form

$$\mathbf{A} = \tfrac{1}{2}\mathcal{B}\mathbf{V} + \lambda\mathbf{V}' \tag{27}$$

correspond to the same uniform induction **B** regardless of the value of λ (Fig. 13.6). Later we shall make use of the fact that it is always possible to select a gauge (that is, choose the gradient of a scalar function to add to a given vector potential) which ensures that the vector potential has zero divergence. In the present case, **V** has zero divergence already, so we do not need to make any gauge transformation to it. A gauge that corresponds to zero divergence of a vector potential is called the **Coulomb gauge**.

Magnetic perturbations

The whole point of introducing the vector potential was to enable us to set up the perturbation hamiltonian for molecules exposed to magnetic fields. With the form of the perturbation established, we shall be able to develop expressions for the magnetic susceptibility and related properties.

13.6 The perturbation hamiltonian

We show in *Further information 2* that there is a simple rule for constructing the hamiltonian of a system in the presence of a magnetic field from its hamiltonian in the absence of the field: wherever $\mathbf{p}$ occurs in the hamiltonian, it should be replaced by $\mathbf{p} + e\mathbf{A}$, where $\mathbf{A}$ is the vector potential for the field. This prescription is valid in classical and quantum mechanics: in the latter we have to be careful to take into account the possible non-commutation of operators.

To see the rule in action, consider a hamiltonian for an electron with a potential energy V (which may vary with position):

$$H^{(0)} = \frac{p^2}{2m_e} + V$$

In the presence of a magnetic field described by a vector potential $\mathbf{A}$, the term $p^2 = \mathbf{p}\cdot\mathbf{p}$ is replaced by

$$(\mathbf{p} + e\mathbf{A})\cdot(\mathbf{p} + e\mathbf{A}) = p^2 + e(\mathbf{p}\cdot\mathbf{A} + \mathbf{A}\cdot\mathbf{p}) + e^2A^2 \tag{28}$$

Some care is needed with the term $\mathbf{p}\cdot\mathbf{A}$ because in the position representation the linear momentum is a differential operator and it operates on the function $\mathbf{A}$ and the unwritten wavefunction on which the hamiltonian operates. When that wavefunction is included, we have

$$\mathbf{p}\cdot\mathbf{A}\psi = \left(\frac{\hbar}{\mathrm{i}}\right)\nabla\cdot\mathbf{A}\psi = \left(\frac{\hbar}{\mathrm{i}}\right)\{(\nabla\cdot\mathbf{A})\psi + \mathbf{A}\cdot(\nabla\psi)\}$$

However, if we adopt the Coulomb gauge, then the term $(\nabla\cdot\mathbf{A})$ is zero, and

$$\mathbf{p}\cdot\mathbf{A}\psi = \left(\frac{\hbar}{\mathrm{i}}\right)\mathbf{A}\cdot(\nabla\psi) = \mathbf{A}\cdot\mathbf{p}\psi$$

In this gauge, the vector potential and the linear momentum commute, and eqn 13.28 can be written

$$(\mathbf{p} + e\mathbf{A})\cdot(\mathbf{p} + e\mathbf{A}) = p^2 + 2e\mathbf{A}\cdot\mathbf{p} + e^2A^2 \tag{29}$$

It follows that the hamiltonian in the presence of the field is

$$H = \frac{p^2}{2m_e} + V + \frac{e}{m_e}\mathbf{A}\cdot\mathbf{p} + \left(\frac{e^2}{2m_e}\right)A^2 \tag{30}$$

This hamiltonian differs from the original hamiltonian by the presence of two terms, one of which is first order in the magnetic induction (via $\mathbf{A}$, which is proportional to $\mathcal{B}$), and the other of which is second order (via A^2). We shall therefore write

$$H = H^{(0)} + H^{(1)} + H^{(2)} \qquad \begin{cases} H^{(1)} = (e/m_e)\mathbf{A}\cdot\mathbf{p} \\ H^{(2)} = (e^2/2m_e)A^2 \end{cases} \tag{31}$$

The first-order term can be written in a more familiar form by considering a uniform magnetic field and replacing the vector potential by eqn 13.25:

$$H^{(1)} = \frac{e}{2m_e}\mathbf{B}\times\mathbf{r}\cdot\mathbf{p} = \frac{e}{2m_e}\mathbf{B}\cdot\mathbf{r}\times\mathbf{p} = \frac{e}{2m_e}\mathbf{B}\cdot\mathbf{l}$$

For the second equality we have used the vector identity $\mathbf{a} \times \mathbf{b} \cdot \mathbf{c} = \mathbf{a} \cdot \mathbf{b} \times \mathbf{c}$, and in the final step we have recognized the orbital angular momentum operator $\mathbf{l} = \mathbf{r} \times \mathbf{p}$. Finally, because the magnetogyric ratio (Section 7.3) is defined as $\gamma_e = -e/2m_e$, we can conclude that

$$H^{(1)} = -\gamma_e \mathbf{B} \cdot \mathbf{l} = -\mathbf{B} \cdot \mathbf{m} \tag{32}$$

where $\mathbf{m} = \gamma_e \mathbf{l}$. It should be noted that spin does not appear in this expression: for spin to appear naturally, we would need to work from the Dirac equation.

The second-order perturbation hamiltonian can also be expressed very simply when the field is uniform. Suppose it lies in the z-direction, then we can use the vector potential in eqn 13.23 and obtain

$$A^2 = \tfrac{1}{4}\mathcal{B}^2(-y\hat{\mathbf{i}} + x\hat{\mathbf{j}}) \cdot (-y\hat{\mathbf{i}} + x\hat{\mathbf{j}}) = \tfrac{1}{4}\mathcal{B}^2(x^2 + y^2)$$

Therefore, for such a field,

$$H^{(2)} = \left(\frac{e^2\mathcal{B}^2}{8m_e}\right)(x^2 + y^2) \tag{33}$$

For a uniform field in a general direction, it follows from eqn 13.25 and the identity $(\mathbf{a} \times \mathbf{b}) \cdot (\mathbf{a} \times \mathbf{b}) = a^2b^2 - (\mathbf{a} \cdot \mathbf{b})^2$ that

$$H^{(2)} = \left(\frac{e^2}{8m_e}\right)\{\mathcal{B}^2 r^2 - (\mathbf{B} \cdot \mathbf{r})^2\} \tag{34}$$

13.7 The magnetic susceptibility

Because the total hamiltonian has both first- and second-order contributions, we must use the full expression given in Section 6.5 to calculate properties to second order in the field:

$$E^{(2)} = \langle 0|H^{(2)}|0\rangle + \sum_n{}' \frac{\langle 0|H^{(1)}|n\rangle\langle n|H^{(1)}|0\rangle}{E_0 - E_n} \tag{35}$$

The first-order contribution is zero for a species in a non-degenerate state. For a uniform field in the z-direction, this expression becomes

$$\begin{aligned} E^{(2)} &= \left(\frac{e^2}{8m_e}\right)\langle 0|x^2 + y^2|0\rangle\mathcal{B}^2 + \left(\frac{e\mathcal{B}}{2m_e}\right)^2 \sum_n{}' \frac{\langle 0|l_z|n\rangle\langle n|l_z|0\rangle}{E_0 - E_n} \\ &= \left\{\left(\frac{e^2}{8m_e}\right)\langle x^2 + y^2\rangle - \left(\frac{e}{2m_e}\right)^2 \sum_n{}' \frac{l_{z;0n}l_{z;n0}}{\Delta E_{n0}}\right\}\mathcal{B}^2 \end{aligned} \tag{36}$$

where $l_{z;n0} = \langle n|l_z|0\rangle$ and $\Delta E_{n0} = E_n - E_0$. It should be noted that the first term is positive, and increases the energy of the molecule as the field is increased; the second term is negative, and decreases the energy.

We now construct the relation between the energy in the presence of a field and molecular properties. The energy of a magnetic dipole in a region of magnetic flux is $-m_z\mathcal{B}$, but we cannot simply write $E^{(2)} = -\langle m_z\rangle\mathcal{B}$ because $\langle m_z\rangle$ changes as the field is increased from zero. This variation is expressed in terms of the magnetizability, ξ, through

$$\langle m_z \rangle = \xi_{zz}\mathcal{H} + \ldots = \xi_{zz}\mathcal{B}/\mu + \ldots \tag{37}$$

where the unwritten terms are of higher order in the field strength. For the second equality, we have used eqn 13.3 to introduce the magnetic induction in place of the magnetic field strength $\mathcal{H}$ which occurs in the definition of the magnetizability. The quantity μ is the permeability (Section 13.1); to a good approximation, it can be replaced by its vacuum value, μ_0. Because an infinitesimal increase in induction, $\mathrm{d}\mathcal{B}$, results in an infinitesimal increase in energy $\mathrm{d}E^{(2)} = -\langle m_z \rangle \mathrm{d}\mathcal{B}$, the total change in energy when the induction is increased from 0 to its final value $\mathcal{B}$ is

$$\begin{aligned} E^{(2)} &= -\int_0^{\mathcal{B}} \langle m_z \rangle \,\mathrm{d}\mathcal{B} = -\int_0^{\mathcal{B}} (\xi_{zz}\mathcal{B}/\mu_0 + \ldots)\,\mathrm{d}\mathcal{B} \\ &= -\tfrac{1}{2}\xi_{zz}\mathcal{B}^2/\mu_0 + \ldots \end{aligned} \tag{38}$$

All we need now do is to compare this result with eqn 13.36, which gives

$$\xi_{zz} = -\left(\frac{e^2\mu_0}{4m_\mathrm{e}}\right)\langle x^2 + y^2 \rangle + \left(\frac{e^2\mu_0}{2m_\mathrm{e}^2}\right)\sum_n{}' \frac{l_{z;0n}l_{z;n0}}{\Delta E_{n0}} \tag{39}$$

The mean magnetizability of a freely rotating molecule is

$$\xi = \tfrac{1}{3}(\xi_{xx} + \xi_{yy} + \xi_{zz})$$

To evaluate this mean from the expression in eqn 13.39 we use

$$\langle (x^2 + y^2) + (y^2 + z^2) + (z^2 + x^2) \rangle = 2\langle r^2 \rangle$$

$$l_{x;0n}l_{x;n0} + l_{y;0n}l_{y;n0} + l_{z;0n}l_{z;n0} = \mathbf{l}_{0n} \cdot \mathbf{l}_{n0} = |l_{0n}|^2$$

It then follows that

$$\xi = -\left(\frac{e^2\mu_0}{6m_\mathrm{e}}\right)\langle r^2 \rangle + \left(\frac{e^2\mu_0}{6m_\mathrm{e}^2}\right)\sum_n{}' \frac{|l_{0n}|^2}{\Delta E_{n0}} \tag{40}$$

The connection between the magnetizability and the molar magnetic susceptibility is the analogue of the expression for polarizability and electric susceptibility modified by multiplication by the molar volume:

$$\chi_\mathrm{m} = \mathcal{N}\xi V_\mathrm{m} = N_\mathrm{A}\xi \tag{41}$$

It then follows that

$$\chi_\mathrm{m} = -\left(\frac{N_\mathrm{A}e^2\mu_0}{6m_\mathrm{e}}\right)\langle r^2 \rangle + \left(\frac{N_\mathrm{A}e^2\mu_0}{6m_\mathrm{e}^2}\right)\sum_n{}' \frac{|l_{0n}|^2}{\Delta E_{n0}} \tag{42}$$

The expression for the molar susceptibility apparently (we shall say why 'apparently' shortly) falls into two contributions, one positive and the other negative. Therefore, we express it as the sum of a negative **diamagnetic susceptibility**, $\chi_\mathrm{m}^\mathrm{d}$, and a positive **paramagnetic susceptibility**, $\chi_\mathrm{m}^\mathrm{p}$:

$$\chi_\mathrm{m} = \chi_\mathrm{m}^\mathrm{d} + \chi_\mathrm{m}^\mathrm{p} \qquad \begin{cases} \chi_\mathrm{m}^\mathrm{d} = -(N_\mathrm{A}e^2\mu_0/6m_\mathrm{e})\langle r^2 \rangle \\ \chi_\mathrm{m}^\mathrm{p} = (N_\mathrm{A}e^2\mu_0/6m_\mathrm{e}^2)\sum_n' |l_{0n}|^2/\Delta E_{n0} \end{cases} \tag{43}$$

It should be emphasized that this paramagnetic contribution to the susceptibility has nothing to do with electron spin and, unlike spin paramagnetism, is independent of the temperature. Hence, it is known as **temperature-independent paramagnetism** (TIP). The diamagnetic contribution is often called the **Langevin term**.

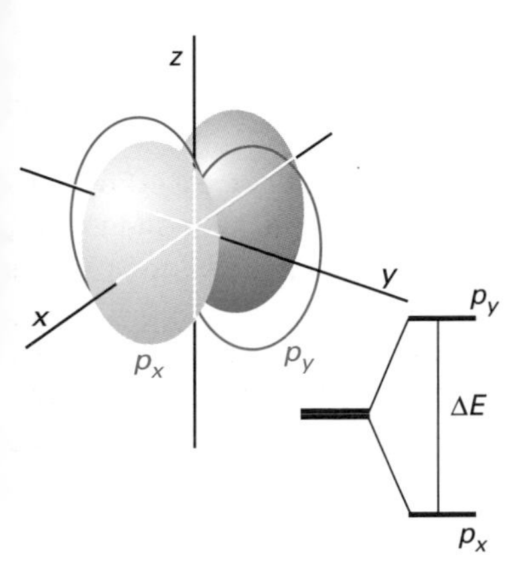

Fig. 13.7 The model system used for a number of illustrative calculations in this chapter: the degeneracy of the p-orbitals is removed (for instance, by the presence of neighbouring atoms).

Example 13.2 The calculation of magnetic susceptibility

Consider a model system in which one electron occupies a $2p_x$-orbital and where the $2p_y$-orbital lies at an energy ΔE above it (Fig. 13.7). Calculate the molar magnetic susceptibility in the z-direction.

Method. Use eqn 13.39 with the expression for ξ_{zz} rather than the mean magnetizability, eqn 13.40. For the expectation value $\langle x^2+y^2\rangle$, use Slater type orbitals as specified in Example 12.4. There is only one nonzero term in the sum for the paramagnetic contribution, and the matrix elements of l_z may be evaluated as in Example 12.4.

Answer. It follows from $l_z \cos\phi = \mathrm{i}\hbar \sin\phi$ that $l_z p_x = \mathrm{i}\hbar p_y$. Therefore,

$$\sum_n{}' \frac{l_{z;0n} l_{z;n0}}{\Delta E_{n0}} = \frac{|\langle p_y|l_z|p_x\rangle|^2}{\Delta E} = \frac{\hbar^2}{\Delta E}$$

$$\begin{aligned}\langle x^2+y^2\rangle &= \int (x^2+y^2)|\psi(2p_x)|^2\,\mathrm{d}\tau = \int r^2|\psi(2p_x)|^2 \sin^2\theta\,\mathrm{d}\tau \\ &= \left(\frac{3}{4\pi}\right)\left(\frac{2^5\zeta_\mathrm{p}^5}{4!}\right)\int_0^{2\pi}\cos^2\phi\,\mathrm{d}\phi \int_0^{\pi}\sin^5\theta\,\mathrm{d}\theta \int_0^{\infty} r^6 \mathrm{e}^{-2\zeta_\mathrm{p} r}\,\mathrm{d}r \\ &= \frac{6a_0^2 n^2}{Z_p^{*2}}\end{aligned}$$

It follows that

$$\chi_\mathrm{m} = -\frac{3N_\mathrm{A}e^2\mu_0 a_0^2 n_p^2}{2m_\mathrm{e} Z_p^{*2}} + \frac{N_\mathrm{A}e^2\mu_0\hbar^2}{2m_\mathrm{e}^2\Delta E}$$

Comment. The susceptibility is paramagnetic if $\Delta E < Z_p^{*2}\hbar^2/3n_p^2 m_\mathrm{e} a_0^2$.

The observed susceptibility of a sample depends on the competition between the diamagnetic and paramagnetic contributions. In free atoms, the paramagnetic contribution is zero because we are free to choose the z-direction as the axis of quantization of the z-component of magnetization; as a result, $|0\rangle$ and $|n\rangle$ are eigenstates of l_z, and hence all off-diagonal elements of l_z are zero. The total molar susceptibility of a sample of atoms is therefore

$$\chi_\mathrm{m} = -\left(\frac{N_\mathrm{A}e^2\mu_0}{6m_\mathrm{e}}\right)\langle r^2\rangle \qquad (44)$$

provided that there are no unpaired spins. For a typical atom with $\langle r^2\rangle \approx R^2$, where R is the radius of the atom, and $R \approx 0.15$ nm, $\chi_\mathrm{m} \approx -8\times10^{-11}\,\mathrm{m^3\,mol^{-1}}$. If an unpaired spin is present on each atom, the spin-only molar susceptibility at

300 K is $1.6 \times 10^{-8}\,\mathrm{m^3\,mol^{-1}}$ (Section 13.2), which overwhelms the diamagnetic contribution. In the absence of spin all atoms have a nonzero but small net diamagnetic susceptibility.

In the case of molecules, the axis of quantization of the orbital angular momentum is no longer necessarily the direction of the applied field (unless the two happen to align). Now the susceptibility is the sum of diamagnetic and paramagnetic (TIP) terms.[4] In most molecules the former dominates, and most molecules without unpaired electron spins are diamagnetic, with molar susceptibilities proportional to $\langle r^2 \rangle$. Only when there are low-lying excited electronic states may the orbital paramagnetic term dominate the Langevin term and the molecule be weakly paramagnetic. If the closure approximation (Section 6.7) is used in eqn 13.42, we obtain

$$\chi_\mathrm{m} \approx -\left(\frac{N_\mathrm{A} e^2 \mu_0}{6m_\mathrm{e}}\right)\langle r^2 \rangle + \left(\frac{N_\mathrm{A} e^2 \mu_0}{6m_\mathrm{e}^2 \Delta E}\right) l(l+1)\hbar^2$$
$$\approx -\left(\frac{N_\mathrm{A} e^2 \mu_0}{6m_\mathrm{e}}\right)\left(\langle r^2 \rangle - \frac{l(l+1)\hbar^2}{m_\mathrm{e}\Delta E}\right)$$

where ΔE is the mean excitation energy, and we have used the fact that $\langle 0|l_q|0\rangle = 0$. The paramagnetic term dominates the diamagnetic when

$$\Delta E < \frac{l(l+1)\hbar^2}{m_\mathrm{e} R^2} \qquad (45)$$

where $R^2 = \langle r^2 \rangle$. With $l \approx 1$ and $R \approx 0.3\,\mathrm{nm}$ the right-hand side evaluates to about 2 eV (16 000 $\mathrm{cm^{-1}}$), which corresponds to very low-lying energy levels.

One of the pitfalls in the interpretation of magnetic susceptibilities in terms of diamagnetic and paramagnetic (TIP) contributions is that the division of the total susceptibility into two contributions depends on the gauge of the vector potential. It is even possible to choose a gauge that eliminates the paramagnetic term completely! The only physically meaningful quantity is the *total* magnetic susceptibility, which remains constant as the gauge is changed. It follows that, because the gauge of the vector potential is arbitrary, so is the division of the susceptibility into two components. The choice of gauge, which is effectively the choice of origin of a coordinate system, is less arbitrary in atoms, where the nucleus is the natural centre. However, there is no such natural centre in molecules, and so the discussion of the individual contributions must be treated with great caution.

13.8 The current density

More insight into the nature of the two contributions to the magnetic susceptibility can be obtained by investigating the electronic currents that are induced by the applied field. Here we shall build the discussion on the concept of **current density**, **j**, which is essentially the flux density introduced in Section 2.7 for the flow of particles in scattering processes (eqn 2.11), but multiplied by

[4] Classical mechanics cannot account for the magnetic susceptibilities of molecules. This is the content of a theorem courteously referred to by Van Vleck as 'Miss van Leeuwen's theorem', which demonstrates that the diamagnetic and paramagnetic contributions cancel in a classical mechanical calculation. This is a late but interesting illustration of the inadequacy of classical physics.

the electric charge:

$$\mathbf{j}_0 = -\left(\frac{e}{2m_e}\right)(\psi^*\mathbf{p}\psi + \psi\mathbf{p}^*\psi^*) \tag{46}$$

(The subscript 0 signifies zero magnetic field.) To recapitulate the justification in Section 2.7: the velocity of an electron is related to its linear momentum by $\mathbf{v} = \mathbf{p}/m_e$, and the current is $-e$ times this velocity, or $-e\mathbf{p}/m_e$. The current density is obtained by weighting this expression by the probability density of the electron at each point in space, which results in terms of the form $-e\psi^*\mathbf{p}\psi/m_e$; the addition of the complex conjugate ensures that the current density is real. The precise definition in eqn 13.46 ensures that (as demonstrated for flux in Example 2.1) the current density obeys a continuity equation characteristic of an incompressible fluid.

In the presence of a magnetic field, the linear momentum $\mathbf{p}$ is replaced wherever it occurs by $\mathbf{p} + e\mathbf{A}$, where $\mathbf{A}$ is the (real) vector potential corresponding to the field. Then the appropriate expression for the current density, with $(\mathbf{p} + e\mathbf{A})^* = -\mathbf{p} + e\mathbf{A}$, is

$$\mathbf{j} = -\left(\frac{e}{2m_e}\right)(\psi^*\mathbf{p}\psi - \psi\mathbf{p}\psi^*) - \left(\frac{e^2}{m_e}\right)\mathbf{A}\psi^*\psi \tag{47}$$

We shall analyse this expression for various cases.

Consider first the current density in a molecule in which the single electron of interest is described by a real wavefunction and there is no magnetic field present. In this case eqn 13.46 becomes

$$\mathbf{j}_0 = -\left(\frac{e}{2m_e}\right)(\psi\mathbf{p}\psi - \psi\mathbf{p}\psi) = 0$$

There is zero current density at every point in the molecule. It will be recalled that we have already seen that a molecule in an orbitally non-degenerate state is described by a real wavefunction (or, at least, by a wavefunction that may be chosen to be real), and that its electrons have zero orbital angular momentum. This current-density result is another way of visualizing that lack of motion.

Now consider an electron in an orbitally degenerate state, but still with no magnetic field applied. In this case, the wavefunction is not necessarily real. For example, suppose the electron occupies a π-orbital in a linear molecule; then if it has a definite component of orbital angular momentum about the z-axis, its wavefunction has the form $f(r,z)e^{i\lambda\phi}$ in the cylindrical coordinates shown in Fig. 13.8, with f a real function. For the state with $\lambda = -1$, the current density is

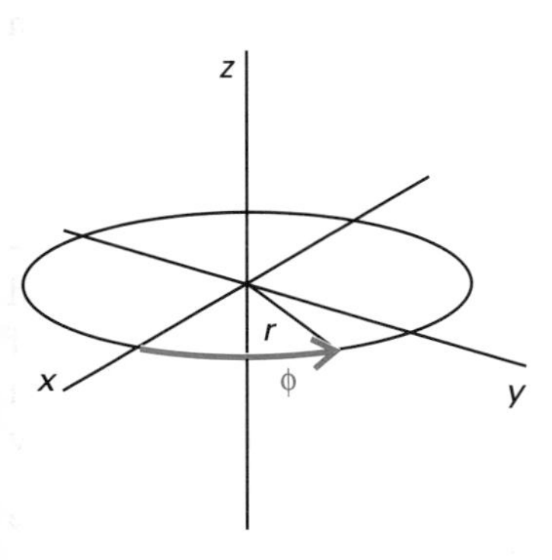

Fig. 13.8 The cylindrical coordinates used to discuss the current density in a molecule.

$$\begin{aligned}\mathbf{j}_0 &= i\left(\frac{e\hbar}{2m_e}\right)\{fe^{i\phi}\nabla fe^{-i\phi} - fe^{-i\phi}\nabla fe^{i\phi}\} \\ &= i\left(\frac{e\hbar}{2m_e}\right)\{fe^{i\phi}(\nabla f)e^{-i\phi} - ife^{i\phi}(\nabla\phi)fe^{-i\phi} - fe^{-i\phi}(\nabla f)e^{i\phi} \\ &\quad - ife^{-i\phi}(\nabla\phi)fe^{i\phi}\} \\ &= \left(\frac{e\hbar}{m_e}\right)f^2(\nabla\phi)\end{aligned}$$

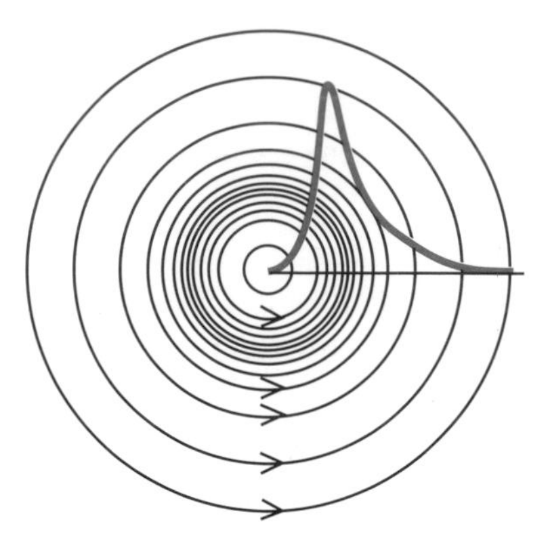

Fig. 13.9 The current density in the xy-plane for a system like that shown in Fig. 13.7 but for degenerate orbitals.

The gradient of ϕ is evaluated most easily by noting that $\phi = \arctan(y/x)$, fo[r] then

$$\nabla\phi = \left(\frac{\partial\phi}{\partial x}\right)\hat{\mathbf{i}} + \left(\frac{\partial\phi}{\partial y}\right)\hat{\mathbf{j}} + \left(\frac{\partial\phi}{\partial z}\right)\hat{\mathbf{k}}$$
$$= \frac{-y\hat{\mathbf{i}} + x\hat{\mathbf{j}}}{x^2 + y^2} = \frac{\mathbf{V}}{x^2 + y^2} \qquad (48$$

where $\mathbf{V}$ is the swirling vector function of eqn 13.16 and Fig. 13.2. It follow[s] that the current density has the form

$$\mathbf{j}_0 = \left(\frac{e\hbar}{m_e}\right)\left(\frac{f^2}{x^2 + y^2}\right)\mathbf{V} \qquad (49$$

This current density is proportional to $\mathbf{V}$, but it varies in a more complicate[d] way with distance from the origin (Fig. 13.9). The flow lines of the curren[t] density are obvious from the illustration, and they are closest together in th[e] region of greatest density of the orbital (after allowing for the $x^2 + y^2$ in th[e] denominator). The flow lines are clockwise seen from below (from a poin[t] $z < 0$), opposite in sense to the orbital angular momentum: the differenc[e] reflects the negative charge of the electron, so charge and mass flow in opposit[e] directions.

Finally, we consider an orbitally non-degenerate molecule in a uniform mag[-]netic field. Because the vector potential is nonzero and the wavefunctions ar[e] distorted by the applied field (and, as we shall see, are no longer real), th[e] current density is in general nonzero. We shall carry out the perturbation t[o] first order in $\mathcal{B}$.

In the presence of a field, the wavefunctions are distorted from ψ_0 t[o] $\psi_0 + \psi^{(1)}$, where

$$\psi^{(1)} = \sum_n{}' a_n\psi_n \qquad a_n = -\frac{H^{(1)}_{n0}}{\Delta E_{n0}} \qquad (50$$

as we deduced in Section 6.4 (see eqn 6.22). The coefficients a_n are now pro[-]portional to the off-diagonal matrix elements of l_z, which are imaginary, and s[o] the overall wavefunction is now complex, which is what we need for a nonzer[o] current density. (This acquisition of an imaginary component to the wavefunc[-]tion is another example of how the character of the perturbation is impressed o[n] the system.) To calculate the first-order correction to the current density, w[e] need the distortion of the wavefunction only to first order in the perturbation and so for this calculation we do not need to trouble about the role of $H^{(2)}$[.] Similarly, because the vector potential is already first order in the inductio[n] $\mathcal{B}$, in the expression $\mathbf{A}\psi^*\psi$ we can replace the wavefunctions by ψ_0. I[t] follows that to first order,

$$\begin{aligned}\mathbf{j} &= -\left(\frac{e}{2m_e}\right)\{(\psi_0+\psi^{(1)})^*\mathbf{p}(\psi_0+\psi^{(1)}) - (\psi_0+\psi^{(1)})\mathbf{p}(\psi_0+\psi^{(1)})^*\} \\ &\quad - \left(\frac{e^2}{m_e}\right)\mathbf{A}\psi_0^2 \\ &= -\left(\frac{e}{2m_e}\right)\{\psi_0\mathbf{p}\psi^{(1)} + \psi^{(1)*}\mathbf{p}\psi_0 - \psi_0\mathbf{p}\psi^{(1)*} - \psi^{(1)}\mathbf{p}\psi_0\} \\ &\quad - \left(\frac{e^2}{m_e}\right)\mathbf{A}\psi_0^2\end{aligned}$$

In the final line, we have used the fact that ψ_0 is real ($\psi_0^* = \psi_0$) and have retained only first-order terms. Because the ψ_n are also real (but the coefficients a_n are not), this expression becomes

$$\mathbf{j} = -\mathrm{i}\left(\frac{e\hbar}{2m_e}\right)\sum_n{}'(a_n - a_n^*)(\psi_n\nabla\psi_0 - \psi_0\nabla\psi_n) - \left(\frac{e^2}{m_e}\right)\mathbf{A}\psi_0^2 \tag{51}$$

The natural apparent division of this expression is into a **diamagnetic current density**, $\mathbf{j}^{\mathrm{d}}$, which depends only on the ground-state wavefunction, and a **paramagnetic current density**, $\mathbf{j}^{\mathrm{p}}$, which depends on the admixture of excited states:

$$\mathbf{j} = \mathbf{j}^{\mathrm{d}} + \mathbf{j}^{\mathrm{p}} \qquad \begin{cases}\mathbf{j}^{\mathrm{d}} = -(e^2/m_e)\mathbf{A}\psi_0^2 \\ \mathbf{j}^{\mathrm{p}} = -\mathrm{i}(e\hbar/2m_e)\sum_n'(a_n - a_n^*)(\psi_n\nabla\psi_0 - \psi_0\nabla\psi_n)\end{cases} \tag{52}$$

However, we stress again, that while this might seem a natural division of the current density, it is only natural for the gauge of the vector potential that we happen to have chosen. A gauge transformation of the kind specified in eqn 13.27 will result in a change of the diamagnetic current density by the addition of a term proportional to $\lambda(\nabla f)\psi_0^2$, and so this contribution to the current density can be varied almost at will. Only the overall current density has a real physical significance, and any division of it into contributions, while convenient, is arbitrary.

13.9 The diamagnetic current density

When the applied field lies in the z-direction, the vector potential in the Coulomb gauge is given by eqn 13.23, so

$$\mathbf{j}^{\mathrm{d}} = -\left(\frac{e^2\mathcal{B}}{2m_e}\right)\psi_0^2\mathbf{V} \tag{53}$$

Although this current density has the characteristic swirling form of $\mathbf{V}$, it is swirling in the opposite direction (the negative sign in this expression) and its shape is modified by the presence of the factor ψ_0^2. If the ground-state wavefunction is a hydrogen 1s-orbital, the explicit form of this current density is

$$\mathbf{j}^{\mathrm{d}} = -\left(\frac{e^2\mathcal{B}}{2\pi m_e a_0^3}\right)(-y\hat{\mathbf{i}} + x\hat{\mathbf{j}})\mathrm{e}^{-2r/a_0} \tag{54}$$

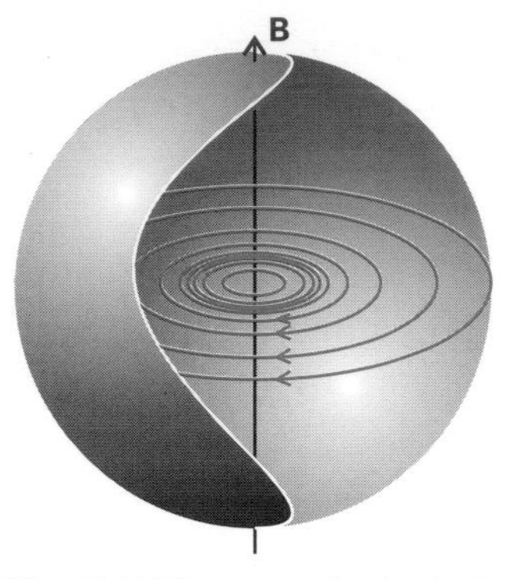

Fig. 13.10 The current density in the xy-plane for a ground-state hydrogen atom in a magnetic field.

This current density is sketched in Fig. 13.10. The magnitude of the current density is proportional to the magnetic induction, and its magnitude is greatest in the equatorial plane of the atom, and then at a radius of $\frac{1}{2}a_0$. On this circle even a field as small as 1×10^{-4} T produces a current density of 80 MA m^{-2}. This enormous current density is brought into perspective when expressed on an atomic scale, for it corresponds to about 0.5 electrons pm^{-2} μs^{-1}.

When the magnetic field is applied perpendicular to the axis of a p-orbital, the shapes of the contours are more complicated. For a hydrogenic $2p_x$-orbital of the form $\psi_0 = Nr \sin\theta \cos\phi\, e^{-r/2a_0}$, the diamagnetic current density is

$$\mathbf{j}^{\mathrm{d}} = -\left(\frac{e^2\mathcal{B}}{2\pi m_e a_0^3}\right)(-y\hat{\mathbf{i}} + x\hat{\mathbf{j}})N^2 r^2 \sin^2\theta \cos^2\phi\, e^{-r/a_0} \tag{55}$$

The direction of the current density at each point is still determined by the factor $(-y\hat{\mathbf{i}} + x\hat{\mathbf{j}})$, but the details are much more complicated (Fig. 13.11). The point to note is that the current density is zero at the angular node of the wavefunction, so there is no flow from one lobe of the orbital to the other.

In summary, the central feature of the diamagnetic current density is that it is a circulating distortion confined to the zone occupied by the orbital, and it vanishes where the orbital amplitude vanishes (at its nodes).

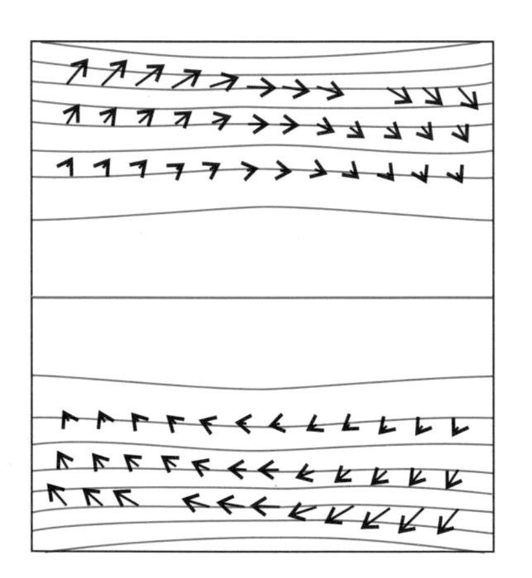

Fig. 13.11 The diamagnetic current density in the xy-plane for an electron in a hydrogenic $2p_x$-orbital with a magnetic field applied in the z-direction.

13.10 The paramagnetic current density

We can discover the principal features of the paramagnetic current density by concentrating on a simple model system consisting of two p-orbitals with their degeneracy removed (as in Fig. 13.7). The magnetic field is applied along the z-axis, and we shall need the following matrix elements (see Example 13.2):

$$\langle p_y|l_z|p_x\rangle = i\hbar \qquad \langle p_x|l_z|p_y\rangle = -i\hbar \tag{56}$$

The coefficient in the perturbation expression for the admixture of the $2p_y$-orbital into the $2p_x$-orbital is therefore

$$a(p_y) = \frac{\gamma_e l_{z;n0}\mathcal{B}}{\Delta E} = -i\frac{\mu_B \mathcal{B}}{\Delta E} \tag{57}$$

The paramagnetic current density therefore consists of a single term:

$$\begin{aligned}\mathbf{j}^{\mathrm{p}} &= -i\left(\frac{e\hbar}{2m_e}\right)\{a(p_y) - a(p_y)^*\}(p_y\nabla p_x - p_x\nabla p_y)\\ &= -\left(\frac{e\hbar\mu_B\mathcal{B}}{m_e\Delta E}\right)(p_y\nabla p_x - p_x\nabla p_y)\end{aligned}$$

The remaining work is to evaluate the gradients:

$$\begin{aligned}p_y\nabla p_x - p_x\nabla p_y &= f\sin\theta\sin\phi\,\nabla f\sin\theta\cos\phi - f\sin\theta\cos\phi\,\nabla f\sin\theta\sin\phi\\ &= f^2\sin^2\theta(\sin\phi\,\nabla\cos\phi - \cos\phi\,\nabla\sin\phi)\\ &= f^2\sin^2\theta(-\sin^2\phi\,\nabla\phi - \cos^2\phi\,\nabla\phi) = -f^2\sin^2\theta\nabla\phi\end{aligned}$$

Therefore, because we have already evaluated $\nabla\phi$ (eqn 13.48), the current density is

$$\mathbf{j}^{\mathrm{p}} = \left(\frac{e\hbar\mu_{\mathrm{B}}\mathcal{B}}{m_{\mathrm{e}}\Delta E}\right)\left(\frac{f^2\sin^2\theta}{x^2+y^2}\right)\mathbf{V} \tag{58}$$

and the ubiquitous swirling vector function $\mathbf{V}$ is back on stage again. This expression is the same as that for the current density in the degenerate case, eqn 13.49, apart from the presence of the factor $\mu_{\mathrm{B}}\mathcal{B}/\Delta E$ and $\sin^2\theta$. Therefore, for the xy-plane we can write

$$\mathbf{j}^{\mathrm{p}} = \left(\frac{\mu_{\mathrm{B}}\mathcal{B}}{\Delta E}\right)\mathbf{j}_0 \tag{59}$$

We can now construct a picture of the induced paramagnetic current density. Its form is exactly the same as the current density that exists when the orbitals are degenerate and the electron is in a state of well-defined orbital angular momentum, the only difference being the magnitude of the current density. The factor $\mu_{\mathrm{B}}\mathcal{B}/\Delta E$ represents the degree of success of the perturbation (of strength $\mu_{\mathrm{B}}\mathcal{B}$) in overcoming the energy separation (ΔE), which tends to lock the electron in its original location. For $\mathcal{B} \approx 1$ T, the ratio works out to be about $0.5/(\tilde{\nu}/\mathrm{cm}^{-1})$, where the energy separation has been expressed as a wavenumber. Hence, the ratio is very small for most systems and the paramagnetic current density is very much smaller than the diamagnetic current density.

It should be noted that the diamagnetic and paramagnetic current densities are in opposite directions around the direction of the applied field. This difference accounts for the opposite signs of the corresponding susceptibilities. The diamagnetic current acts as a source of magnetic field that opposes the applied field and so reduces the induction within the sample. The paramagnetic current generates a magnetic field that augments the applied field.

Magnetic resonance parameters

Most interest in the magnetic properties of molecules now centres on the parameters encountered in magnetic resonance, especially nuclear magnetic resonance (NMR) and to a lesser extent electron spin resonance (ESR or EPR). In this section we shall indicate how these parameters, which include shielding constants, g-values, spin–spin coupling constants, and hyperfine coupling constants, are related to a variety of molecular characteristics and, to some extent, can be rationalized in terms of the currents induced in the electron distributions of molecules. This will not be an introduction to magnetic resonance, which has become a highly specialized field of spectroscopy, and the basic principles will be assumed to be known.

13.11 Shielding constants

Different groups of nuclei in a molecule have resonance frequencies that reflect the fact that they experience different local magnetic fields, $\mathcal{B}_{\mathrm{loc}}$. To a good approximation, the difference between the local and applied fields is proportional to the applied field, and we write

$$\mathcal{B}_{\mathrm{loc}} = \mathcal{B} - \sigma\mathcal{B} \tag{60}$$

where σ is called the **shielding constant**. Our task in this section is to see how the currents induced by the applied field modify the local field and hence give rise to the chemical shift. The strategy is to set up the perturbation hamiltonian that describes a system in which there are two sources of magnetic field, the applied field and the field arising from the magnetic nucleus of interest, to calculate the energy of the system in the presence of both fields, and then to express the energy in terms of a local field.

Consider a molecule containing a single magnetic nucleus (and any number of other nuclei). The uniform, applied magnetic field is described by the vector potential $\mathbf{A}_{\mathrm{ex}} = \frac{1}{2}\mathbf{B} \times \mathbf{r}$ (where the subscript 'ex' denotes an externally applied field). The magnetic field arising from the nucleus is described by a vector potential $\mathbf{A}_{\mathrm{nuc}}$. Our first task is to determine the latter's form.

The classical expression for the magnetic field generated by a magnetic dipole is[5]

$$\mathbf{B} = -\left(\frac{\mu_0}{4\pi r^3}\right)\left\{\mathbf{m} - \frac{3\mathbf{r}(\mathbf{r}\cdot\mathbf{m})}{r^2}\right\} \tag{61}$$

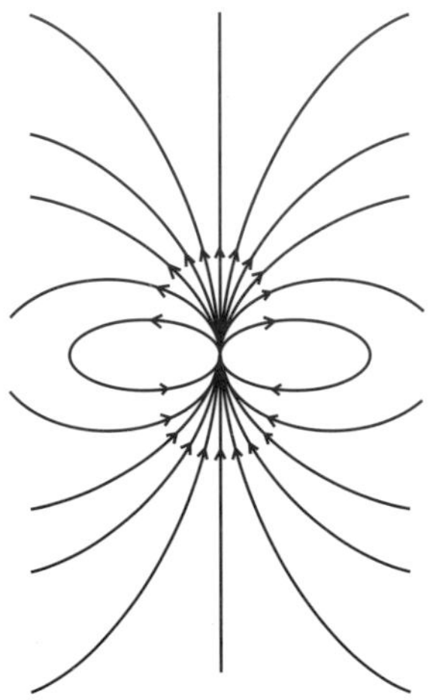

Fig. 13.12 The magnetic field arising from a point magnetic dipole.

This field is not uniform (Fig. 13.12). We can therefore expect the corresponding vector potential to be more complex than those we have considered so far. Nevertheless, it is not much more complicated, and we confirm in *Further information 21* that

$$\mathbf{A}_{\mathrm{nuc}} = \left(\frac{\mu_0}{4\pi r^3}\right)\mathbf{m} \times \mathbf{r} \tag{62}$$

The magnetic moment of a nucleus is related to its spin angular momentum $\mathbf{I}$ by $\mathbf{m} = \gamma_{\mathrm{N}}\mathbf{I}$, where γ_{N} is its magnetogyric ratio. Therefore, the vector potential for a nuclear dipole field is

$$\mathbf{A}_{\mathrm{nuc}} = \left(\frac{\gamma_{\mathrm{N}}\mu_0}{4\pi r^3}\right)\mathbf{I} \times \mathbf{r} \tag{63}$$

It can be confirmed that the divergence of this vector potential is zero.

The hamiltonian for the molecule in a magnetic field is constructed in the normal way by replacing $\mathbf{p}$ wherever it occurs by $\mathbf{p} + e\mathbf{A}$, where now $\mathbf{A} = \mathbf{A}_{\mathrm{ex}} + \mathbf{A}_{\mathrm{nuc}}$ because the electrons are exposed to both sources of magnetic field. It proves sensible to proceed in two stages, first to consider the molecule with no applied field, and then to switch on the field. Therefore, we begin by replacing $\mathbf{p}$ by $\mathbf{p} + e\mathbf{A}_{\mathrm{nuc}}$. The hamiltonian becomes (by analogy with eqn 13.31)

$$H = H^{(0)} + H^{(1)} + H^{(2)} \qquad \begin{cases} H^{(1)} = (e/2m_{\mathrm{e}})(\mathbf{p}\cdot\mathbf{A}_{\mathrm{nuc}} + \mathbf{A}_{\mathrm{nuc}}\cdot\mathbf{p}) \\ H^{(2)} = (e^2/2m_{\mathrm{e}})A_{\mathrm{nuc}}^2 \end{cases} \tag{64}$$

We shall disregard the contributions to the energy that are quadratic in the nuclear magnetic moment, and therefore ignore $H^{(2)}$. Moreover, because the

[5] See *Further reading* for references.

vector potential has zero divergence, $\mathbf{p}\cdot\mathbf{A}_{\text{nuc}} = \mathbf{A}_{\text{nuc}}\cdot\mathbf{p}$; so the first-order hamiltonian is

$$H^{(1)} = \left(\frac{e}{m_{\text{e}}}\right)\mathbf{A}_{\text{nuc}}\cdot\mathbf{p} \tag{65}$$

Now we calculate the first-order correction to the energy:

$$E^{(1)} = \langle 0|H^{(1)}|0\rangle = \left(\frac{e}{m_{\text{e}}}\right)\int \psi^*\mathbf{A}_{\text{nuc}}\cdot\mathbf{p}\psi\,\mathrm{d}\tau$$

For reasons that will shortly become clear, we shall express the integral as the sum of two identical terms:

$$\begin{aligned}\int \psi^*\mathbf{A}_{\text{nuc}}\cdot\mathbf{p}\psi\,\mathrm{d}\tau &= \tfrac{1}{2}\int \psi^*\mathbf{A}_{\text{nuc}}\cdot\mathbf{p}\psi\,\mathrm{d}\tau + \tfrac{1}{2}\int \psi^*\mathbf{A}_{\text{nuc}}\cdot\mathbf{p}\psi\,\mathrm{d}\tau\\ &= \tfrac{1}{2}\int \mathbf{A}_{\text{nuc}}\cdot\psi^*\mathbf{p}\psi\,\mathrm{d}\tau + \tfrac{1}{2}\int \mathbf{A}_{\text{nuc}}\cdot\psi^*\mathbf{p}\psi\,\mathrm{d}\tau\end{aligned}$$

The second term can be manipulated by invoking the hermiticity of the linear momentum operator and the zero divergence of the vector potential:

$$\begin{aligned}\int \mathbf{A}_{\text{nuc}}\cdot\psi^*\mathbf{p}\psi\,\mathrm{d}\tau &= \int (\mathbf{p}^*\cdot\mathbf{A}_{\text{nuc}}\psi^*)\psi\,\mathrm{d}\tau \qquad \text{(hermiticity of } \mathbf{p})\\ &= \int (\mathbf{p}^*\cdot\mathbf{A}_{\text{nuc}})\psi^*\psi\,\mathrm{d}\tau + \int \mathbf{A}_{\text{nuc}}\cdot(\mathbf{p}^*\psi^*)\psi\,\mathrm{d}\tau\\ &\qquad (\mathbf{p}\text{ is a differential operator})\\ &= \int \mathbf{A}_{\text{nuc}}\cdot(\mathbf{p}^*\psi^*)\psi\,\mathrm{d}\tau \qquad (\nabla\cdot\mathbf{A}=0)\end{aligned}$$

It follows that the first-order energy is

$$E^{(1)} = \left(\frac{e}{2m_{\text{e}}}\right)\int \mathbf{A}_{\text{nuc}}\cdot(\psi^*\mathbf{p}\psi + \psi\mathbf{p}^*\psi^*)\,\mathrm{d}\tau$$

However, we can now recognize the current density (eqn 13.46), and so we can write this expression in the very succinct form

$$E^{(1)} = -\int \mathbf{A}_{\text{nuc}}\cdot\mathbf{j}_0\,\mathrm{d}\tau \tag{66}$$

When the external field is applied, the prescription to replace $\mathbf{p}$ by $\mathbf{p} + e\mathbf{A}_{\text{ex}}$ results in the conversion of $\mathbf{j}_0$ into $\mathbf{j}$, the current density in the presence of the applied field. Then

$$E^{(1)} = -\int \mathbf{A}_{\text{nuc}}\cdot\mathbf{j}\,\mathrm{d}\tau \tag{67}$$

This result shows very plainly how shifts in the energy of a magnetic nucleus arise from the coupling of its magnetic dipole (which occurs in the vector potential) with the currents that may exist in the electron distribution (which may have been induced by an applied magnetic field).

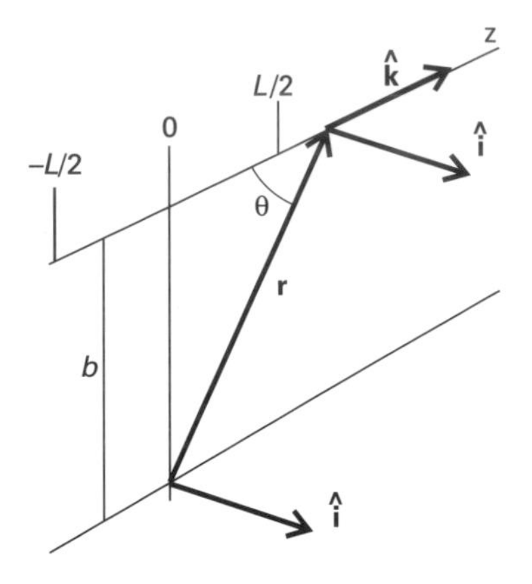

Fig. 13.13 The coordinates used in the calculation in Example 13.3 in which a beam of electrons travels from the left to the right.

Insertion of the explicit form for the nuclear vector potential and use of the vector identity $(\mathbf{a} \times \mathbf{b}) \cdot \mathbf{c} = \mathbf{a} \cdot (\mathbf{b} \times \mathbf{c})$ turns the last equation into

$$E^{(1)} = -\left(\frac{\mu_0 \gamma_N}{4\pi}\right) \int \frac{(\mathbf{I} \times \mathbf{r}) \cdot \mathbf{j}}{r^3} \mathrm{d}\tau = -\left(\frac{\mu_0 \gamma_N}{4\pi}\right) \mathbf{I} \cdot \int \frac{\mathbf{r} \times \mathbf{j}}{r^3} \mathrm{d}\tau \qquad (68)$$

Because the energy of a magnetic dipole in a magnetic field $\mathbf{B}$ is $-\mathbf{m} \cdot \mathbf{B}$, we can interpret this energy as the interaction of a nuclear dipole $\gamma_N \mathbf{I}$ with a local contribution to the magnetic field

$$\mathbf{B}_{\text{loc}} = \left(\frac{\mu_0}{4\pi}\right) \int \frac{\mathbf{r} \times \mathbf{j}}{r^3} \mathrm{d}\tau \qquad (69)$$

Example 13.3 The evaluation of a coupling energy

A beam of electrons of number density $\mathcal{N}$ travels in the z-direction with linear momentum $k\hbar$ at a perpendicular distance b from a neutron (Fig. 13.13). Calculate the energy of interaction between the neutron magnetic moment and the electron beam.

Method. This is a one-dimensional problem, so $\mathrm{d}\tau = \mathrm{d}z$. The flux density of a particle beam was calculated in Section 2.7, and it may readily be converted into a current density by multiplication by $-e$. For normalization, suppose that the beam lies in the range $-\frac{1}{2}L < z < \frac{1}{2}L$, and let $L \to \infty$ at the end of the calculation. The number density of electrons is related to their actual number by $\mathcal{N} = N_e/L$. Note from Fig. 13.13 that only the x-component of $\mathbf{r} \times \mathbf{k}$ is nonzero.

Answer. The flux density is $N_e k\hbar |A|^2/m_e$; for the normalization in a region of length L, $|A|^2 = 1/L$. The current density is therefore

$$j_z = -\frac{eN_e k\hbar}{m_e L} = -\frac{\mathcal{N} ek\hbar}{m_e}$$

Then, by making use of the relation $\hat{\mathbf{k}} \times \mathbf{r} = -\hat{\mathbf{i}} r \sin\theta$ (see the illustration), we find

$$E^{(1)} = -\left(\frac{\mathcal{N} ek\hbar \gamma_N \mu_0}{4\pi m_e}\right) I_x \int \frac{\sin\theta}{r^2} \mathrm{d}z$$

To evaluate the integral we write $\sin\theta = b/r$ and $r = (b^2 + z^2)^{1/2}$, which implies that

$$\begin{aligned} E^{(1)} &= -\left(\frac{\mathcal{N} ek\hbar \gamma_N \mu_0}{4\pi m_e}\right) I_x b \int_{-\frac{1}{2}L}^{\frac{1}{2}L} (b^2 + z^2)^{-3/2} \mathrm{d}z \\ &= -\left(\frac{\mathcal{N} ek\hbar \gamma_N \mu_0}{4\pi m_e}\right) \frac{I_x b}{b^2} \frac{L}{(b^2 + \frac{1}{4}L^2)^{1/2}} \end{aligned}$$

Finally, we take the limit $L \to \infty$, when the last factor becomes 2 and the final result is

$$E^{(1)} = -\left(\frac{\mathcal{N} ek\hbar \gamma_N \mu_0}{2\pi m_e b}\right) I_x$$

Comment. If the energy of interaction is written as $E^{(1)} = -\gamma_{\mathrm{N}}\mathbf{I}\cdot\mathbf{B}$, then we can interpret the interaction as arising between the magnetic moment of the neutron and a field of induction $\mathcal{B}$ in the x-direction, where

$$\mathcal{B} = \frac{\mathcal{N}ek\hbar\mu_0}{2\pi m_{\mathrm{e}}b}$$

A neutron has figured in the calculation to avoid the effects of charge on the path of the electron beam.

The effect of the external magnetic field is to induce a current density in the electron distribution which is given by

$$\mathbf{j} = -\left(\frac{e}{2m_{\mathrm{e}}}\right)(\psi^*\mathbf{p}\psi - \psi\mathbf{p}\psi^*) - \left(\frac{e^2}{m_{\mathrm{e}}}\right)\mathbf{A}_{\mathrm{ex}}\psi^*\psi \tag{70}$$

where the wavefunctions are those in the presence of the external field, and this current density is the source of the local magnetic field in eqn 13.69. If we identify the contribution to the local field with $-\sigma\mathbf{B}$, as in eqn 13.60, and identify a term proportional to the applied field $\mathbf{B}$, then we shall be able to identify an expression for the shielding constant σ.

To make progress with this programme, we decompose the current density into diamagnetic and paramagnetic contributions (this division is arbitrary, on account of the arbitrary character of the gauge, as explained earlier), and make a corresponding (arbitrary) division of the shielding constant:

$$\sigma = \sigma^{\mathrm{d}} + \sigma^{\mathrm{p}} \qquad \begin{cases} \sigma^{\mathrm{d}}\mathbf{B} = -(\mu_0/4\pi)\int(\mathbf{r}\times\mathbf{j}^{\mathrm{d}}/r^3)\,\mathrm{d}\tau \\ \sigma^{\mathrm{p}}\mathbf{B} = -(\mu_0/4\pi)\int(\mathbf{r}\times\mathbf{j}^{\mathrm{p}}/r^3)\,\mathrm{d}\tau \end{cases} \tag{71}$$

We have seen that the two components of current density travel in opposite directions, and so the two components of the shielding constant will have opposite signs.

13.12 The diamagnetic contribution to shielding

The diamagnetic contribution to the current density is given by eqn 13.52. For a field applied in the z-direction,

$$\begin{aligned}
\mathbf{r}\times\mathbf{j}^{\mathrm{d}} &= -\left(\frac{e^2}{m_{\mathrm{e}}}\right)\mathbf{r}\times\mathbf{A}_{\mathrm{ex}}\psi_0^2 \\
&= -\left(\frac{e^2\mathcal{B}}{2m_{\mathrm{e}}}\right)\psi_0^2\begin{vmatrix}\hat{\mathbf{i}} & \hat{\mathbf{j}} & \hat{\mathbf{k}} \\ x & y & z \\ -y & x & 0\end{vmatrix} \\
&= -\left(\frac{e^2\mathcal{B}}{2m_{\mathrm{e}}}\right)\psi_0^2\{-xz\hat{\mathbf{i}} - yz\hat{\mathbf{j}} + (x^2+y^2)\hat{\mathbf{k}}\}
\end{aligned}$$

The local field therefore has components in all three directions. We are interested only in the component along the z-direction (the coefficient of $\hat{\mathbf{k}}$), and so

$$\sigma_{zz}^{\mathrm{d}}\mathcal{B} = \left(\frac{e^2\mu_0\mathcal{B}}{8\pi m_{\mathrm{e}}}\right)\int\left(\frac{x^2+y^2}{r^3}\right)\psi_0^2\,\mathrm{d}\tau$$

It follows that we can identify the shielding constant in the z-direction as

$$\sigma_{zz}^{\mathrm{d}} = \left(\frac{e^2\mu_0}{8\pi m_{\mathrm{e}}}\right)\int\left(\frac{x^2+y^2}{r^3}\right)\psi_0^2\,\mathrm{d}\tau \tag{72}$$

The mean shielding constant for a freely rotating molecule is $\sigma = \frac{1}{3}(\sigma_{xx}+\sigma_{yy}+\sigma_{zz})$, and because $(x^2+y^2)+(y^2+z^2)+(z^2+x^2)=2r^2$, we arrive at the **Lamb formula**:

$$\sigma^{\mathrm{d}} = \left(\frac{e^2\mu_0}{12\pi m_{\mathrm{e}}}\right)\left\langle\frac{1}{r}\right\rangle \tag{73}$$

The magnitude of the diamagnetic contribution to the shielding therefore depends on the average distance of the electrons from the nucleus in question. This is an easy quantity to estimate for atoms. For the ground state of the hydrogen atom, for instance, $\langle 1/r\rangle = 1/a_0$, and insertion of the numerical values gives $\sigma^{\mathrm{d}} = 1.8\times10^{-5}$.

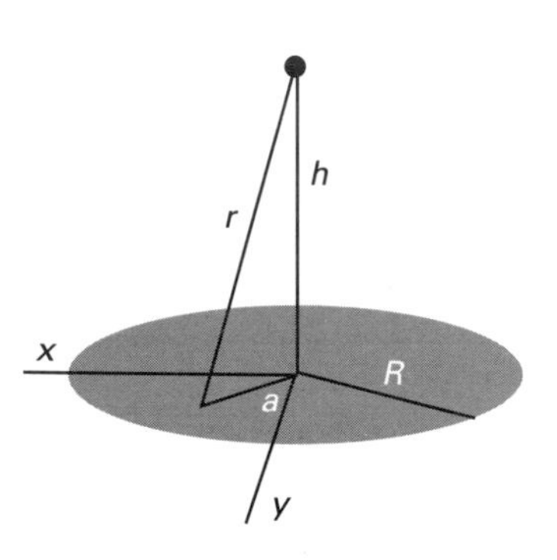

Fig. 13.14 The model used in the calculation in Example 13.4: the tinted disk represents the region of uniform electron density.

Example 13.4 The calculation of shielding constants

In a model of a certain molecule, an electron is confined to a two-dimensional disk-like region of radius R with a uniform probability distribution. A magnetic dipole lies vertically above the centre of the disk at a height h (Fig. 13.14). Calculate the diamagnetic contribution to the shielding constant when the field is applied perpendicular to the disk.

Method. For a field in the z-direction, we use eqn 13.72 with $x^2+y^2=a^2$ and $r^2=h^2+a^2$. The probability density is uniform, so $\psi_0^2 = 1/\pi R^2$.

Answer. Substitution of these relations into eqn 13.72 gives

$$\begin{aligned}\sigma_{zz}^{\mathrm{d}} &= \left(\frac{e^2\mu_0}{8\pi m_{\mathrm{e}}}\right)\left(\frac{1}{\pi R^2}\right)\int_0^{2\pi}\mathrm{d}\phi\int_0^R\left\{\frac{a^2}{(a^2+h^2)^{3/2}}\right\}a\,\mathrm{d}a\\ &= \left(\frac{e^2\mu_0}{8\pi m_{\mathrm{e}}}\right)\left(\frac{1}{\pi R^2}\right)(2\pi)\left\{\frac{R^2+2h^2}{(R^2+h^2)^{1/2}}-2h\right\}\\ &= \left(\frac{e^2\mu_0}{4\pi m_{\mathrm{e}}R^2}\right)\left\{\frac{R^2+2h^2}{(R^2+h^2)^{1/2}}-2h\right\}\end{aligned}$$

The behaviour of this function as h increases is shown in Fig. 13.15.

Comment. When $h=0$ and the nucleus lies in the centre of the disk, the shielding constant is

$$\sigma_{zz}^{\mathrm{d}} = \frac{e^2\mu_0}{4\pi m_{\mathrm{e}}R}$$

Substitution of numerical values gives $\sigma_{zz}^{\mathrm{d}} = 2.8\times10^{-6}/(R/\mathrm{nm})$, so a disk the radius of an atom (about 0.1 nm) gives a shielding constant of about 3×10^{-5}. Note that the shielding constant decreases as R increases because the currents induced by the applied field are increasingly far from the nucleus.

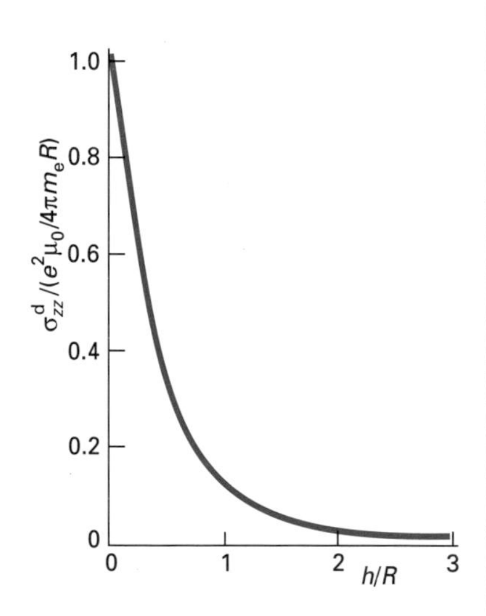

Fig. 13.15 The diamagnetic shielding constant for the system illustrated in the preceding diagram and its variation with height above the plane of the disk.

13.13 The paramagnetic contribution to shielding

The paramagnetic contribution to the shielding constant arises from the interaction of the nucleus with the field generated by the paramagnetic currents like those illustrated in Fig. 13.9, and therefore it depends on the ability of the applied field to mix excited states into the ground state. It follows from the earlier discussion, that in free atoms and ions there will be no paramagnetic contribution because the angular momentum operator l_z is diagonal in the eigenstates of the atom. In molecules, however, there can be a paramagnetic contribution (except parallel to the axis of linear molecules), and in many cases it is dominant.

The strategy for a model calculation involves substituting an expression for the paramagnetic current density into eqn 13.71 for the shielding constant, and then extracting the term that is both linear in the applied field and parallel to it. The coefficient of $\mathcal{B}$ is then identified as $-\sigma^{\mathrm{p}}_{zz}$. For instance, if we use the first-order perturbation expression derived in eqn 13.52, we would obtain

$$-\sigma^{\mathrm{p}}\mathbf{B} = \left(\frac{e\mu_0}{8\pi m_{\mathrm{e}}}\right)\sum_n{}'(a_n - a_n^*)\int \frac{\mathbf{r}\times(\psi_n\mathbf{p}\psi_0 - \psi_0\mathbf{p}\psi_n)}{r^3}\,\mathrm{d}\tau$$

We recognize that $\mathbf{r}\times\mathbf{p} = \mathbf{l}$ occurs in the integrand, so a simpler version of this expression is

$$-\sigma^{\mathrm{p}}\mathbf{B} = \left(\frac{e\mu_0}{8\pi m_{\mathrm{e}}}\right)\sum_n{}'(a_n - a_n^*)\left(\langle n|\frac{\mathbf{l}}{r^3}|0\rangle - \langle 0|\frac{\mathbf{l}}{r^3}|n\rangle\right)$$

At this point we make use of the fact that the orbital angular momentum operator is hermitian and its off-diagonal elements between real states are imaginary. Then[6]

$$\langle 0|\frac{\mathbf{l}}{r^3}|n\rangle = \langle n|\frac{\mathbf{l}}{r^3}|0\rangle^* = -\langle n|\frac{\mathbf{l}}{r^3}|0\rangle$$

and so

$$\sigma^{\mathrm{p}}\mathbf{B} = -\left(\frac{e\mu_0}{4\pi m_{\mathrm{e}}}\right)\sum_n{}'(a_n - a_n^*)\langle n|\frac{\mathbf{l}}{r^3}|0\rangle \tag{74}$$

Now we introduce the mixing coefficients $a_n = -H^{(1)}_{n0}/\Delta E_{n0}$, where $H^{(1)}$ is the perturbation due to the applied field. Because the field lies in the z-direction, we have $H^{(1)}_{n0} = -\gamma_{\mathrm{e}} l_z \mathcal{B}$, so

$$a_n - a_n^* = \left(\frac{\gamma_{\mathrm{e}}\mathcal{B}}{\Delta E_{n0}}\right)(l_{z;n0} - l^*_{z;n0}) = -\left(\frac{2\gamma_{\mathrm{e}}\mathcal{B}}{\Delta E_{n0}}\right)l_{z;0n}$$

We have used hermiticity to write $l_{z;n0} = l^*_{z;0n}$ and then $l^*_{z;0n} = -l_{z;0n}$.

Finally, we can tie everything together. We require the z-component of the local field, so we can write

$$\sigma^{\mathrm{p}}_{zz} = \left(\frac{e\gamma_{\mathrm{e}}\mu_0}{2\pi m_{\mathrm{e}}}\right)\sum_n{}'\frac{l_{z;0n}(r^{-3}l_z)_{n0}}{\Delta E_{n0}}$$

[6] You might worry about the possible lack of commutation of r^{-3} and l_q, and hence the ambiguity in the meaning of l_q/r^3: is it $l_q r^{-3}$ or $r^{-3}l_q$? However, we have seen that l_q is a generator of infinitesimal rotations about the q-axis, and as r is invariant under rotations, it follows that l_q commutes with r and consequently with any function of r. If you do not believe that argument, evaluate $[l_q, r]$ explicitly.

Then, with $\gamma_e = -e/2m_e$, this expression becomes

$$\sigma^{p}_{zz} = -\left(\frac{e^2\mu_0}{4\pi m_e^2}\right)\sum_n{}' \frac{l_{z;0n}(r^{-3}l_z)_{n0}}{\Delta E_{n0}} \tag{75}$$

and the mean value for a freely tumbling molecule is

$$\sigma^{p} = -\left(\frac{e^2\mu_0}{12\pi m_e^2}\right)\sum_n{}' \frac{\mathbf{l}_{0n}\cdot(r^{-3}\mathbf{l})_{n0}}{\Delta E_{n0}} \tag{76}$$

This, at last, is the expression we have been seeking.

The sign of σ^p is negative, which reflects an increase of flux density at the nucleus ($\mathcal{B}_{\text{local}} > \mathcal{B}$). A simple interpretation of the form of the expression is that the factor $\gamma_e l_z \mathcal{B}/\Delta E$ represents the extent to which a current is induced by the applied field, and the other factor, l_z/r^3, represents the transmission of the current magnetically to a dipole at a distance r away. If we apply the closure approximation, with $\mathbf{l}\cdot\mathbf{l}$ replaced by $l(l+1)\hbar^2$ and $l \approx 1$, a very approximate form of eqn 13.76 is

$$\sigma^{p} \approx -\frac{e^2\mu_0\hbar^2}{6\pi m_e^2\Delta E}\left\langle\frac{1}{r^3}\right\rangle \approx -\frac{e^2\mu_0\hbar^2}{6\pi m_e^2 R^3\Delta E} \tag{77}$$

where we have replaced $\langle 1/r^3\rangle$ by $1/R^3$. It follows that

$$|\sigma^{p}/\sigma^{d}| \approx \frac{2\hbar^2}{m_e R^2 \Delta E}$$

With ΔE equivalent to about 30 000 cm^{-1} and $R \approx 0.5$ nm, this ratio works out to about 16, which suggests that paramagnetic contributions to shielding are of greater importance than diamagnetic contributions when low-lying excited states are available. This is because the $1/r^3$ term magnifies the effects of currents when they lie close to the nucleus, but there is no such magnification effect for an external observer measuring magnetic susceptibility.

13.14 The *g*-value

The *g*-value in ESR (EPR) plays a similar role to the shielding constant in NMR, for it takes into account the presence of local fields induced by the applied field by modifying the 'vacuum' value of the interaction from $-g_e\gamma_e\mathbf{s}\cdot\mathbf{B}$ to

$$H^{(1)} = -g\gamma_e\mathbf{s}\cdot\mathbf{B} \tag{78}$$

Although we could proceed in much the same way as for the shielding constant, it is instructive to take a different route to find the relation between g and molecular parameters, and to introduce the concept of a **spin hamiltonian**, a concept widely used in ESR. The thought behind the spin hamiltonian is that whereas the true hamiltonian for an electron involves a lot of different operators, it may be possible to express it as an *effective* hamiltonian in which the effect of all the operators other than the spin have been collected into several parameters. For example, the true hamiltonian for a radical in a magnetic field includes the following terms:

$$H^{(1)} = -g_e\gamma_e\mathbf{s}\cdot\mathbf{B} + \lambda\mathbf{l}\cdot\mathbf{s} - \gamma_e\mathbf{l}\cdot\mathbf{B} \tag{79}$$

representing the effect of the applied field on the spin and orbital angular momenta (the first and third terms) and the spin–orbit coupling (the second term). The spin hamiltonian absorbs the effects of the second and third terms into the single parameter g, and eqn 13.78 is an example of a spin hamiltonian.

To see how this idea works in practice, suppose that the eigenstates of the unperturbed hamiltonian $H^{(0)}$ are $|n\rangle$, with $n = 0$ the ground state. The first-order correction to the energy is the expectation value of $H^{(1)}$ within the orbitally non-degenerate (real) ground state with the field parallel to z:

$$\begin{aligned} E^{(1)} &= -g_e\gamma_e\langle 0|s_z|0\rangle\mathcal{B} + \lambda\langle 0|\mathbf{l}\cdot\mathbf{s}|0\rangle - \gamma_e\langle 0|l_z|0\rangle\mathcal{B} \\ &= -g_e\gamma_e m_s\hbar\mathcal{B} \end{aligned} \tag{80}$$

The second two terms are zero because the expectation value of l_q is zero for real states. The same expression can be obtained for the first-order correction to the energy by introducing a hamiltonian

$$H_1^{(\text{spin})} = -g_e\gamma_e s_z\mathcal{B}$$

and operating on the spin states alone. The spin hamiltonian is starting to emerge.

Now consider the energy correction to second order in the perturbation. The starting point is the perturbation expression

$$E^{(2)} = {\sum_n}' \frac{\langle 0|H^{(1)}|n\rangle\langle n|H^{(1)}|0\rangle}{E_0 - E_n}$$

When the three-term perturbation hamiltonian (eqn 13.79) is inserted, there will be nine terms. However, we are looking for a contribution that can be expressed like eqn 13.78, and therefore are interested only in terms that are bilinear in the spin and applied field (that is, of the form **s** . . . **B**). Only the cross terms between the spin–orbit coupling and the orbital interaction with the applied field have the right form, and so we can confine attention to the following expression:

$$E^{(2)} = (-\lambda\gamma_e\mathcal{B}){\sum_n}' \frac{\langle 0|l_z|n\rangle\langle n|\mathbf{l}\cdot\mathbf{s}|0\rangle + \langle 0|\mathbf{l}\cdot\mathbf{s}|n\rangle\langle n|l_z|0\rangle}{E_0 - E_n}$$

Furthermore, in this simple introduction, we are interested only in an effective hamiltonian containing the operator s_z for the spin, because we are assuming that the local field is parallel to the applied field (in advanced work that assumption is not made). Therefore, with $\Delta E_{n0} = E_n - E_0$, this expression simplifies to

$$\begin{aligned} E^{(2)} &= \lambda\gamma_e\mathcal{B}{\sum_n}' \frac{l_{z;0n}l_{z;n0}m_s\hbar + m_s\hbar l_{z;0n}l_{z;n0}}{\Delta E_{n0}} \\ &= 2\lambda\gamma_e\mathcal{B}m_s\hbar{\sum_n}' \frac{l_{z;0n}l_{z;n0}}{\Delta E_{n0}} \end{aligned}$$

Exactly the same contribution to the energy is obtained if we use the following operator on the spin states:

$$H^{(\text{spin})} = 2\lambda\gamma_e\mathcal{B}\left({\sum_n}' \frac{l_{z;0n}l_{z;n0}}{\Delta E_{n0}}\right)s_z \tag{81}$$

This is the second-order contribution to the spin hamiltonian.

It follows from the preceding discussion that the total spin hamiltonian is the effective operator

$$\begin{aligned} H^{(\mathrm{spin})} &= H_1^{(\mathrm{spin})} + H_2^{(\mathrm{spin})} + \dots \\ &= -g_{\mathrm{e}}\gamma_{\mathrm{e}}\mathcal{B}s_z + 2\gamma_{\mathrm{e}}\lambda\mathcal{B}\left(\sum_n{}' \frac{l_{z;0n}l_{z;n0}}{\Delta E_{n0}}\right)s_z + \dots \\ &= -g_{zz}\gamma_{\mathrm{e}}\mathcal{B}s_z \end{aligned} \tag{82}$$

with

$$g_{zz} = g_{\mathrm{e}} - 2\lambda\sum_n{}' \frac{l_{z;0n}l_{z;n0}}{\Delta E_{n0}} \tag{83}$$

The quantity of interest for rapidly tumbling species in fluid solution is the mean value $g = \frac{1}{3}(g_{xx} + g_{yy} + g_{zz})$, which is

$$g = g_{\mathrm{e}} + \delta g \qquad \delta g = -\tfrac{2}{3}\lambda\sum_n{}' \frac{\mathbf{l}_{0n}\cdot\mathbf{l}_{n0}}{\Delta E_{n0}} \tag{84}$$

Example 13.5 The estimation of a g-value

Consider the model system illustrated in Fig. 13.7, in which a single unpaired electron occupies a p_x-orbital and there is an unoccupied p_y-orbital an energy ΔE above it. Calculate the g-value when the magnetic field is applied in the z-direction.

Method. We use eqn 13.83. The matrix elements required have already been evaluated (in Example 13.2): they are $\langle p_y|l_z|p_x\rangle = i\hbar$ and its hermitian conjugate.

Answer. Substitution of the matrix elements into eqn 13.83 gives

$$\begin{aligned} g_{zz} &= g_{\mathrm{e}} - 2\lambda\frac{\langle p_x|l_z|p_y\rangle\langle p_y|l_z|p_x\rangle}{\Delta E} \\ &= g_{\mathrm{e}} - \frac{2\lambda\hbar^2}{\Delta E} \end{aligned}$$

Comment. When the field is applied along the x-axis, the off-diagonal matrix elements of l_x are zero, and the g-value has its free-spin value.

Exercise 13.5. Calculate the shift when an electron occupies a d_{xy}-orbital with an empty $d_{x^2-y^2}$-orbital at an energy ΔE above it. What difference would there be if there were also p-orbitals at a similar energy above the ground-state orbital?

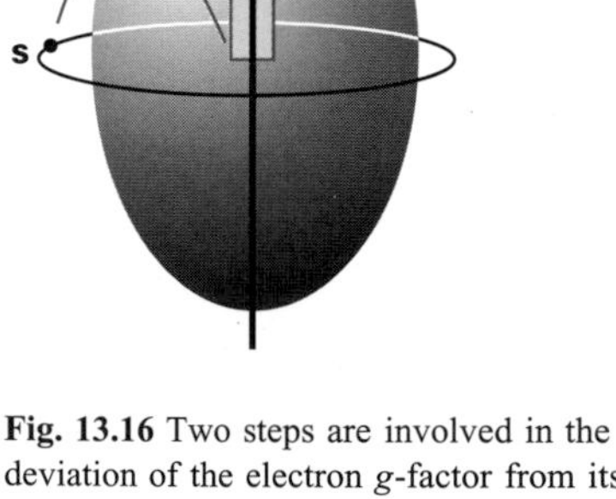

Fig. 13.16 Two steps are involved in the deviation of the electron g-factor from its free-spin value: the applied magnetic field induces orbital angular momentum in the electron, and that orbital angular momentum is transmitted to the spin by the spin–orbit coupling (denoted λ here).

The extent of the deviation of the g-value from the free-spin value increases with increasing spin–orbit coupling constant and with decreasing excitation energy. The factor $\mathcal{B}/\Delta E$ (in eqn 13.81) represents the ease with which the applied field can mix in excited states and so provide a pathway for the electron to circulate through the molecule and hence acquire orbital angular momentum (Fig. 13.16). This orbital angular momentum is then transmitted to the spin as an effective magnetic field through the agency of spin–orbit coupling (the term λ in eqn 13.81). As the excitation energy decreases, the currents can be stirred up more effectively by a given magnetic field, and as the spin–orbit coupling increases, so a given current is experienced by the spin as a stronger magnetic field.

13.15 Spin–spin coupling

There are three types of spin–spin coupling in molecules:

1. **Electron–electron coupling**, which gives rise to the **fine structure** of triplet-state ESR spectra.
2. **Electron–nucleus coupling**, which gives rise to the **hyperfine structure** of ESR and (much less importantly) electronic spectra.
3. **Nucleus–nucleus coupling**, which gives rise to the fine structure of NMR spectra.

We shall deal briefly with the first of these topics, and then introduce electron–nucleus coupling, largely as a foundation for the principal topic of this section, which is spin–spin coupling in NMR.

One mechanism for the interaction between electron spins is the direct dipole–dipole interaction of their magnetic moments. The hamiltonian for the interaction has the form $-\mathbf{m}\cdot\mathbf{B}$ with $\mathbf{B}$ given in eqn 13.61. If the electron spins are aligned along the z-direction, the interaction simplifies to

$$H = \left(\frac{\mu_0 g_e^2 \gamma_e^2}{4\pi r^3}\right)(1 - 3\cos^2\theta)s_{1z}s_{2z} \tag{85}$$

where r is the separation of the electrons. The energy of their interaction is therefore

$$E = \left(\frac{\mu_0 g_e^2 \mu_B^2}{4\pi r^3}\right)(1 - 3\cos^2\theta)m_{s1}m_{s2} \tag{86}$$

In a rapidly tumbling molecule in fluid solution, only the average value of this expression would be observed. However, the average value of $1 - 3\cos^2\theta$ over a sphere is zero, and so there is no net dipole–dipole interaction energy in a rapidly tumbling molecule. The interaction energy does not average to zero in a solid, and so investigating the energy of interaction by solid-state triplet ESR is a way of exploring the distribution of two electrons.

Another mechanism of interaction between electron spins has the same directional dependence as the dipolar interaction. Each of the two electrons interacts with its own orbital angular momentum through a spin–orbit coupling term of the form $\zeta_i \mathbf{s}_i \cdot \mathbf{l}_i$. When these perturbations are used in second-order perturbation theory, they give rise to a second-order contribution that can be modelled by a term in the spin-hamiltonian that is bilinear in the two spin operators $(\mathbf{s}_1 \ldots \mathbf{s}_2)$. This term turns out to have the form $\mathbf{s}_1 \cdot \mathbf{s}_2 - 3(\mathbf{s}_1 \cdot \mathbf{r})(\mathbf{s}_2 \cdot \mathbf{r})/r^2$, exactly as for the direct magnetic dipole interaction (but not with the latter's simple $1/r^3$ dependence). It can be thought of as expressing the energy of interaction of two electron spins that are communicating via their orbital angular momenta: a spin stirs up its own orbital angular momentum, which is experienced by the other electron, which in turn transmits its induced orbital angular

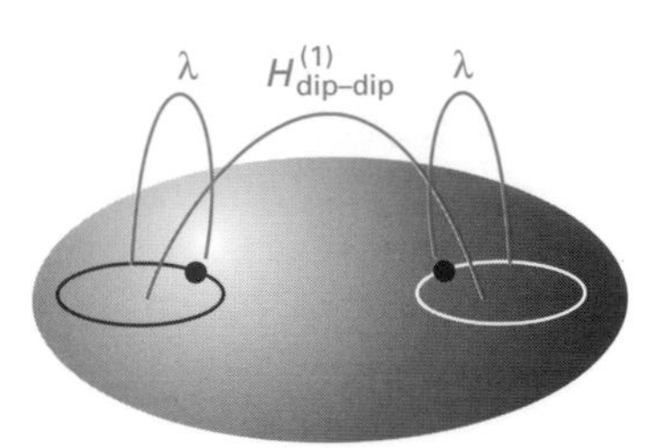

Fig. 13.17 One contribution to electron spin–spin coupling in triplet molecules arises from the spin–orbit coupling, which converts spin angular momentum into orbital angular momentum, and the coupling of these two orbital moments by a dipole–dipole interaction.

momentum to its spin via its own spin–orbit coupling (Fig. 13.17). The direct dipole–dipole mechanism dominates this indirect route in most inorganic species.

13.16 Hyperfine interactions

The term 'hyperfine interaction' denotes any interaction between electrons and nuclei other than the Coulombic interaction between their point electric charges. The interaction may be electric or magnetic. The former includes the interaction between an electric quadrupole moment of the nucleus and the electric field gradient arising from anisotropies in the electron distribution in the molecule. The latter includes the magnetic interactions, such as that between the magnetic dipole moments of the nucleus and the surrounding electrons. We shall concentrate on these magnetic interactions.

There are two types of magnetic interaction between electron and nuclear spins. One is a direct dipolar interaction between the two magnetic moments (Fig. 13.18). The hamiltonian describing this interaction has the form that by now should be familiar:

$$H_{\text{hf}} = \left(\frac{\mu_0 g_{\text{e}}\gamma_{\text{e}}\gamma_{\text{N}}}{4\pi r^3}\right)\left(\mathbf{s}\cdot\mathbf{I} - \frac{3(\mathbf{s}\cdot\mathbf{r})(\mathbf{r}\cdot\mathbf{I})}{r^2}\right) \tag{87}$$

where r is the electron–nucleus distance. When the electron and nuclear spins are so strongly aligned by an external field that only their z-components are of interest, this expression simplifies to

$$\begin{aligned} H_{\text{hf}} &= \left(\frac{\mu_0 g_{\text{e}}\gamma_{\text{e}}\gamma_{\text{N}}}{4\pi r^3}\right)(1-3\cos^2\theta)s_zI_z \\ &= -\left(\frac{g_{\text{e}}g_{\text{N}}\mu_{\text{B}}\mu_{\text{N}}\mu_0}{\hbar^2 4\pi r^3}\right)(1-3\cos^2\theta)s_zI_z \end{aligned}$$

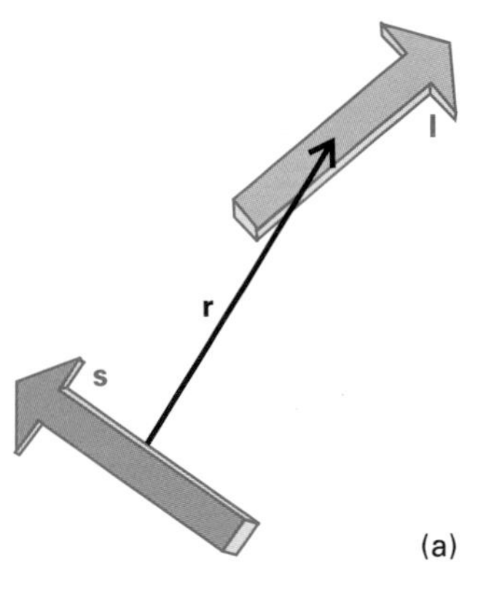

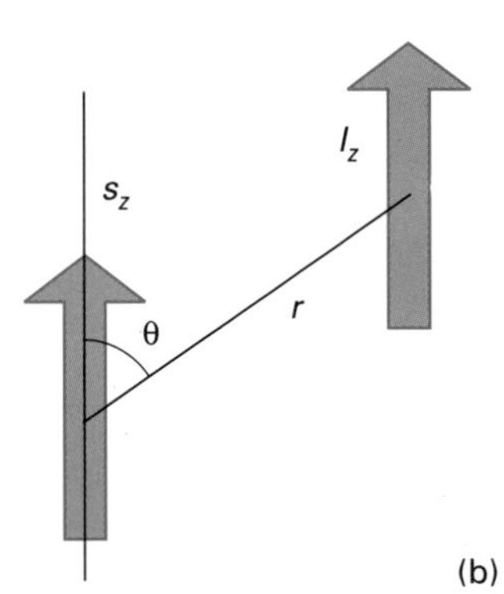

Fig. 13.18 (a) The general relative orientation of two spin angular momenta (and their associated magnetic moments) used in the formulation of the dipolar interaction hamiltonian and (b) the simplified version when the two angular momenta are parallel.

where the Bohr magneton μ_{B} is

$$\mu_{\text{B}} = \frac{e\hbar}{2m_{\text{e}}} = -\gamma_{\text{e}}\hbar$$

and the analogous nuclear magneton μ_{N} is

$$\mu_{\text{N}} = \frac{e\hbar}{2m_{\text{p}}}$$

(Both are positive quantities.) The first-order contribution to the energy is the expectation value of this hamiltonian for the ground-state wavefunction:

$$E_{\text{hf}} = -\left(\frac{g_{\text{e}}g_{\text{N}}\mu_{\text{B}}\mu_{\text{N}}\mu_0}{4\pi}\right)\left\langle\frac{1-3\cos^2\theta}{r^3}\right\rangle m_s m_I \tag{88}$$

and $\hbar\gamma_{\text{N}} = g_{\text{N}}\mu_{\text{N}}$. If the orbital occupied by the electron is an s-orbital, the angular integration can be performed immediately, with the result that the integral over $1-3\cos^2\theta$ vanishes:

$$\int_0^{\pi}(1-3\cos^2\theta)\sin\theta\,\mathrm{d}\theta = \int_{-1}^{1}(1-3x^2)\,\mathrm{d}x = (x-x^3)\Big|_{-1}^{1} = 0$$

We can conclude that an electron in an s-orbital has no net magnetic interaction with its nucleus. However, if the electron occupies some other type of orbital, then its interaction is nonzero. For instance, if it occupies a p_z-orbital, then it has the form

$$\psi = \left(\frac{3}{4\pi}\right)^{1/2} R(r)\cos\theta$$

In this case,

$$\left\langle\frac{1-3\cos^2\theta}{r^3}\right\rangle = \left(\frac{3}{4\pi}\right)\int_0^{2\pi}\mathrm{d}\phi\int_0^{\pi}(1-3\cos^2\theta)\cos^2\theta\sin\theta\,\mathrm{d}\theta \times\int_0^{\infty}\left(\frac{1}{r^3}\right)R^2r^2\mathrm{d}r = -\tfrac{4}{5}\left\langle\frac{1}{r^3}\right\rangle \tag{89}$$

where

$$\left\langle\frac{1}{r^3}\right\rangle = \int_0^{\infty}\left(\frac{1}{r^3}\right)R^2r^2\mathrm{d}r \tag{90}$$

The radial integral, the expectation value of $1/r^3$, can be evaluated by substituting the appropriate expressions for the atomic orbitals (see, for example, Table 3.2 for hydrogenic orbitals or Table 7.1 for STOs).

Although the dipolar interaction is nonzero for a specific orientation of the field with respect to the orbital, when the molecule is tumbling we have to take an orientational average to obtain the mean interaction energy. This mean is zero. So, for rapidly tumbling radicals in solution, there is no net dipolar hyperfine interaction energy.

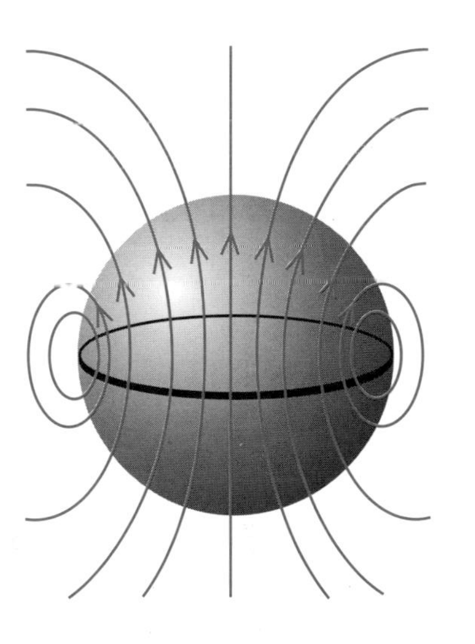

Fig. 13.19 The origin of the Fermi contact interaction is the deviation of the magnetic field pattern from the form it takes on the assumption that the moment can be treated as a point. Note that within the spherical region, loosely denoting the extent of the nucleus, all the lines of force run in the same direction and the angular average is nonzero.

The second hyperfine interaction mechanism we should consider is the **Fermi contact interaction**. It is only an approximation that the magnetic field arising from a nucleus is that of a *point* magnetic dipole. In reality, the nucleus has a finite extent, and it can be treated as a point dipole only when the point of observation is far away. This approximation is valid for all orbitals other than s-orbitals, for electrons in such orbitals are never found at the nucleus itself. However, an electron in an s-orbital can be found at the nucleus, and consequently the point dipole approximation is invalid. That there is a nonzero average magnetic field in this case is illustrated in Fig. 13.19. A quantitative demonstration that there is a nonzero field is developed in *Further information 21*, which takes the vector potential in eqn 13.62 and shows that it implies that the hamiltonian contains the term

$$H_{\mathrm{hf}} = -\tfrac{2}{3}g_{\mathrm{e}}\gamma_{\mathrm{e}}\gamma_{\mathrm{N}}\mu_0\delta(\mathbf{r}_{\mathrm{N}})\mathbf{s}\cdot\mathbf{I} \tag{91}$$

where $\delta(\mathbf{r}_{\mathrm{N}})$ is the **δ-function**, a function that has the following property:

$$\int f(x)\delta(x)\,\mathrm{d}x = f(0) \tag{92}$$

That is, it picks out of $f(x)$ its value at $x = 0$. It follows that when we evaluate the first-order correction to the energy, we find

$$E^{(1)} = -\tfrac{2}{3}g_e\gamma_e\gamma_N\mu_0\left(\int \psi^*\delta(\mathbf{r}_N)\psi\,d\tau\right)\langle 0|\mathbf{s}\cdot\mathbf{I}|0\rangle$$
$$= -\tfrac{2}{3}g_e\gamma_e\gamma_N\mu_0|\psi(0)|^2 m_s m_I\hbar^2$$

where $|\psi(0)|^2$ is the probability density for finding the electron at the nucleus and the ket $|0\rangle$ now refers only to the spin state. The same first-order energy is obtained by adding to the spin hamiltonian a term

$$H^{(\text{spin})} = -\tfrac{2}{3}g_e\gamma_e\gamma_N\mu_0|\psi(0)|^2\mathbf{s}\cdot\mathbf{I} \quad (93)$$

For a $1s$-orbital of hydrogen, $|\psi(0)|^2 = 1/\pi a_0^3$. If the external field is also so strong that only the z-components of the spins are important (which is the case if the off-diagonal matrix elements of this term are much smaller than the energy separations of the spin states in a strong externally applied field), this term becomes

$$H^{(\text{spin})} = -\left(\frac{2}{3\pi a_0^3}\right)g_e\gamma_e\gamma_N\mu_0 I_z s_z \quad (94)$$

The eigenvalues of this effective operator (in the sense that it operates only on the spin states of the system) are

$$E^{(1)} = \left(\frac{2}{3\pi a_0^3}\right)g_e g_N\mu_B\mu_N\mu_0 m_s m_I$$

Insertion of the numerical values gives $E^{(1)}/h \approx (1423\text{ MHz})m_s m_I$. This energy contribution corresponds to the electron experiencing a magnetic field of about 0.5 mT.

At this point we have arrived at the stage where we can express the total spin hamiltonian as

$$H^{(\text{spin})} = -g\gamma_e\mathcal{B}s_z + (A/\hbar^2)I_z s_z + (C/\hbar^2)I_z s_z \quad (95)$$

where g is given by eqn 13.84 and

$$A = -\left(\frac{g_e g_N\mu_B\mu_N\mu_0}{4\pi}\right)\left\langle\frac{1-3\cos^2\theta}{r^3}\right\rangle$$
$$C = \tfrac{2}{3}g_e g_N\mu_B\mu_N\mu_0|\psi(0)|^2 \quad (96)$$

The first-order energies are therefore

$$E^{(1)} = g\mu_B\mathcal{B}m_s + (A+C)m_s m_I \quad (97)$$

The anisotropic term (A) averages to zero if the molecule is tumbling rapidly in solution.

Example 13.6 The estimation of the magnitude of the anisotropic hyperfine coupling

Use STOs to estimate the magnitude of the dipolar hyperfine interaction between an ^{14}N nucleus and an electron in a N$2p_z$-orbital when the spins are (a) parallel, (b) perpendicular to the orbital's axis.

Method. The STOs are specified in Section 7.14 and Table 7.1; according to that table, $Z^* = 3.8340$. Nuclear data are given in Table 13.1. We need to evaluate eqn 13.96 with $g_N = 0.403\,56$. The only tricky point is to ensure that the angle θ is defined appropriately. When the field lies parallel to the orbital's axis, the θ in $1 - 3\cos^2\theta$ is the same as the θ in $p_z \propto \cos\theta$. When the field lies perpendicular to the axis, we can let the form of the interaction remain the same, but we need to interpret the orbital as a p_x orbital instead, in which case we use $p_x \propto \sin\theta\cos\phi$.

Answer. (a) The STO to use is

$$\psi = \left(\frac{Z^{*5}}{32\pi a_0^5}\right)^{1/2} r\cos\theta\, \mathrm{e}^{-Z^* r/2a_0}$$

The expectation value in eqn 13.96 is therefore

$$\left\langle \frac{1 - 3\cos^2\theta}{r^3} \right\rangle = \left(\frac{Z^{*5}}{32\pi a_0^5}\right) \int_0^{2\pi} \mathrm{d}\phi \int_0^{\pi} (1 - 3\cos^2\theta)\cos^2\theta \sin\theta\, \mathrm{d}\theta \times \int_0^{\infty} \left(\frac{1}{r^3}\right) r^2 \mathrm{e}^{-Z^* r/a_0} r^2\, \mathrm{d}r = -\frac{Z^{*3}}{30a_0^3}$$

(b) In this case we use

$$\psi = \left(\frac{Z^{*5}}{32\pi a_0^5}\right)^{1/2} r\sin\theta\cos\phi\, \mathrm{e}^{-Z^* r/2a_0}$$

The same integration as before gives

$$\left\langle \frac{1 - 3\cos^2\theta}{r^3} \right\rangle = \frac{Z^{*3}}{60a_0^3}$$

Therefore,

$$\text{(a) } A = \frac{g_e g_N \mu_B \mu_N \mu_0 Z^{*3}}{120\pi a_0^3} \qquad \text{(b) } A = -\frac{g_e g_N \mu_B \mu_N \mu_0 Z^{*3}}{240\pi a_0^3}$$

The numerical values (expressed as frequencies by dividing by h) are (a) 72 MHz and (b) -36 MHz.

Comment. The values obtained by using SCF orbitals are (a) 134 MHz and (b) -67 MHz. Slater orbitals are not very accurate close to the nucleus, where $1/r^3$ is important.

Exercise 13.6. Show analytically that the magnitude of the hyperfine interaction parallel to the axis of a *p*-orbital is exactly twice the value perpendicular to the axis.

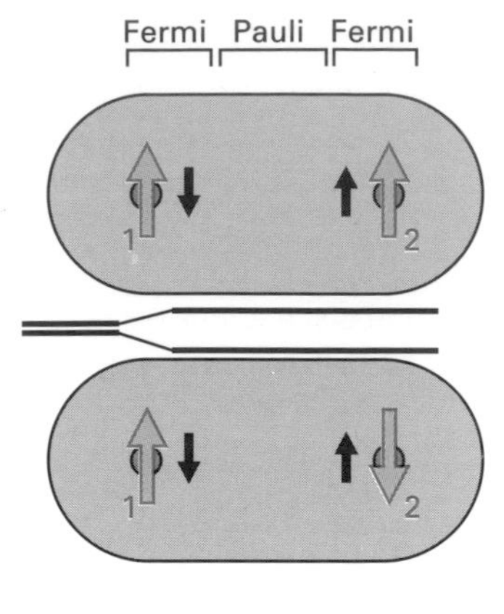

Fig. 13.20 The chain of interactions responsible for nuclear spin–spin coupling.

Table 13.1 Nuclear spin properties

Nuclide	Natural abundance, per cent	Spin I	Magnetic moment, μ/μ_N	g-value g_N
^{1}n*		$\frac{1}{2}$	−1.9130	−3.8260
^{1}H	99.9844	$\frac{1}{2}$	2.792 85	5.5857
^{2}H	0.0156	1	0.857 45	0.857 45
^{3}H*		$\frac{1}{2}$	−2.127 65	−4.2553
^{13}C	1.108	$\frac{1}{2}$	0.7023	1.4046
^{14}N	99.635	1	0.403 56	0.403 56
^{17}O	0.037	$\frac{5}{2}$	−1.893	−0.757 20
^{19}F	100	$\frac{1}{2}$	2.628 35	5.2567
^{31}P	100	$\frac{1}{2}$	1.1317	2.2634
^{33}S	0.74	$\frac{3}{2}$	0.6434	0.4289
^{35}Cl	75.4	$\frac{3}{2}$	0.8218	0.5479
^{37}Cl	24.6	$\frac{3}{2}$	0.6841	0.4561

* Radioactive

13.17 Nuclear spin–spin coupling

There are several interactions in molecules that can contribute to the coupling of nuclear spins. One mechanism is the direct magnetic dipole–dipole interaction of the kind discussed for electrons. This interaction is important for solid samples, but in mobile liquids it averages to zero as a result of the rapid tumbling of the molecules. The mechanisms of importance in fluid samples are those stemming from indirect coupling mediated by the electrons. We shall concentrate on these mechanisms in this section.

One indirect mechanism is illustrated in Fig. 13.20. The first step is a hyperfine interaction between one nucleus and an electron. This interaction has the effect of favouring one orientation of the electron spin rather than the other. The other electron in the bond must have the opposite spin (by the Pauli principle), and is most likely to be found near the other nucleus (because it tends to keep well away from its partner in the bond to minimize electron–electron repulsion). This second electron has a hyperfine interaction with the second nucleus, and consequently one orientation of that nucleus is favoured over the other orientation. As a result, there is an energy difference between the *relative* orientations of the two nuclear spins. Intuitively, we can suspect that there will be a contribution to the spin hamiltonian of the form $\mathbf{I}_1 \cdot \mathbf{I}_2$, because the scalar product is a measure of the angle between the two spins.

Example 13.7 The evaluation of the expectation value of a scalar product

Evaluate the expectation value of the operator $\mathbf{I}_1 \cdot \mathbf{I}_2$ for the triplet ($I = 1$) and singlet ($I = 0$) states of two spin-$\frac{1}{2}$ nuclei and hence find the angles between the spins in the two states.

Method. The scalar product of the two operators should first be expressed in terms of operators with known expectation values: a good starting point is to express it in terms of $\mathbf{I} = \mathbf{I}_1 + \mathbf{I}_2$, because the expectation values of the magnitudes of $\mathbf{I}$, $\mathbf{I}_1$, and $\mathbf{I}_2$ are known. For the second part, use the expression for a scalar product in terms of the angle (θ) between two vectors, $\mathbf{a} \cdot \mathbf{b} = ab \cos\theta$.

Answer. We first note that

$$\mathbf{I}_1 \cdot \mathbf{I}_2 = \tfrac{1}{2}(\mathbf{I}_1 + \mathbf{I}_2) \cdot (\mathbf{I}_1 + \mathbf{I}_2) - \tfrac{1}{2}\mathbf{I}_1 \cdot \mathbf{I}_1 - \tfrac{1}{2}\mathbf{I}_2 \cdot \mathbf{I}_2 = \tfrac{1}{2}I^2 - \tfrac{1}{2}I_1^2 - \tfrac{1}{2}I_2^2$$

The expectation values we require can therefore be calculated from

$$\langle IM_I|\mathbf{I}_1 \cdot \mathbf{I}_2|IM_I\rangle = \tfrac{1}{2}\{I(I+1) - I_1(I_1+1) - I_2(I_2+1)\}\hbar^2$$

Note that the expectation values are independent of M_I. It follows that with $I_1 = I_2 = \frac{1}{2}$,

$$\langle 1M_I|\mathbf{I}_1 \cdot \mathbf{I}_2|1M_I\rangle = \tfrac{1}{4}\hbar^2 \qquad \langle 00|\mathbf{I}_1 \cdot \mathbf{I}_2|00\rangle = -\tfrac{3}{4}\hbar^2$$

To calculate the angles, we use

$$\mathbf{I}_1 \cdot \mathbf{I}_2 = |\mathbf{I}_1||\mathbf{I}_2| \cos\theta = \tfrac{3}{4}\hbar^2 \cos\theta$$

It follows that for the triplet state,

$$\theta = \arccos\frac{(1/4)}{(3/4)} = \arccos\tfrac{1}{3} = 70.5^\circ$$

and for the singlet state,

$$\theta = \arccos\frac{(-3/4)}{(3/4)} = \arccos(-1) = 180^\circ$$

Comment. Note that the expectation values of the scalar products have opposite signs in each case, so if the energy is written as proportional to the scalar product, in one case it rises and in the other case it falls.

The explicit calculation runs as follows. First, we need to decide which hyperfine mechanism to use. In many cases the contact interaction is the most important, and we shall confine our attention to it.[7] The contact interaction for the two nuclei is

$$H^{(1)} = -\tfrac{2}{3}\mu_0 g_e \gamma_e \left\{ \gamma_A \sum_i \mathbf{I}_A \cdot \mathbf{s}_i \delta(\mathbf{r}_{iA}) + \gamma_B \sum_i \mathbf{I}_B \cdot \mathbf{s}_i \delta(\mathbf{r}_{iB}) \right\} \qquad (98)$$

where the sum over i is over all the electrons in the molecule and $\mathbf{r}_{iA}$ is the vector from nucleus A to electron i (and likewise for B). When this operator is integrated over all the spatial variables of the electrons, the δ-functions pick out the value of $|\psi(0)|^2$ at each nucleus for each electron, and so we get the familiar form of the spin hamiltonian for the contact interactions of the electrons with each nucleus. For simplicity of notation we write

$$\mathbf{A} = \sum_i \mathbf{s}_i \delta(\mathbf{r}_{iA}) \qquad \mathbf{B} = \sum_i \mathbf{s}_i \delta(\mathbf{r}_{iB})$$

[7] The dipolar interaction can make a contribution, even in fluids, because it occurs as its square in second-order perturbation theory, and $(1 - 3\cos^2\theta)^2$ does not vanish when averaged over a sphere.

Then the hamiltonian becomes

$$H^{(1)} = -\tfrac{2}{3}\mu_0 g_e \gamma_e \{\gamma_A \mathbf{I}_A \cdot \mathbf{A} + \gamma_B \mathbf{I}_B \cdot \mathbf{B}\} \tag{99}$$

The first-order correction is zero because in a singlet-state molecule the expectation values of the electron spin operators are zero.

When the first-order perturbation hamiltonian is used to calculate the second-order correction to the energy, it gives rise to four terms of the form $\mathbf{I}\ldots\mathbf{I}$. We are interested in the contribution to the spin hamiltonian of the form $J\mathbf{I}_A \cdot \mathbf{I}_B$ and so we need retain only two of these four terms, those proportional to $\mathbf{I}_A \ldots \mathbf{I}_B$ and $\mathbf{I}_B \ldots \mathbf{I}_A$. The second-order contribution to the spin hamiltonian then has the form

$$H^{(\text{spin})} = -\tfrac{8}{9}\mu_0^2 g_e^2 \gamma_e^2 \gamma_A \gamma_B {\sum_n}' \frac{\mathbf{I}_A \cdot \langle 0|\mathbf{A}|n\rangle\langle n|\mathbf{B}|0\rangle \cdot \mathbf{I}_B}{E_n - E_0}$$

We make the usual replacement $\Delta E_{n0} = E_n - E_0$.

Under the operation of taking the spherical average, we can write[8]

$$\langle(\mathbf{I}_A \cdot \mathbf{A})(\mathbf{B} \cdot \mathbf{I}_B)\rangle = \tfrac{1}{3}(\mathbf{I}_A \cdot \mathbf{I}_B)(\mathbf{A} \cdot \mathbf{B})$$

Consequently

$$H^{(\text{spin})} = J\mathbf{I}_A \cdot \mathbf{I}_B \tag{100}$$

with

$$J = -\tfrac{8}{27}\mu_0^2 g_e^2 \gamma_e^2 \gamma_A \gamma_B {\sum_n}' \frac{\langle 0|\mathbf{A}|n\rangle \cdot \langle n|\mathbf{B}|0\rangle}{\Delta E_{n0}} \tag{101}$$

Equation 13.101 is the basic expression for the calculation of the spin–spin coupling constant J, but it obviously needs to be simplified if we are to interpret it physically. The major difficulty lies in the effects of the operators $\mathbf{A}$ and $\mathbf{B}$. If we confine our attention to a two-electron system (such as a chemical bond between the two nuclei), the operator $\mathbf{A}$ would be

$$\begin{aligned}\mathbf{A} &= \mathbf{s}_1\delta(\mathbf{r}_{1A}) + \mathbf{s}_2\delta(\mathbf{r}_{2A}) \\ &= \tfrac{1}{2}(\mathbf{s}_1 + \mathbf{s}_2)\{\delta(\mathbf{r}_{1A}) + \delta(\mathbf{r}_{2A})\} + \tfrac{1}{2}(\mathbf{s}_1 - \mathbf{s}_2)\{\delta(\mathbf{r}_{1A}) - \delta(\mathbf{r}_{2A})\}\end{aligned} \tag{102}$$

and likewise for the operator $\mathbf{B}$. The antisymmetric parts of these operators (the ones with minus signs) have the same general form as the operators introduced in Section 11.9, where we saw that their effect was to mix in *triplet* excited states into a singlet ground state. Because the triplet state of an excited configuration can be expected to lie lower in energy than the corresponding singlet, we can expect the triplet to dominate in the perturbation expression. That implies that in an application of the closure approximation, we should use the mean *triplet* excitation energy $\Delta E^{(T)}$. In that case, under closure we obtain

$$J \approx -\tfrac{8}{27}\mu_0^2 g_e^2 \gamma_e^2 \gamma_A \gamma_B \frac{\langle 0|\mathbf{A} \cdot \mathbf{B}|0\rangle}{\Delta E^{(T)}} \tag{103}$$

[8] See *Further reading*.

For two electrons,

$$\begin{aligned}\mathbf{A}\cdot\mathbf{B} &= \{\mathbf{s}_1\delta(\mathbf{r}_{1A})+\mathbf{s}_2\delta(\mathbf{r}_{2A})\}\cdot\{\mathbf{s}_1\delta(\mathbf{r}_{1B})+\mathbf{s}_2\delta(\mathbf{r}_{2B})\}\\ &= \mathbf{s}_1\cdot\mathbf{s}_1\delta(\mathbf{r}_{1A})\delta(\mathbf{r}_{1B})\\ &\quad+\mathbf{s}_2\cdot\mathbf{s}_2\delta(\mathbf{r}_{2A})\delta(\mathbf{r}_{2B})+\mathbf{s}_1\cdot\mathbf{s}_2\delta(\mathbf{r}_{1A})\delta(\mathbf{r}_{2B})\\ &\qquad+\mathbf{s}_2\cdot\mathbf{s}_1\delta(\mathbf{r}_{2A})\delta(\mathbf{r}_{1B})\end{aligned}$$

The first two terms give zero when integrated over the wavefunction, because an electron cannot simultaneously be at two different nuclei. The action of $\mathbf{s}_1\cdot\mathbf{s}_2$ has already been established in Example 13.7, where we saw that (with change of detail, writing the operator for electrons rather than nuclei)

$$\mathbf{s}_1\cdot\mathbf{s}_2=\tfrac{1}{2}(S^2-s_1^2-s_2^2)$$

The expectation value of this operator in the singlet ground state of the molecule is $-\frac{3}{4}\hbar^2$. It follows that (introducing the Bohr magneton $\mu_B=-\gamma_e\hbar$)

$$J\approx\tfrac{2}{9}g_e^2\mu_0^2\mu_B^2\gamma_A\gamma_B\left(\frac{\langle 0|\delta(\mathbf{r}_{1A})\delta(\mathbf{r}_{2B})+\delta(\mathbf{r}_{2A})\delta(\mathbf{r}_{1B})|0\rangle}{\Delta E^{(T)}}\right)$$

At this point we shall suppose that the electrons occupy an orbital of the form $\psi=c_A\phi_A+c_B\phi_B$ where the ϕs are atomic orbitals on the two nuclei and the coefficients are real. It follows that

$$J\approx\tfrac{4}{9}g_e^2\mu_0^2\mu_B^2\gamma_A\gamma_B|\phi_A(0)|^2|\phi_B(0)|^2\left(\frac{c_A^2c_B^2}{\Delta E^{(T)}}\right)\qquad(104)$$

Because only s-orbitals have nonzero amplitudes at their nucleus, the coefficients that appear in this expression must be those of s-orbitals in the molecular orbital.

Problems

13.1 Calculate the spin contribution to the molar magnetic susceptibility of hydrogen atoms at 298 K.

13.2 Consider a molecule in which there is an excited state at an energy ΔE above the non-degenerate ground state. Show that the angular momentum is no longer completely quenched when a magnetic field is present. *Hint.* Review the argument in Section 13.2 and consider how it is modified when the ground state is perturbed.

13.3 Calculate the average values of S_z^2, S_xS_z, and S_z^4 for a state with spin quantum number S and with all M_S states equally occupied. *Hint.* Use

$$\sum_{r=1}^{n}r^2=\tfrac{1}{6}n(n+1)(2n+1)$$

For the sum over higher powers, see M. Abramowitz and I.A. Stegun, *Handbook of mathematical functions*, Dover (1965), Chapter 23, especially Section 23.1.4.

13.4 The average value of S_z^2 can also be evaluated very simply by noting that in the absence of fields $\langle S_x^2\rangle=\langle S_y^2\rangle=\langle S_z^2\rangle$. Find the average value of S_z^2 in this way for a system with spin quantum number S and with all M_S states equally occupied.

13.5 Sketch the form of the vector function $\mathbf{V}=x\hat{\mathbf{k}}-z\hat{\mathbf{i}}$ and calculate its divergence and curl.

13.6 Confirm that the vector potential $\mathbf{A}=\frac{1}{2}\mathbf{B}\times\mathbf{r}$ describes a uniform magnetic field $\mathbf{B}$ and show that it has zero divergence.

13.7 Find expressions for vector potentials corresponding to a uniform magnetic field (a) parallel to the x-axis, (b) along the direction of the unit vector (1,1,1). Find an expression for A^2 for the vector potential $\mathbf{A}$, and evaluate it for the two special cases.

13.8 Take a vector potential of the form in eqn 13.27 and find expressions for the hamiltonian in the presence of the corresponding magnetic field but for general values of the gauge transformation parameter λ. Is it possible to choose a value of λ such that $H^{(2)}$ is absent?

13.9 Show that the Schrödinger equation is invariant under the gauge transfomation $\mathbf{A} \to \mathbf{A} + \nabla\chi$, $\phi \to \phi - (1/c)\partial\chi/\partial t$, where χ is an arbitrary scalar function, provided that the wavefunction is also multiplied by a factor $e^{-i\chi/\hbar c}$.

13.10 Calculate the contribution to the molar susceptibility of (a) a $1s$-electron, (b) a $2s$-electron, taking Slater orbitals. Specialize to (i) the hydrogen atom, (ii) the carbon atom.

13.11 Estimate the contribution to the molar diamagnetic susceptibility of a $2p$-electron when the field is (a) parallel, (b) perpendicular to the axis. Use Slater orbitals, and then specialize to the carbon atom. What is the mean value?

13.12 An electron occupies one of a doubly degenerate pair of d-orbitals, and its orbital angular momentum corresponds to $\Lambda = +2$. Compute an expression for the current density and plot it for a $3d$ STO on carbon (take $Z^* \approx 1$).

13.13 Plot contour diagrams of the type shown in Fig. 13.10 for planes parallel to the equatorial plane of the hydrogen atom at heights 0, a_0, and $2a_0$ above the nucleus.

13.14 Calculate the form of the diamagnetic and paramagnetic contributions to the current density induced by a magnetic field in the z-direction when the electron occupies (a) a $3d_{xy}$-orbital, (b) a $3d_{x^2-y^2}$-orbital. Suppose that all the degeneracies have been removed by a crystal field. Sketch the form of the current density in the equatorial plane. *Hint*. For the diamagnetic contribution, follow Section 13.9, and for the paramagnetic, follow Section 13.10.

13.15 Sketch the form of the diamagnetic and paramagnetic current densities for an electron in (a) a $2s$-orbital, (b) a $3p_z$-orbital.

13.16 Consider a nitrogen monoxide molecule (nitric oxide, NO) in which the unpaired electron occupies a $2p\pi^*$-orbital formed from a linear combination of the nitrogen and oxygen $2p$-orbitals. For simplicity, take the molecular orbital to be $(1/\sqrt{2})(\psi_N - \psi_O)$; we have ignored the overlap integral. Consider a plane containing both nuclei. Plot contours of the magnitude of the diamagnetic current density taking the p-orbitals to be Slater atomic orbitals: note that this procedure produces a broadside view of the current density.

13.17 Suppose that the NO molecule treated in Problem 13.16 is trapped in a matrix that removes the degeneracy of the π^*-orbitals and separates them by 1.0 eV. What magnetic flux density is needed to restore 10 per cent of the original current density?

13.18 Find an expression for the energy of interaction of the current density computed in Problem 13.16 with the magnetic moment of the nitrogen nucleus. What magnetic flux density does the current give rise to? *Hint*. Use eqn 13.68; $g(^{14}N) = 0.403\,56$ and $I(^{14}N) = 1$.

13.19 Calculate the diamagnetic contribution to the mean shielding constant of an electron in (a) a $2s$-orbital, (b) a $2p$-orbital. Take Slater orbitals, and then specialize to an electron of a carbon atom.

13.20 Calculate the magnitude of the paramagnetic contribution to the mean shielding constant for the same species as in Problem 13.19. Assume that the field mixes in an orbital lying about 5.0 eV above the orbital of interest.

13.21 The ground state of the NO_2 molecule is $^2A_1$, and that of the ClO_2 molecule is $^2B_1$. What states contribute to the deviations of the g-value of the radicals from g_e? *Hint*. The perturbation transforms as a rotation; both molecules are C_{2v}.

13.22 Long ago, in Problem 8.11, the structure of H_2O was investigated. Take the same molecular orbitals for the molecular ion H_2O^+ and estimate its g-values.

13.23 In tetrahedral complexes of Ti^{3+} (configuration d^1), a tetragonal distortion removes the degeneracy of the d-orbitals almost completely. The lowest orbital is d_{z^2}, and the d_{xz}- and d_{yz}-orbitals, which remain degenerate, are at an energy Δ above it. Find an expression for the g-values when the field is applied along the x-, y-, and z-axes of the complex, and estimate their values. Take $\Delta \approx 1.0 \times 10^4\,\mathrm{cm}^{-1}$ and $\zeta = 154\,\mathrm{cm}^{-1}$.

13.24 Show that the energy of dipolar interaction of two electron spin magnetic moments may be expressed as $\mathbf{S} \cdot \mathbf{D} \cdot \mathbf{S}$, where $\mathbf{S} = \mathbf{s}_1 + \mathbf{s}_2$. *Hint*. The energy is proportional to $\mathbf{s}_1 \cdot \mathbf{s}_2 - 3\mathbf{s}_1 \cdot \hat{\mathbf{r}}\hat{\mathbf{r}} \cdot \mathbf{s}_2$. Expand this expression in terms of its Cartesian components and employ relations such as $s_{1x}^2 = \frac{1}{4}\hbar^2$, $S_x^2 = 2s_{1x}s_{2x} + \frac{1}{2}\hbar^2$, etc.

13.25 A Slater $2s$-orbital is zero at the nucleus. Adopt the orthogonalization procedure mentioned in Problem 7.14, which also changes the amplitude from zero, and find a relation for the Fermi contact interaction first for general Z^*, and then for ^{14}N.

13.26 Find an expression for the dipolar hyperfine interaction constant for an electron in a Slater $3d_{z^2}$-orbital when the field is (a) parallel, (b) perpendicular to the axis.

13.27 Estimate the spin–spin coupling constant for the molecule $^1H^2H$. *Hint*. Use eqn 13.104 with a simple LCAO-MO. Take $\Delta E^{(T)} = 10\,\mathrm{eV}$. Express J as a frequency. The experimental value is 40 Hz.

13.28 Write the NMR spin hamiltonian for a molecule containing two protons, one in an environment with chemical shift δ_A and the other with chemical shift δ_B. Let them be coupled through a constant J. Evaluate the matrix elements of the hamiltonian for the states $|m_{IA}m_{IB}\rangle$, and construct and solve the 4×4 secular determinant for the eigenvalues and eigenstates. Determine the allowed magnetic dipole transitions (they correspond to matrix elements of $I_{Ax} + I_{Bx}$), and find their relative intensities. Draw a diagram of the spectrum expected when (a) $J = 0$, (b) $J \ll (\delta_A - \delta_B)\nu_0$, (c) $J = (\delta_A - \delta_B)\nu_0$, (d) $\delta_A = \delta_B$, where ν_0 is the spectrometer frequency. *Hint*. Construct the matrix of the hamiltonian and evaluate its eigenvalues and eigenvectors. Intensities are proportional to the squares of the matrix elements of $I_{Ax} + I_{Bx}$.

Further reading

Classical electrodynamics. J.D. Jackson; Wiley, New York (1975).
Electromagnetic fields. R.K. Wangsness; Wiley, New York (1986).
The theory of electric and magnetic susceptibilities. J.H. van Vleck; Oxford University Press, Oxford (1948).
Magnetochemistry. R.L. Carlin; Springer, Berlin (1986).
Principles of magnetic resonance. C.P. Slichter; Springer, New York (1990).
Hyperfine interactions. A.J. Freeman and R.B. Frankel (ed.); Academic Press, New York (1967).
Quantum theory of magnetic resonance parameters. J.D. Memory; McGraw–Hill, New York (1968).
Nuclear magnetic resonance spectroscopy: a physicochemical view. R.K. Harris; Longman, London (1986).
Ab initio determination of molecular properties. A. Hinchcliffe; Adam Hilger, Bristol (1987).
Molecular electromagnetism. A. Hinchcliffe and R.W. Munn; Wiley, Chichester (1985).
Nuclear magnetic resonance. P.J. Hore; Oxford University Press, Oxford (1995).
Nuclear magnetic resonance spectroscopy. F.A. Bovey, L. Jelinski, and P.A. Mirau; Academic Press, New York (1988).

14 Scattering theory

Scattering experiments have been and continue to be the focus of many experimental and theoretical studies in chemical physics. An early example is the formulation by Rutherford of his nuclear model of the atom, which resulted from his famous experiments with α-particles scattered by gold foil. One reason for their importance is that these experiments can provide a wealth of information on the nature of the interactions between a variety of particles. In addition, the techniques presented here enable us to compute rate constants for chemical reactions from potential surfaces that have been computed by using techniques like those described in Chapter 9.

Collision events come in a variety of forms. In **elastic scattering**, the total translational kinetic energy of the particles remains unchanged as translational kinetic energy is transferred between the two particles. In **inelastic scattering**, the total translational kinetic energy changes and some of it is used to excite internal modes of the projectile or the target. In both cases the composition of the particles remains unchanged, so they are both examples of **non-reactive scattering**. In **reactive scattering**, the composition of the particles does change as old bonds are broken and new bonds form. For example, the collision of A with BC may result in the formation of AC and the release of B.

The formulation of scattering events

We shall direct most of our attention at elastic scattering, which will introduce many of the basic ideas important for understanding the more complex inelastic and reactive scattering events. Although classical and semi-classical treatments of collision events are often quite informative, we shall describe only the quantum mechanical treatment of collision problems and confine the discussion to nonrelativistic processes based on the Schrödinger equation.

14.1 The scattering cross-section

One of the most important quantities determined in any type of scattering experiment is the **scattering cross-section**. The cross-section comes in two varieties, differential and integral, and we define them in this section.

Consider the arrangement shown in Fig. 14.1 in which a beam of incident particles is directed towards the target particles. A detector far from the area of interaction of the incident and target particles presents an 'eye' of area $r^2\mathrm{d}\Omega$ at the orientation (θ, ϕ), where $\mathrm{d}\Omega = \sin\theta\mathrm{d}\theta\mathrm{d}\phi$ is the solid angle subtended by the 'eye'. Suppose that the incident flux of particles, the number of particles per unit area per unit time, is J_i, and that the number of particles per unit time falling on the detector is $\mathrm{d}N_s$, then we can write

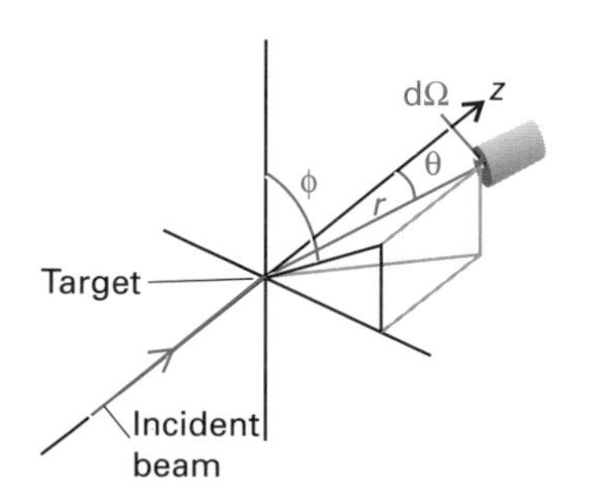

Fig. 14.1 The definition of the scattering cross-section.

$$\mathrm{d}N_s \propto J_i\mathrm{d}\Omega$$

The constant of proportionality, which is called the **differential cross-section**, is written σ, so

$$dN_s = \sigma J_i \, d\Omega \tag{1}$$

The differential cross-section varies with the orientation of the detector to the incident beam. Because we cannot distinguish experimentally between a particle that is not scattered and a particle scattered in the forward direction ($\theta = 0$), the differential cross-section in the forward direction is not a directly experimentally observable quantity. However, in many cases, it can be determined mathematically by extrapolation of experimental results in the vicinity of the forward direction.

The **integral scattering cross-section**, σ_{tot}, is the total cross-section for scattering regardless of angle. It is obtained by integrating the differential cross-section:

$$\sigma_{\text{tot}} = \int_0^{\pi} \int_0^{2\pi} \sigma \sin\theta \, d\theta d\phi \tag{2}$$

The integral scattering cross-section is the constant of proportionality between the total number of scattered particles per unit time, N_s, and the flux of incident particles:

$$N_s = \sigma_{\text{tot}} J_i$$

These definitions of cross-sections are applicable to all varieties of scattering, including elastic, inelastic, and reactive scattering. Cross-sections have dimensions of area, and are often expressed as multiples of a_0^2: they represent the effective area presented to the incident beam for a particular kind of scattering. The non-SI, and faintly jocular unit, the 'barn', with 1 barn $= 10^{-28}$ m^2, is also encountered, particularly in particle physics.

There are a number of hidden assumptions in the interpretation of the cross-sections which need to be brought into the open. One assumption is that the collisions are independent events between a given incident particle and a single target particle. For this condition to be realized experimentally, the incident beam must not be so intense that the incident particles interact with one another. Another assumption is that multiple scattering of one incident particle by several target particles does not occur. A third is that there is no interference between the waves scattered by each of the target particles. In many experiments, these assumptions are valid, though there are notable exceptions. One such exception is the Davisson–Germer experiment in which an electron beam is scattered off a nickel crystal and an extensive interference pattern is observed.

14.2 Stationary scattering states

At the outset we need to represent the incident beam of particles in some way. A rigorous quantum mechanical description uses wavepackets, in which a single incident particle is described by a superposition of plane waves of different momenta. It will be easier, though, to use single-momentum plane waves rather than wavepackets because that greatly simplifies the mathematics yet gives almost identical results.[1]

[1] For a discussion of the rigorous quantal treatment invoking wavepackets, see R.G. Newton, *Scattering theory of waves and particles*, Springer, New York (1982).

We consider first the case of elastic scattering between two structureless particles of masses m_1 and m_2. We assume that their interaction is described by a time-independent potential energy, $V(\mathbf{r})$, that depends only on the separation $\mathbf{r} = (r, \theta, \phi)$ of the two particles. As demonstrated in *Further information 4*, which was first used for the discussion of the hydrogen atom, a two-particle problem can be expressed in terms of the motion of the centre of mass (which does not concern us here) and the relative motion of a particle of reduced mass μ, where

$$\frac{1}{\mu} = \frac{1}{m_1} + \frac{1}{m_2} \tag{3}$$

Furthermore, we shall limit consideration to potentials that approach zero more rapidly than $1/r$ as $r \to \infty$. This restriction rules out scattering by a Coulomb potential.[2]

Because collision events are time-dependent phenomena, it is natural to discuss the problem in terms of the time-dependent Schrödinger equation:

$$H\Psi(\mathbf{r}, t) = \mathrm{i}\hbar\frac{\partial \Psi(\mathbf{r}, t)}{\partial t} \qquad H = -\frac{\hbar^2}{2\mu}\nabla^2 + V(\mathbf{r}) \tag{4}$$

where $\Psi(\mathbf{r}, t)$ is the wavefunction describing the evolution in time of the particle of mass μ. However, because the potential energy is independent of time, the equation can be separated in the usual way and written in terms of solutions of the form

$$\Psi(\mathbf{r}, t) = \psi(\mathbf{r})\mathrm{e}^{-\mathrm{i}Et/\hbar}$$

where the time-independent wavefunction is the solution of

$$H\psi(\mathbf{r}) = E\psi(\mathbf{r}) \tag{5}$$

We are concerned with the solution of the time-independent Schrödinger equation for a positive energy E equal to the relative translational energy of the particles (which for elastic scattering is conserved during the collision). Because there are an infinite number of solutions $\psi(\mathbf{r})$ with $E > 0$, we must find the *particular* solution that satisfies the boundary conditions for the problem of interest. The so-called **asymptotic form** of the solutions, the form of the functions as $r \to \infty$, is thus a very important quantity because it enables us to pin down the solutions by referring to their form when the particles are far apart. At large distances from the target, the wavefunction consists of three components. One we can recognize as a plane wave of definite momentum towards the target. Another is a wave that corresponds to transmission through the target without scattering; this component is a plane wave corresponding to linear momentum away from the target, a continuation of the incident wave. The third component is a scattered wave (Fig. 14.2).

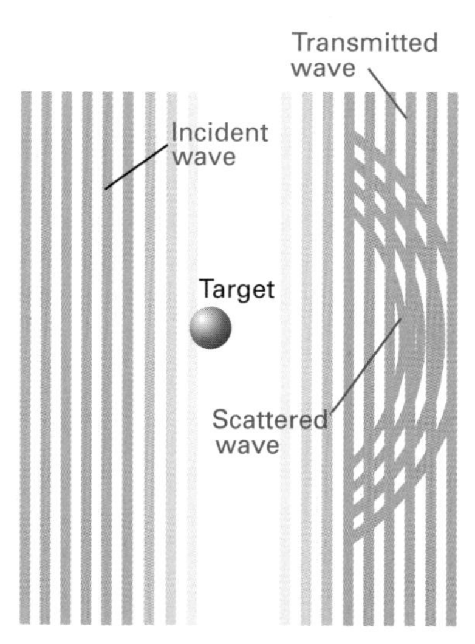

Fig. 14.2 In a scattering experiment, an incident plane wave gives rise to a transmitted wave in the same direction (with the same linear momentum) and a scattered wave.

We now construct these various components. The wavefunction for a particle with linear momentum $\mathbf{p} = \mathbf{k}\hbar$, where $\mathbf{k}$ is the **wavevector** of the motion, is

$$\psi(\mathbf{r}) = \mathrm{e}^{\mathrm{i}\mathbf{k}\cdot\mathbf{r}}$$

[2] For a discussion of Coulomb scattering, see A. Messiah, *Quantum mechanics*, Vol. 1, North Holland, Amsterdam (1961), which is still an excellent source.

(We ignore normalization questions at this stage.) This form of wavefunction was discussed in Section 2.6. The magnitude of the wavevector, k, is related to the kinetic energy of the projectile by

$$E = \frac{k^2\hbar^2}{2\mu} \tag{6}$$

Because the transmitted wave is a continuation of the incident wave, the plane incident wave is also the form of the component corresponding to transmission. In the region of the target molecule, the wavefunction is distorted from this simple form, perhaps in a very complicated manner. However, we are interested only in the shape of the function far from the target where $V \simeq 0$. At these great distances, it will have the form of an outgoing wave with an amplitude that varies with angle; such a function has the following form:

$$f_k(\theta, \phi)\frac{e^{ikr}}{r}$$

The r in the denominator ensures that this outgoing wave is square-integrable (recall that the volume element increases as r^2). The angle-dependent function will also depend on the energy, and so we have included an index k on f. It follows that the asymptotic form of the total wavefunction, allowing for incident, transmitted, and scattered components, will be of the form

$$\psi(\mathbf{r}) \simeq e^{i\mathbf{k}\cdot\mathbf{r}} + f_k(\theta, \phi)\frac{e^{ikr}}{r} \tag{7}$$

The function $\psi(\mathbf{r})$ that has this asymptotic form is called a **stationary scattering state**. The **scattering amplitude**, $f_k(\theta, \phi)$, reflects the anisotropy of the scattering event. We will show its explicit relation to the interaction potential $V(\mathbf{r})$ in the next section. Note that f_k has the dimensions of length.

We now need to establish the link between the asymptotic form of the wavefunction and the outcome of observations as expressed by the scattering cross-section. Indeed, we shall now show that the differential cross-section is related to the scattering amplitude by

$$\sigma = |f_k(\theta, \phi)|^2 \tag{8}$$

To confirm this relation, we need to consider the **flux density**, **J**, which was first introduced in Section 2.7 and is defined as

$$\mathbf{J} = \frac{\hbar}{2\mu i}(\psi^*\nabla\psi - \psi\nabla\psi^*) \tag{9}$$

where the operator ∇ ('grad') can be written in either cartesian or spherical polar coordinates:

$$\begin{aligned}\nabla &= \hat{\mathbf{i}}\frac{\partial}{\partial x} + \hat{\mathbf{j}}\frac{\partial}{\partial y} + \hat{\mathbf{k}}\frac{\partial}{\partial z} \\ &= \hat{\mathbf{r}}\frac{\partial}{\partial r} + \hat{\boldsymbol{\theta}}\frac{1}{r}\frac{\partial}{\partial\theta} + \hat{\boldsymbol{\phi}}\frac{1}{r\sin\theta}\frac{\partial}{\partial\phi}\end{aligned} \tag{10}$$

The flux density corresponding to a wave in the z-direction is

$$J_z = \frac{\hbar}{2\mu\mathrm{i}}\left(\mathrm{e}^{-\mathrm{i}kz}\frac{\partial}{\partial z}\mathrm{e}^{\mathrm{i}kz} - \mathrm{e}^{\mathrm{i}kz}\frac{\partial}{\partial z}\mathrm{e}^{-\mathrm{i}kz}\right) = \frac{k\hbar}{\mu} \tag{11}$$

Therefore, for a general wavevector $\mathbf{k}$,

$$\mathbf{J} = \frac{\mathbf{k}\hbar}{\mu} \tag{12}$$

and the flux density is essentially the velocity of the particle (its momentum divided by its mass). For a plane wave, which corresponds to a uniform probability density throughout space, the flux density is also uniform everywhere. For the scattered wave, it is easier to use the spherical polar form of the operator, and to consider each component separately. The radial component of the flux density is

$$\begin{aligned} J_r &= \frac{\hbar}{2\mu\mathrm{i}} f_k^* f_k \left(\frac{\mathrm{e}^{-\mathrm{i}kr}}{r}\frac{\partial}{\partial r}\frac{\mathrm{e}^{\mathrm{i}kr}}{r} - \frac{\mathrm{e}^{\mathrm{i}kr}}{r}\frac{\partial}{\partial r}\frac{\mathrm{e}^{-\mathrm{i}kr}}{r}\right) \\ &= \frac{k\hbar|f_k|^2}{\mu r^2} \end{aligned} \tag{13}$$

When the same calculation is performed for the angular components (see Problem 14.5), the resulting expressions are proportional to r^{-3}. As we are interested only in the asymptotic contributions, we need retain only the radial component, which survives out to greater distances than the angular components of the flux density.

The differential cross-section is now evaluated by using eqn 14.1. The magnitude of the incident flux is $\hbar k/\mu$. The number of particles scattered into the detector of area $r^2\mathrm{d}\Omega$ per unit time is the scattered radial flux times the area:

$$\mathrm{d}N_s = J_r r^2 \mathrm{d}\Omega = \frac{k\hbar|f_k|^2}{\mu}\mathrm{d}\Omega$$

Then, by substituting these expressions into eqn 14.1,

$$\frac{k\hbar|f_k|^2}{\mu}\mathrm{d}\Omega = \frac{\sigma\hbar k}{\mu}\mathrm{d}\Omega$$

from which eqn 14.8 follows. The integrated cross-section is obtained from

$$\sigma_{\mathrm{tot}} = \int |f_k(\theta,\phi)|^2 \sin\theta \, \mathrm{d}\theta \mathrm{d}\phi \tag{14}$$

where the integration is over the surface of a sphere.

In this analysis we have neglected the contribution to the flux that results from the interference between the transmitted wave $\mathrm{e}^{\mathrm{i}\mathbf{k}\cdot\mathbf{r}}$ and the scattered wave $f_k\mathrm{e}^{\mathrm{i}kr}/r$. For example, we have ignored terms in eqn 14.9 such as

$$f_k^*(\theta,\phi)\frac{\mathrm{e}^{-\mathrm{i}kr}}{r}\nabla\mathrm{e}^{\mathrm{i}\mathbf{k}\cdot\mathbf{r}}$$

These interference terms are only important for scattering in the forward direction; as long as the detector is positioned elsewhere than at $\theta = 0$ we do not need to consider them.

14.3 The integral scattering equation

We shall now show that the Schrödinger equation and its asymptotic boundary conditions can be combined into a single equation which, although difficult to solve, is ideally suited to the formulation of approximations. The combination of a differential equation with its boundary conditions into a single equation introduces the powerful technique of 'Green's functions', which are widely used throughout scattering theory.

Equation 14.5 can be rewritten by using eqn 14.6 as

$$(\nabla^2 + k^2)\psi(\mathbf{r}) = \frac{2\mu}{\hbar^2} V(\mathbf{r})\psi(\mathbf{r}) \tag{15}$$

It will prove convenient to simplify the appearance of this equation by introducing

$$U(\mathbf{r}) = \frac{2\mu}{\hbar^2} V(\mathbf{r}) \tag{16}$$

for it then becomes

$$(\nabla^2 + k^2)\psi(\mathbf{r}) = U(\mathbf{r})\psi(\mathbf{r}) \tag{17}$$

A **Green's function**, $G(\mathbf{r}, \mathbf{r}')$, is a solution of the following equation:

$$(\nabla^2 + k^2)G(\mathbf{r}, \mathbf{r}') = -4\pi\delta(\mathbf{r} - \mathbf{r}') \tag{18}$$

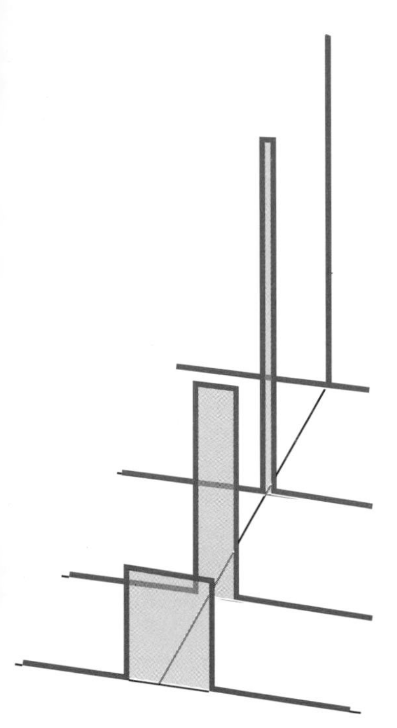

Fig. 14.3 A Dirac δ-function can be regarded as the limit of a rectangular function that shrinks in width and increases in height in such a way as to preserve unit area.

The term $\delta(\mathbf{r} - \mathbf{r}')$ is the **Dirac δ-function** (Fig. 14.3). It can be pictured as being zero everywhere except at $\mathbf{r}' = \mathbf{r}$, and has the following effect:

$$\int g(\mathbf{r})\delta(\mathbf{r} - \mathbf{r}')\,\mathrm{d}\mathbf{r} = g(\mathbf{r}')$$

That is, it picks out of a function g its value at one particular point. It follows from this property that a contribution to the solution of eqn 14.17 is

$$\psi(\mathbf{r}) = -\frac{1}{4\pi}\int G(\mathbf{r}, \mathbf{r}')U(\mathbf{r}')\psi(\mathbf{r}')\,\mathrm{d}\mathbf{r}' \tag{19}$$

To verify that this function is a solution, we proceed as follows:

$$\begin{aligned}(\nabla^2 + k^2)\psi(\mathbf{r}) &= -\frac{1}{4\pi}\int (\nabla^2 + k^2)G(\mathbf{r}, \mathbf{r}')U(\mathbf{r}')\psi(\mathbf{r}')\,\mathrm{d}\mathbf{r}' \\ &= \int \delta(\mathbf{r} - \mathbf{r}')U(\mathbf{r}')\psi(\mathbf{r}')\,\mathrm{d}\mathbf{r}' = U(\mathbf{r})\psi(\mathbf{r})\end{aligned}$$

as required. However, it is not the complete solution, because if there is a function $\psi^{(0)}(\mathbf{r})$ that satisfies

$$(\nabla^2 + k^2)\psi^{(0)}(\mathbf{r}) = 0 \tag{20}$$

then this function could be added to the previous solution and the sum would still satisfy eqn 14.17. Therefore, the complete general solution of eqn 14.17 is

$$\psi(\mathbf{r}) = \psi^{(0)}(\mathbf{r}) - \frac{1}{4\pi}\int G(\mathbf{r}, \mathbf{r}')U(\mathbf{r}')\psi(\mathbf{r}')\,\mathrm{d}\mathbf{r}' \tag{21}$$

There are several different Green's functions (that is, solutions of eqn 14.18), and the choice of which one to use depends on the boundary conditions that we have imposed. We shall adopt the Green's function that results in an asymptotic solution of the form given in eqn 14.7. Such a function is called an **outgoing Green's function** and denoted $G^{(+)}(\mathbf{r}, \mathbf{r}')$. It is demonstrated in *Further information 12* that

$$G^{(+)}(\mathbf{r}, \mathbf{r}') = \frac{e^{ik|\mathbf{r}-\mathbf{r}'|}}{|\mathbf{r}-\mathbf{r}'|} \tag{22}$$

It then follows that a formal solution of the Schrödinger equation for this scattering problem is

$$\psi(\mathbf{r}) = \psi^{(0)}(\mathbf{r}) - \frac{1}{4\pi}\int \frac{e^{ik|\mathbf{r}-\mathbf{r}'|}}{|\mathbf{r}-\mathbf{r}'|} U(\mathbf{r}')\psi(\mathbf{r}')\,d\mathbf{r}' \tag{23}$$

Notice that ψ appears on both sides of this expression, and it is therefore an example of an integral equation: it is called the **integral scattering equation**. What we have achieved is the conversion of a differential equation (the Schrödinger equation) and its boundary conditions into an integral equation that contains, implicitly, the boundary conditions. Now, as integral equations are in general much harder to solve than differential equations, it may appear that we are moving away from finding solutions. However, we shall soon see that integral equations are well-formed for finding *approximate* solutions.

Example 14.1 Green's functions and boundary conditions

Confirm that $\psi(\mathbf{r})$ as given by eqn 14.23 is consistent with the asymptotic form of the solution established in eqn 14.7.

Method. Whenever dealing with asymptotic expressions, we need retain terms that decay most slowly (as the smallest power of $1/r$). We can also make use of the fact (as asserted earlier) that V decays more rapidly than $1/r$, and that it may therefore be considered to be negligibly small for r' larger than a certain small value. That in turn implies that the only terms in the integrand that contribute have $r' \ll r$, so $|\mathbf{r}-\mathbf{r}'|$ in the denominator can be approximated by $|\mathbf{r}-\mathbf{r}'| \approx r - \hat{\mathbf{r}}\cdot\mathbf{r}'$, where $\hat{\mathbf{r}}$ is the unit vector in the $\mathbf{r}$ direction.

Answer. In the denominator of eqn 14.23, we keep only the first term in the expansion of $|\mathbf{r}-\mathbf{r}'|$ since we are considering the asymptotic form of the stationary scattering state, but we keep both terms in the exponent. We then obtain

$$\psi(\mathbf{r}) \simeq \psi^{(0)}(\mathbf{r}) - \frac{1}{4\pi}\frac{e^{ikr}}{r}\int e^{-ik\hat{\mathbf{r}}\cdot\mathbf{r}'} U(\mathbf{r}')\psi(\mathbf{r}')\,d\mathbf{r}'$$

This expression is identical to eqn 14.7 if we take the free particle state $\psi^{(0)}(\mathbf{r})$ to be $e^{i\mathbf{k}\cdot\mathbf{r}}$ and equate the scattering amplitude to

$$f_k(\theta,\phi) = -\frac{1}{4\pi}\int e^{-ik\hat{\mathbf{r}}\cdot\mathbf{r}'} U(\mathbf{r}')\psi(\mathbf{r}')\,d\mathbf{r}' \tag{24}$$

Comment. Notice that, as desired, integration yields a function of only the angles (θ, ϕ) via the unit vector $\hat{\mathbf{r}}$.

Exercise 14.1. Evaluate the scattering amplitude if the stationary scattering state $\psi(\mathbf{r})$ is approximated by $e^{ik\hat{\mathbf{r}}\cdot\mathbf{r}'}$ and the potential $V(\mathbf{r})$ is given simply by $e^{-\alpha r^3}$, where α is a constant.

$[f_k = -2\mu/3\alpha\hbar^2]$

14.4 The Born approximation

We promised that the integral scattering equation would be easier to solve by approximation than the Schrödinger equation itself. To see that this is indeed the case, we shall now begin to solve eqn 14.21 iteratively. The equation itself is

$$\psi(\mathbf{r}) = \psi^{(0)}(\mathbf{r}) - \frac{1}{4\pi}\int G(\mathbf{r},\mathbf{r}')U(\mathbf{r}')\psi(\mathbf{r}')\,\mathrm{d}\mathbf{r}' \tag{25}$$

The problem with this equation is that we do not know the value of $\psi(\mathbf{r}')$, so we cannot evaluate the integral to find $\psi(\mathbf{r})$. However, we can form an equation for $\psi(\mathbf{r}')$ by changing $\mathbf{r}'$ to $\mathbf{r}''$ and $\mathbf{r}$ to $\mathbf{r}'$, for the equation then becomes

$$\psi(\mathbf{r}') = \psi^{(0)}(\mathbf{r}') - \frac{1}{4\pi}\int G(\mathbf{r}',\mathbf{r}'')U(\mathbf{r}'')\psi(\mathbf{r}'')\,\mathrm{d}\mathbf{r}''$$

This expression can now be substituted into the integrand in eqn 14.25, which results in

$$\begin{aligned}\psi(\mathbf{r}) = \psi^{(0)}(\mathbf{r}) &- \frac{1}{4\pi}\int G(\mathbf{r},\mathbf{r}')U(\mathbf{r}')\psi^{(0)}(\mathbf{r}')\,\mathrm{d}\mathbf{r}' \\ &+ \left(\frac{1}{4\pi}\right)^2\int\int G(\mathbf{r},\mathbf{r}')U(\mathbf{r}')G(\mathbf{r}',\mathbf{r}'')U(\mathbf{r}'')\psi(\mathbf{r}'')\,\mathrm{d}\mathbf{r}'\mathrm{d}\mathbf{r}''\end{aligned} \tag{26}$$

Now the first *two* terms on the right-hand side of this equation are known and only the third (final) term contains the unknown function ψ. We can repeat this procedure, and substitute the equation for $\psi(\mathbf{r}'')$ into the integrand of the third term, and so successively generate terms of the **Born expansion** of $\psi(\mathbf{r})$.

It may be helpful to see the structure of the Born expansion by writing it symbolically. Thus, if we write eqn 14.25 as

$$\psi = \psi^{(0)} + GU\psi \tag{27}$$

where we have not shown the integrations explicitly (or the numerical factor), then eqn 14.26 would be

$$\psi = \psi^{(0)} + GU\psi^{(0)} + GUGU\psi$$

and continuation of this series gives

$$\psi = \psi^{(0)} + GU\psi^{(0)} + GUGU\psi^{(0)} + \ldots \tag{28}$$

The utility of the Born expansion is that each successive term in the expansion has one higher power in U and so, if the potential is weak, successive terms get smaller and smaller. We can presume that the expansion converges and that for very weak scattering potentials it does so quite rapidly.[3]

[3] See E. Merzbacher, *Quantum mechanics*, Wiley, New York (1970), p229 and M. Rotenberg, *Ann. Phys.*, 21, 579 (1963) for a discussion of the convergence of the Born expansion.

We are now in a position to round off the calculation by substituting the Bor expansion for the stationary scattering state into eqn 14.24 and so obtain th Born expansion of the scattering amplitude. The so-called **Born approxima tion** is the result of keeping only the first term, and neglecting all terms highe than first order in U:

$$f_k(\theta,\phi) = -\frac{1}{4\pi}\int \mathrm{e}^{-\mathrm{i}k\hat{\mathbf{r}}\cdot\mathbf{r}'}U(\mathbf{r}')\mathrm{e}^{\mathrm{i}\mathbf{k}\cdot\mathbf{r}'}\,\mathrm{d}\mathbf{r}' \quad (29$$

In short, the Born approximation replaces the stationary scattering state by plane wave in the expression for the scattering amplitude.

Example 14.2 How to use the Born approximation

Calculate the differential cross-section for scattering from a Yukawa potential, which is a potential of the form:

$$V(r) = V_0\frac{\mathrm{e}^{-\alpha r}}{r}$$

where V_0 and α are constants. The Yukawa potential is an example of a central potential, one that depends only on r and not (θ,ϕ), and was originally introduced to represent the interaction between fundamental particles.

Method. The first step is to insert the Yukawa potential into the Born approximation for the scattering amplitude, and then to evaluate the integral. We shall need the following definite integral:

$$\int_0^\infty \mathrm{e}^{-ax}\sin bx\,\mathrm{d}x = \frac{b}{a^2+b^2}$$

Then, with f_k determined, the differential scattering cross-section can be determined by using eqn 14.8.

Answer. Within the Born approximation, we have

$$f_k(\theta,\phi) = -\frac{1}{4\pi}\left(\frac{2\mu V_0}{\hbar^2}\right)\int \mathrm{e}^{-\mathrm{i}k\hat{\mathbf{r}}\cdot\mathbf{r}'}\frac{\mathrm{e}^{-\alpha r'}}{r'}\mathrm{e}^{\mathrm{i}\mathbf{k}\cdot\mathbf{r}'}\,\mathrm{d}\mathbf{r}'$$

Integration over the angles associated with $\mathbf{r}'$ gives

$$\begin{aligned} f_k(\theta,\phi) &= -\frac{1}{4\pi}\left(\frac{2\mu V_0}{\hbar^2}\right)\int_0^\infty\int_{-1}^{1}\int_0^{2\pi}\frac{\mathrm{e}^{-\alpha r'}}{r'}\mathrm{e}^{\mathrm{i}(|\mathbf{k}-k\hat{\mathbf{r}}|r'\cos\theta')} \\ &\quad\times r'^2\mathrm{d}r'\,\mathrm{d}\cos\theta'\,\mathrm{d}\phi' \\ &= -\left(\frac{2\mu V_0}{\hbar^2}\right)\int_0^\infty \frac{\mathrm{e}^{-\alpha r'}}{r'}\frac{\sin(|\mathbf{k}-k\hat{\mathbf{r}}|r')}{|\mathbf{k}-k\hat{\mathbf{r}}|r'}r'^2\,\mathrm{d}r' \end{aligned}$$

The definite integral quoted above then gives

$$f_k(\theta,\phi) = -\left(\frac{2\mu V_0}{\hbar^2}\right)\frac{1}{\alpha^2+|\mathbf{k}-k\hat{\mathbf{r}}|^2}$$

Because θ is the angle between the incident wavevector $\mathbf{k}$ and the unit vector $\hat{\mathbf{r}}$ in the scattered direction (see Fig. 14.4), it follows that

$$|\mathbf{k}-k\hat{\mathbf{r}}| = 2k\sin\tfrac{1}{2}\theta$$

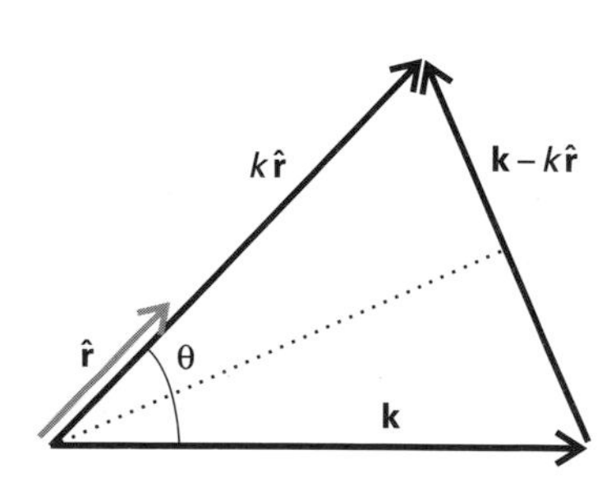

Fig. 14.4 The vectors used in the calculation in Example 14.2.

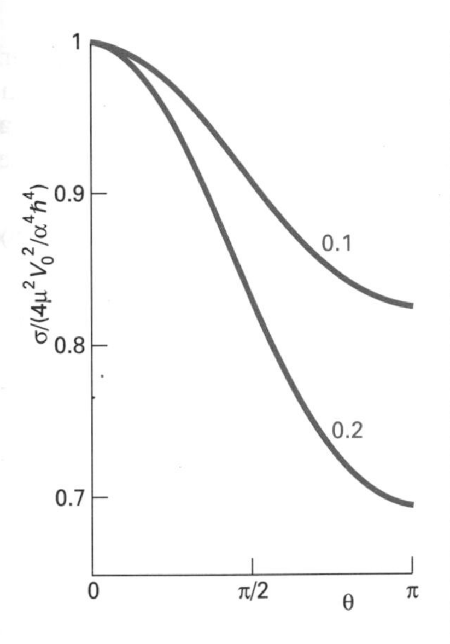

Fig. 14.5 The differential scattering cross-section for a Yukawa potential as a function of scattering angle. The numbers labelling the curves are the values of $4k^2/\alpha^2$.

and therefore that

$$f_k(\theta, \phi) = -\frac{2\mu V_0/\hbar^2}{\alpha^2 + 4k^2 \sin^2 \frac{1}{2}\theta}$$

Note that the scattering amplitude (and consequently the differential cross-section) is independent of the angle ϕ. This independence is a general result for elastic scattering by a central potential. It follows from eqn 14.8 that the differential cross-section is

$$\sigma = \frac{4\mu^2 V_0^2/\hbar^4}{\left(\alpha^2 + 4k^2 \sin^2 \frac{1}{2}\theta\right)^2}$$

Comment. The differential cross-section varies with k and so it also varies with energy E. In the limit of zero energy ($k \rightarrow 0$), the differential cross-section is

$$\sigma = \frac{4\mu^2 V_0^2}{\hbar^4 \alpha^4}$$

and is independent of θ as well as of ϕ. Except at zero energy, σ peaks in the forward direction ($\theta = 0$) and decreases monotonically as θ varies from 0 to π (Fig. 14.5). Notice that, within the Born approximation, the differential cross-section is independent of the sign of V_0, and gives the same result if the Yukawa potential is attractive ($V_0 < 0$) or repulsive ($V_0 > 0$).

Exercise 14.2. Use the Born approximation to calculate the differential cross-section for scattering from the central potential $V(r) = \alpha/r^2$, where α is a constant. A useful definite integral is

$$\int_0^\infty \frac{\sin x}{x}\,dx = \frac{\pi}{2}$$

$[\sigma = (\pi^2\mu^2\alpha^2)/(4k^2\hbar^4 \sin^2 \frac{1}{2}\theta)]$

Partial-wave stationary scattering states

We now focus on scattering by a central potential, $V(r)$, a potential that depends only on the distance, r, between the incident and target particles. It then follows by the same argument that we have used in the example, that the scattering amplitude and the asymptotic form of the stationary scattering state depend only on k, r, and the angle between the incident wavevector $\mathbf{k}$ and scattering direction $\hat{\mathbf{r}}$. Without loss of generality, we choose $\mathbf{k}$ to lie along the z-axis, in which case the scattering amplitude is independent of ϕ.

14.5 Partial waves

The asymptotic form, eqn 14.7, of the stationary scattering state can be written as

$$\psi(r, \theta) \simeq e^{ikr\cos\theta} + f_k(\theta)\frac{e^{ikr}}{r} \tag{30}$$

For elastic scattering by a central potential, the orbital angular momentum l of the incident particle relative to the target particle is conserved during the collision because there are no torques present to accelerate it. Therefore, we should be able to decompose the scattering problem into a set of smaller problems, each characterized by a unique value of l. This separation is accomplished by expanding the stationary scattering state $\psi(r,\theta)$ and scattering amplitude $f_k(\theta)$ in a complete set of basis functions; the natural choice for this central-field problem are the spherical harmonics, but because the states are independent of ϕ and have cylindrical symmetry about the direction $\mathbf{k}$, we need consider only the functions with $m_l = 0$. These functions are proportional to the Legendre polynomials, $P_l(\cos\theta)$. In the equations that follow, all sums over l range over the complete set of values, from 0 to ∞. It then follows that we can expand the scattering amplitude and the wavefunction as

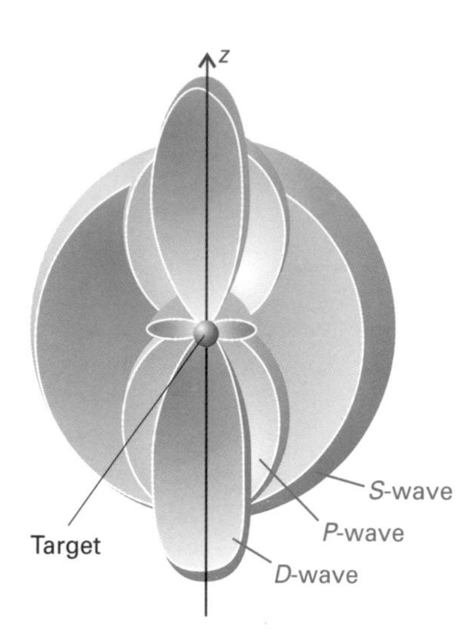

Fig. 14.6 A representation of the S-, P-, and D-wave contributions to the total scattered wave. Note that they differ in their angular distribution but have cylindrical symmetry around the direction of propagation of the incident particles (the z-direction).

$$f_k(\theta) = \sum_l f_l P_l(\cos\theta) \qquad \psi(r,\theta) = \sum_l R_l(r) P_l(\cos\theta)$$

where $R_l(r)$ is a radial function and f_l is a constant, both dependent on the value of l and the wavevector k. Each product $R_l P_l$, which we shall denote ψ_l, is called the **partial-wave stationary scattering state**, and our first task is to find the equation these products satisfy. Each one is the solution of a Schrödinger equation, and after we have solved these individual equations, we can reconstruct the overall wavefunction by adding their individual solutions together. This approach is called a **partial-wave analysis** of the stationary scattering state. Likewise, the decomposition of f_k is a partial-wave analysis of the scattering amplitude. The contribution with $l = 0$ is called ***S*-wave scattering**, that with $l = 1$ is called ***P*-wave scattering**, and so on by analogy with atomic orbitals (Fig. 14.6).

14.6 The partial-wave equation

We shall now derive the differential equation and boundary conditions satisfied by each $\psi_l(r,\theta)$. This analysis will lead us to the concept of the scattering 'phase shift', and, by making use of that concept, to an expression for the scattering amplitude and cross-section.

To construct the partial-wave equation we insert the partial-wave expansion into eqn 14.5. The effect of the laplacian, ∇^2, in spherical coordinates is

$$\nabla^2 R_l P_l = P_l \frac{1}{r}\frac{\mathrm{d}^2}{\mathrm{d}r^2} rR_l + R_l \frac{1}{r^2}\Lambda^2 P_l \tag{31}$$

where Λ^2 is the legendrian operator (eqn 3.19). It should be recalled from Section 3.5 that

$$\Lambda^2 Y_{lm_l} = -l(l+1)Y_{lm_l}, \text{ which implies that } \Lambda^2 P_l = -l(l+1)P_l$$

because Y_{l0} is proportional to P_l. Then, by using eqns 14.31 and 14.4, eqn 14.5 becomes

$$\sum_l \left\{ -\frac{\hbar^2}{2\mu} P_l \frac{1}{r}\frac{\mathrm{d}^2}{\mathrm{d}r^2} rR_l + \frac{\hbar^2 l(l+1)}{2\mu r^2} R_l P_l + V R_l P_l \right\} = E \sum_l R_l P_l$$

For this equation to hold, the differential equation must be satisfied for each value of l individually. Therefore, for each value of l, we need

$$-\frac{\hbar^2}{2\mu}\frac{1}{r}\frac{\mathrm{d}^2}{\mathrm{d}r^2}rR_l + \frac{\hbar^2 l(l+1)}{2\mu r^2}R_l + VR_l = ER_l$$

To simplify the appearance of this equation we introduce the function $u_l = rR_l$, which turns it into

$$-\frac{\hbar^2}{2\mu}\frac{\mathrm{d}^2 u_l}{\mathrm{d}r^2} + \frac{\hbar^2 l(l+1)}{2\mu r^2}u_l + Vu_l = Eu_l \tag{32}$$

When we need to reconstruct the partial-wave scattering state, we use

$$\psi_l(r,\theta) = \frac{1}{r}u_l(r)P_l(\cos\theta) \tag{33}$$

We now need to consider the boundary conditions on u_l. First, consider its value at $r = 0$. Because $R_l(0)$ is finite, it follows that $u_l(0) = 0$. Secondly, we need to consider in detail the asymptotic behaviour as $r \to \infty$. This step requires a discussion of states of the free particle, which is provided in the following section.

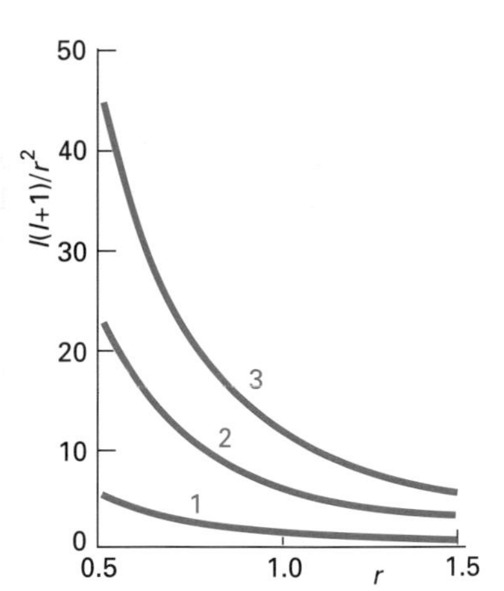

Fig. 14.7 A fragment of the repulsive potential energy arising from the centrifugal effect of orbital angular momentum. The numbers labelling the curves are the values of l.

14.7 Free-particle radial wavefunctions

As $r \to \infty$, $V(r) \to 0$ faster than $1/r$, and eqn 14.32 reduces to the equation for a free particle with orbital angular momentum l:

$$-\frac{\hbar^2}{2\mu}\frac{\mathrm{d}^2 u_l^0}{\mathrm{d}r^2} + \frac{\hbar^2 l(l+1)}{2\mu r^2}u_l^0 = Eu_l^0$$

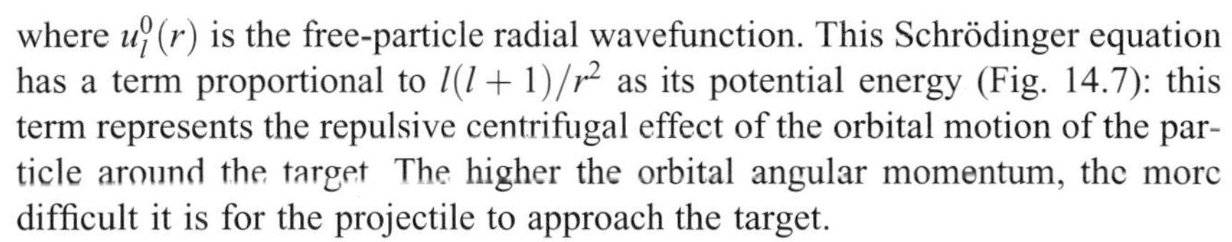

where $u_l^0(r)$ is the free-particle radial wavefunction. This Schrödinger equation has a term proportional to $l(l+1)/r^2$ as its potential energy (Fig. 14.7): this term represents the repulsive centrifugal effect of the orbital motion of the particle around the target. The higher the orbital angular momentum, the more difficult it is for the projectile to approach the target.

A small manipulation of the last equation turns it into

$$\frac{\mathrm{d}^2 u_l^0}{\mathrm{d}r^2} + \left\{k^2 - \frac{l(l+1)}{r^2}\right\}u_l^0 = 0 \tag{34}$$

This differential equation is well-known to mathematicians. Its general solution is a linear combination of a **Riccati–Bessel function**, $\hat{j}_l(kr)$, and a **Riccati–Neumann function**, $\hat{n}_l(kr)$:

$$u_l^0 = A_l\hat{j}_l(kr) + B_l\hat{n}_l(kr) \tag{35}$$

Although these functions (which we shall refer to jointly as the 'Riccati functions') might be unfamiliar, there is nothing particularly mysterious about them, and a few of them are listed in Table 14.1 and plotted in Fig. 14.8. Any properties that we need we shall develop.[4] In particular, we shall make use of their

[4] For a summary of their more important properties, see Newton, *Scattering theory of waves and particles*, Springer, New York (1982). Some authors, including Newton, define $\hat{n}_l(kr)$ with the opposite sign.

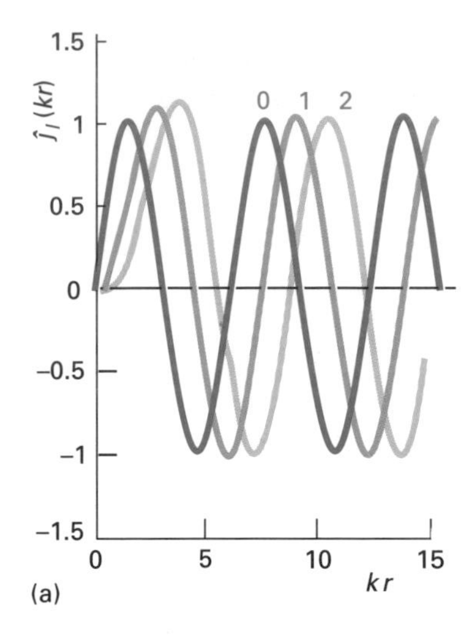

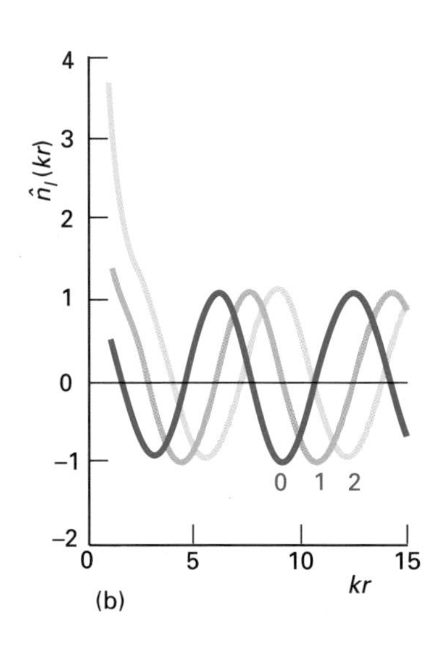

Fig. 14.8 Three examples of (a) Riccati–Bessel functions and (b) Riccati–Neumann functions.

Table 14.1 Riccati functions

$\hat{j}_0(z) = \sin z$	$\hat{n}_0(z) = \cos z$
$\hat{j}_1(z) = \frac{1}{z}\sin z - \cos z$	$\hat{n}_1(z) = \sin z + \frac{1}{z}\cos z$
$\hat{j}_2(z) = \left(\frac{3}{z^2} - 1\right)\sin z - \frac{3}{z}\cos z$	$\hat{n}_2(z) = \frac{3}{z}\sin z + \left(\frac{3}{z^2} - 1\right)\cos z$

relationship to the spherical Bessel functions $j_l(kr)$ and spherical Neumann functions $n_l(kr)$ through

$$\hat{j}_l(kr) = krj_l(kr) \qquad \hat{n}_l(kr) = krn_l(kr)$$

We also need their asymptotic behaviour as $r \to \infty$:

$$\hat{j}_l(kr) \simeq \sin\left(kr - \tfrac{1}{2}l\pi\right) \qquad \hat{n}_l(kr) \simeq \cos\left(kr - \tfrac{1}{2}l\pi\right)$$

That is, far from the origin, the functions are simple sine and cosine functions that are shifted in phase by $\frac{1}{2}l\pi$. It follows that the asymptotic form of the free-particle radial wavefunction is

$$u_l^0 \simeq A_l \sin\left(kr - \tfrac{1}{2}l\pi\right) + B_l \cos\left(kr - \tfrac{1}{2}l\pi\right)$$

14.8 The scattering phase shift

The asymptotic form of the free-particle radial wavefunction suggests that it will be sensible to write the two constants A_l and B_l as

$$A_l = C_l \cos\delta_l \qquad B_l = C_l \sin\delta_l$$

because then

$$\begin{aligned} u_l^0 &\simeq C_l\{\cos\delta_l \sin\left(kr - \tfrac{1}{2}l\pi\right) + \sin\delta_l \cos\left(kr - \tfrac{1}{2}l\pi\right)\} \\ &\simeq C_l \sin\left(kr - \tfrac{1}{2}l\pi + \delta_l\right) \end{aligned}$$

Therefore, all the information necessary to discuss the asymptotic form is carried by the **scattering phase shift**, δ_l.

The free-particle asymptotic form is also the asymptotic form of the scattered wave, because the potential does not influence the particle as $r \to \infty$, so we now know that

$$u_l \simeq C_l \sin\left(kr - \tfrac{1}{2}l\pi + \delta_l\right) \tag{36}$$

The scattering phase shift, δ_l, will prove to be of critical importance, for we shall see that it contains all the information necessary to compute cross-sections for elastic scattering. The phase shift can be obtained from the coefficients A_l and B_l by using the relation

$$\tan\delta_l = \frac{B_l}{A_l}$$

However, this relation leaves the phase shift unspecified up to the addition of an arbitrary integral multiple of π; this ambiguity is referred to as the **modulo-π ambiguity** in δ_l. The modulo-π ambiguity affects neither the differential nor the integrated cross-section computed from the phase shift (see below).

At this point we also know the asymptotic form of the stationary state $\psi(r,\theta)$ as $r \to \infty$, because substitution of eqn 14.36 into eqn 14.33 gives

$$\psi(r,\theta) \simeq \sum_l \frac{C_l}{r} P_l(\cos\theta)\sin(kr - \tfrac{1}{2}l\pi + \delta_l) \tag{37}$$

However, this expression must be consistent with eqn 14.30, which in terms of partial waves is

$$\psi(r,\theta) \simeq e^{ikr\cos\theta} + \sum_l f_l \frac{e^{ikr}}{r} P_l(\cos\theta)$$

To bring the two expressions into a form in which they can be compared, we need the following expansion:[5]

$$e^{ikr\cos\theta} = \sum_l i^l(2l+1)j_l(kr)P_l(\cos\theta) \tag{38}$$

This expansion expresses a wave corresponding to linear momentum along the z-axis (the wavefunction e^{ikz}) as an infinite superposition of orbital angular momentum states with $m_l = 0$. The asymptotic form of this expansion as $r \to \infty$ is obtained by substituting the asymptotic form of the spherical Bessel functions, and is

$$e^{ikr\cos\theta} \simeq \sum_l i^l(2l+1)\frac{\sin(kr - \frac{1}{2}l\pi)}{kr} P_l(\cos\theta)$$

It follows that the asymptotic form of the scattering state is

$$\psi(r,\theta) \simeq \sum_l \left\{ i^l(2l+1)\frac{\sin(kr - \frac{1}{2}l\pi)}{kr} + f_l \frac{e^{ikr}}{r} \right\} P_l(\cos\theta)$$

By comparing this equation with the same expression in terms of phase shifts, we see that the two are equivalent if

$$\frac{C_l}{r}\sin(kr - \tfrac{1}{2}l\pi + \delta_l) = i^l(2l+1)\frac{\sin(kr - \frac{1}{2}l\pi)}{kr} + f_l\frac{e^{ikr}}{r}$$

It is now possible to manipulate this expression into something much simpler by making use of the relations

$$\sin x = \frac{e^{ix} - e^{-ix}}{2i} \qquad \text{and } i^l = e^{il\pi/2}$$

and collecting terms with a common factor of e^{-ikr} (see Problem 14.10). This procedure leads to the identification

$$C_l = \frac{i^l(2l+1)}{k} e^{i\delta_l} \tag{39}$$

Similarly, when we equate terms with a common factor of e^{ikr} we find

$$f_l = \frac{(2l+1)}{2ik}\left(e^{2i\delta_l} - 1\right) = \frac{(2l+1)}{k} e^{i\delta_l}\sin\delta_l \tag{40}$$

[5] See C. Cohen-Tannoudji, B. Diu, and F. Laloë, *Quantum mechanics*, Vol. 2, Wiley, New York (1977).

14.9 Phase shifts and scattering cross-sections

Now at last we can find an expression for the scattering amplitude in terms of the phase shift, the quantity that is central to the computation of physical observables. All we need is to add together the partial-wave contributions ($f_k = \sum_l f_l P_l$):

$$f_k(\theta) = \frac{1}{k}\sum_l (2l+1)e^{i\delta_l} \sin\delta_l P_l(\cos\theta) \tag{41}$$

This important formula is exactly what we need to relate the scattering cross-section to the phase shift, which is simply

$$\begin{aligned}\sigma &= \frac{1}{k^2}\left|\sum_l (2l+1)e^{i\delta_l} \sin\delta_l \, P_l(\cos\theta)\right|^2 \\ &= \frac{1}{k^2}\sum_{l,l'} (2l+1)(2l'+1)e^{i(\delta_l-\delta_{l'})} \sin\delta_l \sin\delta_{l'} P_l(\cos\theta)P_{l'}(\cos\theta)\end{aligned} \tag{42}$$

Because $P_0(\cos\theta) = 1$, the contribution with $l = l' = 0$ to the differential cross-section is isotropic:

$$\sigma = \frac{\sin^2\delta_0}{k^2}$$

This equation implies that if the experimental differential cross-section is anisotropic, then partial waves with $l > 0$ are almost certainly important in the scattering.

To obtain the expression for the integral cross-section we make use of the following orthogonality property of the Legendre polynomials:

$$\int_0^\pi P_l(\cos\theta)P_{l'}(\cos\theta)\sin\theta \, d\theta = \frac{2}{2l+1}\delta_{ll'} \tag{43}$$

where $\delta_{ll'}$ is the Kronecker delta (Section 1.6). This integration, when applied to eqn 14.42, eliminates all terms for which $l' \neq l$, and enables us to write the integrated cross-section as a sum of **partial-wave cross-sections**, σ_l:

$$\sigma_{\text{tot}} = \sum_l \sigma_l \qquad \text{with } \sigma_l = \frac{4\pi}{k^2}(2l+1)\sin^2\delta_l \tag{44}$$

Notice that the σ_l are in fact *integrated* cross-sections for the partial wave l. The k in the denominator shows that σ_l is small for very high energies. The proportionality of σ_l to $\sin^2\delta_l$ shows that as the phase shift increases from zero the cross-section increases; the factor $2l+1$ plays the role of a degeneracy factor that magnifies this effect.

We see that the phase shifts δ_l and their variation with angular momentum l and energy (effectively k) are indispensible for a calculation of elastic scattering cross-sections. For example, at an energy at which $\delta_l = (n+\frac{1}{2})\pi$, where n is an integer, the partial-wave cross-section σ_l reaches its maximum value

$$\sigma_{l,\max} = \frac{4\pi}{k^2}(2l+1)$$

Depending on the behaviour of the other phase shifts at this energy, this maximization of σ_l may result in a maximum in the integral cross-section σ. This behaviour is often associated with a phenomenon known as resonance, and we shall encounter it again in Section 14.11.

We can obtain some insight into the significance of the phase shift δ_l by comparing the asymptotic forms of the stationary scattering state ψ and the wave $e^{ikr\cos\theta}$. For the former, eqns 14.37 and 14.39 imply that

$$\psi(r,\theta) \simeq \sum_l \frac{i^{l-1}(2l+1)}{2kr} P_l(\cos\theta)\left(-e^{-i(kr-l\pi/2)} + e^{i(kr-l\pi/2)}e^{2i\delta_l}\right) \quad (45)$$

For the latter, eqn 14.38 implies that

$$e^{ikr\cos\theta} \simeq \sum_l \frac{i^{l-1}(2l+1)}{2kr} P_l(\cos\theta)\left(-e^{-i(kr-l\pi/2)} + e^{i(kr-l\pi/2)}\right)$$

Comparison of the components with the same angular momentum l shows that both this wave and the stationary scattering state are superpositions of incoming and outgoing spherical waves. The incoming waves are identical in each case, but the outgoing waves differ by a factor of $e^{2i\delta_l}$. Thus, the effect of the potential is to shift the phase of each outgoing partial wave. This effect is the origin of the name 'scattering phase shift' for δ_l.

For the sum $\sigma = \sum_l \sigma_l$ to converge, σ_l must vanish in the limit of large l. However, σ_l is proportional to $2l+1$, so the sine function must vanish as l grows large. It will do so if $\delta_l \to n\pi$, where n is an integer, as $l \to \infty$. We can understand this decrease in partial-wave scattering cross-section by recalling the interpretation of $\hbar^2 l(l+1)/(2\mu r^2)$ in eqn 14.32 as a repulsive centrifugal potential. As l gets large, this centrifugal potential dominates $V(r)$ and the incident particle does not interact with the target particle. Thus, for large l, the stationary scattering state is described by $e^{ikr\cos\theta}$ itself, which implies that $\delta_l = 0$ (to a modulo-π ambiguity).[6]

Example 14.3 Scattering by a hard sphere

As an example of determining scattering phase shifts, consider the case of S-wave scattering by a hard sphere, where $V(r)$ takes the form

$$V(r) = \begin{cases} \infty & \text{if } r \le a \\ 0 & \text{if } r > a \end{cases}$$

The constant a represents the distance of closest approach. Calculate σ_0 for this system.

Method. Classically, we would expect a collision to occur if the incident particle (treated as structureless) approaches the target particle to within a distance a. The classical cross-section is thus πa^2 (Fig. 14.9). For S-wave scattering the centrifugal potential is zero at all distances, so the equations are easier to solve than when it is nonzero. We must first establish the

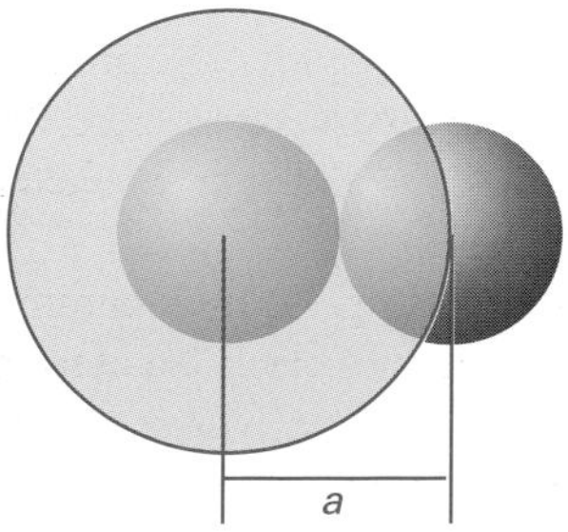

a

Fig. 14.9 The classical collision cross-section for two hard spheres of diameter a.

[6] This ambiguity is eliminated by requiring that δ_l be a smooth function of the energy that vanishes as the energy becomes infinite.

appropriate boundary conditions for the problem. Because the potential energy is infinite for $r \leq a$, the radial wavefunction u_0 must vanish for $r \leq a$.

Answer. The potential energy is zero for $r > a$, so the solution u_0 is of the form

$$u_0 = A \sin kr + B \cos kr = C \sin(kr + \delta_0)$$

This equation is consistent with the asymptotic form given in eqn 14.36. Because the radial wavefunction must be continuous, the following condition must be satisfied at $r = a$:

$$C \sin(ka + \delta_0) = 0$$

This condition implies that $\delta_0 = -ka$ (to within a modulo-π ambiguity). The partial-wave scattering amplitude f_0 is given by eqn 14.40 as

$$f_0 = -\frac{1}{k} e^{-ika} \sin ka$$

It then follows from eqn 14.42 that the S-wave differential cross-section is $(\sin^2 ka)/k^2$; as expected, it is isotropic. The partial-wave cross-section from eqn 14.44 is

$$\sigma_0 = \frac{4\pi}{k^2} \sin^2 ka$$

For $ka \ll 1$, which corresponds to very low energies, we can write $(\sin x)/x \approx 1$, so σ_0 is to an excellent approximation $4\pi a^2$.

Comment. At low energies, the S-wave scattering cross-section is independent of energy and four times the classical cross-section.

Exercise 14.3. Consider the case of P-wave scattering by a hard sphere. By imposing the condition that the radial wavefunction is continuous at $r = a$, find an expression for the $l = 1$ phase shift δ_1.

[$\tan \delta_1 = -\hat{j}_1(ka)/\hat{n}_1(ka)$]

14.10 Scattering by a spherical square well

We shall now consider as an example a central potential with the characteristics of a spherical square well:

$$V(r) = \begin{cases} -V_0 & \text{if } r \leq a \\ 0 & \text{if } r > a \end{cases}$$

We note that this potential might be able to support bound states; that is, there may be solutions of eqn 14.32 for discrete energies $E < 0$. The ability of the potential well to possess quantized energy levels will depend on the values of V_0, a, μ, and l. Here, however, we shall consider the solutions u_l corresponding to continuum or scattering states with $E > 0$.

We solve this problem by writing down the solution u_l of eqn 14.32 in the two regions inside and outside the well and then require that the radial wavefunction and its first derivative be continuous at $r = a$.

First, we consider the region $r \leq a$. The equation we have to solve is

$$\frac{d^2u_l}{dr^2} + \left\{K^2 - \frac{l(l+1)}{r^2}\right\}u_l = 0$$

where

$$\hbar^2K^2 = 2\mu(E + V_0)$$

By analogy with eqn 14.34, the general solution u_l is a linear combination of a Riccati–Bessel function and a Riccati–Neumann function:

$$u_l = A_l\hat{j}_l(Kr) + A_l'\hat{n}_l(Kr)$$

To ensure that ψ is not infinite at the origin, we require $u_l(0) = 0$. Therefore, we need to know the behaviour of the Riccati functions at the origin. The Riccati–Bessel function $\hat{j}_l(kr)$ behaves like $(kr)^{l+1}$ as $kr \to 0$, whereas the Riccati–Neumann function $\hat{n}_l(kr)$ behaves like $(kr)^{-l}$. It follows that for $u_l(r)$ to vanish at the origin, we must have $A_l' = 0$. Therefore, inside the well the solution is of the form

$$u_l = A_l\hat{j}_l(Kr) \tag{46}$$

In the outer region, $r > a$, the potential has no direct influence and eqn 14.32 reduces to the free-particle equation, eqn 14.34, and we can immediately write down

$$u_l = C_l\hat{j}_l(kr) + D_l\hat{n}_l(kr)$$

where, as usual, k is related to the energy by $E = k^2\hbar^2/2\mu$. As in the general case, we introduce the constant δ_l by the relations

$$C_l = B_l\cos\delta_l \qquad D_l = B_l\sin\delta_l$$

and write

$$u_l = B_l\cos\delta_l\hat{j}_l(kr) + B_l\sin\delta_l\,\hat{n}_l(kr) \tag{47}$$

The asymptotic forms of the Riccati functions can quickly be used to show that u_l has the asymptotic form given in eqn 14.36 and that δ_l is the desired scattering phase shift for partial wave l.

To determine the solution u_l we require continuity of the wavefunction and its first derivative at $r = a$. From eqns 14.46 to 14.47 we obtain from the continuity of u_l,

$$A_l\hat{j}_l(Ka) = B_l\cos\delta_l\hat{j}_l(ka) + B_l\sin\delta_l\,\hat{n}_l(ka) \tag{48}$$

and for the continuity of the slope

$$KA_l\hat{j}_l'(Ka) = kB_l\cos\delta_l\hat{j}_l'(ka) + kB_l\sin\delta_l\,\hat{n}_l'(ka)$$

where the prime denotes the derivative with respect to r. Division of the second of these equations by the first results in a complicated expression for the phase shift:

$$K\frac{\hat{j}_l'(Ka)}{\hat{j}_l(Ka)} = k\frac{\hat{j}_l'(ka) + \tan\delta_l\,\hat{n}_l'(ka)}{\hat{j}_l(ka) + \tan\delta_l\,\hat{n}_l(ka)} \tag{49}$$

Therefore, for a given energy (and corresponding K and k), we can, in principle, determine the phase shift δ_l and subsequently, the scattering amplitude, differential cross-section, and partial-wave cross-section.

We now focus on S-wave ($l = 0$) scattering, in which case eqn 14.49 takes on a simpler form. Using the relations in Table 14.1 (together with a fair amount of trigonometry) reduces this condition to

$$K \cot Ka = k \cot (ka + \delta_0)$$

and solving for δ_0 yields

$$\delta_0 = -ka + \arctan\left(\frac{k}{K} \tan Ka\right)$$

If we set $B_0 = 1$ (the solution u_0 is determined uniquely to within an arbitrary 'normalization' constant), then it follows from this result and eqn 14.48 that

$$\begin{aligned} A_0 &= \frac{k}{\left(k^2 \sin^2 Ka + K^2 \cos^2 Ka\right)^{1/2}} \\ &= \frac{k}{\left(k^2 + K_0^2 \cos^2 Ka\right)^{1/2}} \end{aligned} \tag{50}$$

where

$$K_0^2 = K^2 - k^2 = \frac{2\mu V_0}{\hbar^2}$$

Therefore, the solution u_0 is given by

$$\begin{aligned} u_0(r) &= A_0 \sin Kr \ \text{ for } r \leq a \\ u_0(r) &= \sin (kr + \delta_0) \ \text{ for } r > a \end{aligned} \tag{51}$$

Inside the well, the solutions are harmonic with a wavelength determined by K; outside the well, they are also harmonic, but their wavelength depends on k and they have undergone a phase shift.

14.11 Backgrounds and resonances

We need to consider in some detail the scattering phase shift δ_0. If we write

$$\delta_0(E) = \delta_{\text{bg}}(E) + \delta_{\text{res}}(E)$$

with

$$\delta_{\text{bg}}(E) = -ka \qquad \delta_{\text{res}}(E) = \arctan\left(\frac{k}{K} \tan Ka\right)$$

then the partial-wave cross-section (eqn 14.44) can be expressed as

$$\sigma_0(E) = \frac{4\pi}{k^2} \sin^2(\delta_{\text{bg}} + \delta_{\text{res}}) \tag{52}$$

(The significance of the subscripts bg and res will become apparent shortly.) Notice that $\delta_{\text{bg}}(E)$ is identical to the scattering phase shift for S-wave scattering by a hard sphere of radius a; it is a monotonically decreasing function of energy.

When analysing $\delta_{\text{res}}(E)$, we first note that for energies in the range $0 \leq E \ll V_0$, then $k/K \ll 1$, which implies that $\delta_{\text{res}}(E) \approx 0$ (to within the

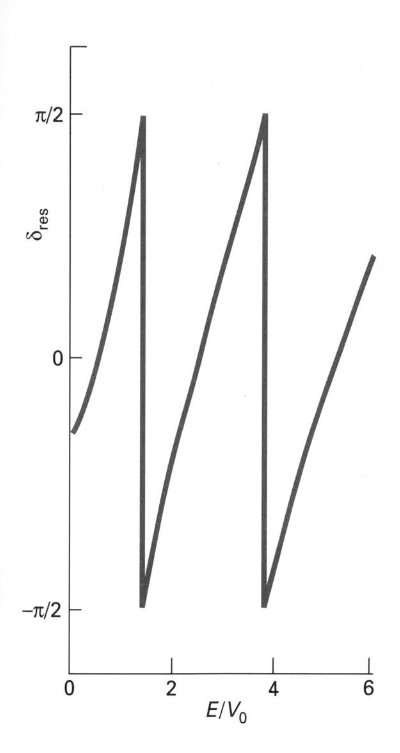

Fig. 14.10 Resonance phase shifts as a function of the energy of the incident particles for S-wave scattering by a spherical square-well potential. We have chosen $a(2\nu V_0)/\hbar = 1$.

modulo-π ambiguity) for most of the energy range. (There will be certain energies and corresponding values of K in the energy range where $\delta_{res} \approx 0$ is not true; see below.) Therefore the partial-wave cross-section $\sigma_0(E)$ is dominated by the contribution from δ_{bg}. In other words, over most of the energy range, the spherical square well potential has virtually the same effect as a hard-sphere potential. From eqns 14.50 and 14.51, we see that the incident particle penetrates very little into the region $r \leq a$ for most of the energy range; A_0 is very small on account of K_0 being large.

However, even for a very deep square well, there are particular energies at which σ_0 is not dominated by the hard-sphere scattering phase shift. If we look more closely at δ_{res} (Fig. 14.10), we see that it 'jumps' by π as the energy traverses $E = E_{res}$, where E_{res} is the energy correponding to

$$K_{res} = \frac{(2n+1)\pi}{2a} \tag{53}$$

with n a non-negative integer. At these energies, which explicitly are

$$E_{res} = \frac{(2n+1)^2\pi^2\hbar^2}{8\mu a^2} - V_0 \tag{54}$$

the phase shift is

$$\delta_{res}(E_{res}) = \tfrac{1}{2}\pi$$

(to modulo π) and this shift contributes to the partial-wave cross-section σ_0 of eqn 14.52. Note that it is only at energies E close to E_{res} that δ_{res} will contribute, but when it does, the increase by π in $\delta_{res}(E)$ can be responsible for a rapid variation in σ_0. Furthermore, note that A_0 reaches its maximum value of 1 at an energy $E = E_{res}$. We conclude that it is only in the vicinity of E_{res} that the incident particle will have appreciable intensity in the region $r \leq a$. These observations are characteristically associated with resonance phenomena (Section 14.9), which we shall introduce more formally shortly. For now, we note that E_{res} is referred to as the (real part of the) **resonance energy** and δ_{res} as the **resonant component** of the phase shift. The contribution δ_{bg} is called the **background phase shift**. The background phase shift makes the dominant contribution to δ over most of the energy range; however, superimposed on this background may be contributions from δ_{res} due to resonances.

We now have to admit that the discussion has been somewhat misleading for the following reason. For S-wave scattering by a spherical square well, the background phase shift $\delta_{bg} = -ka$ in fact decreases with energy at least as fast as the resonant phase shift δ_{res} increases as the energy passes across E_{res}. The net result is rapid variation in neither δ_0 nor σ_0! However, in many cases it is in fact the case that δ_{bg} is either very close to zero or very slowly varying in energy. In these cases (for which the eqn 14.32 is typically solved numerically), rapid changes in the resonant phase shift $\delta_{res}(E)$ do produce structure (that is, rapid variations) in the partial-wave cross-section.

Because the *concepts* introduced by considering S-wave scattering off a spherical square well are commonly observed even though they are not in fact observed for that actual case, we continue our discussion. We define a

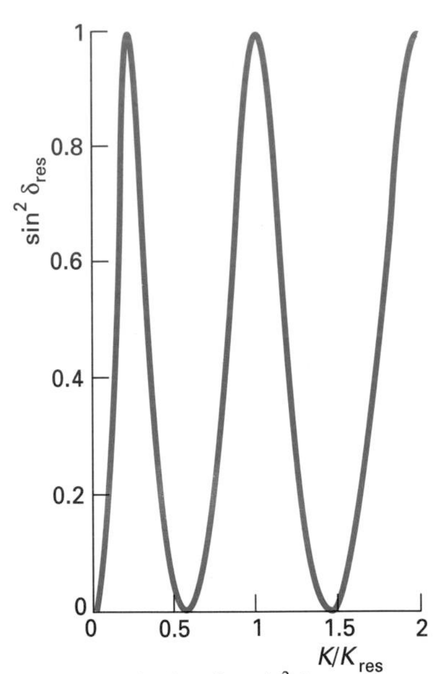

Fig. 14.11 The function $\sin^2 \delta_{res}$ as a function of energy for S-wave scattering from a spherical square well. The illustration shows the effects of three resonances. For the purposes of the plot, K_{res} has been taken to be $3\pi/a$.

resonant part of σ_l analogous to eqn 14.44:

$$\sigma_{res} = \frac{4\pi}{k^2}(2l+1)\sin^2 \delta_{res}$$

It is then easy to show that, for S-wave scattering by a spherical square well,

$$\sin^2 \delta_{res} = \frac{k^2}{k^2 + K^2 \cot^2 Ka} \tag{55}$$

We see that $\sin^2 \delta_{res}$ will be a maximum at $K = K_{res}$ and will increase from 0 to 1 as the energy approaches E_{res} and then decrease from 1 to 0 as we move out of the vicinity of the resonance energy (Fig. 14.11).

14.12 The Breit–Wigner formula

We are now in a position to make a connection between the expression for $\sin^2 \delta_{res}$ in eqn 14.55 and a formula originally derived by G. Breit and E.P. Wigner. For energies E close to E_{res} (such that $|E - E_{res}| \ll E_{res} + V_0$) the expression

$$\tan \delta_{res}(E) = \frac{k}{K}\tan Ka \tag{56}$$

can be written in the simplified form

$$\tan \delta_{res}(E) = \frac{b}{E_{res} - E} \tag{57}$$

where

$$b = \frac{\hbar^2 k_{res}}{\mu a} \tag{58}$$

and k_{res} is related to E_{res} in the normal way through $E = k^2\hbar^2/2\mu$.

Example 14.4 The approximation of expressions near resonance

Confirm that eqn 14.57 is equivalent to eqn 14.56 close to resonance.

Method. This is an exercise in making approximations when the energy lies close to resonance. We can expect to use ΔE as a parameter, where $\Delta E = E - E_{res}$ and $|\Delta E/E_{res}| \ll 1$. The art of approximation is to express all factors in terms of ΔE and then to expand them to first order in ΔE. The following relations will be helpful:

$$\sin(a+b) = \sin a \cos b + \cos a \sin b$$
$$\cos(a+b) = \cos a \cos b - \sin a \sin b$$
$$\sin a = a + \dots \qquad \cos a = 1 - \dots \qquad (1+x)^{1/2} = 1 + \tfrac{1}{2}x + \dots$$

Answer. Close to resonance we can set

$$k\hbar = (2\mu E_{\text{res}})^{1/2}\left(1+\frac{\Delta E}{E_{\text{res}}}\right)^{1/2} \approx \hbar k_{\text{res}}\left(1+\frac{\Delta E}{2E_{\text{res}}}\right)$$

$$K\hbar = \{2\mu(E_{\text{res}}+V_0)\}^{1/2}\left(1+\frac{\Delta E}{E_{\text{res}}+V_0}\right)^{1/2}$$

$$\approx \hbar K_{\text{res}}\left\{1+\frac{\Delta E}{2(E_{\text{res}}+V_0)}\right\}$$

We now address $\tan Ka = \sin Ka/\cos Ka$. First, we expand the sine and cosine terms, and use eqn 14.53, which implies that $\cos K_{\text{res}}a = 0$ and $\sin K_{\text{res}}a = \pm 1$:

$$\tan Ka \approx \frac{\overbrace{\sin K_{\text{res}}a}^{\pm 1}\cos\left(\frac{K_{\text{res}}a\Delta E}{2(E_{\text{res}}+V_0)}\right)+\overbrace{\cos K_{\text{res}}a}^{0}\sin\left(\frac{K_{\text{res}}a\Delta E}{2(E_{\text{res}}+V_0)}\right)}{\underbrace{\cos K_{\text{res}}a}_{0}\cos\left(\frac{K_{\text{res}}a\Delta E}{2(E_{\text{res}}+V_0)}\right)-\underbrace{\sin K_{\text{res}}a}_{\pm 1}\sin\left(\frac{K_{\text{res}}a\Delta E}{2(E_{\text{res}}+V_0)}\right)}$$

$$\approx -\frac{\cos\left(\frac{K_{\text{res}}a\Delta E}{2(E_{\text{res}}+V_0)}\right)}{\sin\left(\frac{K_{\text{res}}a\Delta E}{2(E_{\text{res}}+V_0)}\right)} \approx -\frac{2(E_{\text{res}}+V_0)}{K_{\text{res}}a\Delta E}$$

Then, with $k/K \approx k_{\text{res}}/K_{\text{res}}$ (any correction to this expression results in a ΔE in the numerator, which is close to zero), we obtain

$$\tan\delta_{\text{res}}(E) \approx \left(\frac{k_{\text{res}}}{K_{\text{res}}}\right)\left(-\frac{2(E_{\text{res}}+V_0)}{K_{\text{res}}a\Delta E}\right) = -\frac{k_{\text{res}}\hbar^2}{\mu a\Delta E}$$

This result has the form of eqn 14.57.

Comment. Whoever first derived this relation clearly deserves our admiration for sticking to a sequence of steps that to a normal mortal might seem to be leading nowhere. There is a lesson to be learned here about application.

Exercise 14.4. Confirm that $\delta_{\text{res}}(E_{\text{res}}) = \pi/2$ to modulo π.

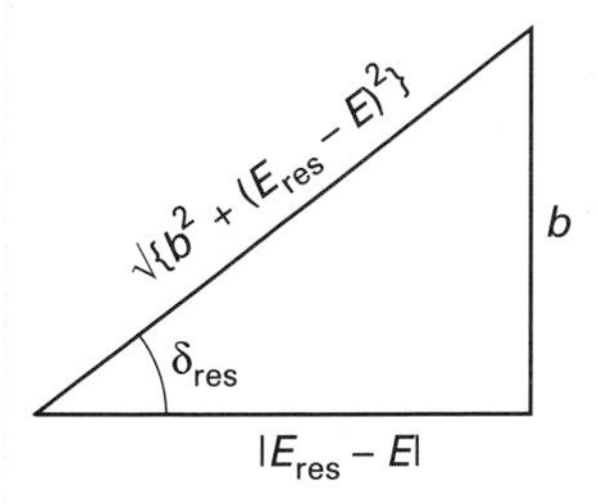

Fig. 14.12 The construction required for the derivation of eqn 14.59.

As can be seen by reference to Fig. 14.12, eqn 14.57 can be rewritten as

$$\sin^2\delta_{\text{res}} = \frac{b^2}{b^2+(E_{\text{res}}-E)^2}$$

If we replace b by $\Gamma/2$, with

$$\Gamma = \frac{2\hbar^2 k_{\text{res}}}{\mu a}$$

this expression becomes identical to the **Breit–Wigner formula** for the resonance phase shift:

$$\sin^2\delta_{\text{res}} - \frac{(\Gamma/2)^2}{(\Gamma/2)^2+(E_{\text{res}}-E)^2} \tag{59}$$

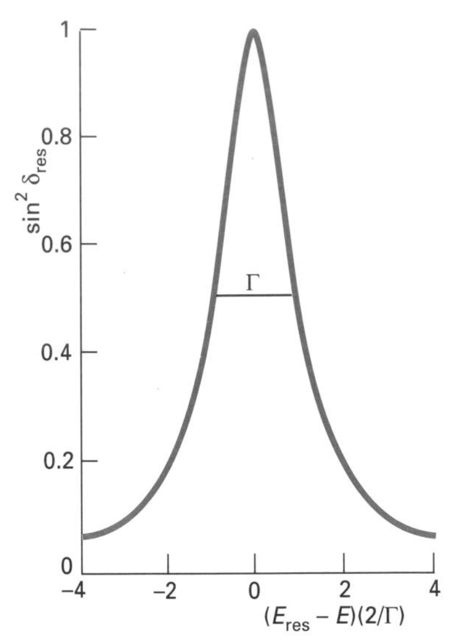

Fig. 14.13 The phase shift according to the Breit–Wigner formula close to resonance.

where Γ is referred to as the **resonance width**. Equation 14.59 is a general expression for the behaviour of the resonant phase shift near E_{res}. Although we have derived it for S-wave scattering by a square-well potential, it is in fact applicable to partial-wave scattering from a wide variety of potentials. It shows that $\sin^2 \delta_{res}$ peaks at $E = E_{res}$, is zero at energies where $|E_{res} - E| \gg \Gamma$, and has a full width at half-maximum given by Γ (Fig. 14.13).

The Breit–Wigner formula has very important implications for scattering experiments. If the form of the potential energy $V(r)$ is such that the background phase shift δ_{bg} is insignificant, then the partial-wave cross-section will be dominated by δ_{res} (that is, by σ_{res}) and will be given by

$$\sigma_l(E) = \frac{4\pi}{k^2}(2l+1)\frac{(\Gamma/2)^2}{(\Gamma/2)^2 + (E_{res} - E)^2} \tag{60}$$

Thus, the partial-wave cross-section will show a peak in the vicinity of E_{res}. A resonance that produces this kind of behaviour for $\sigma_l(E)$ is called a **Breit–Wigner resonance**. Furthermore, if there is a resonance of angular momentum l with its associated E_{res} and all other phase shifts (of all other angular momenta) are slowly varying in the neighbourhood of E_{res}, then the peak in σ_l will also result in a rapid variation of the total integral cross-section σ. Of course, we should be mindful of the fact that if δ_{bg} is nonzero or is varying significantly with energy, then a Breit–Wigner resonance will not be apparent and should not expect the partial-wave cross-section to vary in the simple way given by eqn 14.60.

14.13 The scattering matrix element

We need to consider eqn 14.45 again. For convenience we reproduce it here:

$$\psi(r,\theta) \simeq \sum_l \frac{\mathrm{i}^{l-1}(2l+1)}{2kr} P_l(\cos\theta)\left(-\mathrm{e}^{-\mathrm{i}(kr-l\pi/2)} + \mathrm{e}^{\mathrm{i}(kr-l\pi/2)}\mathrm{e}^{2\mathrm{i}\delta_l}\right)$$

We noted earlier that this equation shows that each component $\psi_l(r,\theta)$ of the stationary scattering state $\psi(r,\theta)$ is asymptotically a superposition of an incoming spherical wave $(1/r)\mathrm{e}^{-\mathrm{i}(kr-l\pi/2)}$ and an outgoing spherical wave $(1/r)\mathrm{e}^{\mathrm{i}(kr-l\pi/2)}$ that has been shifted in phase by $2\delta_l$ relative to the incoming wave with the same angular momentum l. We shall now write this phase shift as

$$S_l = \mathrm{e}^{2\mathrm{i}\delta_l} \tag{61}$$

The quantity S_l is a special case of a **scattering matrix element**, which we first encountered at the end of Chapter 2. However, as we shall soon see, it takes on its full generality when we discuss inelastic and reactive scattering. In terms of the scattering matrix element, the scattering amplitude and cross-sections for elastic scattering are

$$
\begin{aligned}
f_l &= \frac{(2l+1)}{2ik}(S_l - 1) \\
f_k(\theta) &= \frac{1}{2ik}\sum_l (2l+1)(S_l - 1)P_l(\cos\theta) \\
\sigma_l &= \frac{\pi}{k^2}(2l+1)|S_l - 1|^2
\end{aligned}
\tag{62}
$$

We shall now explore the relation between S_l and a resonance of angular momentum l. We have already seen that a resonance is characterized by E_{res} and a width Γ, and that the phase shift δ_l can be written in general as a sum of two contributions:

$$
\delta_l(E) = \delta_{l,\text{bg}}(E) + \delta_{l,\text{res}}(E) \tag{63}
$$

where $\delta_{l,\text{bg}}$ and $\delta_{l,\text{res}}$ are, respectively, the background and resonant phase shifts for the partial wave with orbital angular momentum l. It follows that we can write the scattering matrix element as a product of factors:

$$
S_l = e^{2i\delta_{l,\text{bg}}} e^{2i\delta_{l,\text{res}}} \tag{64}
$$

and hence that

$$
S_l = e^{2i\delta_{l,\text{bg}}}\left(\cos^2\delta_{l,\text{res}} - \sin^2\delta_{l,\text{res}} + 2i\sin\delta_{l,\text{res}}\cos\delta_{l,\text{res}}\right) \tag{65}
$$

We now utilize eqn 14.57 (as illustrated in Fig. 14.12), which is valid for E close to E_{res}, and the relation $\Gamma = 2b$, to obtain

$$
\begin{aligned}
S_l &= e^{2i\delta_{l,\text{bg}}}\left\{\frac{(E_{\text{res}} - E)^2 - (\Gamma/2)^2 + 2i(E_{\text{res}} - E)(\Gamma/2)}{(E_{\text{res}} - E)^2 + (\Gamma/2)^2}\right\} \\
&= e^{2i\delta_{l,\text{bg}}}\left\{\frac{E - E_{\text{res}} - i\Gamma/2}{E - E_{\text{res}} + i\Gamma/2}\right\}
\end{aligned}
\tag{66}
$$

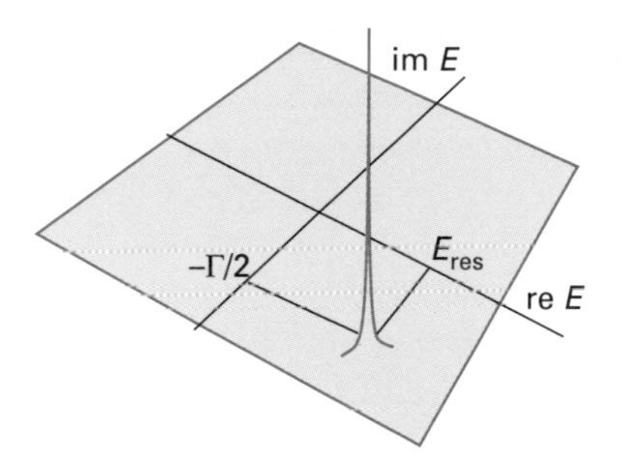

Fig. 14.14 The interpretation of a pole in the complex energy plane. The real coordinate is the real part of the resonance energy and the complex coordinate is proportional to the width of the resonance.

In a scattering experiment, we are interested in measurements and calculations at real scattering energies E. However, this expression for S_l is valid mathematically at both real and *complex* energies in the vicinity of E_{res}. If we think of it as extended into real and imaginary energies (Fig. 14.14), then we see that S_l has a pole (a point at which a function becomes infinite) at the complex energy $E = \bar{E}$ given by

$$
\bar{E} = E_{\text{res}} - i\Gamma/2 \tag{67}
$$

The complex energy $\bar{E}$ is called the **resonance energy**; its real component is E_{res} and its negative imaginary part is $\Gamma/2$. From now on we shall take the existence of a pole in the scattering matrix element as the *definition* of a resonance. From what we have already seen about the physical significance of resonances, we now know that a pole in the scattering matrix element for the orbital angular momentum l signifies the likelihood of a rapid variation of partial-wave cross-section close to the real part of $\bar{E}$, and the rapidity of the variation is determined by the imaginary part of $\bar{E}$ (provided the background phase shift is well-behaved).

Consider now the nature of the state associated with the resonance energy. From eqn 14.45 (which we reproduced at the start of this section), we see that at

$E \approx \bar{E}$, because S_l is so large close to a resonance,

$$\psi_l(r,\theta) \simeq \frac{\mathrm{i}^{l-1}(2l+1)}{2kr} P_l(\cos\theta) S_l(\bar{E}) \mathrm{e}^{\mathrm{i}(kr - l\pi/2)} \tag{68}$$

and $\psi_l(r,\theta)$ has no incoming wave; it is purely an outgoing wave. The function $\psi_l(r,\theta)$ violates the asymptotic expression for a *stationary* state. Instead, we can think of the scattering state $\psi_l(r,\theta)$ associated with the resonance of angular momentum l as the solution of the time-independent Schrödinger equation for a particular value l with complex energy $E_{\mathrm{res}} - \mathrm{i}\Gamma/2$ and an asymptotic boundary condition of a purely outgoing spherical wave. Furthermore, the analogous time-dependent wavefunction is a *decaying* state because

$$\mathrm{e}^{-\mathrm{i}\bar{E}t/\hbar} = \mathrm{e}^{-\mathrm{i}E_{\mathrm{res}}t/\hbar}\mathrm{e}^{-\Gamma t/(2\hbar)} \tag{69}$$

Thus, the intensity of the resonance state wavefunction which is given by $|\Psi(r,\theta,t)|^2$ (for a particular l) decays exponentially with time as $\mathrm{e}^{-\Gamma t/\hbar}$. If we define the mean lifetime τ of the resonance state as the time t at which its intensity has decreased to $1/\mathrm{e}$ of its intensity at $t = 0$, then we immediately see that

$$\tau = \frac{\hbar}{\Gamma} \tag{70}$$

It follows that the mean lifetime of the resonance state is inversely proportional to its width Γ, in accord with the general principles of lifetime broadening (Section 6.18). At this point, we can make a connection with the discussion of predissociation in Section 11.5. In the language of scattering theory, a predissociating state of finite lifetime τ is a resonance of finite width Γ.

At this point we have collected several equivalent descriptions of the resonance. We can characterize the resonance by a state at the complex energy $\bar{E} = E_{\mathrm{res}} - \mathrm{i}\Gamma/2$ with a mean lifetime $\hbar/\Gamma$. To characterize the resonance at real physical energies E, we must regard it as having an imprecise energy. This imprecision is associated with the resonance width Γ. In a range of real energies about E_{res} the resonance may have physically observable effects. These descriptions of the resonance are also applicable to inelastic and reactive scattering though we shall need to generalize our definition of a resonance.

Multichannel scattering

In the previous sections, we have limited ourselves to a discussion of elastic scattering between two structureless particles. We shall now consider some of the fundamental concepts pertinent to inelastic and reactive scattering. Rather than presenting derivations in full, we shall present many results by generalizing the findings for elastic scattering.

14.14 Channels for scattering

First, we must define what is meant by a 'channel', which in a general sense refers to each possible grouping of the various particles involved in the scattering event.[7] This concept is best illustrated by an example. Consider the

[7] We give a more precise definition in Section 14.16.

scattering of an incident atom A off a target diatomic molecule BC. There are many possible results of the collision depending on the energy of the reaction; we shall restrict ourselves to the electronic ground states of all atomic and diatomic species. For example, some possibilities are

$$
\begin{aligned}
&(1) \quad \mathrm{A + BC \longrightarrow A + BC} \\
&(2) \quad \mathrm{A + BC \longrightarrow A + (BC)^*} \\
&(3) \quad \mathrm{A + BC \longrightarrow AB + C} \\
&(4) \quad \mathrm{A + BC \longrightarrow (AB)^* + C} \\
&(5) \quad \mathrm{A + BC \longrightarrow AC + B} \\
&(6) \quad \mathrm{A + BC \longrightarrow (AC)^* + B} \\
&(7) \quad \mathrm{A + BC \longrightarrow A + B + C}
\end{aligned}
$$

where a superscript * refers to a vibrationally or rotationally excited state of the molecule. Each grouping on the right side of the arrow represents a channel; in this case we can have seven channels, so the collision of A and BC would be referred to as a **multichannel process**.

Process 1 represents the elastic scattering event, in which the relative translational energy of A and BC is unchanged by the collision, and the energy of the internal modes of motion of the diatomic molecule (its vibration and rotation) remains the same. Process 2 represents an inelastic scattering process in which the internal state of BC is changed by the process. Because the total energy is conserved during the collision, the relative translational energy also changes in this process. Of course, there may be many possible internal states of BC; each final internal state $(\mathrm{BC})^{**}$ could be associated with a channel $\mathrm{A} + (\mathrm{BC})^{**}$ and there may be many inelastic processes. Processes 3–6 represent reactive scattering events. For simplicity, we have supposed that AB and AC are each limited to only two internal states. Process 7 involves the dissociative channel A + B + C.

Which processes are actually possible depends on the total energy E. Channels that are energetically accessible are called **open channels** and channels that are not energetically accessible are called **closed channels**. For example, if the total energy E is less than the energy of the internal state $(\mathrm{AB})^*$, then the channel $(\mathrm{AB})^* + \mathrm{C}$ is closed and process 4 cannot occur. If we define the zero of energy as the energy of the channel $\mathrm{A + B + C}$ with all three atoms infinitely far apart and stationary, then at scattering energies $E \geq 0$, the bound state of all three particles (the species ABC) is always associated with closed channels.

In theory, if the scattering event begins with the particles in some incident channel, and if there are N open channels, then there are N possible processes. However, in principle, each of the N channels can be the incident channel and thus there are N^2 qualitatively different processes that can be considered, one of the N possible incident channels resulting in each of the N final channels. The multichannel process will be described by an $N \times N$ matrix, called the **scattering matrix**, or S matrix. Each element of the S matrix, S_{ji}, conveys information about the process connecting incident channel i and final channel j. The S matrix will play an important role in the remainder of this chapter.

14.15 Multichannel stationary scattering states

A discussion of multichannel scattering gets quite complicated for a number of reasons. Depending on the total collision energy, as discussed above, there may be many open channels and in a complete treatment we may need to consider electronic, vibrational, and rotational states. This wide range of different species and states of excitation then clearly affects the angular momenta (and other quantum numbers) that must be considered in the problem. In the case of elastic scattering, we saw that the orbital angular momentum was a conserved quantity and we could therefore decompose the scattering problem into a set of simpler problems, each characterized by a partial wave l. For multichannel scattering, we may have to consider quantum numbers relating to vibration and rotation as well as orbital motion. However, one simplification is that we know that the *total* angular momentum, represented by the quantum number J, is conserved during the collision (provided there are no external fields present that can exercise torques on the system), and the scattering problem can be decomposed into sets of smaller problems, each one referring to one value of J.

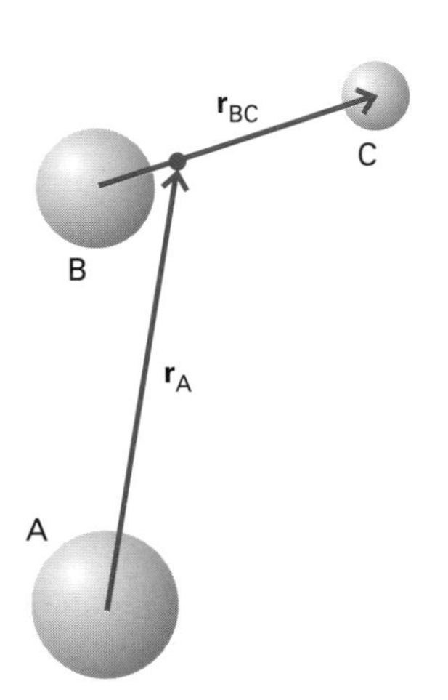

Fig. 14.15 The vectors used to specify the location of particles in an example of multichannel scattering.

It is not our goal here to give the complete and rigorous relations between J and the various quantum numbers of the modes that contribute to J; we are more interested in conveying a sense of what is involved in treating multichannel scattering processes. We shall therefore restrict consideration to the specific case of an atom A that collides with a diatomic molecule BC. The initial internal state of the molecule is designated by a set of quantum numbers α_0. The multichannel stationary scattering state (the solution to eqn 14.5) is denoted $\psi_{\alpha_0}(\mathbf{r}_A, \mathbf{r}_{BC})$, where $\mathbf{r}_A$ is the vector from A to the centre of mass of BC and where $\mathbf{r}_{BC}$ is the vector from B to C (Fig. 14.15). As in elastic scattering, we are interested in the asymptotic behaviour of $\psi_{\alpha_0}(\mathbf{r}_A, \mathbf{r}_{BC})$; however, do we want this behaviour for $r_A \to \infty$ or for $r_{BC} \to \infty$? Actually we want both! The asymptotic behaviour as $r_A \to \infty$ will correspond to final channels in which A is moving infinitely far away from B and C; that is, elastic, inelastic, and dissociative processes (for example, processes 1, 2, and 7 in the preceding section). On the other hand, as $r_{BC} \to \infty$, the asymptotic behaviour will correspond to channels in which B is moving infinitely far away from AC or likewise C is moving infinitely far away from AB; that is, it corresponds to the reactive and dissociative processes (processes 3–7).

14.16 Inelastic collisions

We focus first on the much simpler case of inelastic scattering. The various internal states of the diatomic molecule BC are labelled by the index α with a corresponding wavefunction $\chi_\alpha(\mathbf{r}_{BC})$ and energy E_α. For example, χ_α may represent a vibrational-rotational state of the diatomic molecule. Because the total collisional energy is E, if the diatomic molecule has energy E_α, then the relative translational energy of atom A with respect to BC will be $E - E_\alpha$ with a wavevector of magnitude k_α, for which

$$E - E_\alpha = \frac{k_\alpha^2 \hbar^2}{2\mu} \tag{71}$$

where μ is the reduced mass of A+BC:

$$\frac{1}{\mu} = \frac{1}{m_{\mathrm{A}}} + \frac{1}{m_{\mathrm{BC}}}.$$

where $m_{\mathrm{BC}} = m_{\mathrm{B}} + m_{\mathrm{C}}$. Therefore, the wavefunction that describes the system well before the collision is

$$\mathrm{e}^{\mathrm{i}\mathbf{k}_{\alpha_0}\cdot\mathbf{r}_{\mathrm{A}}}\chi_{\alpha_0}(\mathbf{r}_{\mathrm{BC}}) = \sum_{l_0,m_{l0}} \mathrm{i}^{l_0}(2l_0+1)j_{l_0}(k_{\alpha_0}r_{\mathrm{A}})Y_{l_0m_{l0}}(\theta,\phi)\chi_{\alpha_0}(\mathbf{r}_{\mathrm{BC}}) \qquad (72)$$

where the coordinates are defined in Fig. 14.15. This equation resembles eqn 14.38, but is more general because the direction of $\mathbf{k}_{\alpha_0}$ is no longer fixed and the relative orientation of the diatomic species needs to be taken into account (through θ and ϕ). We define a **channel**, labelled λ, by a particular set of labels l, m_l, and α; the incident channel λ_0 is specified by l_0, m_{l0}, and α_0. Each term in the sum in the equation represents the free wavefunction of a unique incident channel. It is imperative to recognize that the initial state α_0, which is experimentally accessible, is different from the incident channel λ_0, which includes the orbital angular momentum l_0 and its component m_{l0} and therefore is not experimentally well-defined. We can observe transitions experimentally between initial and final *states*. Scattering theory will allow us to analyse transitions between initial and final *channels* which then must be related to state-to-state processes.

The asymptotic expression for the multichannel stationary scattering state as $r_{\mathrm{A}} \to \infty$ is the analogue of the scattering states that we have already considered:

$$\psi_{\alpha_0}(\mathbf{r}_{\mathrm{A}},\mathbf{r}_{\mathrm{BC}}) \simeq \mathrm{e}^{\mathrm{i}\mathbf{k}_{\alpha_0}\cdot\mathbf{r}_{\mathrm{A}}}\chi_{\alpha_0}(\mathbf{r}_{\mathrm{BC}}) + \sum_{\alpha} f_{\alpha\alpha_0}(\hat{\mathbf{r}}_{\mathrm{A}})\frac{\mathrm{e}^{\mathrm{i}k_\alpha r_{\mathrm{A}}}}{r_{\mathrm{A}}}\chi_\alpha(\mathbf{r}_{\mathrm{BC}}) \qquad (73)$$

where $f_{\alpha\alpha_0}$ is the scattering amplitude for final state α (with incident state α_0) and $\hat{\mathbf{r}}_{\mathrm{A}}$ is the unit vector in the direction of $\mathbf{r}_{\mathrm{A}}$. (Note that it is essential to continue the distinction between states α and channels λ.) This result is nothing but a straightforward generalization of the elastic scattering result, eqn 14.7. The first term contains the plane-wave initial state α_0 and the second term is a sum over all states α of BC (including the state α_0 and the possibility of elastic scattering). Each term in the sum in this expansion is the product of a scattering amplitude, an outgoing spherical wave for A with wavevector of magnitude k_α, and an internal state wavefunction for BC with energy E_α. If the total collision energy E is less than E_α, then k_α is imaginary. In this case, $\mathrm{e}^{\mathrm{i}k_\alpha r_{\mathrm{A}}}$ is an exponentially decreasing function and will vanish as $r_{\mathrm{A}} \to \infty$. Therefore states of BC that are closed (that is, energetically inaccessible) do not contribute to the sum in eqn 14.73 and we need consider only the open states α and their scattering amplitudes $f_{\alpha\alpha_0}$.

Differential cross-sections can be found in terms of the scattering amplitudes in a manner entirely analogous to that in Section 14.2 by using the flux operator **J**. The differential cross-section for scattering into the solid angle dΩ in the direction $\hat{\mathbf{r}}_{\mathrm{A}}$ for an initial state α_0 and final state α is given by

$$\sigma_{\alpha\alpha_0} = \frac{k_\alpha}{k_{\alpha_0}}|f_{\alpha\alpha_0}(\hat{\mathbf{r}}_{\mathrm{A}})|^2 \qquad (74)$$

The integral cross-section is obtained from this differential cross-section by integration in the normal way (recall eqn 14.2).

The determination of the scattering amplitudes requires solution of the Schrödinger equation. For elastic scattering, the stationary scattering state and the scattering amplitude were expanded in a basis of Legendre polynomials P_l, the unknown quantities being the radial functions R_l and the partial-wave scattering amplitudes f_l. We then determined the relation between the radial function and the scattering phase shift δ_l, expressed the partial-wave scattering amplitude in terms of the phase shift, and obtained an equation for the partial-wave cross-section σ_l in terms of δ_l. We follow a similar type of procedure here, giving only the highlights and leaving out the details.

The first step is to expand the scattering amplitude $f_{\alpha\alpha_0}$ as a sum over partial waves l_0, m_{l0}, which introduces the partial-wave scattering amplitudes. Then the multichannel stationary scattering state ψ_{α_0} is expanded as a sum over partial waves:

$$\psi_{\alpha_0}(\mathbf{r}_\mathrm{A}, \mathbf{r}_\mathrm{BC}) = \sum_{l_0, m_{l0}} \psi_{\alpha_0 l_0 m_{l0}}(\mathbf{r}_\mathrm{A}, \mathbf{r}_\mathrm{BC}) \tag{75}$$

where $\psi_{\alpha_0 l_0 m_{l0}}$ is the **partial-wave multichannel stationary scattering state**. Our goal is to find the asymptotic form of $\psi_{\alpha_0 l_0 m_{l0}}$ and, by using eqn 14.73, relate it to the partial-wave scattering amplitudes. This procedure will yield expressions for $f_{\alpha\alpha_0}$ and ultimately, through eqns 14.74 and 14.2, for differential and integral state-to-state cross-sections. The latter quantities make contact with experimental measurements.

If, at this point, we were to expand each partial-wave multichannel stationary scattering state in a complete basis set and then substitute eqn 14.75 into the Schrödinger equation, we would get an infinite set of coupled differential equations, which would be computationally difficult. The calculation can be significantly simplified by taking advantage of the fact that during the scattering process, both the total angular momentum J and its z-component M_J are conserved. As we have already remarked, this conservation of angular momentum allows us to decompose the problem into independent equations, each one relating to one value of J. Therefore, before expanding $\psi_{\alpha_0 l_0 m_{l0}}$ in a complete basis set, we first expand it as

$$\psi_{\alpha_0 l_0 m_{l0}}(\mathbf{r}_\mathrm{A}, \mathbf{r}_\mathrm{BC}) = \sum_{J, M_J} c(JM_J\alpha_0 l_0 m_{l0})\psi^{JM_J}_{\alpha_0 l_0 m_{l0}}(\mathbf{r}_\mathrm{A}, \mathbf{r}_\mathrm{BC}) \tag{76}$$

where J ranges from 0 to infinity and $M_J = J, J-1, \ldots, -J$. The $c(JM_J\alpha_0 l_0 m_{l0})$ are vector coupling coefficients (Section 4.12) for the construction of the coupled angular momentum state (J, M_J) from its component angular momenta. These coefficients are known from tables, and we do not have to calculate them separately in this problem. It is the $\psi^{JM_J}_{\alpha_0 l_0 m_{l0}}$ that we now expand in a complete set of basis functions and do the work to determine.

The ψ^{JM_J} are labelled by $(\alpha_0 l_0 m_{l0})$; that is, by the channel λ_0. Our complete set of basis functions will also be labelled by channels λ. We shall denote the complete set of basis functions $\{\Phi^{JM_J}_\lambda(\mathbf{r}_\mathrm{A}, \mathbf{r}_\mathrm{BC})\}$, where, to ensure completeness, we must have an infinite number of channels λ. To demonstrate the form of these basis functions, we consider a specific example of inelastic scattering between A and BC in which only the rotational state (j, m_j), the significance of α_0, of the diatomic molecule is allowed to change. Channels for the scattering process are then labelled (j, m_j, l, m_l), restricted to those values for which the vector coupling of j and l yield J and for which $M_J = m_j + m_l$. However, for the basis to be complete, we must have an infinite number of sets (j, l) for each value of J. For example, for $J = 0$, we require $j = l$ $(j = 0, 1, \ldots, \infty)$; all rotational states (j, m_j) must be included in the basis. The basis function $\Phi_\lambda^{JM_J}$, in this rotationally inelastic case, can be taken as a product of a BC rotational wavefunction (a function of $\mathbf{r}_{\mathrm{BC}}$) and a spherical harmonic $Y_{lm_l}(\hat{\mathbf{r}}_{\mathrm{A}})$.

In general, for inelastic scattering the expansion of $\psi_{\lambda_0}^{JM_J}$ takes the form

$$\psi_{\lambda_0}^{JM_J}(\mathbf{r}_{\mathrm{A}}, \mathbf{r}_{\mathrm{BC}}) = \frac{1}{r_{\mathrm{A}}} \sum_\lambda R_{\lambda\lambda_0}^J(r_{\mathrm{A}}) \Phi_\lambda^{JM_J}(\hat{\mathbf{r}}_{\mathrm{A}}, \mathbf{r}_{\mathrm{BC}}) \qquad (77)$$

where $R_{\lambda\lambda_0}^J(r_{\mathrm{A}})$ is an as yet unknown radial function which varies with J but is independent of M_J due to the isotropy of space. It is important to reiterate that the sum in this expression runs over an infinite number of channels; that is, it includes channels in which the bound state of BC is energetically open, channels in which the bound state of BC is energetically closed, and also channels involving continuum states of BC. In the example of the rotationally inelastic process, these continuum states might be large j states for which the repulsive centrifugal potential proportional to $j(j+1)/r_{\mathrm{BC}}^2$ results in dissociation of the molecule.

Substitution of the expansions in eqns 14.75–77 into eqn 14.5 results in an infinite set of coupled differential equations for the radial functions $R_{\lambda\lambda_0}^J(r_{\mathrm{A}})$. There is still an infinite set of equations for each value of J; however, radial functions with different values of J are decoupled. We do not intend to derive or even to show the infinite set of coupled equations, but mention simply that the coupling terms in the differential equation for $R_{\lambda\lambda_0}^J$ take the form

$$\sum_{\lambda' \neq \lambda} R_{\lambda'\lambda_0}^J(r_{\mathrm{A}}) \int \left(\Phi_\lambda^{JM_J}(\hat{\mathbf{r}}_{\mathrm{A}}, \mathbf{r}_{\mathrm{BC}})\right)^* V(\mathbf{r}_{\mathrm{A}}, \mathbf{r}_{\mathrm{BC}}) \Phi_{\lambda'}^{JM_J}(\hat{\mathbf{r}}_{\mathrm{A}}, \mathbf{r}_{\mathrm{BC}})\, \mathrm{d}\hat{\mathbf{r}}_{\mathrm{A}} \mathrm{d}\mathbf{r}_{\mathrm{BC}}$$

The infinite set of coupled differential equations is of little practical utility. However, in many cases it is possible, to a very good approximation, to retain only a small finite set of channels λ in the expansion in eqn 14.77. If we let P be the number of channels retained in the expansion, we will have a set of P coupled differential equations for each value of J. This truncation is known as the **close-coupling approximation** or the **coupled-channel approximation**. It is common to take $P \geq N$, where N is the number of open channels, the expectation being that although the closed states do not play a role in the asymptotic form of the multichannel stationary scattering state, some closed channels may be necessary to represent the multichannel scattering state accurately at all values of $(\mathbf{r}_{\mathrm{A}}, \mathbf{r}_{\mathrm{BC}})$.

To obtain expressions for the scattering amplitudes and cross-sections we need the $r_{\mathrm{A}} \to \infty$ asymptotic form of the radial functions $R_{\lambda\lambda_0}^J(r_{\mathrm{A}})$ which is given as

$$R_{\lambda\lambda_0}^J \simeq \delta_{\lambda\lambda_0} \mathrm{e}^{-\mathrm{i}(k_{\alpha_0} r_{\mathrm{A}} - l_0\pi/2)} - \left(\frac{k_{\alpha_0}}{k_\alpha}\right)^{1/2} S_{\lambda\lambda_0}^J \mathrm{e}^{\mathrm{i}(k_\alpha r_{\mathrm{A}} - l\pi/2)} \qquad (78)$$

where $(k_\alpha, l, m_l, \alpha)$ corresponds to channel λ, $(k_{\alpha_0}, l_0, m_{l0}, \alpha_0)$ corresponds to channel λ_0, and $S^J_{\lambda\lambda_0}$ is an element of the scattering matrix $\mathbf{S}^J$. You should notice the similarity to eqns 14.45 and 14.61 of the elastic scattering case. If λ is a closed channel, then the second term on the right of this expression vanishes on account of k_α being imaginary and hence of the exponential term becoming zero as $r_A \to \infty$. Thus, the only physically meaningful scattering matrix elements are $S^J_{\lambda\lambda_0}$ with $\lambda = 0, 1, \ldots, N$. Because, in theory, *any* of the open channels can be the incident channel, the scattering matrix $\mathbf{S}^J$ is in fact an $N \times N$ matrix, the element S^J_{ji} referring to incident channel i and final channel j. In principle, it is possible to derive an expression relating the scattering amplitude $f_{\alpha\alpha_0}$ and cross-section $\sigma_{\alpha\alpha_0}$ to the scattering matrix elements $S^J_{\lambda\lambda_0}$ by comparing the asymptotic form in eqn 14.73 of the multichannel stationary scattering state with the expansions in eqns 14.75–77 and the asymptotic form in eqn 14.78.

14.17 Reactive scattering

A discussion of reactive scattering proceeds in a similar fashion to that of inelastic scattering given above. The multichannel stationary scattering state can be expanded in a complete basis as above, and a set of coupled differential equations produced. At first sight, it might appear that if we used eqn 14.77, we would not allow for the possibility of reactive scattering. It is clear that inelastic and elastic scattering are reflected in the asymptotic form of $\psi_{\alpha_0}(\mathbf{r}_A, \mathbf{r}_{BC})$ as $r_A \to \infty$. However, the effect of the reactive scattering channels (which appear only when r_{BC} goes to infinity) is concealed in the infinite sum in eqn 14.77. Therefore, if we were to use this expansion for reactive scattering, then we would probably need a very large number of basis functions within the coupled-channel approximation to get accurate results. Thus, it is often of great advantage to use a variety of basis set functions in the expansion; for example, basis set functions of the form $\Phi^{JM_J}_\beta(\hat{\mathbf{r}}_B, \mathbf{r}_{AC})$ or $\Phi^{JM_J}_\gamma(\hat{\mathbf{r}}_C, \mathbf{r}_{AB})$ for which the effect of reactive scattering into channels B + AC or C + AB is more transparent. In any event, the scattering problem will still be characterized by an $N \times N$ scattering matrix $\mathbf{S}^J$, where N is the total number of open channels.[8]

Finally, a concept of the greatest significance in chemistry is the rate constant for a chemical reaction. To establish the relation of this quantity and the S matrix elements we need to find an expression for the probability P_{ji} that a transition occurs from the incident channel i to final channel j. By using the asymptotic form in eqn 14.78, it is possible to show (for a given value of J) that the ratio of the outgoing flux in channel j to the incoming flux in channel i is $|S^J_{ji}|^2$. Therefore

$$P_{ji} = \sum_J (2J+1) P^J_{ji} = \sum_J (2J+1) |S^J_{ji}|^2 \qquad (79)$$

where P^J_{ji} is the probability for a given total angular momentum J. In the simplest possible case of purely elastic scattering, for which the incident and final channels are identical and the scattering matrix is a simple number $S_l = e^{2i\delta_l}$,

[8] When reactive scattering can occur, it can be accompanied by elastic and inelastic processes; N includes elastic, inelastic, and reactive scattering channels that are open.

the probability is

$$P^{l}_{\lambda_0\lambda_0} = |S_l|^2 = 1 \tag{80}$$

as it should be, because there is only one possible result for the outcome of the scattering event.

Now consider the general chemical reaction

$$\mathrm{A + BC \rightarrow AB + C}$$

for which we are interested in the ordinary temperature-dependent rate constant $k(T)$ (as in the expression *rate* $= k(T)[\mathrm{A}][\mathrm{BC}]$). We let λ be an open channel in the reactants A + BC and γ be an open channel in the products AB + C. Then we define the **cumulative reaction probability**, $N(E)$, as

$$N(E) = \sum_{\gamma,\lambda} P_{\gamma\lambda}(E) \tag{81}$$

The cumulative reaction probability represents the sum at a fixed total energy E over all possible channel-to-channel reactive transition probabilities. The rate constant is then given by

$$k(T) \doteq \frac{\int_0^\infty N(E)\mathrm{e}^{-E/kT}\,\mathrm{d}E}{hQ_\mathrm{r}(T)} \tag{82}$$

where $Q_\mathrm{r}(T)$ is the partition function density (the partition function divided by the volume) of the reactants at the temperature T. We do not intend to go into the details of this equation, but its importance should be clear. It provides a critical link between an experimentally measurable quantity (a rate constant) and a theoretically calculable quantity, $N(E)$.

14.18 The *S* matrix and multichannel resonances

In this section, we mention some of the properties of the scattering matrix and give its connection to resonances.[9] One extremely important property of **S** is that it is *unitary*, which means that

$$\mathbf{S}^\dagger\mathbf{S} = \mathbf{S}\mathbf{S}^\dagger = \mathbf{1}$$

where **1** is the unit matrix and $\mathbf{S}^\dagger$ is the adjoint of **S**, the complex conjugate of its transpose (see *Further information 23* for information on the properties of matrices). Therefore,

$$\sum_k S^\dagger_{ik}S_{kl} = \delta_{il} \qquad \sum_k S^*_{ki}S_{kl} = \delta_{il} \qquad \sum_k |S_{ki}|^2 = 1$$

This set of equations is the mathematical expression for the conservation of flux during the scattering event. Furthermore, the scattering matrix is often also symmetric:

$$S_{ij} = S_{ji}$$

[9] The full *S* matrix is a block-diagonal matrix, with blocks consisting of the smaller matrices $\mathbf{S}^J$ and S_{ji} being identically zero for channels j and i that correspond to different total angular momenta.

(To be precise, if the scattering system is time-reversal invariant, then the S matrix is symmetric.)

We conclude by giving the relation between the S matrix and resonances. Recall for elastic scattering, resonances of partial wave l correspond to poles in S_l at complex energies $E_{\text{res}} - \frac{1}{2}\text{i}\Gamma$. Resonances do not have to occur at the same resonance energies for different partial waves. In fact, resonances may be found only for certain partial waves and not for others. For multichannel scattering, a general form of the scattering matrix is

$$\mathbf{S}^J = \mathbf{S}^J_{\text{bg}} - \frac{\text{i}\mathbf{C}^J}{E - E^J_{\text{res}} + \text{i}\Gamma^J/2} \tag{83}$$

where $\mathbf{S}^J_{\text{bg}}$ is a unitary background scattering matrix and $\mathbf{C}^J$ is another $N \times N$ matrix with properties that do not concern us here. Thus, a pole will occur in the scattering matrix $\mathbf{S}^J$ at complex energies $\bar{E}^J = E^J_{\text{res}} - \frac{1}{2}\text{i}\Gamma^J$. These poles correspond to resonances, the properties of which we have already described. Notice that at $\bar{E}^J$, a pole will appear in *each* scattering matrix element S^J_{ji}. However, resonances need not occur for different values of J. For example, there may be a scattering resonance for $J = 0$ but none for $J > 0$. Therefore, the effect of a resonance may not be observed experimentally, because experiments reflect averages over all total angular momenta.

When resonances are found, they usually occur in multichannel systems (as opposed to the one-channel elastic case). They can have very important effects on cross-sections and state-to-state transition probabilities at real energies in the vicinity of the real part E_{res} of the resonance energy. As such, they play a very important role in understanding the dynamics of scattering processes, and their study is one of the current growth points of modern molecular quantum mechanics.

Problems

14.1 Characterize each of the following scattering processes as either elastic, inelastic, or reactive.

(i) $\text{O}(^3\text{P}) + \text{O}(^3\text{P}) \longrightarrow \text{O}(^3\text{P}) + \text{O}(^1\text{D})$
(ii) $\text{O}(^3\text{P}) + \text{O}(^3\text{P}) \longrightarrow \text{O}(^3\text{P}) + \text{O}(^3\text{P})$
(iii) $\text{Cl}(^2\text{P}) + \text{HF}(v = 0, j = 0) \longrightarrow \text{Cl}(^2\text{P}) + \text{HF}(v = 1, j = 0)$
(iv) $\text{Cl}(^2\text{P}) + \text{HF}(v = 0, j = 0) \longrightarrow \text{F}(^2\text{P}) + \text{HCl}(v = 0, j = 0)$
(v) $\text{Cl}(^2\text{P}) + \text{HF}(v = 0, j = 0) \longrightarrow \text{Cl}(^2\text{P}) + \text{HF}(v = 0, j = 0)$

14.2 Given that the scattering amplitude has the simple analytical form

$$f_k(\theta, \phi) = \sin\theta\cos\phi$$

find an expression for the differential cross-section.

14.3 Evaluate the integral scattering cross-section for a case in which the differential cross-section is a constant C independent of the angles θ and ϕ.

14.4 The first two ($l = 0, 1$) Riccati–Bessel functions are

$$\hat{j}_0(kr) = \sin(kr) \qquad \text{and} \qquad \hat{j}_1(kr) = \frac{\sin(kr)}{kr} - \cos(kr)$$

Confirm that they are solutions of the free-particle radial wave equation, eqn 14.34.

14.5 Calculate the angular components of the flux density, J_θ and J_ϕ, for the scattered wave

$$\psi = f_k(\theta, \phi)\frac{e^{ikr}}{r}$$

and confirm that in the limit $r \to \infty$, only the radial component J_r given in eqn 14.13 needs to be retained.

14.6 The incoming Green's function is given by

$$G^{(-)}(\mathbf{r}, \mathbf{r}') = \frac{e^{-ik|\mathbf{r}-\mathbf{r}'|}}{|\mathbf{r}-\mathbf{r}'|}$$

Show that $G^{(-)}$ is a solution of eqn 14.18. (*Hint*. Use an analysis similar to that given in *Further information 12*.) Although the incoming Green's function does not yield the desired asymptotic form of the stationary scattering state (eqn 14.7), $G^{(-)}$ appears in some of the formal expressions of scattering theory. (See Chapter 19 of A. Messiah, *Quantum mechanics*, Vol. II, North-Holland, Amsterdam (1965).)

14.7 The differential cross-section for the Yukawa potential using the Born approximation is given in Example 14.2. Plot it as a function of the angle θ for (i) zero energy, (ii) moderate energy ($k \approx \alpha$), and (iii) high energy ($k \gg \alpha$). For the plots, choose the range of the y-axis to be 0 to $\{(2\mu V_0)/(\hbar^2\alpha^2)\}^2$. For moderate energy, take $k = \alpha/2$; for high energy, take $k = 10\alpha$.

14.8 Use the Born approximation to calculate the differential cross-section for scattering from the spherical square-well potential (Section 14.10). *Hint*. Use integration by parts to determine the scattering amplitude.

14.9 Consider the scattering of an electron by an atom of atomic number Z. The interaction potential energy can be approximated by the screened Coulomb potential energy $V(r) = -(Ze^2/4\pi\varepsilon_0 r)e^{-r/a}$, where a is the screening length. (a) Use the Born approximation to calculate the differential cross-section for scattering from the screened Coulomb potential. Go on to evaluate the integral scattering cross-section. (b) In the limit $a \to \infty$, $V(r)$ becomes exactly the Coulomb potential energy. Evaluate the differential cross-section obtained in part (a) in this limit. The expression obtained is the same as the Coulomb scattering cross-section and is the celebrated *Rutherford formula*. It is interesting that, although the Born approximation gives only an approximate differential cross-section and, in addition, does not apply to the Coulomb $(1/r)$ potential, the result obtained here is precisely Rutherford's formula.

14.10 Derive the expressions given in eqns 14.39 and 14.40.

14.11 Consider the differential cross-section for elastic scattering given in eqn 14.42. At a given energy, sketch its dependence on the scattering angle θ when the $l = 1$ partial wave dominates the scattering. Do the same for the $l = 0$ and $l = 2$ partial waves.

14.12 Show for the elastic scattering of a particle by a central potential $V(r)$ that approaches zero more rapidly than $1/r$ as $r \to \infty$, that the integrated cross-section can be written as

$$\sigma = \frac{4\pi}{k}\,\mathrm{Im}\, f_k(0)$$

where $\mathrm{Im}\, f_k(0)$ is the imaginary part of the forward scattering amplitude ($\theta = 0$). This is the so-called *optical theorem*. (Note: The Legendre polynomials are required to satisfy $P_l(1) = 1$ for all values of l.)

14.13 For elastic scattering off a central potential, it is possible to show analytically that if the potential is repulsive, with $V(r) > 0$ for all r, then the scattering phase shift $\delta_l(E)$ is negative; likewise, if the potential is attractive, with $V(r) < 0$ for all r, then the phase shift δ_l is positive. (The interested reader is referred to pp 404–5 of A. Messiah, *Quantum mechanics*, Vol. I, North-Holland, Amsterdam (1965).) Explain this result qualitatively by considering the effect of a repulsive (or attractive) potential on the wavelength of the scattered particle.

14.14 Derive an expression for the scattering phase shift δ_l for l-wave scattering by a hard sphere, where $V(r)$ is given in Example 14.3.

14.15 A particle of mass m is scattered off a central potential $V(r)$ of the form

$$V(r) = \begin{cases} \infty & \text{if } r = 0 \\ V_0 & \text{if } 0 < r < a \\ 0 & \text{if } r \geq a \end{cases}$$

where V_0 is a positive constant. For energies $E > V_0$, find an expression for the S-wave scattering phase shift δ_0. *Hint.* Require that the wavefunction and its first derivative be continuous at $r = a$.

14.16 A particle of mass m is scattered off a central potential $V(r)$ of the form

$$V(r) = \begin{cases} \infty & \text{if } r = 0 \\ 0 & \text{if } 0 < r < a \\ V_0 & \text{if } a < r < b \\ 0 & \text{if } r \geq b \end{cases}$$

where V_0 is a positive constant. For energies $E > V_0$, find an expression for the S-wave scattering phase shift δ_0. *Hint.* Require that the wavefunction and its first derivative be continuous at $r = a$ and at $r = b$.

14.17 For scattering by a spherical square-well potential (Section 14.10), show that the S-wave cross-section can be written at low energies (that is, $ka \ll 1$) as

$$\sigma_0 = 4\pi a^2 \left(\frac{\tan Ka}{Ka} - 1 \right)^2$$

14.18 In the Ramsauer–Townsend effect it is observed that when electrons are scattered off some inert gas atoms, there is a nearly complete transmission of the bombarding electrons at low energies around 0.7 eV. For energies above and below 0.7 eV, the scattering cross-section is significantly greater than zero. Model the interaction between the bombarding electrons and the inert gas atom as a spherical square-well potential, and give an explanation for the Ramsauer–Townsend effect on the basis of the expression given in Problem 14.17. Suggest a reason why this effect is not observed for non-inert gas atoms.

14.19 For elastic scattering off a central potential, the scattering phase shift for partial wave l can be written as

$$\delta_l(E) = \delta_{\text{bg}}(E) + \delta_{\text{res}}(E)$$

where the resonant part of the phase shift is given by

$$\tan \delta_{\text{res}}(E) = \frac{\Gamma}{2(E_{\text{res}} - E)}$$

and the background phase shift is often a slowly varying function of energy. (a) Sketch the behavior of δ_l as a function of energy in the vicinity of E_{res} if δ_{bg} is taken to be independent of energy with a constant value of (i) 0; (ii) $\pi/4$; (iii) $\pi/2$; (iv) $3\pi/4$. (b) The partial wave cross-section $\sigma_l(E)$ is proportional to $\sin^2 \delta_l(E)$. Sketch the dependence of the latter on energy in the vicinity of E_{res} for the four values of δ_{bg} given in part (a). Note that for $\delta_{bg} = 0$, $\sin^2 \delta_l(E)$ has the Breit–Wigner form (eqn 14.59).

14.20 At a total collision energy E_1, the products of the scattering process involving atom A and diatomic molecule BC include A + BC, AB + C, and AC + B. It is known that there are eleven A + BC channels, six AB + C channels, and sixteen AC + B channels that are energetically accessible at energy E_1. What is the dimension of the scattering matrix at this scattering energy?

14.21 Explain the appearance of the factor k_α/k_{α_0} in eqn 14.74 for the differential cross-section for scattering from an initial state α_0 to a final state α.

14.22 The reactance matrix, $\mathbf{K}$, defined in relation to the scattering matrix through $\mathbf{K} = \mathrm{i}(\mathbf{1} - \mathbf{S})(\mathbf{1} + \mathbf{S})^{-1}$ also appears in scattering theory. Show for elastic scattering off a central potential with partial wave l that $\mathbf{K}$ is a 1×1 matrix with element $K_l = \tan \delta_l$.

14.23 Consider a scattering process in which there are two possible channels, denoted 1 and 2. According to the principle of microscopic reversibility (also called the principle of detailed balance), the probability of being incident in channel 1 and ending up in channel 2 is equal to the probability of being incident in channel 2 and ending up in channel 1. Discuss this principle in light of the properties of the scattering matrix.

14.24 During a scattering process, a system in incident channel i undergoes a transition to final channel j. It is possible to define a channel-to-channel delay time Δt_{ji} in terms of the scattering matrix element S_{ji} by

$$\Delta t_{ji} = \mathrm{Im}\left\{ \frac{\hbar}{S_{ji}} \frac{\mathrm{d}S_{ji}}{\mathrm{d}E} \right\}$$

The delay time represents the time difference between starting in channel i and ending in channel j, relative to the time difference in the absence of the potential V. Because a resonance represents a metastable state with a finite lifetime, one would expect that at real energies near E_{res}, the scattered particle should experience a significant delay time. Confirm the latter statement by showing that the maximum in Δt_{ji} occurs precisely at $E = E_{\mathrm{res}}$ and, in addition, that the product of the maximum Δt_{ji} and the resonance width equals $2\hbar$, reminiscent of the lifetime broadening relation (Section 1.18). To demonstrate the above, begin with eqn 14.83 for the scattering matrix element S_{ji} and assume that the background contribution $S_{ji,\mathrm{bg}}$ is negligible and that C_{ji} is independent of energy. [Interested readers are encouraged to see R.S. Friedman, V.D. Hullinger, and D.G. Truhlar, *J. Phys. Chem.*, 3184, **99** (1995).]

Further reading

Quantum mechanics in chemistry. G.C. Schatz and M.A. Ratner; Prentice–Hall, Englewood Cliffs (1993).

Quantum mechanics: foundations and applications. A. Bohm; Springer, New York (1993).

The theory of atomic and molecular collisions. J.N. Murrell and S.D. Bosanac; *Chem. Soc. Rev.*, **21**, 17 (1992).

A computational approach to chemistry. D.M. Hirst; Blackwell Scientific, Oxford (1990).

Introduction to the theory of atomic and molecular collisions. J.N. Murrell and S.D. Bosanac; Wiley, New York (1989).

Scattering theory: the quantum theory of nonrelativistic collisions. J.R. Taylor; Krieger Publishing Company, Malabar, Florida (1987).

Molecular reaction dynamics and chemical reactivity. R.D. Levine and R.B. Bernstein; Oxford University Press, Oxford (1987).

The theory of chemical reaction dynamics. M. Baer (ed.); CRC Press, Boca Raton, Florida (1985).

The theory of chemical reaction dynamics. D.C. Clary (ed.); Reidel, Dordrecht (1986).

Scattering theory of waves and particles. R.G. Newton; Springer, New York (1982).

Quantum mechanics. Vols I and II., A. Messiah; North-Holland, Amsterdam (1961).

Quantum mechanics. L.I. Schiff; McGraw–Hill, New York (1968).

Molecular collision theory. M.S. Child; Dover, New York (1997).

Further information

Equation numbers without a prefix refer to the current sections; those of the form eqn $X.x$ refer to eqn x of Chapter X in the text; cross-references to equations in other *Further information* sections are prefixed by FI.

Classical mechanics

1 Action

Hamilton's principle asserts the following:

The path taken by a particle is the one that involves the least action.

The **action**, S, can be expressed as an integral over the **lagrangian**, L, which depends on the position (x) and speed ($\dot{x}$) of a particle at each point along its path:

$$S = \int_{t_1}^{t_2} L(x, \dot{x})\,\mathrm{d}t \tag{1.1}$$

The function L is formulated so that Hamilton's principle results in a path that agrees with observation. We illustrate what this means in the following paragraphs.

Suppose that x and $\dot{x}$ are varied a little at each point of the particle's path except the end points. As a result of this modification, the lagrangian changes by δL and the action changes by

$$\delta S = \int_{t_1}^{t_2} \delta L(x, \dot{x})\,\mathrm{d}t$$

Because L is a function of x and $\dot{x}$, changes in these quantities result in a change in L given by

$$\delta L = \left(\frac{\partial L}{\partial x}\right)\delta x + \left(\frac{\partial L}{\partial \dot{x}}\right)\delta \dot{x}$$

Therefore, on integration by parts,

$$\begin{aligned}\delta S &= \int_{t_1}^{t_2} \left(\frac{\partial L}{\partial x}\right)\delta x\,\mathrm{d}t + \int_{t_1}^{t_2} \left(\frac{\partial L}{\partial \dot{x}}\right)\left(\frac{\mathrm{d}\delta x}{\mathrm{d}t}\right)\mathrm{d}t \\ &= \int_{t_1}^{t_2} \left(\frac{\partial L}{\partial x}\right)\delta x\,\mathrm{d}t + \left\{\left(\frac{\partial L}{\partial \dot{x}}\right)\delta x\bigg|_{t_1}^{t_2} - \int_{t_1}^{t_2} \frac{\mathrm{d}}{\mathrm{d}t}\left(\frac{\partial L}{\partial \dot{x}}\right)\delta x \mathrm{d}t\right\}\end{aligned}$$

Because the end points of the path are fixed, the middle term in this expression is zero (δx is zero at the end points). Therefore, the change in S is

$$\delta S = \int_{t_1}^{t_2} \left\{\left(\frac{\partial L}{\partial x}\right) - \frac{\mathrm{d}}{\mathrm{d}t}\left(\frac{\partial L}{\partial \dot{x}}\right)\right\}\delta x\,\mathrm{d}t$$

According to Hamilton's principle, the action is a minimum for the actual path; hence any small variation of the path corresponds to $\delta S = 0$, the usual condition for an extremum (a maximum or minimum) of a function. In this case, $\delta S = 0$ is achieved for small but otherwise arbitrary variations δx only if the factor multiplying δx vanishes everywhere. Thus, we arrive at the **Euler–Lagrange equation of motion**:

$$\left(\frac{\partial L}{\partial x}\right) - \frac{\mathrm{d}}{\mathrm{d}t}\left(\frac{\partial L}{\partial \dot{x}}\right) = 0 \tag{1.2}$$

This equation should correspond to the equations of motion obeyed by the particle, so the lagrangian should be modified until that is so.

As an illustration, suppose we propose that the lagrangian for a particle of mass m is

$$L = \tfrac{1}{2}m(\dot{x})^2 - V(x) \tag{1.3}$$

then the Euler–Lagrange equation is constructed from

$$\left(\frac{\partial L}{\partial x}\right) = -\frac{\mathrm{d}V}{\mathrm{d}x} = F$$

$$\frac{\mathrm{d}}{\mathrm{d}t}\left(\frac{\partial L}{\partial \dot{x}}\right) = \frac{\mathrm{d}(m\dot{x})}{\mathrm{d}t} = m\ddot{x}$$

It follows that the Euler–Lagrange equation is

$$F - m\ddot{x} = 0 \tag{1.4}$$

which should be recognized as Newton's second law of motion. Hence, for a particle that obeys Newtonian mechanics, the lagrangian given above is appropriate. It should be noted that the lagrangian has the form

$$L = T - V \tag{1.5}$$

where T is the kinetic energy and V is the potential energy.

Consider now the total derivative of the action with respect to the time t_2. This calculation tells us how the action changes as the end point is varied:

$$\frac{\mathrm{d}S}{\mathrm{d}t_2} = \frac{\partial S}{\partial t_2} + \left(\frac{\partial S}{\partial x_2}\right)\left(\frac{\partial x_2}{\partial t_2}\right)$$

It follows from eqn 1 that

$$\frac{\mathrm{d}S}{\mathrm{d}t_2} = L \text{ evaluated at the end point}$$

Moreover, by using eqn 2 we can write

$$\frac{\partial S}{\partial x_2} = \int_{t_1}^{t_2}\left(\frac{\partial L}{\partial x_2}\right)\mathrm{d}t = \int_{t_1}^{t_2}\frac{\mathrm{d}}{\mathrm{d}t}\left(\frac{\partial L}{\partial \dot{x}_2}\right)\mathrm{d}t = \frac{\partial L}{\partial \dot{x}_2}$$

evaluated at the same end point. Because we know that $\partial L/\partial \dot{x} = m\dot{x}$, we arrive at

$$L = \left(\frac{\partial S}{\partial t_2}\right) + m\dot{x}_2{}^2 = \left(\frac{\partial S}{\partial t_2}\right) + 2T$$

Then, as $T - V = L$, it follows that

$$T + V = -\left(\frac{\partial S}{\partial t_2}\right)$$

The sum of T and V is the total energy of the system, E; and so we can conclude that

$$E = -\left(\frac{\partial S}{\partial t_2}\right) \tag{1.6}$$

as used in the text.

2 The canonical momentum

'Canon' means rule. The following is a rule for finding the momentum of any system of particles. As an example of each step, we shall construct the expression for the linear momentum in the presence of electric and magnetic fields.

1. Choose a lagrangian L such that the Euler–Lagrange equations (*Further information 1*) correspond to the known equations of motion.

The equation of motion of an electron in the presence of electric and magnetic fields is given by the **Lorentz force law**:

$$m_e\ddot{\mathbf{r}} = -e(\mathbf{E} + \dot{\mathbf{r}} \times \mathbf{B}) \tag{2.1}$$

This equation of motion is reproduced by the Euler–Lagrange equations if we take as the lagrangian the expression

$$L = \tfrac{1}{2}m_e\dot{r}^2 + e\phi - e\dot{\mathbf{r}} \cdot \mathbf{A} \tag{2.2}$$

where ϕ is the scalar potential and $\mathbf{A}$ is the vector potential describing the fields (see *Further information 20*). This expression may be confirmed by noting that

$$\frac{\partial L}{\partial x} = e\frac{\partial \phi}{\partial x} - e\frac{\partial}{\partial x}(\dot{\mathbf{r}} \cdot \mathbf{A})$$

so that in three dimensions

$$\nabla L = e\nabla\phi - e\nabla(\dot{\mathbf{r}} \cdot \mathbf{A}) = e\nabla\phi - e\dot{\mathbf{r}} \cdot \nabla\mathbf{A} - e\dot{\mathbf{r}} \times (\nabla \times \mathbf{A})$$

To derive this result, we have used the vector relations listed in *Further information 22*; note that $\mathbf{F} \cdot \nabla\mathbf{G}$ should be interpreted as $(\mathbf{F} \cdot \nabla)\mathbf{G}$. Likewise,

$$\begin{aligned}\frac{\mathrm{d}}{\mathrm{d}t}\left(\frac{\partial L}{\partial \dot{x}}\right) &= \frac{\mathrm{d}}{\mathrm{d}t}(m_e\dot{x} - eA_x) \\ &= m_e\ddot{x} - e\frac{\mathrm{d}A_x}{\mathrm{d}t} \\ &= m_e\ddot{x} - e\left(\frac{\partial A_x}{\partial t}\right) - e\left\{\left(\frac{\partial x}{\partial t}\right)\left(\frac{\partial A_x}{\partial x}\right) + \ldots\right\}\end{aligned}$$

where the dots indicate the analogous terms with y and z in place of x, and in three dimensions (in a notation that should be self-explanatory by comparison with the expression above)

$$\frac{\mathrm{d}}{\mathrm{d}t}\left(\frac{\partial L}{\partial \dot{\mathbf{r}}}\right) = m_{\mathrm{e}}\ddot{\mathbf{r}} - e\dot{\mathbf{A}} - e(\dot{\mathbf{r}} \cdot \nabla)\mathbf{A}$$

It follows that the Euler–Lagrange equation is

$$e\nabla\phi - e\dot{\mathbf{r}} \cdot \nabla\mathbf{A} - e\dot{\mathbf{r}} \times (\nabla \times \mathbf{A}) = m_{\mathrm{e}}\ddot{\mathbf{r}} - e\dot{\mathbf{A}} - e(\dot{\mathbf{r}} \cdot \nabla)\mathbf{A}$$

which reduces to the Lorentz expression by using eqns FI20.3 and FI20.5. Hence, the lagrangian in eqn 2 is acceptable.

2. Form the **canonical momentum**, which is defined as

$$p_q = \frac{\partial L}{\partial \dot{q}} \tag{2.3}$$

From the lagrangian developed above, it follows that

$$p_x = m_{\mathrm{e}}\dot{x} - eA_x$$

and hence in three dimensions

$$\mathbf{p} = m_{\mathrm{e}}\dot{\mathbf{r}} - e\mathbf{A} \tag{2.4}$$

3. Form the **hamiltonian**, which is defined as

$$H = \mathbf{p} \cdot \dot{\mathbf{r}} - L \tag{2.5}$$

and express it in terms of **p** and **r** as variables.

In the present example, because $\dot{\mathbf{r}} = (\mathbf{p} + e\mathbf{A})/m_{\mathrm{e}}$, we obtain

$$\begin{aligned} H &= \frac{1}{m_{\mathrm{e}}}\mathbf{p} \cdot (\mathbf{p} + e\mathbf{A}) - \frac{1}{2m_{\mathrm{e}}}(\mathbf{p} + e\mathbf{A})^2 - e\phi + \frac{e}{m_{\mathrm{e}}}(\mathbf{p} + e\mathbf{A}) \cdot \mathbf{A} \\ &= \frac{1}{2m_{\mathrm{e}}}(\mathbf{p} + e\mathbf{A})^2 - e\phi \end{aligned} \tag{2.6}$$

The same expression would be obtained by replacing **p**, wherever it occurs in the hamiltonian, by $\mathbf{p} + e\mathbf{A}$, which is the rule used in the text.

3 The virial theorem

In classical mechanics, the proof of the virial theorem (Section 2.17) is based on the disappearance of the time average of the time derivative of the product $\mathbf{p} \cdot \mathbf{r}$, where **p** is the linear momentum and **r** is the position of a particle. The proof is similar in quantum mechanics, but it makes use of the time derivative of the expectation value of the operator $\mathbf{p} \cdot \mathbf{r}$.

From the equation

$$\frac{\mathrm{d}\langle\Omega\rangle}{\mathrm{d}t} = \frac{\mathrm{i}}{\hbar}\langle[H, \Omega]\rangle$$

(this is eqn 1.35), we can write

$$\frac{\mathrm{d}}{\mathrm{d}t}\langle\mathbf{p} \cdot \mathbf{r}\rangle = \frac{\mathrm{i}}{\hbar}\langle[H, \mathbf{p} \cdot \mathbf{r}]\rangle \tag{3.1}$$

For simplicity, we shall consider only a one-dimensional system, but the extension to more dimensions is straightforward. We need the following relations:

$$[H, p_x x] = [H, p_x]x + p_x[H, x]$$

$$[H, p_x] = [-\frac{\hbar^2}{2m}\frac{d^2}{dx^2} + V, \frac{\hbar}{i}\frac{d}{dx}] = i\hbar\frac{dV}{dx}$$

$$[H, x] = [-\frac{\hbar^2}{2m}\frac{d^2}{dx^2}, x] = -2\left(\frac{\hbar^2}{2m}\right)\frac{d}{dx} = -\frac{i\hbar}{m}p_x$$

The first of these relations is a special case of the general result that

$$[A, BC] = [A, B]C + B[A, C]$$

Then, because the kinetic energy operator, T, can be identified with the operator $p_x^2/2m$, it follows that

$$\frac{d}{dt}\langle p_x x\rangle = 2\langle T\rangle - \left\langle x\frac{dV}{dx}\right\rangle$$

The time average of this expression is

$$\frac{1}{\tau}\int_0^\tau \frac{d}{dt}\langle p_x x\rangle\, dt = \frac{1}{\tau}\int_0^\tau \left\{2\langle T\rangle - \left\langle x\frac{dV}{dx}\right\rangle\right\} dt$$

Therefore, because the expectation values on the right are independent of time,

$$\frac{1}{\tau}\langle p_x x\rangle\Big|_0^\tau = 2\langle T\rangle - \left\langle x\frac{dV}{dx}\right\rangle$$

The term on the left is zero, for if the motion is periodic we may choose τ to be the period, and if the motion is not periodic, then we may choose τ to be infinite. In the latter case, the value of $\langle p_x x\rangle_\tau - \langle p_x x\rangle_0$ is finite in a bounded system and τ is infinite. Therefore,

$$2\langle T\rangle = \left\langle x\frac{dV}{dx}\right\rangle \tag{3.2}$$

The force experienced by the particle is $F_x = -dV/dx$, so this equation may be written

$$2\langle T\rangle = -\langle xF_x\rangle \tag{3.3}$$

and in three dimensions this expression is the **virial theorem**:

$$2\langle T\rangle = -\langle \mathbf{r}\cdot\mathbf{F}\rangle \tag{3.4}$$

If the potential energy of the particle has the form $V = ax^s$, then eqn 2 gives

$$\langle T\rangle = \tfrac{1}{2}s\langle V\rangle \tag{3.5}$$

as used in the text. The theorem may be extended to operators other than $\mathbf{p}\cdot\mathbf{r}$; then, different choices lead to a variety of **hypervirial theorems**.

4 Reduced mass

Here we show that the motion of two particles may be separated into the motion of their centre of mass and their relative motion. Let the masses be m_1 and m_2, their locations $\mathbf{r}_1$ and $\mathbf{r}_2$, and the total mass be $m = m_1 + m_2$.Their separation is

$$\mathbf{r} = \mathbf{r}_1 - \mathbf{r}_2 \tag{4.1}$$

and the location of the centre of mass is at

$$\mathbf{R} = \left(\frac{m_1}{m}\right)\mathbf{r}_1 + \left(\frac{m_2}{m}\right)\mathbf{r}_2 \tag{4.2}$$

The hamiltonian for a system in which the potential energy depends only on their separation is

$$H = -\frac{\hbar^2}{2m_1}\nabla_1^2 - \frac{\hbar^2}{2m_2}\nabla_2^2 + V(|\mathbf{r}_1 - \mathbf{r}_2|) \tag{4.3}$$

We want to show that this operator can be transformed into

$$H = -\frac{\hbar^2}{2m}\nabla_{\mathbf{R}}^2 - \frac{\hbar^2}{2\mu}\nabla_{\mathbf{r}}^2 + V(r) \tag{4.4}$$

in what should be an obvious notation. If this is so, then it follows that the wavefunction can be expressed as the product $\Psi(\mathbf{R})\psi(\mathbf{r})$.

The transformation of the potential energy contribution is trivial; the work we have to do resides in the derivatives. To analyse them, we consider the x components, which are

$$x = x_1 - x_2 \qquad X = \left(\frac{m_1}{m}\right)x_1 + \left(\frac{m_2}{m}\right)x_2 \tag{4.5}$$

It follows that

$$\frac{\partial}{\partial x_1} = \left(\frac{\partial X}{\partial x_1}\right)\frac{\partial}{\partial X} + \left(\frac{\partial x}{\partial x_1}\right)\frac{\partial}{\partial x} = \frac{m_1}{m}\frac{\partial}{\partial X} + \frac{\partial}{\partial x}$$
$$\frac{\partial}{\partial x_2} = \left(\frac{\partial X}{\partial x_2}\right)\frac{\partial}{\partial X} + \left(\frac{\partial x}{\partial x_2}\right)\frac{\partial}{\partial x} = \frac{m_2}{m}\frac{\partial}{\partial X} - \frac{\partial}{\partial x}$$

Therefore, the x-component of the sum of the two laplacians is

$$\begin{aligned}\frac{1}{m_1}\frac{\partial^2}{\partial x_1^2} + \frac{1}{m_2}\frac{\partial^2}{\partial x_2^2} &= \frac{1}{m_1}\left(\frac{m_1}{m}\frac{\partial}{\partial X} + \frac{\partial}{\partial x}\right)^2 + \frac{1}{m_2}\left(\frac{m_2}{m}\frac{\partial}{\partial X} - \frac{\partial}{\partial x}\right)^2 \\ &= \frac{1}{m}\frac{\partial^2}{\partial X^2} + \left(\frac{1}{m_1} + \frac{1}{m_2}\right)\frac{\partial^2}{\partial x^2}\end{aligned}$$

The y- and z-components can be dealt with similarly, and when they are added together, we obtain

$$\frac{1}{m_1}\nabla_1^2 + \frac{1}{m_2}\nabla_2^2 = \frac{1}{m}\nabla_{\mathbf{R}}^2 + \frac{1}{\mu}\nabla_{\mathbf{r}}^2 \tag{4.6}$$

with μ as defined earlier. Substitution of this expression into eqn 3 gives eqn 4, as required.

Solutions of the Schrödinger equation

5 The motion of wavepackets

The time-dependent form of the wavefunction of a particle of mass m in a state of linear momentum $p = k\hbar$ is given by eqn 2.12 in Chapter 2 as

$$\Psi_k(x,t) = A\mathrm{e}^{\mathrm{i}kx - \mathrm{i}E(k)t/\hbar} \qquad E(k) = \frac{k^2\hbar^2}{2m}$$

Such a particle is regarded as having a **phase velocity**, v_p, given by

$$v_p = \frac{p}{m} = \frac{k\hbar}{m} = \frac{h}{\lambda m} \tag{5.1}$$

The wavefunction of an imprecisely prepared system is the superposition

$$\Psi(x,t) = \int g(k)\Psi_k(x,t)\,\mathrm{d}k$$

We suppose that the **shape function**, $g(k)$, peaks sharply around k_0 and falls to zero for values of $|k - k_0| \gg \Gamma$, the width parameter. For example, $g(k)$ might be the normalized Gaussian function

$$g(k) = N\mathrm{e}^{-(k-k_0)^2/2\Gamma^2} \qquad N = \frac{1}{\Gamma\sqrt{2\pi}} \tag{5.2}$$

If we write

$$G(x) = N\int \mathrm{e}^{-(k-k_0)^2/2\Gamma^2 + \mathrm{i}(k-k_0)x}\,\mathrm{d}k = \mathrm{e}^{-x^2\Gamma^2/2}$$

then it follows that the probability density for finding the particle at $t = 0$ is

$$|\Psi(x,0)|^2 = |AG(x)\mathrm{e}^{\mathrm{i}k_0x}|^2 = |A|^2\mathrm{e}^{-x^2\Gamma^2} \tag{5.3}$$

which is a Gaussian function centred on $x = 0$ with a width $\delta x \approx 1/\Gamma$.

Now consider the shape of the packet at later times. Because g peaks sharply around k_0, the only values of $E(k)$ that contribute significantly to the integral are those near $E(k_0)$. Therefore, we expand $E(k)$ as a Taylor series and discard all but the first few terms:

$$\begin{aligned} E(k) &= E(k_0) + (k-k_0)\left(\frac{\mathrm{d}E}{\mathrm{d}k}\right)_{k_0} + \tfrac{1}{2}(k-k_0)^2\left(\frac{\mathrm{d}^2E}{\mathrm{d}k^2}\right)_{k_0} + \ldots \\ &= E(k_0) + (k-k_0)v_\mathrm{g}\hbar + \tfrac{1}{2}(k-k_0)^2 w_\mathrm{g}\hbar + \ldots \end{aligned}$$

where the **group velocity**, v_g, is

$$v_\mathrm{g} = \frac{1}{\hbar}\left(\frac{\mathrm{d}E}{\mathrm{d}k}\right)_{k_0} = \frac{k_0\hbar}{m} = \frac{p_0}{m} \tag{5.4}$$

and

$$w_\mathrm{g} = \frac{1}{\hbar}\left(\frac{\mathrm{d}^2E}{\mathrm{d}k^2}\right)_{k_0} = \frac{\hbar}{m} \tag{5.5}$$

The wavepacket therefore has the form

$$\Psi(x,t) = AN\int \mathrm{e}^{-(k-k_0)^2/2\Gamma^2 + \mathrm{i}kx - \mathrm{i}E(k_0)t/\hbar - \mathrm{i}(k-k_0)v_{\mathrm{g}}t - \frac{1}{2}\mathrm{i}(k-k_0)^2 w_{\mathrm{g}}t + \cdots}\,\mathrm{d}k$$

If we wait for only short times, in the sense $\Gamma^2 w_{\mathrm{g}}t \ll 1$, we can neglect the term $\frac{1}{2}(k-k_0)^2 w_{\mathrm{g}}t$ relative to $(k-k_0)^2/2\Gamma^2$, and write

$$\begin{aligned}\Psi(x,t) &\approx AN\mathrm{e}^{\mathrm{i}k_0x - \mathrm{i}E(k_0)t/\hbar}\int \mathrm{e}^{-(k-k_0)^2/2\Gamma^2 + \mathrm{i}(k-k_0)(x-v_{\mathrm{g}}t)}\,\mathrm{d}k \\ &\approx A\mathrm{e}^{\mathrm{i}k_0x - \mathrm{i}E(k_0)t/\hbar}G(x - v_{\mathrm{g}}t)\end{aligned}$$

and its probability density is

$$|\Psi(x,t)|^2 \approx |AG(x-v_{\mathrm{g}}t)|^2 = |A|^2\mathrm{e}^{-(x-v_{\mathrm{g}}t)^2\Gamma^2} \tag{5.6}$$

This is the same function as in eqn 3, but centred on $x = v_{\mathrm{g}}t$. That is, the packet has moved without change of shape from $x = 0$ to $x = v_{\mathrm{g}}t$, and is therefore travelling uniformly with the group velocity $v_{\mathrm{g}} = p_0/m$, the classical velocity of the particle.

The conclusion is valid provided that $\Gamma^2 w_{\mathrm{g}}t \ll 1$, or $\hbar\Gamma^2 t/m \ll 1$. When sufficient time has elapsed for this condition to be invalid, we may no longer neglect terms in w_{g}. These additional terms result in the spreading of the packet. For example, $G(x)$ becomes

$$G(x) = N\int \mathrm{e}^{-(k-k_0)^2/2\Gamma^2 + \mathrm{i}(k-k_0)(x-v_{\mathrm{g}}t) - \frac{1}{2}\mathrm{i}(k-k_0)^2 w_{\mathrm{g}}t}\,\mathrm{d}k$$

so that

$$|G(x)|^2 = \frac{1}{\gamma}\mathrm{e}^{-(x-v_{\mathrm{g}}t)^2\Gamma^2/\gamma^2} \qquad \gamma^2 = 1 + w_{\mathrm{g}}^2t^2\Gamma^4 \tag{5.7}$$

This function has the same exponential dependence as before with Γ replaced by Γ/γ. Therefore, the width of the packet increases as time passes (as Γ/γ decreases), and if its initial uncertainty in location is $\delta x_0 = 1/\Gamma$, then at a time t its spread has become

$$\delta x = \frac{\gamma}{\Gamma} = \frac{1}{\Gamma}(1 + w_{\mathrm{g}}^2t^2\Gamma^4)^{1/2} = \left(1 + \frac{w_{\mathrm{g}}^2t^2}{\delta x_0^4}\right)^{1/2}\delta x_0 \tag{5.8}$$

Because $w_{\mathrm{g}} = \hbar/m$, we find

$$\delta x = \left(1 + \frac{t^2\hbar^2}{m^2\delta x_0^4}\right)^{1/2}\delta x_0 \tag{5.9}$$

It follows that the time for the uncertainty in location to spread from δx_0 to δx is

$$t = \frac{m}{\hbar}\delta x_0\{(\delta x)^2 - (\delta x_0)^2\}^{1/2} \tag{5.10}$$

If $\delta x \gg \delta x_0$, this expression simplifies to

$$t \approx \frac{m}{\hbar}\delta x_0\delta x \tag{5.11}$$

This result means that the location even of an apparently stationary particle spreads with time, but the effect is negligible for most macroscopic objects.

For instance, if $m = 1$ g and $\delta x_0 = 1$ nm, then the uncertainty in location reaches 1 μm after an interval of 10^{16} s, or about 300 million years. On the other hand, for an electron localized to within 1 pm initially, an uncertainty in position of 0.1 nm (the radius of an atom) is reached in only 1 as (1 as $= 10^{-18}$ s).

6 The harmonic oscillator: solution by factorization

The Schrödinger equation for the harmonic oscillator is specified in eqn 2.41. Its appearance is greatly simplified by making the following substitutions:

$$\lambda = \frac{E}{(\hbar\omega/2)} \qquad y = \left(\frac{m\omega}{\hbar}\right)^{1/2} x \qquad \omega = \left(\frac{k}{m}\right)^{1/2} \tag{6.1}$$

The equation then becomes

$$\left(\frac{d^2}{dy^2} - y^2\right)\psi = -\lambda\psi \tag{6.2}$$

The left-hand side of this equation can be factorized by noting that

$$\left(\frac{d}{dy} + y\right)\left(\frac{d}{dy} - y\right)\psi = \left(\frac{d^2}{dy^2} - y^2 - 1\right)\psi$$
$$\left(\frac{d}{dy} - y\right)\left(\frac{d}{dy} + y\right)\psi = \left(\frac{d^2}{dy^2} - y^2 + 1\right)\psi$$

For convenience, we introduce the following operators:

$$a = \frac{1}{\sqrt{2}}\left(\frac{d}{dy} + y\right) \qquad a^+ = \frac{1}{\sqrt{2}}\left(\frac{d}{dy} - y\right) \tag{6.3}$$

Then the last pair of equations have the following form:

$$\begin{aligned} \left(\frac{d^2}{dy^2} - y^2\right)\psi &= (2aa^+ + 1)\psi \\ \left(\frac{d^2}{dy^2} - y^2\right)\psi &= (2a^+a - 1)\psi \end{aligned} \tag{6.4}$$

Therefore, the Schrödinger equation may be written in either of the following forms:

$$\begin{aligned} aa^+\psi_\lambda &= -\tfrac{1}{2}(\lambda + 1)\psi_\lambda \\ a^+a\psi_\lambda &= -\tfrac{1}{2}(\lambda - 1)\psi_\lambda \end{aligned} \tag{6.5}$$

where ψ_λ is the wavefunction corresponding to the energy equivalent to λ. It follows that

$$(a^+a - aa^+)\psi_\lambda = \psi_\lambda$$

This equation is true for any value of λ, and so it can be expressed as the operator identity

$$a^+a - aa^+ = 1 \tag{6.6}$$

or, equivalently, $[a^+, a] = 1$. We shall now see what can be developed from this commutation relation. From the second line of eqn 5, multiplication from the left with a produces

$$aa^+a\psi_\lambda = -\tfrac{1}{2}(\lambda - 1)a\psi_\lambda$$

Use of eqn 6 in the form $aa^+ = a^+a - 1$ turns this equation into

$$a^+aa\psi_\lambda = -\tfrac{1}{2}(\lambda - 3)a\psi_\lambda$$

However, it follows from the second line of eqn 5 that the Schrödinger equation for a state of energy equivalent to $\lambda - 2$ is

$$a^+a\psi_{\lambda-2} = -\tfrac{1}{2}(\lambda - 3)\psi_{\lambda-2}$$

Hence, by comparison of the last two equations,

$$a\psi_\lambda \propto \psi_{\lambda-2} \tag{6.7}$$

This proportionality is valid only for non-degenerate states; but by symmetry, all the states of a one-dimensional oscillator are non-degenerate. We conclude that the wavefunction corresponding to the energy $\lambda - 2$ can be generated from the wavefunction for energy λ by operating on the latter with a. This process may be continued, and wavefunctions corresponding to the energies $\lambda - 4, \lambda - 6\ldots$ may be constructed similarly. In the same way, it is easy to show that successive operations with a^+ generate the wavefunctions corresponding to $\lambda + 2,\ \lambda + 4, \ldots$ from the wavefunction ψ_λ. The fact that a steps wavefunctions down a ladder of energy levels is the origin of its name, the **annihilation operator**. Similarly, a^+ is called a **creation operator**.

The generation of states of lower energy by repeated application of a cannot be continued indefinitely because the energy of a harmonic oscillator is non-negative (it is the eigenvalue of a hamiltonian which is the sum of the squares of two hermitian operators, Example 1.6). Therefore, there must exist a certain minimum value of λ, which we shall call λ_{min}. Because there is no wavefunction corresponding to a lower energy, it follows that $a\psi_{\lambda_{\text{min}}} = 0$. If both sides of this equation are operated on by a^+ we obtain $a^+a\psi_{\lambda_{\text{min}}} = 0$. Then, by using eqn 5, this expression becomes

$$0 = a^+a\psi_{\lambda_{\text{min}}} = -\tfrac{1}{2}(\lambda_{\text{min}} - 1)\psi_{\lambda_{\text{min}}}$$

Consequently, $\lambda_{\text{min}} = 1$ and the allowed values of λ are $1, 3, 5, \ldots$, or $\lambda = 2v + 1$ with $v = 0, 1, 2, \ldots$. We conclude that the allowed energy levels are

$$E = (2v + 1)(\hbar\omega/2) = (v + \tfrac{1}{2})\hbar\omega \qquad v = 0, 1, 2, \ldots \tag{6.8}$$

which are the values quoted in Section 2.16.

The wavefunction for a state of any energy can be found by applying a^+ the appropriate number of times to the wavefunction corresponding to $v = 0$ (that is, $\lambda = 1$). From now on, we shall label the wavefunctions with v in place of λ. It follows that we need to determine the form of ψ_0. Because we know that $a\psi_0 = 0$, it follows that

$$\frac{1}{\sqrt{2}}\left(\frac{\mathrm{d}}{\mathrm{d}y} + y\right)\psi_0 = 0$$

This first-order differential equation rearranges into

$$\frac{d\psi_0}{\psi_0} = -y dy$$

The solution is

$$\psi_0 = N_0 e^{-y^2/2} \tag{6.9}$$

where N_0 is a normalization constant. Successive applications of a^+ (with the constants of proportionality absorbed into normalization constants) then yield

$$\psi_1 = N_1 y e^{-y^2/2} \qquad \psi_2 = N_2(2y^2 - 1)e^{-y^2/2}$$

and so on. Each wavefunction has the form of a Gaussian function multiplied by a Hermite polynomial, as was asserted in Section 2.16 and as is demonstrated explicitly in the following *Further information* section.

The following matrix elements are consistent with the commutation rules in eqn 6:

$$\langle v+1|a^+|v\rangle = -(v+1)^{1/2} \qquad \langle v-1|a|v\rangle = v^{1/2} \tag{6.10}$$

All other matrix elements are zero.

7 The harmonic oscillator: the standard solution

The Schrödinger equation for the harmonic oscillator is

$$\frac{d^2\psi}{dy^2} - y^2\psi = -\lambda\psi \tag{7.1}$$

where we have made the substitutions described at the start of the preceding section. In the conventional approach to solving such an equation, we first establish the solutions for $y \to \infty$. (This approach will be used a number of times in the text: see particularly Chapter 14.) In such a limit, the term in y^2 dominates the term in λ, so the asymptotic form of the equation is

$$\frac{d^2\psi}{dy^2} - y^2\psi \simeq 0$$

where the symbol $\simeq$ means 'asymptotically equal to' in the limit of a variable (in this case y) becoming infinitely large. The solutions have the form

$$\psi \simeq e^{\pm y^2/2}$$

as may be verified as follows:

$$\frac{d^2\psi}{dy^2} - y^2\psi \simeq y^2 e^{\pm y^2/2} - y^2 e^{\pm y^2/2} \simeq 0$$

The solution with + in the exponent is not square-integrable, so we discard it and write

$$\psi \simeq e^{-y^2/2}$$

The next stage is to set up an equation for the entire function by writing

$$\psi = f(y)e^{-y^2/2} \tag{7.2}$$

where $f(y)$ is a polynomial in y. Substitution of this solution into the full Schrödinger equation produces the following equation:

$$(f'' - 2yf' + y^2 f - f)\mathrm{e}^{-y^2/2} - y^2 f \mathrm{e}^{-y^2/2} = (f'' - 2yf' - f)\mathrm{e}^{-y^2/2}$$
$$= -\lambda f \mathrm{e}^{-y^2/2}$$

That is, we need to solve

$$f'' - 2yf' + (\lambda - 1)f = 0 \tag{7.3}$$

To solve eqn 3 we suppose that the polynomial f has the form

$$f = \sum_n a_n y^n$$

and is such as to ensure that the wavefunction is square-integrable. For that to be the case, the polynomial cannot go to infinity more rapidly than $\mathrm{e}^{y^2/2}$, for otherwise the wavefunction would become infinite as $|y| \to \infty$. Substitution of this polynomial solution into the preceding differential equation for f produces the following expression:

$$\sum_n a_n \{n(n-1)y^{n-2} - 2ny^n + (\lambda - 1)y^n\} = 0$$

Inspection of the expression on the left shows that the coefficient of y^n is

$$(n+1)(n+2)a_{n+2} + (\lambda - 2n - 1)a_n$$

Therefore, for the sum to vanish for any value of y, this coefficient must itself be equal to zero for all values of n. It follows that

$$a_{n+2} = \left\{\frac{2n+1-\lambda}{(n+1)(n+2)}\right\} a_n \tag{7.4}$$

This expression is a **recursion formula** for the coefficients, for it enables all a_n with n even to be expressed in terms of a_0 and all a_n with n odd to be expressed in terms of a_1. Notice that it implies that all the powers of y that appear in f are either even or odd, not a mixture (symmetry considerations also require the same conclusion).

For the function f to be a polynomial rather than an infinite series, the coefficients must vanish after some value of n, which we shall call v. By eqn 4, termination is ensured if $\lambda = 2v + 1$. It follows from eqn FI6.1 that the allowed values of the energy are

$$E = \lambda(\hbar\omega/2) = (2v+1)\hbar\omega/2 = (v + \tfrac{1}{2})\hbar\omega \qquad v = 0, 1, 2\ldots \tag{7.5}$$

which is the result quoted in the text and derived in the preceding section. The recursion relation in eqn 4 enables us to write down the polynomial for any value of v, and the procedure develops the polynomials in Table 2.1, which are termed the Hermite polynomials.

The following definition of the Hermite polynomials is sometimes more useful than their definition in terms of the recursion relation:

$$H_v(y) = (-1)^v \mathrm{e}^{y^2} \frac{\mathrm{d}^v}{\mathrm{d}y^v} \mathrm{e}^{-y^2} \tag{7.6}$$

8 The radial wave equation

The radial component of the Schrödinger equation for a hydrogenic atom of atomic number Z and reduced mass μ was given in eqn 3.40 as

$$\frac{\mathrm{d}^2\Pi}{\mathrm{d}r^2}+\left(\frac{2\mu}{\hbar^2}\right)\left(\frac{Ze^2}{4\pi\varepsilon_0 r}-\frac{l(l+1)\hbar^2}{2\mu r^2}\right)\Pi=-\left(\frac{2\mu E}{\hbar^2}\right)\Pi \tag{8.1}$$

where $\Pi = rR$, with R the radial wavefunction. We can simplify the appearance of this equation by introducing the following parameters:

$$a=\left(\frac{2\mu}{\hbar^2}\right)\frac{Ze^2}{4\pi\varepsilon_0} \qquad b=l(l+1)\frac{\hbar^2}{2\mu} \qquad \lambda^2=\frac{2\mu|E|}{\hbar^2} \tag{8.2}$$

and henceforth consider only bound states ($E < 0$). Then

$$\Pi''+\left(\frac{a}{r}-\frac{b}{r^2}\right)\Pi=\lambda^2\Pi \tag{8.3}$$

where $\Pi''=\mathrm{d}^2\Pi/\mathrm{d}r^2$. Guidance towards the solutions is obtained, as for the harmonic oscillator in *Further information 7*, by considering the asymptotic form of the equation as $r\to\infty$. When r is large, eqn 3 becomes

$$\Pi''\simeq\lambda^2\Pi \tag{8.4}$$

The solutions of this equation are

$$\Pi\simeq \mathrm{e}^{\pm\lambda r} \tag{8.5}$$

and we can eliminate the positive exponential because it gives a function that is not square-integrable. Hence, $\Pi\simeq\mathrm{e}^{-\lambda r}$.

To find the full solution, we write

$$\Pi=L(r)\mathrm{e}^{-\lambda r} \tag{8.6}$$

where $L(r)$ is a polynomial in r. Substitution of this expression into eqn 3 gives

$$L''-2\lambda L'+\left(\frac{a}{r}-\frac{b}{r^2}\right)L=0 \tag{8.7}$$

To solve this equation, we write

$$L(r)=\sum_n c_n r^n \tag{8.8}$$

which implies that

$$\sum_n c_n\{(n(n-1)-b)r^{n-2}-(2n\lambda-a)r^{n-1}\}=0$$

For this sum to be zero for all values of r, each coefficient of r^n must be zero, so

$$\{(n+2)(n+1)-b\}c_{n+2}-\{2(n+1)\lambda-a\}c_{n+1}=0$$

or, equivalently,

$$\{n(n+1)-b\}c_{n+1}-\{2n\lambda-a\}c_n=0$$

This expression gives a recursion relation for the coefficients:

$$c_{n+1} = \left\{ \frac{2n\lambda - a}{n(n+1) - b} \right\} c_n \tag{8.9}$$

For this series to terminate at a given value n (so that Π is square integrable), it must be the case that

$$2n\lambda = a$$

which rearranges to

$$|E| = \frac{Z^2 e^4 \mu}{32 n^2 \pi^2 \varepsilon_0^2 \hbar^2} \tag{8.10}$$

which is the expression given in the text (with the identification of E as a negative quantity).The polynomials developed by applying the recursion formula for the coefficients are the associated Laguerre functions which are used to construct the hydrogenic radial functions listed in Table 3.2.

9 The angular wavefunction

The wavefunctions for rotation in three dimensions are solutions of

$$\Lambda^2 \psi = -k\psi \qquad k = 2IE/\hbar^2 \tag{9.1}$$

We have indicated (Section 3.5) that the equation is separable with solutions of the form $\psi = \Theta(\theta)\Phi(\phi)$, and that the latter factor has the form

$$\Phi_{m_l}(\phi) = \left(\frac{1}{2\pi}\right)^{1/2} \mathrm{e}^{\mathrm{i}m_l\phi} \qquad m_l = 0, \pm 1, \pm 2, \ldots \tag{9.2}$$

Our concern in this section is to determine the solutions Θ, which satisfy

$$\frac{1}{\sin\theta} \frac{\mathrm{d}}{\mathrm{d}\theta} \sin\theta \frac{\mathrm{d}\Theta}{\mathrm{d}\theta} - \frac{m_l^2 \Theta}{\sin^2\theta} + k\Theta = 0 \tag{9.3}$$

To solve this equation, we introduce $z = \cos\theta$, with $-1 \leq z \leq 1$ and henceforth (to bring the equations into line with standard notation) denote $\Theta(\theta)$ by the function $P(z)$. Because $\sin^2\theta = 1 - \cos^2\theta = 1 - z^2$ and

$$\frac{\mathrm{d}\Theta}{\mathrm{d}\theta} = \frac{\mathrm{d}P}{\mathrm{d}z}\frac{\mathrm{d}z}{\mathrm{d}\theta} = -\frac{\mathrm{d}P}{\mathrm{d}z}\sin\theta$$

the equation to solve is

$$\frac{\mathrm{d}}{\mathrm{d}z}\left\{(1 - z^2)\frac{\mathrm{d}P}{\mathrm{d}z}\right\} + \left\{k - \frac{m_l^2}{1 - z^2}\right\} P = 0 \tag{9.4}$$

It turns out to be fruitful to try a substitution of the form

$$P(z) = (1 - z^2)^{|m_l|/2} G(z) \tag{9.5}$$

which leads to the following equation for G:

$$(1 - z^2)G'' - 2(|m_l| + 1)zG' + \{k - |m_l|(|m_l| + 1)\}G = 0 \tag{9.6}$$

with $G' = \mathrm{d}G/\mathrm{d}z$ and $G'' = \mathrm{d}^2G/\mathrm{d}z^2$. We try a polynomial solution of the form

$$G = \sum_n a_n z^n \tag{9.7}$$

and after substitution into the differential equation, collect coefficients of z^n. For a general value of n,

$$(n+1)(n+2)a_{n+2} + \{[k - |m_l|(|m_l|+1)] - 2n(|m_l|+1) - n(n-1)\}a_n = 0$$

which implies the following recursion formula:

$$a_{n+2} = \left\{\frac{(n+|m_l|)(n+|m_l|+1) - k}{(n+1)(n+2)}\right\}a_n \tag{9.8}$$

An infinite series based on this relation between coefficients diverges for $z = \pm 1$, so there must be a restriction that ensures that the series terminates after a finite number of terms. This restriction implies that there must be a value of $n = 0, 1, 2\ldots$ for which

$$k = (n+|m_l|)(n+|m_l|+1)$$

We now introduce the quantum number $l = n + |m_l|$, and write this restriction as

$$k = l(l+1) \qquad \text{with } l = |m_l|, |m_l|+1, \ldots \tag{9.9}$$

At this stage we have demonstrated that the original equation may be written

$$\Lambda^2\psi = -l(l+1)\psi \tag{9.10}$$

as claimed in eqn 3.22 and know the coefficients in the expansion of G, which identify $P(z)$ as the **associated Legendre functions**. The specific relation between the normalized functions Θ and the associated Legendre functions is

$$\Theta(\theta) = \left\{\left(\frac{2l+1}{2}\right)\frac{(l-|m_l|)!}{(l+|m_l|)!}\right\}^{1/2} P_l^{|m_l|}(\cos\theta) \tag{9.11}$$

The products of Θ in eqn 11 and Φ in eqn 2 are called **spherical harmonics** and denoted $Y_{lm_l}(\theta, \phi)$.

10 Molecular integrals

In the case of the MO description of H_2^+, the energy is given by the expression quoted in eqn 8.24, with

$$\begin{aligned} j'/j_0 &= \int \frac{a^2(1)}{r_{1b}}\,\mathrm{d}\tau_1 = \frac{1}{R}\{1-(1+s)\mathrm{e}^{-2s}\} \\ k'/j_0 &= \int \frac{a(1)b(1)}{r_{1b}}\,\mathrm{d}\tau_1 = \frac{1}{a_0}\{1+s\}\mathrm{e}^{-s} \\ S &= \int a(1)b(1)\,\mathrm{d}\tau_1 = \{1+s+\tfrac{1}{3}s^2\}\mathrm{e}^{-s} \end{aligned} \tag{10.1}$$

where $j_0 = e^2/4\pi\varepsilon_0$ and $s = R/a_0$. We have taken $1s$-orbitals on each atom, and have denoted them a and b.

For the MO description of H_2, the energy is given by eqn 8.28. The following integrals are required in addition to those given above:

$$\begin{aligned}
j/j_0 &= \int \frac{a^2(1)b^2(2)}{r_{12}}\,d\tau_1 d\tau_2 \\
&= \frac{1}{R} - \frac{1}{2a_0}\left\{\frac{2}{s} + \tfrac{11}{4} + \tfrac{3}{2}s + \tfrac{1}{3}s^2\right\}e^{-2s} \\
k/j_0 &= \int \frac{a(1)b(1)a(2)b(2)}{r_{12}}\,d\tau_1 d\tau_2 = \frac{A-B}{5a_0} \\
l/j_0 &= \int \frac{a^2(1)a(2)b(2)}{r_{12}}\,d\tau_1 d\tau_2 \\
&= \frac{1}{2a_0}\left\{\left(2s + \tfrac{1}{4} + \frac{5}{8s}\right)e^{-s} - \left(\tfrac{1}{4} + \frac{5}{8s}\right)e^{-3s}\right\} \\
m/j_0 &= \int \frac{a^2(1)a^2(2)}{r_{12}}\,d\tau_1 d\tau_2 = \frac{5}{8a_0}
\end{aligned} \tag{10.2}$$

with

$$\begin{aligned}
A &= \frac{6}{s}\{(\gamma + \ln s)S^2 - E_1(4s)S'^2 + 2E_1(2s)SS'\} \\
B &= \{-\tfrac{25}{8} + \tfrac{23}{4}s + 3s^2 + \tfrac{1}{3}s^3\}e^{-2s} \\
S'(s) &- S(-s)
\end{aligned} \tag{10.3}$$

where γ is Euler's constant ($\gamma = 0.577\,22\ldots$) and $E_1(x)$ is the tabulated function known as the exponential integral.[1] These equations give some of the idea of the complexity of integrals that arise in analytical treatments of molecules.

11 The Hartree–Fock equations

The normalized Hartree–Fock (HF) ground-state wavefunction Φ_0 is given by the n-electron Slater determinant

$$\Phi_0 = (n!)^{-1/2}\det|\phi_a(1)\phi_b(2)\ldots\phi_z(n)| \tag{11.1}$$

and the ground-state HF electronic energy is given by

$$E = \langle\Phi_0|H|\Phi_0\rangle \tag{11.2}$$

where H is given in eqn 7.43. We seek the set of orthonormal spinorbitals ϕ that yield a minimum energy E: this condition leads to the HF equations.

As a first step, we derive an expression for E in terms of the spinorbitals. From eqns 2 and 7.43 (of Chapter 7),

$$E = \langle\Phi_0|\sum_i h_i + \tfrac{1}{2}\sum_{i,j}{}'\left(\frac{e^2}{4\pi\varepsilon_0 r_{ij}}\right)|\Phi_0\rangle \tag{11.3}$$

For the first term, we can write

$$\langle\Phi_0|\sum_i h_i|\Phi_0\rangle = \langle\Phi_0|h_1 + h_2 + \ldots + h_n|\Phi_0\rangle = n\langle\Phi_0|h_1|\Phi_0\rangle \tag{11.4}$$

[1] See M. Abramowitz and I.A. Stegun, *Handbook of mathematical functions*, Dover, New York (1965).

The second equality follows from the fact that all the electrons are indistinguishable in the determinant and therefore the matrix elements of all the h_i are equal. Expansion of the Slater determinant Φ_0 in eqn 4, using eqn 1 and the orthonormality of the spinorbitals, gives

$$\langle \Phi_0 | \sum_i h_i | \Phi_0 \rangle = \sum_{i=1}^{n} \langle \phi_i(1) | h_1 | \phi_i(1) \rangle \tag{11.5}$$

which can be rewritten using the one-electron notation

$$[\phi_i | h | \phi_i] = \langle \phi_i(1) | h_1 | \phi_i(1) \rangle \tag{11.6}$$

as

$$\langle \Phi_0 | \sum_i h_i | \Phi_0 \rangle = \sum_{i=1}^{n} [\phi_i | h | \phi_i] \tag{11.7}$$

The second sum in H is over all $\frac{1}{2}n(n-1)$ unique pairs of electrons. Each term in the sum gives the same result because the electrons are indistinguishable, so

$$\langle \Phi_0 | \tfrac{1}{2} \sum_{i,j}{}' \frac{e^2}{4\pi\varepsilon_0 r_{ij}} | \Phi_0 \rangle = \tfrac{1}{2}n(n-1) \langle \Phi_0 | \frac{e^2}{4\pi\varepsilon_0 r_{12}} | \Phi_0 \rangle$$

The expansion of Φ_0 in terms of its spinorbitals in eqn 1 turns this expression into

$$\tfrac{1}{2} \sum_{i,j}{}' \int \phi_i^*(1)\phi_j^*(2) \left(\frac{e^2}{4\pi\varepsilon_0 r_{12}} \right) \{ \phi_i(1)\phi_j(2) - \phi_j(1)\phi_i(2) \} \, \mathrm{d}\mathbf{x}_1 \mathrm{d}\mathbf{x}_2 \tag{11.8}$$

Then, with

$$[\phi_a \phi_b | \phi_c \phi_d] = \int \phi_a^*(1)\phi_b(1) \left(\frac{e^2}{4\pi\varepsilon_0 r_{12}} \right) \phi_c^*(2)\phi_d(2) \, \mathrm{d}\mathbf{x}_1 \mathrm{d}\mathbf{x}_2 \tag{11.9}$$

and eqns 3, 7, and 8, we have

$$E = \sum_{i=1}^{n} [\phi_i | h | \phi_i] + \tfrac{1}{2} \sum_{i,j}^{n} \{ [\phi_i \phi_i | \phi_j \phi_j] - [\phi_i \phi_j | \phi_j \phi_i] \} \tag{11.10}$$

(Note that we do not have to exclude $i \neq j$ in the second sum because the term with $i = j$ is identically zero.) Equation 10 is an expression for the energy as a functional of the spinorbitals; that is, for every set of functions ϕ, there is associated a single value E. We shall use this equation shortly.

To derive the HF equations, we introduce and use the technique of functional variation. The energy E is a functional of the Slater determinant Φ in eqn 2. To find the particular determinant Φ for which E is a minimum (that is, Φ_0), we find Φ for which a small change $\Phi \to \Phi + \delta\Phi$ yields no change in the value of E to first order in $\delta\Phi$.[2] For an infinitesimal change $\delta\Phi$, we have

$$E[\Phi + \delta\Phi] = \langle \Phi + \delta\Phi | H | \Phi + \delta\Phi \rangle = \langle \Phi | H | \Phi \rangle + \delta \langle \Phi | H | \Phi \rangle \tag{11.11}$$

[2] This step is analogous to what is done in finding a minimum in a one-dimensional function $f(x)$: we seek the value of x such that $f'(x) = 0$; in other words an infinitesimally small change dx yields no change df.

where

$$\delta\langle\Phi|H|\Phi\rangle = \langle\delta\Phi|H|\Phi\rangle + \langle\Phi|H|\delta\Phi\rangle = \delta E$$

We seek the determinant Φ for which $\delta E = 0$. We use the expression for E as given in terms of the spinorbitals in eqn 10. However, because we have an additional constraint that the spinorbitals be orthonormal, we must use the technique of undetermined multipliers.[3]

We must satisfy the condition

$$\int \phi_i^*(1)\phi_j(1)\,\mathrm{d}\mathbf{x}_1 = \delta_{ij} \tag{11.12}$$

where δ_{ij} is the Kronecker delta. The constraint is of the form

$$g = \sum_{i,j=1}^{n}\{\langle\phi_i|\phi_j\rangle - \delta_{ij}\} = 0$$

When the spinorbitals are changed by an arbitrary infinitesimal amount $\delta\phi$, then because δ_{ij} is a constant, g changes by

$$\delta g = \sum_{i,j=1}^{n}\delta\langle\phi_i|\phi_j\rangle = \sum_{i,j=1}^{n}\{\langle\delta\phi_i|\phi_j\rangle + \langle\phi_i|\delta\phi_j\rangle\}$$

At this stage we take the constraint into account by introducing the set of undetermined multipliers λ_{ji}, and then look for the condition for which

$$\delta E - \sum_{i,j=1}^{n}\lambda_{ji}\{\langle\delta\phi_i|\phi_j\rangle + \langle\phi_i|\delta\phi_j\rangle\} = 0 \tag{11.13}$$

We now examine this condition.

From eqn 10, we get the following expression for δE:

$$\begin{aligned}\delta E = \sum_{i=1}^{n}\{[\delta\phi_i|h|\phi_i] + [\phi_i|h|\delta\phi_i]\} + \tfrac{1}{2}\sum_{i,j}^{n}\{&[(\delta\phi_i)\phi_i|\phi_j\phi_j] + [\phi_i(\delta\phi_i)|\phi_j\phi_j] \\ &+ [\phi_i\phi_i|(\delta\phi_j)\phi_j] + [\phi_i\phi_i|\phi_j(\delta\phi_j)] - [(\delta\phi_i)\phi_j|\phi_j\phi_i] \\ &- [\phi_i(\delta\phi_j)|\phi_j\phi_i] - [\phi_i\phi_j|(\delta\phi_j)\phi_i] - [\phi_i\phi_j|\phi_j(\delta\phi_i)]\}\end{aligned} \tag{11.14}$$

At this point we substitute eqn 14 into eqn 13, recognize complex conjugates of terms, and obtain

$$\sum_{i=1}^{n}[\delta\phi_i|h|\phi_i] + \sum_{i,j}^{n}\{[(\delta\phi_i)\phi_i|\phi_j\phi_j] - [(\delta\phi_i)\phi_j|\phi_j\phi_i] - \lambda_{ji}\langle\delta\phi_i|\phi_j\rangle\} + \mathrm{cc} = 0 \tag{11.15}$$

where cc stands for the complex conjugate of all the terms explicitly shown in eqn 15. Now we factor out the common term $\delta\phi_i^*$ and use eqns 6 and 9 and the definitions in eqns 9.9 and 9.10 (of Chapter 9) for the Coulomb and exchange

[3] See, for example, *Further information 14* in P.W. Atkins, *Physical chemistry*, 5th Edn., Oxford University Press and W.H. Freeman, New York (1994).

operators, and obtain

$$\sum_{i=1}^{n}\int \delta\phi_i^*(1)\left(h_1\phi_i(1)+\sum_{j=1}^{n}\{J_j(1)\phi_i(1)-K_j(1)\phi_i(1)-\lambda_{ji}\phi_j(1)\}\right)\mathrm{d}\mathbf{x}_1+\mathrm{cc}=$$

As the variation $\delta\phi_i^*$ is arbitrary, each term in the parentheses must be identically zero. Therefore, for each spinorbital,

$$h_1\phi_i(1)+\sum_{j=1}^{n}\{J_j(1)\phi_i(1)-K_j(1)\phi_i(1)\}=\sum_{i=1}^{n}\lambda_{ji}\phi_j(1) \qquad (11.16)$$

When the definition of the Fock operator (eqn 9.8) is used in eqn 16, the equations for the spinorbitals become

$$f_1\phi_i(1)=\sum_{i=1}^{n}\lambda_{ji}\phi_j(1) \qquad (11.17)$$

Equation 17 is not quite the standard form of the HF equations as given in eqn 9.7 because the set of spinorbitals ϕ is not unique; it is possible to form a new set of spinorbitals, each a linear combination of the ϕ, without changing the minimum energy E. In particular, it is possible to transform the original set into a new set of orthonormal **canonical spinorbitals**, ϕ', such that the transformed Fock operator f_1' is the same as f_1 and the matrix composed of the multipliers λ_{ji} is transformed into a diagonal matrix with elements ε_i'.[4] The canonical spinorbital ϕ_i' solves the equation

$$f_1\phi_i'(1)=\varepsilon_i'\phi_i'(1) \qquad (11.18)$$

At this point, we discard the primes and obtain the HF equations as given in eqn 9.7. The HF equation, eqn 18, for the spinorbital ϕ_i can be converted to an equation for the spatial function ψ_i by writing it as a product of a spatial and spin function and using the orthonormality of the latter. For the closed-shell case, where all spatial functions are doubly occupied, this procedure results in the set of equations given in eqn 7.44.

12 Green's functions

We saw in Example 14.1 that insertion of the outgoing Green's function

$$G^{(+)}(\mathbf{r},\mathbf{r}')=\frac{\mathrm{e}^{\mathrm{i}k|\mathbf{r}-\mathbf{r}'|}}{|\mathbf{r}-\mathbf{r}'|} \qquad (12.1)$$

into eqn 14.21 results in the correct asymptotic form (eqn 14.7) for the stationary scattering state $\psi(\mathbf{r})$. In this section, we show that $G^{(+)}(\mathbf{r},\mathbf{r}')$ is a solution of the equation

$$(\nabla^2+k^2)G(\mathbf{r},\mathbf{r}')=-4\pi\delta(\mathbf{r}-\mathbf{r}') \qquad (12.2)$$

where $\delta(\mathbf{r}-\mathbf{r}')$ is the Dirac δ-function. Equivalently, we demonstrate that

$$G^{(+)}(\mathbf{r})=\frac{\mathrm{e}^{\mathrm{i}kr}}{r} \qquad (12.3)$$

[4] See A. Szabo and N.S. Ostlund, *Modern quantum chemistry: introduction to advanced electronic structure*, Macmillan, New York (1982).

is a solution of

$$(\nabla^2 + k^2)G(\mathbf{r}) = -4\pi\delta(\mathbf{r}) \tag{12.4}$$

First, consider the term $\nabla^2 G^{(+)}(\mathbf{r})$. Because $\nabla^2 = \nabla \cdot \nabla$, we can use the properties of differential operators acting on a product of functions to obtain

$$\begin{aligned}
\nabla^2 G^{(+)} &= \nabla \cdot (\nabla G^{(+)}) \\
&= \nabla \cdot \{\frac{1}{r}\nabla(\mathrm{e}^{ikr})\} + \nabla \cdot \{\mathrm{e}^{ikr}\nabla\frac{1}{r}\} \\
&= \frac{1}{r}\nabla^2\mathrm{e}^{ikr} + \mathrm{e}^{ikr}\nabla^2\frac{1}{r} + 2\nabla\frac{1}{r}\cdot\nabla(\mathrm{e}^{ikr})
\end{aligned} \tag{12.5}$$

Two standard properties introduced in *Further information 21*, which will be very handy here, are

$$\nabla\frac{1}{r} = -\frac{\mathbf{r}}{r^3} \qquad \nabla^2\frac{1}{r} = -4\pi\delta(\mathbf{r}) \tag{12.6}$$

We need to evaluate the effects of ∇ and ∇^2 on e^{ikr}. As a first step we write

$$\nabla\mathrm{e}^{ikr} = \mathrm{i}k\mathrm{e}^{ikr}\nabla r \tag{12.7}$$

and we are forced to evaluate ∇r where, as usual, $\mathbf{r} = x\hat{\mathbf{i}} + y\hat{\mathbf{j}} + z\hat{\mathbf{k}}$, $r = (x^2 + y^2 + z^2)^{1/2}$, and

$$\begin{aligned}
\nabla r &= \hat{\mathbf{i}}\frac{\partial r}{\partial x} + \hat{\mathbf{j}}\frac{\partial r}{\partial y} + \hat{\mathbf{k}}\frac{\partial r}{\partial z} \\
&= \tfrac{1}{2}(x^2 + y^2 + z^2)^{-1/2}(2x\hat{\mathbf{i}} + 2y\hat{\mathbf{j}} + 2z\hat{\mathbf{k}}) = \frac{\mathbf{r}}{r}
\end{aligned} \tag{12.8}$$

We obtain from eqns 7 and 8

$$\nabla\mathrm{e}^{ikr} = \frac{\mathrm{i}k\mathbf{r}\mathrm{e}^{ikr}}{r} \tag{12.9}$$

We must now evaluate $\nabla^2\mathrm{e}^{ikr}$. From $\nabla^2 = \nabla \cdot \nabla$ and eqn 9, we obtain

$$\nabla^2\mathrm{e}^{ikr} = \nabla \cdot \left(\frac{\mathrm{i}k\mathbf{r}\mathrm{e}^{ikr}}{r}\right) = \frac{\mathrm{i}k\mathbf{r}\cdot\nabla(\mathrm{e}^{ikr})}{r} + \mathrm{i}k\mathrm{e}^{ikr}\nabla\cdot\frac{\mathbf{r}}{r} \tag{12.10}$$

The type of analysis that led to eqn 8 leads to

$$\nabla\cdot\frac{\mathbf{r}}{r} = \frac{2}{r} \tag{12.11}$$

and therefore, from eqns 9–11,

$$\nabla^2\mathrm{e}^{ikr} = -k^2\mathrm{e}^{ikr} + \frac{2\mathrm{i}k\mathrm{e}^{ikr}}{r} \tag{12.12}$$

Then, by using eqns 3, 5, 6, 9, and 12, we obtain

$$(\nabla^2 + k^2)G^{(+)} = -4\pi\mathrm{e}^{ikr}\delta(\mathbf{r}) = -4\pi\delta(\mathbf{r})$$

The last equality follows from the fact that because $\delta(\mathbf{r})$ is nonvanishing only at the origin $(\mathbf{r} = 0)$, then $\mathrm{e}^{ikr} = 1$.

13 The unitarity of the S matrix

Here, we demonstrate that the S matrix for scattering off an arbitrary one-dimensional potential, $V(x)$, is unitary. The only requirement is that $V \to 0$ as $x \to \pm\infty$. The finite-width barrier of Section 2.10 is an example of such a potential.

As $x \to \pm\infty$, the general time-independent solution $\psi(x)$ approaches the solution of the free-particle, $V = 0$, Schrödinger equation. Therefore, the asymptotic form of $\psi(x)$ can immediately be written as

$$\psi \simeq A\mathrm{e}^{\mathrm{i}kx} + B\mathrm{e}^{-\mathrm{i}kx} \qquad \text{as } x \to -\infty$$

$$\psi \simeq A''\mathrm{e}^{\mathrm{i}kx} + B''\mathrm{e}^{-\mathrm{i}kx} \qquad \text{as } x \to +\infty$$

where $k\hbar = (2mE)^{1/2}$. The scattering matrix is given by eqn 2.52 and relates the coefficients A'' and B of the outgoing waves to the coefficients A and B'' of the incoming waves.

We now use the result of Example 2.1, that the flux density, J_x (eqn 2.11), associated with a wavefunction of definite energy is independent of location. Evaluation of J_x as $x \to \infty$ yields $(\hbar k/m)(|A''|^2 - |B''|^2)$ whereas evaluation of J_x as $x \to -\infty$ yields $(\hbar k/m)(|A|^2 - |B|^2)$. Because J_x must be independent of x, we have

$$|A''|^2 - |B''|^2 = |A|^2 - |B|^2$$

or

$$|A''|^2 + |B|^2 = |B''|^2 + |A|^2$$

This equation can be put into the matrix form

$$\begin{pmatrix} A''^* & B^* \end{pmatrix} \begin{pmatrix} A'' \\ B \end{pmatrix} = \begin{pmatrix} B''^* & A^* \end{pmatrix} \begin{pmatrix} B'' \\ A \end{pmatrix}$$

We can now develop the left-hand side of this equation by introducing the S matrix (eqn 2.52) and using the matrix properties described in *Further information 23*:

$$\begin{aligned} \begin{pmatrix} B''^* & A^* \end{pmatrix} \begin{pmatrix} B'' \\ A \end{pmatrix} &= \{\begin{pmatrix} B'' & A \end{pmatrix}(\mathbf{S}^{\mathrm{T}})\}^* \mathbf{S} \begin{pmatrix} B'' \\ A \end{pmatrix} \\ &= \{\begin{pmatrix} B'' & A \end{pmatrix}^* (\mathbf{S}^{\mathrm{T}})^*\} \mathbf{S} \begin{pmatrix} B'' \\ A \end{pmatrix} \\ &= \begin{pmatrix} B''^* & A^* \end{pmatrix} \{(\mathbf{S}^{\mathrm{T}})^* \mathbf{S}\} \begin{pmatrix} B'' \\ A \end{pmatrix} \end{aligned}$$

In the first line, we have used the matrix property that if $\mathbf{X} = \mathbf{YZ}$ then $\mathbf{X}^{\mathrm{T}} = \mathbf{Z}^{\mathrm{T}}\mathbf{Y}^{\mathrm{T}}$.We see that $(\mathbf{S}^{\mathrm{T}})^*\mathbf{S} = \mathbf{1}$, so $\mathbf{S}$ is unitary.

Group theory and angular momentum

14 The orthogonality of basis functions

In this section we prove the following two theorems:

Theorem 1
Two functions are orthogonal if they are basis functions for different irreducible representations of a group, or they are members of a basis of a particular irreducible representation but are in different positions in the row.

Theorem 2
The integral $\langle f_i^{(l)}|f_{i'}^{(l')}\rangle$ is independent of the index i.

Proof of Theorem 1. Let the set of functions $f_i^{(l)}$ with $i = 1, 2, \ldots, d_i$ be the basis of an irreducible representation of symmetry species $\Gamma^{(l)}$, and the set $f_{i'}^{(l')}$ with $i' = 1, 2, \ldots, d_{i'}$ be the basis of an irreducible representation of symmetry species $\Gamma^{(l')}$. Then for all the operations R of the group

$$Rf_i^{(l)} = \sum_j f_j^{(l)} D_{ji}^{(l)}(R) \qquad Rf_{i'}^{(l')} = \sum_{j'} f_{j'}^{(l')} D_{j'i'}^{(l')}(R) \tag{14.1}$$

The value of an integral is independent of any symmetry operation, and so

$$\langle f_i^{(l)}|f_{i'}^{(l')}\rangle = \langle Rf_i^{(l)}|Rf_{i'}^{(l')}\rangle \tag{14.2}$$

for all operations R. Therefore, because there are h elements in the group and each one leaves the integral unchanged

$$\begin{aligned}\langle f_i^{(l)}|f_{i'}^{(l')}\rangle &= \frac{1}{h}\sum_R \langle Rf_i^{(l)}|Rf_{i'}^{(l')}\rangle \\ &= \frac{1}{h}\sum_R \sum_{jj'} D_{ji}^{(l)}(R)^* D_{j'i'}^{(l')}(R)\langle f_j^{(l)}|f_{j'}^{(l')}\rangle\end{aligned} \tag{14.3}$$

The great orthogonality theorem (Section 5.10) may be used to write this relation as

$$\begin{aligned}\langle f_i^{(l)}|f_{i'}^{(l')}\rangle &= \frac{1}{d_l}\delta_{ll'}\sum_{jj'}\delta_{jj'}\delta_{ii'}\langle f_j^{(l)}|f_{j'}^{(l')}\rangle \\ &= \delta_{ll'}\delta_{ii'}\frac{1}{d_l}\sum_j \langle f_j^{(l)}|f_j^{(l')}\rangle\end{aligned} \tag{14.4}$$

Therefore, $\langle f_i^{(l)}|f_{i'}^{(l')}\rangle \propto \delta_{ll'}\delta_{ii'}$, which completes the proof of the theorem.

Proof of Theorem 2. From the preceding theorem we have

$$\langle f_i^{(l)}|f_i^{(l)}\rangle = \frac{1}{d_l}\sum_j \langle f_j^{(l)}|f_j^{(l)}\rangle \tag{14.5}$$

and the sum on the right is independent of i.

15 Vector coupling coefficients

Vector coupling coefficients are listed in Appendix 2. As an example of their application, consider the determination of the energy of an atom in a magnetic field **B** of magnitude B in the z-direction, its single p-electron having a spin–orbit coupling constant ζ. The hamiltonian is

$$H = (\mu_B B/\hbar)(l_z + 2s_z) + (hc\zeta/\hbar^2)\mathbf{l}\cdot\mathbf{s}$$

The matrix elements of this hamiltonian may be expressed in the coupled or the uncoupled representations. For the latter, it is convenient to express H in the form

$$H = (\mu_B B/\hbar)(l_z + 2s_z) + (hc\zeta/\hbar^2)\{l_z s_z + \tfrac{1}{2}(l_+ s_- + l_- s_+)\}$$

When all the matrix elements are calculated we obtain

	$\lvert +1, +\frac{1}{2}\rangle$	$\lvert 0, +\frac{1}{2}\rangle$	$\lvert -1, +\frac{1}{2}\rangle$	$\lvert +1, -\frac{1}{2}\rangle$	$\lvert 0, -\frac{1}{2}\rangle$	$\lvert -1, -\frac{1}{2}\rangle$
$\langle +1, +\frac{1}{2}\rvert$	$2\mu_B B + \frac{1}{2}hc\zeta$	0	0	0	0	0
$\langle 0, +\frac{1}{2}\rvert$	0	$\mu_B B$	0	$hc\zeta/\sqrt{2}$	0	0
$\langle -1, +\frac{1}{2}\rvert$	0	0	$-\frac{1}{2}hc\zeta$	0	$hc\zeta/\sqrt{2}$	0
$\langle +1, -\frac{1}{2}\rvert$	0	$hc\zeta/\sqrt{2}$	0	$-\frac{1}{2}hc\zeta$	0	0
$\langle 0, -\frac{1}{2}\rvert$	0	0	$hc\zeta/\sqrt{2}$	0	$-\mu_B B$	0
$\langle -1, -\frac{1}{2}\rvert$	0	0	0	0	0	$-2\mu_B B + \frac{1}{2}hc\zeta$

where the states are described by the notation $\lvert m_l m_s\rangle$. For the coupled representation it is sensible to write H in the form

$$H = (\mu_B B/\hbar)(l_z + s_z) + (\mu_B B/\hbar)s_z + \left(\frac{hc\zeta}{2\hbar^2}\right)(j^2 - l^2 - s^2)$$

In the coupled representation, the states are eigenstates of j^2, j_z, l^2, and s^2, and so most of the elements of the hamiltonian can be calculated very simply. The difficulty is associated with the effect of s_z, for the coupled states are not eigenstates of this operator. The effect of s_z may be determined by expanding the coupled states in terms of the uncoupled states by using the vector coupling coefficients. As an example, consider the element $(\frac{1}{2}, +\frac{1}{2}\lvert s_z \rvert \frac{3}{2}, +\frac{1}{2})$, where the notation $\lvert jm_j)$ implies that we are working in the coupled representation. From the table of coefficients, we expand both coupled states as linear combinations of uncoupled states $\lvert m_l m_s\rangle$. For instance:

$$\lvert\tfrac{3}{2}, +\tfrac{1}{2}) = (\tfrac{1}{3})^{1/2}\lvert +1, -\tfrac{1}{2}\rangle + (\tfrac{2}{3})^{1/2}\lvert 0, +\tfrac{1}{2}\rangle$$

The effect of the operator s_z on this state is

$$s_z\lvert\tfrac{3}{2}, +\tfrac{1}{2}) = -\tfrac{1}{2}\hbar(\tfrac{1}{3})^{1/2}\lvert +1, -\tfrac{1}{2}\rangle + \tfrac{1}{2}\hbar(\tfrac{2}{3})^{1/2}\lvert 0, +\tfrac{1}{2}\rangle$$

The state $\lvert\frac{1}{2}, +\frac{1}{2})$ can be expressed as the linear combination

$$\lvert\tfrac{1}{2}, +\tfrac{1}{2}) = (\tfrac{2}{3})^{1/2}\lvert +1, -\tfrac{1}{2}\rangle - (\tfrac{1}{3})^{1/2}\lvert 0, +\tfrac{1}{2}\rangle$$

The matrix element we require is therefore

$$(\tfrac{1}{2}, +\tfrac{1}{2}\lvert(\mu_B/\hbar)Bs_z\rvert\tfrac{3}{2}, +\tfrac{1}{2}) = -\tfrac{1}{3}\sqrt{2}\mu_B B$$

The entire matrix can be constructed in this way, and we obtain

	$\lvert\frac{3}{2},+\frac{3}{2})$	$\lvert\frac{3}{2},+\frac{1}{2})$	$\lvert\frac{3}{2},-\frac{1}{2})$	$\lvert\frac{3}{2},-\frac{3}{2})$	$\lvert\frac{1}{2},+\frac{1}{2})$	$\lvert\frac{1}{2},-\frac{1}{2})$
$(\frac{3}{2},+\frac{3}{2}\rvert$	$2\mu_B B+\frac{1}{2}hc\zeta$	0	0	0	0	0
$(\frac{3}{2},+\frac{1}{2}\rvert$	0	$\frac{2}{3}\mu_B B+\frac{1}{2}hc\zeta$	0	0	$-\frac{1}{3}\sqrt{2}\mu_B B$	0
$(\frac{3}{2},-\frac{1}{2}\rvert$	0	0	$-\frac{2}{3}\mu_B B+\frac{1}{2}hc\zeta$	0	0	$-\frac{1}{3}\sqrt{2}\mu_B B$
$(\frac{3}{2},-\frac{3}{2}\rvert$	0	0	0	$-2\mu_B B+\frac{1}{2}hc\zeta$	0	0
$(\frac{1}{2},+\frac{1}{2}\rvert$	0	$-\frac{1}{3}\sqrt{2}\mu_B B$	0	0	$\frac{1}{3}\mu_B B-hc\zeta$	0
$(\frac{1}{2},-\frac{1}{2}\rvert$	0	0	$-\frac{1}{3}\sqrt{2}\mu_B B$	0	0	$-\frac{1}{3}\mu_B B-hc\zeta$

The point of the calculation now becomes clear. To determine the energy levels of the electron, we need to diagonalize the matrix. If the externally applied field is very weak (in the sense $\mu_B B \ll hc\zeta$), then the matrix of H has much smaller off-diagonal elements in the coupled representation than in the uncoupled representation: only B occurs in the off-diagonal elements in the coupled representation whereas ζ occurs in them in the uncoupled representation. Conversely, if the field is so strong that $\mu_B B \gg hc\zeta$, then the uncoupled representation has smaller off-diagonal elements and the matrix is more closely diagonal. Therefore, for practical convenience, it is better to set up the matrix in a representation that reflects the physics of the problem because then it is much easier to diagonalize. When the spin–orbit coupling is strong, the coupled representation should be used. When the applied field is strong, the uncoupled representation is more appropriate. The representation that most nearly diagonalizes the hamiltonian is the closest to the 'true' description of the system, and so we conclude that the coupled representation, with vectors adopting precise relative orientations, is better when the spin–orbit coupling is strong. The uncoupled representation, in which the vectors make precise angles with respect to the applied field but not to one another, is better when the external field is strong.

Spectroscopic properties

16 Electric dipole transitions

Consider a molecule exposed to light with its electric vector lying in the z-direction and oscillating at a frequency $\omega = 2\pi\nu$. The perturbation is

$$H^{(1)}(t) = -\mu_z \mathcal{E}(t), \qquad \mathcal{E}(t) = 2\mathcal{E}\cos\omega t \tag{16.1}$$

The transition rate from an initial state i to a continuum of final states f due to a perturbation of this form is given by eqn 6.78

$$W_{i\to f} = 2\pi\hbar\lvert V_{fi}\rvert^2 \rho(E_{fi}) \tag{16.2}$$

and in this instance is

$$W_{i\to f} = \frac{2\pi}{\hbar}\lvert\mu_{z,fi}\rvert^2 \mathcal{E}^2 \rho(E_{fi}) \tag{16.3}$$

where $\rho(E_{fi})$ is the density of the continuum states at an energy $E_{fi} = \hbar\omega_{fi}$, with ω_{fi} the transition frequency. In a fluid sample, the z-direction corresponds to all possible directions in the molecules, and so in such a case we should replace

$|\mu_{z,\text{fi}}|^2$ by its mean value $\frac{1}{3}|\mu_{\text{fi}}|^2$. The energy of a classical electromagnetic field is

$$E = \tfrac{1}{2}\int \left\{\varepsilon_0\langle\mathcal{E}(t)^2\rangle + \mu_0\langle\mathcal{H}(t)^2\rangle\right\}\mathrm{d}\tau \tag{16.4}$$

where $\langle\mathcal{E}(t)^2\rangle$ and $\langle\mathcal{H}(t)^2\rangle$ are the time-averages of the squared field strengths and, as usual, $\mathrm{d}\tau$ is the volume element. In the present case, because the period is $2\pi/\omega$,

$$\langle\mathcal{E}(t)^2\rangle = \frac{4\mathcal{E}^2}{2\pi/\omega}\left\{\int_0^{2\pi/\omega}\cos^2\omega t\,\mathrm{d}t\right\} = 2\mathcal{E}^2 \tag{16.5}$$

From electromagnetic theory, $\mu_0\langle\mathcal{H}(t)^2\rangle = \varepsilon_0\langle\mathcal{E}(t)^2\rangle$. Therefore, for a field in a region of volume V,

$$E = 2\varepsilon_0\mathcal{E}^2 V, \qquad \text{or } \mathcal{E}^2 = \frac{E}{2\varepsilon_0 V}$$

It follows that

$$W_{\text{i}\to\text{f}} = \frac{\pi}{3\varepsilon_0\hbar}|\mu_{\text{fi}}|^2\left(\frac{E\rho(E_{\text{fi}})}{V}\right) \tag{16.6}$$

The expression just derived is for the transition rate from an initial state i to a continuum of states f. Up to this point, we have treated the radiation as effectively monochromatic, with $\omega = \omega_{\text{fi}}$. Now we are interested in the transition rate from a discrete initial state i to a *discrete* final state f under the influence of non-monochromatic radiation. To proceed, we need to introduce the density of radiation states, $p_{\text{rad}}(E)$, where $p_{\text{rad}}(E)\mathrm{d}E$ is the number of waves with photon energies in the range E to $E + \mathrm{d}E$. The same analysis employed in Section 6.16 that led to eqn 6 above can be used here to sum (integrate) over all the waves present, and results in

$$W_{\text{i}\to\text{f}} = \frac{1}{6\varepsilon_0\hbar^2}|\mu_{\text{fi}}|^2\left(\frac{E_{\text{fi}}p_{\text{rad}}(E_{\text{fi}})}{V}\right) \tag{16.7}$$

We now note that the term $Ep_{\text{rad}}(E)/V$ is the product of the energy of a monochromatic wave and the density of radiation states per unit volume; hence it is the **energy density of radiation states**, and we write it $\rho_{\text{rad}}(E)$. Therefore,

$$W_{\text{i}\to\text{f}} = \left(\frac{|\mu_{fi}|^2}{6\varepsilon_0\hbar^2}\right)\rho_{\text{rad}}(E_{\text{fi}}) \tag{16.8}$$

and we can identify the coefficient of stimulated absorption as

$$B = \frac{|\mu_{fi}|^2}{6\varepsilon_0\hbar^2} \tag{16.9}$$

It then follows from eqn 6.88 that the coefficient of spontaneous emission is

$$A = 8\pi h\left(\frac{\nu_{fi}}{c}\right)^3\left(\frac{|\mu_{fi}|^2}{6\varepsilon_0\hbar^2}\right) = \left(\frac{8\pi^2\nu_{fi}^3}{3\varepsilon_0\hbar c^3}\right)|\mu_{fi}|^2 \tag{16.10}$$

17 Oscillator strength

In this section, we establish the relation between the integrated absorption coefficient ($\mathcal{A}$) of a band and the transition dipole moment (μ_{fi}). To do so, consider a plane of area A at x with radiation incident from the left. All the photons within a distance $c\Delta t$, and hence in a volume $Ac\Delta t$, will pass through the plane in an interval Δt. If the energy density of the field is $\mathcal{U}$, then the total electromagentic energy passing through the plane in that interval is $\mathcal{U}Ac\Delta t$. The **energy flux**, J, is the energy per unit time per unit area, and so $J = \mathcal{U}Ac\Delta t/A\Delta t = c\mathcal{U}$. The energy density in the range ν to $\nu + \mathrm{d}\nu$ is $\mathrm{d}\mathcal{U} = \rho_{\mathrm{rad}}(E)\mathrm{d}\nu$, and so the energy flux in the same range is $\mathrm{d}J = c\rho_{\mathrm{rad}}(E)\mathrm{d}\nu$. We write $\mathrm{d}J = I(\nu)\mathrm{d}\nu$, where I is the **intensity** of the radiation; hence $I = c\rho_{\mathrm{rad}}$.

Now consider the absorption that occurs within a slab of thickness $\mathrm{d}l$. Let the number density of molecules able to absorb light of frequency in the range ν to $\nu + \mathrm{d}\nu$ be $n(\nu)\mathrm{d}\nu$, so the total number density of absorbers is $\mathcal{N} = \int n(\nu)\,\mathrm{d}\nu$. The rate at which any one molecule absorbs a photon is $W = B\rho_{\mathrm{rad}}(E)$, and as each photon has an energy $h\nu$, the rate of change of energy density is

$$\frac{\mathrm{d}\mathcal{U}}{\mathrm{d}t} = -h\nu W n(\nu)\mathrm{d}\nu = -n(\nu)h\nu B\rho_{\mathrm{rad}}(E)\mathrm{d}\nu \tag{17.1}$$

The energy entering the slab at x from the left during the interval $\mathrm{d}t$ is $J(x)A\mathrm{d}t$, and the energy leaving the slab on the right at $x + \mathrm{d}l$ is $J(x + \mathrm{d}l)A\mathrm{d}t$. By the conservation of energy, the difference is the rate of change of energy in the slab:

$$\frac{\mathrm{d}(\mathcal{U}A\mathrm{d}l)}{\mathrm{d}t} = J(x + \mathrm{d}l)A - J(x)A$$

This expression rearranges into

$$\frac{\mathrm{d}\mathcal{U}}{\mathrm{d}t} = \frac{J(x + \mathrm{d}l) - J(x)}{\mathrm{d}l} = \frac{\mathrm{d}J}{\mathrm{d}l}$$

This conservation expression is valid for each frequency component, and by using eqn 1 and noting that

$$\frac{\mathrm{d}J}{\mathrm{d}l} = \frac{\mathrm{d}I}{\mathrm{d}l}\mathrm{d}\nu$$

we obtain

$$\mathrm{d}I = -n(\nu)h\nu B\rho_{\mathrm{rad}}\mathrm{d}l = -\frac{n(\nu)h\nu}{c}BI\mathrm{d}l$$

The reduction in intensity when a beam passes through a solution of length $\mathrm{d}l$ when the absorbers A are at a molar concentration [A] is

$$\mathrm{d}I = -\varepsilon(\nu)[\mathrm{A}]I\mathrm{d}l$$

where ε is the molar absorption coefficient. Comparison of the two expressions leads to

$$\frac{\varepsilon(\nu)}{\nu} = \frac{hn(\nu)B}{c[\mathrm{A}]}$$

Multiplication of both sides by $\mathrm{d}\nu$ and integration over all the frequencies of the band lead to $Bh\mathcal{N}/c[\mathrm{A}]$ on the right; but $\mathcal{N} = [\mathrm{A}]N_{\mathrm{A}}$, where N_{A} is the Avogadro constant. Hence,

$$\int \frac{\varepsilon(\nu)}{\nu}\,\mathrm{d}\nu = \frac{BhN_{\mathrm{A}}}{c}$$

For typical absorption bands, the frequency is virtually constant over the range for which $\varepsilon(\nu)$ is nonzero, and so we set $\nu \approx \nu_{\mathrm{fi}}$ on the left and recognize $\mathcal{A} = \int \varepsilon(\nu)\mathrm{d}\nu$, the integrated absorption coefficient. It then follows that

$$\mathcal{A} = \left(\frac{h\nu_{\mathrm{fi}}}{c}\right)N_{\mathrm{A}}B \tag{17.2}$$

We show in *Further information 16* that for electric dipole transitions $B = |\mu_{\mathrm{fi}}|^2/6\varepsilon_0\hbar^2$; therefore

$$\mathcal{A} = \frac{\pi\nu_{\mathrm{fi}}N_{\mathrm{A}}|\mu_{\mathrm{fi}}|^2}{3\varepsilon_0\hbar c} \tag{17.3}$$

which is a direct link between a measurable quantity $\mathcal{A}$ and a calculated quantity μ_{fi}.

It turns out to be useful to introduce the dimensionless **oscillator strength**, f, of a transition:

$$f = \left(\frac{4\pi m_{\mathrm{e}}\nu_{\mathrm{fi}}}{3e^2\hbar}\right)|\mu_{\mathrm{fi}}|^2 \tag{17.4}$$

The relation between this quantity and the integrated absorption coefficient is obtained by combining the last two equations, and is

$$f = \left(\frac{4m_{\mathrm{e}}c\varepsilon_0}{N_{\mathrm{A}}e^2}\right)\mathcal{A} \tag{17.5}$$

The practical form of this expression is

$$f = 6.257 \times 10^{-19} \times (\mathcal{A}/\mathrm{m^2\,mol^{-1}\,s^{-1}})$$

For a one-dimensional harmonic oscillator, $f = \frac{1}{3}$. For an electron bound so that it oscillates harmonically in three dimensions (which was an early model of a hydrogen atom), $f = 1$. The observed oscillator strength is therefore the ratio of the intensity of the transition to the intensity of an harmonically oscillating electron (in three dimensions). In practice, $f \approx 1$ for allowed electric dipole transitions and $f \ll 1$ for forbidden transitions.

18 Sum rules

In this section, we establish the **Kuhn–Thomas sum rule**:

$$\sum_n f_{n0} = N_{\mathrm{e}} \tag{18.1}$$

where f_{n0} is the oscillator strength for the transition $n \leftarrow 0$ (with $n = 0$ implicitly excluded from the sum here and throughout), and N_{e} is the number of electrons in the molecule. The first step is to derive the **velocity–dipole relation**:

$$\mathbf{p}_{mn} = -\mathrm{i}\left(\frac{m_{\mathrm{e}}\omega_{mn}}{e}\right)\boldsymbol{\mu}_{mn} \tag{18.2}$$

where $\hbar\omega_{mn}$ is the transition energy and $\boldsymbol{\mu}_{mn}$ its dipole moment. To derive this result, we consider the x-component with $\mu_x = -ex$:

$$p_{x,mn} = \mathrm{i}m_{\mathrm{e}}\omega_{mn}x_{mn} \tag{18.3}$$

The proof hinges on the evaluation of the commutator of the hamiltonian and the position operator:

$$\begin{aligned}\langle m|[H,x]|n\rangle &= \langle m|Hx|n\rangle - \langle m|xH|n\rangle \\ &= (E_m - E_n)\langle m|x|n\rangle = \hbar\omega_{mn}x_{mn}\end{aligned}$$

The commutator may also be written as follows:

$$\begin{aligned}[H,x] &= -\frac{\hbar^2}{2m_{\mathrm{e}}}\left[\frac{\mathrm{d}^2}{\mathrm{d}x^2},x\right] + [V(x),x] \\ &= -\frac{\hbar^2}{2m_{\mathrm{e}}}\left(\frac{\mathrm{d}^2}{\mathrm{d}x^2}x - x\frac{\mathrm{d}^2}{\mathrm{d}x^2}\right) \\ &= -\frac{\hbar^2}{2m_{\mathrm{e}}}\left(2\frac{\mathrm{d}}{\mathrm{d}x} + x\frac{\mathrm{d}^2}{\mathrm{d}x^2} - x\frac{\mathrm{d}^2}{\mathrm{d}x^2}\right) \\ &= -\frac{\hbar^2}{m_{\mathrm{e}}}\frac{\mathrm{d}}{\mathrm{d}x} = \frac{\hbar}{\mathrm{i}m_{\mathrm{e}}}p_x\end{aligned}$$

It follows that

$$\langle m|[H,x]|n\rangle = \frac{\hbar}{\mathrm{i}m_{\mathrm{e}}}p_{x,mn} \tag{18.4}$$

and eqn 3 follows immediately.

At this point, we develop f_{n0} in terms of the linear momentum by using the velocity–dipole relation and $\omega_{n0} = 2\pi\nu_{n0}$:

$$\begin{aligned}f_{n0} &= \frac{4\pi m_{\mathrm{e}}\nu_{n0}|\mu_{n0}|^2}{3e^2\hbar} \\ &= \frac{m_{\mathrm{e}}\omega_{n0}}{3e^2\hbar}(\boldsymbol{\mu}_{0n}\cdot\boldsymbol{\mu}_{n0} + \boldsymbol{\mu}_{0n}\cdot\boldsymbol{\mu}_{n0}) \\ &= \frac{m_{\mathrm{e}}}{3e^2\hbar}\left(\frac{e\mathrm{i}}{m_{\mathrm{e}}}\right)(\boldsymbol{\mu}_{0n}\cdot\mathbf{p}_{n0} - \mathbf{p}_{0n}\cdot\boldsymbol{\mu}_{n0})\end{aligned}$$

The sum over n produces the commutator:

$$\sum_n f_{n0} = -\left(\frac{\mathrm{i}}{3\hbar}\right)\langle 0|\mathbf{r}\cdot\mathbf{p} - \mathbf{p}\cdot\mathbf{r}|0\rangle$$

For each component of the scalar product the commutator is $\mathrm{i}\hbar$, so the outcome is

$$\sum_n f_{n0} = -\left(\frac{\mathrm{i}}{3\hbar}\right)(3\mathrm{i}\hbar)\langle 0|0\rangle = 1$$

This result is for a single electron. If the system consists of N_e electrons, each one gives the same contribution, so overall

$$\sum_n f_{n0} = N_e$$

as was to be proved.

19 Normal modes: an example

Consider a linear triatomic molecule BAB in which the mass of A is m_A and the mass of B is m_B. For simplicity, we shall confine attention to displacement along the axis of the molecule, and the displacement of the atoms B, A, and B will be written ξ_1, ξ_2, and ξ_3, respectively. Because the relative displacements of the bonded pairs of atoms are $\xi_1 - \xi_2$ and $\xi_3 - \xi_2$, and the force constants of the two bonds are the same, the potential energy is

$$V = \tfrac{1}{2}k(\xi_1 - \xi_2)^2 + \tfrac{1}{2}k(\xi_3 - \xi_2)^2 \qquad (19.1)$$

The force constant matrix (with matrix elements specified in eqn 10.68) is

$$\mathbf{k} = \begin{pmatrix} k & -k & 0 \\ -k & 2k & -k \\ 0 & -k & k \end{pmatrix}$$

We shall work with the mass-weighted coordinates q_i:

$$q_i = m_i^{1/2}\xi_i \qquad (19.2)$$

The force constant matrix then turns into $\mathbf{K}$, where

$$K_{ij} = \left(\frac{\partial^2 V}{\partial q_i \partial q_j}\right) = \left(\frac{1}{m_i m_j}\right)^{1/2} k_{ij}$$

Therefore,

$$\mathbf{K} = \begin{pmatrix} k/m_B & -k/(m_A m_B)^{1/2} & 0 \\ -k/(m_A m_B)^{1/2} & 2k/m_A & -k/(m_A m_B)^{1/2} \\ 0 & -k/(m_A m_B)^{1/2} & k/m_B \end{pmatrix} \qquad (19.3)$$

We seek a linear combination of the coordinates that diagonalizes this matrix. According to the procedures set out in *Further information 23*, we need to solve the secular equations

$$|\mathbf{K} - \kappa\mathbf{1}| = \begin{vmatrix} k/m_B - \kappa & -k/(m_A m_B)^{1/2} & 0 \\ -k/(m_A m_B)^{1/2} & 2k/m_A - \kappa & -k/(m_A m_B)^{1/2} \\ 0 & -k/(m_A m_B)^{1/2} & k/m_B - \kappa \end{vmatrix} = 0$$

The roots of this cubic equation for κ are

$$\kappa_1 = 0 \qquad \kappa_2 = \frac{k}{m_B} \qquad \kappa_3 = \frac{k}{\mu} \qquad (19.4)$$

with the effective mass

$$\mu = \frac{m_A m_B}{m_A + 2m_B} \tag{19.5}$$

Note that the effective force constants κ_i depend on the masses of the atoms. The mode with zero force constant (no restoring force) corresponds to the translation of the entire molecule parallel to the axis.

The eigenvectors Q_l of $\mathbf{K}$ are the combinations

$$Q_l = \sum_i c_{il} q_i \tag{19.6}$$

and are found by solving the set of simultaneous equations

$$\sum_j (K_{ij} - \kappa_l \delta_{jl}) c_{jl} = 0 \tag{19.7}$$

with $l = 1, 2, 3$ in turn. As the simplest example, consider the mode Q_1, which corresponds to $\kappa_1 = 0$. The equations for c_{i1} reduce to

$$\sum_j K_{ij} c_{j1} = 0$$

or, specifically,

$$\begin{aligned} K_{11}c_{11} + K_{12}c_{21} + K_{13}c_{31} &= 0 \\ K_{21}c_{11} + K_{22}c_{21} + K_{23}c_{31} &= 0 \\ K_{31}c_{11} + K_{32}c_{21} + K_{33}c_{31} &= 0 \end{aligned}$$

The coefficients are given in eqn 3, so

$$c_{11} = \left(\frac{m_B}{m_A}\right)^{1/2} c_{21} \qquad c_{31} = \left(\frac{m_B}{m_A}\right)^{1/2} c_{21} \tag{19.8}$$

We also require

$$c_{11}^2 + c_{21}^2 + c_{31}^2 = 1 \tag{19.9}$$

It follows that

$$c_{11} = c_{31} = \left(\frac{m_B}{m}\right)^{1/2} \qquad c_{21} = \left(\frac{m_A}{m}\right)^{1/2} \tag{19.10}$$

where $m = m_A + 2m_B$, the total mass of the molecule. Therefore,

$$\begin{aligned} Q_1 &= \frac{1}{m^{1/2}} \left(m_B^{1/2} q_1 + m_A^{1/2} q_2 + m_B^{1/2} q_3 \right) \\ &= \frac{1}{m^{1/2}} (m_B \xi_1 + m_A \xi_2 + m_B \xi_3) \end{aligned} \tag{19.11}$$

The modes corresponding to κ_2 and κ_3 are found in a similar way:

$$\begin{aligned} Q_2 &= \left(\frac{1}{2}\right)^{1/2} (q_1 - q_3) \\ Q_3 &= \left(\frac{1}{2m}\right)^{1/2} \left(m_A^{1/2} q_1 - 2m_B^{1/2} q_2 + m_A^{1/2} q_3 \right) \end{aligned} \tag{19.12}$$

The former is a symmetrical mode (the B atoms move in opposite directions) and involves no motion of the central atom. The second involves the motion of the outer pair of atoms against the central atom, and is the antisymmetric mode.

It may be verified that the kinetic energy may be expressed in the form $\frac{1}{2}\sum_i \dot{Q}_i^2$, so both the kinetic and potential energy contributions are diagonal, as required.

The electromagnetic field

20 The Maxwell equations

The Maxwell equations describe the properties of the electromagnetic field. They are expressed in terms of the following six quantities (with their SI units in parentheses):

E	electric field strength ($\mathrm{V\,m^{-1}}$)
D	electric displacement ($\mathrm{C\,m^{-2}}$)
ρ	charge density ($\mathrm{C\,m^{-3}}$)
H	magnetic field strength ($\mathrm{A\,m^{-1}}$)
B	magnetic flux density (T)
J	current density ($\mathrm{A\,m^{-2}}$)

The electric displacement and the magnetic flux density are related to the magnetic and electric field strengths by the polarization, **P**, and the magnetization, **M**, respectively:

$$\mathbf{D} = \varepsilon_0\mathbf{E} + \mathbf{P} \qquad \mathbf{B} = \mu_0\mathbf{H} + \mu_0\mathbf{M} \tag{20.1}$$

The **Maxwell equations** are then

$$(i)\quad \nabla\cdot\mathbf{D} = \rho \qquad (ii)\quad \nabla\cdot\mathbf{B} = 0$$
$$(iii)\quad \nabla\times\mathbf{E} = -\frac{\partial\mathbf{B}}{\partial t} \qquad (iv)\quad \nabla\times\mathbf{H} = \mathbf{J} + \frac{\partial\mathbf{D}}{\partial t} \tag{20.2}$$

The fields **B** and **E** may be expressed in terms of two potentials, a scalar potential ϕ and a vector potential **A**. Because the divergence of a curl is identically zero, it follows that the second Maxwell equation ($\nabla\cdot\mathbf{B} = 0$) is satisfied by writing

$$\mathbf{B} = \nabla\times\mathbf{A} \tag{20.3}$$

It then follows from the third Maxwell equation that

$$\nabla\times\left(\mathbf{E} + \frac{\partial\mathbf{A}}{\partial t}\right) = 0$$

and hence

$$\mathbf{E} = -\frac{\partial\mathbf{A}}{\partial t} + \mathbf{f} \tag{20.4}$$

where **f** is a vector function with zero curl. Because the curl of a gradient of a scalar function is identically zero, we may write $\mathbf{f} = -\nabla\phi$, and so obtain

$$\mathbf{E} = -\frac{\partial\mathbf{A}}{\partial t} - \nabla\phi \tag{20.5}$$

When the vector potential is independent of time,

$$\mathbf{E} = -\nabla\phi \tag{20.6}$$

The Maxwell equations take on special importance in a vacuum, for which $\mathbf{P} = 0$, $\mathbf{M} = 0$, $\rho = 0$, and $\mathbf{J} = 0$. Under these conditions, $\mathbf{D} = \varepsilon_0\mathbf{E}$ and $\mathbf{B} = \mu_0\mathbf{H}$, and the equations become

$$\begin{aligned} &(i)\ \nabla\cdot\mathbf{E} = 0 \qquad &(ii)\ \nabla\cdot\mathbf{H} = 0 \\ &(iii)\ \nabla\times\mathbf{E} = -\mu_0\frac{\partial\mathbf{H}}{\partial t} \qquad &(iv)\ \nabla\times\mathbf{H} = \varepsilon_0\frac{\partial\mathbf{E}}{\partial t} \end{aligned} \tag{20.7}$$

If the curl of the third Maxwell equation is taken, we obtain

$$\nabla\times(\nabla\times\mathbf{E}) + \frac{\partial(\nabla\times\mathbf{B})}{\partial t} = 0$$

Then, because $\nabla\times(\nabla\times\mathbf{E}) = \nabla(\nabla\cdot\mathbf{E}) - \nabla^2\mathbf{E}$ (see *Further information 22*),

$$\nabla(\nabla\cdot\mathbf{E}) - \nabla^2\mathbf{E} + \frac{\partial(\nabla\times\mathbf{B})}{\partial t} = 0$$

The first term in this expression is zero by eqn 7(*i*), and by eqn 7(*iv*) the third term is

$$\frac{\partial(\nabla\times\mathbf{B})}{\partial t} = \varepsilon_0\mu_0\frac{\partial^2\mathbf{E}}{\partial t^2}$$

Therefore, in free space the electric field satisfies the equation

$$\nabla^2\mathbf{E} - \varepsilon_0\mu_0\ddot{\mathbf{E}} = 0 \tag{20.8}$$

where $\ddot{\mathbf{E}} = \partial^2\mathbf{E}/\partial t^2$, which is the equation of a wave propagating with velocity $c = 1/(\varepsilon_0\mu_0)^{1/2}$.

For electromagnetic radiation propagating in a medium we need to allow for the polarization. Suppose (see eqn 12.81) that

$$\mathbf{P} = \mathcal{N}\alpha\mathbf{E} - \mathcal{N}\beta\dot{\mathbf{B}} \tag{20.9}$$

where $\dot{\mathbf{B}} = \partial\mathbf{B}/\partial t$. In optically inactive media, the second term on the right is absent. From eqns 1, 2, and 9, and setting $\mathbf{B} \approx \mu_0\mathbf{H}$,

$$\nabla\times\mathbf{B} \approx \mu_0\dot{\mathbf{D}} = \varepsilon_0\mu_0\dot{\mathbf{E}} + \mu_0\mathcal{N}\alpha\dot{\mathbf{E}} - \mu_0\mathcal{N}\beta\ddot{\mathbf{B}}$$

When the curl is taken of both sides, and the relation $\nabla\times(\nabla\times\mathbf{B}) = \nabla(\nabla\cdot\mathbf{B}) - \nabla^2\mathbf{B} = -\nabla^2\mathbf{B}$ (where the second equality follows from eqn 2) is used, it follows that

$$\nabla^2\mathbf{B} = \varepsilon_0\mu_0\left(1 + \frac{\alpha\mathcal{N}}{\varepsilon_0}\right)\ddot{\mathbf{B}} + \mu_0\beta\mathcal{N}\nabla\times\ddot{\mathbf{B}} \tag{20.10}$$

Suppose for the moment that $\beta = 0$; then by comparison of this expression with eqn 8 we see that in a medium the magnetic field propagates at

$$v = \left(\frac{1}{\varepsilon_0\mu_0(1 + \alpha\mathcal{N}/\varepsilon_0)}\right)^{1/2}$$

and hence the refractive index is

$$n_r = \frac{c}{v} = \left(1 + \frac{\alpha\mathcal{N}}{\varepsilon_0}\right)^{1/2} \approx 1 + \frac{\alpha\mathcal{N}}{2\varepsilon_0}$$

as in eqn 12.59.

When β is nonzero, we can expect birefringence ($n_+ \neq n_-$). A circularly polarized electric field with propagation direction z has the form (see eqn 12.75)

$$\mathbf{E}^{\pm} = \mathcal{E}\hat{\mathbf{i}}\cos\phi_{\pm} \pm \mathcal{E}\hat{\mathbf{j}}\sin\phi_{\pm} \tag{20.11}$$

where the amplitude $\mathcal{E}$ is assumed here to be time-independent, $\hat{\mathbf{i}}$ and $\hat{\mathbf{j}}$ are unit vectors perpendicular to the propagation direction and

$$\phi_{\pm} = \omega t - k_{\pm}z \tag{20.12}$$

The wavevector of magnitude $k_{\pm}$ depends on the sense of polarization because $k = 2\pi/\lambda$ and λ depends on the refractive index through $\lambda = v/\nu = c/n_r\nu$. It proves convenient to work with the magnetic component of the electromagnetic field, and from Maxwell's equation $\nabla \times \mathbf{E}^{\pm} = -\dot{\mathbf{B}}^{\pm}$, it follows that

$$\mathbf{B}^{\pm} = \left(\frac{\mathcal{E}k_{\pm}}{\omega}\right)\{\hat{\mathbf{j}}\cos\phi_{\pm} \mp \hat{\mathbf{i}}\sin\phi_{\pm}\}$$

has the correct polarization characteristics. Then, because

$$\nabla \times \mathbf{B}^{\pm} = \pm k_{\pm}\mathbf{B}^{\pm}$$

and in addition

$$\nabla^2\mathbf{B}^{\pm} = -k_{\pm}^2\mathbf{B}^{\pm}$$

and

$$\ddot{\mathbf{B}}^{\pm} = -\omega^2\mathbf{B}^{\pm}$$

equation 10 becomes

$$k_{\pm}^2 = \varepsilon_0\mu_0\omega^2\left(1 + \frac{\alpha\mathcal{N}}{\varepsilon_0}\right) \pm \mu_0\beta\mathcal{N}\omega^2 k_{\pm} \tag{20.13}$$

Because $\varepsilon_0\mu_0 = 1/c^2$ and $k_{\pm} = 2\pi\nu n_{\pm}/c = \omega n_{\pm}/c$ (from the remark above), it follows that

$$n_{\pm}^2 = 1 + \frac{\alpha\mathcal{N}}{\varepsilon_0} \pm \frac{\omega\beta\mathcal{N}n_{\pm}}{c\varepsilon_0} \tag{20.14}$$

This is a quadratic equation for $n_{\pm}$. The solution to first order in β is

$$n_{\pm} \approx 1 + \frac{\alpha\mathcal{N}}{2\varepsilon_0} \pm \frac{\omega\beta\mathcal{N}}{2c\varepsilon_0} \tag{20.15}$$

which is the expression used in the text (eqn 12.82).

21 The dipolar vector potential

In this section, we deduce the form of the magnetic field corresponding to the vector potential

$$\mathbf{A} = a\frac{\mathbf{m} \times \mathbf{r}}{r^3} \qquad a = \frac{\mu_0}{4\pi} \tag{21.1}$$

This potential was introduced in Section 13.11 in connection with the discussion of the field of a magnetic dipole.

First, note that as $\nabla(1/r) = -\mathbf{r}/r^3$,

$$\mathbf{A} = -a\mathbf{m} \times \nabla\left(\frac{1}{r}\right) \tag{21.2}$$

Then, from FI22.8 with $\mathbf{F} = \mathbf{m}$ and $\mathbf{G} = (\nabla r^{-1})$,

$$\begin{aligned}\nabla \times \mathbf{A} &= -a\nabla \times \left(\mathbf{m} \times \nabla\frac{1}{r}\right) \\ &= -a\left\{\mathbf{m}\left(\nabla\cdot\nabla\frac{1}{r}\right) - (\nabla\cdot\mathbf{m})\left(\nabla\frac{1}{r}\right) + \left(\nabla\frac{1}{r}\cdot\nabla\right)\mathbf{m} \right. \\ &\qquad \left. - (\mathbf{m}\cdot\nabla)\left(\nabla\frac{1}{r}\right)\right\} \\ &= -a\left\{\mathbf{m}\left(\nabla^2\frac{1}{r}\right) - (\mathbf{m}\cdot\nabla)\left(\nabla\frac{1}{r}\right)\right\}\end{aligned} \tag{21.3}$$

because $\mathbf{m}$ is a constant and $\nabla\cdot\nabla = \nabla^2$. The second term may be written

$$\begin{aligned}(\mathbf{m}\cdot\nabla)\left(\nabla\frac{1}{r}\right) &= -(\mathbf{m}\cdot\nabla)\left(\frac{\mathbf{r}}{r^3}\right) \\ &= -\left(m_x\frac{\partial}{\partial x} + m_y\frac{\partial}{\partial y} + m_z\frac{\partial}{\partial z}\right)\frac{x\hat{\mathbf{i}} + y\hat{\mathbf{j}} + z\hat{\mathbf{k}}}{r^3} \\ &= -\frac{\mathbf{m}}{r^3} - \mathbf{r}\mathbf{m}\cdot\left(\nabla\frac{1}{r^3}\right) = -\frac{\mathbf{m}}{r^3} + 3\frac{\mathbf{r}(\mathbf{m}\cdot\mathbf{r})}{r^5} \\ &= -\frac{\mathbf{m} - 3(\mathbf{m}\cdot\hat{\mathbf{r}})\hat{\mathbf{r}}}{r^3}\end{aligned}$$

Therefore, this part of the vector potential accounts for the contribution

$$\mathbf{B}_{\text{dipolar}} = -\frac{a}{r^3}\{\mathbf{m} - 3(\mathbf{m}\cdot\hat{\mathbf{r}})\hat{\mathbf{r}}\} \tag{21.4}$$

as in eqn 13.61.

When the system is spherically symmetrical, detailed analysis shows that the first term of the last line in eqn 3 does not necessarily vanish when it is averaged over the appropriate wavefunctions. Furthermore, the spherical average of the second term in the last line produces

$$\left\langle(\mathbf{m}\cdot\nabla)\left(\nabla\frac{1}{r}\right)\right\rangle = \tfrac{1}{3}\mathbf{m}\left(\nabla^2\frac{1}{r}\right) \tag{21.5}$$

Therefore, in this case

$$\nabla \times \mathbf{A} = -\tfrac{2}{3}a\mathbf{m}\nabla^2\frac{1}{r} \tag{21.6}$$

A standard property is

$$\nabla^2\frac{1}{r} = -4\pi\delta(\mathbf{r}) \tag{21.7}$$

where $\delta(\mathbf{r})$ is the Dirac delta-function (Section 14.3). Therefore, this term contributes

$$\mathbf{B}_{\text{contact}} = \left(\frac{8\pi}{3}\right) a\mathbf{m}\delta(\mathbf{r}) = \tfrac{2}{3}\mu_0\mathbf{m}\delta(\mathbf{r}) \tag{21.8}$$

Mathematical relations

22 Vector properties

We shall consider the properties of vectors written as

$$\mathbf{F} = F_x\hat{\mathbf{i}} + F_y\hat{\mathbf{j}} + F_z\hat{\mathbf{k}} \tag{22.1}$$

and likewise for **G, H**, and **I**.

1. *Vector multiplication*

The **scalar product** of two vectors is defined as

$$\mathbf{F}\cdot\mathbf{G} = F_xG_x + F_yG_y + F_zG_z = \sum_r F_rG_r \tag{22.2}$$

The scalar product is a scalar quantity. The **vector product** of two vectors is defined as

$$\mathbf{F}\times\mathbf{G} = \begin{vmatrix} \hat{\mathbf{i}} & \hat{\mathbf{j}} & \hat{\mathbf{k}} \\ F_x & F_y & F_z \\ G_x & G_y & G_z \end{vmatrix} \tag{22.3}$$

A vector product, a vector quantity, is also often denoted $\mathbf{F}\wedge\mathbf{G}$.

The following relations are useful:

$$\begin{aligned} \mathbf{F}\times\mathbf{G} &= -\mathbf{G}\times\mathbf{F} \\ \mathbf{F}\cdot(\mathbf{G}\times\mathbf{H}) &= \mathbf{G}\cdot(\mathbf{H}\times\mathbf{F}) = \mathbf{H}\cdot(\mathbf{F}\times\mathbf{G}) = (\mathbf{F}\times\mathbf{G})\cdot\mathbf{H} \\ \mathbf{F}\times(\mathbf{G}\times\mathbf{H}) &= \mathbf{G}(\mathbf{F}\cdot\mathbf{H}) - \mathbf{H}(\mathbf{F}\cdot\mathbf{G}) \\ (\mathbf{F}\times\mathbf{G})\cdot(\mathbf{H}\times\mathbf{I}) &= (\mathbf{F}\cdot\mathbf{H})(\mathbf{G}\cdot\mathbf{I}) - (\mathbf{F}\cdot\mathbf{I})(\mathbf{G}\cdot\mathbf{H}) \end{aligned} \tag{22.4}$$

2. *Vector differentiation*

The vector differentiation of a scalar function f is denoted ∇ or grad, and is called the **gradient** of the function:

$$\nabla f = \left(\frac{\partial f}{\partial x}\right)\hat{\mathbf{i}} + \left(\frac{\partial f}{\partial y}\right)\hat{\mathbf{j}} + \left(\frac{\partial f}{\partial z}\right)\hat{\mathbf{k}} \tag{22.5}$$

The quantity ∇f is a vector. There are two versions of the differentiation of a vector. The **divergence** of a vector is defined as

$$\nabla\cdot\mathbf{F} = \left(\frac{\partial F_x}{\partial x}\right) + \left(\frac{\partial F_y}{\partial y}\right) + \left(\frac{\partial F_z}{\partial z}\right) = \sum_r\left(\frac{\partial F_r}{\partial r}\right) \tag{22.6}$$

The **curl** of a vector is defined as

$$\nabla\times\mathbf{F} = \begin{vmatrix} \hat{\mathbf{i}} & \hat{\mathbf{j}} & \hat{\mathbf{k}} \\ \frac{\partial}{\partial x} & \frac{\partial}{\partial y} & \frac{\partial}{\partial z} \\ F_x & F_y & F_z \end{vmatrix} \tag{22.7}$$

The divergence $\nabla \cdot \mathbf{F}$ is a scalar and the curl $\nabla \times \mathbf{F}$ (which is also often denoted $\nabla \wedge \mathbf{F}$) is a vector.

The following relations are useful:

$$\begin{aligned}
\nabla(fg) &= f\nabla g + g\nabla f \\
\nabla^2 f &= \nabla \cdot \nabla f \\
\nabla \times (\nabla f) &= 0 \\
\nabla \cdot (f\mathbf{F}) &= f(\nabla \cdot \mathbf{F}) + \mathbf{F} \cdot (\nabla f) \\
\nabla \times (f\mathbf{F}) &= f(\nabla \times \mathbf{F}) + (\nabla f) \times \mathbf{F} \\
\nabla \cdot (\mathbf{F} \times \mathbf{G}) &= \mathbf{G} \cdot (\nabla \times \mathbf{F}) - \mathbf{F} \cdot (\nabla \times \mathbf{G}) \\
\nabla \times (\nabla \times \mathbf{F}) &= \nabla(\nabla \cdot \mathbf{F}) - \nabla^2 \mathbf{F} \\
\nabla \times (\mathbf{F} \times \mathbf{G}) &= \mathbf{F}(\nabla \cdot \mathbf{G}) - \mathbf{G}(\nabla \cdot \mathbf{F}) + (\mathbf{G} \cdot \nabla)\mathbf{F} - (\mathbf{F} \cdot \nabla)\mathbf{G} \\
\nabla(\mathbf{F} \cdot \mathbf{G}) &= (\mathbf{F} \cdot \nabla)\mathbf{G} + (\mathbf{G} \cdot \nabla)\mathbf{F} + \mathbf{F} \times (\nabla \times \mathbf{G}) + \mathbf{G} \times (\nabla \times \mathbf{F})
\end{aligned} \tag{22.8}$$

23 Matrices

A **matrix** is an array of numbers. Matrices may be combined together by addition or multiplication according to generalizations of the rules for ordinary numbers (which are 1×1 matrices). Most numerical matrix manipulations are now carried out computationally.

Consider a square matrix $\mathbf{M}$ of n^2 numbers arranged in n columns and n rows.These n^2 numbers are the **elements** of the matrix, and may be specified by stating the row, r, and column, c, at which they occur. Each element is therefore denoted M_{rc}. For example, in the matrix

$$\mathbf{M} = \begin{pmatrix} 1 & 2 \\ 3 & 4 \end{pmatrix}$$

the elements are $M_{11} = 1$, $M_{12} = 2$, $M_{21} = 3$, and $M_{22} = 4$. The **determinant**, $|\mathbf{M}|$, of this matrix is

$$|\mathbf{M}| = \begin{vmatrix} 1 & 2 \\ 3 & 4 \end{vmatrix} = 1 \times 4 - 2 \times 3 = -2$$

Note that if $\mathbf{P} = \mathbf{MN}$, then $|\mathbf{P}| = |\mathbf{M}||\mathbf{N}|$.

Two matrices $\mathbf{M}$ and $\mathbf{N}$ may be added to give the sum $\mathbf{S} = \mathbf{M} + \mathbf{N}$, according to the rule

$$S_{rc} = M_{rc} + N_{rc} \tag{23.1}$$

(that is, corresponding elements are added). Thus, with $\mathbf{M}$ given above and

$$\mathbf{N} = \begin{pmatrix} 5 & 6 \\ 7 & 8 \end{pmatrix}$$

the sum is

$$\mathbf{S} = \begin{pmatrix} 1 & 2 \\ 3 & 4 \end{pmatrix} + \begin{pmatrix} 5 & 6 \\ 7 & 8 \end{pmatrix} = \begin{pmatrix} 6 & 8 \\ 10 & 12 \end{pmatrix}$$

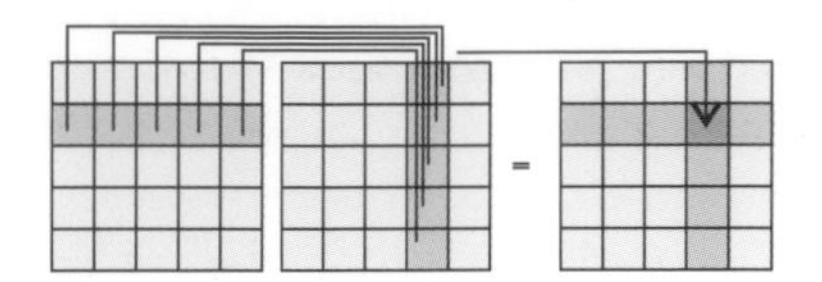

Fig. FI23.1 A schematic illustration of matrix multiplication. The product of the elements linked by lines is taken, and then the sum of these products is placed at the intersection of the row and column.

Two matrices may also be multiplied to give the product $\mathbf{P} = \mathbf{MN}$ according to the rule

$$P_{rc} = \sum_n M_{rn} N_{nc} \tag{23.2}$$

This rule is illustrated in Fig. FI23.1. For example, with the matrices given above,

$$\mathbf{P} = \begin{pmatrix} 1 & 2 \\ 3 & 4 \end{pmatrix}\begin{pmatrix} 5 & 6 \\ 7 & 8 \end{pmatrix} = \begin{pmatrix} 1\times 5 + 2\times 7 & 1\times 6 + 2\times 8 \\ 3\times 5 + 4\times 7 & 3\times 6 + 4\times 8 \end{pmatrix} = \begin{pmatrix} 19 & 22 \\ 43 & 50 \end{pmatrix}$$

It should be noticed that in general $\mathbf{MN} \neq \mathbf{NM}$, and matrix multiplication is in general non-commutative.

A **diagonal matrix** is a matrix in which the only nonzero elements lie on the major diagonal (the diagonal from M_{11} to M_{nn}). Thus, the matrix

$$\mathbf{D} = \begin{pmatrix} 1 & 0 & 0 \\ 0 & 2 & 0 \\ 0 & 0 & 1 \end{pmatrix}$$

is diagonal. The condition may be written

$$M_{rc} = m_r \delta_{rc} \tag{23.3}$$

where δ_{rc} is the **Kronecker delta**, which is equal to 1 for $r = c$ and to 0 for $r \neq c$. In the above example, $m_1 = 1$, $m_2 = 2$, and $m_3 = 1$. The **unit matrix, 1** (and occasionally **I**), is a special case of a diagonal matrix in which all nonzero elements are 1.

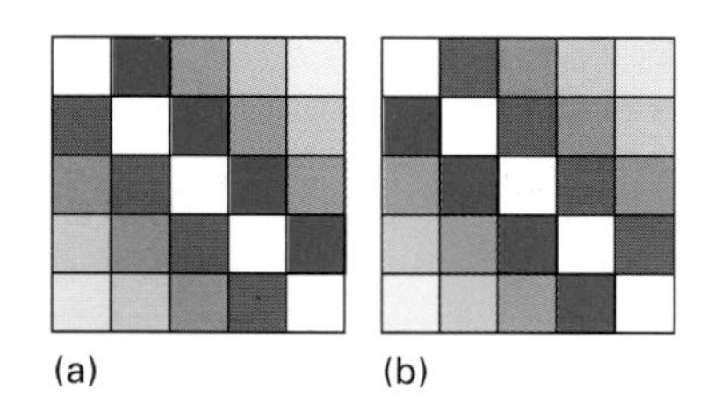

Fig. FI23.2 The transpose of a matrix is formed by reflecting the elements across the principal diagonal.

The **transpose** of a matrix **M** is denoted $\mathbf{M}^{\mathrm{T}}$ and is defined by

$$M^{\mathrm{T}}_{mn} = M_{nm} \tag{23.4}$$

(See Fig. FI23.2.) Thus, for the matrix **M** we have been using,

$$\mathbf{M}^{\mathrm{T}} = \begin{pmatrix} 1 & 3 \\ 2 & 4 \end{pmatrix}$$

If $\mathbf{A} = \mathbf{BC}$, then $\mathbf{A}^{\mathrm{T}} = \mathbf{C}^{\mathrm{T}}\mathbf{B}^{\mathrm{T}}$.

The **complex conjugate** of a matrix, $\mathbf{M}^*$, with complex elements is the matrix obtained by taking the complex conjugate of each element:

$$M^*_{rc} = (M_{rc})^* \tag{23.5}$$

If $\mathbf{A} = \mathbf{BC}$, then $\mathbf{A}^* = \mathbf{B}^*\mathbf{C}^*$. The **adjoint** of a matrix , $\mathbf{M}^\dagger$, is the complex conjugate of the transpose:

$$M^\dagger_{mn} = M^*_{nm} \tag{23.6}$$

A matrix is **hermitian** or **self-adjoint** if it is equal to its own adjoint:

$$M^\dagger_{mn} = M_{mn} \qquad \text{that is, } M^*_{nm} = M_{mn}$$

(See the discussion of hermiticity in Section 1.13.)

The **inverse** of a matrix **M** is denoted $\mathbf{M}^{-1}$, and is defined so that

$$\mathbf{MM}^{-1} = \mathbf{M}^{-1}\mathbf{M} = \mathbf{1} \tag{23.7}$$

A matrix is **unitary** if $\mathbf{M}^{-1} = \mathbf{M}^\dagger$.

The inverse of a matrix can be constructed using a mathematical software program, but in simple cases the following procedure can be carried through without much effort:

1. Form the determinant of the matrix. For example, for our matrix **M**, $|\mathbf{M}| = -2$.
2. Form the transpose of the matrix. For example, $\mathbf{M}^{\mathrm{T}} = \begin{pmatrix} 1 & 3 \\ 2 & 4 \end{pmatrix}$.
3. Form $\tilde{\mathbf{M}}'$, where $\tilde{M}'_{rc}$ is the **cofactor** of the element M_{rc}; that is, it is the determinant formed from **M** with the row r and column c struck out. For example, $\tilde{\mathbf{M}}' = \begin{pmatrix} 4 & -2 \\ -3 & 1 \end{pmatrix}$.
4. Construct the inverse as $\mathbf{M}^{-1} = \tilde{\mathbf{M}}'/|\mathbf{M}|$. For example,

$$\mathbf{M}^{-1} = \left(\frac{1}{-2}\right)\begin{pmatrix} 4 & -2 \\ -3 & 1 \end{pmatrix} = \begin{pmatrix} -2 & 1 \\ \frac{3}{2} & -\frac{1}{2} \end{pmatrix}$$

A set of n **simultaneous equations**

$$\begin{aligned} a_{11}x_1 + a_{12}x_2 + \ldots a_{1n}x_n &= b_1 \\ a_{21}x_1 + a_{22}x_2 + \ldots a_{2n}x_n &= b_2 \\ \ldots \quad \ldots \quad \ldots & \\ a_{n1}x_1 + a_{n2}x_2 + \ldots a_{nn}x_n &= b_n \end{aligned} \tag{23.8}$$

can be written in matrix notation if we introduce the **column vectors x** and **b**:

$$\mathbf{x} = \begin{bmatrix} x_1 \\ x_2 \\ \vdots \\ x_n \end{bmatrix} \qquad \mathbf{b} = \begin{bmatrix} b_1 \\ b_2 \\ \vdots \\ b_n \end{bmatrix}$$

For then, with **a** the matrix of coefficients a_{rc}, the n equations are

$$\mathbf{ax} = \mathbf{b} \tag{23.9}$$

The formal solution is obtained by multiplying both sides of this matrix equation by $\mathbf{a}^{-1}$, for then

$$\mathbf{x} = \mathbf{a}^{-1}\mathbf{b} \tag{23.10}$$

In the special case that $\mathbf{b} = \lambda\mathbf{x}$, eqn 9 is an **eigenvalue equation**:

$$\mathbf{ax} = \lambda\mathbf{x} \tag{23.11}$$

where λ is the **eigenvalue** and **x** is the **eigenvector**. In general, there are n eigenvalues $\lambda^{(i)}$, and they satisfy the n simultaneous equations

$$(\mathbf{a} - \lambda\mathbf{1})\mathbf{x} = 0 \tag{23.12}$$

There are n corresponding eigenvectors $\mathbf{x}^{(i)}$. The matrix equation, eqn 12, is equivalent to a set of n simultaneous equations, and they have a solution only if the determinant of the coefficients is zero. However, this determinant is just $|\mathbf{a} - \lambda\mathbf{1}|$, and so the n eigenvalues may be found from the solution of the **secular equation**

$$|\mathbf{a} - \lambda\mathbf{1}| = 0 \tag{23.13}$$

The n eigenvalues the secular equation yields may be used to find the n eigenvectors. These eigenvectors (which are $n \times 1$ matrices) may be used to form an $n \times n$ matrix $\mathbf{X}$. Thus, since

$$\mathbf{x}^{(1)} = \begin{bmatrix} x_1^{(1)} \\ x_2^{(1)} \\ \vdots \\ x_n^{(1)} \end{bmatrix} \qquad \mathbf{x}^{(2)} = \begin{bmatrix} x_1^{(2)} \\ x_2^{(2)} \\ \vdots \\ x_n^{(2)} \end{bmatrix} \qquad \text{etc.}$$

we may form the matrix

$$\mathbf{X} = (\mathbf{x}^{(1)}, \mathbf{x}^{(2)}, \ldots, \mathbf{x}^{(n)}) = \begin{pmatrix} x_1^{(1)} & x_1^{(2)} & \ldots & x_1^{(n)} \\ x_2^{(1)} & x_2^{(2)} & \ldots & x_2^{(n)} \\ \vdots & \vdots & & \vdots \\ x_n^{(1)} & x_n^{(2)} & \ldots & x_n^{(n)} \end{pmatrix}$$

so that $X_{rc} = x_r^{(c)}$. If further we write $\Lambda_{rc} = \lambda_r \delta_{rc}$, so that $\mathbf{\Lambda}$ is a diagonal matrix with the elements $\lambda_1, \lambda_2, \ldots, \lambda_n$ along the diagonal, then all the eigenvalue equations $\mathbf{a}\mathbf{x}^{(i)} = \lambda_i \mathbf{x}^{(i)}$ may be confined into the single equation

$$\mathbf{aX} = \mathbf{X\Lambda} \tag{23.14}$$

because this expression is equal to

$$\sum_n a_{rn} X_{nc} = \sum_n X_{rn} \Lambda_{nc}$$

or

$$\sum_n a_{rn} x_n^{(c)} = \sum_n x_r^{(n)} \lambda_n \delta_{nc} = \lambda_c x_r^{(c)}$$

as required. Therefore, if we form $\mathbf{X}^{-1}$ from $\mathbf{X}$, we construct a **similarity transformation**

$$\mathbf{\Lambda} = \mathbf{X}^{-1}\mathbf{aX} \tag{23.15}$$

that makes $\mathbf{a}$ diagonal (because $\mathbf{\Lambda}$ is diagonal). It follows that if the matrix $\mathbf{X}$ that causes $\mathbf{X}^{-1}\mathbf{aX}$ to be diagonal is known, then the problem is solved: the diagonal matrix so produced has the eigenvalues as its only nonzero elements, and the matrix $\mathbf{X}$ used to bring about the transformation has the corresponding eigenvectors as its columns.

Appendix 1 Character tables and direct products

Character tables

C_{2v}, $2mm$	E	C_2	σ_v	σ'_v	$h = 4$	
A_1	1	1	1	1	z, z^2, x^2, y^2	
A_2	1	1	−1	−1	xy	R_z
B_1	1	−1	1	−1	x, xz	R_y
B_2	1	−1	−1	1	y, yz	R_x

C_{3v}, $3m$	E	$2C_3$	$3\sigma_v$	$h = 6$	
A_1	1	1	1	z, z^2, $x^2 + y^2$	
A_2	1	1	−1		R_z
E	2	−1	0	(x, y), $(xy, x^2 - y^2)(xz, yz)$	(R_x, R_y)

C_{4v}, $4mm$	E	C_2	$2C_4$	$2\sigma_v$	$2\sigma_d$	$h = 8$	
A_1	1	1	1	1	1	z, z^2, $x^2 + y^2$	
A_2	1	1	1	−1	−1		R_z
B_1	1	1	−1	1	−1	$x^2 - y^2$	
B_2	1	1	−1	−1	1	xy	
E	2	−2	0	0	0	(x, y), (xz, yz)	(R_x, R_y)

C_{5v}	E	$2C_5$	$2C_5^2$	$5\sigma_v$	$h = 10$, $\alpha = 72°$	
A_1	1	1	1	1	z, z^2, $x^2 + y^2$	
A_2	1	1	1	−1		R_z
E_1	2	$2\cos\alpha$	$2\cos 2\alpha$	0	(x, y), (xz, yz)	(R_x, R_y)
E_2	2	$2\cos 2\alpha$	$2\cos\alpha$	0	$(xy, x^2 - y^2)$	

$C_{6v}, 6mm$	E	C_2	$2C_3$	$2C_6$	$3\sigma_d$	$3\sigma_v$	$h = 12$	
A_1	1	1	1	1	1	1	z, z^2, x^2+y^2	
A_2	1	1	1	1	−1	−1		R_z
B_1	1	−1	1	−1	−1	1		
B_2	1	−1	1	−1	1	−1		
E_1	2	−2	1	1	0	0	$(x, y), (xz, yz)$	(R_x, R_y)
E_2	2	2	−1	−1	0	0	$(xy, x^2 - y^2)$	

$C_{\infty v}$	E	$2C_\phi$ †	$\infty\sigma_v$	$h = \infty$	
$A_1(\Sigma^+)$	1	1	1	$z, z^2, x^2 + y^2$	
$A_2(\Sigma^-)$	1	1	−1		R_z
$E_1(\Pi)$	2	$2\cos\phi$	0	$(x, y), (xz, yz)$	(R_x, R_y)
$E_2(\Delta)$	2	$2\cos 2\phi$	0	$(xy, x^2 - y^2)$	
⋮					

† There is only one member of this class if $\phi = \pi$.

$D_2, 222$	E	C_2^z	C_2^y	C_2^x	$h = 4$	
A_1	1	1	1	1	x^2, y^2, z^2	
B_1	1	1	−1	−1	z, xy	R_z
B_2	1	−1	1	−1	y, xz	R_y
B_3	1	−1	−1	1	x, yz	R_x

$D_3, 32$	E	$2C_3$	$3C_2'$	$h = 6$	
A_1	1	1	1	$z^2, x^2 + y^2$	
A_2	1	1	−1	z	R_z
E	2	−1	0	$(x, y), (xz, yz)(xy, x^2 - y)$	(R_x, R_y)

D_{4h}, 422	E	C_2	$2C_4$	$2C_2'$	$2C_2''$	$h=8$	
A_1	1	1	1	1	1	z^2, x^2+y^2	
A_2	1	1	1	−1	−1	z	R_z
B_1	1	1	−1	1	−1	x^2-y^2	
B_2	1	1	−1	−1	1	xy	
E	2	−2	0	0	0	$(x,y), (xz,yz)$	(R_x, R_y)

D_{3h}, $\bar{6}2m$	E	σ_h	$2C_3$	$2S_3$	$3C_2'$	$3\sigma_v$	$h=12$	
A_1'	1	1	1	1	1	1	$z^2,\ x^2+y^2$	
A_2'	1	1	1	1	−1	−1		R_z
A_1''	1	−1	1	−1	1	−1		
A_2''	1	−1	1	−1	−1	1	z	
E'	2	2	−1	−1	0	0	$(x,\ y),\ (xy,\ x^2-y^2)$	
E''	2	−2	−1	1	0	0	$(xz,\ yz)$	$(R_x,\ R_y)$

$D_{\infty h}$	E	$2C_\phi$	$\infty C_2'$	i	$2iC_\phi$	iC_2'	$h=\infty$	
$A_{1g}(\Sigma_g^+)$	1	1	1	1	1	1	$z^2,\ x^2+y^2$	
$A_{1u}(\Sigma_u^+)$	1	1	1	−1	−1	−1	z	
$A_{2g}(\Sigma_g^-)$	1	1	−1	1	1	−1		R_z
$A_{2u}(\Sigma_u^-)$	1	1	−1	−1	1	1		
$E_{1g}(\Pi_g)$	2	$2\cos\phi$	0	0	$-2\cos\phi$	0	$(xz,\ yz)$	$(R_x,\ R_y)$
$E_{1u}(\Pi_u)$	2	$2\cos\phi$	0	0	$2\cos\phi$	0	$(x,\ y)$	
$E_{2g}(\Delta_g)$	2	$2\cos 2\phi$	0	0	$2\cos 2\phi$	0	$(xy,\ x^2-y^2)$	
$E_{2u}(\Delta_u)$	2	$2\cos 2\phi$	0	0	$-2\cos 2\phi$	0		
⋮								

T_d, $\bar{4}3m$	E	$8C_3$	$3C_2$	$6\sigma_d$	$6S_4$	$h=24$	
A_1	1	1	1	1	1	$x^2+y^2+z^2$	
A_2	1	1	1	−1	−1		
E	2	−1	2	0	0	$(3z^2-r^2,\ x^2-y^2)$	
T_1	3	0	−1	−1	1		$(R_x,\ R_y,\ R_z)$
T_2	3	0	−1	1	−1	$(x,\ y,\ z),\ (xy,\ xz,\ yz)$	

O, 432	E	$8C_3$	$3C_2$	$6C_2'$	$6C_4$	$h = 24$	
A_1	1	1	1	1	1	$x^2 + y^2 + z^2$	
A_2	1	1	1	−1	−1		
E	2	−1	2	0	0	$(x^2 - y^2, 3z^2 - r^2)$	
T_1	3	0	−1	−1	1	(x, y, z)	(R_x, R_y, R_z)
T_2	3	0	−1	1	−1	(xy, yz, zx)	

Direct products

In general g × g = g, g × u = u, u × u = g;

$$\Gamma' \times \Gamma' = \Gamma', \qquad \Gamma' \times \Gamma'' = \Gamma'', \qquad \Gamma'' \times \Gamma'' = \Gamma'$$

For C_2, C_{2v}, C_{2h}; C_3, C_{3v}, C_{3h}; D_3, D_{3h}, D_{3d}; C_6, C_{6v}, C_{6h}, D_6, S_6

	A_1	A_2	B_1	B_2	E_1	E_2
A_1	A_1	A_2	B_1	B_2	E_1	E_2
A_2		A_1	B_2	B_1	E_1	E_2
B_1			A_1	A_2	E_2	E_1
B_2				A_1	E_2	E_1
E_1					$A_1 + [A_2] + E_2$	$B_1 + B_2 + E_1$
E_2						$A_1 + [A_2] + E_2$

For T, T_h, T_d; O, O_h:

	A_1	A_2	E	T_1	T_2
A_1	A_1	A_2	E	T_1	T_2
A_2		A_1	E	T_2	T_1
E			$A_1 + [A_2] + E$	$T_1 + T_2$	$T_1 + T_2$
T_1				$A_1 + E + [T_1] + T_2$	$A_2 + E + T_1 + T_2$
T_2					$A_1 + E + [T_1] + T_2$

For $C_{\infty v}$, $D_{\infty h}$:

	Σ^+	Σ^-	Π	Δ	…
Σ^+	Σ^+	Σ^-	Π	Δ	…
Σ^-		Σ^+	Π	Δ	…
Π			$\Sigma^+ + [\Sigma^-] + \Delta$	$\Pi + \Phi$	…
Δ				$\Sigma^+ + [\Sigma^-] + \Gamma$	…
⋮					…

Appendix 2 Vector coupling coefficients

1. $j_1 = j_2 = \frac{1}{2}$. $|jm_j\rangle$

m_{j1}	m_{j2}	$\|1, 1\rangle$	$\|1, 0\rangle$	$\|0, 0\rangle$	$\|1, -1\rangle$
$\frac{1}{2}$	$\frac{1}{2}$	1			
$\frac{1}{2}$	$-\frac{1}{2}$		$\sqrt{\frac{1}{2}}$	$\sqrt{\frac{1}{2}}$	
$-\frac{1}{2}$	$\frac{1}{2}$		$\sqrt{\frac{1}{2}}$	$-\sqrt{\frac{1}{2}}$	
$-\frac{1}{2}$	$-\frac{1}{2}$				1

2. $j_1 = 1,\ j_2 = \frac{1}{2}$. $|jm_j\rangle$

m_{j1}	m_{j2}	$\|\frac{3}{2}, \frac{3}{2}\rangle$	$\|\frac{3}{2}, \frac{1}{2}\rangle$	$\|\frac{1}{2}, \frac{1}{2}\rangle$	$\|\frac{3}{2}, -\frac{1}{2}\rangle$	$\|\frac{1}{2}, -\frac{1}{2}\rangle$	$\|\frac{3}{2}, -\frac{3}{2}\rangle$
1	$\frac{1}{2}$	1					
1	$-\frac{1}{2}$		$\sqrt{\frac{1}{3}}$	$\sqrt{\frac{2}{3}}$			
0	$\frac{1}{2}$		$\sqrt{\frac{2}{3}}$	$-\sqrt{\frac{1}{3}}$			
0	$-\frac{1}{2}$				$\sqrt{\frac{2}{3}}$	$\sqrt{\frac{1}{3}}$	
-1	$\frac{1}{2}$				$\sqrt{\frac{1}{3}}$	$-\sqrt{\frac{2}{3}}$	
-1	$-\frac{1}{2}$						1

3. $j_1 = 1\ j_2 = 1$. $|jm_j\rangle$

m_{j1}	m_{j2}	$\|2,2\rangle$	$\|2,1\rangle$	$\|1,1\rangle$	$\|2,0\rangle$	$\|1,0\rangle$	$\|0,0\rangle$	$\|2,-1\rangle$	$\|1,-1\rangle$	$\|2,-2\rangle$
1	1	1								
1	0		$\sqrt{\frac{1}{2}}$	$\sqrt{\frac{1}{2}}$						
0	1		$\sqrt{\frac{1}{2}}$	$-\sqrt{\frac{1}{2}}$						
1	-1				$\sqrt{\frac{1}{6}}$	$\sqrt{\frac{1}{2}}$	$\sqrt{\frac{1}{3}}$			
0	0				$\sqrt{\frac{2}{3}}$	0	$-\sqrt{\frac{1}{3}}$			
-1	1				$\sqrt{\frac{1}{6}}$	$-\sqrt{\frac{1}{2}}$	$\sqrt{\frac{1}{3}}$			
-1	0							$\sqrt{\frac{1}{2}}$	$-\sqrt{\frac{1}{2}}$	
0	-1							$\sqrt{\frac{1}{2}}$	$\sqrt{\frac{1}{2}}$	
-1	-1									1

Answers to selected problems

0.1 (a) 6.626×10^{-19} J, (b) 6.626×10^{-20} J, (c) 6.626×10^{-34} J. **0.4** 6000 K. **0.6** (a) $(\theta_E/T)^2 e^{-\theta_E/T}$, (b) $1 + (\theta_E/T)$. **0.8** $2.93R$, $0.24R$. **0.9** $3.144R$. **0.10** 2.97×10^{20}. **0.11** (a) 8.0×10^5 m s^{-1}, (b) no electrons emitted. **0.14** $\mathcal{R}_H = 1.097 \times 10^5$ cm^{-1}, $I = 13.6$ eV $= hc\mathcal{R}_H$. **0.15** (a) 6.6×10^{-29} m, (b) 7.3×10^{-40} m, (c) 0.145 nm, (d)(i) 1.23 nm, (ii) 12.3 pm.

1.3 (a) Eigenfunction is (a); (b) eigenfunctions are (a), (c), (e), (f). **1.4** (a) $-(\hbar^2/2m)(d^2/dx^2)$ in one dimension, $-(\hbar^2/2m)\nabla^2$ in three dimensions; (b) multiplication by $(1/x)$; (c) multiplication by $\sum_i e_i \mathbf{r}_i$; (d) $(\hbar/i)\{x(\partial/\partial y) - y(\partial/\partial x)\}$; (e) multiplication by $x^2 - \langle x\rangle^2$, $-\hbar^2(\partial^2/\partial x^2) - \langle p\rangle^2$. **1.6** $-\hbar^2 c^2(\partial^2\Psi/\partial x^2) + m^2c^4\Psi = -\hbar^2(\partial^2\Psi/\partial t^2)$, probability is not conserved. **1.10** No. **1.11** (a) 0, (b) 0, (c) $i\hbar$, (d) $2i\hbar x$, (e) $ni\hbar x^{n-1}$. **1.12** (a) $\hbar/(ix^2)$, (b) $(2\hbar/x^3)(\hbar - ixp_x)$, (c) $i\hbar(zp_x - xp_z)$, (d) $2x^2(\partial^2/\partial x\partial y) - 2xy(\partial^2/\partial y^2)$. **1.14** $\hbar^2 l_z$. **1.17** (a) $i\hbar(\partial V/\partial x)$, (b) $(\hbar/im)p_x$; for (i) (a) 0, (b) $(\hbar/im)p_x$; for (ii) (a) $i\hbar kx$, (b) $(\hbar/im)p_x$; for (iii) (a) $-(i\hbar e^2/4\pi\varepsilon_0)(x/r^3)$, (b) $(\hbar/im)p_x$. **1.19** $(d/dt)\langle x\rangle = (1/m)\langle p\rangle$, $(d/dt)\langle p\rangle = -k\langle x\rangle$. **1.21** Eigenvalues $(v + \frac{1}{2})\hbar\omega$. **1.22** $(\hbar/2)^2(2v^2 + 2v + 1)$. **1.25** $N = (a^3/\pi)^{1/2}$. **1.26** $N = (1/\Gamma\sqrt{\pi})^{1/2}$, 0.8427.... **1.27** $\pm\Gamma$. **1.28** (a) 2.1×10^{-6} pm^{-3}, (b) 2.9×10^{-7} pm^{-3}; 2.1×10^{-6}, 2.9×10^{-7}. **1.29** (a) 0.459, (b) 135 pm.

2.1 (a) (i) $A\ e^{\{5.123i(x/\text{nm})\}}$, (ii) $A\ e^{\{512.3i(x/\text{nm})\}}$; (b) $A \exp\{9.48\times 10^{31} i(x/\text{m})\}$. **2.2** $A^2 = 1/L; L \to \infty$. **2.8** $4\gamma/\{4\gamma^2 + (1-\gamma^2)\sin^2 k'L\}$ where $\gamma = k/k'$ with $k^2 = 2mE/\hbar^2$ and $k'^2 = 2m(E-V)/\hbar^2$. **2.9** (a) 1, (b) $\{(\sqrt{E-V} - \sqrt{E})/(\sqrt{E-V} + \sqrt{E})\}^2$. **2.10** (a) 1/2 for all n; (b) $(1/4)\{1 - (2/\pi n)\sin(n\pi/2)\}$, 0.09085 for $n = 1$; (c) $(2/L)\{\delta - (-1)^n(L/2\pi n)\sin(2n\pi\delta/L)\}$, $(2/L)\{\delta + (L/2\pi)\sin(2\pi\delta/L)\}$ for $n = 1$. **2.11** $\lambda_C/\sqrt{8}$. **2.12** (a) $n^2h^2/4mL^3$, (b) 0.49 pm. **2.14** $\Delta x = (L/2\sqrt{3})\{1 - (6/n^2\pi^2)\}^{1/2}$; as $n \to \infty, \Delta x \to L/2\sqrt{3}$. **2.15** $\langle p\rangle = 0$, $\langle p^2\rangle = n^2h^2/4L^2$, $\Delta p = nh/2L$. For general n, we have $\Delta x\Delta p = (n\pi/\sqrt{3})\{1 - (6/n^2\pi^2)\}^{1/2}(\hbar/2)$; For $n = 1$, we get $\Delta x\Delta p = 1.1357(\hbar/2)$. **2.19** $E_n = n^2h^2/8m_eL^2$, $\lambda/\text{nm} = 3.297 \times 10^{-3}(R_{CC}/\text{pm})^2(N-1)^2/(N+1)$. **2.20** (b) $\Psi = (2/L)^{3/2}\sin(n_x\pi x/L)\sin(n_y\pi y/L)\sin(n_z\pi z/L)$, $E = (n_x^2 + n_y^2 + n_z^2)(h^2/8mL^2)$, (d) 6. **2.27** 4.57×10^{-20} J, 4.35×10^{-6} m. **2.28** (a) 0.171, (b) 0.620. **2.29** (a) 0, (b) $(1/2)\hbar\omega/k$, (c) 0, (d) $(1/2)\hbar k/\omega$, $\Delta x\Delta p = \hbar/2$.

3.1 $E = (1.30 \times 10^{-22}\,\text{J})m_l^2$, 1.53 mm. **3.5** $E = (2.2 \times 10^{-65}\,\text{J})m_l^2$, -1.5×10^{33}. **3.7** (a) $N = 1/(2\pi I_0(2))^{1/2} = 0.2642$; (b) $0, 0, 0.698\hbar$. **3.8** $N = 1/(2\pi I_0(2\alpha))^{1/2}$, $\langle l_z\rangle = \alpha\hbar\{I_1(2\alpha)/I_0(2\alpha)\}$. **3.9** $\Theta''\sin^2\theta + \Theta'\sin\theta\cos\theta - \{m_l^2 - (2IE/\hbar^2)\}\,\Theta = 0$. **3.14** $l = 0, E = 0$, nondegenerate; $l = 1, E = 2.60 \times 10^{-22}$ J, triply degenerate; $l = 2, E = 7.80 \times 10^{-22}$J, fivefold degenerate; 0.764 nm. **3.15** $\arccos m_l/\{l(l+1)\}^{1/2}$; With angles in degrees: for $l = 1$, 45, 90, 135; for

$l = 2$, 35.3, 65.9, 90, 114.1, 144.7; for $l = 3$, 30, 54.7, 73.2, 90, 106.8, 125.3, 150. **3.17** $\langle T \rangle = -E_{1s}$, $\langle V \rangle = 2E_{1s}$, $\langle T \rangle = (-1/2)\langle V \rangle$. **3.19** (a) $2a$, (b) $(3 \pm \sqrt{3})(3a/2)$. **3.20** For 1s, (a) $3a/2Z$, (b) $3(a/Z)^2$, (c) a/Z; for 2s, (a) $6a/Z$, (b) $42(a/Z)^2$, (c) $5.24a/Z$; for 3s, (a) $27a/2Z$, (b) $207(a/Z)^2$, (c) $13.07a/Z$. **3.24** For 1s, $(1/\pi)(Z/a)^3$; for 2s, $(1/8\pi)(Z/a)^3$; for 3s, $(1/27\pi)(Z/a)^3$. **3.25** $(1/24)(Z/a)^3$. **3.26** $-0.357\,\text{kJ mol}^{-1}$.

4.1 $i\hbar l_z$. **4.2** (a) $-i\hbar(l_z l_y + l_y l_z)$, (b) $-i\hbar(l_x l_z l_y + l_x l_y l_z + l_z l_y l_x + l_y l_z l_x)$, (c) $\hbar^2 l_y$. **4.3** Upon expansion of the determinant, $\mathbf{l} \times \mathbf{l} = \hat{\mathbf{i}}(l_y l_z - l_z l_y) - \hat{\mathbf{j}}(l_x l_z - l_z l_x) + \hat{\mathbf{k}}(l_x l_y - l_y l_x)$, which is then identified (term by term) with $i\hbar\mathbf{l}$. **4.4** (a) $[s_x, s_y] = i\hbar s_z$, (b) eigenvalues of $s^2 = s_x^2 + s_y^2 + s_z^2$ are $\frac{3}{4}\hbar^2 = s(s+1)\hbar^2$. **4.5** (a) $i\hbar s_z/2$, (b) $(\hbar/2)^4 s_x$, (c) $(\hbar/2)^6$. **4.7** l_+ would be a lowering operator and l_- a raising operator. **4.8** (a) 0, (b) $\hbar\sqrt{6}$, (c) $2\hbar^2\sqrt{6}$, (d) $6\hbar^2$, (e) $6\hbar^2$, (f) $48\hbar^5$. **4.10** (a) $-i\hbar$, (b) 0, (c) $-i\hbar$, (d) $i\hbar$, (e) 0. **4.16** (a) 7, 6, ..., 1; (b)(i) 2, 1, 0, (ii) 4, 3, 2, 1, 0, (iii) 3, 2, 1; (c) 2, 1, 1, 1, 0, 0. **4.22** $\langle \text{G}, M_L | l_{1z} | \text{G}, M_L \rangle = M_L\hbar/2$.

5.1 (a) C_{2v}, (b) $D_{\infty h}$, (c) D_{2h}, (d) C_{2v}, (e) C_{2h}, (f) D_{6h}, (g) D_{2h}, (h) C_1, (i) C_{3h}. **5.2** (a), (d), and (h). **5.5** $3A_1 + B_1 + 2B_2$; for A_1: $\frac{1}{2}(\text{H}1s_A + \text{H}1s_B)$, O2$s$, O2$p_z$; for B_1: O2p_x; for B_2: $\frac{1}{2}(\text{H}1s_A - \text{H}1s_B)$, O2$p_y$. **5.8** $A_1 + T_2$; for A_1: $\text{H}1s_A + \text{H}1s_B + \text{H}1s_C + \text{H}1s_D$; for T_2: $\text{H}1s_A - \text{H}1s_B - \text{H}1s_C + \text{H}1s_D$, $\text{H}1s_A + \text{H}1s_B - \text{H}1s_C - \text{H}1s_D$, $\text{H}1s_A - \text{H}1s_B + \text{H}1s_C - \text{H}1s_D$. **5.9** (a) A_1, (b) E, (c) E_2, (d) $A_1 + A_2 + E_2$, (e) $A_1 + A_2 + 2E + 2T_1 + 2T_2$. **5.11** $A_1 + A_2 + B_1 + B_2$. **5.13** $3A_1 + 2A_2 + 2B_1 + 3B_2$. **5.14** $A_1 + B_1 + E_1 + E_2$; in D_{6h}, it is $A_{2u} + B_{2g} + E_{1g} + E_{2u}$. **5.16** (a) ${}^1A_2, {}^3A_2$; (b)(i) ${}^1E, {}^3E$, (ii) ${}^1A_1, {}^3A_2, {}^1E$; (c)(i) ${}^1E, {}^3E$, (ii) ${}^1T_1, {}^3T_1, {}^1T_2, {}^3T_2$, (iii) ${}^1A_2, {}^3A_2, {}^1E, {}^3E, {}^1T_1, {}^3T_1, {}^1T_2, {}^3T_2$, (iv) ${}^1A_1, {}^1E, {}^1T_2, {}^3T_1$, (v) ${}^1A_1, {}^1E, {}^1T_2, {}^3T_1$; (d)(i) ${}^1A_1, {}^3A_2, {}^1E$, (ii) ${}^1T_1, {}^3T_1, {}^1T_2, {}^3T_2$, (iii) ${}^1A_1, {}^1E, {}^3T_1, {}^1T_2$. **5.20** 3 (can be increased by accidental degeneracies). **5.22** $A_1 + E$.

6.1 (a) 25 739.45 cm^{-1} (99.998% Ψ_1), 50 267.29 cm^{-1} (99.998% Ψ_2); (b) 25 699.16 cm^{-1} (99.835% Ψ_1), 50 307.58 cm^{-1} (99.835% Ψ_2); (c) 25 759.74 cm^{-1} (84.278% Ψ_1), 51 246.99 cm^{-1} (84.278% Ψ_2). **6.2** (a) -74.8 eV, (b) 20.4 eV. **6.4** $E^{(1)} = mgL/2$, $E^{(1)}/L = 4.47 \times 10^{-30}\,\text{J m}^{-1}$. **6.5** With $a = mgL/(h^2/8mL^2)$, we have $E^{(2)} = -0.1013amgL$, $\Psi^{(1)} = a\{0.1801\Psi_2 + 0.0144\Psi_4 + 0.0040\Psi_6 + \ldots\}$. **6.7** (a) d_{xz}, (b) d_{z^2}, (c) f_{xyz}. **6.8** (a) B_1, (b) B_2. **6.10** $E^{(2)} = -0.02949\varepsilon^2/\Delta E$; $\Delta E = 8.12(h^2/8mL^2)$. **6.12** (a) $k = \pi/L, E = h^2/8mL^2$; (b) $kL = 0.5147, E = h^2/6.0489mL^2$; (c) trial function inadmissible since first derivative is discontinuous. **6.13** $\frac{1}{2}(s_A - \sqrt{2}s_B + s_C)$, $E = \alpha - \beta\sqrt{2}$; $\frac{1}{\sqrt{2}}(s_A - s_C)$, $E = \alpha$; $\frac{1}{2}(s_A + \sqrt{2}s_B + s_C)$, $E = \alpha + \beta\sqrt{2}$. **6.15** 0 at all times. **6.16** $P(t) \approx \sin^2(\mu_B Bt/2000\hbar)$, 36 ns. **6.19** 0. **6.20** $A = (2^9/3^7)(\pi\alpha^5 c/\lambda_C)Z^4$, $\rho B = (2^{12}/3^7)(\pi\alpha^5 c/\lambda_C)Z^4\exp(-3hc\mathcal{R}Z^2/4kT)$. **6.21** $A \propto 1/L^4$, $B \propto L^2$.

7.1 $(4.3889 \times 10^5\,\text{cm}^{-1})(1/4 - 1/n^2)$. **7.2** $(1.092 \times 10^5\,\text{cm}^{-1})(1/n_1^2 - 1/n_2^2)$. **7.3** (b), (c), and (e). **7.4** B^{4+}, $3.283 \times 10^4\,\text{kJ mol}^{-1}$. **7.8** $E_{so}(j) - E_{so}(j-1) = jhc\zeta_{nl}$. **7.9** $\zeta_{3d,\text{mean}} = 95.5\,\text{cm}^{-1}$. **7.11** Li ${}^2S_{1/2}$; Be ${}^1S_0$; B $E({}^2P_{1/2}) < E({}^2P_{3/2})$; C $E({}^3P_0) < E({}^3P_1) < E({}^3P_2) < E({}^1D_2) < E({}^1S_0)$; N $E({}^4S) < E({}^2D) < E({}^2P)$; O $E({}^3P_2) < E({}^3P_1) < E({}^3P_0) < E({}^1D_2) < E({}^1S_0)$; F $E({}^2P_{3/2}) < E({}^2P_{1/2})$. **7.12** For 1$s$, $E^{(1)}/hc = 7.305\,\text{cm}^{-1}$. **7.15** $-(3^6/2^7)hc\mathcal{R}$; $\zeta = 1.69a_0$; 23.1 eV and 54.4 eV. **7.18** 3211 and 814 cm^{-1}, respectively. **7.19** $^1S_0$; $^2P_{3/2,1/2}$; $^3P_{2,1,0}$; $^3D_{3,2,1}$; $^2D_{5/2,3/2}$; $^1D_2$; $^4D_{7/2,5/2,3/2,1/2}$; $^3P_0 < {}^3P_1 < {}^3P_2 < {}^1P_1$; $^3D_1 < {}^3D_2 < {}^3D_3 < {}^3P_0 < {}^3P_1 < {}^3P_2 < {}^3S_1 < {}^1D_2 < {}^1P_1 < {}^1S_0$; $^3F_2 < {}^3F_3 < {}^3F_4 < {}^3D_1 < {}^3D_2 < {}^3D_3 < {}^3P_0 < {}^3P_1 < {}^3P_2 < {}^1F_3 < {}^1D_2 < {}^1P_1$. (a) $^1G, {}^3F, {}^1D, {}^3P, {}^1S$; (b) ^1I, ^3H, ^1G, ^3F, ^1D, ^3P, ^1S. **7.21** 2.14 T; (a) $1 + S/(L+S)$, (b) $1 - S/(L - S + 1)$.

8.1 130 pm, 170 kJ mol^{-1}. **8.3** $\rho_+ = (1.462 \times 10^{-6}\,\text{pm}^{-3})\{e^{-(r_a+r_b)/a} - 0.235(e^{-2r_a/a} + e^{-2r_b/a})\}$; $\rho_- = -2.767\rho_+$. **8.5** $2E_{1s} + (j_0/R) + \frac{1}{2}(j + 2k + 4l + m)/(1+S)^2 - 2(j' + k')/(1+S)$. **8.9** (a) $^1\Sigma_g^+$, (b) $^2\Pi_u$, (c) $^2\Sigma_g^+$, (d) $^2\Sigma_g^+$, (e) $^2\Pi_g$, (f) $^2\Pi_g$, (g) $^2\Sigma_u^+$. **8.10** (a) $^1\Sigma$, (b) $^2\Pi$. **8.11** Let $\Delta^2 = (\alpha_O - \alpha_H)^2 + 4\beta^2$; for A_1: $E_\pm = \frac{1}{2}(\alpha_O + \alpha_H) \pm \frac{1}{2}\Delta$; for B_1: $E = \alpha_O$; for B_2: same as for A_1. **8.17** For A_2: $E = \alpha_C \pm \beta$; for B_2: $E = \alpha_C + 1.9337\beta, \alpha_C + 0.8410\beta, \alpha_C - 1.1672\beta, \alpha_C - 2.1074\beta$; π-electron energy is $6\alpha_C + 7.5494\beta$. **8.18** $\alpha, \alpha, (\alpha \pm \beta)/(1 \pm 2S)$. **8.19** For six equivalent bond lengths, delocalization energy is $-0.26\alpha + 1.1\beta$; for alternating bond lengths, delocalization energy is $-0.26\alpha - 0.31\beta$. **8.20** $E(d_{z^2}, d_{x^2-y^2})$ and $T_2(d_{xy}, d_{xz}, d_{yz})$. **8.21** Free ion $\rightarrow$ complex: $^1I \rightarrow {}^1A_1 + {}^1A_2 + {}^1E + {}^1T_1 + 2{}^1T_2$: $^3H \rightarrow {}^3E + 2{}^3T_1 + {}^3T_2$: $^1G \rightarrow {}^1A_1 + {}^1E + {}^1T_1 + {}^1T_2$: $^3F \rightarrow {}^3A_2 + {}^3T_1 + {}^3T_2$: $^1D \rightarrow {}^1E + {}^1T_2$: $^3P \rightarrow {}^3T_1$: $^1S \rightarrow {}^1A_1$. **8.22** (a) For e_g^2 : $^1A_{1g}, {}^3A_{2g}, {}^1E_g$; for $t_{2g}^1e_g^1$: $^1T_{1g}, {}^3T_{1g}, {}^1T_{2g}, {}^3T_{2g}$; for t_{2g}^2 : $^1A_{1g}, {}^1E_g, {}^3T_{1g}, {}^1T_{2g}$; (b) $^1G \rightarrow {}^1A_{1g} + {}^1E_g + {}^1T_{1g} + {}^1T_{2g}$: $^1D \rightarrow {}^1E_g + {}^1T_{2g}$: $^1S \rightarrow {}^1A_{1g}$; (c) $^1G \rightarrow {}^1A + {}^1E + 2{}^1T$: $^3F \rightarrow {}^3A + 2{}^3T$: $^1D \rightarrow {}^1E + {}^1T$: $^3P \rightarrow {}^3T$: $^1S \rightarrow {}^1A$.

9.1 $E_a^0 + E_b^0 + E_c^0 + \ldots + E_z^0$. **9.3** Explicitly expanding the 3×3 determinant, one obtains a value of identically zero. **9.6** 23. **9.7** $(ab|cd) = (ba|cd) = (ab|dc) = (ba|dc) = (cd|ab) = (cd|ba) = (dc|ab) = (dc|ba)$. **9.9** (i) 3H1*s*, C1*s*, C2*s*, 3C2*p*, Cl1*s*, Cl2*s*, 3Cl2*p*, Cl3*s*, 3Cl3*p*; 17 functions; (ii) 6H1*s*, C1*s*, 2C2*s*, 6C2*p*, Cl1*s*, Cl2*s*, 3Cl2*p*, 2Cl3*s*, 6Cl3*p*; 28 functions; (iii) 6H1*s*, 9H2*p*, 2C1*s*, 2C2*s*, 6C2*p*, 6C3*d*, 2Cl1*s*, 2Cl2*s*, 6Cl2*p*, 2Cl3*s*, 6Cl3*p*, 6Cl3*d*; 55 functions. **9.11** (i) 39 basis functions, 90 primitives; (ii) 57 basis functions, 108 primitives; (iii) 75 basis functions, 126 primitives. **9.12** 9.41×10^{28}. **9.14** (a), (b), (d), and (e). **9.15** (a), (d), and (e). **9.17** Use the result of Exercise 9.18. **9.20** Inactive: $1\sigma_g, 1\sigma_u$; active: $2\sigma_g, 2\sigma_u$, $1\pi_u, 3\sigma_g, 3\sigma_u, 1\pi_g$; virtual: $4\sigma_g, 4\sigma_u, \ldots$; 4 inactive and 8 active electrons. **9.23** (a) Includes (iv) and (vi); (b) includes (i), (iii), and (v). **9.24** Need second derivatives of one- and two-electron integrals, second derivatives of non-variationally determined c_{ji} (eqn 9.14), and first derivatives of variationally determined C_{Js} (eqn 9.32). **9.27** (i) (a) and (f); (ii) (a), (b), and (f); (iii) (a), (b), (d), (e), and (f).

10.2 See Table 10.1. **10.3** See Table 10.1 **10.6** $E = hcB\{J(J+1) + K^2\}$ as $A = 2B$. **10.7** $\mu(t) = \alpha\mathcal{E}_0 \cos\omega t + \frac{1}{4}\beta\mathcal{E}_0^2(1 + \cos 2\omega t)$. **10.8** (a) 4.718×10^{-48} kg m^2, (b) 9.429×10^{-48} kg m^2, (c) 2.644×10^{-47} kg m^2. **10.9** 162 pm, 6.45 cm^{-1}. **10.10** $^{16}O^{12}C^{32}S$: 1.37996×10^{-45} kg m^2; $^{16}O^{12}C^{34}S$: 1.41450×10^{-45} kg m^2; $R_{CO} = 116.3$ pm, $R_{CS} = 160.0$ pm. **10.12** (a) Most intense transition would be $4 \leftarrow 3$, (b) most intense transition is $3 \leftarrow 2$. **10.15** Effective masses (in atomic mass units, u); force constants (in N m^{-1}); and wavenumbers (cm^{-1}) of deuterated compounds: (a) 0.5039; 574.9; 3811; (b) 0.9570; 965.7; 3000; (c) 0.9796; 516.3; 2145;

(d) 0.9954; 411.5; 1885; (e) 0.9999; 313.8; 1639. **10.16** $\mu_{v-1,v} = -e(\hbar/2m\omega)^{1/2}v^{1/2}$; $\mu_{v+1,v} = -e(\hbar/2m\omega)^{1/2}(v+1)^{1/2}$. **10.18** $\langle 1/R^2\rangle \approx (1/R_e^2)\{1+(\delta R/2R_e)^2\}$; with $B_e = \hbar/4\pi c\mu R_e^2$ and $\gamma_v = 12\pi c(v+\frac{1}{2})B_e/\omega$, we obtain $B_v \approx (1+\gamma_v)B_e$. **10.19** Mean value $1505\,\mathrm{N\,m^{-1}}$. **10.20** (a) All 3 modes, (b) all 3 modes. **10.21** $3A_g + 2B_{1g} + B_{2g} + A_u + B_{1u} + 2B_{2u} + 2B_{3u}$; B_{1u}, B_{2u}, B_{3u} are infrared active while A_g, B_{1g}, B_{2g} are Raman active.

11.2 (a), (b), (e), and (f). **11.5** In H_2CO: ${}^1A_2 \leftarrow {}^1A_1$ is allowed only if it is vibronic with possible singly excited B_1 and B_2 vibronic states of the upper electronic state. In ethene, ${}^1B_{2u} \leftarrow {}^1A_g$ is allowed for y-polarized radiation, and, in addition, singly excited B_{2u} and B_{3u} vibronic states of the upper electronic state are electric-dipole allowed. **11.6** Both are forbidden; for the first: $T_{1u}, T_{2u}, E_u, A_{1u}$; for the second: $T_{1u}, T_{2u}, E_u, A_{2u}$; for both: p-orbital admixtures can account for intensity. **11.7** $|S_{00}|^2 = \exp\{-(m\omega/2\hbar)\Delta R^2\}$, $|S_{10}|^2 = (m\omega/2\hbar)\Delta R^2\exp\{-(m\omega/2\hbar)\Delta R^2\}$, $|S_{20}|^2 = \frac{1}{8}(m\omega/\hbar)^2\Delta R^4 \exp\{-(m\omega/2\hbar)\Delta R^2\}$; $\sqrt{(2\hbar/m\omega)} < \Delta R < \sqrt{(4\hbar/m\omega)}$. **11.10** ${}^1B_{2u}, {}^1E_{2u}$ into ${}^3B_{1u}$; ${}^1B_{1u}, {}^1E_{2u}$ into ${}^3B_{2u}$. **11.11** $B_{2u} \longrightarrow A_{1g}$ (y-polarized), $B_{3u} \longrightarrow A_{1g}$ (x-polarized).

12.1 3.5×10^{-6} D, $-3.52 \times 10^{-5}\,\mathrm{kJ\,mol^{-1}}$. **12.2** $0.2026e^2L^2/(h^2/8mL^2)$. **12.4** $5.966 \times 10^{-41}\,\mathrm{J^{-1}\,C^2\,m^2}$; $5.36 \times 10^{-25}\,\mathrm{cm^3}$. **12.5** $\alpha_{zz} = 2.44\times 10^{-41}\,\mathrm{J^{-1}\,C^2\,m^2}$. **12.8** $\beta_{xxx} = 0$. **12.10** (a) $1.5 \times 10^{-40}\,\mathrm{J^{-1}\,C^2\,m^2}$, (b) $4.9\times 10^{-41}\,\mathrm{J^{-1}\,C^2\,m^2}$; $1s$ contribution is $6.4 \times 10^{-44}\,\mathrm{J^{-1}\,C^2\,m^2}$; $2s$ contribution is $4.2 \times 10^{-41}\,\mathrm{J^{-1}\,C^2\,m^2}$; total mean polarizibility for configuration $1s^22s2p_x2p_y2p_z$ is $2.9 \times 10^{-40}\,\mathrm{J^{-1}\,C^2\,m^2}$. **12.11** $5 \times 10^{-31}\mathrm{m^3}$. **12.13** $n_r - 1 \approx 1.7 \times 10^{-24}$. **12.14** (a) $(-34.80\,\mathrm{kJ\,mol^{-1}})(L/R)^6$, (b) $(-8.70\,\mathrm{kJ\,mol^{-1}})(L/R)^6$. **12.16** $(-4.29 \times 10^{-4}\,\mathrm{kJ\,mol^{-1}})/(R/\mathrm{nm})^6$. **12.18** With the upper orbital wavefunction denoted $2p_z\cos\zeta + 3d_{xy}\sin\zeta$, one obtains $R = (-2.108\times 10^{-54}\,\mathrm{C^2\,m^3\,s^{-1}})\sin 2\zeta$; $\Delta\theta \approx -0.13\sin 2\zeta$.

13.1 1.57×10^{-5}. **13.3** $\langle S_z^2\rangle = \frac{1}{3}S(S+1)\hbar^2$, $\langle S_xS_z\rangle = 0$, $\langle S_z^4\rangle = \frac{1}{15}\hbar^4\{(S+1)/(2S+1)\}\{6(S+1)^4 - 15(S+1)^3 + 10(S+1)^2 - 1\}$. **13.4** $\langle S_z^2\rangle = \frac{1}{3}S(S+1)\hbar^2$. **13.5** $\nabla\cdot\mathbf{V} = 0$, $\nabla\times\mathbf{V} = -2\mathbf{j}$. **13.7** (a) $\mathbf{A} = -\frac{1}{2}B(-z\mathbf{j} + y\mathbf{k})$, $A^2 = \frac{1}{4}B^2(x^2+y^2)$; (b) $\mathbf{A} = (B/2\sqrt{3})\{(z-y)\hat{\mathbf{i}} - (z-x)\hat{\mathbf{j}} + (y-x)\hat{\mathbf{k}}\}$, $A^2 = \frac{1}{4}B^2\{r^2 - \frac{1}{3}(x+y+z)^2\}$. **13.10** (a) $\chi(\mathrm{H}) = -2.96 \times 10^{-8}$, $\chi(\mathrm{C}) = -7.73\times 10^{-11}$; (b) $\chi(\mathrm{C}) = -2.40 \times 10^{-9}$. **13.12** With $s = r/a_0$, $(-5.738\times 10^{13}\,\mathrm{A\,m^{-2}})s^3\sin^3\theta(-\hat{\mathbf{i}}\sin\phi + \hat{\mathbf{j}}\cos\phi)\mathrm{e}^{-2s/3}$. **13.15** For a field along z, there is no paramagnetic contribution. With $s = r/a_0$, (a) $\mathbf{j}^{\mathrm{d}}(2s) = -(Z^3e^2B/16\pi m_e a_0^2)s(1-Zs)^2(-\hat{\mathbf{i}}\sin\phi + \hat{\mathbf{j}}\cos\phi)\mathrm{e}^{-2Zs}\sin\theta$; (b) $\mathbf{j}^{\mathrm{d}}(3p_z) = -(2Z^5e^2B/729\pi a_0^2)s^3(1-\frac{1}{3}Zs)^2(-\hat{\mathbf{i}}\sin\phi + \hat{\mathbf{j}}\cos\phi)\mathrm{e}^{-2Zs/3}\cos^2\theta\sin\theta$. **13.17** 17 kT. **13.19** For each type of orbital, $\sigma^{\mathrm{d}} = e^2\mu_0 Z^*/48\pi m_e a_0$. **13.20** The $2s$-orbital gives zero paramagnetic contribution. For a $2p$-electron, $\sigma^{\mathrm{p}} = -(e^2\mu_0\hbar^2/288\pi m_e^2 a_0^3)(Z^{*3}/\Delta)$. **13.21** For NO_2: ${}^2B_2, {}^2B_1, {}^2A_2$; for ClO_2: ${}^2A_2, {}^2A_1, {}^2B_2$. **13.23** $g_{zz} = g_e = 2.002$; $g_{xx} = g_{yy} = g_e - 6\zeta/\Delta = 1.910$. **13.26** Let $\hbar^2A^0 = g_e g_N\mu_B\mu_N\mu_0/4\pi a_0^3$; (a) $A_\parallel/A^0 = 1.41 \times 10^{-3}Z^{*3}$, (b) $A_\perp = -A_\parallel/5$. **13.27** 32 Hz.

14.1 (i) Inelastic, (ii) elastic, (iii) inelastic, (iv) reactive, (v) elastic. **14.2** $\sin^2\theta\cos^2\phi$. **14.3** $4\pi C$. **14.5** $J_\theta = (\hbar/\mu r^3)\mathrm{Re}\{-\mathrm{i}f_k^*(\partial f_k/\partial\theta)\}$; $J_\phi = (\hbar/\mu r^3\sin\theta)\mathrm{Re}\{-\mathrm{i}f_k^*(\partial f_k/\partial\phi)\}$. **14.8** $(2\mu V_0/\hbar^2q^3)^2\{\sin qa - qa\cos qa\}^2$

with $q = 2k\sin\frac{1}{2}\theta$. **14.9** (a) Let $q = 2k\sin\frac{1}{2}\theta$; $\sigma = (2\mu Ze^2/4\pi\varepsilon_0\hbar^2 q)^2\ \{qa^2/(q^2a^2+1)\}^2$, $\sigma_{\text{tot}} = 16\pi\mu^2Z^2e^4a^4/\{\hbar(4k^2a^2+1)\}$; (b) $\sigma = (\mu^2Z^2e^4/4\pi^2\varepsilon_0^2\hbar^4q^4)$. **14.11** For $l=0$: independent of θ; for $l=1$: as $\cos^2\theta$; for $l=2$: as $\frac{1}{4}(3\cos^2\theta - 1)^2$. **14.14** $\tan\delta_l = -\hat{j}_l(ka)/\hat{n}_l(ka)$. **14.15** With $k^2 = 2mE/\hbar^2, k_1^2 = 2m(E - V_0)/\hbar^2$, $\tan\delta_0 = \{(k/k_1)\ \tan k_1a \cos ka - \sin k_1a\}/\{(k/k_1)\tan k_1a \sin ka + \cos k_1a\}$. **14.18** $\sigma = 0$ at energies E such that $\tan(Ka) = Ka$, where $K^2 = 2\mu(E+V_0)/\hbar^2$. Non-inert gas atoms have a much greater effective range a and thus the condition $ka \ll 1$ is not satisfied. **14.20** 33. **14.23** If the scattering system is time-reversal invariant, then the scattering matrix is symmetric; $S_{12} = S_{21}$ then implies $P_{12} = P_{21}$.

Index

T signifies a Table